PERIODIC TABLE OF THE ELEMENTS

Main groups

Main groups

1 / 1A[a][b]	2 / 2A	3 / 3B	4 / 4B	5 / 5B	6 / 6B	7 / 7B	8 / 8B	9 / 8B	10 / 8B	11 / 1B	12 / 2B	13 / 3A	14 / 4A	15 / 5A	16 / 6A	17 / 7A	18 / 8A
1 H Hydrogen 1.00794																	**2 He** Helium 4.002602
3 Li Lithium 6.941	**4 Be** Beryllium 9.012182											**5 B** Boron 10.811	**6 C** Carbon 12.0107	**7 N** Nitrogen 14.00674	**8 O** Oxygen 15.9994	**9 F** Fluorine 18.998403	**10 Ne** Neon 20.1797
11 Na Sodium 22.989770	**12 Mg** Magnesium 24.3050											**13 Al** Aluminum 26.981538	**14 Si** Silicon 28.0855	**15 P** Phosphorus 30.973762	**16 S** Sulfur 32.066	**17 Cl** Chlorine 35.4527	**18 Ar** Argon 39.948
19 K Potassium 39.0983	**20 Ca** Calcium 40.078	**21 Sc** Scandium 44.95591	**22 Ti** Titanium 47.867	**23 V** Vanadium 50.9415	**24 Cr** Chromium 51.9961	**25 Mn** Manganese 54.938049	**26 Fe** Iron 55.845	**27 Co** Cobalt 58.933200	**28 Ni** Nickel 58.6934	**29 Cu** Copper 63.546	**30 Zn** Zinc 65.39	**31 Ga** Gallium 69.723	**32 Ge** Germanium 72.61	**33 As** Arsenic 74.92160	**34 Se** Selenium 78.96	**35 Br** Bromine 79.904	**36 Kr** Krypton 83.80
37 Rb Rubidium 85.4678	**38 Sr** Strontium 87.62	**39 Y** Yttrium 88.90585	**40 Zr** Zirconium 91.224	**41 Nb** Niobium 92.90638	**42 Mo** Molybdenum 95.94	**43 Tc** Technetium [98]	**44 Ru** Ruthenium 101.07	**45 Rh** Rhodium 102.90550	**46 Pd** Palladium 106.42	**47 Ag** Silver 107.8682	**48 Cd** Cadmium 112.411	**49 In** Indium 114.818	**50 Sn** Tin 118.710	**51 Sb** Antimony 121.760	**52 Te** Tellurium 127.60	**53 I** Iodine 126.90447	**54 Xe** Xenon 131.29
55 Cs Cesium 132.90545	**56 Ba** Barium 137.327	**57 *La** Lanthanum 138.9055	**72 Hf** Hafnium 178.49	**73 Ta** Tantalum 180.9479	**74 W** Tungsten 183.84	**75 Re** Rhenium 186.207	**76 Os** Osmium 190.23	**77 Ir** Iridium 192.217	**78 Pt** Platinum 195.078	**79 Au** Gold 196.96655	**80 Hg** Mercury 200.59	**81 Tl** Thallium 204.3833	**82 Pb** Lead 207.2	**83 Bi** Bismuth 208.98038	**84 Po** Polonium [209]	**85 At** Astatine [210]	**86 Rn** Radon [222]
87 Fr Francium [223]	**88 Ra** Radium 226.025	**89 †Ac** Actinium 227.028	**104 Rf** Rutherfordium [261]	**105 Db** Dubnium [262]	**106 Sg** Seaborgium [266]	**107 Bh** Bohrium [267]	**108 Hs** Hassium [269]	**109 Mt** Meitnerium [268]	**110 Ds** Darmstadtium [281]	**111[c] Rg** Roentgenium [272]	**112** [277]	**113** [284]	**114** [285]	**115** [288]	**116** [289]	**117**	**118** [294]

Transition metals

Metals   Nonmetals   Noble gases

*Lanthanide series

58 Ce Cerium 140.116	59 Pr Praseodymium 140.90765	60 Nd Neodymium 144.24	61 Pm Promethium [145]	62 Sm Samarium 150.36	63 Eu Europium 151.964	64 Gd Gadolinium 157.25	65 Tb Terbium 158.92534	66 Dy Dysprosium 162.50	67 Ho Holmium 164.93032	68 Er Erbium 167.26	69 Tm Thulium 168.93421	70 Yb Ytterbium 173.04	71 Lu Lutetium 174.967

†Actinide series

90 Th Thorium 232.0381	91 Pa Protactinium 231.03588	92 U Uranium 238.0289	93 Np Neptunium 237.048	94 Pu Plutonium [244]	95 Am Americium [243]	96 Cm Curium [247]	97 Bk Berkelium [247]	98 Cf Californium [251]	99 Es Einsteinium [252]	100 Fm Fermium [257]	101 Md Mendelevium [258]	102 No Nobelium [259]	103 Lr Lawrencium [262]

Atomic masses in brackets are the masses of the longest-lived or most important isotope of certain radioactive elements.

[a] The labels on top (1, 2, 3 … 18) are the group numbers recommended by the International Union of Pure and Applied Chemistry.

[b] The labels on the bottom (1A, 2A, … 8A) are the group numbers commonly used in the United States and the ones we use in this text.

[c] The names and symbols of elements 112 and above have not been assigned.

Further information is available at the Web site of WebElements™ .

TABLE OF ATOMIC MASSES BASED ON CARBON-12

Name	Symbol	Atomic Number	Atomic Mass	Name	Symbol	Atomic Number	Atomic Mass
Actinium	Ac	89	227.028	Meitnerium	Mt	109	(268)
Aluminum	Al	13	26.9815	Mendelevium	Md	101	(258)
Americium	Am	95	(243)	Mercury	Hg	80	200.59
Antimony	Sb	51	121.760	Molybdenum	Mo	42	95.94
Argon	Ar	18	39.948	Neodymium	Nd	60	144.24
Arsenic	As	33	74.9216	Neon	Ne	10	20.1797
Astatine	At	85	(210)	Neptunium	Np	93	237.048
Barium	Ba	56	137.327	Nickel	Ni	28	58.6934
Berkelium	Bk	97	(247)	Niobium	Nb	41	92.9064
Beryllium	Be	4	9.01218	Nitrogen	N	7	14.0067
Bismuth	Bi	83	208.980	Nobelium	No	102	(259)
Bohrium	Bh	107	(267)	Osmium	Os	76	190.23
Boron	B	5	10.811	Oxygen	O	8	15.9994
Bromine	Br	35	79.904	Palladium	Pd	46	106.42
Cadmium	Cd	48	112.411	Phosphorus	P	15	30.9738
Calcium	Ca	20	40.078	Platinum	Pt	78	195.078
Californium	Cf	98	(251)	Plutonium	Pu	94	(244)
Carbon	C	6	12.0107	Polonium	Po	84	(209)
Cerium	Ce	58	140.116	Potassium	K	19	39.0983
Cesium	Cs	55	132.905	Praseodymium	Pr	59	140.908
Chlorine	Cl	17	35.4527	Promethium	Pm	61	(145)
Chromium	Cr	24	51.9961	Protactinium	Pa	91	231.036
Cobalt	Co	27	58.9332	Radium	Ra	88	226.025
Copper	Cu	29	63.546	Radon	Rn	86	(222)
Curium	Cm	96	(247)	Rhenium	Re	75	186.207
Darmstadtium	Ds	110	(281)	Rhodium	Rh	45	102.906
Dubnium	Db	105	(262)	Roentgenium	Rg	111	(272)
Dysprosium	Dy	66	162.50	Rubidium	Rb	37	85.4678
Einsteinium	Es	99	(252)	Ruthenium	Ru	44	101.07
Erbium	Er	68	167.26	Rutherfordium	Rf	104	(261)
Europium	Eu	63	151.964	Samarium	Sm	62	150.36
Fermium	Fm	100	(257)	Scandium	Sc	21	44.9559
Fluorine	F	9	18.9984	Seaborgium	Sg	106	(266)
Francium	Fr	87	(223)	Selenium	Se	34	78.96
Gadolinium	Gd	64	157.25	Silicon	Si	14	28.0855
Gallium	Ga	31	69.723	Silver	Ag	47	107.868
Germanium	Ge	32	72.61	Sodium	Na	11	22.9898
Gold	Au	79	196.967	Strontium	Sr	38	87.62
Hafnium	Hf	72	178.49	Sulfur	S	16	32.066
Hassium	Hs	108	(269)	Tantalum	Ta	73	180.948
Helium	He	2	4.00260	Technetium	Tc	43	(98)
Holmium	Ho	67	164.930	Tellurium	Te	52	127.60
Hydrogen	H	1	1.00794	Terbium	Tb	65	158.925
Indium	In	49	114.818	Thallium	Tl	81	204.383
Iodine	I	53	126.904	Thorium	Th	90	232.038
Iridium	Ir	77	192.217	Thulium	Tm	69	168.934
Iron	Fe	26	55.845	Tin	Sn	50	118.710
Krypton	Kr	36	83.80	Titanium	Ti	22	47.867
Lanthanum	La	57	138.906	Tungsten	W	74	183.84
Lawrencium	Lr	103	(262)	Uranium	U	92	238.029
Lead	Pb	82	207.2	Vanadium	V	23	50.9415
Lithium	Li	3	6.941	Xenon	Xe	54	131.29
Lutetium	Lu	71	174.967	Ytterbium	Yb	70	173.04
Magnesium	Mg	12	24.3050	Yttrium	Y	39	88.9059
Manganese	Mn	25	54.9380	Zinc	Zn	30	65.39
				Zirconium	Zr	40	91.224

Atomic masses in this table are relative to carbon-12 and limited to six significant figures, although some atomic masses are known more precisely. For certain radioactive elements the numbers listed (in parentheses) are the mass numbers of the most stable isotopes.

CHEMISTRY
FOR CHANGING TIMES

TWELFTH EDITION

John W. Hill
University of Wisconsin–River Falls

Terry W. McCreary
Murray State University

Doris K. Kolb

Prentice Hall

New York   Boston   San Francisco
London   Toronto   Sydney   Tokyo   Singapore   Madrid
Mexico City   Munich   Paris   Cape Town   Hong Kong   Montreal

Library of Congress Cataloging-in-Publication-Data

Hill, John William
        Chemistry for changing times / John W. Hill. — 12th ed. / Terry W. McCreary, Doris K. Kolb.
            p. cm.
        Includes index.
        ISBN 0-13-605449-8
        1. Chemistry—Textbooks. I. McCreary, Terry Wade, 1955– II. Kolb, Doris K. (1927–2005) III. Title.
    QD33.2.H54 2010
    540–dc22

2008053839

Acquisitions Editor: Dawn Giovanniello
Assistant Editor: Jessica Neumann
Editor in Chief, Science: Nicole Folchetti
Marketing Manager: Elizabeth Averbeck
Editorial Assistant: Lisa Tarabokjia
Editor in Chief, Development: Ray Mullaney
Development Editor: Carol Pritchard-Martinez
Managing Editor, Chemistry and Geosciences: Gina M. Cheselka
Project Manager, Science: Beth Sweeten
Associate Media Producer: Kristin Mayo
Art Editor: Connie Long
Art Studio: Precision
Art Director: Maureen Eide
Interior Design: Yellow Dog Designs
Cover Design: Maureen Eide
Senior Operations Supervisor: Alan Fischer
Image Permissions Coordinator: Elaine Soares
Photo Researcher: Eric Shrader
Production Supervision/Composition: Prepare, Inc.
Cover Credit: BLOOMimage/Getty Images, Inc

Printed in the United States of America
10 9 8 7 6 5 4 3 2 1

ISBN-10: 0-13-605449-8
ISBN-13: 978-0-13-605449-8

Prentice Hall
is an imprint of

www.pearsonhighered.com

BRIEF CONTENTS

ON-LINE
eCHAPTERS

GREEN CHEMISTRY

The twelfth edition of *Chemistry for Changing Times* is pleased to present the Green Chemistry essays listed below. The topics have been carefully chosen to introduce students to the concepts of Green Chemistry—a new approach to designing chemicals and chemical transformations that are beneficial for human health and the environment. The Green Chemistry essays in this edition highlight cutting-edge research by chemists, molecular scientists, and engineers to explore the fundamental science and practical applications of chemistry that is "benign by design." These examples emphasize the responsibility of chemists for the consequences of the new materials they create and the importance of building a sustainable chemical enterprise.

CONTENTS

1 CHEMISTRY 1

2 ATOMS 41

3 ATOMIC STRUCTURE 61

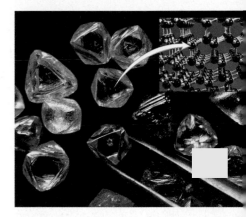

4 CHEMICAL BONDS 89

5 CHEMICAL ACCOUNTING 123

6 GASES, LIQUIDS, SOLIDS, AND INTERMOLECULAR FORCES 149

7 ACIDS AND BASES 171

8 OXIDATION AND REDUCTION 195

9 ORGANIC CHEMISTRY 223

10 POLYMERS 261

11 NUCLEAR CHEMISTRY 291

12 CHEMISTRY OF EARTH   325

13 AIR   345

14 WATER   381

15 ENERGY 407

16 BIOCHEMISTRY 453

17 FOOD 491

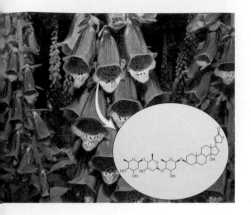

18 DRUGS 527

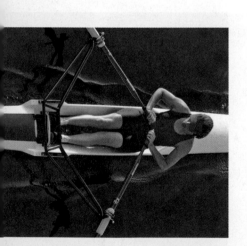

19 FITNESS AND HEALTH 583

20 CHEMISTRY DOWN ON THE FARM . . . 611

21 HOUSEHOLD CHEMICALS   639

22 POISONS   679

PREFACE

Chemistry for Changing Times is now in its twelfth edition. Times have changed immensely since the first edition appeared in 1972 and continue to change more rapidly than ever—especially in the vital areas of biochemistry, the environment, energy, drugs, and health and nutrition—and we have thoroughly updated the material accordingly. The emphasis on the concept of green chemistry throughout the book, especially in the Green Chemistry essays (see page v), is one of the major changes to the twelfth edition.

A Special Kind of Course

Our knowledge base has expanded enormously since the first edition, never more so than in the last few years. We have faced tough choices in deciding what to include and what to leave out. We now live in what has been called the "information age." Unfortunately, information is not knowledge; the information may or may not be valid. Our focus, more than ever, is on helping students evaluate information. May we all someday gain the gift of wisdom.

A major premise is that a chemistry course for students who are not majoring in science should be quite different from the course we offer our science majors. It must present basic chemical concepts with intellectual honesty, but it need not—and probably should not—focus on esoteric theories or rigorous mathematics. It should include lots of modern everyday applications. The textbook should be appealing to look at, easy to understand, and interesting to read.

Three-fourths of the legislation considered by the U.S. Congress involves questions having to do with science or technology, yet only rarely does a scientist or engineer enter politics. Most of the people who make important decisions regarding our health and our environment are not trained in science, but it is critical that these decision makers be scientifically literate. In the judicial system, decisions often depend on scientific evidence, but judges and jurors frequently have little education in the sciences. A chemistry course for students who are not science majors should emphasize practical applications of chemistry to problems involving such things as environmental pollution, radioactivity, energy sources, and human health. The students who take our liberal arts chemistry courses include future teachers, business leaders, lawyers, legislators, accountants, artists, journalists, jurors, and judges.

Objectives

Our main objectives in a chemistry course for students who are not majoring in science are as follows:

- To attract lots of students from a variety of disciplines. If students do not enroll in the course, we can't teach them.
- To help students learn so that they may become productive, creative, ethical, and engaged citizens.
- To use topics of current interest to illustrate chemical principles. We want students to appreciate the importance of chemistry in the real world.
- To relate chemical problems to the everyday lives of our students. Chemical problems become more significant to students when they can see a personal connection.

- To instill in students an appreciation for chemistry as an open-ended learning experience. We hope that our students will develop a curiosity about science and will want to continue learning throughout their lives.
- To acquaint students with scientific methods. We want students to be able to read about science and technology with some degree of critical judgment. This is especially important because many of the scientific problems discussed are complex and controversial.
- To show students—through Green Chemistry essays and exercises—that better, safer, and more environmentally friendly processes and products are being developed. We want students to know that "green chemistry" is not a buzzword but a goal to be pursued.
- To help students become literate in science. We want our students to develop a comfortable knowledge of science so that they find news articles relating to science interesting rather than intimidating.

New Features in the Twelfth Edition

In preparing this new edition, we have responded to suggestions from users and reviewers of the eleventh edition, and we have used our own writing and teaching experience. The text is fully revised and updated to reflect the latest scientific developments in a fast-changing world.

A very visible change to the twelfth edition is the reorganization of the chapters. At most institutions, the course for nonscience majors brings together a group of students who are quite heterogeneous both in their science backgrounds and in their academic interests. A major challenge to the instructor is to find the balance between these needs and interests. As authors we have tried to create a text that is flexible and that can be used in a variety of ways. One goal of the reorganization was to make it easier for the instructor to skip sections or (in some cases) whole chapters.

- Chapters 1 through 8 deal with some of the basic chemistry, including atoms and elements, structures of ionic and molecular compounds, acids and bases, and oxidation–reduction reactions. Most of the numerical manipulations are concentrated in Chapters 5 and 6.
- Chapters 9 and 10 introduce organic chemistry and polymers.
- Chapter 11 is concerned with nuclear chemistry. This was Chapter 4 in previous editions; it was moved to this new location in response to user feedback to present a more concise core. For those who so desire, Chapter 11 can still be taught after Chapter 3.
- Chapters 12 through 15 (which deal with Earth, air, water, and energy, respectively) focus on environmental issues. An environmental focus permeates the entire book, especially in the incorporation of the concept of green chemistry in the Green Chemistry essays and exercises.
- Chapter 16 introduces biochemistry and provides a background for the some of the material in subsequent chapters, especially Chapter 17 (Food) and eChapter 19 (Fitness and Health).
- Chapter 18 deals with various aspects of drugs: legal and illegal, life-sustaining and death-dealing.
- Another dramatic change is seen in Chapters 19 through 22, which are presented online. Most of today's students use the Web routinely; for them this mode of delivery will probably be welcomed! By doing this, we also save paper—the book is now greener both in content and in fact—and the cost has been reduced. These four chapters delve into fitness and health, agriculture and gardening, household chemicals, and poisons, respectively.

Clearly, for a one-semester course, choices will have to be made. In some places, the text refers to later chapters where further details of a particular topic are to be found, but it is fairly easy for instructors to select the parts most important and useful to their students. If you have specific questions about skipping sections or chapters, please contact John Hill (jwhill602@comcast.net) or Terry McCreary (terry.mccreary@murraystate.edu).

Improvements in Pedagogy

The following changes have been made to strengthen and improve the pedagogy in this edition.

- We have incorporated Self-Assessment Questions at the end of most sections in the text. These multiple-choice items provide immediate feedback to test the student's understanding of the material.
- We extensively revised the Review Questions to account for the fact that many of them were superseded by the Self-Assessment Questions. We also revised about 25% of the end-of-chapter matched-set Problems and many of the Additional Problems.
- Several of the worked-out Examples and their accompanying Exercises have been modified or revised extensively.
- Each chapter starts with a set of *Questions You May Have Asked Yourself*. We have attempted to select real-life questions that may have arisen in an inquiring student's mind and are related to the chapter content. Answers to those questions are found in the margins of the chapter near the text content associated with the question.
- Critical Thinking Exercises have been added to or modified in certain chapters. These exercises require the student to apply information and learning from the chapter in both concrete and abstract fashion, in the evaluation of claims both hypothetical and real.
- Each chapter has a category of Collaborative Group Projects as a part of the end-of-chapter exercises. This will make it easy for instructors who want to encourage collaborative work and to make group assignments. In this way, the students' learning of chemistry can be extended far beyond the textbook.
- Some of the figures use voice balloons to point out important features. We also use voice balloons in text displays and problem solving to guide the student through the learning process and thus improve the pedagogy.
- The chapter summaries are organized by sections with key terms highlighted in red for easy recognition.
- We have added *More to Explore* notes in the margins, which suggest references for those who want to learn more about a particular topic.
- Keeping in mind that today's student is more visually oriented than ever, we have made extensive changes in the photographs, figures, graphs, and other illustrations. Figures that present numeric data have been updated where possible.

Applications

Periodically through the text, the reader will find box features. These essays focus on interesting, relevant applications of the chemistry covered in a particular chapter or section. Box features include the following:

- Nanoworld (Chapter 1)
- A Compound by Any Other Name Would Smell As Sweet. . . (Chapter 4)
- Photochromic Glass (Chapter 8)
- Cell Phones and Microwaves and Power Lines, Oh My! (Chapter 11)

- A Closed Ecosystem? (Chapter 13)
- Energy Return on Energy Invested (Chapter 15)
- Infectious Prions: Deadly Protein (Chapter 16)
- Enzymes and Green Chemistry (Chapter 16)
- It's a Drug! No, It's a Food! No, It's...a Dietary Supplement! (Chapter 17)
- Cisplatin: The Platinum Standard of Cancer Treatment (Chapter 18)
- Some Chemistry of Love, Trust, and Sexual Fidelity (Chapter 18)
- Chemistry and Athletic Performance (eChapter 19)
- Chemistry of Sports Materials (eChapter 19)
- Renaissance Poisoners and Chemistry and Counterterrorism (eChapter 22)

In addition, you will find short *It DOES Matter!* items in the margin. These features briefly discuss a specific application or phenomenon and show the student that the material being studied does relate to the world in which we live.

Visualization

Visual material adds greatly to the general appeal of a textbook. For that reason new color photographs and diagrams have been added throughout the text. Color diagrams can also be highly instructive, and colorful photographs relating to descriptive chemistry do much to enhance the learning process. We have added and revised illustrations that use both microscopic (molecular) and macroscopic (visual) views to help students visualize chemical phenomena. Some of the figure captions feature questions to focus attention on the concept illustrated in the figure.

Readability

Over the years, students have told us that they have found this textbook easy to read. The language is simple, and the style is conversational. Explanations are clear and easy to understand. The friendly tone of the book has been maintained in this edition. Since the format and the amount of open space on a page also contribute to readability, we have made conscious improvements in the design of this edition. For example, many of the margin notes of previous editions have been incorporated directly into the text to ensure that pages don't appear to be crowded.

Units of Measurement

The United States continues to use the traditional English system for many kinds of measurement even though the metric system has long been used internationally. A modern version of the metric system, the Système International (SI), is now widely used, especially by scientists. So what units should be used in a text for liberal arts students? In presenting chemical principles, we use primarily metric units. In other parts of the book we use those units that the students are most likely to encounter elsewhere in the same context.

Chemical Structures

The structures of many complicated molecules are presented in the text, especially in the later chapters. These structures are presented mainly to emphasize that they are actually known and to illustrate the fact that substances with similar properties often have similar structures. In many of the structures, functional groups or specific molecular features have been emphasized with color. For example, basic functional groups are usually shown in blue, acidic ones in red, and neutral in green. Students should not feel that they must learn all these structures, but they should take the time to look at them. We hope that they will come to recognize familiar features in these molecules.

Chapter Summaries and Glossary

The chapter summaries, with key terms highlighted in red, provide a quick review. The Glossary gives definitions of terms that appear in boldface throughout the text as well as some other terms frequently encountered. These terms include all key terms highlighted in the chapter summaries.

Questions and Problems

Worked-out Examples and accompanying Exercises are given within almost all of the chapters. Each Example carefully guides students through the process for solving a particular type of problem. It is then usually followed by one or more Exercises that allow students to check their comprehension right away. Many Examples are now followed by two Exercises, labeled A and B. The goal in an A Exercise is to apply to a similar situation the method outlined in the Example. In a B Exercise, students often must combine that method with other ideas previously learned. Because many of the B Exercises provide a context closer to that in which chemical knowledge is applied, they thus serve as a bridge between the worked Examples and the more challenging problems at the end of the chapter. The A and B Exercises provide a simple way for the instructor to assign homework that is closely related to the Examples. Answers to all the in-chapter Exercises are given in the Answers section at the back of the book.

The end-of-chapter exercises include the following:

- Review Questions that for the most part simply ask for a recall of material in the chapter. Answers to Review Questions identified by red numbers are given in the Answers section at the back of the book.

- A set of matched-pair Problems, arranged according to subject matter from each chapter, with answers to the odd-numbered Problems given in the Answers section.

- Additional Problems that are not grouped by type. Some of these are more challenging than the matched-pair Problems and often require a synthesis of ideas from more than one chapter. Other Additional Problems pursue further an idea in the text or introduce new ideas. Answers to odd-numbered Additional Problems are also given in the Answers section.

- Many colleges and universities now emphasize group learning as well as individual assignments. The Collaborative Group Projects permit an instructor to easily assign group work in an open-ended context; most of these Projects can have multiple directions and multiple focus points.

Supplementary Materials

The most important learning aid is the teacher. In order to make the instructor's job easier and enrich the education of students, we have provided a variety of supplementary materials.

Print Resources for Students

Study Guide and Selected Solutions Manual (0321612434). Prepared by Richard Jones of Sinclair Community College. This book assists students through the text material and contains learning objectives, chapter outlines, key terms, additional problems along with self-tests and answers, and answers to the odd-numbered problems from the text.

Chemical Investigations for Changing Times, Twelfth Edition (0321612450). Prepared by C. Alton Hassell and Paula Marshall. Contains 59 laboratory experiments and is specifically referenced to *Chemistry for Changing Times.*

Print Resources for Instructors

Transparency Set (0321615093). Selected by Terry McCreary and John W. Hill. This set contains 150 full-color acetates.

Printed Test Bank (0321612442). Prepared by Rill Ann Reuter, Winona State University. The *Test Item File* now contains over 2400 test questions that are referenced to the text.

Media Resources for Instructors

Online Instructor Manual for Chemical Investigations for Changing Times (0321615085). Prepared by Paula Marshall and C. Alton Hassell. This laboratory manual reference includes notes for experiments, safety regulations, procedural instructions, and specifications for equipment and supplies.

Instructor Resource Center on CD/DVD (0321612477). This integrated collection of resources includes everything you need organized in one easy-to-access place. It is designed to help you make efficient and effective use of your lecture preparation time as well as to enhance your classroom presentations and assessment efforts. This package features all of the art from the text, including tables; three prebuilt PowerPoint™ presentations; PDF files of the art for high-resolution printing; the *Instructor's Resource Manual,* and a set of "clicker" questions for use with Classroom Response Systems. Also included is the TestGen test-generation software and a TestGen version of the *Test Item File* that enables you to create and tailor exams to your needs or to create online quizzes for delivery in WebCT, Blackboard, or CourseCompass.

Online Instructor Resource Manual (0321612469). Prepared by Paul Karr and David Pietz of Wayne State College. This useful guide describes all the different resources available to instructors and shows how to integrate them into your course. Organized by chapter, this manual offers lecture outlines, answers and solutions to all questions and problems that are not answered by the authors in Appendix C, suggested in-class demonstrations recommended by Doris Kolb, and other suggested resources.

Course Management Options Pearson Prentice Hall offers three content cartridges for online, text-specific course management systems depending on your preferred platform. Hundreds of text-specific problems are provided.

Visit *www.pearsonhighered.com/elearning* for details on how to communicate with your students online, customize content to meet your course needs, create online quizzes and tests, track grades, and much more.

Acknowledgments

Through the last four decades we have greatly benefited from hundreds of helpful reviews. It would take far too many pages to list all of those reviewers here. Many of you have contributed to the flavor of the book and helped us minimize our errors. Please know that your contributions are deeply appreciated. For the twelfth edition, we are grateful for challenging reviews from

Michelle Boucher, *Utica College*

Roxanne Finney, *Skagit Valley College*

Luther Giddings, *Salt Lake Community College*

Todd Hamilton, *Georgetown College*

Alton Hassell, *Baylor University*

Scott Hewitt, *California State University, Fullerton*

Sherell Hickman, *Brevard Community College, Cocoa Beach*

Beth Hixon, *Tulsa Community College*

James L. Klino, *SUNY-Agriculture-Technical College, Cobleskill*

Meghan Knapp, *Georgetown College*

Kim Loomis, *Century College*

Jeremy Mason, *Texas Technical University*

Lois Schadenwald, *Normandale Community College*

Joseph Sinski, *Bellarmine University*

Kelli M. Slunt, *University of Mary Washington*

Jie Song, *University of Michigan, Flint*

Wayne M. Stalick, *University of Central Missouri*

Durwin Striplin, *Davidson College*

Shashi Unnithan, *Front Range Community College*

Edward Vitz, *Kutztown University*

Dan Wacks, *University of Redlands*

Elizabeth Wallace, *Western Oklahoma State College*

Matthew E. Wise, *University of Colorado, Boulder*

We also appreciate the many people who have called, written, or e-mailed with corrections and other helpful suggestions. Cynthia S. Hill prepared much of the original material on biochemistry, food, and health and fitness. For this edition, we are especially indebted to Jan William Simek, California Polytechnic State University, who shared several helpful suggestions, including his course outline and exams.

We owe a special debt of gratitude to Doris K. Kolb (1927–2005), who was an esteemed coauthor of the seventh through the eleventh editions. Doris and her husband Ken were friends and helpful supporters long before Doris joined the author team. She provided much to the spirit and flavor of the book. Doris's contributions to *Chemistry for Changing Times*—and indeed to all of chemistry and chemical education—will live on for many years to come, not only in her publications but also in the hearts and minds of her many students, colleagues, and friends. Doris was a poet of some note. Her main interest was humorous poetry, and she consistently added a touch of fun to *Chemistry for Changing Times.* Four of her verses that appear in this volume were first published in the *Journal of Chemical Education.* We acknowledge, with thanks, the permission to reprint them here. Doris Kolb wrote those verses plus all of the others.

Throughout her career as a teacher, scientist, community leader, poet, and much more, Doris K. Kolb was blessed with a wonderful spouse, colleague, and companion, Kenneth E. Kolb. Over the years, Ken did chapter reviews, made suggestions, and gave invaluable help for many editions. All who knew her miss Doris greatly. Those of us who had the privilege of working closely with her miss her wisdom and wit most profoundly. Let us all dedicate our lives, as Doris did hers, to making this world a better place.

We also want to thank our colleagues at the University of Wisconsin–River Falls, Murray State University, and Bradley University for all their help and support through the years.

We owe a debt of gratitude to the many creative people at Pearson Prentice Hall who have contributed their talents to this edition. Kent Porter-Hamann and Dawn Giovanniello, our chemistry editors, have provided valuable guidance throughout the project. Development editor Carol Pritchard-Martinez contributed in so many ways to this project, especially in challenging us to be better authors in every way. We treasure her many helpful suggestions of new material and better presentation of all the subject matter. Ray Mullaney, vice president and editor in chief, Science Book Development, has provided exemplary guidance and impeccable judgment throughout this process. As chemistry editorial assistant, Lisa Tarabokjia helped keep the team up-to-date on developments. We are also grateful to assistant editor Jessica Neumann, who coordinated development of the Green Chemistry essays and excelled in the critical role of keeping the project on schedule; to Gina Chesel-ka, Beth Sweeten, and Francesca Monaco in production and art director Maureen Eide for their diligence and patience in bringing all the parts together to yield a finished work; and to our media producer, Kristin Mayo, who has brought new facets to the media set accompanying our text. We are indebted to our copy editor, Donna E. Mulder, whose expertise helped improve the consistency of the text; to proofreader, Donna Young; and accuracy checker, Brad Sieve, whose sharp eyes caught many of our errors and typos. We also salute our photo researchers, Eric Schrader and Travis Amos, who vetted hundreds of images in the search for quality photographic illustrations.

John W. Hill owes a very special kind of thanks to his wonderful spouse, Ina, who over the years has done typing, library research, and so many other tasks. Most of all, he is grateful for her enduring love and boundless patience. Terry W. McCreary would like to thank his wife, Geniece, and their children, Corinne and Yvette, for their unflagging support, understanding, and love.

Finally, we also thank all those many students whose enthusiasm has made teaching such a joy. It is gratifying to have students learn what you are trying to teach them, but it is a supreme pleasure to find that they want to learn even more. And, of course, we are grateful to all who have made so many helpful suggestions. We welcome and appreciate all your comments, corrections, and criticisms.

John W. Hill
jwhill602@comcast.net

Terry W. McCreary
terry.mccreary@murraystate.edu

Doris K. Kolb

To the Student

Welcome to Our Chemical World!

Chemistry is fun. Through this book, we would like to share with you some of the excitement of chemistry and some of the joy of learning about it. You do not need to exclude chemistry from your learning experiences. Learning chemistry will enrich your life—now and long after this course is over—through a better understanding of the natural world, the technological questions now confronting us, and the choices we must face as citizens within a scientific and technological society.

Learning chemistry involves thinking logically, critically, and creatively. Skills gained in this course can be exceptionally useful in many aspects of your life. You will learn how to use the language of chemistry: symbols, formulas, and equations. More important, you will learn how to obtain meaning from information. The most important thing you will learn is how to learn. Memorized material will quickly fade into oblivion unless it is arranged on a framework of understanding.

Chemistry Directly Affects Our Lives

How does the human body work? How does aspirin cure headaches, reduce fevers, and perhaps lessen the chance of a heart attack or stroke? Is ozone a good thing or a threat to our health? Are iron supplement pills poisonous? Is global warming real? If so, did humans contribute to it, and what are some of the possible consequences? Why do most weight-loss diets seem to work in the short run but fail in the long run? Why do our moods swing from happy to sad? Can a chemical test on urine predict possible suicide attempts? How does penicillin kill bacteria without harming our healthy body cells? Chemists have found answers to questions such as these and continue to seek the knowledge that will unlock still other secrets of our universe. As these mysteries are resolved, the direction of our lives often changes—sometimes dramatically. We live in a chemical world—a world of drugs, biocides, food additives, fertilizers, fuels, detergents, cosmetics, and plastics. We live in a world with toxic wastes, polluted air and water, and dwindling petroleum reserves. Knowledge of chemistry will help you better understand the benefits and hazards of this world and will enable you to make intelligent decisions in the future.

Chemical Dependency

We are all chemically dependent. Even in the womb we depend on a constant supply of oxygen, water, glucose, and a multitude of other chemicals.

Our bodies are intricate chemical factories. They are durable but delicate systems. Innumerable chemical reactions that allow our bodies to function properly are constantly taking place within us. Thinking, learning, exercising, feeling happy or sad, putting on too much weight or not gaining enough, and virtually all life processes are made possible by these chemical reactions. Everything that we ingest is part of a complex process that determines whether our bodies work effectively or not. The consumption of some substances can initiate chemical reactions that will stop body functions. Other substances, if consumed, can cause permanent disabilities, and still others can make living less comfortable. A proper balance of the right foods provides the chemicals and generates the reactions we need in order to function at our best. The knowledge of chemistry that you will soon be gaining will help you better understand how your body works so that you will be able to take proper care of it.

Changing Times

We live in a world of increasingly rapid change. It has been said that the only constant is change itself. At present, we are facing some of the greatest problems that humans have ever encountered, and the dilemmas with which we are now confronted seem to have no perfect solutions. We are sometimes forced to make a best choice among only bad alternatives, and our decisions often provide only temporary solutions to our problems. Nevertheless, if we are to choose properly, we must understand what our choices are. Mistakes can be costly, and they cannot always be rectified. It is easy to pollute, but cleaning up pollution once it is there is enormously expensive. We can best avoid mistakes by collecting as much information as possible and evaluating it carefully before making critical decisions. Science is a means of gathering and evaluating information, and chemistry is central to all the sciences.

Chemistry and the Human Condition

Above all else, our hope is that you will learn that the study of chemistry need not be dull and difficult. Rather, it can enrich your life in so many ways—through a better understanding of your body, your mind, your environment, and the world in which you live. After all, the search to understand the universe is an essential part of what it means to be human.

Highlights of the Twelfth Edition

Chemistry for Changing Times, the most successful book in liberal arts chemistry, defined the course in its first edition. With each subsequent revision, this text has reflected the changing times and the changing needs of the market.

Visually appealing, understandable, and interesting to read, the goal of the twelfth edition of *Chemistry for Changing Times* is to help inform students as scientifically literate consumers and decision makers. The authors present basic chemical concepts with abundant everyday applications, personalizing the chemistry experience for today's students. In this way, the text focuses students on evaluating information about real-life issues instead of memorizing rigorous theory and mathematics. Important in this new edition is the use of green chemistry as a theme to show the positive impact of chemistry on the future.

Green Chemistry

The concept of **green chemistry** is used throughout the book and appears prominently in the **Green Chemistry** essays in each chapter.

Critical Thinking

Critical Thinking Exercises. Critical thinking is introduced in Chapter 1 and carried throughout the text. At the end of every chapter is an expanded set of Critical Thinking Exercises that encourages students to think critically about and evaluate the most up-to-date, relevant issues. These exercises require the student to apply information and learning from the chapter in both concrete and abstract ways.

Collaborative Group Projects. These end-of-chapter exercises, which extend student learning of chemistry beyond the text, are available for instructors who want to engage students in collaborative work with group assignments.

Conceptual Problem Solving

Conceptual Examples guide students through the process of learning and understanding important chemical concepts.

- Each Example shows a title indicating the skill being covered.
- Solutions are expanded with more explanation to guide the students through solving the problem.
- **A and B Exercises** follow many of the Examples. The **A** Exercise parallels the Example; the **B** Exercise requires the student to incorporate information from earlier material. The dual Exercises help the students synthesize their learning into a coherent whole rather than just learning isolated facts.
- **Voice balloons** show students the logic of the problem-solving process.

Visualization

Illustrations using both microscopic (molecular) and macroscopic (visual) views help students visualize chemical phenomena.

Questions shown at the end of some of the figure captions direct the student to the things that are particularly important to visualize and to expand on the concept illustrated in the figure or photograph.

Chapter Summaries. The chapter summaries are presented by sections with key terms highlighted in red for easy recognition.

Applications That Focus on the Environment

Applications. Box features and *It DOES Matter!* discussions include many interesting, relevant, and environmental applications.

About Our Sustainability Initiatives

This book is carefully crafted to minimize environmental impact. The materials used to manufacture this book originated from sources committed to responsible forestry practices. The paper is FSC certified. The binding, cover, and paper come from facilities that minimize waste, energy consumption, and the use of harmful chemicals.

Pearson closes the loop by recycling every out-of-date text returned to our warehouse. We pulp the books, and the pulp is used to produce items such as paper coffee cups and shopping bags. In addition, Pearson aims to become the first carbon neutral educational publishing company.

Pearson is also supporting student sustainability efforts, through our Sustainable Solutions Awards, our Student Sustainability Summits, and our Student Activity Fund.

The future holds great promise for reducing our impact on Earth's environment, and Pearson is proud to be leading the way. We strive to publish the best books with the most up-to-date and accurate content, and to do so in ways that minimize our impact on Earth.

Mixed Sources
Product group from well-managed forests and other controlled sources
www.fsc.org Cert no. SW-COC-002550
© 1996 Forest Stewardship Council
FSC

Prentice Hall
is an imprint of

www.pearsonhighered.com

ENGAGING YOU WITH CHEMISTRY

Chemistry for Changing Times engages you by connecting your world and the chemistry that surrounds you. In this edition, John Hill and Terry McCreary continue to make chemistry exciting and interesting by incorporating a robust media package that makes it easier than ever for you to access tutoring, practice problems, and virtual office hours.

my**e**Book

my**e**Book enables you to make virtual notes within the text, as well as enlarge images, charts, and figures for better viewing 24/7.

Within my**e**Book you are also able to access Virtual Lectures (see below), where you can see and hear a professor working out the examples throughout the text.

Virtual Lectures

Directly linked in my**e**Book, these professor-narrated videos provide step-by-step commentary on a virtual whiteboard for all of the examples and conceptual examples throughout the text. These prerecorded videos enable you to pause, rewind, and fast-forward, to individualize your learning for on-demand tutoring.

EXAMPLE 1.8 Unit Conversions

Convert

a. 1.83 kg to grams and   b. 729 μL to milliliters.

Solution

a. We start with the given quantity 1.83 kg and use the equivalence 1 kg = 1000 g to form a conversion factor (Appendix) that allows us to cancel the unit kg and end with the unit g.

$$1.83\ \text{kg} \times \frac{1000\ \text{g}}{1\ \text{kg}} = 1830\ \text{g}$$

The Chemistry Place™ Website

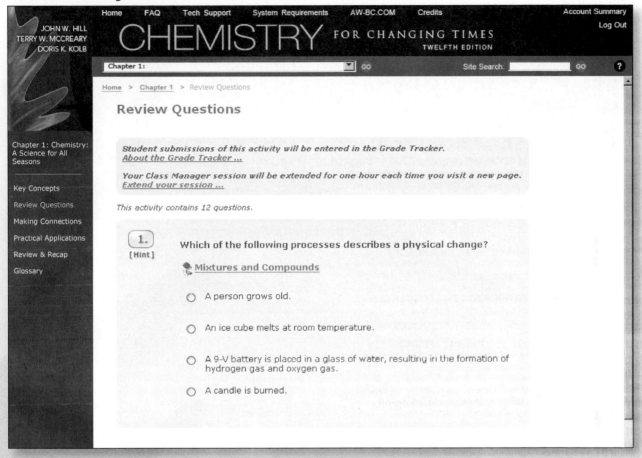

www.chemplace.com

The Chemistry Place Website gives you everything you need to delve deeper into learning chemistry:

- **Key Concepts** are a collection of learning goals that help you identify what you should understand and be able to articulate after reading the chapter.

- **Review Activities** provide you with a variety of assessment opportunities that help you check your understanding of key concepts with the choice to view helpful hints and receive instant feedback on selected answers. Instructors have the option to assign any of these online self-grading activities for homework credit or simply allow you to work at your own pace toward mastery.

- **Media Enhancements** include selected review questions that are enhanced with media—short movies, animations, and 3-D molecules—to help you visualize key concepts.

- **Application and Critical Thinking Activities** are collections of thought-provoking questions designed to involve you in scenarios and independent and collaborative research opportunities that focus on current chemistry issues.

- **Virtual Lectures** are examples from the text that are worked out and describe the method to you.

- **myeBook** features all chapters of the text, including the four eChapters, which will be made available for you in the Chemistry Place. In addition to the content in the print version, myeBook will also include links to Self-Assessment Question quizzes and the Virtual Lectures.

PROMOTING GREEN CHEMISTRY

The concept of "Green Chemistry" is incorporated throughout the text by implementing a framework using the Twelve Principles of Green Chemistry written by experts in the field and tying in the Green Chemistry Essays at the end of each chapter.

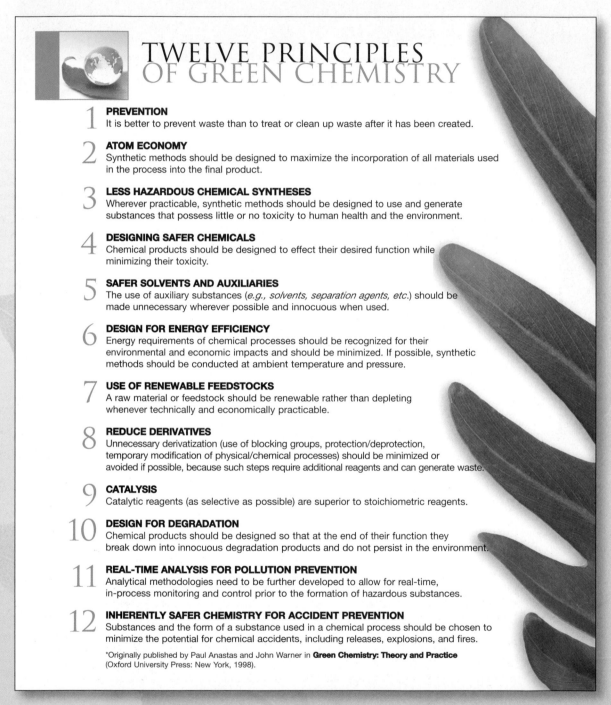

TWELVE PRINCIPLES
OF GREEN CHEMISTRY

1 PREVENTION
It is better to prevent waste than to treat or clean up waste after it has been created.

2 ATOM ECONOMY
Synthetic methods should be designed to maximize the incorporation of all materials used in the process into the final product.

3 LESS HAZARDOUS CHEMICAL SYNTHESES
Wherever practicable, synthetic methods should be designed to use and generate substances that possess little or no toxicity to human health and the environment.

4 DESIGNING SAFER CHEMICALS
Chemical products should be designed to effect their desired function while minimizing their toxicity.

5 SAFER SOLVENTS AND AUXILIARIES
The use of auxiliary substances (*e.g., solvents, separation agents, etc.*) should be made unnecessary wherever possible and innocuous when used.

6 DESIGN FOR ENERGY EFFICIENCY
Energy requirements of chemical processes should be recognized for their environmental and economic impacts and should be minimized. If possible, synthetic methods should be conducted at ambient temperature and pressure.

7 USE OF RENEWABLE FEEDSTOCKS
A raw material or feedstock should be renewable rather than depleting whenever technically and economically practicable.

8 REDUCE DERIVATIVES
Unnecessary derivatization (use of blocking groups, protection/deprotection, temporary modification of physical/chemical processes) should be minimized or avoided if possible, because such steps require additional reagents and can generate waste.

9 CATALYSIS
Catalytic reagents (as selective as possible) are superior to stoichiometric reagents.

10 DESIGN FOR DEGRADATION
Chemical products should be designed so that at the end of their function they break down into innocuous degradation products and do not persist in the environment.

11 REAL-TIME ANALYSIS FOR POLLUTION PREVENTION
Analytical methodologies need to be further developed to allow for real-time, in-process monitoring and control prior to the formation of hazardous substances.

12 INHERENTLY SAFER CHEMISTRY FOR ACCIDENT PREVENTION
Substances and the form of a substance used in a chemical process should be chosen to minimize the potential for chemical accidents, including releases, explosions, and fires.

*Originally published by Paul Anastas and John Warner in **Green Chemistry: Theory and Practice** (Oxford University Press: New York, 1998).

Twelve Principles Page

The Twelve Principles of Green Chemistry page, located at the front of the text, outlines the principles developed by Paul Anastas and John Warner that are used today by many in the chemistry community, including the American Chemical Society and the Environmental Protection Agency.

GREEN CHEMISTRY

Introduction to Green Chemistry

Paul Anastas, Yale University

In seeking a cleaner environment and a more sustainable future, chemists have developed a set of green chemistry principles. These principles are listed below and several are illustrated throughout this text. These principles should help to frame your perspective on the chemistry you learn in this course.

Basically, green chemistry involves the design of products and chemical processes that reduce or eliminate the use or production of hazardous substances. Green chemistry can be applied at all steps in the life cycle of a chemical product, from its design through manufacture and use to its disposal. The many benefits of green chemistry include reduced waste, safer products, and reduced use of energy and resources.

The U.S. Environmental Protection Agency (EPA) promotes green chemistry through Presidential Green Chemistry Challenge Awards, which recognize outstanding green chemistry achievement in academia, industry, and small businesses. The EPA Green Chemistry Program, the ACS Green Chemistry Institute®, and a growing list of organizations, universities, and companies are enabling green chemistry through projects and programs, including educational activities and research and development.

You will see award winning examples and explore many of these concepts in the Green Chemistry essays in future chapters. By using green chemistry, new technologies for a sustainable future are being developed in all kinds of applications, such as energy, biofuels, soap, nanotechnology, medicines, plastics, and much more as you will see in the coming chapters.

As the future leaders and citizens of the world, you will have the power and capability of making a huge impact on the sustainability of the planet and all of its inhabitants. Green chemistry and green technology will be key.

Twelve Principles of Green Chemistry*

1. **Prevention:** It is better to prevent waste than to treat or clean up waste after it has been created.
2. **Atom Economy:** Synthetic methods should be designed to maximize the incorporation of all materials used in the process into the final product.
3. **Less Hazardous Chemical Syntheses:** Wherever practicable, synthetic methods should be designed to use and generate substances that possess little or no toxicity to human health and the environment.

4. **Designing Safer Chemicals:** Chemical products should be designed to effect their desired function while minimizing their toxicity.
5. **Safer Solvents and Auxiliaries:** The use of auxiliary substances (e.g., solvents, separation agents, etc.) should be made unnecessary wherever possible and innocuous when used.
6. **Design for Energy Efficiency:** Energy requirements of chemical processes should be recognized for their environmental and economic impacts and should be minimized. If possible, synthetic methods should be conducted at ambient temperature and pressure.
7. **Use of Renewable Feedstocks:** A raw material or feedstock should be renewable rather than depleting whenever technically and economically practicable.
8. **Reduce Derivatives:** Unnecessary derivatization (use of blocking groups, protection/deprotection, temporary modification of physical/chemical processes) should be minimized or avoided if possible, because such steps require additional reagents and can generate waste.
9. **Catalysis:** Catalytic reagents (as selective as possible) are superior to stoichiometric reagents.
10. **Design for Degradation:** Chemical products should be designed so that at the end of their function they break down into innocuous degradation products and do not persist in the environment.
11. **Real-time Analysis for Pollution Prevention:** Analytical methodologies need to be further developed to allow for real-time, in-process monitoring and control prior to the formation of hazardous substances.
12. **Inherently Safer Chemistry for Accident Prevention:** Substances and the form of a substance used in a chemical process should be chosen to minimize the potential for chemical accidents, including releases, explosions, and fires.

*Originally published by Paul Anas[...] and Practice, New York: Oxford U[...]

Recycling

Because it is an element, iron cannot be created or destroyed, but that does not mean it will always be found in its elemental state. Let's consider the following two different pathways for the recycling of iron.

1. Iron ore (hematite) is mined from the ground and converted into pig iron, which is used to make steel cans. Once discarded, the cans rust. The rust slowly leaches into groundwater as iron ions (ions are explained in Chapter 3) and eventually winds up in the ocean. At this point, the original iron atoms have become dissolved ions that can be absorbed and incorporated by marine plants. Marine creatures that eat plants will in turn absorb and incorporate iron. The original iron atoms are now widely separated in space. Perhaps a few of them might even become part of the hemoglobin in your own bloodstream. The iron has been recycled, but it will never again resemble the original pig iron.

2. Iron ore is mined from the ground and converted to pig iron and then into steel. The steel is used in making an automobile, which is driven for a decade and then sent to the junkyard. The junkyard compresses the automobile and sends it to a recycling plant, where the steel is recovered and ultimately used again in a new automobile. Once the iron was removed from its ore, it has been conserved in its elemental metallic form. In this form of recycling the iron continues to be useful for a long time.

HOW CHEMISTRY APPLIES TO EVERYDAY LIFE

Chemistry for Changing Times presents basic chemical concepts with abundant everyday applications, personalizing the chemistry experience. In this way, the text allows you to focus on evaluating information about real-life issues instead of having to memorize rigorous theory and mathematics. These features help you apply information and learning from the chapter and make real-life connections with chemistry.

QUESTIONS YOU MAY HAVE ASKED YOURSELF

1. If we can't see atoms, how can we discuss even smaller particles?
2. What is radioactivity? Is it contagious?
3. What causes the colors of fireworks?
4. Most lightbulbs produce white light. Why does a neon bulb give off colored light?
5. I see the word *quantum* used; what does it mean?
6. What is "periodic" about the periodic table?

Questions and Answers
Each chapter starts with a set of **Questions You May Have Asked Yourself.** These real-life questions may have arisen in an inquiring student's mind and are related to the chapter content. Answers to those questions are found in the margins of the chapter near the text content associated with the question.

Conceptual Problem-Solving Examples
These examples guide you through the process of learning and understanding important chemical concepts.

CONCEPTUAL EXAMPLE 1.2   Mass and Weight

On the planet Mercury, gravity is 0.376 times that on Earth.

a. What would be the mass on Mercury of a person who has a mass of 62.5 kilograms (kg) on Earth?
b. What would be the weight on Mercury of a person who weighs 124 pounds (lb) on Earth?

Solution

a. The person's mass would be the same (62.5 kg) as on Earth; the quantity of matter has not changed.
b. The person would weigh only $0.376 \times 124 \ lb = 46.6 \ lb$; the force of attraction between planet and person is only 0.376 times that on Earth.

■ EXERCISE 1.2A

At the surface of Venus, the force of gravity is 0.903 times that on Earth's surface.

a. What would be the mass of a standard 1.00-kg object on Venus?
b. A man who weighs 198 lb on Earth would weigh how much on the surface of Venus?

■ EXERCISE 1.2B

On the (liquid) surface of Jupiter, the force of gravity is 2.34 times that on Earth.

a. What would be the mass of a 52.5-kg woman at that location on Jupiter?
b. A man who weighs 212 lb on Earth would weigh how much on Jupiter?

■ COLLABORATIVE GROUP PROJECTS

Prepare a PowerPoint, poster, or other presentation (as directed by your instructor) for presentation to the class.

87. List five chemical activities you have engaged in today. Compare your list with those of the other members of your group.
88. Prepare a brief biographical report on one of the following and share it with your group.
    a. Francis Bacon   b. Rachel Carson
    c. Thomas Malthus   d. George Washington Carver

(The following problem is best done in a group of four students.)

89. Make copies of the following form. Student 1 should write a word from the list in the first column of the form and its definition in the second column. Then she or he should fold the first column under to hide the word and pass the sheet to Student 2, who uses the definition to determine what word was defined and place that word in the third column. Student 2 then folds the second column under to hide it and passes the sheet to Student 3, who writes a definition for the word in the third column and then folds the third column under and passes the form to Student 4. Finally, Student 4 writes the word corresponding to the definition given by Student 3.

Compare the word in the last column with that in the first column. Discuss any differences in the two definitions. If the word in the last column differs from that in the first column, determine what went wrong in the process.
    a. hypothesis   b. theory   c. mixture   d. substance

Text Entry	Student 1	Student 2	Student 3	Student 4
Word	Definition	Word	Definition	Word
___	___	___	___	___
___	___	___	___	___

Collaborative Group Projects
These projects are end-of-chapter exercises that help professors engage you in collaborative work with group assignments.

Self-Assessment Questions

1. When Roentgen took an X-ray photograph of a hand, the bones showed up slightly dark because the X-rays were
   a. absorbed by bone more so than by flesh
   b. light that is less energetic than visible light
   c. made up of fast-moving atomic nuclei
   d. a type of radioactivity

2. Radioactivity arises from elements that
   a. absorb radiofrequency energy and release stronger radiation
   b. are unstable and release radiation
   c. give off X-rays
   d. undergo fission when they absorb energy

Answers: 1, a; 2, b

NEW! Self-Assessment Questions
These questions have been added to the ends of each section within the chapters to allow you to track your understanding of chapter material.

■ SUMMARY

Section 6.1—When a substance is melted or vaporized, the forces that hold the particles (molecules, atoms, or ions) close to one another are overcome. Several properties are related to the strengths of those forces. The **melting point** of a solid is the temperature at which the forces holding the particles together in a regular arrangement are overcome. The reverse of the melting process, a liquid changing to a solid, is called **freezing**. **Vaporization** is the conversion of a liquid to a gas; the reverse process is called **condensation**. The temperature at which a substance vaporizes at ordinary pressure is called its **boiling point**. Some substances undergo **sublimation**, a conversion directly from the solid to the gas state.

Section 6.2— Ionic solids have very strong forces holding the ions to one another, so most ionic solids have high melting points and high boiling points compared to molecular compounds.

Section 6.3—Not all solids are held together by ionic bonds. **Dipole–dipole forces** involve the attraction of the positive end of one dipole for the negative end of another dipole. Nonpolar molecules also have forces between them. These **dispersion forces** result from electron motion of neighboring atoms, which causes tiny, short-lived dipoles. Dispersion forces are generally weak. A special, strong type of dipole force occurs when the positive end is a hydrogen atom attached to F, O, or N, and the negative end is a pair of electrons on an atom of F, O, or N. Such a force is called a **hydrogen bond**.

Section 6.4—A **solution** is a **homogeneous** mixture, a mixture in which all parts have the same composition. The substance

that is dissolved is called the **solute**, and the dissolving substance is called the **solvent**. Nonpolar solutes dissolve in nonpolar solvents. Polar substances and many ionic substances dissolve in polar solvents.

Section 6.5—The behavior of gases is explained using **kinetic–molecular theory**, which describes gas molecules: (1) They move rapidly, constantly, and in straight lines; (2) they are far apart; (3) there is little attraction between them; (4) when they collide, energy is conserved; and (5) temperature is a measure of average kinetic energy of the gas molecules.

Section 6.6—**Boyle's law** says that the volume of a fixed amount of gas varies inversely with pressure at constant temperature, or $P_1V_1 = P_2V_2$. **Charles's law** says that the volume of a fixed amount of gas varies directly with absolute (Kelvin) temperature, or $V_1/T_1 = V_2/T_2$. The **molar volume** of a gas is the volume occupied by 1 mol of any gas. At 0 °C and 1 atm, known as **standard temperature and pressure (STP)**, the molar volume is 22.4 L for any gas. We can use that molar volume to convert from moles of gas to volume and vice versa.

Section 6.7—The simple gas laws can be formed into a single relationship called the **combined gas law**. The **ideal gas law** shows the relationship among pressure (P), volume (V), number of moles (n), and absolute temperature (T): $PV = nRT$. When P = atmospheres, V = liters, n = moles of gas, and T = kelvins, the value of the constant R is 0.0821 liter · atm/ (mol · K).

Chapter Summaries
These summaries are organized by sections, with key terms highlighted to remind you of important concepts.

Chemistry is everywhere, not just in a laboratory.
Chemistry occurs in soil and rocks, in waters, in clouds, and in us.

CHEMISTRY

1

QUESTIONS YOU MAY HAVE ASKED YOURSELF

1. Why do people suggest that chemicals are bad for us?
2. Why should I study chemistry?
3. Why do scientists so often say "more study is needed"?
4. Why do scientists bother with studies that have no immediate application?
5. Can we change lead into gold?

A Science for All Seasons

Try a word-association game with schoolmates or friends. Say the word *chemicals* and ask them for the first single word that they associate with it. Play the game often enough and you'll get responses like "harmful," "toxic," and "risky." The word *chemical* is commonly associated with artificial or processed substances, such as the products of the chemical industry. To a chemist, how-ever, *chemical* is a somewhat technical term referring to *any material with a fixed, specific composition*—a chemical element or compound (more about these later). Look around you. In reality, everything you see is made of chemi-cals: the food we eat, the air we breathe, the clothes we wear, the medicines we take, the vehicles we ride in, and the buildings we live and work in.

▲ This woman proudly asserts, "I don't use chemicals in *my* garden." Is this really true?

Everything we *do* also involves chemistry. Whenever we eat a sandwich, take a shower, drive a car, listen to music, or ride a bicycle, we use chemistry. Even when we are asleep, chemical reactions go on constantly throughout our bodies.

Most developments in health and medicine involve a lot of chem-istry. The astounding advances in biotechnology—decoding the human genome, developing new drugs, improving nutrition, and much more—have a huge chemical component. Understanding environmental prob-lems requires knowledge of chemistry. The great worldwide issues of global warming and ozone depletion involve chemistry.

The media often carry reports about toxic chemicals and cancer-causing chemicals polluting our air, water, and food. Half a century ago the media rarely mentioned such health risks. Rather, they emphasized the wonders of chemical technology: miracle drugs, amazing plastics, fertilizers, detergents, and more. Has chemistry changed over the years or just our perception of it?

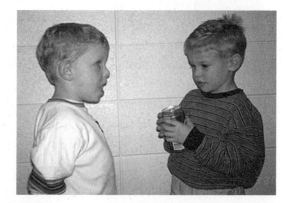

▲ "Mom won't let me drink that stuff. It has chemicals in it."

A N S W E R

1. Why do people suggest that chemicals are bad for us?
Everything you can see, smell, taste, or touch is either a chemical or is made of chemicals. Chemicals are neither good nor bad in themselves; they can be put to good use, bad use, or anything in between.

A N S W E R

2. Why should I study chemistry?
Chemistry is a part of many areas of study. Knowledge of chemistry helps you to understand many facets of modern life.

▲ Aristotle (384–322 B.C.E.), Greek philosopher and tutor of Alexander the Great, believed that we could understand nature through logic. The idea of experimental science did not triumph over Aristotelian logic until about 1500 C.E.

Certain chemicals do indeed cause problems, but many others are extremely helpful. Some chemicals kill bacteria that cause dreadful diseases; some relieve our pain and suffering; some increase our food production; some provide fuel for our heating, cooling, lighting, and transportation; and some provide materials for building our machines, making our clothing, and constructing our houses. Chemistry has provided ordinary people with luxuries that were not available even to the mightiest of kings in ages past. Chemicals are highly important to our lives—so much so that life itself would be impossible without chemicals.

So what is chemistry anyway? The usual definition is that **chemistry** is the study of matter and the changes it undergoes. *Matter* is anything that has *mass*, a property that reflects the quantity of matter in a sample. This means that if you can weigh it, it is matter. Even though you cannot see air, it has mass; therefore, air is matter. Matter is the stuff of which all material things are made. The science of chemistry deals with every kind of matter, from the tiniest parts of atoms to the most complex materials in living plants and animals. Thus, the word *chemical* may sound ominous, but it is nothing more than a name for matter—any and all matter. Gold, water, salt, sugar, coffee, ice cream, a computer, a pencil—all are chemicals or are made entirely of chemicals.

Then there are those *changes* that matter undergoes. Sometimes matter changes on its own—a tool rusts when left out in the wet grass. Often we change matter intentionally to make it more useful, as when we light a candle or cook an egg. Most changes in matter are accompanied by changes in energy. For example, when we burn gasoline, the reaction gives off energy that we can use to propel an automobile.

Your own body is an incredibly marvelous chemical factory. It takes the food you eat and turns it into skin, bones, blood, and muscle, while also generating energy for all your activities. This amazing chemical plant operates continuously 24 hours a day for as long as you live. Chemistry not only affects your own life every moment, but it also transforms society as a whole. Chemistry shapes our civilization.

1.1 Science and Technology: The Roots of Knowledge

Chemistry is a *science,* but what is science? Let's examine the roots of science. Our study of the material universe has two facets: the *technological* (or *factual*) and the *philosophical* (or *theoretical*).

Technology, the application of knowledge for practical purposes, arose long before science and originated in prehistoric times. Early people used fire to bring about chemical changes. For example, they cooked food, baked pottery, and smelted ores to produce metals such as copper. They made beer and wine by fermentation and obtained dyes and drugs from plant materials. These things—and many others—were accomplished without an understanding of the scientific principles involved.

The Greek philosophers, about 2500 years ago, were perhaps the first to formulate rational *theories* of chemistry—explanations of the behavior of matter. They generally did not test their theories by experimentation, however. Nevertheless, their view of nature—attributed mainly to Aristotle—dominated Western thinking about the workings of the material world for 2000 years.

The experimental roots of chemistry are planted in **alchemy,** a mystical mixture of chemistry and magic that flourished in Europe during the Middle Ages, from about 500 to 1500 C.E. Alchemists searched for a "philosopher's stone" that would turn cheaper metals into gold, and they sought an elixir that would confer immortality on those exposed to it. Alchemists never achieved these goals, but they discovered many new chemical substances and perfected techniques such as distillation and extraction that are still used today. Modern chemists inherited from the alchemists an abiding interest in human health and the quality of life.

Technology also developed rapidly during the Middle Ages in Europe even though it was not aided by the Aristotelian philosophy that prevailed. The beginnings of modern science were more recent, however, and coincided with the emergence of the experimental method. What we now call science grew out of **natural philosophy**—that is, out of philosophical speculation about nature. Science had its true beginnings in the seventeenth century, when astronomers, physicists, and physiologists began to rely on experimentation.

▲ Alchemists made many positive contributions to chemistry and life. Here we see a distillation apparatus from the 1512 edition of Hieronymus Brunschwig's *Liber de arte distillandi* (On the art of distilling). *(Source: Donald and Mildred Othmer Collection)*

Modern Alchemy

Modern chemists can transform matter in ways that would astound the alchemists. They can change ordinary salt into lye and laundry bleach. They can convert sand into transistors and computer chips. And they can turn crude oil into plastics, fibers, pesticides, drugs, detergents, and a host of other products. Many products of modern chemistry are much more valuable than the glittering metal that the alchemists vainly sought. For example, a relatively new form of carbon, called *nanotubes* because of their size and shape, has thousands of potential uses, but nanotubes are extremely expensive to make. A pound of purified nanotubes costs more than $50,000, whereas a pound of gold will only set you back about $13,000.

1.2 The Baconian Dream and the Carsonian Nightmare

Francis Bacon was a philosopher who practiced law and served as a judge. Although not a scientist, he was interested in science and argued that it should be experimental. His dream was that science could solve the world's problems and enrich human life with new inventions, thereby increasing happiness and prosperity.

By the middle of the twentieth century, science and technology appeared to have made the Baconian dream come true. Many deadly diseases such as smallpox, polio, and plague had been almost eliminated. The use of fertilizers, pesticides, and scientific animal and plant breeding had increased and enriched our food supply. New materials provided us better clothing and shelter, swifter transportation, and nearly instantaneous communication. Nuclear energy seemed to promise unlimited power for our every need. Science and technology had done much toward creating our "modern" world.

▲ Francis Bacon (1561–1626), English philosopher and Lord Chancellor to King James I.

▲ Rachel Carson (1907–1964) at Woods Hole, Massachusetts, in 1951.

The Baconian dream has lost much of its luster in recent decades. People have learned that the products of science are not an unmitigated good. Some have predicted that science might bring not wealth and happiness, but death and destruction.

Rachel Carson, a biologist, was among the more noteworthy critics of modern technology. Her poetic and polemic book *Silent Spring* was published in 1962. The book's main theme is that through our use of chemicals to control insects, we are threatening the destruction of all life, including ourselves. People in the pesticide industry and their allies roundly denounced Carson as a "propagandist," while other scientists rallied to her support. By the late 1960s, however, massive fish kills, the threatened extinction of several species of birds, and the disappearance of fish from rivers, lakes, and areas of the ocean that had long been productive caused many scientists to move into Carson's camp. Popular support for Carson's views was overwhelming.

Carson was not the first prophet of doom. As early as 1798 in his "Essay upon the Principles of Population," Thomas Malthus predicted that the rapid increase in world population would outpace the increase in the food supply and result in widespread famine. During the nineteenth century and much of the twentieth century, however, technological developments enabled food production to keep up with population growth, at least in developed countries.

Today the picture has changed. In spite of declining birth rates in the developed countries, population growth in much of the rest of the world still threatens to overtake even the most optimistic projections of food production. Some scientists project a dismal future for our world; others confidently predict that science and technology, properly applied, will save us from disaster.

1.3 Science: Reproducible, Testable, Tentative, Predictive, and Explanatory

What *is* science? If scientists disagree about what is and what will be, is science merely a guessing game in which one guess is as good as another? Science is difficult to define precisely, but we will try to describe it.

Scientific Hypotheses

Science is an accumulation of knowledge about nature and our physical world and the theories that we use to explain that knowledge. Science is based on observations and experimental tests of our assumptions. Science is not simply a collection of unalterable facts. We cannot force nature to fit our preconceived ideas. Science is good at correcting errors, but it is not especially good at establishing truths. Science is an unfinished work. The things we have learned from science fill millions of books and scholarly journals, but what we know pales in comparison to what we have yet to discover.

Scientists collect data by making careful observations. Data reported by a scientist must also be observable by other scientists; that is, the data must be *reproducible*. Careful observations and measurements are essential, but scientific work is not fully accepted until it has been verified by other scientists.

Scientists develop *testable* **hypotheses** (educated guesses) as tentative explanations of observed data and test these hypotheses by designing and performing experiments. This is the main thing that distinguishes science from the arts and humanities: The tenets of science are *testable*. In the humanities, people often still argue about some of the same questions that were being debated thousands of years ago: What is truth? What is beauty? These arguments persist because the answers cannot be tested and confirmed objectively. However, experiments can be devised to answer most scientific questions. Ideas can be tested and thereby either verified or rejected. For example, it was long thought that exercise caused muscles to tire and become

What Science Is Not

News media generally try to be fair, presenting both sides of an issue. Science is not fair in that sense; not all ideas are considered to be equal. The idea of a flat Earth is not equal to the view that Earth is (roughly) spherical. Only ideas that have survived experimental testing or that can be tested by experiment are considered valid. Ideas that are beautiful, elegant, or even sacrosanct can be destroyed by experimental data.

Science is not a democratic process. Majority rule does not determine what is sound science. Science does not accept notions that are proven false by experiment.

sore from a buildup of lactic acid. Recent findings suggest instead that lactic acid actually *delays* muscle tiredness and that the cause of tired, sore muscles may be related to other factors, including leakage of calcium ions inside the muscle cells that weakens contractions. With many experiments, scientists have established a firm foundation of knowledge so that each new generation can build on the past.

Scientific Laws

Large amounts of scientific data can sometimes be summarized in brief statements called **scientific laws**. For example, Robert Boyle (1627–1691), an Englishman, conducted many experiments on gases. In each experiment, he found the volume of the gas to decrease when the pressure applied to the gas was increased. Many scientific laws can be stated mathematically. For example, Boyle's law can be written as $PV = k$, where P is the pressure on a gas, V is its volume, and k is a constant number. If P is doubled, V will be cut in half. Scientific laws are *universal*; under the stated conditions they hold everywhere in the observable universe.

Experimental observations are just the beginning of the intellectual processes of science. There are many different paths to scientific discovery, and there is no general set of rules. Science is not just a straightforward logical process for cranking out discoveries.

Scientific Theories

Science is a body of knowledge. Scientists organize this knowledge on a framework of detailed explanations called **theories**. The theories represent the best current explanations for various phenomena, but they are always *tentative*. In the future, a theory may have to be modified or even discarded in the light of new observations. Science is a body of knowledge that is rapidly growing and always changing.

Theories provide organization for scientific knowledge, but they also are useful for their *predictive* value. Predictions based on theories are then tested by further experiments. Theories that make successful predictions generally are widely accepted by the scientific community. A theory developed in one area is often found to apply in others.

Scientific Models

Scientists often use models to help *explain* complicated phenomena. A **scientific model** uses tangible items or pictures to represent invisible processes. For example, the invisible particles of a gas can be visualized as billiard balls or marbles, or as dots or circles on paper. We know that when a glass of water is left standing for a period of time, the water disappears through a process called evaporation (Figure 1.1). Scientists explain evaporation with a theory, specifically the kinetic–molecular theory, based on tiny, invisible particles called *molecules,* which are in constant motion. The kinetic–molecular model pictures the liquid water as molecules.

3. Why do scientists so often say "more study is needed"?
More data help the scientist refine a hypothesis so that it is better defined, clearer, or more applicable.

ANSWER

A Possible Scientific Process

Observation reported
↓
Observation confirmed by others
↓
Hypothesis suggested
↓
Experiments designed to test hypothesis
↓
Experiments do not support hypothesis
↓
Hypothesis rejected
↓
New hypothesis offered
↓
New experiments tried
↓
Experiments support (new) hypothesis
↓
Experiments repeated and results confirmed
↓
Theory formulated
↓
Many further experiments
↓
and so on

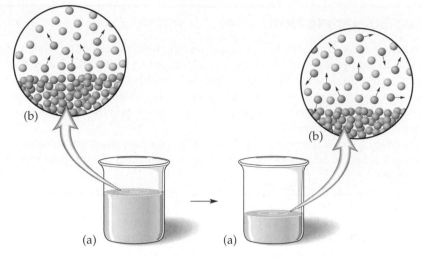

▶ **Figure 1.1** The evaporation of water. (a) When a container of water is left standing open to the air, the water slowly disappears. (b) Scientists explain evaporation in terms of the motion of molecules.

● = Water molecule

● = Air (nitrogen or oxygen) molecule

In the bulk of the liquid, these molecules are held together by forces of attraction. The molecules collide with one another like billiard balls on a playing table. Sometimes a "hard break" of billiard balls causes one ball to fly off the table. Likewise, some of the molecules of a liquid gain enough energy through collisions to break the attraction to their neighbors, escape from the liquid, and disperse among the widely spaced air molecules. The water in the glass gradually disappears. It is much more rewarding to understand evaporation through the use of a model than merely to have a name for it.

When developing theories and constructing models, it is important to note that a *correlation* between two items isn't necessarily evidence that one *causes* the other. For example, many people have allergies in the fall when goldenrod is in bloom. However, research has shown that the main cause of these allergies is ragweed pollen. There is a correlation between the goldenrod blooming and autumnal allergies, but goldenrod pollen is not the cause. Ragweed happens to bloom at the same time.

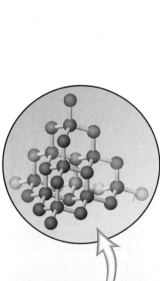

▲ A molecular model of diamond shows the tightly linked, rigid structure that explains why diamonds are so hard.

Molecular Modeling

A *molecule* is a group of two or more atoms in a definite arrangement held together by forces called chemical bonds (Chapter 4). Molecular models are three-dimensional representations of molecules. They can be put together using simple balls to represent atoms and sticks to connect the balls. The models can also be made with various space-filling model kits, or they can be computer generated. Using a computer, one can draw various kinds of molecules and then rotate them on the screen to get a three-dimensional perspective.

Some molecular modeling programs can compute the structures of complicated molecules, even indicating the distribution of electrons. These computational programs can be used to look at individual molecules or to study their interactions with each other. Computer modeling is especially useful in the study of complex biological molecules and drug molecules.

Self-Assessment Questions

Select the best answer or response.

1. The study of matter and the changes it undergoes is called
   a. alchemy   **b.** chemistry
   c. natural philosophy   **d.** technology

2. The ancient Greek philosophers—for example, Aristotle—were not successful as scientists because they
   a. failed to formulate viable theories
   b. failed to test their ideas with experiments
   c. lacked modern instrumentation
   d. lacked the ability to formulate scientific laws

3. Who first argued that science should be experimental rather than just speculative?
   a. Aristotle   b. Francis Bacon
   c. Rachel Carson   d. Robert Boyle

4. The main theme of Rachel Carson's *Silent Spring* was that life on Earth would be destroyed by
   a. botulism   b. nuclear bombs
   c. overpopulation   d. pesticides

5. To gather information to support or discredit a hypothesis, a scientist
   a. conducts experiments   b. consults an authority
   c. consults a psychic   d. forms a theory

6. The declaration that mass is always conserved when chemical changes occur is an example of a chemical
   a. experiment   b. hypothesis
   c. law   d. theory

7. A successful theory
   a. can be used to predict behavior under different conditions
   b. is not subject to further testing
   c. is permanently accepted as true
   d. becomes a scientific law

8. Which of the following is *not* a hypothesis?
   a. All solutions of dubnium are blue.
   b. Cesium can be produced from cesium chloride by electrolysis.
   c. A psychic can locate a corpse when conditions are right.
   d. Water reacts with dubnium to form hydrogen gas.

9. A possible explanation for a collection of observations is called a(n)
   a. experiment   b. law
   c. observation   d. theory

Answers: 1, b; 2, b; 3, b; 4, d; 5, a; 6, c; 7, a; 8, c; 9, d

1.4 The Limitations of Science

We sometimes hear people say that we could solve all our problems if we would only attack them using the scientific method. We have seen already that there is no single scientific method, but why can't the procedures of the scientist be applied to social, political, ethical, and economic problems? Why do scientists disagree over environmental, social, and political issues?

Often the disagreement results from the inability to control *variables*. A **variable** is something that can change over the course of an experiment. If, for example, we wanted to study in the laboratory how the volume of a gas varies with changes in pressure, we would hold constant such factors as temperature and the amount and kind of gas. If, on the other hand, we wanted to determine the health effect of low levels of a particular pollutant on a human population, we would find it difficult, if not impossible, to control such variables as the people's diets, habits, and exposure to other substances.

We can make observations, formulate hypotheses, and conduct experiments, even if most of the variables are not subject to control. Interpretation of the results, however, is much more difficult and more subject to disagreement. Nonscientists sometimes use some of the methods and language of scientists. Artists *experiment*

More to Explore
Angier, Natalie. *The Canon: A Whirligig Tour of the Beautiful Basics of Science*. New York: Houghton Mifflin, 2007.

with new techniques and new materials. Playwrights and novelists *observe* life as it is before trying to express its essence in their writings. Scientific ideas and methods pervade almost every aspect of society.

1.5 Science and Technology: Risks and Benefits

Most people recognize that society has benefited from science and technology, but there are risks associated with every technological advance. How can we determine when the benefits outweigh the risks? One approach, called **risk–benefit analysis,** involves the estimation of a desirability quotient (DQ).

$$DQ = \frac{Benefits}{Risks}$$

A *benefit* is anything that promotes well-being or has a positive effect. Benefits may be economic, social, or psychological. A *risk* is any hazard that leads to loss or injury. Some of the risks in modern technology have led to disease, death, economic loss, and environmental deterioration. Risks may involve one individual, a group, or society as a whole.

Most people recognize the automobile as a technological development that benefits us greatly. But driving a car involves risk—individual risks of injury or death in a traffic accident and risks to society such as pollution and global warming. Most people consider the benefits of driving a car to outweigh the risks.

Weighing the benefits and risks connected with a product is more difficult when one considers a group of people. For example, pasteurized milk is a safe, clean, nutritious beverage for many people of northern European descent. However, a few people in this group can't tolerate lactose, the sugar in milk. And some are allergic to milk proteins. For these groups of people, drinking milk can be harmful. Because milk allergies and lactose intolerance are relatively uncommon among people of northern European descent, the benefits are large and the risks are small, resulting in a large DQ. Skim or low-fat milk is generally beneficial for this group. However, adults in much of the rest of the world are lactose intolerant and would find that milk has a small DQ. Thus, milk is not always suitable for use in programs to relieve malnutrition.

In 1958, the drug thalidomide was introduced in Germany. Many pregnant women took the drug to prevent morning sickness. But soon thalidomide was shown to involve an enormous risk. Many malformed infants were born to women who had taken the drug during pregnancy. For thalidomide, then, there are small benefits, huge risks, and a very small DQ. Thalidomide was eventually judged to present unacceptable risks and was banned. However, there is now a renewed interest in thalidomide. Its use in treating a debilitating skin condition of leprosy was approved in September 1998. Current research is evaluating the use of thalidomide in treating Kaposi's sarcoma (a complication of AIDS), as well as for graft-versus-host reactions in organ transplants and autoimmune diseases, for ulcers in AIDS, and for tuberculosis.

The artificial sweetener aspartame is the most highly studied of all food additives. There is little evidence that use of an artificial sweetener helps one to lose weight. (Aspartame may provide some benefit to diabetics, however, because sugar consumption presents a large risk to them.) There are anecdotal reports of problems with the sweetener, but these have not been confirmed in controlled studies. To most people, the risk involved in using aspartame is small. This leads to an uncertain DQ—and, it seems, to endless debate over the safety of aspartame.

Other technologies provide large benefits and present large risks. For these technologies, too, the DQ is uncertain. An example is the conversion of coal to liquid fuels. Most people find that liquid fuels provide large benefits in the areas of

▲ **It DOES Matter!**

For most people of northern European ancestry, milk is a wholesome food; its benefits far outweigh its risks. Other people have high rates of lactose intolerance among adults, and the desirability quotient for milk is much smaller.

transportation, home heating, and industry. The risks associated with coal conversion are also large, however. These include the risks to mine workers, air and water pollution, and the exposure of conversion plant workers to toxic chemicals. The result, again, is an uncertain DQ and political controversy.

There are yet other problems in risk–benefit analysis. Some technologies benefit one group of people while presenting a risk to another. For example, it may be economically advantageous to a community to spend as little as possible on a sewage treatment plant and to dump raw wastes into a nearby stream. These wastes might present a hazard to downstream communities, however. Difficult political decisions are needed in such cases.

Other technologies provide current benefits but present future risks. For example, although nuclear power now provides useful electricity, wastes from nuclear power plants, if improperly stored, might present hazards for centuries. Thus, the use of nuclear power is controversial.

Science and technology obviously involve *both* risks and benefits. The determination of benefits is almost entirely a social judgment; risk assessment also involves social decisions, but scientific investigation can help considerably in risk evaluation.

According to the National Safety Council, the chemical industry ranks at or near the top each year in worker safety among 42 basic industries. The number of incidents of occupational illness and injury in the chemical industry is only one-third of the average rate for all U.S. industries. This is in sharp contrast to the popular belief that chemicals are so dangerous. (Some are, of course, but even they can be used safely with proper precautions.)

CONCEPTUAL EXAMPLE 1.1   Risk–Benefit Analysis

Heroin is thought by some to be more effective for the relief of severe pain than other medicines. For example, heroin is a legal prescription drug in the United Kingdom. Heroin can become addictive with continued use in as little as three days, and recreational use often renders the addict unable to function in society. Do a risk–benefit analysis for the use of heroin in treating the pain of

a. a young athlete's broken leg and
b. a terminally ill cancer patient.

Solution

a. The heroin would provide the benefit of pain relief, but its use for such purposes has been judged to be too risky by the U.S. Food and Drug Administration. The DQ is low.
b. The heroin would provide the benefit of pain relief. The risk of addiction in a dying person is irrelevant. Heroin is banned for any purpose in the United States. The DQ is uncertain. (Both answers involve judgments that are not clearly scientific; people can differ in their assessments of each.)

■ EXERCISE 1.1A
Chloramphenicol is a powerful antibacterial drug that often destroys bacteria unaffected by other drugs. It is highly dangerous to some individuals, however, causing fatal aplastic anemia in about 1 in 30,000 people. Do a risk–benefit analysis for the use of chloramphenicol in

a. sick farm animals, from which people might consume milk or meat with residues of the drug, and
b. a person with Rocky Mountain spotted fever faced with a high probability of death or permanent disability.

■ EXERCISE 1.1B
A four-year clinical trial (6800 participants) of the drug tamoxifen in healthy women at high risk showed a 49% lower rate of breast cancer than a group of 6800 women taking a placebo. Tamoxifen has a low rate of serious side effects, including potentially fatal blood clots, uterine cancer, hot flashes, loss of libido, and cataracts. Do a risk–benefit analysis for the use of tamoxifen in

a. all women and   b. women at high risk.

Risks of Death

Our perception of risk often differs from the real risks that we face. Some people fear flying but readily assume the risk of an automobile trip. The odds of dying from various causes are listed in the following table.

Approximate Lifetime Risks of Death in the United States

Action	Lifetime Risk[a]	Approximate Lifetime Odds	Details/Assumptions
All causes	1	1 in 1	We all die of something.
Cigarettes	0.25	1 in 4	Cigarette smoking, 1 pack/day
Heart disease	0.20	1 in 5	Heart attacks, congestive heart failure
All cancers	0.14	1 in 7	All cancers
Motor vehicles	0.011	1 in 88	Death in motor vehicle accident
Home accidents	0.010	1 in 100	Home accident death
Radon	0.0030	1 in 300	Radon in homes, cancer deaths
Background radiation	0.0010	1 in 1000	Sea level background radiation, cancers
Peanut butter (aflatoxin)	0.00060	1 in 1700	4 tablespoons peanut butter per day
Airplane accidents	0.0002	1 in 5000	Death in aircraft crashes
Terrorist attack[b]	0.00077	1 in 1300	One 9/11-level attack per year
Terrorist attack[c]	0.000077	1 in 13,000	One 9/11-level attack per 10 years

[a] The odds of dying of a particular cause in a given year are calculated by dividing the population by the number of deaths by that cause in that year. Lifetime odds of dying of a specific cause are calculated by dividing the one-year odds by the life expectancy of a person born in that year.
[b] Unlikely scenario
[c] More reasonable scenario

More to Explore
For more on risks of dying, see "Ways to Go." *National Geographic*, August 2006, p. 21.

1.6 Chemistry: Its Central Role

Science is a unified whole. Common scientific laws apply everywhere and on all levels of organization. The various areas of science interact and support one another. Accordingly, chemistry not only is useful in itself but is also fundamental to other scientific disciplines. The application of chemical principles has revolutionized biology and medicine, has provided materials for powerful computers used in mathematics, and has profoundly influenced other fields such as psychology. The social goals of better health and more and better food, housing, and clothing are dependent to a large extent on the knowledge and techniques of chemists. Many modern materials have been developed by chemists and are used by engineers, and even more amazing materials are on the way. Recycling of basic materials—paper, glass, and metals—involves chemical processes. Chemistry is indeed a central science (Figure 1.2). There is scarcely a single area of our daily lives that is not affected by chemistry.

Chemistry is also important to the *economy* of industrial nations. In the United States, chemical process industries make thousands of consumer products, including personal care products, agricultural products, plastics, coatings, soaps, and detergents. It produces 80% of the materials used to make medicines. The U.S. chemical industry is one of the country's largest industries, with sales of more than $600 billion per year, and it accounts for 10% of all U.S. exports. It employs 80,000 scientists, engineers and technicians, and it generates nearly 11% of all U.S. patents. And despite widespread fear of chemicals, the chemical industry is five times safer for its workers than the average of the U.S. manufacturing sector.

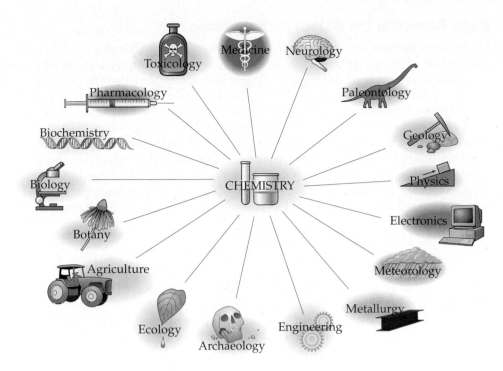

▶ **Figure 1.2** Chemistry has a central role among the sciences.

1.7 Solving Society's Problems: Scientific Research

Chemistry is a powerful force in shaping society today. Chemical research not only plays a pivotal role in other sciences, but it also has a profound influence on society as a whole. Chemists, like other scientists, do research in one of two categories, *applied research* or *basic research*. The two often overlap, and it isn't always possible to label a particular project as one or the other.

Applied Research

Most chemists work in the area of applied research. They test polluted soil, air, and water. They analyze foods, fuels, cosmetics, detergents, and drugs. They synthesize new substances for use as drugs or pesticides and formulate plastics for new applications. These activities are examples of **applied research**—work oriented toward the solution of a particular problem in industry or the environment.

Among the most monumental accomplishments in the realm of applied research were those of George Washington Carver. Born in slavery, Carver attended Simpson College and later graduated from Iowa State University. A botanist and agricultural chemist, Carver taught and did research at Tuskegee Institute. He developed over 300 products from peanuts, ranging from peanut butter to hand cleaner to insulating board. He made other new products from sweet potatoes, pecans, and clay. Carver also taught southern farmers to rotate crops and to use legumes to replenish the nitrogen removed from the soil by cotton crops. Carver's work helped to revitalize the economy of the South.

Another example of applied research is the synthesis of an antifungal drug. Many drugs are effective against bacteria, but few are useful and safe against fungal infections in humans. Seeking an effective fungal antibiotic, Rachel Brown (1898–1980) and Elizabeth Hazen (1885–1975) of the New York State Department of Health discovered nystatin, a compound effective against such fungal infections as thrush and vaginitis. Nystatin also has been used in veterinary medicine, in agriculture (to protect fruit), and in the art world (to prevent the growth of mold on art treasures in Florence, Italy, after the objects were damaged by floods). Brown and Hazen set out to find a fungicide and did so—an excellent example of applied research.

▲ George Washington Carver (1864–1943), research scientist, in his laboratory at Tuskegee Institute.

More to Explore
"What We Don't Know." *Science*, 1 July 2005, pp. 75–102. A special section on the scientific puzzles that drive research.

A N S W E R

4. Why do scientists bother with studies that have no immediate application? Basic research may not have a use now, but because it extends our understanding of the world, it can be considered an investment in the future.

Basic Research: The Search for Knowledge

Many chemists are involved in **basic research**, the search for knowledge for its own sake. Some chemists work out the fine points of atomic and molecular structure. Others measure the intricate energy changes that accompany complex chemical reactions. Chemists synthesize new compounds and determine their properties. These types of investigations are called basic research. Done for the sheer joy of unraveling the secrets of nature and discovering order in our universe, basic research is characterized by the absence of any predictable, marketable product.

This doesn't mean that basic research is useless—far from it! Findings from basic research often *are* applied at some point. This may be the hope but is not the primary goal of the researcher. In fact, most of our modern technology is based on results obtained in basic research. Without this base of factual information, technological innovation would be haphazard and slow.

Applied research is carried out mainly by industries seeking a competitive edge with a novel, better, or more salable product. Its ultimate aim is usually profit for the stockholders. Basic research is conducted mainly at universities and research institutes. Most of its support comes from federal and state governments and foundations, although some larger industries also support it.

An example of basic research later applied to improving human welfare is the work of Gertrude Elion (1918–1999) and George Hitchings (1905–1998), who studied compounds called purines in an attempt to understand their role in the chemistry of the cell. Their basic research at Burroughs Wellcome Research Laboratories in North Carolina led to the discovery of a number of valuable new drugs for carrying out successful organ transplants and for treating various diseases such as gout, malaria, herpes, and cancer. Elion and Hitchings shared the 1988 Nobel Prize in Physiology or Medicine, and in 1991 Elion became the first woman to be inducted into the National Inventors Hall of Fame.

▲ **It DOES Matter!**
Gertrude Elion won a Nobel Prize for her basic research on purines. Her work has led to many applications in pharmaceuticals and medicine. Many important modern applications started out with basic research.

Importance of Basic Research

There are two compelling reasons why society must support basic science. One is substantial: The theoretical physics of yesterday is the nuclear defense of today; the obscure synthetic chemistry of yesterday is curing disease today. The other reason is cultural. The essence of our civilization is to explore and analyze the nature of man and his surroundings. As proclaimed in the Bible in the Book of Proverbs: "Where there is no vision, the people perish."

Arthur Kornberg (1918–2007), American biochemist, Nobel Prize in Physiology or Medicine, 1959

Spinning Nuclei

An example of basic research later applied to improving human welfare is the work of two physicists, Otto Stern (1888–1969) and Isidor Isaac Rabi (1898–1988). In 1930 they determined that certain atomic nuclei (Chapter 3) have a property called *spin*, and that in spinning, the nuclei act as tiny magnets. For their basic research in measuring these minuscule magnetic fields and thoroughly studying this phenomenon, Stern, a German, and Rabi, an American, won the Nobel Prize in Physics in 1943 and 1944, respectively. In the 1940s teams led by Edward M. Purcell (1912–1997) and Felix Bloch (1905–1983) used the fact that nuclear spin is influenced by its chemical environment (that is, by the atoms around the nucleus) to work out the structure of complicated molecules. This technique, called *nuclear magnetic resonance* (NMR), has become a major tool of chemists in determining molecular structure. Purcell and Bloch shared the Nobel Prize in Physics in 1952 for this research. In the 1960s Paul Lauterbur, Peter Mansfield, and other scientists applied the principles of NMR to scans of the human body. This technique, called *magnetic resonance imaging* (MRI), has replaced many exploratory surgical operations with a noninvasive procedure and made other surgery processes much more precise. Lauterbur (1929–2007) and Mansfield (b. 1933) won the Nobel Prize in Physiology or Medicine in 2003 for their work.

1.8 Chemistry: A Study of Matter and Its Changes

Earlier we defined chemistry as the study of matter and the changes it undergoes. Because the entire physical universe is made up of nothing more than matter and energy, the field of chemistry extends from atoms to stars, from rocks to living organisms. Now let's look at matter a little more closely.

Matter is the stuff that makes up all material things; it is anything that occupies space and has mass. Matter has *mass*; you can weigh it. Wood, sand, water, air, and people have mass and are therefore matter. **Mass** is a measure of the quantity of matter that an object contains. The greater the mass of an object, the more difficult it is to change its velocity. You can easily deflect a tennis ball coming toward you at 30 meters per second (m/s), but you would have difficulty stopping a cannonball of the same size moving at the same speed. A cannonball has more mass than a tennis ball of equal size.

The mass of an object does not vary with location. An astronaut has the same mass on the moon as on Earth. In contrast, **weight** measures a force. On Earth, it measures the force of attraction between our planet and the mass in question. On the moon, where gravity is one-sixth that on Earth, an astronaut weighs only one-sixth as much as on Earth (Figure 1.3). Weight varies with gravity; mass does not.

▲ **Figure 1.3** Astronaut John W. Young leaps from the lunar surface, where gravity pulls at him with only one-sixth the force on Earth.

QUESTION: If an astronaut has a mass of 72 kg, what would be his mass on the moon? If an astronaut weighs 180 lb on Earth, what would he weigh on the moon?

> ### CONCEPTUAL EXAMPLE 1.2   Mass and Weight
>
> On the planet Mercury, gravity is 0.376 times that on Earth.
>
> **a.** What would be the mass on Mercury of a person who has a mass of 62.5 kilograms (kg) on Earth?
> **b.** What would be the weight on Mercury of a person who weighs 124 pounds (lb) on Earth?
>
> #### Solution
>
> **a.** The person's mass would be the same (62.5 kg) as on Earth; the quantity of matter has not changed.
> **b.** The person would weigh only 0.376×124 lb $= 46.6$ lb; the force of attraction between planet and person is only 0.376 times that on Earth.
>
> #### ■ EXERCISE 1.2A
>
> At the surface of Venus, the force of gravity is 0.903 times that on Earth's surface.
>
> **a.** What would be the mass of a standard 1.00-kg object on Venus?
> **b.** A man who weighs 198 lb on Earth would weigh how much on the surface of Venus?
>
> #### ■ EXERCISE 1.2B
>
> On the (liquid) surface of Jupiter, the force of gravity is 2.34 times that on Earth.
>
> **a.** What would be the mass of a 52.5-kg woman at that location on Jupiter?
> **b.** A man who weighs 212 lb on Earth would weigh how much on Jupiter?

Physical and Chemical Properties

We can use our knowledge of chemistry to change matter to make it more useful. Chemists can change crude oil into gasoline, plastics, pesticides, drugs, detergents, and thousands of other products. Changes in matter are accompanied by changes in energy. Often we change matter to extract part of its energy. For example, we burn gasoline to get energy to propel our automobiles.

▶ **Figure 1.4** A comparison of the physical properties of two elements. Copper (left) can be hammered into thin foil, drawn into wire, or milled into granules. Iodine (right) consists of brittle gray crystals that crumble into a powder when struck.

QUESTION: What additional physical properties of copper and iodine are apparent from the photographs?

To distinguish between samples of matter, we can compare their properties (Figure 1.4). A **physical property** of a substance is a physical characteristic or behavior, such as color, odor, and hardness (Table 1.1). A **chemical property** describes how a substance reacts with other types of matter—how its atomic building blocks can change (Table 1.2).

A **physical change** involves an alteration in the physical appearance of matter without changing its chemical identity or composition. An ice cube can melt to form a liquid, but it is still water. Melting is a physical change, and the temperature at which it occurs—the melting point—is a physical property. A **chemical change** involves a change in the chemical identity of matter into other substances that are chemically different. In exhibiting a chemical property, matter undergoes a chemical change; the substance(s) in the original matter is replaced by one or more new substances. Iron metal reacts with oxygen from the air to form rust (iron oxide); when sulfur burns in air, sulfur, which is made up of one type of atom, and oxygen (from air), which is made up of another type of atom, combine to form sulfur dioxide, which is made up of molecules that have sulfur and oxygen atoms in the ratio 1:2. (A *molecule* is a group of atoms bound together as a single unit. More about atoms and molecules later.)

It is difficult at times to determine whether a change is physical or chemical, but we can decide on the basis of what happens to the composition or structure of the

Table 1.1 Some Examples of Physical Properties

Property	Examples
Temperature	0 °C for ice water, 100 °C for boiling water.
Mass	A nickel weighs 5 g; a penny weighs 2.5 g.
Structure	Ice is crystalline; glass is amorphous.
Color	Sulfur is yellow; bromine is reddish-brown.
Taste	Acids are sour; bases are bitter.
Odor	Benzyl acetate smells like jasmine; hydrogen sulfide smells like rotten eggs.
Boiling point	Water boils at 100 °C; ethyl alcohol boils at 78.5 °C.
Freezing point	Water freezes at 0 °C; methane freezes at −182 °C.
Specific heat	Water has a high specific heat; iron has a low specific heat.
Hardness	Diamond is exceptionally hard; sodium metal is soft.
Conductivity	Copper conducts electricity; diamond does not. Aluminum is a good conductor of heat; glass is a poor heat conductor.
Solubility	Ethyl alcohol dissolves in water; gasoline does not.
Density	1.00 g/mL for water; 19.3 g/cm^3 for gold.

Table 1.2 Some Examples of Chemical Properties

Substance	Typical Chemical Property
Iron	rusts (combines with oxygen to form iron oxide).
Carbon	burns (combines with oxygen to form carbon dioxide).
Silver	tarnishes (combines with sulfur to form silver sulfide).
Nitroglycerin	explodes (decomposes to produce a mixture of gases).
Carbon monoxide	is toxic (combines with hemoglobin, causing anoxia).
Neon	is inert (does not react with anything).

matter involved. *Composition* refers to the types of atoms that are present and their relative proportions, and *structure* refers to the arrangement of those atoms in particular assemblages or in space. A chemical change results in a change in composition or structure, whereas a physical change does not.

> **CONCEPTUAL EXAMPLE 1.3** Chemical Change and Physical Change

Which of the following events involve chemical changes and which involve physical changes?

a. Your hair is cut.
b. Lemon juice converts milk to curds and whey.
c. Water boils.
d. Water is broken down into hydrogen gas and oxygen gas.

Solution

We examine each change and determine whether there has been a change in composition or structure. In other words, we ask "Have new substances that are chemically different been created?" If so, the change is chemical; if not, it is physical.

a. Physical change: The composition of the hair is not changed by cutting.
b. Chemical change: The compositions of curds and whey are different from the composition of the milk.
c. Physical change: Both the liquid water and the invisible water vapor formed when liquid water boils have the same composition; the water merely changes from a liquid to a gas.
d. Chemical change: New substances, hydrogen and oxygen, are formed.

■ **EXERCISE 1.3A**

Which of the following events involve chemical changes and which involve physical changes?

a. Gasoline vaporizes from an open container.
b. A piece of magnesium metal burns in air to form a white powder called *magnesium oxide*.
c. A dull knife is sharpened with a whetstone.

■ **EXERCISE 1.3B**

Which of the following events involve chemical changes and which involve physical changes?

a. A steel wrench left out in the rain becomes rusty.
b. A stick of butter melts.
c. A wooden log is burned.
d. A piece of wood is ground up into sawdust.

Self-Assessment Questions

1. Which of the following is *not* an example of matter?
   a. automobile exhaust   b. iron
   c. love   d. natural gas

2. Which of the following is an example of matter?
   a. air   b. blue light
   c. an idea   d. a prayer

3. Two identical items on Earth are taken to the moon where they have
   a. both the same mass and the same weight
   b. neither the same mass nor the same weight
   c. the same mass but not the same weight
   d. the same weight but not the same mass

4. What kind of change alters the identity of a material?
   a. chemical   b. physical
   c. physical state   d. temperature

5. Bending glass tubing in a flame involves
   a. applied research   b. a chemical change
   c. an experiment   d. a physical change

6. Transforming liquid water into ice involves
   a. adding energy to the water   b. a change in chemical formula
   c. a chemical change   d. a physical change

7. Which of the following is a chemical property?
   a. Chlorine gas is green.   b. Iron rusts.
   c. Ethanol dissolves in water. d. Wood is easily carved.

8. Which of the following is an example of a physical change?
   a. A cake is baked from flour, baking powder, sugar, eggs, shortening, and milk.
   b. Milk left outside a refrigerator overnight turns sour.
   c. Sheep are sheared and the wool is spun into yarn.
   d. Silkworms eat mulberry leaves and make silk.

9. Which of the following is an example of a chemical change?
   a. An egg is broken and poured into an eggnog mix.
   b. A lawn is mowed, shortening the grass.
   c. A lawn is watered and fertilized, and the grass grows thicker.
   d. Frost forms on a cold window pane.

Answers: 1, c; 2, a; 3, c; 4, a; 5, d; 6, d; 7, b; 8, c; 9, c

1.9 Classification of Matter

In this section, we will examine three of the many ways of classifying matter. First we look at the physical forms or *states of matter*.

The States of Matter

There are three familiar states of matter: solid, liquid, and gas (Figure 1.5). They can be classified by bulk properties (a *macro* view) or by arrangement of the particles that comprise them (a *"molecular"* or *micro* view). A **solid** object ordinarily maintains its shape and volume regardless of its location. A **liquid** occupies a definite volume but assumes the shape of the occupied portion of its container. If you have a 355-milliliter (mL) soft drink, you have 355 mL whether the soft drink is in a can, in a bottle, or, through a mishap, on the floor—which demonstrates another property of liquids. Unlike solids, liquids flow readily. A **gas** maintains neither shape nor volume. It expands to fill completely whatever container it occupies. Gases flow and are easily compressed. For example, enough air for many minutes of breathing can be compressed into a steel tank for underwater diving.

Bulk properties of *solids, liquids,* and *gases* are explained using the kinetic–molecular theory. In solids, the particles are close together and in fixed positions.

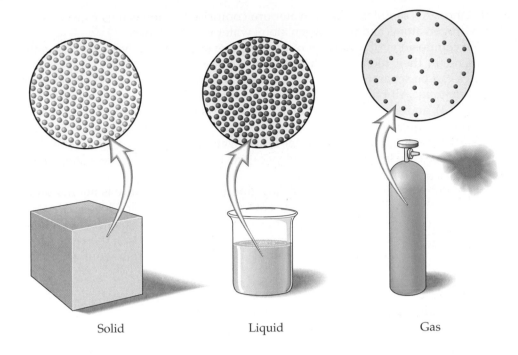

◀ **Figure 1.5** The kinetic–molecular theory can be used to interpret (or explain) the bulk properties of *solids, liquids,* and *gases.* In solids, the particles are close together and in fixed positions. In liquids, the particles are close together, but they are free to move about. In gases, the particles are far apart and are in rapid random motion.

QUESTION: Based on this figure, why does a quart of water vapor weigh so much less than a quart of liquid water?

In liquids, the particles are close together, but they are free to move about. In gases, the particles are far apart and are in rapid random motion. We discuss the states of matter in more detail in Chapter 5.

Substances and Mixtures

Matter can be either pure or mixed (Figure 1.6). Pure matter is considered to be a substance. A **substance** has a definite, or fixed, composition that does not vary from one sample to another. Pure gold (24-karat gold) consists entirely of gold atoms; it is

▼ **Figure 1.6** A scheme for classifying matter. The "molecular-level" views are of gold—an *element*; water—a *compound*; 12-karat gold—a *homogeneous mixture* of silver and gold; and a miner's pan containing a *heterogeneous mixture* of gold flakes in water.

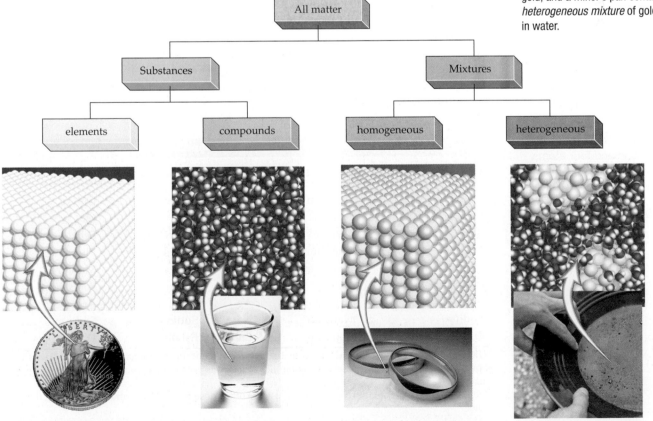

a substance. All samples of pure water are comprised of molecules consisting of two hydrogen atoms and one oxygen atom; water is a substance.

The composition of a **mixture** of two or more substances is variable. The substances retain their identities. They do not change chemically; they simply mix. Mixtures can be separated by physical means. Mixtures can be either *homogeneous* or *heterogeneous*. All parts of a homogeneous mixture have the same composition (Section 6.4) and the same appearance. A saline solution—a solution of salt in water—is a homogeneous mixture. The proportions of salt and water can vary from one solution to another, but the water is still water and the salt remains salt. The two substances can be separated by physical means. For example, the water can be boiled away, leaving the salt behind.

Sand and water form a heterogeneous mixture; the appearance is not the same throughout. The grains of sand are different from the liquid water. Most of what we deal with on a day-to-day basis are heterogeneous mixtures. All you need do is look at a computer, a book, a pen, or a person to see that different parts of those mixtures have different compositions (your hair is quite different from your skin, for example).

Elements and Compounds

A *substance* is either an element or a compound. An **element** is one of the fundamental substances from which all material things are constructed. Elements cannot be broken down into simpler substances by any chemical process. A **compound** is a substance made up of two or more elements chemically combined. Compounds have a fixed composition. For example, water has fixed proportions, by mass, of hydrogen (H) and oxygen (O). There are more than 100 known elements, of which only about a third will be dealt with in this book. A list of these elements is provided on the inside front cover. Oxygen, carbon, sulfur, aluminum, and iron are familiar elements. Aluminum oxide ("sand" on sandpaper), carbon dioxide, and iron disulfide (FeS_2, "fool's gold") are compounds.

Because elements are so fundamental to our study of chemistry, we find it useful to refer to them in a shorthand form. Each element can be represented by a **chemical symbol** made up of one or two letters derived from the name of the element. Symbols for all the elements are listed in the Table of Atomic Masses (inside front cover). The first letter of the symbol is always capitalized; the second is always lowercase. (It makes a difference. For example, Co is the symbol for cobalt, an element, but CO is the formula for carbon monoxide, a compound.)

Symbols are the alphabet of chemistry. Most are based on the English names of the elements, but a few are based on Latin names (Table 1.3).

Table 1.3	Some Elements with Symbols Derived from Latin Names			
Usual English Name	**Latin Name**	**Symbol**	**Spanish Name**[b]	**French Name**[b]
Copper	Cuprum	Cu	Cobre	Cuivre
Gold	Aurum	Au	Oro	Or
Iron	Ferrum	Fe	Hierro	Fer
Lead	Plumbum	Pb	Plomo	Plomb
Mercury	Hydrargyrum	Hg	Mercurio	Mercure
Potassium	Kalium[a]	K	Potasio	Potassium
Silver	Argentum	Ag	Plata	Argent
Sodium	Natrium[a]	Na	Sodio	Sodium
Tin	Stannum	Sn	Estaño	Étain

[a]The elements potassium and sodium were unknown in ancient times. The names kalium and natrium shown here are the names that were used for the compounds potassium carbonate and sodium carbonate.

[b]Element names in languages other than English are often close to the Latin names as shown in this table.

---

5. Can we change lead into gold?
One element cannot be changed into another element by chemical reactions. Although an element cannot be created or destroyed, it can be extracted from a mixture or compound containing the element.

A chemical symbol in a formula stands for one atom of the element. If more than one atom is to be indicated in a formula, a subscript number is used after the symbol. For example, the formula H_2 represents two atoms of hydrogen, and the formula CH_4 stands for one atom of carbon and four atoms of hydrogen.

> **CONCEPTUAL EXAMPLE 1.4** Elements and Compounds
>
> Which of the following represent elements and which represent compounds?
>
> C   Ca   HI   BN   In   HBr
>
> ### Solution
>
> C, Ca, and In represent elements (each is a single symbol). HI, BN, and HBr are composed of two symbols each and represent compounds.
>
> ■ **EXERCISE 1.4A**
>
> Which of the following represent elements and which represent compounds?
>
> He   CuO   No   NO   KI   Os
>
> ■ **EXERCISE 1.4B**
>
> How many *different* elements are represented in the entire list of Exercise 1.4A?

Atoms and Molecules

An **atom** is the smallest characteristic part of an element. Each element is composed of atoms of a particular kind. For example, the element copper is made up of copper atoms, and gold is made up of gold atoms. All copper atoms are alike in a fundamental way and are different from gold atoms. The smallest characteristic part of most compounds is a molecule. A **molecule** is a group of atoms bound together as a unit. Each molecule of a given compound has the same atoms in the same proportions as all the other molecules of the compound. For example, all water molecules have two hydrogen (H) atoms and one oxygen (O) atom, as indicated by the formula (H_2O). We will discuss atoms in some detail in Chapter 2, and much of the focus of many of the remaining chapters is on molecules.

Self-Assessment Questions

1. Which state of matter has neither a definite volume nor a definite shape?
   a. homogeneous   **b.** gas   **c.** liquid   **d.** solid

2. Which of the following is a mixture?
   a. carbon   **b.** carbonated water   **c.** lead   **d.** silver

3. Which of the following is *not* an element?
   a. aluminum   **b.** bronze   **c.** lead   **d.** sodium

4. A compound can be separated into its elements
   a. by chemical means   **b.** by mechanical means
   c. by physical means   **d.** using a magnet

5. The smallest part of an element is a(n)
   a. atom   **b.** ion   **c.** mass   **d.** molecule

6. The symbol for potassium is
   a. K   **b.** Kr   **c.** P   **d.** Pm

7. The symbol for calcium is
   a. C   **b.** Ca   **c.** Cm   **d.** K

Answers: 1, b; 2, b; 3, b; 4, a; 5, a; 6, a; 7, b

1.10 The Measurement of Matter

Accurate measurements of such quantities as mass, volume, time, and temperature are essential to the compilation of dependable scientific data. These data are of critical importance in all science-related fields. Measurements of temperature and blood pressure are routinely made in medicine, and modern medical diagnosis depends on a whole battery of other measurements, including careful chemical analyses of blood and urine.

▶ **Figure 1.7** Comparisons of metric and customary units of measure. The meter stick is 100 centimeters long and is slightly longer than the yardstick below it (1 m = 1.0936 yd).

QUESTION: From this figure, approximately how many pounds are in a kilogram? Which two units are closer in size, the inch and centimeter or the quart and the liter?

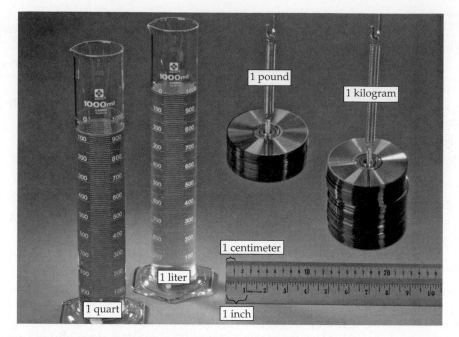

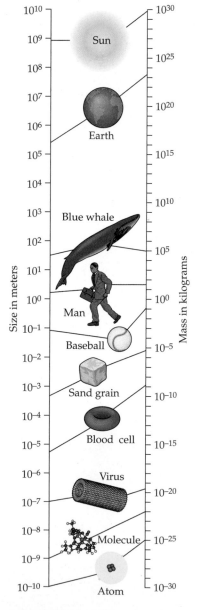

▲ Comparison of very large and very small objects is much easier with exponential notation.

The measurement system agreed on since 1960 is the *International System of Units* or **SI units** (from the French *Système Internationale*), a modernized version of the metric system established in France in 1791. Most countries use metric measures in everyday life, but in the United States SI units are used mainly in science laboratories. However, metric measures are increasingly being used in commerce, especially in businesses with an international component. The contents of most bottled beverages are now given in metric units, and metric measurements are also common in sporting events. Figure 1.7 compares some metric and customary units.

Because SI units are based on the decimal system, it is easy to convert from one unit to another. All measured quantities can be expressed in terms of the seven base units listed in Table 1.4. We use the first six in this text.

Exponential Numbers: Powers of Ten

Scientists deal with objects smaller than atoms and as large as the universe. We usually use exponential notation to describe the sizes of such objects. A number is in *exponential notation* when it is written as the product of a coefficient and a power of ten, such as 1.6×10^{-19} or 35×10^6. A number is said to be expressed in *scientific notation* when the coefficient has a value between 1 and 10. The Appendix provides a more detailed discussion of scientific notation.

An electron has a diameter of about 1×10^{-15} meter (m) and a mass of about 1×10^{-30} kilogram (kg). At the other extreme, our galaxy measures about 1×10^{23} m across and has a mass of about 1×10^{41} kg. It is difficult even to imagine numbers so small or so large. The accompanying figure offers some perspectives on size.

Table 1.4	The Seven SI Base Units	
Physical Quantity	**Name of Unit**	**Symbol of Unit**
Length	meter[a]	m
Mass	kilogram	kg
Time	second	s
Temperature	kelvin	K
Amount of substance	mole	mol
Electric current	ampere	A
Luminous intensity	candela	cd

[a]Spelled *metre* in most countries.

Table 1.5 Approved Numerical Prefixes[a]

Exponential Expression	Decimal Equivalent	Prefix	Pronunciation	Symbol
10^{12}	1,000,000,000,000	tera-	TER-uh	T
10^{9}	1,000,000,000	giga-	GIG-uh	G
10^{6}	1,000,000	mega-	MEG-uh	M
10^{3}	1,000	kilo-	KIL-oh	k
10^{2}	100	hecto-	HEK-toe	h
10^{1}	10	deka-	DEK-uh	da
10^{-1}	0.1	deci-	DES-ee	d
10^{-2}	0.01	centi-	SEN-tee	c
10^{-3}	0.001	milli-	MIL-ee	m
10^{-6}	0.000,001	micro-	MY-kro	μ
10^{-9}	0.000,000,001	nano-	NAN-oh	n
10^{-12}	0.000,000,000,001	pico-	PEE-koh	p
10^{-15}	0.000,000,000,000,001	femto-	FEM-toe	f

[a]The most commonly used prefixes are shown in color.

Measurements using the basic SI units are often of awkward magnitude. As we have just seen, we can use exponential numbers. However, in many cases it is more convenient to use prefixes (Table 1.5) to indicate units larger and smaller than the base unit. The following examples show how prefixes and powers of ten are interconverted.

EXAMPLE 1.5   Prefixes and Powers of Ten

Convert each of the following measurements to a unit that replaces the power of ten by a prefix.

a. 2.89×10^{-3} g   **b.** 4.30×10^{3} m

Solution

Our goal is to replace each power of ten with the appropriate prefix from Table 1.5. For example, $10^{-3} = 0.001$, corresponding to milli(unit). (It doesn't matter what the unit is; here we are dealing only with the prefixes.)

a. 10^{-3} corresponds to the prefix *milli-*; 2.89 mg; that is,
   $10^{-3} \times$ (unit) = milli(unit).

b. 10^{3} corresponds to the prefix *kilo-*; 4.30 km; that is,
   $10^{3} \times$ (unit) = kilo(unit).

■ EXERCISE 1.5

Convert each of the following measurements to a unit that replaces the power of ten by a prefix.

a. 7.24×10^{3} g   **b.** 4.29×10^{-6} m   **c.** 7.91×10^{-3} s
d. 2.29×10^{-2} g   **e.** 7.90×10^{6} m

EXAMPLE 1.6   Prefixes and Powers of Ten

Use exponential notation to express each of the following measurements in terms of an SI base unit.

a. 4.12 cm   **b.** 947 μs   **c.** 3.17 nm

Solution

a. Our goal is to find the power of ten that relates the given unit to the SI base unit. That is, centi(base unit) = $10^{-2} \times$ (base unit)

$$4.12 \text{ centimeter} = 4.12 \times 10^{-2} \text{ m}$$

b. To change microsecond to the base unit second, we replace the prefix *micro-* by 10^{-6}. To obtain an answer in conventional scientific notation, we also need to replace the coefficient 947 by 9.47×10^2. The result of these two changes is

$$947 \, \mu s = 947 \times 10^{-6} \, s = 9.47 \times 10^2 \times 10^{-6} \, s = 9.47 \times 10^{-4} \, s$$

c. To change nanometer to the base unit meter, we replace the prefix *nano-* by 10^{-9}. The answer in exponential form is 3.17×10^{-9} m.

■ EXERCISE 1.6A

Use scientific notation to express each of the following measurements in terms of an SI base unit.

a. 7.45 nm   **b.** 5.25 ms   **c.** 1.415 km   **d.** 2.06 mm

■ EXERCISE 1.6B

Use scientific notation to express each of the following measurements in terms of an SI base unit.

a. 284 nm   **b.** 119 ms   **c.** 754 km   **d.** 6.19×10^6 mm

Mass

The SI base quantity of mass is the **kilogram (kg),** about 2.2 pounds (lb). This base quantity is unusual in that it already has a prefix. A more convenient mass unit for most laboratory work is the gram (g).

$$1 \, kg = 10^3 \, g = 1000 \, g \qquad or \qquad 1 \, g = 0.001 \, kg = 10^{-3} \, kg$$

The milligram (mg) is a suitable unit for small quantities of materials, such as some drug dosages.

$$1 \, mg = 10^{-3} \, g = 0.001 \, g$$

Chemists can now detect masses in the microgram (μg), nanogram (ng), picogram (pg), and even smaller ranges.

Length, Area, and Volume

The SI base unit of length is the **meter (m),** a unit about 10% longer than 1 yard (yd). The kilometer (km) is used to measure distances along a highway.

$$1 \, km = 1000 \, m$$

In the laboratory, we usually find lengths smaller than the meter to be more convenient. For example, we use the centimeter (cm), which is about the width of a typical calculator button—and the millimeter (mm), which is about the thickness of the cardboard backing in a notepad.

$$1 \, cm = 0.01 \, m \qquad 1 \, mm = 0.001 \, m$$

For measurements at the atomic and molecular levels, we use the micrometer (μm), the nanometer (nm), and the picometer (pm). For example, a hemoglobin molecule, which is nearly spherical, has a diameter of 5.5 nm or 5500 pm, and the diameter of a sodium atom is 372 pm.

The units for area and volume are derived from the base unit of length. The SI unit of area is the square meter (m^2), but we often find square centimeters (cm^2) or square millimeters (mm^2) more convenient for laboratory work.

$$1 \, cm^2 = (10^{-2} \, m)^2 = 10^{-4} \, m^2 \qquad 1 \, mm^2 = (10^{-3} \, m)^2 = 10^{-6} \, m^2$$

Similarly, the SI unit of volume is the cubic meter (m^3), but two units more likely to be used in the laboratory are the cubic centimeter (cm^3 or cc) and the cubic decimeter (dm^3). A cubic centimeter is about the volume of a sugar cube, and a cubic decimeter is slightly larger than 1 quart (qt).

$$1 \, cm^3 = (10^{-2} \, m)^3 = 10^{-6} \, m^3 \qquad 1 \, dm^3 = (10^{-1} \, m)^3 = 10^{-3} \, m^3$$

The cubic decimeter is commonly called a liter. A **liter (L)** is 1 cubic decimeter or 1000 cubic centimeters.

$$1 \, L = 1 \, dm^3 = 1000 \, cm^3$$

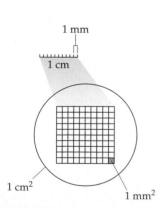

▲ Area units such as mm^2 and cm^2 are derived from units of length.

QUESTION: How many mm are in 1 cm? How many mm^2 are in 1 cm^2?

The milliliter (mL) and/or cubic centimeter are frequently used in laboratories.

$$1 \text{ mL} = 1 \text{ cm}^3$$

Time

The SI base unit for measuring intervals of time is the second (s). Extremely short time periods are expressed through the usual SI prefixes: milliseconds (ms), microseconds (μs), nanoseconds (ns), and picoseconds (ps).

$$1 \text{ ms} = 10^{-3} \text{ s} \qquad 1 \, \mu\text{s} = 10^{-6} \text{ s} \qquad 1 \text{ ns} = 10^{-9} \text{ s} \qquad 1 \text{ ps} = 10^{-12} \text{ s}$$

Long time intervals, in contrast, are usually expressed in traditional, non-SI units: minutes (min), hours (h), days (d), and years (y).

$$1 \text{ min} = 60 \text{ s} \qquad 1 \text{ h} = 60 \text{ min} \qquad 1 \text{ d} = 24 \text{ h} \qquad 1 \text{ y} = 365 \text{ d}$$

Problem Solving: Estimation

Many chemistry problems require calculations that yield numerical answers. When you use a calculator, it will always give you an answer—but the answer may not be correct. You may have punched a wrong number or used the wrong function. You can learn to make *estimations* of answers that enable you to know whether your answer is reasonable. At times an estimated answer is good enough, as the following example and exercises illustrate. This ability to estimate answers can be important in everyday life as well as in chemistry. Note that estimation does *not* require a detailed calculation. Only a rough calculation—or none at all—is required.

EXAMPLE 1.7 Mass, Length, Area, Volume

Without doing a detailed calculation, determine which of the following is a reasonable **(a)** mass (weight) and **(b)** height for a typical two-year-old child.

Mass:	10 mg	10 g	10 kg	100 g
Height:	85 mm	85 cm	850 cm	8.5 m

Solution

a. In customary units, a two-year-old child should weigh about 20–25 lb. Knowing that 1 kg is about 2.2 lb, the only reasonable answer is 10 kg.

b. In customary units, a two-year-old child should be about 30–36 in. tall. Knowing that 1 cm is a little less than 0.5 in., the answer must be a little *more* than twice 30–36, or a bit more than 60–72 cm. Thus, the only reasonable answer is 85 cm.

■ EXERCISE 1.7A

Without doing a detailed calculation, determine which of the following is a reasonable area for the front cover of your textbook.

$$500 \text{ mm}^2 \qquad 50 \text{ cm}^2 \qquad 500 \text{ cm}^2 \qquad 50 \text{ m}^2$$

■ EXERCISE 1.7B

Without doing a detailed calculation, determine which of the following is a reasonable volume for your textbook.

$$1600 \text{ mm}^3 \qquad 16 \text{ cm}^3 \qquad 1600 \text{ cm}^3 \qquad 1.6 \text{ m}^3$$

Problem Solving: Unit Conversions

It is easy to convert from one metric unit to another. We use the unit conversion method of problem solving. If you are not familiar with that method, you should study the topic in the Appendix before proceeding. We also discuss the conversion of common units to metric units in the Appendix. You will also find a discussion of *significant figures,* a way of indicating the precision of measurements, in the Appendix. In the following problems and throughout the text, we simply will carry three digits in most calculations.

Nanoworld

For over two centuries, chemists have been able to rearrange atoms to make molecules. These molecules generally have dimensions in the *picometer* (10^{-12} m) range. This ability has led to revolutions in the design of drugs, plastics, and many other materials. Over the last several decades, scientists have made vast strides in handling materials in the *micrometer* (10^{-6} m) range. The revolution in microelectronic devices—computers, cellular phones, and so on—was spurred by the ability of scientists to produce computer chips by photolithography on a micrometer scale.

Recently the news has focused on *nanotechnology*. Just what are the pundits talking about? The prefix *nano-* means one-billionth (10^{-9}). With nanotechnology it is possible to bridge the gap between picometer-sized molecules and micrometer-sized electronics. By tailoring the structure of materials in the range from about 1 nm to 100 nm, a scientist can systematically change the properties of materials. A nanometer-sized object contains just a few hundred to a few thousand atoms or molecules, and such objects often have entirely new properties. For example, bulk gold is yellow, but a gold ring made up of nanometer-sized particles appears red. Carbon *nanotubes* are made of the same element as the graphite in pencil lead, but nanotubes are much, much stronger. Scientists can now control matter on every scale, from nanometers to meters, giving us great new power in materials design. We will examine some of the many practical applications of nanotechnology in subsequent chapters.

Bulk gold = Yellow

Nanogold = Red

EXAMPLE 1.8   Unit Conversions

Convert

a. 1.83 kg to grams and   **b.** 729 μL to milliliters.

Solution

a. We start with the given quantity 1.83 kg and use the equivalence 1 kg = 1000 g to form a conversion factor (Appendix) that allows us to cancel the unit kg and end with the unit g.

$$1.83 \ \cancel{kg} \times \frac{1000 \ g}{1 \ \cancel{kg}} = 1830 \ g$$

b. Here we start with the given quantity 729 μL and use the equivalences $10^6 \ \mu$L = 1 L and 10^3 mL = 1 L to form conversion factors that allow us to cancel the unit μL and end with the unit mL.

$$729 \ \cancel{\mu L} \times \frac{1 \ \cancel{L}}{10^6 \ \cancel{\mu L}} \times \frac{1000 \ mL}{1 \ \cancel{L}} = 0.729 \ mL$$

■ **EXERCISE 1.8**

Convert

a. 7.5 m to millimeters,   **b.** 2056 mL to liters,
c. 2.06 g to micrograms, and   **d.** 0.738 cm to millimeters.

Self-Assessment Questions

1. The SI unit for length is the
   a. foot   **b.** kelvin   **c.** kilometer   **d.** meter

2. The SI unit for mass is the
   a. dram   **b.** gram   **c.** grain   **d.** kilogram

3. The prefix that means 10^{-3} is
   a. *centi-*   **b.** *deci-*   **c.** *micro-*   **d.** *milli-*

4. A meter is about 10% longer than a(n)
   a. foot   **b.** inch   **c.** mile   **d.** yard

5. One cubic centimeter is equal to one
   a. deciliter   **b.** dram   **c.** liter   **d.** milliliter

6. One quart is slightly less than one
  a. deciliter **b.** kiloliter **c.** liter **d.** milliliter

7. A rope is 5775 cm long. What is its length in kilometers?
  a. 0.05775 km **b.** 0.5775 km **c.** 5.775 km **d.** 57.75 km

8. Which of the following has a mass of roughly 1 g?
  a. an ant **b.** an orange **c.** a peanut **d.** this textbook

9. Which of the following has a thickness of roughly 1 cm?
  a. a concrete block **b.** a knife blade
  c. a writing tablet **d.** this textbook

10. Which of the following has a volume of roughly 250 mL (0.250 L)?
  a. 1 cup **b.** 1 pint
  c. 1 tablespoon **d.** 1 teaspoon

Answers: 1, d; 2, d; 3, d; 4, d; 5, d; 6, c; 7, a; 8, c; 9, c; 10, a

1.11 Density

In everyday life, we might speak of lead as "heavy" or aluminum as "light," but such descriptions are imprecise at best. Scientists use the term *density* to describe this important property. The **density, *d***, of a substance is the quantity of mass, *m*, per unit of volume, *V*.

$$d = \frac{m}{V}$$

For substances that don't mix, such as oil and water, the concept of density allows us to predict which will float on the other. It isn't just the mass of the materials, but the *mass per unit volume*—the density—that determines the result. Density is one of the properties that helps us to know that in an oil-and-vinegar salad dressing, oil (lower density) floats on water (higher density). Figure 1.8 gives more examples.

We can rearrange the equation for density to give

$$m = d \times V \quad \text{and} \quad V = \frac{m}{d}$$

These equations are useful for calculations. Densities of some common substances are listed in Table 1.6. They are reported, as is customary, in grams per milliliter (g/mL) for liquids and grams per cubic centimeter (g/cm³) for solids. Values listed in the table are used in some of the following examples and exercises and in some of the end-of-chapter problems.

▲ One cubic centimeter of copper weighs 8.94 g, so the density of copper is 8.94 g/cm³.

QUESTION: What would be the mass of 2 cm³ of copper? What would be the density of 2 cm³ of copper?

Table 1.6 Densities of Some Common Substances at Specified Temperatures

Substance[a]	Density	Temperature
Solids		
Copper (Cu)	8.94 g/cm³	25 °C
Gold (Au)	19.3 g/cm³	25 °C
Magnesium (Mg)	1.738 g/cm³	20 °C
Water (ice) (H_2O)	0.917 g/cm³	0 °C
Liquids		
Ethyl alcohol (CH_3CH_2OH)	0.789 g/mL	20 °C
Hexane ($CH_3CH_2CH_2CH_2CH_2CH_3$)	0.660 g/mL	20 °C
Mercury (Hg)	13.534 g/mL	25 °C
Urine (a mixture)	1.003–1.030 g/mL	25 °C
Water (H_2O)	0.997 g/mL	0 °C
	1.000 g/mL	4 °C
	0.998 g/mL	20 °C

[a]Formulas are provided for possible future reference.

▲ **Figure 1.8** At room temperature, the density of water is 1.00 g/mL. A coin sinks in water, but a cork floats on water.

QUESTION: Is the density of the coin less than, equal to, or greater than 1.00 g/mL? Is the density of the cork less than, equal to, or greater than 1.00 g/mL?

As with mass, length, area, and volume (page 23), it is often sufficient to estimate answers to problems involving densities. For example, an estimated answer is usually good enough to determine relative volumes of materials or whether or not a material will float or sink in water. These ideas are illustrated in the following example and exercises. Note again that estimation does *not* require a detailed calculation.

EXAMPLE 1.9   Mass, Volume, and Density

Answer the following without doing a detailed calculation.

a. Which has the greater volume, a 50.0-g block of copper or a 50.0-g block of gold?
b. Which has the greater mass, 225 mL of ethyl alcohol or 225 mL of hexane?

Solution

a. From Table 1.6 we see that the density of gold ($19.3 \ \text{g/cm}^3$) is greater than that of copper ($8.94 \ \text{g/cm}^3$). It takes a larger block of copper to have a mass of 50.0 g than of gold. A 50.0-g block of copper has a greater volume than a 50.0-g block of gold.

b. From Table 1.6 we see that the density of ethyl alcohol (0.789 g/mL) is greater than that of hexane (0.660 g/mL). Because it is more dense (that is, it has more mass in each unit of volume), 225 mL of ethyl alcohol has a greater mass than does 225 mL of hexane.

■ EXERCISE 1.9A

Without doing a detailed calculation, determine which has the greater volume, a 500.0-g block of ice or a 500.0-g block of magnesium.

■ EXERCISE 1.9B

Wood does not dissolve in water and will float on water if its density is less than that of water. Padouk ($d = 0.86 \ \text{g/cm}^3$) and ebony ($d = 1.2 \ \text{g/cm}^3$) are tropical woods. Which will float on water and which will sink in water?

When we need a numerical answer, we can use the relationship of mass and volume to density to form conversion factors. Then we use the unit conversion method of problem solving.

EXAMPLE 1.10   Density from Mass and Volume

What is the density of iron if 156 g of iron occupies a volume of 20.0 cm³?

Solution

The given quantities are

$$m = 156 \ \text{g} \quad \text{and} \quad V = 20.0 \ \text{cm}^3$$

We can use the equation that defines density.

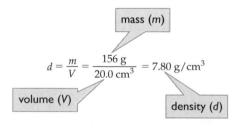

$$d = \frac{m}{V} = \frac{156 \ \text{g}}{20.0 \ \text{cm}^3} = 7.80 \ \text{g/cm}^3$$

■ EXERCISE 1.10A

What is the density, in grams per milliliter, of a salt solution if 52.5 mL has a mass of 58.5 g?

■ EXERCISE 1.10B

What is the density, in grams per cubic centimeter, of a metal alloy if a cube that measures 2.00 cm on an edge has a mass of 89.2 g?

EXAMPLE 1.11 Mass from Density and Volume

What is the mass of 1.00 L of gasoline if its density is 0.703 g/mL?

Solution

We can express density as a ratio, 0.703 g/1.00 mL, and use it as a conversion factor. We also need the factor 1000 mL/1 L to convert liters to milliliters.

$$m = d \times V = \frac{0.703\ g}{1\ mL} \times 1\ L \times \frac{1000\ mL}{1\ L} = 703\ g$$

■ **EXERCISE 1.11A**

What is the mass of 200.0 mL of glycerol, which has a density of 1.264 g/mL at 20 °C?

■ **EXERCISE 1.11B**

What is the mass of a lead brick that is 2.50 cm × 3.50 cm × 8.25 cm? (The density of lead at 20 °C is 11.34 g/cm³.)

EXAMPLE 1.12 Volume from Mass and Density

What volume is occupied by 461 g of mercury? (See Table 1.6.)

Solution

Here we can express the inverse of density as a ratio, 1.00 mL/13.6 g, and use it as a conversion factor.

$$V = 461\ g \times \frac{1\ mL}{13.534\ g} = 34.1\ mL$$

■ **EXERCISE 1.12A**

What volume is occupied by an 845-g piece of magnesium? (See Table 1.6.)

■ **EXERCISE 1.12B**

What volume is occupied by 225 g of hexane? (See Table 1.6.)

Self-Assessment Questions

1. The density of a wood plank floating in a lake is
   a. less than that of air
   b. less than that of water
   c. more than that of water
   d. the same as that of water

2. A pebble and a lead sinker both sink when dropped into a pond. This shows that the two objects
   a. are both less dense than water
   b. are both more dense than water
   c. have different densities
   d. have the same density

3. Ice floats on liquid water. A reasonable value for the density of ice is
   a. 0.92 g/cm³
   b. 1.08 g/cm³
   c. 1.98 g/cm³
   d. 4.90 g/cm³

4. A 49er panning for gold shakes a mixture of mud, gravel, and water in a pan. He looks for gold at the bottom because gold
   a. dissolves in water and sinks
   b. has a high density and sinks
   c. has the same density as rocks
   d. is repelled by the other minerals

5. Which of the following metals will sink in a pool of mercury (d = 13.534 g/cm³)?
   a. aluminum (d = 2.6 g/cm³)
   b. iron (d = 7.9 g/cm³)
   c. lead (d = 11.34 g/cm³)
   d. uranium (d = 19.5 g/cm³)

6. If the volume of a rock is 8.0 cm³ and its mass is 16.0 g, its density is
   a. 0.03 g/cm³
   b. 0.50 g/cm³
   c. 2.0 g/cm³
   d. 128 g/cm³

Answers: 1, b; 2, b; 3, a; 4, b; 5, d; 6, c

1.12 Energy: Heat and Temperature

The physical and chemical changes that matter undergoes are almost always accompanied by changes in energy. **Energy** is required to make something happen that wouldn't happen by itself; it is the ability to change matter, either physically or chemically.

Two important concepts in science that are related, and sometimes confused, are *temperature* and *heat*. When two objects at different temperatures are brought together, heat flows from the warmer to the cooler object until both are at the same temperature. **Heat** is energy on the move, the energy that flows from a warmer object to a cooler one. **Temperature** is a measure of how hot or cold an object is. Temperature tells us the direction in which heat will flow. Heat flows from more energetic (higher-temperature) to less energetic (lower-temperature) atoms or molecules. For example, if you touch a hot test tube, heat will flow from the tube to your hand. If the tube is hot enough, your hand will be burned.

The SI base unit of temperature is the **kelvin (K)**. For laboratory work, we often use the more familiar **Celsius scale**. On this temperature scale, the freezing point of water is zero degrees Celsius (°C) and the boiling point is 100 °C. The interval between these two reference points is divided into 100 equal parts, each a *degree Celsius*. The Kelvin scale is called an *absolute scale* because its zero point is the coldest temperature possible, or absolute zero. (This fact was determined by theoretical considerations and has been confirmed by experiment, as we will see in Chapter 6.) The zero point on the Kelvin scale, 0 K, is equal to -273.15 °C, often rounded to -273 °C. A kelvin is the same size as a degree Celsius, and so the freezing point of water on the Kelvin scale is 273 K. The Kelvin scale has no negative temperatures, and we don't use a degree sign with the K. To convert from degrees Celsius to kelvins, simply add 273.15 to the Celsius temperature.

$$K = °C + 273.15$$

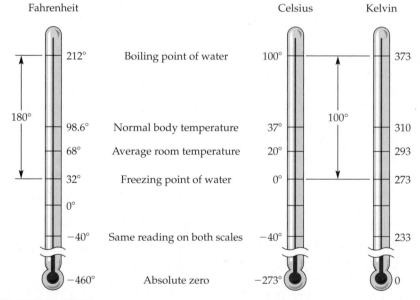

▶ **Figure 1.9** A comparison of the Fahrenheit, Celsius, and Kelvin temperature scales.

EXAMPLE 1.13   Temperature Conversions

Ether boils at 36 °C. What is the boiling point of ether on the Kelvin scale?

Solution

$$K = °C + 273.15 \qquad K = 36 + 273.15 = 309 \text{ K}$$

■ **EXERCISE 1.13A**

What is the boiling point of water (100 °C) expressed in kelvins?

■ **EXERCISE 1.13B**

Express a temperature of -78 °C in kelvins.

The Fahrenheit temperature scale is widely used in the United States. Figure 1.9 compares the three temperature scales. Note that on the Fahrenheit scale, the freezing point of water is 32 °F and the boiling point is 212 °F, so that a 10-degree temperature interval on the Celsius scale equals an 18-degree interval on the Fahrenheit scale. Conversion between Fahrenheit and Celsius temperatures is discussed in the Appendix.

The SI-derived unit of energy is the **joule (J),** but the **calorie (cal)** is the more familiar unit in everyday life.

$$1 \text{ cal} = 4.184 \text{ J}$$

A calorie is the amount of heat required to raise the temperature of 1 g of water 1 °C.

The "calorie" used for measuring the energy content of foods is actually a **kilocalorie (kcal).**

$$1000 \text{ cal} = 1 \text{ kcal} = 4184 \text{ J}$$

A dieter might be aware that a banana split contains 1500 "calories." If the same dieter realized that this was really 1500 *kilo*calories or 1,500,000 calories, giving up a banana split might be easier! The concept of heat is discussed further in the Appendix. (See Figure 1.10.)

◀ **Figure 1.10** Temperature and heat are different phenomena. Both the tub and the pool are at roughly are at the same temperature—about 40 °C—but it took much more heat to warm the pool to that temperature than it did the tub of water.

Body Temperature, Hypothermia, and Hyperthermia (Fever)

Carl Wunderlich (1815–1877), a German physician, first recognized fever as a symptom of disease. He averaged thousands of human temperature measurements and reported the value as 37 °C. When this value was converted to the Fahrenheit scale, it somehow (improperly) acquired an extra significant figure, and 98.6 °F became widely (and incorrectly) known as the *normal body temperature*. In recent years millions of measurements have revealed that *average* normal body temperature is actually 98.2 °F and that body temperatures in healthy people range from 97.7 °F to 99.5 °F.

The healthy human body maintains a fairly constant temperature. Heat is a by-product of metabolism, and we must constantly get rid of some of it. Because heat always flows spontaneously from a hot object to a cold one, we can get rid of excess heat by simple conduction only if the surrounding temperature is less than 37 °C. Above that temperature, the body depends on air movement, blood vessel dilation, and perspiration to keep its temperature normal. If the outside temperature is above 40 °C, the U.S. National Weather Service usually issues a heat advisory because the body can *gain* heat from the environment. The body temperature then rises above the normal range, a condition known as *hyperthermia*. Certain diseases also cause abnormally high

(continued)

body temperatures or *fevers*. Mild fevers (up to 102 °F) are usually not dangerous and may even help the body fight off an infection. Prolonged high fevers (104 °F or more in adults) can be fatal.

When exposed to prolonged cold, the body can lose too much heat to the environment. The body temperature drops below the normal range, a condition known as *hypothermia*. A drop of only 2 or 3 °F leads to shivering, a condition in which muscles contract in an attempt to generate more heat. Prolonged or severe hypothermia (body temperature below 93 °F) can lead to unconsciousness and death.

EXAMPLE 1.14   Energy Conversions

When 1.00 g of gasoline burns, it yields about 10.3 kcal of energy. What is the quantity of energy in kilojoules (kJ)?

Solution

$$10.3 \ \cancel{kcal} \times \frac{1000 \ \cancel{cal}}{1 \ \cancel{kcal}} \times \frac{4.184 \ \cancel{J}}{1 \ \cancel{cal}} \times \frac{1 \ kJ}{1000 \ \cancel{J}} = 43.1 \ kJ$$

■ **EXERCISE 1.14A**

A European woman on average consumes food with an energy content of 7525 kJ per day. What is her daily intake in kilocalories (food calories)?

■ **EXERCISE 1.14B**

It takes about 12 kJ of energy to melt a single ice cube. How much energy in kilocalories does it take to melt three ice cubes?

Self-Assessment Questions

1. Heat is
   a. an element
   b. energy flow from a hot object to a cold one
   c. a measure of energy intensity
   d. a temperature above room temperature

2. The SI scale for temperature is known as the
   a. Celsius scale   b. Fahrenheit scale
   c. Kelvin scale   d. metric scale

3. Water freezes at
   a. 0 °C   b. 0 °F
   c. 273 °C   d. 373 K

4. To convert a Celsius temperature to kelvins, we
   a. add 273
   b. divide by 273
   c. multiply by 273
   d. subtract 273

5. The SI unit of energy is the
   a. einstein
   b. joule
   c. newton
   d. watt

GREEN CHEMISTRY

Introduction to Green Chemistry

Paul Anastas, Yale University

In seeking a cleaner environment and a more sustainable future, chemists have developed a set of green chemistry principles. These principles are listed below and several are illustrated throughout this text. These principles should help to frame your perspective on the chemistry you learn in this course.

Basically, green chemistry involves the design of products and chemical processes that reduce or eliminate the use or production of hazardous substances. Green chemistry can be applied at all steps in the life cycle of a chemical product, from its design through manufacture and use to its disposal. The many benefits of green chemistry include reduced waste, safer products, and reduced use of energy and resources.

The U.S. Environmental Protection Agency (EPA) promotes green chemistry through Presidential Green Chemistry Challenge Awards, which recognize outstanding green chemistry achievement in academia, industry, and small businesses. The EPA Green Chemistry Program, the ACS Green Chemistry Institute®, and a growing list of organizations, universities, and companies are enabling green chemistry through projects and programs, including educational activities and research and development.

You will see award winning examples and explore many of these concepts in the Green Chemistry essays in future chapters. By using green chemistry, new technologies for a sustainable future are being developed in all kinds of applications, such as energy, biofuels, soap, nanotechnology, medicines, plastics, and much more as you will see in the coming chapters.

As the future leaders and citizens of the world, you will have the power and capability of making a huge impact on the sustainability of the planet and all of its inhabitants. Green chemistry and green technology will be key.

Twelve Principles of Green Chemistry*

1. **Prevention:** It is better to prevent waste than to treat or clean up waste after it has been created.
2. **Atom Economy:** Synthetic methods should be designed to maximize the incorporation of all materials used in the process into the final product.
3. **Less Hazardous Chemical Syntheses:** Wherever practicable, synthetic methods should be designed to use and generate substances that possess little or no toxicity to human health and the environment.
4. **Designing Safer Chemicals:** Chemical products should be designed to effect their desired function while minimizing their toxicity.
5. **Safer Solvents and Auxiliaries:** The use of auxiliary substances (e.g., solvents, separation agents, etc.) should be made unnecessary wherever possible and innocuous when used.
6. **Design for Energy Efficiency:** Energy requirements of chemical processes should be recognized for their environmental and economic impacts and should be minimized. If possible, synthetic methods should be conducted at ambient temperature and pressure.
7. **Use of Renewable Feedstocks:** A raw material or feedstock should be renewable rather than depleting whenever technically and economically practicable.
8. **Reduce Derivatives:** Unnecessary derivatization (use of blocking groups, protection/deprotection, temporary modification of physical/chemical processes) should be minimized or avoided if possible, because such steps require additional reagents and can generate waste.
9. **Catalysis:** Catalytic reagents (as selective as possible) are superior to stoichiometric reagents.
10. **Design for Degradation:** Chemical products should be designed so that at the end of their function they break down into innocuous degradation products and do not persist in the environment.
11. **Real-time Analysis for Pollution Prevention:** Analytical methodologies need to be further developed to allow for real-time, in-process monitoring and control prior to the formation of hazardous substances.
12. **Inherently Safer Chemistry for Accident Prevention:** Substances and the form of a substance used in a chemical process should be chosen to minimize the potential for chemical accidents, including releases, explosions, and fires.

*Originally published by Paul Anastas and John Warner in *Green Chemistry: Theory and Practice*, New York: Oxford University Press, 1998.

1.13 Critical Thinking

One of the hallmarks of science is the ability to think critically—to evaluate statements and claims in a rational, objective fashion. This ability can be learned, and it will serve you well in everyday life as well as in science courses. Critical thinking is important for workers in all types of professions and employees in all kinds of industries. We will use an approach adapted from one developed by James Lett,[1] first outlining the approach and then working through some simple examples.

You can use the acronym **FLaReS** (ignore the vowels) to remember four rules used to test a claim: *f*alsifiability, *l*ogic, *r*eplicability, and *s*ufficiency. We will first list the rules and describe them briefly, and then we will illustrate them with examples.

The FLaReS method is a good starting point for evaluating the myriad claims that you will find on the Internet and elsewhere.

- **Falsifiability:** Can any conceivable evidence show the claim to be false? It must be possible to think of evidence that would prove the claim false. Falsifiability is an essential component of the scientific method. A hypothesis that cannot be falsified is of no value. Science cannot prove anything true in an absolute sense, although it can provide overwhelming evidence. Science *can* prove something false.

- **Logic:** Any argument offered as evidence in support of any claim must be sound. An argument is sound if its conclusion follows inevitably from its premises and if its premises are true. It is unsound if a premise is false, and it is unsound if there is a single exception in which the conclusion does not necessarily follow from the premises.

- **Replicability:** If the evidence for any claim is based on an experimental result, it is necessary for the evidence to be replicable in subsequent experiments or trials. Scientific research is almost always reviewed by other qualified scientists. The peer-reviewed research is then published in a form that enables others to repeat the experiment. Sometimes bad science slips through the peer-review process, but such science eventually fails the test of replicability. For example, in May 2005 the journal *Science* published research by Korean biomedical scientist Hwang Woo-suk in which he claimed to have tailored embryonic stem cells so that every patient could receive custom treatment. Others were unable to reproduce Hwang's findings, and evidence emerged that he had intentionally fabricated results. *Science* retracted the article in January 2006. Research results published in media that are not peer reviewed have little or no standing in the scientific community.

- **Sufficiency:** The evidence offered in support of any claim must be adequate to establish the truth of that claim, with these stipulations: (1) The burden of evidence for any claim rests on the claimant, (2) extraordinary claims demand extraordinary evidence, and (3) evidence based on authority and/or testimony is never adequate.

If a claim passes all four FLaReS tests, then it *might* be true. On the other hand, it could still be proven false. However, if a claim fails even one of the FLaReS tests it is likely to be false.

[1]Lett used the acronym FiLCHeRS (ignore the vowels) as a mnemonic for six rules: *f*alsifiability, *l*ogic, *c*omprehensiveness, *h*onesty, *r*eplicability, and *s*ufficiency. See Lett, James. "A Field Guide to Critical Thinking." *The Skeptical Inquirer,* Winter 1990, pp. 153–160.

Critical Thinking Examples

1.1 A psychic claims he can bend a spoon using only the powers of his mind. However, he says he can do so only when the conditions are right; there must be no one with negative energy present. Apply the FLaReS tests to evaluate the psychic's claim.

Solution

1. Is the claim falsifiable? No. If the psychic fails, he can always claim that someone present had negative energy.

2. Is the claim logical? No. What kind of matter or energy could move from the psychic's mind to the spoon with enough force to bend it? Just what is "negative energy"?

3. Is the claim reproducible? No. The psychic can do it only when "conditions are right." (Actually, one such psychic was caught cheating; he was seen bending the spoon with his hands. Now his proponents claim he cheats only sometimes!)

4. Is the claim sufficient? No. Any such claim is extraordinary; it would require extraordinary evidence. The burden of proof for any such claim rests on the claimant, and only flimsy evidence is provided.

1.2 Some people claim that repetitive biorhythm cycles that date from a person's birth date can predict airplane crashes. They say that more crashes occur when the pilot, copilot, and/or navigator are experiencing critically low points in their intellectual, emotional, and/or physical cycles. Several studies show no such correlation. Apply the FLaReS tests to evaluate this claim.

Solution

1. Is the claim falsifiable? Yes. Crash data and birth dates of the crew members can be analyzed to see whether or not there is a correlation.

2. Is the claim logical? No. Hundreds of thousands of people are born each day. It is highly unlikely that all of them would share the same up-and-down cycles day by day for a lifetime.

3. Is the claim reproducible? No. It has not held up in other studies.

4. Is the claim sufficient? No. It does not address the real causes of airplane crashes.

1.3 Many "psychics" claim to be able to predict the future. These predictions are often made near the end of one year for the next year. Look up several such predictions and apply the FLaReS tests to evaluate the claim.

Solution

1. Is the claim falsifiable? Yes. You can write down a claim and then check it yourself at the end of the year.

2. Is the claim logical? No. If psychics could really predict the future, they could readily win lotteries, ward off all kinds of calamities by providing advance warning, and so on.

3. Is the claim reproducible? No. Psychics usually make broad, general claims. They later make claim-specific references to "hits" while ignoring "misses." (Some past predictions: World War III will begin in 1958; Kennedy will *not* be elected in 1960; Fidel Castro will die in 1969; George H. W. Bush will be elected in 1992. None of the well-known psychics, including James Van Praagh, John Edward, Sylvia Browne, or the Jamison sisters, Terry and Linda, predicted the terrorist attacks on the United States on September 11, 2001.)

4. Is the claim sufficient? No. Any such claim is extraordinary; it would require extraordinary evidence. The burden of proof for any such claim rests on the claimant, and only flimsy evidence is provided.

Critical Thinking Exercises

Apply knowledge that you have gained in this chapter and one or more of the FLaReS principles to evaluate the following statements or claims.

1.1 An alternative health practitioner claims that a nuclear power plant releases radiation at a level so low that it cannot be measured but that is harmful to the thyroid gland. He sells a thyroid extract that he claims can fix the problem.

1.2 A doctor claims that she can cure a patient of arthritis by simply massaging the affected joints. If she has the patient's complete trust, she can cure the arthritis within a year. Several of her patients have testified that the doctor has cured their arthritis.

1.3 Some people claim that crystals have special powers. Crystal therapists claim that they can use quartz to restore balance and harmony to a person's spiritual energy.

1.4 A sixth-grade student has a beautiful tigereye stone that her grandfather gave her. When she holds it in her hand, she can think more clearly. She claims that the stone really works because one day when she left the stone at home, she got the worst grade she had ever received on an exam. She believes that her stone has magic power. What do you think?

1.5 A woman claims that she has memorized the New Testament. She offers to quote any chapter of any book entirely from memory.

1.6 A television advertisement extols the virtue of a pill called "Diamondine." The pill is claimed to "improve masculinity" and "extend endurance." The advertisement is accompanied by this fine print: "These statements have not been evaluated by the Food and Drug Administration. This product is not intended to treat or cure any disease." A phone call to the company reveals that the statement about the FDA is required by law. Evaluate the advertisement's claims.

■ SUMMARY

Sections 1.1–1.3—**Chemistry** is the study of matter and the changes it undergoes. The roots of chemistry lie in **natural philosophy**—philosophical speculation about nature—and in **alchemy,** a mystical investigation practiced in the Middle Ages. **Technology** is the practical application of knowledge, whereas **science** is an accumulation of knowledge about nature and our physical world based on observations and experimental tests of our assumptions. Science is *testable, reproducible, explanatory, predictive,* and always *tentative.* In one common scientific method, a confirmed observation about nature may lead to a **hypothesis**—a tentative explanation of observations. If a hypothesis stands up to testing and further experimentation, it may become a **theory**—the best current explanation for a phenomenon. A theory is always tentative and can be rejected if it does not continue to stand up to further testing. A valid theory can be used to predict new scientific facts. A **scientific law** is a brief statement that summarizes large amounts of data. A **scientific model** uses tangible items or pictures to represent invisible processes and help explain complex phenomena.

Sections 1.4–1.6—Scientists disagree over social and political issues partly because of the inability to control **variables,** those things that can change during an experiment. Science and technology have provided many benefits for our world, but they have also introduced new risks. A **risk–benefit analysis** can help us decide which outweighs the other. Chemistry is fundamental to other scientific disciplines and plays a central role in science.

Section 1.7—Chemists are usually involved in some kind of research. The purpose of **applied research** is to make particular kinds of useful products, whereas **basic research** is carried out simply to obtain new knowledge or to answer fundamental questions. Basic research often may result in a useful product or process.

Section 1.8—**Matter** is anything that has mass and takes up space. **Mass** is a measure of the amount of matter in an object, whereas **weight** represents the gravitational force of attraction for an object. **Physical properties** of matter can be observed without making new substances. When **chemical properties** are observed, new substances are formed. A **physical change** does not entail a change in chemical composition. A **chemical change** does involve such a change.

Section 1.9—Matter can be classified according to its physical state. A **solid** has definite shape and volume; a **liquid** has definite volume but takes the shape of its container; and a **gas** takes both the shape and the volume of its container. Matter also can be classified according to composition. A pure **substance** always has the same composition, no matter how it is made or found. A **mixture** may have different compositions depending on how it is prepared. Pure substances are either elements or compounds. An **element** is composed of atoms all of one type, an **atom** being the smallest particle possible of an element. There are only about a hundred elements, each of which is represented by a **chemical symbol,** which is made up of one or two letters derived from the element's name. A **compound** is made of two or more elements, chemically combined in fixed proportions. Many compounds exist as **molecules,** groups of atoms bound together as a unit.

Section 1.10—Scientific measurements are made using **SI units,** an agreed-upon standard version of the metric system. Base units are the **meter (m)** for length, the **kilogram (kg)** for mass, and the **kelvin (K)** for temperature. Although it is not truly SI, the **liter (L)** is a common unit of volume. Prefixes make basic units larger or smaller by factors of ten.

Section 1.11—**Density** can be thought of as "how heavy something is for its size." It is the amount of mass per unit volume:

$$d = \frac{m}{V}$$

Density can be used as a conversion factor. The density of water is almost exactly 1 g/mL.

Section 1.12—When matter undergoes a physical or chemical change, there is also a change in **energy,** the ability to change matter physically or chemically. Energy is either given off or absorbed in each process. **Heat** is energy flow from a hot object to a cold one. **Temperature** is a measure of how hot or cold an object is. The **kelvin (K)** is the SI unit of temperature, but the **Celsius scale (°C)** is more commonly used. On the Celsius scale water boils at 100 °C and freezes at 0 °C. The SI unit of heat is the **joule (J).** The **calorie (cal)** is a more familiar unit, the amount of heat needed to raise the temperature of 1 g of water by 1 °C. The "food calorie" is actually a **kilocalorie (kcal)** or 1000 calories.

Section 1.13—Claims may be tested with critical thinking. The acronym **FLaReS** can be useful in such exercises. The letters FLRS represent rules used to test a claim: *f*alsifiability, *l*ogic, *r*eplicability, and *s*ufficiency.

∎ REVIEW QUESTIONS

1. State five distinguishing characteristics of science. Which characteristic best serves to distinguish science from other disciplines?
2. Why have Thomas Malthus's predictions not been fulfilled in developed countries?
3. Why can't scientific methods always be used to solve social, political, ethical, and economic problems?
4. How does technology differ from science?
5. What is risk–benefit analysis?
6. What sorts of judgments go into (a) the evaluation of benefits and (b) the evaluation of risks?
7. What is a DQ? What does a large DQ mean? Why is it often difficult to estimate a DQ?
8. What derived units of (a) mass and (a) length are commonly used in the laboratory?
9. Following is an incomplete table of SI prefixes, their symbols, and their meanings. Fill in the blank cells. The first row is completed as an example.

Prefix	Symbol	Definition
tera-	T	10^{12}
	M	
centi-		
	μ	
milli-		
		10^{-1}
	K	
nano-		

10. What is the SI-derived unit for volume? What volume units are more often used in the laboratory?
11. Identify the following research projects as either applied or basic.
    a. A Virginia Tech chemist examines a method for analyzing coal powder in less than a minute before the powder goes into a coal-fired energy plant.
    b. A Purdue engineer develops a method for causing aluminum to react with water to generate hydrogen for automobiles.
    c. A worker at University of Illinois Urbana-Champaign examines the behavior of atoms at high temperatures and low pressures.
12. Identify the following work as either applied research or basic research.
    a. An engineer determines the strength of a titanium alloy that will be used to construct notebook-computer cases.
    b. A biochemist runs experiments to determine the way in which oxygen binds to the blood cells of lobsters.
    c. A biologist determines the number of eagles that nest annually in Kentucky's Land Between the Lakes area over a six-year period.

∎ PROBLEMS

A word of advice

You cannot learn to work problems by reading them or watching your instructor work them, just as you cannot become a piano player solely by reading about piano-playing skills or attending a performance. Working through problems will help you to improve your understanding of the ideas presented in the chapter and to practice your estimation skills and your ability to synthesize concepts. Plan to work through the great majority of these problems.

Risk–Benefit Analysis

13. Penicillin kills bacteria, thus saving the lives of thousands of people who otherwise might die of infectious diseases. Penicillin causes allergic reactions in some people; in extreme cases the allergic reaction can lead to death if the resulting condition is not treated. Do a risk–benefit analysis of the use of penicillin for society as a whole.

14. Do a risk–benefit analysis of the use of penicillin for a person who is allergic to it. (See Problem 13.)

15. X-rays are used in medicine to diagnose injuries and illnesses, in industry to detect unseen flaws in metal equipment, in airport security to detect hidden weapons, in science to determine the structure of crystals, and in many other ways. X-rays can cause cancer, but for exposures from ordinary medical procedures, the risk is quite low. What is the DQ for use of X-rays in (a) the detection of flaws in the structural steel of a bridge, (b) detecting a cancerous growth in the brain, and (c) determining the fit of a new pair of shoes?

16. In his 1970 book, *Vitamin C and the Common Cold*, Linus Pauling, who won two Nobel Prizes (for chemistry in 1954 and for peace in 1962), advocated the use of large doses of vitamin C to protect against the common cold. A review of many vitamin C studies shows that the vitamin doesn't help much except for people undergoing prolonged periods of high stress, such as running marathons. What is the DQ for large daily doses of vitamin C for the average person?

17. Synthetic food colors make food more attractive and increase sales. A few such dyes are suspected carcinogens (cancer inducers). Who derives most of the benefits from the use of food colors? Who assumes most of the risk associated with use of these dyes?

18. A virus called HIV causes AIDS, a devastating and often deadly disease. Several drugs are available to treat HIV, but all are expensive. Used separately and in combination, these drugs have resulted in a huge drop in AIDS deaths. An expensive new drug shows promise in treating HIV/AIDS patients. It is especially promising in preventing passage of the HIV virus from a pregnant woman to her fetus. What is the DQ for (a) a man who thinks he may be infected with HIV, (b) for a pregnant woman who is HIV positive, and (c) an unborn child whose mother has AIDS?

Mass and Weight

19. Which of the following is a realistic mass for your textbook?

20 mg   20 g   200 g   2 kg

20. At home in the United States, you usually work out with 12-lb weights. While in Europe you frequent the exercise center in your hotel. What weights should you select from the rack?

1 kg   5 kg   10 kg   20 kg

21. Two samples are weighed under identical conditions in a laboratory. Sample A weighs 1.00 lb and Sample B weighs 2.00 lb. Does Sample B have twice the mass of Sample A?

22. Sample A on the moon has exactly the same mass as Sample B on Earth. Do the two samples weigh the same? Explain.

Length, Area, and Volume

23. Which of the following is a reasonable volume for a teacup?

25 mL   250 mL   2.5 L   25 L

24. Which of the following is a reasonable value for the length of a pickup truck?

3.5 mm   3.5 cm   3.5 m   3.5 km

25. Earth's oceans contain $3.50 \times 10^8 \text{ mi}^3$ of water and cover an area of $1.40 \times 10^8 \text{ mi}^2$. What is the volume of ocean water in cubic kilometers?

26. What is the area of the oceans in square kilometers? See Problem 25.

27. Consider the two tubes shown here. The aluminum tube has an outside diameter of 0.998" and an inside diameter of 0.782". Which one(s) of the following could be the inside diameter of the paper tube?

19.9 mm   24.9 mm   18.7 mm

28. Without doing a detailed calculation, determine whether the aluminum tube in Problem 27 will fit inside another aluminum tube with an inside diameter of 26.3 mm.

Physical and Chemical Properties and Change

29. Identify the following as physical or chemical properties.
   a. Glass is transparent.
   b. Methanol, made from wood, burns in air with a pale blue flame.
   c. Argon is a gas at room temperature.
   d. Ethanol is a liquid with a density of 0.789 g/mL.

30. Identify the following as physical or chemical properties.
   a. Zinc metal can be shaped into many different forms by hammering.
   b. Silver metal tarnishes, forming black silver sulfide.
   c. Sodium metal is soft enough to cut with a butter knife.
   d. Gold melts at 1064 °C.

31. Identify the following changes as physical or chemical.
   a. Bacteria are killed by the chlorine in a swimming pool.
   b. Brown print from an inkjet printer is made by mixing cyan, magenta, and yellow inks.
   c. Sugar is dissolved in water to make syrup.

32. Identify the following changes as physical or chemical.
   a. A machine makes a nail from wire by stretching the wire, then cutting it at angles to form the point, then flattening the other end to form a head.
   b. Pulp-free orange juice is prepared by filtering most of the pulp from fresh orange juice.
   c. The compound glycerin, used in lotions, is formed as a by-product during the preparation of "biodiesel" fuel from waste restaurant oils.

Substances and Mixtures

33. Identify each of the following as a substance or a mixture.
   a. carbon dioxide   **b.** oxygen
   c. smog   **d.** a carrot
   e. blueberry pancakes **f.** cellophane tape

34. Identify each of the following as a substance or a mixture.
   a. helium gas used to fill a balloon
   b. the juice squeezed from an orange
   c. distilled water   **d.** carbon dioxide gas

35. Which of the following mixtures are homogeneous and which are heterogeneous?
   a. gasoline   **b.** Italian salad dressing
   c. raisin pudding   **d.** an intravenous glucose solution

36. Which of the following mixtures are homogeneous and which are heterogeneous?
   a. maple syrup   **b.** distilled water
   c. liquid oxygen   **d.** chicken noodle soup

37. Every sample of the sugar glucose (collected anywhere on Earth) consists of 8 parts (by mass) oxygen, 6 parts carbon, and 1 part hydrogen. Is glucose a substance or a mixture? Explain.

38. An advertisement for shampoo says, "Pure shampoo, with nothing artificial added." Is this shampoo a substance or a mixture? Explain.

Elements and Compounds

39. Which of the following represent elements and which represent compounds?
   a. H   **b.** He   **c.** HF   **d.** Hf

40. Which of the following represent elements and which represent compounds?
   a. Li   **b.** CO   **c.** Cf   **d.** CF_4

41. Without consulting tables, write symbols for each of the following.
   a. chlorine   **b.** iron
   c. phosphorus   **d.** plutonium

42. Without consulting tables, name each of the following.
   a. H   **b.** N   **c.** S   **d.** Na

43. In his 1789 textbook, *Traité élementaire de Chimie*, Antoine Lavoisier (Chapter 2) listed 33 known elements, one of which was baryte. Which of the following observations best shows that baryte cannot be an element?
   a. Baryte is insoluble in water.
   b. Baryte melts at 1580 °C.
   c. Baryte has a density of 4.48 g/cm^3.
   d. Baryte is formed in hydrothermal veins and around hot springs.

e. Baryte is formed as a solid when sulfuric acid and barium hydroxide are mixed.

f. Baryte is formed as the sole product when a particular metal is burned in oxygen.

◀ Baryte

44. In 1774 Joseph Priestley isolated a gas that he called "dephlogisticated air." Which of the following observations best shows that dephlogisticated air is an element?
   a. Dephlogisticated air combines with charcoal to form "fixed air."
   b. Dephlogisticated air combines with "inflammable air" to form water.
   c. Dephlogisticated air combines with a metal to form a solid called a "calx."
   d. Dephlogisticated air has never been separated into simpler substances.
   e. Dephlogisticated air and "mephitic air" are the main components of the atmosphere.

The Metric System: Measurement and Unit Conversion

45. Change the unit used to report each of the following measurements by replacing the power of ten with an appropriate SI prefix.
   a. 8.01×10^{-6} g   **b.** 7.9×10^{-3} L
   c. 1.05×10^{3} m

46. Use exponential notation to express each of the following measurements in terms of an SI base unit.
   a. 45 mg   **b.** 125 ns
   c. 10.7 μL

47. Carry out the following conversions.
   a. 37.4 mL to L   **b.** 1.55×10^{2} km to m
   c. 0.198 g to mg   **d.** 1.19 m^2 to cm^2
   e. 78 μs to ms

48. Carry out the following conversions.
   a. 546 mm to m   **b.** 65 ns to μs
   c. 87.6 mg to kg   **d.** 46.3 dm^3 to L
   e. 181 pm to μm

49. For each of the following, indicate which is the larger unit.
   a. mm or cm   **b.** kg or g
   c. dL or μL

50. For each of the following, indicate which is the larger unit.
   a. L or cm^3 **b.** dm^3 or mL
   c. μs or ps

51. How many milliliters are there in 1.00 cm^3? In 15.3 cm^3?

52. How many millimeters are there in 1.00 cm? In 1.83 m?

Density

(You may need data from Table 1.6 for some of these problems.)

53. What is the density of **(a)** a salt solution if 37.5 mL has a mass of 43.75 g? **(b)** 2.75 L of the liquid glycerol, which has a mass of 3465 g?

54. What is the density of **(a)** a sulfuric acid solution if 10.00 mL has a mass of 15.04 g? **(b)** a 10.0-cm^3 block of plastic with a mass of 9.23 g?

55. What is the mass, in grams, of **(a)** 125 mL of castor oil, a laxative, which has a density of 0.962 g/mL? **(b)** 477 mL of blood plasma, $d = 1.027$ g/mL?

56. What is the mass, in grams, of **(a)** 30.0 mL of the liquid propylene glycol, a moisturizing agent for foods, which has a density of 1.036 g/mL at 25 °C? **(b)** 1.000 L of mercury at 25 °C?

57. What is the volume of **(a)** a 475-g piece of copper (in cubic centimeters)? **(b)** a 253-g sample of mercury (in milliliters)?

58. What is the volume of **(a)** 227 g of hexane (in milliliters)? **(b)** a 454-g block of ice (in cubic centimeters)?

59. A 40-mL quantity each of mercury and hexane, and 80 mL of water are placed in the 250-mL container shown here. The three liquids do not mix with one another. Provide a sketch that shows the relative location of the three liquids in the container.

60. A piece of maple wood ($d = 0.68$ g/cm^3), a piece of balsa wood ($d = 0.11$ g/cm^3), a copper coin, a gold coin, and a piece of ice are dropped into the mixture from Problem 59. Where will each solid end up in the cylinder?

61. A metal stand specifies a maximum load of 450 lb. Will it support an aquarium that weighs 59.5 lb and is filled with 37.9 L of seawater ($d = 1.03$ g/mL)?

62. E85 fuel is a mixture of 85% ethanol and 15% gasoline by volume. What mass in kilograms of E85 ($d = 0.758$ g/mL) can be contained in a 14.0-gal tank?

63. An artist wants to produce a pot with a mass of 2.45 kg. She starts with wet clay that loses 10.9% of its mass when fired. The wet clay has a density of 1.60 g/cm^3. What diameter sphere of wet clay should she take? (*Hint:* The volume of a sphere is $4\pi r^3/3$ where r is the radius.)

64. Air has an average density of about 1.29 g/L. What is the mass in grams of air inside a car that has a volume of about 2550 liters?

Energy: Heat and Temperature

65. Convert 37 °C to kelvins.

66. Temperature-dependent scientific data often are recorded at 298 K. What is that value in degrees Celsius?

67. The label on a 100-mL container of orange juice packaged in New Zealand reads in part, "energy...161 kJ." What is that value in kilocalories ("food calories")?

68. To vaporize 1.00 g of sweat (water) from your skin, the water must absorb 584 calories. What is that value in kilojoules?

■ ADDITIONAL PROBLEMS

69. The brass weight and the pillow have the same mass. Which has the greater density? Explain.

70. Arrange the following in order of increasing length (shortest first): (1) a 1.21-m chain, (2) a 75-in. board, (3) a 3-ft 5-in. rattlesnake, (4) a yardstick.

71. Arrange the following in order of increasing mass (lightest first): (1) a 5-lb bag of potatoes, (2) a 1.65-kg cabbage, (3) 2500 g sugar.

72. One of the people in the photo has a mass of 47.2 kg and a height of 1.53 m. Which one is it likely to be?

73. Some metal chips having a volume of 3.29 cm³ are placed on a piece of paper and weighed. The combined mass is found to be 18.43 g. The paper itself weighs 1.21 g. Calculate the density of the metal.

74. A glass container weighs 48.462 g. A sample of 4.00 mL of antifreeze solution is added, and the container plus the antifreeze weighs 54.51 g. Calculate the density of the antifreeze solution.

75. A rectangular block of balsa wood ($d = 0.11$ g/cm³) is 7.6 cm × 7.6 cm × 94 cm. What is the mass of the block in grams?

76. A collection of gold-colored metal beads has a mass of 425 g. The volume of the beads was found to be 48.0 cm³. Using the given densities and those in Table 1.6, identify the metal: **(a)** bronze ($d = 9.87$ g/cm³); **(b)** copper, **(c)** gold, **(d)** nickel silver ($d = 8.86$ g/cm³).

77. A 5.79-mg piece of gold is hammered into gold leaf of uniform thickness with an area of 44.6 cm². What is the thickness of the gold leaf?

78. A block of wood measures 3.00 cm by 6.00 cm by 4.00 cm and has a mass of 80.0 g. What is the density of the wood? Would the piece of wood float in water?

79. What is the mass, in metric tons, of a cube of gold that is 36.1 cm on each side? (1 metric ton = 1000 kg)

80. A 10.5-inch (26.7-cm) iron skillet has a mass of 7.00 lb (3180 g). How many cubic centimeters of iron does it contain?

81. Refer to Problems 25 and 26. What is the average depth of Earth's oceans in kilometers?

82. In July 2007 gasoline **(a)** in the Netherlands cost 1.51 euros per liter (€1.51/L) and **(b)** in the United Kingdom it cost 0.97 pound per liter (£ 0.97/L). Calculate the cost of each in U.S. dollars per gallon ($/gal). The exchange rates were 1 € = \$1.37889 and 1 £ = \$2.02929. Compare the cost to that in the United States, where the average price was US\$3.10/gal.

83. The density of a planet can be approximated from its average radius and estimated mass. Calculate the approximate average density, in grams per cubic centimeter, of **(a)** Jupiter, which has a radius of about 70,000 km and a mass of 1.9×10^{27} kg; **(b)** Earth, radius 6.4×10^3 km and mass 5.98×10^{24} kg; and **(c)** Saturn, radius 5.82×10^4 km and mass 5.68×10^{26} kg. The volume of a sphere is $4\pi r^3/3$ where r = radius. How does the density of each compare to that of water?

84. The extra-solar planet HAT-P-1 orbits a star 450 light-years from Earth. The planet has about half the mass of Jupiter and its radius is about 1.38 times that of Jupiter (see Problem 83). Calculate the approximate average density, in grams per cubic centimeter, of HAT-P-1. How does this density compare to that of water?

85. The Canadian Forces Snowbirds are a precision aerobatics team similar to the USAF Thunderbirds and the USN Blue Angels. The Snowbirds fly CT-114 Tutor jets, for which the operating manual lists a fuel tank capacity of "1173 litres (2012 lb) 310 U.S. gals." What is the density of the fuel in grams per milliliter?

86. The density of ice at 0 °C is 0.9167 g/cm³ and that of liquid water at 0 °C is 0.99984 g/mL. Calculate the percent expansion of water on freezing. (*Hint:* Assume 1.0000 L of $H_2O(l)$ at 0 °C.)

▪ COLLABORATIVE GROUP PROJECTS

Prepare a PowerPoint, poster, or other presentation (as directed by your instructor) for presentation to the class.

87. List five chemical activities you have engaged in today. Compare your list with those of the other members of your group.

88. Prepare a brief biographical report on one of the following and share it with your group.

    a. Francis Bacon   **b.** Rachel Carson
    c. Thomas Malthus   **d.** George Washington Carver

(The following problem is best done in a group of four students.)

89. Make copies of the following form. Student 1 should write a word from the list in the first column of the form and its definition in the second column. Then she or he should fold the first column under to hide the word and pass the sheet to Student 2, who uses the definition to determine what word was defined and place that word in the third column. Student 2 then folds the second column under to hide it and passes the sheet to Student 3, who writes a definition for the word in the third column and then folds the third column under and passes the form to Student 4. Finally, Student 4 writes the word corresponding to the definition given by Student 3.

Compare the word in the last column with that in the first column. Discuss any differences in the two definitions. If the word in the last column differs from that in the first column, determine what went wrong in the process.

    a. hypothesis   **b.** theory   **c.** mixture   **d.** substance

Text Entry	Student 1	Student 2	Student 3	Student 4
Word	Definition	Word	Definition	Word
___	___	___	___	___

The lump of metal is a large nugget of platinum (Pt), one of the few metals that can be found in its native or uncombined state. The inset shows a highly magnified false-color image of platinum, in which the hexagonal arrangement of its atoms is clearly visible. The image was made by a technique called scanning tunneling microscopy. The hypothesis that all matter is composed of atoms is over 2000 years old, but it was only in the latter part of the twentieth century that scientists were able to obtain images of atoms.

ATOMS

1. How small are atoms?
2. If we can't see atoms, how do we know they exist?
3. Why is it often difficult to destroy hazardous waste?
4. What's the difference between atoms and molecules?

Are They for Real?

We hear something about atoms almost every day. The twentieth century brought us into the Atomic Age. The terms *atomic power*, *atomic energy*, and *atomic bomb* are a part of our ordinary vocabulary. But just what are atoms?

Every material thing in the world is made up of atoms, tiny particles that are much, much too small to see even with the finest optical microscope. The smallest speck of matter that can be detected by the human eye is made up of many billions of atoms. There are more than 10^{22} atoms in a penny. That is 10,000,000,000,000,000,000,000 atoms. Imagine that the atoms in a penny were enlarged until they were barely visible, like tiny grains of sand. The atoms in a single penny would make enough "sand" to cover the entire state of Texas several feet deep. Comparing an atom to a penny is like comparing a grain of sand to a Texas-size sandbox.

Why should we care about something as tiny as an atom? Because our world is made up of atoms, and atoms are a part of all we do. *Everything* is made of atoms, including you and me. And because chemistry is the study of the behavior of matter, chemistry studies the behavior of atoms. The behavior and interactions of atoms ultimately determine the behavior and interactions of matter.

Atoms are not all alike. Each element has its own kind of atoms. On Earth, about 90 elements occur in nature. About two dozen more have been synthesized by scientists. As far as we know, the entire universe is made up of these same few elements. An **atom** is the smallest particle that is characteristic of a given element.

ANSWER

1. How small are atoms?
Atoms are so incredibly small that the smallest particle of matter visible to your eye contains more atoms than you could count in a lifetime!

▶ **Figure 2.1** A sandy beach.
QUESTION: Sand looks continuous—infinitely divisible—when you look at a beach from a distance. Is it really continuous? Water looks continuous, even when viewed up close. Is it really continuous? Is a cloud continuous? Is air?

▲ This 1983 postage stamp from Greece has an image of Democritus.

2.1 Atoms: The Greek Idea

A pool of water can be separated into drops, and then each drop can be split into smaller and smaller drops. Suppose you could keep splitting these drops into still smaller ones even after they became much too small to see. Would you ever reach a point at which the tiny drop could no longer be separated into smaller droplets of water? That is, is water infinitely divisible, or would you eventually come to a particle which, if divided, would no longer be water?

The Greek philosopher Leucippus, who lived in the fifth century B.C.E., and his pupil Democritus (ca. 460–ca. 370 B.C.E.) might well have discussed this question as they strolled along the beach of the Aegean Sea. Based only on intuition, Leucippus thought that there must ultimately be tiny particles of water that could not be subdivided. After all, from a distance the sand on the beach looked continuous, but closer inspection showed it to be made up of tiny grains (Figure 2.1).

Democritus expanded on Leucippus's idea. He called the particles *atomos* (meaning "cannot be cut"), from which we derive the modern name *atom* for the tiny unit particle of an element. Democritus thought that each kind of atom was distinct in shape and size (Figure 2.2). He thought that real substances were mixtures of various kinds of atoms. The Greeks at that time believed that there were four basic elements: earth, air, fire, and water. These "elements" were connected through four "principles"—hot, moist, dry, and cold. The relationships among these are shown in Figure 2.3. These ideas seem strange to us today, but they persisted for two millennia.

Four centuries after Democritus, the Roman poet Lucretius (ca. 95–ca. 55 B.C.E.) wrote a long didactic poem (a poem meant to teach), "On the Nature of Things," in which he presented strong arguments for the atomic nature of matter. Unfortunately, a few centuries earlier, Aristotle (ca. 384–ca. 322 B.C.E.), considered the greatest of the ancient Greek philosophers, had declared that matter was continuous—infinitely divisible—rather

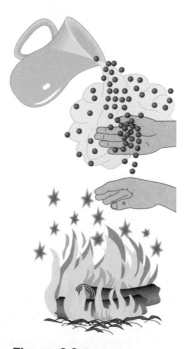

▲ **Figure 2.2** Democritus imagined that "atoms" of water might be smooth, round balls and that atoms of fire could have sharp edges.

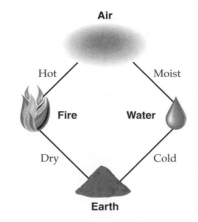

◀ **Figure 2.3** The Greek view of matter was that there were only four elements (in bold) connected by four "principles."

than atomistic—consisting of fundamental particles that are themselves indivisible—and the people of that time had no way to determine which view was correct. To most of them, the continuous view of matter seemed more logical and reasonable, and so Aristotle's view prevailed for 2000 years, even though it was wrong.

More to Explore
Clagett, Marshall. *Greek Science in Antiquity*. Mineola, NY: Courier Dover, 2001.

Self-Assessment Questions

1. Consider the ideas of *discrete* (atomic) and *continuous* as applied to materials at the macroscopic level (visible to the unaided eye). Which of the following would be continuous?
   a. apple juice   **b.** a ream of paper
   c. a bowl of cherries **d.** chocolate chips
2. The view that matter was continuous rather than atomic prevailed for centuries because it was
   a. essentially correct   **b.** not considered important
   c. not tested by experiment **d.** tested and found to be true

Answers: 1, a; 2, c

2.2 Lavoisier: The Law of Conservation of Mass

The eighteenth century saw the triumph of careful observation and measurement. Antoine Laurent Lavoisier (1743–1794) perhaps did more than anyone to establish chemistry as a quantitative science. He found that when a chemical reaction was carried out in a closed container, the total mass of the system was not changed. Perhaps the most important chemical reaction that Lavoisier performed was decomposition of the red oxide of mercury to form metallic mercury and a gas he named oxygen. Karl Wilhelm Scheele (1742–1786), a Swedish apothecary, and Joseph Priestley (1733–1804), a Unitarian minister who later fled England and eventually settled in America, had carried out the same reaction earlier, but Lavoisier was the first to weigh all the substances present before and after the reaction. He was also the first to interpret the reaction correctly.

Lavoisier carried out many quantitative experiments. He found that when coal was burned, it united with oxygen to form carbon dioxide. He experimented with animals, observing that when a guinea pig breathed, oxygen was consumed and carbon dioxide was formed. Lavoisier therefore concluded that respiration was related to combustion. In each of these reactions, he found that matter was *conserved*—its amount remained constant.

Lavoisier summarized his findings in a scientific law. The **law of conservation of mass** states that matter is neither created nor destroyed during a chemical change (Figure 2.4). The total mass of the reaction products is always equal to the total mass of the reactants (starting materials). We will discuss chemical reactions in more detail in Chapter 5. For now, some simple examples and discussion will suffice to illustrate the conservation of mass.

▲ The work of Antoine Lavoisier marked the beginnings of chemistry as a quantitative science.

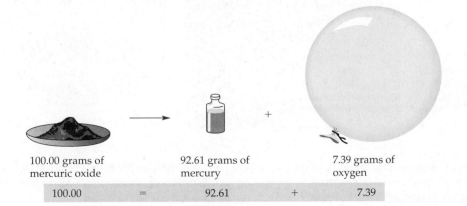

100.00 grams of mercuric oxide		92.61 grams of mercury		7.39 grams of oxygen
100.00	=	92.61	+	7.39

◄ **Figure 2.4** Although mercuric oxide (a red solid) has none of the properties of mercury (a silver liquid) or oxygen (a colorless gas), when 100.00 g of mercuric oxide is decomposed by heating, the products are 92.61 g of mercury and 7.39 g of oxygen. Properties are completely changed in this reaction, but there is no change in mass.

QUESTION: When 10.00 g of mercuric oxide decomposes, 0.739 g of oxygen forms. What mass of mercury forms?

2. If we can't see atoms, how do we know they exist?
If matter had no "smallest particles" the law of conservation of mass, the law of definite proportions, and the law of multiple proportions would be almost impossible to explain. Atomic theory provides a simple explanation for these laws and for many others.

▲ Jöns Jakob Berzelius (1779–1848) was the first person to prepare an extensive list of atomic weights. Published in 1828, it agrees remarkably well with most of our accepted values today.

Scientists had by this time abandoned the Greek idea of the four elements and were almost universally using Robert Boyle's operational definition put forth over a century earlier. In his book *The Sceptical Chymist* (published in 1661), Boyle said that a supposed *element* must be tested to see if it really was simple. If a substance could be broken down into simpler substances, it was not an element. The simpler substances might be elements and would be so regarded until such time (if it ever came) as they in turn could be broken down into still simpler substances. On the other hand, two or more elements might combine to form a complex substance called a *compound*.

Using Boyle's definition, Lavoisier included a table of elements in his book *Elementary Treatise on Chemistry*. The table included some substances we now know to be compounds. (It also included light and heat, which he called "caloric.") Lavoisier was the first to use systematic names for chemical elements. He is often called the "father of modern chemistry," and his book is usually regarded as the first chemistry textbook.

The law of conservation of mass is the basis for many chemical calculations. For example, we can calculate the mass of iron ore needed to produce a ton of iron metal. (Such calculations are discussed in detail in Chapter 5.) This law is not just a matter of academic interest. It states that we cannot create materials from nothing; we can make new materials only by changing the way atoms are combined. Nor can we get rid of wastes by the destruction of matter. We must put wastes somewhere. However, through chemical reactions, we can change some kinds of potentially hazardous wastes to less harmful forms. Such transformations of matter from one form to another are what chemistry is all about.

2.3 Proust: The Law of Definite Proportions

By the end of the eighteenth century, Lavoisier and other scientists noted that many substances were composed of two or more elements. Each compound had the same elements in the same proportions, regardless of where it came from or who prepared it. The painstaking work of Joseph Louis Proust (1754–1826) convinced most chemists of the general validity of these observations. In one set of experiments, for example, Proust found that basic copper carbonate (Figure 2.5), whether prepared in the laboratory or obtained from natural sources, was always composed of 57.48% by mass copper, 5.43% carbon, 0.91% hydrogen, and 36.18% oxygen. To summarize these and many other experiments, Proust in 1799 formulated a new scientific law. The **law of definite proportions** states that a compound always contains the same elements in certain definite proportions and in no other combinations. (This generalization is also sometimes called the *law of constant composition*.)

(a) (b) (c)

▲ **Figure 2.5** The compound known as *basic copper carbonate* [$Cu_2(OH)_2CO_3$] occurs in nature as the mineral *malachite* (a). It is formed as a patina on copper roofs (b). It can also be synthesized in the laboratory (c). Regardless of its source, basic copper carbonate always has the same composition. Analysis of this compound led Proust to formulate the law of definite proportions.

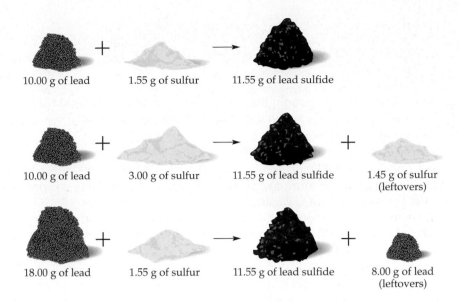

QUESTION: What mass of sulfur can react completely with 20.00 g of lead? What mass of lead sulfide forms?

An early illustration of the law of definite proportions is found in the work of the noted Swedish chemist J. J. Berzelius, represented in Figure 2.6. Berzelius heated a quantity (say, 10.00 g) of lead with various amounts of sulfur to form lead sulfide. Lead is a soft, grayish metal, and sulfur is a yellow solid. Lead sulfide is a shiny, black solid. Therefore, it was easy to tell when all the lead had reacted. Excess sulfur was washed away with carbon disulfide, a solvent that dissolves sulfur but not lead sulfide. As long as he used at least 1.55 g of sulfur with 10.00 g of lead, Berzelius got exactly 11.55 g of lead sulfide. Any sulfur in excess of 1.55 g was left over, unreacted. If he used more than 10.00 g of lead with 1.55 g of sulfur, he got 11.55 g of lead sulfide, with lead left over.

The law of definite proportions is further illustrated by the electrolysis of water. In 1783 Henry Cavendish (1731–1810), a wealthy, eccentric English nobleman, found that water forms when hydrogen burns in oxygen. (It was Lavoisier, however, who correctly interpreted the experiment and who first used the names *hydrogen* and *oxygen*.) Later, in 1800, two English chemists, William Nicholson and Anthony Carlisle, decomposed water into hydrogen and oxygen gases by passing an electric current through the water (Figure 2.7). (The Italian scientist Alessandro Volta had invented the chemical battery only six weeks earlier.) The two gases are always produced in a 2:1 volume ratio. Although this is a volume ratio and Berzelius's experiment is a mass ratio, both substantiate the law of definite proportions. This scientific law led to rapid developments in chemistry and dealt a death blow to the ancient Greek idea of water as an element.

The law of definite proportions is the basis for chemical formulas (Chapter 5) such as H_2O. It also has wider meaning. Not only do compounds have constant composition, but they also have constant properties. Pure water always dissolves salt or sugar, and at normal pressure it always freezes at 0 °C and boils at 100 °C at sea level.

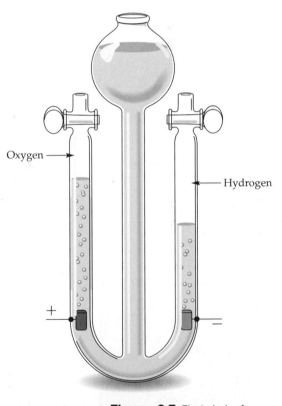

▲ **Figure 2.7** Electrolysis of water. Hydrogen and oxygen are always produced in a volume ratio of 2:1.

QUESTION: Under the same conditions of temperature and pressure, molecules of hydrogen occupy the same volume as an equal number of oxygen molecules. What ratio of hydrogen molecules to oxygen molecules is produced by the electrolysis of water?

Self-Assessment Questions

1. For any chemical change, compared to the mass of the reactants, the mass of the products is
   a. always equal
   b. always greater
   c. always less
   d. often different

2. In an experiment in which 36.04 g of liquid water is decomposed into hydrogen gas and oxygen gas, the total mass of the products is
   a. 0 g
   b. 18.02 g
   c. 36.04 g
   d. uncertain

3. The ancient Greeks thought that water was an element. In 1800 Nicholson and Carlisle decomposed water into hydrogen and oxygen. Their experiment proved that
   a. electricity causes decomposition
   b. hydrogen is an element
   c. oxygen is an element
   d. water is not an element

4. The fact that water from any place on Earth is always 89% oxygen and 11% hydrogen by mass illustrates
   a. Dalton's atomic theory
   b. Democritus's atomic theory
   c. the law of definite proportions
   d. the law of conservation of mass

5. Ammonia produced in an industrial plant contains 3.0 kg of hydrogen for every 14.0 kg of nitrogen. The ammonia dissolved in a window-cleaning preparation contains 30.0 g of hydrogen for every 140.0 g of nitrogen. What law does this illustrate?
   a. Dalton's atomic theory
   b. Democritus's atomic theory
   c. the law of definite proportions
   d. the law of conservation of mass

6. When 6.00 g of carbon is burned in 16.00 g of oxygen, 22.00 g of carbon dioxide is formed. What mass of carbon dioxide is formed when 6.00 g of carbon is burned in 75.00 g of oxygen?
   a. 6.00 g
   b. 16.00 g
   c. 22.00 g
   d. 81.00 g

Answers: 1, a; 2, c; 3, d; 4, c; 5, c; 6, c

2.4 John Dalton: The Atomic Theory of Matter

Lavoisier's law of conservation of mass and Proust's law of definite proportions were repeatedly verified by experiment. This work led to attempts to develop theories to explain these laws.

In 1803 John Dalton, an English schoolteacher, proposed a model to explain the accumulating experimental data. By this time, the composition of a number of substances was known with a fair degree of accuracy. (To avoid confusion, we will use modern values, terms, and examples rather than those actually used by Dalton.) For example, all samples of water have an oxygen-to-hydrogen mass ratio of 7.94:1.00. Similarly, all samples of ammonia have a nitrogen-to-hydrogen mass ratio of 4.63:1.00 (14:3). Dalton explained these unvarying ratios by assuming that matter is made of atoms.

As Dalton refined his model, he discovered another law that his theory would have to explain. Proust had stated that a compound contains elements in certain proportions and only those proportions. Dalton's new law, called the **law of multiple proportions,** stated that elements might combine in *more* than one set of proportions, with each set corresponding to a different compound. For example, carbon combines with oxygen in a mass ratio of 1.00:2.66 (or 3.00:8.00) to form carbon dioxide, a gas familiar as a product of respiration and of the burning of coal and wood. But Dalton found that carbon also combines with oxygen in a mass ratio of 1.00:1.33 (or 3.00:4.00) to form carbon monoxide, a poisonous gas produced when a fuel is burned in the presence of a limited air supply.

Dalton then used his **atomic theory** to explain the various laws. Following are the important points of Dalton's atomic theory, with some modern modifications that we will consider later.

▲ John Dalton (1766–1848). In addition to atomic theory, Dalton carried out important investigations of the behavior of gases. All of his contributions to science were made in spite of the fact that he was color-blind.

Dalton's Atomic Theory	Modern Modifications
1. All matter is composed of extremely small particles called atoms.	1. Dalton assumed atoms to be indivisible. This isn't quite true, as we will see in the next chapter.
2. All atoms of a given element are alike, but atoms of one element differ from the atoms of any other element.	2. Dalton assumed that all the atoms of a given element were identical in all respects, including mass. We now know this to be incorrect, as we will see on page 48.
3. Compounds are formed when atoms of different elements combine in fixed proportions.	3. Unmodified. The numbers of each kind of atom in simple compounds usually form a simple ratio. For example, the ratio of carbon atoms to oxygen atoms is 1:1 in carbon monoxide and 1:2 in carbon dioxide.
4. A chemical reaction involves a *rearrangement* of atoms. No atoms are created, destroyed, or broken apart in a chemical reaction.	4. Unmodified for *chemical* reactions. Atoms are broken apart in *nuclear* reactions. We discuss chemical reactions in Chapter 5 and nuclear reactions in Chapter 11.

A N S W E R

3. Why is it often difficult to destroy hazardous waste?
Hazardous wastes that are compounds or mixtures can be converted to other compounds or mixtures—but the elements from those compounds are still present and may be hazardous as well. For example, insecticide containing arsenic oxide can be broken down into the elements arsenic and oxygen—but arsenic is an element and can't be broken down into something else.

Explanations Using Atomic Theory

Dalton's theory clearly explains the difference between elements and compounds. *Elements* are composed of only one kind of atom. (We will explain more precisely what we mean by *kind* when we talk about protons in Section 3.5.) For example, a sample of the element phosphorus contains only phosphorus atoms. *Compounds* are made up of two or more kinds of atoms chemically combined in definite proportions.

Dalton set up a table of relative atomic masses based on hydrogen as 1. Many of Dalton's atomic masses were inaccurate, as we might expect because of the equipment available at that time. These relative atomic masses are usually simply called *atomic masses*. Historically, the relative masses were usually determined by comparison with a standard mass, a technique called *weighing*. For this historical reason, we often refer to these relative masses as *atomic weights*. You will find a table of atomic masses on the inside front cover of this book. We will use these modern values to show how Dalton's atomic theory explains the various laws.

To explain the law of definite proportions, Dalton's reasoning went something like this: Why should 1 g of hydrogen always combine with 19 g of fluorine? Why shouldn't 1 g of hydrogen also combine with 18 g of fluorine? Or 20 g of fluorine? Or any other mass of fluorine? If an atom of fluorine has a mass 19 times that of a hydrogen atom, the compound formed by the union of one atom of each element would have to consist of 1 part by mass of hydrogen and 19 parts by mass of fluorine. Matter must be atomic for the law of definite proportions to be valid (Figure 2.8).

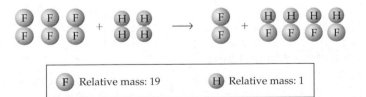

| F | Relative mass: 19 | H | Relative mass: 1 |

◄ **Figure 2.8** The law of definite proportions and the law of conservation of mass interpreted in terms of Dalton's atomic theory.

QUESTION: If 20 molecules of fluorine and 28 molecules of hydrogen react, how many molecules of HF can form? Which element is left over? How many molecules of the leftover element are there?

Atomic theory also explains the law of conservation of mass. When fluorine atoms combine with hydrogen atoms to form hydrogen fluoride, the atoms are merely rearranged. Matter is neither lost nor gained; the mass does not change.

Finally, atomic theory is consistent with the law of multiple proportions. For example, 1.00 g of carbon combines with 1.33 g of oxygen to form carbon monoxide, and with 2.66 g of oxygen to form carbon dioxide. Carbon dioxide has twice the mass of oxygen per gram of carbon as does carbon monoxide. This is because one atom of carbon

▲ It DOES Matter!
A molecule with one carbon (C) atom and two oxygen (O) atoms is carbon dioxide (CO_2), the gas you exhale and the "fizz" in soft drinks. Cooled to about −80 °C it becomes dry ice (shown above), used to keep items frozen during shipping. But a molecule with one less O atom is deadly carbon monoxide (CO). Just 0.2% CO in the air is enough to kill. Unfortunately, most fuels produce a little CO when they burn, which is why a car engine is never left running in a closed garage.

Table 2.1	The Law of Multiple Proportions			
	Compound	Representation[a]	Mass of N per 1.000 g of O	Ratio of the Masses of N[b]
	Nitrous oxide		1.750 g	$(1.750 \div 0.4375) = 4.000$
	Nitric oxide		0.8750 g	$(0.8750 \div 0.4375) = 2.000$
	Nitrogen dioxide		0.4375 g	$(0.4375 \div 0.4375) = 1.000$

[a] ⚪ = nitrogen atom and ⚫ = oxygen atom

[b] We obtain the ratio of the masses of N that combine with a given mass of O by dividing each quantity in the third column by the smallest (0.4375 g).

combines with *one* atom of oxygen to form carbon monoxide, whereas one atom of carbon combines with *two* atoms of oxygen to form carbon dioxide. Using modern values, we assign an oxygen atom a relative mass of 16.0 and a carbon atom a relative mass of 12.0. On this scale carbon monoxide is seen to be made up of one atom of carbon combined with one atom of oxygen to give a mass ratio of 12.0 parts carbon to 16.0 parts oxygen (or 3.00:4.00). Carbon dioxide is composed of one atom of carbon combined with two atoms of oxygen to give a mass ratio of 12.0 parts carbon to $2 \times 16.0 = 32.0$ parts oxygen (or 3.00:8.00). Because oxygen atoms all have the same average mass, and all carbon atoms have the same average mass, the ratio of carbon in these two compounds is 2:1. The same law holds true for other compounds formed from the same two elements. Table 2.1 shows another example of the law of multiple proportions, involving nitrogen and oxygen.

Dalton also invented a set of symbols (Figure 2.9) to represent the different kinds of atoms. These symbols have since been replaced by modern symbols of one or two letters (inside front cover).

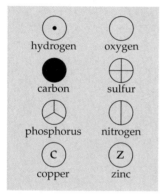

▲ Figure 2.9 Some of Dalton's symbols for the elements.

QUESTION: Using Dalton's symbols, draw diagrams for sulfur dioxide and sulfur trioxide. (Place the sulfur atom in the center and arrange the oxygen atoms around it.) What law do these two compounds illustrate?

Isotopes

As we have noted, Dalton's second assumption, that all atoms of an element are alike, has been modified. Not all atoms of an element have the same mass. Atoms of an element with different masses are called *isotopes*. For example, although most carbon atoms have a relative atomic mass of 12 (carbon-12), 1.1% of carbon atoms have a relative atomic mass of 13 (carbon-13). We discuss isotopes in more detail in Chapter 3.

Problem Solving: Mass and Atom Ratios

We can use proportions, such as those determined by Dalton, to calculate the amount of one substance needed to combine with a given quantity of another substance. To learn how to do this, let's look at some examples.

EXAMPLE 2.1 Mass Ratios

The gas methane (CH_4), the main component of natural gas, can be decomposed to give carbon (C) and hydrogen (H) in a ratio of 3.00 parts by mass of carbon to 1.00 part by mass of hydrogen. How much hydrogen can be made from 90.0 g of methane?

Solution
We can express the parts ratio in any units we choose—pounds, grams, kilograms— as long as it is the same for both elements. Using grams as the units, we see that 3.00 g of C and 1.00 g of H would mean 4.00 g CH_4 at the start. To convert g CH_4 to g H, we need a conversion factor that includes 1.00 g H and 4.00 g CH_4.

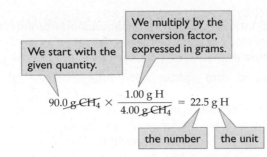

$$90.0 \text{ g } \cancel{CH_4} \times \frac{1.00 \text{ g H}}{4.00 \text{ g } \cancel{CH_4}} = 22.5 \text{ g H}$$

■ **EXERCISE 2.1A**

The gas ammonia can be decomposed to give 3.00 parts by mass of hydrogen and 14.0 parts by mass of nitrogen. What mass of nitrogen is obtained if 1.27 g of ammonia is decomposed?

■ **EXERCISE 2.1B**

Nitrous oxide, sometimes called "laughing gas," can be decomposed to give 7.00 parts by mass of nitrogen and 4.00 parts by mass of oxygen. What mass of nitrogen is obtained if enough nitrous oxide is decomposed to yield 36.0 g of oxygen?

EXAMPLE 2.2 Atom Ratios

Hydrogen sulfide gas can be decomposed to give sulfur and hydrogen in a mass ratio of 16.0:1.00. If the relative mass of sulfur is 32.0 when the mass of hydrogen is taken to be 1.00, how many hydrogen atoms are combined with each sulfur atom in the gas?

Solution

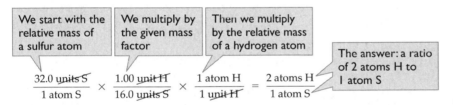

$$\frac{32.0 \text{ units S}}{1 \text{ atom S}} \times \frac{1.00 \text{ unit H}}{16.0 \text{ units S}} \times \frac{1 \text{ atom H}}{1 \text{ unit H}} = \frac{2 \text{ atoms H}}{1 \text{ atom S}}$$

■ **EXERCISE 2.2**

Arsine gas can be decomposed to give arsenic and hydrogen in a mass ratio of 25.0:1.00. If the relative mass of arsenic is 74.9 when the mass of hydrogen is taken to be 1.00, how many hydrogen atoms are combined with each arsenic atom in the gas?

Despite some inaccuracies, Dalton's atomic theory was a great success. Why? Because it served—and still serves—to explain a large amount of experimental data. It also successfully predicted how matter would behave under a wide variety of circumstances. Dalton arrived at his atomic theory by reasoning based on experimental facts, and with modest modification it has stood the test of time and modern, highly sophisticated instrumentation. Formulation of so successful a theory was quite a triumph for a Quaker schoolteacher in 1803.

More to Explore

Kolb, Doris. "Chemical Principles Revisited: But If Atoms Are So Tiny . . . " *Journal of Chemical Education*, September 1977, pp. 543–547.

Self-Assessment Questions

1. Two compounds, hydrazine and diimide, both contain nitrogen and hydrogen. Hydrazine contains 3.00 g of N and 0.432 g H. Diimide contains 3.00 g of N and 0.216 g H. If the formula of diimide is N_2H_2, the formula of hydrazine is

   a. NH
   b. N_2H
   c. NH_3
   d. N_2H_4

2. According to Dalton, elements are distinguished from each other by the
   a. density in the solid state   **b.** nuclear charge
   c. shapes of their atoms   **d.** weights of their atoms

3. Dalton postulated that atoms were indestructible and immutable to explain why
   a. the same two elements can form more than one compound
   b. no two elements have the same atomic mass
   c. mass is conserved in chemical reactions
   d. nuclear fission is impossible

4. Dalton viewed chemical change as
   a. a change of atoms from one type into another
   b. creation and destruction of atoms
   c. a rearrangement of atoms
   d. a transfer of electrons

5. A substance that only contains one kind of atom is called a(n)
   a. compound   **b.** element
   c. mixture   **d.** natural product

6. How many types of atoms are present in a given compound?
   a. at least two   **b.** hundreds
   c. three or more   **d.** a variable number

7. Hydrogen and nitrogen combine in a 3.0:14.0 mass ratio to form ammonia. If every molecule of ammonia contains 3 H atoms and 1 N atom, an atom of nitrogen must have a mass of
   a. 3/14 times the mass of a hydrogen atom
   b. 3 times the mass of a hydrogen atom
   c. 14/3 times the mass of a hydrogen atom
   d. 14 times the mass of a hydrogen atom

8. By experiment, we find that 1.008 g of hydrogen (H) combines with 35.453 g of chlorine (Cl) to form 36.461 g of the compound hydrogen chloride. Using Dalton's atomic theory, the best explanation for this is that
   a. H and Cl atoms are neither created nor destroyed in the process, which means the mass of reactants is equal to the mass of the product
   b. H and Cl atoms combine in a 1:35 ratio
   c. H and Cl atoms combine in more than one small-whole-number ratio
   d. one atom of H combines with 35.453 atoms of Cl in the reaction
   e. the product is a mixture because the ratio of elements is not a whole-number ratio.

Items 9–11 refer to the following sketches:

9. Which figure represents an element?

10. Which figure represents a compound?

11. Which figure represents a mixture?

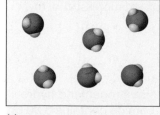 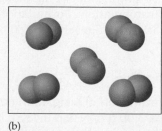 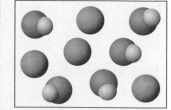

(a)   (b)   (c)

Answers: 1, d; 2, d; 3, c; 4, c; 5, b; 6, a; 7, d; 8, a; 9, b; 10, a; 11, c

2.5 Out of Chaos: The Periodic Table

During the eighteenth century new elements were discovered with surprising frequency, and by 1830 there were 55 known elements, all with different properties and with no apparent order in these properties. John Dalton had set up a table of relative atomic masses in his book *A New System of Chemical Philosophy* in 1808. Dalton's rough values were improved in subsequent years, notably by Berzelius, who published a table of atomic weights in 1828 containing 54 elements. Most of Berzelius's values agree well with modern values.

Relative Atomic Masses

Although it was impossible to determine actual masses of atoms in the 1800s, chemists were able to determine relative atomic masses by measuring the amounts of various elements that combined with a given mass of another element. Dalton's atomic masses were based on an atomic mass of 1 for hydrogen. As more accurate atomic weights were determined, this standard was replaced by one in which the atomic mass of naturally occurring oxygen was assigned a value of 16.0000. Because the isotopic composition of oxygen varies a bit depending on its source, the oxygen standard was replaced in 1961 by a more logical one based on a single isotope of carbon, carbon-12. Adoption of this new standard caused little change in atomic masses. These relative atomic masses are usually expressed in **atomic mass units (amu)**, commonly referred to today simply as *units (u)*.

Mendeleev's Periodic Table

Various attempts were made to arrange the elements in some sort of systematic fashion. The most successful arrangement—one that soon became widely accepted by chemists—was published in 1869 by Dmitri Ivanovich Mendeleev (1834–1907), a Russian chemist. Mendeleev's **periodic table** arranged the elements primarily in order of increasing atomic mass, although in a few cases he put a slightly heavier element before a lighter one in order to place elements with similar chemical properties in the same column (Figure 2.10). For example, he put tellurium, with an atomic mass of 127.6 u, ahead of iodine, which has an atomic mass of 126.9 u. He did this in order to place tellurium in the same column as sulfur and selenium, which it resembles in chemical properties. This rearrangement also put iodine in the same column as chlorine and bromine, which it resembles chemically.

To place elements in groups with similar properties, Mendeleev also had to leave gaps in his table. Instead of considering these blank spaces as defects, he boldly predicted the existence of elements yet undiscovered. Further, he even predicted the properties of some of the missing elements. For example, the missing elements he called eka-boron, eka-aluminum, and eka-silicon were soon discovered and named scandium, gallium, and germanium.[1] As can be seen in Table 2.2, Mendeleev's predictions were amazingly successful. This remarkable predictive value led to wide acceptance of Mendeleev's table.

The modern periodic table (inside front cover) contains 114 elements. Each element is represented by a "box" in the periodic table, as shown in Figure 2.11. We will discuss the periodic table and its theoretical basis in Chapter 3.

▲ Dmitri Mendeleev, the Russian chemist who invented the periodic table of the elements, continues to be honored in his native land. Element 101 (Md) is named mendelevium in his honor.

More to Explore

Strathern, Paul. *Mendeleyev's Dream: The Quest for the Elements*. London: Hamish Hamilton, 2000, and Gordin, Michael D. *A Well-Ordered Thing: Dmitrii Mendeleev and the Shadow of the Periodic Table*. New York: Basic Books, 2004.

---

[1]Gallium was discovered in 1875 by P. E. Lecoq de Boisbaudran, who named it for his native land, Gaul (France). Swedish chemist Lars F. Nilson discovered scandium in 1879 and named it for Scandinavia. Finally, in 1886 C. A. Winkler discovered germanium and named it for his country, Germany.

▶ **Figure 2.10** Mendeleev's original periodic table.

Reihen	Gruppe I. — R²O	Gruppe II. — RO	Gruppe III. — R²O³	Gruppe IV. RH⁴ RO²	Gruppe V. RH³ R²O⁵	Gruppe VI. RH² RO³	Gruppe VII. RH R²O⁷	Gruppe VIII. — RO⁴
1	H=1							
2	Li=7	Be=9,4	B=11	C=12	N=14	O=16	F=19	
3	Na=23	Mg=24	Al=27,3	Si=28	P=31	S=32	Cl=35,5	
4	K=39	Ca=40	—=44	Ti=48	V=51	Cr=52	Mn=55	Fe=56, Co=59, Ni=59, Cu=63.
5	(Cu=63)	Zn=65	—=68	—=72	As=75	Se=78	Br=80	
6	Rb=85	Sr=87	?Yt=88	Zr=90	Nb=94	Mo=96	—=100	Ru=104, Rh=104, Pd=106, Ag=108.
7	(Ag=108)	Cd=112	In=113	Sn=118	Sb=122	Te=125	J=127	
8	Cs=133	Ba=137	?Di=138	?Ce=140	—	—	—	— — —
9	(—)							
10	—	—	?Er=178	?La=180	Ta=182	W=184	—	Os=195, Ir=197, Pt=198, Au=199.
11	(Au=199)	Hg=200	Tl=204	Pb=207	Bi=208			
12	—	—	—	Th=231		U=240		— — —

Tabelle II.

der chemischen Elemente.

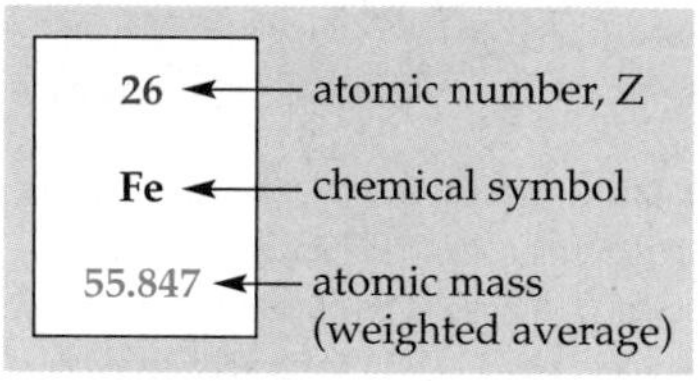

26 ◀—— atomic number, Z

Fe ◀—— chemical symbol

55.847 ◀—— atomic mass (weighted average)

▲ **Figure 2.11** Representation of an element on the periodic table.

Table 2.2 Properties of Germanium: Predicted and Observed

Property	Predicted by Mendeleev for Eka-Silikon (1871)	Observed by Winkler for Germanium (1886)
Atomic mass	72	72.6
Density (g/cm³)	5.5	5.47
Color	Dirty gray	Grayish white
Density of oxide (g/cm³)	EsO_2: 4.7	GeO_2: 4.703
Boiling point of chloride	$EsCl_4$: below 100 °C	$GeCl_4$: 86 °C
Density of chloride (g/cm³)	$EsCl_4$: 1.9	$GeCl_4$: 1.887

Precursors of the Periodic Table

Dobereiner's "Triads"

As early as 1816 Johann Dobereiner, a German chemist, noticed that there were several groups of three elements that were very similar (lithium, sodium, potassium; calcium, strontium, barium; sulfur, selenium, tellurium; chlorine, bromine, iodine). In each case, the middle element seemed to be halfway between the other two in atomic mass, reactivity, and other properties. Dobereiner published his observations in 1829.

De Chancourtois's "Telluric Helix"

In 1862 Beguyer de Chancourtois, a French geologist, arranged the elements in order of atomic mass. When he wound the list spirally around a cylinder, he found that similar elements fell along the same vertical lines.

Newlands's "Law of Octaves"

In 1863 John Newlands, an English chemist, noted that when elements were listed in order of atomic mass, every eighth element had similar properties; however, the rule seemed to break down for elements past calcium.

Meyer's System of Elements

In 1868 Lothar Meyer in Germany came up independently with an arrangement of elements similar to that of Mendeleev. Many believe that he should share the credit for the periodic table. Unfortunately, Meyer did not write up his table until December 1869, and it was not published until March 1870. Mendeleev had published his table in 1869.

Self-Assessment Questions

1. Mendeleev's periodic table had all the following features *except* that in it he
   a. announced the discovery of new elements
   b. arranged the elements in order of increasing atomic weight
   c. left gaps for predicted new elements
   d. placed some lighter elements before heavier ones

2. Relative masses in the modern periodic table are based on
   a. a carbon-12 isotope value   **b.** hydrogen value of 1.000
   c. oxygen value of 16.000   **d.** oxygen-16 value of 16.000

3. The approximate number of elements known today is
   a. 12   **b.** 100
   c. 1000   **d.** 30,000,000

Answers: 1, a; 2, a; 3, b

2.6 Atoms and Molecules: Real and Relevant

Are atoms real? Certainly they are real as a concept, a highly useful concept at that. Scientists can even observe computer-enhanced *images* of individual atoms. These portraits reveal little detail of atoms, but they provide powerful (though still indirect) evidence that atoms exist.

Are atoms relevant? Much of modern science and technology—including the production of new materials and the technology of pollution control—is ultimately based on the concept of atoms. We have seen that atoms are conserved in chemical reactions. Thus, material things—things made of atoms—can be recycled, for the atoms are not destroyed no matter how we use them. The one way we might "lose" a material from a practical standpoint is to spread the atoms so thinly that it would take too much time and energy to put them back together again. The *Recycling* essay on page 54 gives a real-world example of such loss.

Now back to Leucippus and his musings by the seashore. We now know that if we keep dividing drops of water into smaller drops, we will ultimately obtain a small particle—called a *molecule*—that is still water. A **molecule** is a group of atoms, chemically bonded or connected together. Molecules are represented by chemical formulas. The symbol H represents an *atom* of hydrogen; the formula H_2 represents a *molecule* of hydrogen, which is composed of two hydrogen atoms. The formula H_2O represents a molecule of water, which is composed of two hydrogen atoms and one oxygen atom. If we divide a water molecule, we will obtain those *two atoms* of hydrogen and *one atom* of oxygen.

And if we divide these atoms ... but that is a story for another time.

Dalton regarded the atom as indivisible, as did his successors up until the discovery of radioactivity in 1895. We examine the evolving concept of the atom in the next chapter.

A N S W E R

4. What's the difference between atoms and molecules?
A molecule is made of atoms that are combined in fixed proportions. A carbon atom is the smallest particle of the element carbon, and a carbon dioxide molecule is the smallest particle of the compound carbon dioxide. As you have probably figured out, molecules can be broken down into their atoms.

Self-Assessment Questions

1. The recycling of materials is possible because atoms
   a. always combine in the same way   **b.** are conserved
   c. are indivisible   **d.** combine in multiple portions

2. A molecule is
   a. a collection of like atoms
   b. conserved in chemical reactions
   c. a group of atoms chemically bonded together
   d. one of 100 or so simple substances

Answers: 1, b; 2, c

Recycling

Because it is an element, iron cannot be created or destroyed, but that does not mean it will always be found in its elemental state. Let's consider the following two different pathways for the recycling of iron.

1. Iron ore (hematite) is mined from the ground and converted into pig iron, which is used to make steel cans. Once discarded, the cans rust. The rust slowly leaches into groundwater as iron ions (ions are explained in Chapter 3) and eventually winds up in the ocean. At this point, the original iron atoms have become dissolved ions that can be absorbed and incorporated by marine plants. Marine creatures that eat plants will in turn absorb and incorporate iron. The original iron atoms are now widely separated in space. Perhaps a few of them might even become part of the hemoglobin in your own bloodstream. The iron has been recycled, but it will never again resemble the original pig iron.

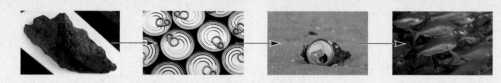

2. Iron ore is mined from the ground and converted to pig iron and then into steel. The steel is used in making an automobile, which is driven for a decade and then sent to the junkyard. The junkyard compresses the automobile and sends it to a recycling plant, where the steel is recovered and ultimately used again in a new automobile. Once the iron was removed from its ore, it has been conserved in its elemental metallic form. In this form of recycling the iron continues to be useful for a long time.

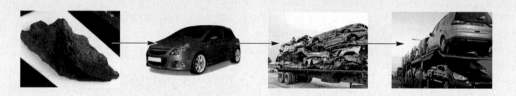

Critical Thinking Exercises

Apply knowledge that you have gained in this chapter and one or more of the FLaReS principles (Chapter 1) to evaluate the following statements or claims.

2.1 A doctor claims that, by giving patients small amounts of selenium, he can cure some types of cancer. He says that selenium atoms have the ability to get inside certain kinds of cancer cells and kill them. He has a list of patients that he says he has cured with this treatment.

2.2 A health food store has a large display of bracelets made of copper metal. Some people claim that wearing a copper bracelet will protect the wearer against arthritis or rheumatoid diseases.

2.3 Mary has just learned that her red blood cell count is low, and her doctor has given her some pills that contain iron. The doctor says that the pills should raise the level of hemoglobin, a substance that has iron atoms as oxygen-carrying entities in her bloodstream, and should keep Mary from becoming anemic. Should Mary take the pills or should she seek another opinion?

2.4 A company markets a device it calls a "water energoxizer." The device is claimed to supply so much energy to drinking water that the mass of oxygen in the water is increased, thereby providing more oxygen to the body.

GREEN CHEMISTRY

Designing Molecules

John Thompson, Lane Community College

For more than a century, chemists have determined the structures of molecules and synthesized new molecules. Now, with the development of the concept of green chemistry and the field of nanotechnology (see the essay *Nanoworld* on p. 24), chemists are gradually becoming *designers* of molecules. In addition to making new substances, chemists must not only think about yield and purity, but they must also consider the wider impact of the materials on sustainability—the pattern of resource use that aims to meet our needs without harming the environment. In doing this, we use the principles of green chemistry as a guide.

People have intentionally changed matter since prehistoric times. With the discovery or development of chemistry, these processes became more scientific, and chemists began to recognize that the changes in matter were actually reactions of atoms and molecules. Chemical synthesis—the creation of desired materials by controlled chemical reactions—played an important role in the Industrial Revolution. New materials and products were made on a large scale for the first time. Industrial chemists sought to maximize the amount of the desired product while minimizing material and production costs. However, this goal generally did not take environmental impacts into account. Pollution strained ecosystems and impacted human health. As these problems increased, governments produced regulations to protect our air, water, and soil. These regulations often increased the cost of doing business.

Many chemists now work to solve these challenges. The principles of green chemistry provide tools for evaluating the sustainability of chemical processes. Using these principles, chemists design greener molecules in much the same way as architects design greener buildings. Green chemistry plays an increasingly significant role in the challenges we face as an industrial society.

The Presidential Green Chemistry Challenge Awards were created in 1995 to promote the use of green chemistry for pollution prevention. The first award for Greener Reaction Conditions was given in 1996 to the Dow Chemical Company. Dow used 100% carbon dioxide (CO_2) as a blowing agent for polystyrene. By replacing traditional blowing agents with CO_2, Dow Chemical eliminated 3.5 million pounds per year of chemicals that either depleted the ozone layer or contributed to ground-level smog. The CO_2 used is a by-product of existing commercial or natural processes; thus, no net additional CO_2 is released to the atmosphere.

The 2007 Greener Synthetic Pathways Award was given to Kaichang Li and colleagues of Oregon State University, Columbia Forest Products, and Hercules Incorporated for the development of a soy-based adhesive that can replace formaldehyde-based adhesives in plywood and particleboard. Formaldehyde-based resins, used since the 1940s, release toxic formaldehyde, a probable human carcinogen. In 2006 Columbia Forest Products used the new soy-based adhesive to replace more than 47 million pounds of formaldehyde-based adhesives, reducing the emission of hazardous air pollutants at each plant by 50 to 90%. This adhesive is not only better for the environment, but it is also cost-competitive!

The number of potential applications for properly designed molecules is almost infinite. Substances might be designed to remove plaque from teeth before it can lead to tooth decay and gum disease. A carefully designed molecule could remove lead or PCBs (polychlorinated biphenyls) or other toxic substances from the environment, substances that are too dilute to remove by conventional means. Lubricating molecules might be constructed to adhere to the moving surfaces of an automobile engine, so that oil would no longer be necessary. The design of molecules for specific purposes is truly an exciting field for chemists and chemistry—and for you!

Li's research group at Oregon State University was inspired to develop a new soy-based adhesive by mussels, which attach themselves so strongly to rocks that they are not moved by water or waves. Here mussels of the genus *Mytilidae* adhere to rocks at low tide.

SUMMARY

Section 2.1—The concept of atoms was first suggested in ancient Greece by Leucippus and Democritus. However, it was rejected for almost 2000 years in favor of Aristotle's view of matter, which declared that matter was continuous in nature.

Section 2.2—The **law of conservation of mass** resulted from careful experiments by Lavoisier and others, who weighed all the reactants and all the products for a number of chemical reactions and found that no change in mass occurred. Boyle said that a supposed element must be tested; if it could be broken down into simpler substances, it was not really an element.

Section 2.3—The **law of definite proportions** (or the law of constant composition) was formulated by Proust, based on his experiments and on those of Berzelius. It states that a given compound always contains the same elements in exactly the same proportions by mass.

Section 2.4—In 1803 John Dalton explained the laws of definite proportions and of conservation of mass with his **atomic theory**, which had four main points: (1) Matter is made up of tiny particles called atoms; (2) atoms of the same element are alike; (3) compounds are formed when atoms of different elements combine in certain proportions; and (4) during chemical reactions atoms are rearranged, not destroyed. In his studies Dalton also discovered the **law of multiple**

proportions, which states that different elements might combine in two or more different sets of proportions, each set corresponding to a different compound. His atomic theory also explained this new law. These laws can be used to perform calculations involving the amounts of elements that combine or are present in a compound or reaction. In the two centuries following, atomic theory has undergone only minor modification.

Section 2.5—Berzelius published a table of atomic weights in 1828, which agrees well with modern values. In 1869 Mendeleev published his version of the **periodic table**, a systematic arrangement of the elements that allowed him to predict the existence and properties of undiscovered elements. The modern periodic table contains over 110 elements. For each element is listed its symbol and the average mass of an atom of that element in **atomic mass units**, which are very tiny units of mass.

Section 2.6—Since atoms are conserved in chemical reactions, matter (which is made of atoms) always can be recycled. If we hope to recycle a particular kind of matter, however, we should take care not to let it spread too thinly throughout nature, or recycling will not be practical. A **molecule** is a group of atoms chemically bonded together. Just as an atom is the smallest unit particle of an element, the smallest unit particle of most compounds is a molecule.

REVIEW QUESTIONS

1. Distinguish between **(a)** the atomic view and the continuous view of matter and **(b)** the ancient Greek definition of an element and the modern one.

2. What was Democritus' contribution to atomic theory? Why did the idea that matter was continuous (rather than atomic) prevail for so long? What discoveries finally refuted the idea?

3. Consider the ideas of *discrete* (atomic) and *continuous* as applied to foods at the macroscopic level (visible to the unaided eye).
   a. Would blending of 1 tablespoon (tbsp) of butter into 1 cup of mashed potatoes verify the law of definite proportions?
   b. Would also blending of 2 tbsp of butter into 1 cup of mashed potatoes verify the law of multiple proportions?
   c. Do these observations confirm the atomic theory?

4. Describe Lavoisier's contribution to the development of modern chemistry.

5. How did Robert Boyle define an element?

6. State the law of definite proportions and illustrate it using the compound zinc sulfide.

7. State the law of multiple proportions. For a fixed mass of chlorine (Cl) in each of the following compounds, what is the relationship between (ratio of)
   a. masses of oxygen in ClO_2 and in ClO?
   b. masses of fluorine in ClF_3 and ClF?

8. Outline the main points of Dalton's atomic theory.

9. In the figure, the blue spheres represent phosphorus atoms, the red ones represent oxygen atoms, and "Initial" represents a mixture. Which one of the three other rectangles could *not* represent that mixture after chemical reaction(s) occur? Explain briefly.

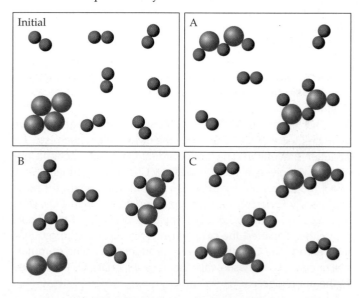

10. Consider the following set of compounds. What principle does this group illustrate?

$$P_4S_3 \quad P_4S_4 \quad P_4S_6 \quad P_4S_{10}$$

11. Use Dalton's atomic theory to explain each of the following laws and give an example that illustrates each law.
    a. conservation of mass
    b. definite proportions
    c. multiple proportions

12. Lavoisier considered *alumina* an element. In 1825 the Danish chemist Hans Christian Oersted isolated aluminum metal by reacting aluminum chloride with potassium, and alumina was later shown to be formed by reacting aluminum metal with oxygen. What did Oersted's experiment prove?

13. Heptane is always composed of 84.0% carbon and 16.0% hydrogen. What law does this illustrate?

14. What did each of the following individuals contribute to the development of modern chemistry?
    a. J. J. Berzelius
    b. Henry Cavendish
    c. Joseph Proust
    d. Dmitri Mendeleev

■ PROBLEMS

Conservation of Mass

15. A balloon filled with helium floats near the ceiling. After several days, the balloon is deflated and lying on the floor. Have the helium atoms been destroyed? If so, how? If not, where are they?

16. A piece of iron is dissolved in a solution of hydrochloric acid. Have the iron atoms been destroyed? If so, how? If not, where are they?

17. If 45.0 g of vinegar is added to 5.0 g of baking soda in an open beaker, the total mass after reaction is less than 50 g. After reaction in a closed system, the total mass of the products is 50.0 g. Is the law of conservation of mass violated in the open-vessel reaction? Explain.

18. Many reactions seem to violate the law of conservation of mass. For example, **(a)** when a metal object rusts, the rusty object has a greater mass than before, and **(b)** when a piece of wood burns, the remaining material (ash) has less mass than the wood. Explain these observations and in each case, suggest an experiment that would demonstrate that mass is actually conserved in these reactions.

19. Methane consists of carbon and hydrogen atoms. When methane is burned in a Bunsen burner, the only visible product is water vapor condensed on a cold beaker held above the flame. Have the carbon atoms of the methane been destroyed? If so, how? If not, where are they?

20. Acetaminophen (generic name for Tylenol®) consists of carbon, hydrogen, nitrogen, and oxygen atoms. A student in an organic chemistry laboratory prepares impure acetaminophen and dissolves it in hot water in order to purify it by recrystallization. No crystals form on cooling, and so the student pours out the solution and starts the experiment over. Have the carbon, hydrogen, nitrogen, and oxygen atoms of the acetaminophen been destroyed? If so, how? If not, where are they?

21. A student heats 1.0000 g of zinc powder with 0.2000 g of sulfur. He reports that he obtains 0.6080 g of zinc sulfide and recovers 0.5920 g of unreacted zinc. Show by calculation whether or not his results obey the law of conservation of mass.

22. A student heats 0.5585 g of iron with 0.3550 g of sulfur. She reports that she obtains 0.8792 g of iron sulfide and recovers 0.0433 g of unreacted sulfur. Show by calculation whether or not her results obey the law of conservation of mass.

23. When 1.00 g zinc and 0.80 g sulfur are allowed to react, all the zinc is used up, 1.50 g of zinc sulfide is formed, and some unreacted sulfur remains. What is the mass of *unreacted* sulfur (choose one)?
    a. 0.20 g   **b.** 0.30 g   **c.** 0.50 g
    d. impossible to determine from this information alone

24. A city has to come up with a plan to solve its solid waste problem. The solid wastes consist of many different kinds of materials, which are comprised of many different kinds of atoms. The options for disposal include burying the wastes in a landfill, incinerating them, or dumping them at sea. Which method, if any, will get rid of the atoms that make up the waste? Which method, if any, will immediately change the chemical form of the waste?

Definite Proportions

25. When 18.02 g of water is decomposed by electrolysis, 16.00 g of oxygen and 2.02 g of hydrogen are formed. According to the law of definite proportions, what mass in grams of hydrogen is formed by the electrolysis of 275 g of water?

26. Hydrogen from the decomposition of water has been promoted as the fuel of the future (Chapter 15). What mass of water, in kilograms, would have to be electrolyzed to produce 125 kg of hydrogen? (See Problem 25.)

27. With a plentiful supply of air, 3.0 parts carbon react with 8.0 parts oxygen to produce carbon dioxide. Use this mass ratio to calculate how much carbon is required to produce 960 g of carbon dioxide.

28. When 31 g of phosphorus reacts with oxygen, 71 g of an oxide of phosphorus is the product. What mass of oxygen is needed to produce 13 g of this product?

Multiple Proportions

29. Carbon combines with oxygen in a mass ratio of 1.000:2.664 to form carbon dioxide (CO_2) and with oxygen in a mass ratio of 1.000:1.332 to form carbon monoxide (CO). A third carbon-oxygen compound, called carbon suboxide, is 52.96% C by mass and 47.04% O by mass. Show that the law of multiple proportions is followed.

30. A compound containing only oxygen and rubidium has 0.187 g O per gram Rb. The relative atomic masses are O = 16.0 and Rb = 85.5. What is a possible O-to-Rb mass ratio for a different oxide of rubidium (choose one)?
    a. 8.0:85.5   **b.** 16.0:85.5   **c.** 32.0:85.5   **d.** 16.0:171

31. A sample of an oxide of tin with the formula SnO consists of 0.742 g of tin and 0.100 g of oxygen. A sample of a second oxide of tin consists of 0.555 g of tin and 0.150 g of oxygen. What is the formula of this second oxide?

32. Consider three oxides of nitrogen, X, Y, and Z. Oxide X has an oxygen-to-nitrogen mass ratio of 2.28:1.00, oxide Y has an oxygen-to-nitrogen mass ratio of 1.14:1.00, and oxide Z has an oxygen-to-nitrogen mass ratio 0.57:1.00. What are the ratios of masses of oxygen in these compounds when we have a fixed amount of nitrogen?

33. Sulfur forms two compounds with fluorine, T and U. Compound T has 0.447 g of sulfur combined with 1.06 g of fluorine while U has 0.438 g of sulfur combined with 1.56 g of fluorine. Show that these data illustrate the law of multiple proportions.

34. Two compounds, V and W, are composed only of hydrogen and carbon. Compound V is 80.0% carbon by mass and 20.0% hydrogen by mass. Compound W is 83.3% carbon by mass and 16.7% hydrogen by mass. What are the ratios of masses of hydrogen in these compounds when we have a fixed amount of carbon?

Dalton's Atomic Theory

35. Are the following findings, expressed to the nearest atomic mass unit, in agreement with Dalton's atomic theory? Explain your answers.
   a. An atom of calcium has a mass of 40 u and one of vanadium, 50 u.
   b. An atom of calcium has a mass of 40 u and one of potassium, 40 u.

36. To the nearest atomic mass unit, one atom of calcium has a mass of 40 u and another calcium atom has a mass of 44 u. Do these findings support or contradict Dalton's atomic theory? Explain.

37. A neutron strikes an atom of uranium and splits it into two smaller atoms. Do these findings support or contradict Dalton's atomic theory? Explain.

38. According to Dalton's atomic theory, when elements react, their atoms combine in (choose one)
   a. a simple whole-number ratio that is unique for each set of elements.
   b. exactly a 1:1 ratio.
   c. one or more simple whole-number ratios.
   d. pairs.
   e. random proportions.

39. Hydrogen and oxygen combine in a mass ratio of 1:8 to form water. If every water molecule consists of two atoms of hydrogen and one atom of oxygen, the mass of an oxygen atom must be what fraction or multiple of that of a hydrogen atom?
   a. $\frac{1}{16}$ **b.** $\frac{1}{8}$
   c. 8 times   **d.** 16 times

40. Fluorine and nitrogen combine in a mass ratio of 57:14 to form a compound. If every molecule of the compound consists of three atoms of fluorine and one atom of nitrogen, the mass of a nitrogen atom must be what fraction or multiple of that of a fluorine atom?
   a. $\frac{19}{14}$ **b.** 3 times
   c. $\frac{14}{19}$ **d.** 14 times

Chemical Compounds

41. A colorless liquid is thought to be a pure compound. Analyses of three samples of the material yield the following results.

	Mass of Sample	Mass of Carbon	Mass of Hydrogen
Sample 1	1.000 g	0.862 g	0.138 g
Sample 2	1.549 g	1.295 g	0.254 g
Sample 3	0.988 g	0.826 g	0.162 g

Could the material be a pure compound? Explain.

42. A blue solid called azulene is thought to be a pure compound. Analyses of three samples of the material yield the following results.

	Mass of Sample	Mass of Carbon	Mass of Hydrogen
Sample 1	1.000 g	0.937 g	0.0629 g
Sample 2	0.244 g	0.229 g	0.0153 g
Sample 3	0.100 g	0.094 g	0.0063 g

Could the material be a pure compound?

■ ADDITIONAL PROBLEMS

43. When 0.2250 g of magnesium is heated with 0.5331 g of nitrogen in a closed container, the magnesium is completely converted to 0.3114 g of magnesium nitride. What mass of unreacted nitrogen must remain?

44. Gasoline can be approximated by the formula C_8H_{18}. An environmental advocate claims that burning 1 gallon (about 7 lb) of gasoline produces about 19 lb of carbon dioxide, a greenhouse gas. Explain this apparent contradiction of the law of conservation of mass.

45. The gas silane can be decomposed to yield silicon and hydrogen in a ratio of 7 parts by mass of silicon to 1 part by mass of hydrogen. If the relative mass of silicon atoms is 28 and the mass of hydrogen atoms is taken to be 1, how many hydrogen atoms are combined with each silicon atom?

46. Two experiments were performed in which sulfur was burned completely in pure oxygen gas, producing sulfur dioxide and leaving some unreacted oxygen. In the first experiment, 0.312 g of sulfur produced 0.623 g of sulfur dioxide. In the second experiment, 1.305 g of sulfur was burned. What mass of sulfur dioxide was produced?

47. In an experiment illustrated in the figure below, about 15 mL of hydrochloric acid solution was placed in a flask and approximately 3 g of sodium carbonate was placed into a balloon. The opening of the balloon was then carefully stretched over the top of the flask, taking care not to allow the sodium carbonate to fall into the acid in the flask. The flask was placed on an electronic balance, and the mass of the flask and its contents was found to be 38.61 g. The sodium carbonate was then slowly shaken into the acid. The balloon began to fill with gas. When the reaction was complete, the mass of the flask and its contents, including the gas in the balloon, was found to be 38.61 g. What law does this experiment illustrate? Explain.

48. In one experiment, 3.06 g hydrogen was allowed to react with an excess of oxygen to form 27.35 g water. In a second experiment, electric current broke down a sample of water into 1.45 g hydrogen and 11.51 g oxygen. Are these results consistent with the law of constant composition? Explain why or why not.

49. Use Figure 2.4 to calculate the mass of mercury oxide that would be needed to produce 100.0 g of mercury metal.

50. Gold chloride, $AuCl_3$, is formed when gold metal is dissolved in *aqua regia*, a highly corrosive acid mixture. Consult Figure 2.10, and determine the mass ratio of gold to chlorine in gold chloride
    a. based on Mendeleev's values from his periodic table.
    b. based on values in the modern periodic table in the inside front cover of this book.

51. Recall Table 2.1. Another compound of nitrogen and oxygen contains 0.5836 g of nitrogen per 1.000 gram of oxygen. Calculate the ratio of mass of N in this compound to mass of N in nitrogen dioxide. What *whole-number* ratio does this value represent?

◼ COLLABORATIVE GROUP PROJECTS

Prepare a PowerPoint, poster, or other presentation (as directed by your instructor) for presentation to the class.

52. Prepare a brief biographical report on one of the following individuals.
    a. Henry Cavendish   b. Joseph Proust
    c. John Newlands   d. Lothar Meyer
    e. Dmitri Mendeleev   f. John Dalton
    g. Antoine Lavoisier

53. Prepare a brief report on early Greek contributions and ideas in the field of science, focusing on the work of one of the following individuals: Aristotle, Leucippus, Democritus, Thales, Anaximander, Anaximenes, Heraclitus, Empedicles, or other Greek philosopher of the time before 300 B.C.E.

54. Prepare a brief report on *phlogiston theory*. Describe what the theory was, how it explained changes in mass when something is burned, and why it was finally abandoned.

55. Write a brief essay on the recycling of one of the following: metals, paper, plastics, glass, food wastes, or grass clippings. Contrast a recycling method that maintains the properties of an element with one that changes them.

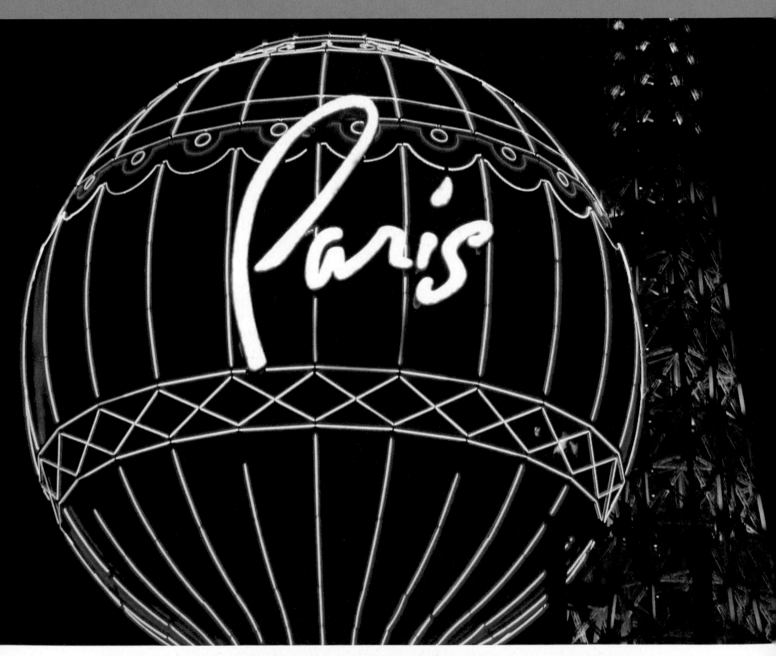

Brilliant neon lights illuminate the Eiffel Tower Restaurant in Las Vegas, Nevada. The colors you see here are caused by electrons in the atoms of the gases that fill the glass and plastic tubes. Those electrons move back and forth between different energy levels in the atoms, causing light to be given off. Different gases have different *energy levels* available, resulting in different colors. The specific energy levels available and the colors they produce are the result of atomic structure, the subject of this chapter.

ATOMIC STRUCTURE

Images of the Invisible

Atoms are exceedingly tiny particles, much too small to see. Yet in this chapter, we discuss their inner structure. If atoms are so small that we cannot see them, even with an optical microscope, how can we possibly know anything about their structures?

It turns out that we really know quite a lot about the structure of atoms. Although scientists have never examined the interior of an atom directly, they have been able to obtain a great deal of *indirect* information. By designing some clever experiments and exercising their powers of deduction, they have been able to construct an amazingly detailed model of what an atom's interior must be like.

Atoms are much too small to be seen with an ordinary light microscope. However, since 1970 scientists have been able to obtain images of individual atoms. In order to see even rough images of atoms, they must use special kinds of instruments such as the scanning tunneling microscope (STM). In this way they obtain pictures such as the one shown on page 40. We can see outlines of atoms in such photographs, and we can tell quite a bit about how they are arranged, but these pictures tell us nothing about the inner structure of the atoms. Why do we care about the structure of particles as tiny as atoms? It is the arrangement of various parts of atoms that determines the properties of different kinds of matter. Only by understanding atomic structure can we learn how atoms combine to make millions of different substances. With such knowledge, we can modify and synthesize materials to meet our needs more precisely. Knowledge of atomic structure is even essential to our health. Many medical diagnoses are based on chemical analyses that have been developed from our understanding of atomic structure.

Perhaps of greater interest to you is the fact that your understanding of chemistry (as well as much of biology and other sciences) depends, at least in part, on your knowledge of atomic structure. Let's start our study of atomic structure by going back to the time of John Dalton.

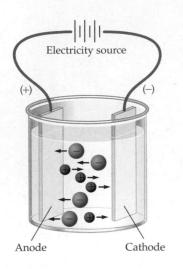

▲ **Figure 3.1** An electrolysis apparatus. The electricity source (for example, a battery) directs electrons through wires from the anode to the cathode. Cations (+) are attracted to the cathode (−), and anions (−) are attracted to the anode (+). This migration of ions is the flow of electricity through the solution.

▲ In addition to Michael Faraday's work in electrochemistry, he also devised methods for liquefying gases, and invented the first electrical transformer.

ANSWER

1. If we can't see atoms, how can we discuss even smaller particles?
The existence of electrons and other subatomic particles was deduced from the behavior of electricity. The existence and behavior of those subatomic particles make atoms behave the way they do.

3.1 Electricity and the Atom

John Dalton, who set forth his atomic theory in 1803, regarded the atom as hard and indivisible. However, it wasn't long before evidence accumulated to show that matter has electrical properties. Indeed, the electrolytic decomposition of water by Nicholson and Carlisle in 1800 (Section 2.3) had already indicated this. Electricity played an important role in unraveling the structure of the atom.

Static electricity has been known since ancient times, but the notion of continuous electric current developed in the nineteenth century. In 1800 Alessandro Volta invented an electrochemical cell much like a modern battery. If the poles of a cell are connected by a wire, current flows through the wire. The current is sustained by chemical reactions inside the cell. Volta's invention soon was applied in many areas of science and everyday life.

Electrolysis

Soon after Volta's invention, Humphry Davy (1778–1829), a British chemist, built a powerful battery that he used to pass electricity through molten (melted) salts. Davy quickly discovered several new elements. In 1807 he liberated highly reactive potassium metal from molten potassium hydroxide. Shortly thereafter he produced sodium metal by passing electricity through molten sodium hydroxide. Within a year, Davy had also produced magnesium, strontium, barium, and calcium metals for the first time. The science of electrochemistry was born. (We will study electrochemistry in more detail in Chapter 8.)

Davy's protégé Michael Faraday (1791–1867) greatly extended this new science. Faraday defined many of the terms we still use today, such as **electrolysis**, the splitting of compounds by electricity (Figure 3.1). Lacking in his own formal education, he consulted English classical scholar William Whewell (1794–1866), who suggested the name **electrolyte** for a compound that conducts electricity when melted or dissolved in water. In the electrolysis apparatus, **electrodes**, carbon rods or metal strips inserted into a molten compound or solution, carry the electric current. The electrode that bears a positive charge is the **anode**, and the negatively charged electrode is the **cathode**. The entities that carry the electric current through a melted compound or solution are called *ions*. An **ion** is an atom or a group of atoms bonded together that has an electric charge. An ion with a negative charge is an **anion**; it travels toward the anode. A positively charged ion is a **cation**; it moves toward the cathode.

Faraday's work established that atoms are electrical in nature, but further details of atomic structure had to wait several decades for more powerful sources of electrical voltage and better vacuum pumps.

Cathode Ray Tubes

In 1875 William Crookes (1832–1919), an English chemist, was able to construct a low-pressure gas discharge tube that would allow electricity to pass through it. His experiment is shown in Figure 3.2. Metal electrodes are sealed in the tube. It is connected to a vacuum pump, and most of the air is removed. A beam of current is seen as a green fluorescence, observed when the beam strikes a screen coated with zinc sulfide. This beam, which seems to leave the cathode and travel to the anode, is called a **cathode ray**.

Thomson's Experiment: Mass-to-Charge Ratio

Considerable speculation arose as to the nature of cathode rays and many experiments were undertaken. Were these rays actually beams of particles, or did they consist of a wavelike form of energy much like visible light? The answer came (as scientific answers should) from experiment performed by the English physicist Joseph John Thomson in 1897. Thomson showed that cathode rays were deflected

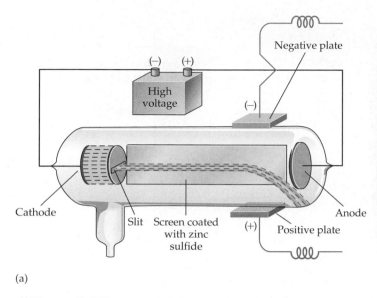

▲ **Figure 3.2** Thomson's apparatus, showing deflection of cathode rays (a beam of electrons). Cathode rays are themselves invisible but are observed through the green fluorescence produced when they strike a zinc sulfide–coated screen. The diagram (a) shows deflection of the beam in an electric field. The photograph (b) shows the deflection in a magnetic field. The magnetic field is created by the magnet to the right of and slightly behind the screen. Cathode rays travel in straight lines unless some kind of external field is applied.

in an electric field (see again Figure 3.2). The beam was attracted to the positive plate and repelled by the negative plate. Thomson therefore concluded that cathode rays consisted of negatively charged particles. His experiments also showed that the particles were the same regardless of the materials from which the electrodes were made or the type of gas in the tube. He concluded that these negative particles are part of all kinds of atoms. Thomson named these negatively charged units **electrons**. *Cathode rays,* then, are beams of electrons emanating from the cathode of a gas discharge tube.

Cathode rays are deflected in magnetic fields as well as in electric fields. The greater the charge on a particle, the more it is deflected in an electric (or magnetic) field. The heavier the particle, the less it is deflected by a force. By measuring the amount of deflection in fields of known strength, Thomson was able to calculate the *ratio* of the mass of the electron to its charge. He could not measure either the mass or the charge separately. This is like knowing that each 1-ft length of a steel beam has a mass of 25 lb. With this information alone, we cannot find either the total mass or length of a beam. Once the beam's mass or its length is known, it is easy to calculate the other from the known value and the 25 lb/1 ft ratio. Thomson was awarded the Nobel Prize in Physics in 1906.

Goldstein's Experiment: Positive Particles

In 1886 German scientist Eugen Goldstein performed experiments with gas discharge tubes that had perforated cathodes (Figure 3.3). He found that although electrons were formed and sped off toward the anode as usual, positive particles were also formed and shot in the opposite direction toward the cathode. Some of these positive particles went through the holes in the cathode. In 1907 a study of the deflection of these particles in a magnetic field indicated that they were of varying mass. The lightest particles, formed when there was a little hydrogen gas in the tube, were later shown to have a mass 1837 times that of an electron.

Millikan's Oil-Drop Experiment: Electron Charge

The charge on the electron was determined in 1909 by Robert A. Millikan (1868–1953), a physicist at the University of Chicago. Millikan observed electrically charged oil drops in an electric field. A diagram of his apparatus is shown in

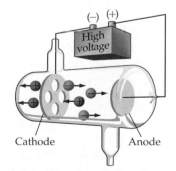

▲ **Figure 3.3** Goldstein's apparatus for the study of positive particles. Some positive ions, attracted toward the cathode, pass through the holes in the cathode. The deflection of these particles (not shown) in a magnetic field can be studied in the region to the left of the cathode.

▶ **Figure 3.4** The Millikan oil-drop experiment. Oil drops irradiated with X-rays pick up electrons and become negatively charged. Their fall due to gravity can be balanced by adjusting the voltage of the electric field. The charge on the oil drop can be determined from the applied voltage and the mass of the oil drop. The charge on each drop is that of some whole number of electrons.

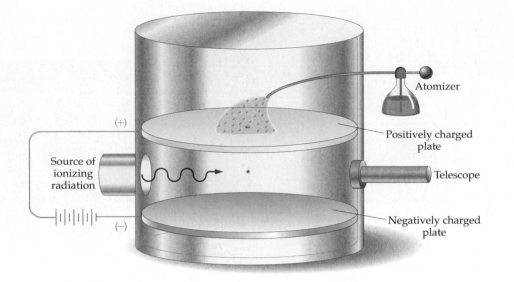

Figure 3.4. A spray bottle is used to form tiny droplets of oil. Some acquire negative charges by picking up electrons from the friction generated as the particles rub against the opening of the spray nozzle and against each other. (The charge is static electricity, just like the charge you get from walking across a nylon carpet.) The negative droplets can acquire one or more extra electrons by this process. Charges can also be produced by irradiation with X-rays.

Some of the oil droplets pass into a chamber where they can be viewed through a microscope. The negative plate at the bottom of the chamber repels the negatively charged droplets; the positive plate attracts them. By manipulating the charge on each plate and observing the behavior of the droplets, the charge on each droplet can be determined. Millikan took the smallest possible difference in charge between two droplets to be the charge of an individual electron. For his research, he received the Nobel Prize in Physics in 1923.

From Millikan's value for the charge and Thomson's value for the mass-to-charge ratio, the mass of the electron was readily calculated. That mass was found to be only 9.1×10^{-28} g. It would take more than 1×10^{27} electrons to weigh one gram—that's a 1 followed by 27 zeros, or a billion billion billion electrons. More important, that mass is much *smaller* than that of the lightest atom. This means that electrons are much smaller than atoms.

You can find out more about numbers such as 9.1×10^{-28} and 1×10^{27} in the Appendix. However, in practice we seldom use the actual mass and charge of the electron. For many purposes, the electron is considered the unit of electrical charge. The charge is shown as a superscript minus sign (meaning $^{1-}$). In indicating charges on ions, we use $^-$ to indicate a net charge of one electron, $^{2-}$ to indicate a charge of two electrons, and so on.

Self-Assessment Questions

1. The use of electricity to cause a chemical reaction is called
   a. anodizing
   b. corrosion
   c. electrolysis
   d. hydrolysis

2. During electrolysis of molten sodium hydroxide, sodium cations move to the
   a. anode, which is negatively charged
   b. anode, which is positively charged
   c. cathode, which is negatively charged
   d. cathode, which is positively charged

3. Cathode rays are *not*
   a. composed of negatively charged particles
   b. composed of particles with a mass of 1 u
   c. deflected by electric fields
   d. the same from element to element

4. Thomson used a cathode ray tube to discover that cathode rays
   a. consist of radiation similar to X-rays
   b. could be deflected by an electric field
   c. could ionize air inside the tube
   d. could ionize the air outside the far end of the tube

5. When J. J. Thomson discovered the electron, he measured its
   a. atomic number
   b. charge
   c. mass
   d. mass-to-charge ratio

6. In his oil-drop experiment, Millikan determined the charge of an electron by
   a. measuring the diameter of the tiniest oil drops
   b. noting that the drops had either a certain minimum charge or multiples of that charge
   c. using an atomizer that could spray out individual electrons
   d. viewing the smallest charged drops microscopically and noting they had one extra electron

Answers: 1, c; 2, c; 3, b; 4, b; 5, d; 6, b

3.2 Serendipity in Science: X-Rays and Radioactivity

Let's return now to the structure of the atom and look at a little scientific *serendipity*—scientific discoveries that are "happy accidents." Have you ever wondered why these accidents often seem to happen to scientists? It is probably because scientists are trained observers. The same accident could happen right before the eyes of an untrained person and go unnoticed (though often even scientists miss important finds at first). Or, if noticed, its significance might not be grasped.

Roentgen: The Discovery of X-Rays

Two serendipitous discoveries in the last years of the nineteenth century profoundly changed the world. In 1895 German scientist Wilhelm Conrad Roentgen (1845–1923) was working in a dark room, studying the glow produced in certain substances by cathode rays. To his surprise, he noted this glow on a chemically treated piece of paper some distance from the cathode ray tube. The paper even glowed when taken into the next room. Roentgen had discovered a new type of ray that could travel through walls. When he waved his hand between the radiation source and the glowing paper, he suddenly was able to see the bones of his own hand through the paper. He called these mysterious rays, which seemed to make his flesh disappear, **X-rays**.

X-rays are a form of electromagnetic radiation—energy with electric and magnetic components. Among other types, electromagnetic radiation also includes visible light, radio waves, microwaves, infrared radiation, ultraviolet light, and gamma rays (Section 3.3). These forms differ in energy, with radio waves lowest in energy, followed by microwaves, infrared radiation, visible light, ultraviolet light, X-rays, and gamma rays.

Today X-rays are one of the most widely used tools in the world for medical diagnosis. Not only are they employed for examining decayed teeth, broken bones, and diseased lungs, but they are also the basis for such procedures as mammography and computerized tomography (Chapter 11). In the United States alone, payment for various radiological procedures totals more than $20 billion each year. How ironic that Roentgen himself made no profit at all from his discovery. He considered X-rays a "gift to humanity" and refused to patent any part of the discovery. However, he did receive much popular acclaim and in 1901 was awarded the first Nobel Prize in Physics.

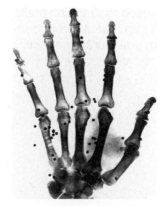

▲ X-rays were used in medicine shortly after they were discovered by Wilhelm Roentgen (1845–1923) in 1895. Michael Purpin of Columbia University made this X-ray in 1896 to aid in the surgical removal of gunshot pellets (the dark spots) from the hand of a patient.

▲ Marie Sklodowska Curie in her laboratory and Marie and Pierre Curie on a French postage stamp.

The Discovery of Radioactivity

Certain chemicals exhibit *fluorescence* after exposure to strong sunlight; they continue to glow even when taken into a dark room. In 1895 Antoine Henri Becquerel (1852–1908), a French physicist, was studying fluorescence by wrapping photographic film in black paper, placing a few crystals of the fluorescing chemical on top of the paper, and then placing the package in strong sunlight. If the glow was like ordinary light, it would not pass through the paper. On the other hand, if similar to X-rays, it would pass through the black paper and fog the film.

While working with a uranium compound, Becquerel made an important accidental discovery. When placed in sunlight, the compound fluoresced and fogged the film. On several cloudy days when exposure to sunlight was not possible, he prepared samples and placed them in a drawer. To his great surprise, the photographic film was fogged even though the uranium compound had not been exposed to sunlight. Further experiments showed that the radiation coming from the uranium compound was unrelated to fluorescence but was a characteristic of the element uranium.

Other scientists immediately began to study this new radiation. Becquerel had a graduate student from Poland, Marie Sklodowska, who gave the phenomenon a name: radioactivity. **Radioactivity** is the spontaneous emission of radiation from certain unstable elements. Marie later married Pierre Curie, a French physicist. Together they discovered the radioactive elements polonium and radium, and with Becquerel they shared the 1903 Nobel Prize in Physics.

After her husband's death in 1906, Marie Curie continued to work with radioactive substances, winning the Nobel Prize in Chemistry in 1911. For more than 50 years she was the only person ever to have received two Nobel Prizes.

Self-Assessment Questions

1. When Roentgen took an X-ray photograph of a hand, the bones showed up slightly dark because the X-rays were
   a. absorbed by bone more so than by flesh
   b. light that is less energetic than visible light
   c. made up of fast-moving atomic nuclei
   d. a type of radioactivity
2. Radioactivity arises from elements that
   a. absorb radiofrequency energy and release stronger radiation
   b. are unstable and release radiation
   c. give off X-rays
   d. undergo fission when they absorb energy

Answers: 1, a; 2, b

3.3 Three Types of Radioactivity

Scientists soon showed that three types of radiation emanated from various radioactive elements. Ernest Rutherford (1871–1937), a New Zealander who spent his career in Canada and Great Britain, chose the names alpha, beta, and gamma for the three types of radiation. When passed through a strong magnetic or electric field, the alpha form was deflected in a manner indicating that it consisted of a beam of positive particles (Figure 3.5). Later experiments showed that an **alpha particle** has a mass four times that of a hydrogen atom and a charge twice the magnitude of, but opposite in sign to, that of an electron. An alpha particle is in fact identical to the nucleus (Section 3.4) of a helium atom and is often symbolized by He^{2+}.

The beta radiation was shown to be made up of negatively charged particles identical to those of cathode rays. Therefore, a **beta particle** is an electron.

Gamma rays are not deflected by a magnetic field. They are a form of electromagnetic radiation, much like the X-rays used in medical work but even more energetic and more penetrating. The three types of radioactivity are summarized in Table 3.1.

The discoveries of the late nineteenth century paved the way for an entirely new picture of the atom, which developed rapidly during the early years of the twentieth century.

Table 3.1	Types of Radioactivity		
Name	**Symbol**	**Mass (u)**	**Charge**
Alpha	α	4	2+
Beta	β	$\dfrac{1}{1837}$	1−
Gamma	γ	0	0

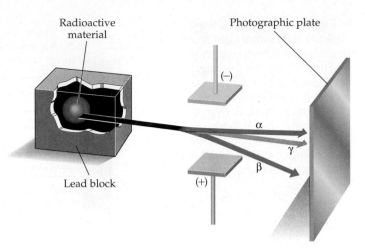

▲ **Figure 3.5** Behavior of radioactive rays in an electric field.

Self-Assessment Questions

1. The mass of an alpha particle is about
   a. four times that of a hydrogen atom
   b. the same as that of an electron
   c. the same as that of a hydrogen atom
   d. the same as that of a beta particle

2. Compared to the charge on an electron (designated −1), that on an alpha particle is about
   a. −4   b. −1   c. +1   d. +2   e. +4

3. A beta particle is a(n)
   a. boron nucleus
   b. electron
   c. helium nucleus
   d. proton

4. The mass of a gamma ray is
   a. −1 u   b. 0 u   c. 1 u   d. 4 u

Answers: 1, a; 2, d; 3, b; 4, b

3.4 Rutherford's Experiment: The Nuclear Model of the Atom

At Rutherford's suggestion, two of his coworkers, Hans Geiger (1882–1945), a German physicist, and Ernest Marsden (1889–1970), an English undergraduate student, bombarded very thin metal foils with alpha particles from a radioactive source (Figure 3.6). In an experiment with gold foil, most of the particles behaved as Rutherford expected, going right through the foil with little or no scattering. However, a few particles were deflected sharply. Occasionally one was sent right back in the direction from which it had come! Rutherford had assumed the positive charge to be spread evenly over all the space occupied by the atom, but obviously it was not. To explain the experiment, Rutherford concluded that all the positive charge

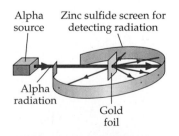

▲ **Figure 3.6** Rutherford's gold-foil experiment. Most alpha particles passed right through the gold foil, but now and then a particle was deflected.

▶ **Figure 3.7** Model explaining the results of Rutherford's gold-foil experiment. Most of the alpha particles pass right through the foil because it is mainly empty space. But some alpha particles are deflected as they pass close to a dense, positively charged atomic nucleus. Once in a while an alpha particle approaches an atomic nucleus head-on and is knocked back in the direction from which it came.

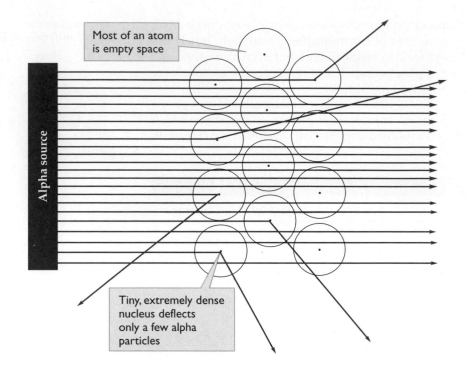

and nearly all the mass of an atom are concentrated at the center of the atom in a tiny core called the **nucleus**.

When an alpha particle, which is positively charged, directly approached the positively charged nucleus, it was strongly repelled and, therefore, sharply deflected (Figure 3.7). Because only a few alpha particles were deflected, Rutherford concluded that the nucleus must occupy only a tiny fraction of the volume of an atom. Most of the alpha particles passed right through because most of an atom is empty space. The space outside the nucleus isn't completely empty, however. It contains the negatively charged electrons. Rutherford concluded that the electrons had so little mass that they were no match for the alpha "bullets." It would be analogous to a mouse trying to stop the charge of a bull elephant.

Rutherford's nuclear theory of the atom, set forth in 1911, was revolutionary. He postulated that all the positive charge and nearly all the mass of an atom are concentrated in a tiny, tiny nucleus. The negatively charged electrons have almost no mass, yet they occupy nearly all the volume of an atom. To picture Rutherford's model, visualize a sphere as big as a giant indoor football stadium. The nucleus at the middle of the sphere is as small as a pea but weighs several million tons. A few flies flitting here and there throughout the sphere represent the electrons.

Self-Assessment Questions

1. The most persuasive evidence found in Rutherford's gold foil experiment was that
   a. most of the alpha particles went right through the foil, but a tiny fraction was deflected right back toward the source
   b. some alpha particles were deflected a bit
   c. some gold atoms were converted to lead atoms
   d. some gold atoms were split

2. Rutherford's gold foil experiment showed that
   a. the gold atoms consisted of protons in a "pudding" of electrons
   b. the nucleus was huge and was positively charged
   c. the nucleus was tiny and positively charged
   d. some gold atoms were shattered by alpha particles

3. From his gold foil experiment, Rutherford concluded that atoms consist mainly of
   a. electrons
   b. empty space
   c. neutrons
   d. protons

4. The positively charged part of the atom is its
   a. cation
   b. electron cloud
   c. nucleus
   d. valence shell

Answers: 1, a; 2, c; 3, b; 4, c

3.5 The Atomic Nucleus

In 1914 Rutherford suggested that the smallest positive particle (the one formed when there is hydrogen gas in the Goldstein apparatus—see Figure 3.3) is the unit of positive charge in the nucleus. This particle, called a **proton**, has a charge equal in magnitude to that of the electron and has nearly the same mass as a hydrogen atom. Rutherford's suggestion was that protons constitute the positively charged matter in all atoms. The nucleus of a hydrogen atom consists of one proton, and the nuclei of larger atoms contain greater numbers of protons.

Except for hydrogen atoms, atomic nuclei are heavier than indicated by the number of positive charges (number of protons). For example, the helium nucleus has a charge of 2+ (and, therefore, two protons, according to Rutherford's theory), but its mass is *four* times that of hydrogen. This excess mass puzzled scientists at first. But in 1932 English physicist James Chadwick (1891–1974) discovered a particle with about the same mass as a proton but with no electric charge. It was called a **neutron**, and its existence explains the unexpectedly high mass of the helium nucleus. Whereas the hydrogen nucleus contains only one proton of mass 1 u, the helium nucleus contains not only two protons (2 u) but also two neutrons (2 u), giving the nucleus a total mass of 4 u.

With the discovery of the neutron, the list of "building blocks" we will need for "constructing" atoms is complete. The properties of these particles are summarized in Table 3.2. (There are dozens of other subatomic particles, but most exist only momentarily and are not important in our discussion here.)

The number of protons in the nucleus of an atom of any element is the **atomic number (Z)** of that element. This number determines the kind of atom—that is, the identity of the element—and it is found on any periodic table or list of the elements. An *element*, then, is a substance in which all the atoms have the same atomic number. Dalton had said that the mass of an atom determines the element. We now know it is not the mass but the number of protons that determines the identity of an element. For example, an atom with 26 protons (one whose atomic number $Z = 26$) is an atom of iron (Fe). An atom with 50 protons ($Z = 50$) is an atom of tin (Sn). In a neutral atom (without an electric charge) the positive charge of the protons is exactly neutralized by the negative charge of the electrons. The attractive forces between the unlike charges help hold the atom together.

Table 3.2 Subatomic Particles

Particle	Symbol	Mass (u)	Charge	Location in Atom
Proton	p^+	1	1+	Nucleus
Neutron	n	1	0	Nucleus
Electron	e^-	$\dfrac{1}{1837}$	1−	Outside nucleus

A proton and a neutron have almost the same mass, 1.0073 u and 1.0087 u, respectively. This is equivalent to saying that two different people weigh 100.7 kg and 100.9 kg. The difference is so small that it usually can be ignored. Thus, for many purposes, we assume the masses of the proton and the neutron to be the same, 1 u. The proton has a charge equal in magnitude but opposite in sign to that of an electron. This charge on a proton is written as $1+$. The electron has a charge of $1-$ and a mass of 0.00055 u. The electrons in an atom contribute so little to its total mass that their mass is usually disregarded and treated as if it were 0.

Isotopes

Atoms of a given element can have different numbers of neutrons in their nuclei. For example, most hydrogen atoms have a nucleus consisting of a single proton and no neutrons. However, about 1 hydrogen atom in 6700 has a neutron as well as a proton in the nucleus. This heavier hydrogen atom is called *deuterium*. Whether it has one neutron, two neutrons, or none, any atom with $Z = 1$—that is, with one proton—is a hydrogen atom. Atoms that have this sort of relationship—having the same number of protons but different numbers of neutrons—are called **isotopes** (Figure 3.8). A third, rare isotope of hydrogen is *tritium,* which has two neutrons and one proton in the nucleus.

Most, but not all, elements exist in nature in isotopic forms. For example, tin (Sn) is present in nature in 10 different isotopic forms. It also has 15 radioactive isotopes that do not occur in nature. The existence of isotopes requires a modification of Dalton's original theory. He said that all atoms of the same element are alike. We now say that all atoms of the same element have the same number of protons. Different isotopes of an element have atoms with the same number of protons but with different numbers of neutrons (and, therefore, different masses).

Isotopes usually are of little importance in ordinary chemical reactions. All three hydrogen isotopes react with oxygen to form water. Because the isotopes differ in mass, compounds formed with different hydrogen isotopes have different physical properties, but such differences are usually slight. For example, water in which both hydrogen atoms are deuterium is called *heavy water,* often represented as D_2O. Heavy water boils at 101.4 °C (instead of 100 °C) and freezes at 3.8 °C (instead of 0 °C). In nuclear reactions, however, isotopes are of utmost importance, as we shall see in Chapter 11.

Symbols for Isotopes

Collectively, the two principal nuclear particles, protons and neutrons, are called **nucleons**. Isotopes are represented by symbols with subscripts and superscripts.

$$^{A}_{Z}X$$

In this general symbol, Z is the nuclear charge (atomic number or number of protons), and A is the **mass number** or the **nucleon number** because it is the num-

▶ **Figure 3.8** The three isotopes of hydrogen. Each has one proton and one electron, but they differ in the number of neutrons in the nucleus.

QUESTION: What is the atomic number and mass number of each of the isotopes? What is the mass of each, in atomic mass units, to the nearest whole number?

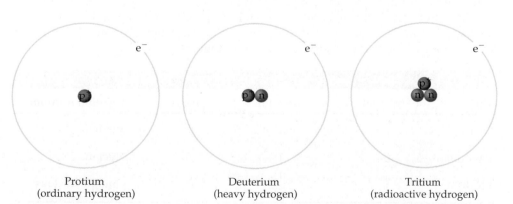

Protium
(ordinary hydrogen)

Deuterium
(heavy hydrogen)

Tritium
(radioactive hydrogen)

ber of nucleons. (*Nucleon number* is the term recommended by the International Union of Pure and Applied Chemistry.) As an example, the isotope with the symbol

$$^{35}_{17}\text{Cl}$$

has 17 protons and 35 nucleons. The number of neutrons in this atom is, therefore, $35 - 17 = 18$.

Isotopes often are named by placing the mass number as a suffix to the name of the element. The three hydrogen isotopes are therefore represented as

$$^1_1\text{H} \quad ^2_1\text{H} \quad ^3_1\text{H}$$

and named as hydrogen-1, hydrogen-2, and hydrogen-3.

EXAMPLE 3.1   Number of Neutrons

How many neutrons are there in the $^{235}_{92}\text{U}$ nucleus?

Solution

Simply subtract the atomic number Z (number of protons) from the mass number A (number of protons plus neutrons).

$$A - Z = \text{number of neutrons}$$
$$235 - 92 = 143$$

There are 143 neutrons in the nucleus.

■ **EXERCISE 3.1A**

How many neutrons are there in the $^{131}_{53}\text{I}$ nucleus?

■ **EXERCISE 3.1B**

A potassium isotope has 21 neutrons in its nucleus. What is the mass number and name of the isotope?

CONCEPTUAL EXAMPLE 3.2   Isotopes

Refer to the following isotope symbols, in which we use the letter X as the symbol for all elements so that the symbol will not identify the elements. **(a)** Which are isotopes of the same element? **(b)** Which have the same mass number? **(c)** Which have the same number of neutrons?

$$^{16}_8\text{X} \quad ^{16}_7\text{X} \quad ^{14}_7\text{X} \quad ^{14}_6\text{X} \quad ^{12}_6\text{X}$$

Solution

a. Isotopes of the same element will have the same atomic number (subscript). Therefore, $^{16}_7\text{X}$ and $^{14}_7\text{X}$ are isotopes of nitrogen (N), and $^{14}_6\text{X}$ and $^{12}_6\text{X}$ are isotopes of carbon (C).
b. The mass number is the superscript, so $^{16}_8\text{X}$ and $^{16}_7\text{X}$ have the same mass number. The first is an isotope of oxygen, and the second an isotope of nitrogen. $^{14}_7\text{X}$ and $^{14}_6\text{X}$ also have the same mass number. The first is an isotope of nitrogen, and the second an isotope of carbon.
c. To determine the number of neutrons, we subtract the atomic number from the mass number. We find that $^{16}_8\text{X}$ and $^{14}_6\text{X}$ each have eight neutrons ($16 - 8 = 8$ and $14 - 6 = 8$, respectively).

■ **EXERCISE 3.2A**

Which of the following are isotopes of the same element?

$$^{90}_{37}\text{X} \quad ^{90}_{35}\text{X} \quad ^{88}_{37}\text{X} \quad ^{88}_{38}\text{X} \quad ^{93}_{38}\text{X}$$

■ **EXERCISE 3.2B**

How many different elements are represented in Exercise 3.2A?

Self-Assessment Questions

1. The three basic components of an atom are
   a. protium, deuterium, and tritium
   b. protons, neutrinos, and ions
   c. protons, neutrons, and electrons
   d. protons, neutrons, and ions

2. The identity of an element is determined by its
   a. atomic mass
   b. number of atoms
   c. number of electrons
   d. number of protons

3. Compared to a proton, the mass of an electron is about
   a. 18 times larger
   b. 1800 times larger
   c. 18 times smaller
   d. 1800 times smaller

4. Which particles have approximately the same mass?
   a. electrons and neutrons
   b. electrons and protons
   c. neutrons and protons
   d. electrons, neutrons, and protons

5. A change in the number of neutrons of an atom produces an
   a. alpha particle   b. element   c. ion   d. isotope

6. The number of protons plus neutrons in an atom is its
   a. atomic mass
   b. ionic charge
   c. isotope number
   d. nucleon number

7. How many protons (p) and how many neutrons (n) are there in a $^{14}_{6}C$ nucleus?
   a. 6 p, 6 n   b. 6 p, 8 n   c. 6 p, 14 n   d. 8 p, 8 n

8. A neutral atom has 11 electrons and 12 neutrons. Its atomic number is
   a. 1   b. 11   c. 12   d. 23

9. The nuclear symbol for barium-138 is
   a. $^{82}_{56}Ba$ b. $^{138}_{56}Ba$ c. $^{138}_{82}Ba$ d. $^{220}_{82}Ba$

Answers: 1, c; 2, d; 3, d; 4, c; 5 d; 6, d; 7, b; 8, b; 9, b

3.6 Electron Arrangement: The Bohr Model

Let us now turn our attention once more to electrons. Rutherford demonstrated that atoms have a tiny, positively charged nucleus with electrons outside the nucleus. Evidence soon accumulated that the electrons were not randomly distributed but were arranged in an ordered fashion. We will examine the evidence soon. But first let's take a side trip into some colorful chemistry and physics to provide a background for our study of the electron structure of atoms.

Fireworks and Flame Tests

Chemists of the eighteenth and nineteenth centuries developed flame tests that used the colors of flames to identify several elements (Figure 3.9). Sodium salts give a persistent yellow flame, potassium salts a fleeting lavender flame, and lithium salts a brilliant red flame. It was not until the structure of atoms was better understood, however, that scientists could explain why each element emits a distinctive color of light.

▲ Flame colors (see Figure 3.9) also are the basis for the brilliant colors of fireworks displays. Strontium compounds produce red, copper compounds produce blue, and sodium compounds produce yellow.

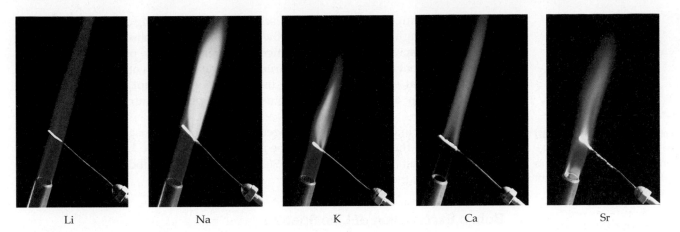

| Li | Na | K | Ca | Sr |

▲ **Figure 3.9** Certain chemical elements can be identified by the characteristic colors their compounds impart to flames. Five examples are shown here.

Fireworks originated earlier than flame tests, in ancient China. The brilliant colors of aerial displays continue to mark our celebrations of patriotic holidays. The colors of fireworks are attributable to specific elements. Brilliant reds are produced by strontium compounds, whereas barium compounds are used to produce green, sodium compounds yield yellow, and copper salts produce blue or violet.

The colors of fireworks and flame tests are not as simple as they seem to the unaided eye. If the light from the flame is passed through a prism, it is separated into light of several different colors.

Continuous and Line Spectra

When white light from an incandescent lamp is passed through a prism, it produces a continuous spectrum, or rainbow of colors (Figure 3.10). A similar phenomenon occurs when sunlight passes through raindrops. The different colors of light represent different wavelengths. Blue light has shorter wavelengths than red light, but there is no sharp transition in moving from one color to the next. All wavelengths are present in a continuous spectrum (Figure 3.11, top). White light is simply a combination of all the various colors.

More to Explore
Conkling, John A. "Pyrotechnics." *Scientific American*, July 1990, pp. 96–102. Discusses the science of fireworks.

▲ **Figure 3.10** A glass prism separates white light into a continuous spectrum or rainbow of colors just as sunlight passing through raindrops causes a rainbow in the sky.

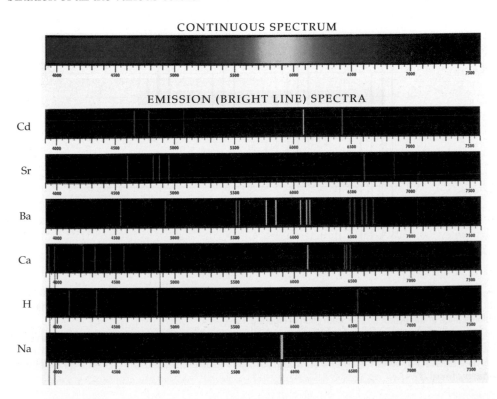

▲ **Figure 3.11** Line spectra of selected elements. Some of the components of the light emitted by excited atoms appear as colored lines. A continuous spectrum is shown at the top for comparison. The numbers are wavelengths of light given in Angstrom units ($1 \text{ Å} = 10^{-10}$ m). Each element has its own characteristic line spectrum that is different from all others and can be used to identify the element.

▲ Danish physicist Niels Bohr (1885–1962) received the 1922 Nobel Prize in Physics for his planetary model of the atom with its quantized electron energy levels. Element 107 is named bohrium in his honor.

3. What causes the colors of fireworks?

The colors of fireworks come from elements that are vaporized and then excited. See Figure 3.11. Sodium gives off bright yellow light and little else. So when sodium is heated a yellow flame is the result. Strontium's strongest emission is a red line, so strontium vapor from strontium compounds gives the beautiful red color we see in a fireworks display.

4. Most lightbulbs produce white light. Why does a neon bulb give off colored light?

When a solid is heated, it goes through a stage of red heat, then yellow, and finally white light is emitted, which consists of all wavelengths. When a gas such as neon is heated, we get a line spectrum (Figure 3.11), which gives a specific color to the light.

5. I see the word *quantum* used; what does it mean?

Just as an atom is the smallest unit of an element, a quantum is the smallest unit of energy. We can have different sizes of quanta, but energy is produced or consumed in discrete rather than continuous units. "Quantum leap" comes from the idea suggested by Figure 3.12, where an electron "leaps" from one energy level to another.

If a thin beam of light from a gas discharge tube containing a particular element is passed through a prism, only narrow colored lines are observed (Figure 3.11). Each line corresponds to light of a particular wavelength. The pattern of lines emitted by an element is called its *line spectrum*. The line spectrum of an element is characteristic of that element and can be used to identify it. Not all the lines in the spectrum of an atom are visible; some lines appear as infrared or ultraviolet radiation.

The visible line spectrum of hydrogen is fairly simple, consisting of four lines in that portion of the electromagnetic spectrum. To explain the hydrogen spectrum, Danish physicist Niels Bohr (1885–1962) worked out a model for the electron arrangement of the hydrogen atom.

Bohr's Explanation of Line Spectra

Niels Bohr presented his explanation of line spectra in 1913. He suggested that electrons cannot have just any amount of energy but can have only certain specified amounts; that is to say, the energy of an electron is *quantized*. A **quantum** (plural: *quanta*) is a tiny unit of energy, the value of which depends on the frequency. A specified energy value for an electron is called its **energy level**.

By absorbing a quantum of energy (for example, when atoms of the element are heated), an electron is elevated to a higher energy level (Figure 3.12). By giving up a quantum of energy, the electron can return to a lower energy level. The energy released shows up as a line spectrum. Each line has a specific wavelength corresponding to a quantum of energy. An electron moves practically instantaneously from one energy level to another with no intermediate stages.

Consider the analogy of a person on a ladder. The person can stand on the first rung, the second rung, the third rung, and so on, but is unable to stand between rungs. As the person moves from one rung to another, the potential energy (energy due to position) changes by definite amounts. As an electron moves from one energy level to another, its total energy (both potential and kinetic) also changes by definite amounts or quanta.

Bohr based his model of the atom on the laws of planetary motion that had been set down by the German astronomer Johannes Kepler (1571–1630) three cen-

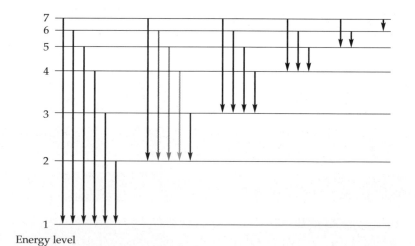

Energy level

▲ **Figure 3.12** Possible electron shifts between energy levels in atoms produce the lines found in spectra. Not all the lines are in the visible portion of the spectrum. The four colored arrows correspond to the four colored lines in the hydrogen spectrum seen in Figure 3.11.

QUESTION: Which transition involves the greater change in energy, a drop from energy level 6 to energy level 2 or a drop from energy level 7 to energy level 6? What color of light is absorbed when an electron in a hydrogen atom is boosted from energy level 2 to energy level 3?

turies before. Bohr imagined the electrons to be orbiting about the nucleus much as planets orbit the sun (Figure 3.13). Different energy levels were pictured as different orbits. The modern picture is different, as we shall see in subsequent sections.

Ground States and Excited States

The electron in a hydrogen atom is usually in the first, or lowest, energy level. Given the choice, electrons usually remain in their lowest possible energy levels (those nearest the nucleus); atoms whose electrons are thus situated are said to be in their **ground state**. When a flame or other source supplies energy to an atom (hydrogen, for example) and an electron jumps from the lowest possible level to a higher level, the atom is said to be in an **excited state**. An atom in an excited state eventually emits a quantum of energy—often a *photon* or "particle" of light—as the electron jumps back down to one of the lower levels and ultimately reaches the ground state.

Bohr's theory was a spectacular success in explaining the line spectrum of hydrogen. It established the important idea of energy levels in atoms. Bohr was awarded the Nobel Prize in Physics in 1922 for this work.

Atoms larger than hydrogen have more than one electron, and Bohr was also able to deduce that a given energy level of an atom could contain at most only a certain number of electrons. We shall simply state Bohr's findings in this regard. The maximum number of electrons that can be in a given level is indicated by the formula

$$\text{Maximum number of electrons} = 2n^2$$

where n is the energy level being considered. For the first energy level ($n = 1$), the maximum population is $2 \times 1^2 = 2 \times 1 = 2$. For the second energy level ($n = 2$), the maximum number of electrons is $2 \times 2^2 = 2 \times 4 = 8$. For the third level, the maximum is $2 \times 3^2 = 2 \times 9 = 18$. (However, the outermost level usually has no more than 8 electrons.)

The various energy levels are often called **shells**. The first energy level ($n - 1$) is the first shell, the second energy level ($n = 2$) is the second shell, and so on.

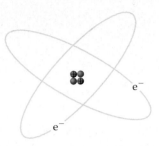

▲ **Figure 3.13** The nuclear atom, as envisioned by Bohr, has most of its mass in an extremely small nucleus. Electrons orbit about the nucleus, occupying most of the volume of the atom but contributing little to its mass.

QUESTION: What element is represented in this drawing?

EXAMPLE 3.3 Electron Shell Capacity

What is the maximum number of electrons in the fifth shell (fifth energy level)?

Solution

The maximum number of electrons in a given shell is given by $2n^2$. For the fifth level, $n = 5$, and so we have

$$2 \times 5^2 = 2 \times 25 = 50$$

■ **EXERCISE 3.3A**

What is the maximum number of electrons in the fourth shell (fourth energy level)?

■ **EXERCISE 3.3B**

What is the lowest shell (value of n) that can hold 18 electrons?

Building Atoms: Main Shells

Imagine building up atoms by adding one electron to the proper shell as *each* proton is added to the nucleus, keeping in mind that electrons will go to the lowest energy level (shell) available. (We can ignore the neutrons in the nucleus; they are not involved in this process.) For hydrogen, with a nucleus of only one proton ($Z = 1$), the single electron goes into the first shell. For helium, with a nucleus having two protons ($Z = 2$), both electrons go into the first shell. According to Bohr, two electrons is the maximum population of the first shell; that level is filled in the helium atom.

With lithium ($Z = 3$), two electrons go into the first shell; the other must go into the second shell. This process of adding electrons is continued until the second shell is filled with eight electrons, as in a neon atom ($Z = 10$), which has two of its 10 electrons in the first shell and the remaining eight in the second shell.

A sodium atom ($Z = 11$) has 11 electrons. Two are in the first shell, the second shell is filled with eight electrons, and the remaining electron is in the third shell. We can indicate the **electron configuration** (or arrangement) of the first 11 elements as follows:

Element	First Shell	Second Shell	Third Shell
H	1		
He	2		
Li	2	1	
⋮	⋮	⋮	
Ne	2	8	
Na	2	8	1

Sometimes the main-shell configuration is abbreviated by simply listing the symbol for the element followed by the number of electrons in each shell, separated by commas, starting with the lowest energy level. Following the color scheme from the table, the configuration for sodium is simply given as

$$\text{Na} \quad 2, 8, 1$$

We could now continue to add electrons to the third shell until we get to argon. The configuration for argon is

$$\text{Ar} \quad 2, 8, 8$$

After argon, the shell model works best if it is enhanced with more detail. The topic of atomic structure is discussed in greater detail in Section 3.8. Meanwhile, Example 3.4 shows how to determine a few main-shell configurations.

EXAMPLE 3.4   Main-Shell Electron Configurations

What are the main-shell electron configurations for **(a)** fluorine and **(b)** aluminum?

Solution

a. Fluorine has the symbol F ($Z = 9$); it has nine electrons. Two of these electrons go into the first shell, and the remaining seven go into the second level.

$$\text{F} \quad 2, 7$$

b. Aluminum ($Z = 13$) has 13 electrons. Two go into the first shell, eight go into the second, and the remaining three go into the third shell.

$$\text{Al} \quad 2, 8, 3$$

Notice that fluorine is in group 7A of the periodic table and that it has seven outer-shell electrons. Aluminum is in group 3A, and it has three electrons in its outermost shell.

■ EXERCISE 3.4

Give the main-shell electron configurations for **(a)** beryllium and **(b)** magnesium. These elements are in the same group of the periodic table. What do you notice about the number of electrons in their outermost shells?

Self-Assessment Questions

1. When electrified in a gas discharge tube, different chemical elements emit light in narrow lines with frequencies that are
   a. characteristic of each different element
   b. multiples of those of the hydrogen spectrum
   c. the same for all elements but with differing intensities
   d. the same for all elements with the exception of one different line

2. When an electron in an atom goes from a lower energy level to a higher one, the electron
   a. absorbs energy
   b. changes spin
   c. changes charge
   d. releases energy

3. An atom in an excited state is one with an electron that has
   a. been moved to a higher energy level
   b. bonded to another atom to make a molecule
   c. combined with a proton to make a neutron
   d. dropped to the lowest energy level

4. The maximum number of electrons in the first shell of an atom is
   a. 2   b. 6   c. 8   d. unlimited

5. The maximum number of electrons in the fourth main shell of an atom is
   a. 6   b. 8   c. 18   d. 32

6. The main-shell electron configuration of silicon is
   a. 2, 8, 2   b. 2, 8, 4   c. 2, 8, 18   d. 2, 8, 8, 8, 2

7. The main-shell electron configuration of sulfur is
   a. 2, 8, 4   b. 2, 8, 6   c. 2, 8, 8, 8, 6   d. 2, 8, 18, 4

Answers: 1, a; 2, a; 3, a; 4, a; 5, d; 6, b; 7, b

3.7 Electron Arrangement: The Quantum Model

The simple planetary Bohr model of the atom has been replaced for many purposes by more sophisticated models in which electrons are treated as *both* particles and waves. The electrons have characteristic wavelengths and their locations are indicated as probabilities. The theory that the electron should have wavelike properties was first suggested in 1924 by Louis de Broglie (1892–1987), a young French physicist. Although it was hard to accept because of Thomson's evidence that electrons are particles, de Broglie's theory was experimentally verified within a few years.

Erwin Schrödinger (1887–1961), an Austrian physicist, used highly mathematical *quantum mechanics* in the 1920s to develop equations that describe the properties of electrons in atoms. Fortunately, we can make use of some of Schrödinger's results without understanding his elaborate equations. The solutions to these equations express the probability of finding an electron in a given volume of space. These shaped volumes of space, called **orbitals**, replace the planetary orbits of the Bohr model.

Suppose you had a camera that could photograph electrons and you left the shutter open while an electron zipped about the nucleus. The developed picture would give a record of where the electron had been. (Doing the same thing with an electric fan would give a blurred image of the rapidly moving blades, a picture resembling a disk.) The electrons in the first shell would appear as a fuzzy ball and are often referred to as a *charge cloud* or an *electron cloud* (Figure 3.14).

Building Atoms by Orbital Filling

Schrödinger concluded that each electron orbital could contain a maximum of two electrons, and that some shells could contain more than one orbital. The first shell contains a single spherical orbital named 1s. The second shell contains four orbitals:

An s orbital

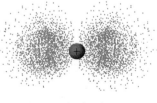

A p orbital

▲ **Figure 3.14** Charge-cloud representations of atomic orbitals.

▶ **Figure 3.15** Electron orbitals of the second main shell. In these drawings, the nucleus of the atom is located at the intersection of the axes. The eight electrons that would be placed in the second shell of Bohr's model are distributed among these four orbitals in the current model of the atom, with two electrons per orbital.

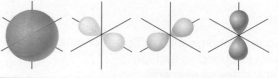

The 4 distinctively shaped or oriented orbitals of the second energy level

Combined in one drawing

The 4 orbitals centered on a single nucleus. (The spherical s orbital is shown only in outline in order that the others can be seen clearly.)

2s is spherical, and the other three orbitals, called 2p, are dumbbell shaped (Figure 3.15). Orbitals in the same shell with the same letter designation make up a **subshell (sublevel)**. The second shell has two subshells: the 2s subshell with only one orbital and the 2p subshell with three orbitals. The third shell contains nine orbitals distributed among three subshells: a spherical 3s orbital in the 3s subshell, three dumbbell-shaped 3p orbitals in the 3p subshell, and five 3d orbitals with more complicated shapes in the 3d subshell.

In building up the electron configuration of atoms of the various elements, the lower subshells are filled first.

Hydrogen ($Z = 1$) has only one electron in the s orbital of the first shell. The electron configuration is

$$\text{H} \quad 1s^1$$

Helium ($Z = 2$) has two electrons and an electron configuration

$$\text{He} \quad 1s^2$$

Lithium ($Z = 3$) has three electrons—two in the first shell and one in the s orbital of the second shell:

$$\text{Li} \quad 1s^2 2s^1$$

Skipping to nitrogen ($Z = 7$), we place the seven electrons as follows:

$$\text{N} \quad 1s^2 2s^2 2p^3$$

Let's review what this notation means.

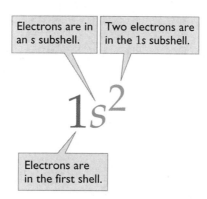

Electrons are in an s subshell.

Two electrons are in the 1s subshell.

Electrons are in the first shell.

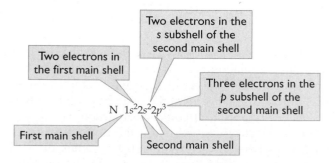

Two electrons in the first main shell

Two electrons in the s subshell of the second main shell

Three electrons in the p subshell of the second main shell

N $1s^2 2s^2 2p^3$

First main shell

Second main shell

Look next at argon ($Z = 18$) with its 18 electrons. The configuration is

$$\text{Ar} \quad 1s^2 2s^2 2p^6 3s^2 3p^6$$

Note that the highest occupied subshell, 3p, is filled. However, the remaining order of subshell filling is not intuitively obvious. When we move to potassium ($Z = 19$), we find that the 4s sublevel fills before the 3d sublevel. The order in which the various electron subshells are filled is shown in Figure 3.16, and Table 3.3 gives the electron configuration for the first 20 elements. The electrons in the outermost main shell, called *valence electrons*, are shown in color.

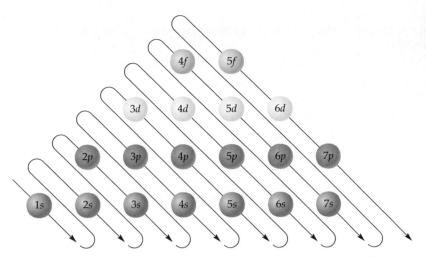

▲ **Figure 3.16** An order-of-filling chart for determining the electron configurations of atoms.

Table 3.3 Electron Structures for Atoms of the First 20 Elements

Name	Atomic Number	Electron Structure
Hydrogen	1	$1s^1$
Helium	2	$1s^2$
Lithium	3	$1s^2 2s^1$
Beryllium	4	$1s^2 2s^2$
Boron	5	$1s^2 2s^2 2p^1$
Carbon	6	$1s^2 2s^2 2p^2$
Nitrogen	7	$1s^2 2s^2 2p^3$
Oxygen	8	$1s^2 2s^2 2p^4$
Fluorine	9	$1s^2 2s^2 2p^5$
Neon	10	$1s^2 2s^2 2p^6$
Sodium	11	$1s^2 2s^2 2p^6 3s^1$
Magnesium	12	$1s^2 2s^2 2p^6 3s^2$
Aluminum	13	$1s^2 2s^2 2p^6 3s^2 3p^1$
Silicon	14	$1s^2 2s^2 2p^6 3s^2 3p^2$
Phosphorus	15	$1s^2 2s^2 2p^6 3s^2 3p^3$
Sulfur	16	$1s^2 2s^2 2p^6 3s^2 3p^4$
Chlorine	17	$1s^2 2s^2 2p^6 3s^2 3p^5$
Argon	18	$1s^2 2s^2 2p^6 3s^2 3p^6$
Potassium	19	$1s^2 2s^2 2p^6 3s^2 3p^6 4s^1$
Calcium	20	$1s^2 2s^2 2p^6 3s^2 3p^6 4s^2$

EXAMPLE 3.5   Subshell Notation

Without referring to Table 3.3, use subshell notation to write out the electron configuration for **(a)** oxygen and **(b)** sulfur. What do the electron configurations of these two elements have in common?

Solution

a. Oxygen ($Z = 8$) has eight electrons. Place them in the lowest unfilled energy sublevels. Two go into the $1s$ orbital and two into the $2s$ orbital. That leaves four electrons to be placed in the $2p$ subshell. The electron configuration is $1s^2 2s^2 2p^4$.
b. Sulfur ($Z = 16$) atoms have 16 electrons each. The electron configuration is $1s^2 2s^2 2p^6 3s^2 3p^4$. Note that the total of the superscripts is 16 and that we have not exceeded the maximum capacity for any sublevel.

Both O and S have electron configurations with four electrons in their highest energy sublevel (outermost subshell).

■ **EXERCISE 3.5A**

Without referring to Table 3.3, use subshell notation to write out the electron configurations for **(a)** fluorine and **(b)** chlorine. What do the electron configurations for these two elements have in common?

■ **EXERCISE 3.5B**

Use Figure 3.16 to write the electron configurations for **(a)** titanium (Ti) and **(b)** gallium (Ga).

Self-Assessment Questions

1. The shape of a $1s$ orbital is best described as
   a. circular   **b.** a dumbbell   **c.** octahedral   **d.** spherical
2. The maximum number of electrons in an atomic orbital
   a. depends on the main shell   **b.** depends on the subshell
   c. is always 2   **d.** is always 8
3. Which of the following is the highest energy subshell?
   a. $2p$ **b.** $2s$ **c.** $3p$ **d.** $3s$
4. What is the lowest-numbered main shell in which d orbitals are found?
   a. 1   **b.** 2   **c.** 3   **d.** 4

5. Which of the following elements has atoms in the ground state with a half-filled subshell?
   a. beryllium   **b.** neon   **c.** nitrogen   **d.** oxygen
6. Which subshell has a total of three orbitals?
   a. d   **b.** f   **c.** p   **d.** s
7. A 3d subshell is located in what main shell?
   a. 1   **b.** 3   **c.** 5   **d.** none
8. The electron capacity of the 4p subshell is
   a. 3   **b.** 4   **c.** 6   **d.** 10

Answers: 1, d; 2, c; 3, c; 4, c; 5, c; 6, c; 7, b; 8, c

3.8 Electron Configurations and the Periodic Table

In general, the properties of elements can be correlated with their electron configurations. (We explore bond formation using electron configurations in Chapter 4.) Because the number of electrons equals the number of protons, the periodic table tells us about electron configuration as well as atomic number.

The modern periodic table (inside front cover) has horizontal rows and vertical columns.

- Each vertical column is a **group** or *family*. Elements in a group have similar chemical properties.
- A horizontal row of the periodic table is called a **period**. Many properties of elements vary in regular fashion across a period.

In the United States, the groups are often indicated by a numeral followed by the letter A or B.

- An element in an A group is a **main group element**.
- An element in a B group is a **transition element**.

The International Union of Pure and Applied Chemistry (IUPAC) recommends numbering the groups from 1 to 18. Both systems are indicated on the periodic table on the inside front cover, but we follow the traditional U.S. method in this book.

Family Features: Outer Electron Configurations

The period in which an element appears in the periodic table tells us how many main electron shells are in that atom. Phosphorus, for example, is in the third period, and so the phosphorus atom has three main shells. The group number (for main group elements) tells us how many electrons are in the outermost shell. An electron in the outermost shell of an atom is called a **valence electron**. From the fact that phosphorus is in group 5A, we can deduce that it has five valence electrons. Two of these electrons are in an *s* orbital, and the other three are in *p* orbitals. We can indicate the outer electron configuration of the phosphorus atom as

$$\text{P} \quad 3s^2 3p^3$$

These valence electrons determine most of the chemistry of an atom. Since all the elements in the same group of the periodic table have the same number of valence electrons, they should have similar chemistry, and they do. Figure 3.17 on page 82 relates the subshell configurations to the groups in the periodic table.

Family Groups

Elements within a group have similar properties. All the elements in group 1A have one valence electron, and all except hydrogen are very reactive metals (hydrogen is a nonmetal). All have the outer electron structure ns^1, where n denotes the number

ANSWER

6. What is "periodic" about the periodic table?
Figure 3.17 highlights some of the periodic or recurring trends of the periodic table. Elements in the same column have similar electron configurations, and often have similar chemical and physical characteristics as well.

GREEN CHEMISTRY

Thinking about Nanotechnology

Dwight Tshudy, Gordon College

Have you ever thought how exciting it would be to build very small objects one atom or one molecule at a time? If you could do that, then you would be doing chemistry on an atomic level. In Chapters 2 and 3 you have learned about the atom and how what we know about atoms has changed over time. Even though the idea of manipulating one atom at a time might sound like science fiction, it is actually being done, and it forms the basis of what we call *nanotechnology*.

Nanotechnology is an exciting and growing field of chemical, physical, and mechanical research that is changing the way chemistry is done. The term *nanotechnology* can be confusing to scientists and nonscientists alike because it is hard to think and imagine in such small terms. The term nano relates to the size of a nanometer (nm). One nanometer is 10^{-9} meter (m) or one billionth of a meter. In very general terms, nanotechnology relates to materials and structures of nanometer dimensions, objects that are 0.1 to 100 nm in size.

What does it mean to be working on this scale? The example of human hair is often used to offer perspective. Human hair is around 50 micrometers (μm) in diameter. That is 50,000 nm—about 500 times larger than what would be included as part of nanotechnology. Another example is a normal human red blood cell that is used to carry oxygen in your blood. These cells are about 6 to 8 μm (6,000 to 8,000 nm) in diameter—still much larger than objects classified within the realm of nanotechnology.

Then what would be nanoscale? A single-cell virus can be in the nanoscale region. If you stretch your imagination, you might even think of an influenza virus as being a biological nanomachine. Nanoscale objects are not really new. The beautiful red colors in centuries-old-stained glass windows are the result of nanosized particles of gold in the glass. Another example is the traditional photography process developed in the 1800s, which relies on nanoscale silver-halide particles to create an image.

Nanotechnology is being developed for a range of consumer and industrial products. Research is being done in developing solar cells, increasing the efficiency fuel cells, designing more targeted drug delivery systems, improving the selectivity of diagnostic medical devices, enabling stronger building materials, manufacturing of electronic components, creating sensors for detecting biological and chemical warfare agents, and purifying drinking water. You might have heard of the terms *buckyballs*, *fullerenes*, *carbon nanotubes* or even *quantum dots*. These nanosized particles are being investigated for their unique chemical and physical properties. Researchers are also working on developing nanomachines that exhibit controlled molecular movement.

Did you know that you can now purchase items in your local store that are the result of applied nanotechnology? Nanosized zinc oxide and titanium oxide are being used in sunscreens, cosmetics, and paints. Silver nanoparticles, which have antibacterial properties, are being added to fabric to help control odors. You now can buy a pair of socks that can help control those smelly feet.

Researchers can produce gold nanoparticles that contain only a few atoms by carefully combining individual atoms of gold together and growing the particle until the desired size is reached. Gold is known for its chemical inertness as well as its intrinsic beauty for jewelry. When it is formed into gold nanoparticles, it has very distinctive properties. These particles are being investigated for their unique chemical catalyst properties as well as for use in electronic and biomedical applications. The ability to make more efficient and selective chemical catalysts is Green Chemistry Principle #9 (Catalysis) in action. The field of nanotechnology fits into green chemistry in many other ways as well. Work on solar energy collectors and fuel cells is directly related to discovering new ways to use renewable resources, which is Green Chemistry Principle #7 (Renewable Feedstocks).

You might have heard scientists and politicians alike who promote the idea that the world will be a different and better place because of nanotechnology. However there are others who warn of the potential consequences of having hordes of uncontrollable nanomachines. Concerns have also been raised about the long- and short-term risks: What are the known and unknown hazards of products produced using nanotechnology? For example, if these nanomaterials released into the environment, what would be the impact on the environment and on ourselves? Some researchers feel it is imperative to develop nanotechnology within the framework of green chemistry and green engineering principles, so that the chemistry and products that are enabled by nanotechnology carry the most benefit with the lowest risk of harm.

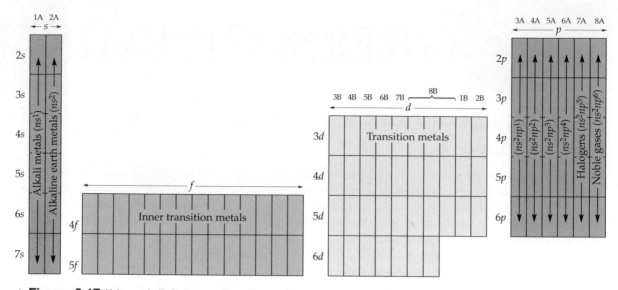

▲ **Figure 3.17** Valence shell electron configurations and the periodic table.

of the outermost main shell. The metals in group 1A are called **alkali metals**; they react vigorously with water to evolve hydrogen gas. Important trends exist within a family. For example, lithium is the hardest metal in the group. Sodium is softer than lithium, potassium is softer still, and so on down the group. Lithium is also the least reactive toward water. Sodium, potassium, rubidium, and cesium are progressively more reactive. Francium is highly radioactive and extremely rare; few of its properties have been measured. Hydrogen is the odd one in group 1A. It is not an alkali metal. Rather, it is a characteristic nonmetal. Based on its properties, hydrogen probably should be put in a group of its own.

Group 2A elements are known as **alkaline earth metals**. The metals in this group have the outer electron structure ns^2. Most are fairly soft and moderately reactive with water. Beryllium is the odd member of the group in that it is rather hard and does not react with water. As in other families, there are trends in properties within the group. For example, magnesium, calcium, strontium, barium, and radium are progressively more reactive toward water.

Group 7A elements, often called **halogens**, also consist of reactive elements. All have seven valence electrons in the configuration ns^2np^5. Halogens react vigorously with alkali metals to form crystalline solids (discussed further in Chapter 4). Trends exist in the halogen family as you move down the group. Fluorine is most reactive toward alkali metals, chlorine is the next most reactive, and so on. Fluorine and chlorine are gases at room temperature, bromine is a liquid, and iodine is a solid. (Astatine, like francium, is highly radioactive and extremely rare; few of its properties have been determined.)

Members of group 8A, to the far right of the periodic table, have a complete set of valence electrons and undergo few, if any, chemical reactions. They are called **noble gases**, known for their lack of chemical reactivity.

(Above) Lithium, an alkali metal, reacts with water to form hydrogen gas. (Below) Potassium, another alkali metal, undergoes the same reaction but much more vigorously.

EXAMPLE 3.6 Valence Shell Electron Configurations

Write out the subshell notation for the electrons in the highest main shell for **(a)** strontium (Sr) and **(b)** arsenic (As).

Solution

a. Strontium is in group 2A with two valence electrons in an s subshell. Because strontium is in the fifth period of the periodic chart, its outer main shell is $n = 5$. Its outer electron configuration is therefore $5s^2$.

b. Arsenic is in group 5A and the fourth period. Its five outer (valence) electrons are in the $n = 4$ shell and have the configuration $4s^24p^3$.

■ **EXERCISE 3.6A**

Write out the subshell notation for the electrons in the highest main shell for **(a)** rubidium (Rb), **(b)** selenium (Se), and **(c)** germanium (Ge).

■ **EXERCISE 3.6B**

The valence electrons in aluminum have the configuration $3s^2 3p^1$. Gallium is directly below aluminum, and indium is directly below gallium in the periodic table. Use only this information to write the configuration for the valence electrons in **(a)** gallium and **(b)** indium.

▲ (Top) Metals such as gold are easily shaped and conduct heat and electricity well. (Bottom) Nonmetals such as sulfur are usually brittle, melt at low temperatures, and often are good insulators.

Metals and Nonmetals

Elements in the periodic table also are divided into two classes by a heavy, stepped diagonal line (see inside front cover). Those to the left of the line are *metals*. A **metal** has a characteristic luster and generally is a good conductor of heat and electricity. Except for mercury, which is a liquid, all metals are solids at room temperature. Metals generally are *malleable*; that is, they can be hammered into thin sheets. Most also are *ductile*; they can be drawn into wires.

Elements to the right of the stepped line are *nonmetals*. A **nonmetal** element lacks metallic properties. Several nonmetals are gases (oxygen, nitrogen, fluorine, chlorine). Others are solids (carbon, sulfur, phosphorus, iodine). Bromine is the only nonmetal that is a liquid at room temperature.

Some of the elements bordering the stepped line are called *semimetals* or *metalloids*, elements that have intermediate properties. Metalloids have properties that resemble those of both metals and nonmetals. There is a lack of agreement on just which elements fit in this category.

More to Explore

Barrett, Jack. *Atomic Structure and Periodicity*. Cambridge, UK: Royal Society of Chemistry, 2002.

Self-Assessment Questions

1. Which pair of elements has the most similar chemical properties?
   a. C and N   **b.** Cl and Br   **c.** K and Ca   **d.** Li and Be

2. Elements within a period of the periodic table have the same number of
   a. electrons in the outermost shell
   b. neutrons in the nucleus
   c. occupied main energy shells
   d. protons in the nucleus

3. The valence shell electron configuration of the halogens is
   a. ns^1 **b.** $ns^2 np^3$ **c.** $ns^2 np^5$ **d.** $ns^2 np^6$

4. The alkaline earth metals are found in group
   a. 1A   **b.** 2A   **c.** 7A   **d.** 8A

5. What is the total number of valence electrons in an atom of iodine?
   a. 1   **b.** 7   **c.** 8   **d.** 17

6. In which of the following sets do the elements exhibit the most similar chemical properties?
   a. Al, Sc, P   **b.** Hg, Br, Kr   **c.** Li, Na, K   **d.** N, O, F

7. Atoms of a fourth-period element in the ground state must have
   a. a 3f sublevel
   b. electrons in the fourth main shell
   c. four valence electrons
   d. properties similar to those of other elements in the fourth period

Answers: 1, b; 2, c; 3, c; 4, b; 5, b; 6, c; 7, b

3.9 Which Model to Choose?

For some purposes in this text, we use only main-shell configurations of atoms to picture the distribution of electrons in atoms. At other times, the electron clouds of the quantum mechanical model are more useful. Even Dalton's model sometimes proves to be the best way to describe certain phenomena (the behavior of gases, for example). The choice of model is always based on which one is most helpful in understanding a particular concept. This, after all, is the whole purpose of scientific models.

Critical Thinking Exercises

Apply knowledge that you have gained in this chapter and one or more of the FLaReS principles (Chapter 1) to evaluate the following statements or claims.

3.1 Suppose you read in the newspaper that a chemist in South America claims to have discovered a new element with an atomic mass of 34. An extremely rare element, it was found in a sample recovered from the Andes Mountains. Unfortunately, the chemist has used up all of the sample in his analyses.

3.2 Some aboriginal tribes have rainmaking ceremonies in which they toss pebbles of gypsum up into the air. (Gypsum is the material used to make plaster of Paris by heating the rock to remove some of its water.) Sometimes it really does rain several days after these rainmaking ceremonies.

■ SUMMARY

Section 3.1—Davy, Faraday, and others showed that matter is electrical in nature. They were able to decompose compounds into elements by **electrolysis** or passing electricity through molten salts. **Electrodes** are the pieces of metal or carbon that carry electricity through the **electrolyte** or solution/compound undergoing electrolysis. The electrolyte contains **ions**—charged atoms or groups of atoms. The **anode** is the positive electrode and **anions** or negatively charged ions move toward it. The **cathode** is the negative electrode and **cations** or positively charged ions move toward it. Experiments with **cathode ray** tubes showed that matter contained negatively charged particles, which were called **electrons**. By deflecting cathode rays with a magnet, Thomson was able to determine the mass-to-charge ratio for the electron. Goldstein's experiment, using gas discharge tubes with perforated cathodes, showed that matter also contained positively charged particles. Millikan's oil-drop experiment measured the charge on the electron, so its mass could then be calculated. That mass was found to be much smaller than that of the lightest atom.

Sections 3.2–3.3—In his studies of cathode rays, Roentgen accidentally discovered **X-rays**, which are a highly penetrating form of radiation used in medical diagnosis. Becquerel accidentally discovered another type of radiation that comes from certain unstable elements. Marie Curie named this new discovery **radioactivity** and studied it extensively, discovering two elements in the process and winning two Nobel Prizes. Radioactivity was soon found to be three different types of radiation: **Alpha particles** have four times the mass of a hydrogen atom and a positive charge twice that of an electron. **Beta particles** are energetic electrons. **Gamma rays** are a form of energy like X-rays but more penetrating.

Section 3.4—Rutherford's later experiments with alpha particles and gold foil showed that most of the alpha particles emitted toward the foil passed through it. A few were deflected or, occasionally, bounced almost directly back to the source. This indicated that all the positive charge and most of the mass of an atom must be in a tiny core, which he called the **nucleus**.

Section 3.5—Rutherford called the unit of positive charge the **proton**, with about the mass of a hydrogen atom and with a charge equal in size but opposite in sign to the electron. In 1932 Chadwick discovered the **neutron**, a nuclear particle as massive as a proton but with no charge. The number of protons in an atom is the **atomic number**. Atoms of the same element have the same number of protons but may have different numbers of neutrons; such atoms are called **isotopes**. Different isotopes of an element are nearly identical chemically. Protons and neutrons collectively are called **nucleons**, and the number of nucleons is called the **mass number** or **nucleon number**. The difference between the mass number and the atomic number is the number of neutrons. The general symbol for an isotope of element X is written $^A_Z X$, where A is the mass number and Z is the atomic number.

Section 3.6—Light from the sun or from an incandescent lamp is a continuous spectrum, containing all colors. Light from a gas discharge tube is a line spectrum, containing only certain colors. Bohr explained line spectra by proposing that an electron in an atom resides only at certain discrete **energy levels**, with these levels differing by a **quantum** or discrete unit of energy. An atom drops to the lowest energy state or **ground state** after being in a higher-energy state or **excited state**. The emitted energy from a change in energy level manifests as line spectra characteristic of each element, corresponding to specific wavelengths or colors of light. Bohr also deduced that the energy levels of an atom could handle at most $2n^2$ electrons, where n is the energy level or **shell** number. A description of the shells occupied by electrons of an atom is one way of giving the atom's **electron configuration** or arrangement of electrons.

Section 3.7—De Broglie theorized that electrons have wave properties. Schrödinger developed equations that described each electron's location in terms of an **orbital**, the volume of space that an electron usually occupies. Each orbital holds at most two electrons. Orbitals in the same shell and with the same energy make up a **subshell**, each of which is designated by a letter. An s orbital is spherical and a p orbital is dumbbell shaped. The d and f orbitals have more complex shapes. The first main shell can hold only one s orbital; the second level can hold one s and three p orbitals; and the third level can hold one s, three p, and five d orbitals. In writing an electron configuration using subshell notation (quantum mechanical notation), the shell number and subshell letter are given, followed by a superscript giving the number of electrons in that subshell. The order of the shells and subshells can be remembered with a chart or by using the periodic table.

Section 3.8—Elements in the periodic table are arranged in columns or **groups** and in rows or **periods**. Electrons in the outermost shell of an atom are called **valence electrons** and are responsible for the reactivity of that atom. Elements in a group usually have the same number of valence electrons and similarities in properties. Groups are designated by a number and by the letter A or B. The A-group elements are called **main group elements** and are on both sides of the periodic table. The B-group elements are **transition elements** in the middle of the table. The first column of the table, or group 1A elements, is known as the **alkali metals**. Except for hydrogen, they are all soft, low-melting, highly reactive metals. The second column, group 2A, is known as the **alkaline earth elements**. They are fairly soft and fairly reactive metals. Group 7A elements are the **halogens** and are reactive nonmetals. Group 8A elements, the **noble gases**, react very little or not at all. A diagonal line divides the periodic table into metals and nonmetals. **Metals** conduct electricity and heat and are shiny, malleable, and ductile. **Nonmetals** tend to lack the properties of metals. Certain elements on the diagonal line are sometimes referred to as metalloids and have properties intermediate between metals and nonmetals.

■ REVIEW QUESTIONS

1. What did each of the following scientists contribute to our knowledge of the atom?
   a. Crookes b. Goldstein

2. What is radioactivity? How did the discovery of radioactivity contradict Dalton's atomic theory?

3. Define or identify each of the following.
   a. deuterium b. tritium
   c. photon

4. How are X-rays and gamma rays similar? How are they different?

5. The following table describes four atoms.

	Atom A	Atom B	Atom C	Atom D
Number of protons	10	11	11	10
Number of neutrons	11	10	11	10
Number of electrons	10	11	11	10

   Are atoms A and B isotopes? A and C? A and D? B and C?

6. What is the mass in u of each atom in Question 5?

7. Compare Dalton's model of the atom with the nuclear model of the atom.

8. What are the symbol, name, and atomic mass of the element with Z = 98? You may use a periodic table and table of atomic masses.

9. What are the symbol, name, and atomic mass of the element that has 18 protons in the nucleus of its atoms?

10. Explain what is meant by the term *atomic mass*.

11. How did Bohr refine the model of the atom?

12. Which atom absorbs more energy, one in which an electron moves from the second shell to the third shell or an otherwise identical atom in which an electron moves from the first to the third shell?

13. Use the periodic table to determine the numbers of protons in atoms of the following elements.
    a. helium b. sodium
    c. chlorine d. oxygen
    e. magnesium f. sulfur

14. How many electrons are there in the neutral atoms of the elements listed in Question 13?

■ PROBLEMS

Nuclear Symbols and Isotopes

15. Give the nuclear symbol and name for **(a)** an isotope with a mass number of 40 and an atomic number of 19 and **(b)** an isotope with 26 neutrons and 22 protons.

16. Fill in the following table:

Element	Mass Number	Protons	Neutrons
Cobalt	60		
	19	9	
		92	143
Lead			124

17. How many different isotopes of silver are listed here? (The X does not necessarily represent silver or any other element.)

$$^{108}_{47}X \qquad ^{108}_{48}X \qquad ^{110}_{47}X \qquad ^{109}_{46}X \qquad ^{107}_{47}X$$

18. How many different elements are listed here?

$$^{23}_{11}X \qquad ^{22}_{10}X \qquad ^{11}_{5}X \qquad ^{24}_{11}X \qquad ^{25}_{12}X$$

Quantum Mechanical Notation for Electron Structure

19. Without referring to the periodic table, give the atomic numbers of the elements with the following electron structures.
    a. $1s^2 2s^2 2p^2$ **b.** $1s^2 2s^2 2p^6 3s^2 3p^4$
    c. $1s^2 2s^2 2p^5$

20. Without referring to the periodic table, give the atomic numbers of the elements with the following electron structures.
    a. $1s^2 2s^2 2p^6 3s^2 3p^1$ **b.** $1s^2 2s^2 2p^6 3s^1$
    c. $1s^2 2s^2 2p^6 3s^2 3p^6 4s^2$

21. Classify the following by indicating whether the electron configuration represents an atom in the ground state, a possible excited state, or is incorrect. In each case, explain why.
    a. $1s^2 2s^2 3s^2$ **b.** $1s^2 2s^2 2p^2 3s^1$
    c. $1s^2 2s^2 2p^6 2d^5$ **d.** $1s^2 2s^4 2p^2$

22. Classify the following by indicating whether the electron configuration represents an atom in the ground state, a possible excited state, or is incorrect. In each case, explain why.
    a. $1s^1 2s^1$ **b.** $1s^2 2s^2 2p^7$
    c. $1s^2 2p^2$ **d.** $1s^2 2s^2 2p^2$

23. Suppose that one electron was added to the outermost shell of a chlorine atom. This new electron structure would resemble that of which element?

24. Suppose that the following changes were made with corresponding proton changes in the nucleus. Give the symbol for the element that would have the new electron structures.
    a. two electrons added to a sulfur atom
    b. one electron removed from a sodium atom
    c. one electron removed from a potassium atom

25. Referring only to a periodic table, tell how the electron structures of magnesium (Mg) and calcium (Ca) are similar. How are they different?

26. Referring only to the periodic table, tell how the electron structures of fluorine (F) and chlorine (Cl) are similar. How are they different? Do you expect the electron structures of bromine (Br) and iodine (I) to be similar to those of fluorine and chlorine?

27. Refer to a periodic table to identify each element as a metal or nonmetal.
    a. manganese **b.** cesium
    c. neon **d.** phosphorus

Use the following list of elements to answer Problems 28–32: Mn, Cs, Ne, P, Kr, K, Ra, N, Ni, Ca, Ba

28. Which element(s) is/are alkali metals?

29. Which element(s) is/are noble gases?

30. Which element(s) is/are transition metals?

31. How many of the elements are nonmetals?

32. All elements with more than 83 protons are *radioactive*, as we shall see in Chapter 11. Are any of the elements in the foregoing list necessarily radioactive? If so, which one(s)?

■ ADDITIONAL PROBLEMS

33. Refer back to Problem 24. Explain how you can answer the three parts of Problem 24 just by looking at a periodic table, without writing electron structures.

34. An atom of an element has two electrons in the first shell, eight electrons in the second shell, and five electrons in the third shell. From this information, give for the element **(a)** its atomic number, **(b)** its name, **(c)** its total number of electrons, **(d)** its total number of *s* electrons, and **(e)** total number of *d* electrons.

35. Without referring to any tables in the text, mark an appropriate location for each of the following in the blank periodic table provided: **(a)** the fourth-period noble gas, **(b)** the third-period alkali metal, **(c)** the fourth-period halogen, and **(d)** a metal in the fourth period and in group 3B.

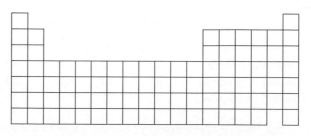

36. Refer to Figure 3.11. A patient is found to be suffering from heavy-metal poisoning. An emission spectrum is obtained from a sample of the patient's blood, by spraying the blood into a hot flame. The brightest emission lines are found at 5520, 5580, 5890, and 5900 Å. What two elements are likely to be present? Which one is more likely to be the culprit?

■ COLLABORATIVE GROUP PROJECTS

Prepare a PowerPoint, poster, or other presentation (as directed by your instructor) for presentation to the class.

37. Prepare a brief biographical report on one of the following individuals.
    a. Humphry Davy
    b. William Crookes
    c. Wilhelm Roentgen
    d. Marie Curie
    e. Robert Millikan
    f. Niels Bohr
    g. Alessandro Volta
    h. Ernest Rutherford
    i. Michael Faraday
    j. J. J. Thomson

38. Draw a grid containing 16 squares, in two rows of 8, representing the 16 elements in the periodic table from lithium to argon. For each of the elements, give **(a)** the main-shell electron configuration and **(b)** the quantum mechanical notation for electron configuration.

39. Prepare a brief report on one of the alkali metals or alkaline earth metals. List sources and commercial uses.

40. Using this book and other references, design a time line of important milestones in the elucidation of atomic structure. Begin with the ancient Greeks and conclude with the present.

41. Prepare a brief report on one of the halogens or noble gases. List sources and commercial uses.

42. Many different forms of the periodic table have been generated. Prepare a brief report showing at least three different versions of the periodic table and comment on their utility.

Atoms combine by forming chemical bonds. The materials so formed have very different properties from the original substance(s). Here we see carbon in the form of industrial diamonds, which are made from carbon in the form of graphite (pencil lead) or soot. The only difference between the forms of carbon are the bonds between the carbon atoms. It is the arrangement and number of these bonds in diamond that make it by far the hardest natural material in the world.

CHEMICAL BONDS

4

QUESTIONS YOU MAY HAVE ASKED YOURSELF

1. We have a negative ion generator in our home. What is a negative ion?
2. Why does pool chlorine have an odor, while chlorine in salt doesn't?
3. If there is iron in blood, why aren't we magnetic?

The Ties That Bind

The element carbon is commonly found as soft, black soot formed by incomplete combustion. By heating that soot under tremendous pressure, the hardest material known—diamond—can be made. We have not changed the carbon in soot to a different element in this process. . . so why is diamond so different from the soot? The answer lies in the bonds that hold the carbon atoms together. We have learned a bit about the structure of the atom. Now we are ready to consider chemical bonds, the ties that bind atoms together.

Chemical bonds are the forces that hold atoms together in molecules and hold ions together in ionic crystals. The vast number and incredible variety of chemical compounds—tens of millions are known—result from the fact that atoms can form bonds with many other types of atoms. In addition, chemical bonds determine the three-dimensional shape of a molecule. Some important consequences of chemical structure and bonding include

- Whether a substance is a solid, liquid, or gas at room temperature
- The strengths of materials (as well as adhesives that hold materials together) used for building bridges, houses, and many other structures
- Whether a liquid is light and volatile (like gasoline) or heavy and viscous (like glycerol)
- The taste, odor, and drug activity of chemical compounds
- The structural integrity of skin, muscles, bones, and teeth
- The toxicity of certain molecules to living organisms

Chemical bonding is related to the arrangement of electrons in compounds. In this chapter, we look at different types of chemical bonds and some of the unusual properties of compounds that result.

4.1 The Art of Deduction: Stable Electron Configurations

In our discussion of the atom and its structure (Chapters 2 and 3), we followed the historical development of some of the more important atomic concepts. We could continue to look at chemistry in this manner, but that would require several volumes of print—and perhaps more of your time than you care to spend. We won't abandon the historical approach entirely, but we will emphasize another important aspect of scientific endeavor: deduction.

The art of deduction works something like this.

- **Fact:** Noble gases, such as helium, neon, and argon, are inert; they undergo few, if any, chemical reactions.
- **Theory:** The inertness of noble gases results from their electron structures; each (except helium) has an octet of electrons in its outermost shell.
- **Deduction:** Other elements that can alter their electron structures to become like those of noble gases would become less reactive by doing so.

We can use an example to illustrate this deductive argument. Sodium has 11 electrons, one of which is in the third shell. Recall that electrons in the outermost shell are called **valence electrons**, while those in all the other shells are lumped together as **core electrons**. If the sodium atom got rid of its valence electron, the remaining core electrons would have the same electron structure as an atom of the noble gas neon. Using main-shell configurations (Chapter 3), we can represent this as

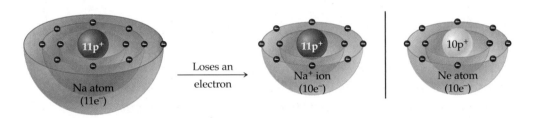

Similarly, if a chlorine atom could gain an electron, it would have the same electron structure as argon.

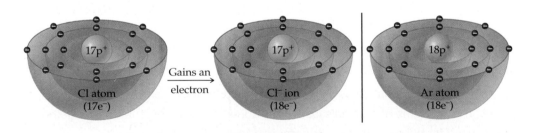

> The outermost shell is "filled" when it contains eight electrons, two *s* and six *p*. The first shell is an exception: It holds only two *s* electrons.

The sodium atom, having lost an electron, becomes positively charged. It has 11 protons (11+) and only 10 electrons (10−). It is written Na$^+$ and is called a *sodium ion*. The chlorine atom, having gained an electron, becomes negatively charged. It has 17 protons (17+) and 18 electrons (18−). It is written Cl$^-$ and is called a *chloride ion*. Note that a positive charge, as in Na$^+$, indicates that one electron has been lost. Similarly, a negative charge, as in Cl$^-$, indicates that one electron has been gained. It is important to note that even though Cl$^-$ and Ar are **isoelectronic** (have the same electron configuration), they are *different* chemical species. In the same way, a sodium atom does not *become* an atom of neon when it loses an electron; the sodium ion is simply isoelectronic with neon.

Self-Assessment Questions

1. Which group of elements in the periodic table is characterized by an especially stable electron arrangement?
   a. 2A **b.** 2B **c.** 8A **d.** 8B

2. The structural difference between a sodium atom and a sodium ion is that the sodium ion has one
   a. less proton than the sodium atom
   b. less electron than the sodium atom
   c. more proton and one less electron than the sodium atom
   d. more proton than the sodium atom

3. Which of the following are isoelectronic?
   a. K^+ and Ar **b.** Mg^{2+} and Ar **c.** Ne and Cl^- **d.** Xe and Kr

4. The anion F^- is isoelectronic with the cation
   a. Ca^{2+} **b.** K^+ **c.** Li^+ **d.** Mg^{2+}

Answers: 1, c; 2, b; 3, a; 4, d

4.2 Lewis (Electron-Dot) Symbols

In forming ions, the cores of sodium atoms and chlorine atoms do not change. It is convenient, therefore, to let the symbol represent the *core* of the atom (nucleus plus inner electrons). The valence electrons are then represented by dots. The equations of the preceding section then can be written as follows:

$$Na\cdot \rightarrow Na^+ + 1\,e^-$$

and

$$\cdot\ddot{\underset{\cdot\cdot}{Cl}}\colon \;+\; 1\,e^- \longrightarrow \colon\!\ddot{\underset{\cdot\cdot}{Cl}}\colon^-$$

In these representations, the symbol of the element represents the core, and dots stand for valence electrons. These **electron-dot symbols** are usually called **Lewis symbols**.

Lewis Symbols and the Periodic Table

It is especially easy to write Lewis symbols for most of the main group elements. The number of valence electrons for most of these elements is equal to the group number (Table 4.1). Because of their more complicated electron configurations,

▲ Electron-dot symbols are called *Lewis symbols* after G. N. Lewis, the famous American chemist (1875–1946) who invented them. Lewis also made important contributions in the fields of thermodynamics, acids and bases, and spectroscopy. His achievements in any of these areas might have merited a Nobel Prize, yet he never received that award.

Table 4.1 Lewis Symbols for Selected Main Group Elements

Group 1A	Group 2A	Group 3A	Group 4A	Group 5A	Group 6A	Group 7A	Noble Gases
H·							He:
Li·	·Be·	·Ḃ·	·Ċ·	:Ṅ·	:Ö·	:Ḟ:	:Ṅe:
Na·	·Mg·	·Ȧl·	·Ṡi·	:Ṗ·	:Ṡ·	:Ċl·	:Ȧr:
K·	·Ca·				:Ṡe·	:Ḃr·	:Ḳr:
Rb·	·Sr·				:Ṫe·	:Ï·	:Ẍe:
Cs·	·Ba·						

In writing a Lewis symbol, only the *number* of dots is important. The dots need not be drawn in any specific positions, except that there should be no more than two dots on any given side of the chemical symbol (right, left, top, or bottom).

elements in the central part of the periodic table (the transition metals) do not easily lend themselves to the electron-dot symbolism.

EXAMPLE 4.1 Lewis Symbols

Without referring to Table 4.1, give Lewis symbols for magnesium, oxygen, and phosphorus. You may use the periodic table.

Solution
Magnesium is in group 2A, oxygen is in group 6A, and phosphorus is in group 5A. The Lewis symbols, therefore, have two, six, and five dots, respectively. They are

·Mg·   :Ö: :P·

■ **EXERCISE 4.1**
Without referring to Table 4.1, give Lewis symbols for each of the following elements. You may use the periodic table.
a. Ar **b.** Ca **c.** F **d.** N **e.** K **f.** S

Self-Assessment Questions

1. How many dots are shown about Be in the Lewis symbol for beryllium?
   a. 2 **b.** 4 **c.** 5 **d.** 9
2. How many dots are shown about F in the Lewis symbol for fluorine?
   a. 1 **b.** 5 **c.** 7 **d.** 9
3. Which of the following Lewis symbols is *incorrect*?
   a. ·C· **b.** :Cl· **c.** Li: **d.** :N·
4. Which of the following Lewis symbols is *incorrect*?
   a. ·As· **b.** :Ö· **c.** Rb· **d.** ·Si·

Answers: 1, a; 2, c; 3, c; 4, a

Chlorine gas is actually composed of Cl_2 molecules, not separate Cl atoms. Each atom of the molecule takes an electron from a sodium atom. Two sodium ions and two chloride ions are formed.

$$Cl_2 + 2\,Na \rightarrow 2\,Cl^- + 2\,Na^+$$

2. Why does pool chlorine have an odor, while chlorine in salt doesn't?
Pool chlorine is the element Cl_2, while the chlorine in salt is Cl^- or chlor*ide* ions. An element and its ions are chemically quite different.

4.3 The Reaction of Sodium and Chlorine

Sodium is a highly reactive metal. It is soft enough to be cut with a knife. When freshly cut, it is bright and silvery, but it dulls rapidly because it reacts with oxygen in the air. In fact, it reacts so readily in air that it is usually stored under oil or kerosene. Sodium reacts violently with water also, becoming so hot that it melts. A small piece forms a spherical bead after melting and races around on the surface of the water as it reacts.

Chlorine is a greenish-yellow gas. It is familiar as a disinfectant for city water supplies and swimming pools. (The actual substance added is often a compound that reacts with water to form chlorine.) Chlorine is extremely irritating to the eyes and nose. In fact, it was used as a poison gas in World War I.

If a piece of sodium is dropped into a flask containing chlorine gas, a violent reaction ensues, producing sodium chloride, beautiful white crystals that you might sprinkle on your food at the dinner table. These white crystals are ordinary table salt. Sodium chloride has very few properties in common with either sodium or chlorine (Figure 4.1).

The Reaction of Sodium and Chlorine: Theory

A sodium atom achieves a filled valence shell by losing one electron. A chlorine atom achieves a filled valence shell by adding one electron. What happens when

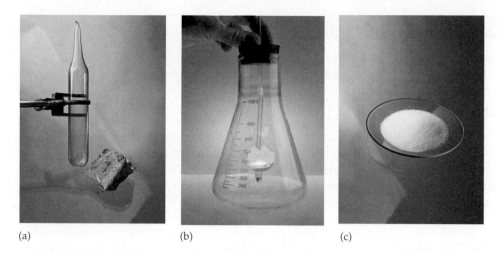

◀ **Figure 4.1** Sodium, a soft, silvery metal, reacts with chlorine, a greenish gas, to form sodium chloride (ordinary table salt), a white crystalline solid.

QUESTION: What particles, ions or molecules, make up sodium chloride?

sodium atoms come into contact with chlorine atoms? The obvious: Chlorine extracts an electron from a sodium atom.

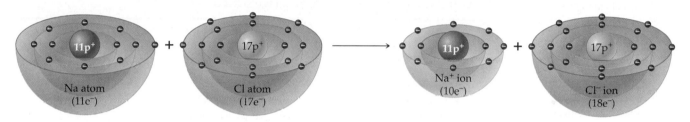

In the abbreviated Lewis form, this reaction is written

$$\text{Na}\cdot \ + \ :\ddot{\text{Cl}}\cdot \longrightarrow \text{Na}^+ \ + \ :\ddot{\text{Cl}}:^-$$

Ionic Bonds

The Na$^+$ and Cl$^-$ ions formed from sodium and chlorine atoms have opposite charges and are strongly attracted to one another. These ions arrange themselves in an orderly fashion. Each sodium ion attracts (and is attracted to) six chloride ions (top and bottom, front and back, left and right) as shown in Figure 4.2a. The arrangement is repeated many times in all directions (Figure 4.2b). The result is a **crystal** of sodium chloride. The orderly microscopic arrangement of the ions is reflected in the macroscopic shape of each salt crystal (Figure 4.2c). Even the tiniest grain of salt has billions and billions of each type of ion. The forces holding the crystal together—the attractive forces between positive and negative ions—are called **ionic bonds**.

Electron loss is known as *oxidation*, and gain of electrons is called *reduction*. We shall explore this concept in Chapter 8.

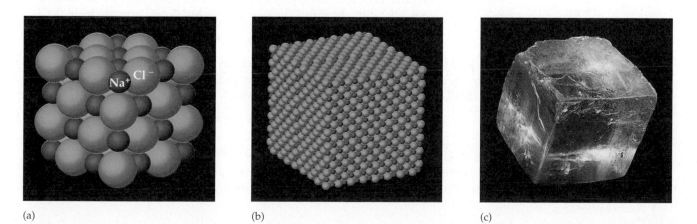

▲ **Figure 4.2** Structure of a sodium chloride crystal. (a) Each Na$^+$ ion (small sphere) is surrounded by six Cl$^-$ ions (large spheres) and each Cl$^-$ ion by six Na$^+$ ions. (b) This arrangement repeats itself many, many times. (c) The highly ordered pattern of alternating Na$^+$ and Cl$^-$ ions is observed in the macroscopic world as a crystal of sodium chloride.

▲ **Figure 4.3** Forms of iron and calcium that are useful to the human body are the ions (usually Fe^{2+} and Ca^{2+}) taken in ionic compounds such as $FeSO_4$ and $CaCO_3$. The elemental forms (Fe and Ca) are chemically very different from forms of the same elements in ionic compounds.

Atoms and Ions: Distinctively Different

Ions are emphatically different from the atoms from which they are made, much as a whole peach (an atom) and a peach pit (a positive ion) are different from one another. The names and symbols of atoms and ions may look a lot alike, but the similarities end there (Figure 4.3). Unfortunately, the situation is confusing because people talk about needing iron to perk up "tired blood" and calcium for healthy teeth and bones. What they really mean is iron(II) *ions* (Fe^{2+}) and calcium *ions* (Ca^{2+}). You wouldn't think of eating iron nails to get iron. (However, some enriched cereals do indeed have powdered iron added; the iron metal is readily converted to Fe^{2+} ions in the stomach.) Nor would you eat highly reactive calcium metal.

Similarly, when you are warned to reduce your sodium intake, your doctor is not concerned that you are eating too much sodium metal—that would be exceedingly unpleasant—but that your intake of Na^+ *ions*—usually as sodium chloride—may be too high. Although the names are similar, the atom and the ion are quite different chemically. It is important that we make careful distinctions and use precise terminology, as we shall do here.

A N S W E R

3. If there is iron in blood, why aren't we magnetic?
Iron metal is magnetic, but compounds—such as hemoglobin—that contain iron usually contain iron(II) ions or iron(III) ions. The ions are chemically and physically very different from the metal.

Self-Assessment Questions

1. A sodium ion and a neon atom have the same
   a. electron configuration   b. net charge
   c. nuclear charge   d. properties

2. The Lewis symbol for chloride ion shows how many dots about the Cl and what charge on it?
   a. 5 dots, no charge   b. 6 dots, + charge
   c. 7 dots, − charge   d. 8 dots, − charge

3. The bonding between Na^+ ions and Cl^- ions in sodium chloride is
   a. covalent   b. ionic
   c. nonpolar   d. polar

Answers: 1, a; 2, d; 3, b

4.4 Using Lewis Symbols: More Ionic Compounds

Potassium, a metal in the same family as sodium and therefore similar to sodium in properties, also reacts with chlorine. The reaction yields potassium chloride (KCl).

$$K\cdot \ + \ \cdot \ddot{\underset{..}{Cl}}: \longrightarrow K^+ + \ :\ddot{\underset{..}{Cl}}:^-$$

Potassium also reacts with bromine, a reddish-brown liquid in the same family as chlorine and therefore similar to chlorine in properties. The product, potassium bromide (KBr), is a stable, white, crystalline solid.

$$K\cdot \ + \ \cdot \ddot{\underset{..}{Br}}: \longrightarrow K^+ + \ :\ddot{\underset{..}{Br}}:^-$$

EXAMPLE 4.2   Electron Transfer to Form Ions

Use Lewis symbols to show the transfer of electrons from sodium atoms to bromine atoms to form ions with noble gas configurations.

Solution

Sodium has one valence electron, and bromine has seven. Transfer of the single electron from sodium to bromine leaves each with a noble gas configuration.

$$Na\cdot \ + \ \cdot \ddot{\underset{..}{Br}}: \longrightarrow Na^+ + \ :\ddot{\underset{..}{Br}}:^-$$

■ **EXERCISE 4.2A**

Use Lewis symbols to show the transfer of electrons from lithium atoms to fluorine atoms to form ions with noble gas configurations.

■ **EXERCISE 4.2B**

Use Lewis symbols to show the transfer of electrons from rubidium atoms to iodine atoms to form ions with noble gas configurations.

Magnesium, a group 2A metal, is harder and less reactive than sodium. Magnesium reacts with oxygen, a group 6A element that is a colorless gas, to form another stable white, crystalline solid called magnesium oxide (MgO).

$$\cdot Mg\cdot \ + \ \cdot \ddot{O}\colon \ \longrightarrow \ Mg^{2+} + \ \colon\ddot{\underset{..}{O}}\colon^{2-}$$

Magnesium must give up two electrons and oxygen must gain two electrons for each to have the same configuration as the noble gas neon.

An atom such as oxygen, which needs two electrons to complete a noble gas configuration, may react with potassium atoms, which have only one electron each to give. In this case, two atoms of potassium are needed for each oxygen atom. The product is potassium oxide (K_2O).

$$\begin{matrix} K\cdot \\ K\cdot \end{matrix} \ + \ \cdot\ddot{O}\colon \ \longrightarrow \ \begin{matrix} K^+ \\ K^+ \end{matrix} \ + \ \colon\ddot{\underset{..}{O}}\colon^{2-}$$

By this process, each potassium atom achieves the argon configuration. As before, oxygen assumes the neon configuration.

When the Mg atom becomes a Mg^{2+} ion, its second electron shell becomes its outermost shell. The Mg^{2+} ion is isoelectronic with the Ne atom and has the same stable electron structure: 2, 8.

EXAMPLE 4.3   Electron Transfer to Form Ions

Use Lewis symbols to show the transfer of electrons from magnesium atoms to nitrogen atoms to form ions with noble gas configurations.

Solution

$$\begin{matrix} \cdot Mg\cdot \\ \cdot Mg\cdot \\ \cdot Mg\cdot \end{matrix} \ + \ \begin{matrix} \cdot\ddot{N}\cdot \\ \cdot\ddot{N}\cdot \end{matrix} \ \longrightarrow \ \begin{matrix} Mg^{2+} \\ Mg^{2+} \\ Mg^{2+} \end{matrix} \ + \ \begin{matrix} \colon\ddot{\underset{..}{N}}\colon^{3-} \\ \colon\ddot{\underset{..}{N}}\colon^{3-} \end{matrix}$$

Each of three magnesium atoms gives up two electrons (a total of six), and each of the two nitrogen atoms acquires three (a total of six). Notice that the total positive and negative charges on the products are equal (6+ and 6−). Magnesium reacts with nitrogen to yield magnesium nitride (Mg_3N_2).

■ **EXERCISE 4.3**

Use Lewis symbols to show the transfer of electrons from aluminum atoms to oxygen atoms to form ions with noble gas configurations.

Generally speaking, metallic elements in groups 1A and 2A (those from the left side of the periodic table) react with nonmetallic elements in groups 6A and 7A (those from the right side) to form ionic compounds. These products are stable crystalline solids.

The Octet Rule

Each atom of a metal tends to give up the electrons in its outer shell, and each atom of a nonmetal tends to take on enough electrons to complete its valence shell. The resulting ions have noble gas configurations. A set of eight valence electrons—an *octet*—is the characteristic arrangement of all noble gases except helium. When atoms react with each other, they often tend to attain this stable noble gas electron configuration. Thus, they are said to follow the **octet rule**, or the "rule of eight." (In the case of helium, a maximum of two electrons can exist in its single electron shell, and so hydrogen follows the "rule of two.")

In following the octet rule, atoms of group 1A metals give up one electron to form 1+ ions, those of group 2A metals give up two electrons to form 2+ ions, and group 3A metals give up three electrons to form 3+ ions. Group 7A nonmetal atoms take on one electron to form 1− ions, and group 6A atoms tend to pick up two electrons to form 2− ions. Atoms of B group metals can give up various numbers of electrons to form positive ions with various charges. These periodic relationships are summarized in Figure 4.4.

Table 4.2 lists symbols and names for some ions formed by the gain or loss of electrons. (Names are explained in the next section.) You can calculate the charge on the negative ions in the table by subtracting 8 from the group number. For example, the charge on the oxide ion (oxygen is in group 6A) is $6 - 8 = -2$. The nitride ion (nitrogen is in group 5A) has a charge of $5 - 8 = -3$.

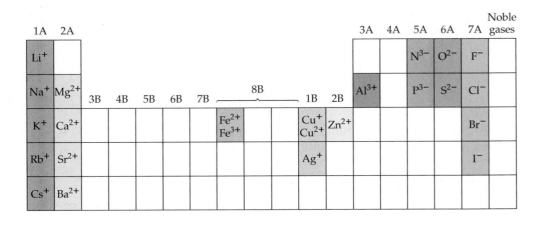

▶ **Figure 4.4** Periodic relationships of some simple ions. Many of the transition elements (B groups) form ions of more than one kind, with different charges.

QUESTION: Can you propose a formula for the simple ion formed from selenium (Se)? For the simple ion formed from germanium (Ge)?

Table 4.2 Symbols and Names for Some Simple (Monatomic) Ions

Group	Element	Name of Ion	Symbol for Ion
1A	Hydrogen	Hydrogen ion	H^+
	Lithium	Lithium ion	Li^+
	Sodium	Sodium ion	Na^+
	Potassium	Potassium ion	K^+
2A	Magnesium	Magnesium ion	Mg^{2+}
	Calcium	Calcium ion	Ca^{2+}
3A	Aluminum	Aluminum ion	Al^{3+}
5A	Nitrogen	Nitride ion	N^{3-}
6A	Oxygen	Oxide ion	O^{2-}
	Sulfur	Sulfide ion	S^{2-}
7A	Fluorine	Fluoride ion	F^-
	Chlorine	Chloride ion	Cl^-
	Bromine	Bromide ion	Br^-
	Iodine	Iodide ion	I^-
1B	Copper	Copper(I) ion (cuprous ion)	Cu^+
		Copper(II) ion (cupric ion)	Cu^{2+}
	Silver	Silver ion	Ag^+
2B	Zinc	Zinc ion	Zn^{2+}
8B	Iron	Iron(II) ion (ferrous ion)	Fe^{2+}
		Iron(III) ion (ferric ion)	Fe^{3+}

EXAMPLE 4.4 Determining Formulas by Electron Transfer

What is the formula of the compound formed by the reaction of sodium and sulfur?

Solution

Sodium is in group 1A; the sodium atom has one valence electron. Sulfur is in group 6A; the sulfur atom has six valence electrons.

$$\text{Na}\cdot \qquad \cdot \ddot{\underset{..}{\text{S}}} \cdot$$

Sulfur needs two electrons to gain an argon configuration, but sodium has only one to give. The sulfur atom, therefore, must react with two sodium atoms.

$$\begin{array}{c} \text{Na}\cdot \\ + \cdot \ddot{\underset{..}{\text{S}}}\cdot \\ \text{Na}\cdot \end{array} \longrightarrow \begin{array}{c} \text{Na}^+ \\ + :\ddot{\underset{..}{\text{S}}}:^{2-} \\ \text{Na}^+ \end{array}$$

The formula of the compound, called sodium sulfide, is Na_2S.

■ EXERCISE 4.4A

What are the formulas of the compounds formed by the reaction of **(a)** calcium with fluorine, and **(b)** lithium with oxygen?

■ EXERCISE 4.4B

Use data in Figure 4.4 to predict the formulas of the two compounds that can be formed from iron (Fe) and chlorine.

Self-Assessment Questions

1. Which of the following pairs of elements would be most likely to form an ionic compound?
   a. Ca and Al   **b.** Cd and Na   **c.** K and O   **d.** Ne and Na

2. Sulfur forms a simple ion with a charge of
   a. 6−   **b.** 2−   **c.** 2+   **d.** 6+

3. Magnesium forms a simple ion with a charge of
   a. 2−   **b.** 2+   **c.** 4+   **d.** 8+

4. Mg and N react to form Mg_3N_2, an ionic compound. How many electrons are there in the N^{3-} valence shell?
   a. 2   **b.** 6   **c.** 8   **d.** 18

5. The formula of the ionic compound formed by lithium and oxygen is
   a. LiO   **b.** Li_2O **c.** LiO_2 **d.** Li_2O_3

6. The formula of the ionic compound formed by magnesium and chlorine is
   a. MgCl   **b.** Mg_2Cl **c.** $MgCl_2$ **d.** Mg_2Cl_3

7. The only one of the following ions likely to be formed in ordinary chemical reactions is
   a. Ar^+ **b.** Br^- **c.** K^{2+} **d.** S^{3-}

Answers: 1, c; 2, b; 3, b; 4, c; 5, b; 6, c; 7, b

4.5 Formulas and Names of Binary Ionic Compounds

Names of simple positive ions (*cations*) are derived from those of their parent elements by the addition of the word *ion*. A sodium atom, on losing an electron, becomes a *sodium ion* (Na^+). A magnesium atom (Mg), on losing two electrons, becomes a *magnesium ion* (Mg^{2+}). When a metal forms more than one ion, the charges on the different ions are denoted by Roman numerals in parentheses. For example, Fe^{2+} is iron(II) ion and Fe^{3+} is iron(III) ion.

Names of simple negative ions (*anions*) are derived from those of their parent elements by changing the usual ending to *-ide* and adding the word *ion*. A chlor*ine* atom, on gaining an electron, becomes a chlor*ide* ion (Cl^-). A sulf*ur* atom gains two electrons, becoming a sulf*ide* ion (S^{2-}).

Simple ions of opposite charge can be combined to form **binary** (two-element) **compounds**. To get the correct **formula** for a binary compound, write each ion with its charge (positive ion to the left), then transpose the charge numbers (but not the plus and minus signs) and write them as subscripts. The process is best learned by practice, as is provided in the following examples and exercises and by problems at the end of the chapter.

EXAMPLE 4.5 Determining Formulas from Ionic Charges

Give the formulas for **(a)** calcium chloride and **(b)** aluminum oxide.

Solution

a. First, write the symbols for the ions. (We write the charge on chloride ion explicitly as "$1-$" to illustrate the method. You may omit the "1" when you are comfortable with the process.)

$$Ca^{2+} \; Cl^{1-}$$

Then cross over the numbers as subscripts.

Then rewrite the formula, dropping the charges. The formula for calcium chloride is

$$Ca_1Cl_2 \text{ or (dropping the "1") simply } CaCl_2$$

b. Write the symbols for the ions.

$$Al^{3+} \; O^{2-}$$

Cross over the numbers as subscripts.

Then rewrite the formula, dropping the charges. The formula for aluminum oxide is

$$Al_2O_3$$

■ **EXERCISE 4.5**

Give the formulas for **(a)** potassium oxide, **(b)** calcium nitride, and **(c)** calcium sulfide.

This method, called the *crossover method*, works because it is based on the transfer of electrons and the conservation of charge. Two Al atoms lose three electrons each (a total of six electrons lost), and three O atoms gain two electrons each (a total of six electrons gained). Electrons lost equal electrons gained. Similarly, two Al^{3+} ions have six positive charges (three each), and three O^{2-} ions have six negative charges (two each). The net charge on Al_2O_3 is 0, just as it should be.

Now you are able to translate the "English," such as aluminum oxide, into the "chemistry," Al_2O_3. You also can translate in the other direction.

EXAMPLE 4.6 Naming Ionic Compounds

What are the names of **(a)** MgS and **(b)** $FeCl_3$?

Solution

a. From Table 4.2 we can determine that MgS is made up of Mg^{2+} (magnesium ion) and S^{2-} (sulfide ion). The name is simply magnesium sulfide.
b. From Table 4.2 we can determine that the ions in $FeCl_3$ are

$$Fe^{3+} \quad Cl^-$$

How do we know the iron ion in $FeCl_3$ is Fe^{3+} and not Fe^{2+}? Because there are three Cl^- ions, each $1-$, the one Fe ion must be $3+$ because the compound $FeCl_3$ is neutral. The names of these ions are iron(III) ion (or ferric ion) and chloride ion. Therefore, the compound is iron(III) chloride (or, by the older system, ferric chloride).

■ **EXERCISE 4.6**

What are the names of **(a)** CaF_2 and **(b)** $CuBr_2$?

A Compound by Any Other Name. . .

As you read labels on foodstuff and pharmaceuticals, you will see some names that are beginning to sound familiar, and some that are confusing or mysterious. For example, why do we have two names for Fe^{2+}? Some names are historical: Iron, copper, gold, silver, and other elements have been known for thousands of years. In naming compounds of these elements, an *-ous* ending came to be used for the ion of smaller charge, and an *-ic* ending for the ion of larger charge. The modern system uses Roman numerals to indicate ionic charge. Although the new system is more logical and easier to apply, the old names persist, especially in everyday life and in some of the biomedical sciences.

Boron 150 mcg
100%
horus 109 mg 11% *Daily Value (DV) n

DIENTS: Dicalcium Phosphate, Potassium Chloride, se, Ascorbic Acid, Ferrous Fumarate, Calcium Carb Tocopheryl Acetate, Croscarmellose Sodium, Niacin sium Stearate, Dextrin, d-Calcium Pantothenat nellose, Manganese Sulfate, Polyethylene Glycol, Si e, Maltodextrin, Cupric Sulfate, Corn Starch, Dex

▲ Dietary supplements often use older, Latin-derived names.

Self-Assessment Questions

1. The formula for the binary ionic compound of barium and sulfur is
   a. BaS   b. Ba_2S c. BaS_2 d. Ba_2S_3
2. Which of the following formulas is incorrect?
   a. AlF_3 b. K_2S c. MgF_2 d. Na_2I
3. Which of the following formulas is incorrect?
   a. $AlCl_3$ b. Al_3P_2 c. CaS   d. Cs_2S
4. The correct name for KCl is
   a. chloride potassium   b. chlorine kryptide
   c. potassium chloride   d. potassium chlorine
5. The correct name for Al_2S_3 is
   a. aluminum sulfide   b. aluminum trisulfide
   c. dialuminum trisulfide   d. antimony sulfide
6. The correct name for LiI is
   a. indium lithide   b. iodine lithide   c. lithium indide   d. lithium iodide
7. The correct name for $ZnCl_2$ is
   a. dichlorozinc   b. zinc chloride
   c. zinc dichloride   d. zirconium chloride

Answers: 1, a; 2, d; 3, b; 4, c; 5, a; 6, d; 7, b

4.6 Covalent Bonds: Shared Electron Pairs

We might expect a hydrogen atom, with its one electron, to acquire another electron and assume the configuration of the noble gas helium. In fact, hydrogen atoms do just that in the presence of atoms of a reactive metal such as lithium—that is, a metal that finds it easy to give up an electron.

$$Li \cdot + H \cdot \longrightarrow Li^+ + H{:}^-$$

But what if there are no other kinds of atoms around, only hydrogen? One atom can't gain an electron from another, for all hydrogen atoms have an equal attraction for electrons. Two hydrogen atoms can compromise, however, by *sharing a pair* of electrons.

$$H \cdot + \cdot H \longrightarrow H{:}H$$

By sharing electrons, the two hydrogen atoms form a hydrogen molecule. The bond formed by a shared pair of electrons is called a **covalent bond**.

H:H
⌐ covalent bond (shared pair of electrons)

Consider next the case of chlorine. A chlorine atom readily takes an extra electron from anything willing to give one up. But again, what if the only things around are other chlorine atoms? Chlorine atoms also can attain a more stable arrangement by sharing a pair of electrons.

$$:\!\ddot{C}l\cdot + \cdot\ddot{C}l\!: \longrightarrow :\!\ddot{C}l\!:\!\ddot{C}l\!:$$

The shared pair of electrons in the chlorine molecule is another example of a covalent bond; they are called a **bonding pair**. The other electrons that stay on one atom and are not shared are called **nonbonding pairs** or **lone pairs**.

$$:\!\ddot{C}l\!:\!\ddot{C}l\!:$$

For simplicity, the hydrogen molecule is often represented as H_2 and the chlorine molecule as Cl_2. In each case, the covalent bond between the atoms is understood. Sometimes the covalent bond is indicated by a dash, H—H and Cl—Cl. Lone pairs of electrons often are not shown. Each chlorine atom in the chlorine molecule has eight electrons around it, an arrangement like that of the noble gas argon. Thus, the atoms in a covalent bond follow the octet rule by sharing electrons, even as those in an ionic bond follow it by giving up or taking on electrons.

Multiple Bonds

One pair of shared electrons is called a **single bond**. In some molecules, atoms must share more than one pair of electrons in order to fulfill the octet rule. In carbon dioxide (CO_2) for example, the carbon atom shares *two* pairs of electrons with each of the two oxygen atoms.

$$:\!\ddot{O}{:}{:}C{:}{:}\ddot{O}\!:$$

Note that each atom has an octet of electrons about it as a result of this sharing. We say that the atoms are joined by a **double bond**, a covalent linkage in which the two atoms share two pairs of electrons. A double bond is indicated by a double dash between atoms: O=C=O.

Atoms also can share three pairs of electrons. In the nitrogen (N_2) molecule, for example, each nitrogen atom shares three pairs of electrons with the other.

$$:\!N{:}{:}{:}N\!:$$

The atoms are joined by a **triple bond**, a covalent linkage in which two atoms share three pairs of electrons (N≡N). Note that each of the nitrogen atoms has an octet of electrons around it.

Covalent bonds are usually represented as dashes. The three kinds of covalent bonds are simply written as follows:

H—Cl   O=C=O   N≡N

Names of Covalent Compounds

Covalent or *molecular* compounds are those in which electrons are shared, not transferred. Molecular compounds generally have molecules that consist of two or more nonmetals. Many molecular compounds have common and widely used names. Examples are water (H_2O), ammonia (NH_3), and methane (CH_4). For other molecular compounds, the naming process is often more systematic. The prefixes *mono-*, *di-*, *tri-*, and so on are used to indicate the number of atoms of each element in the molecule. A list of these prefixes for up to 10 atoms is given in Table 4.3. For example, the compound N_2O_4 is called *dinitrogen tetroxide*. (The *a* often is dropped from *tetra-* and other prefixes when they precede another vowel.) We often leave off the *mono-* prefix (NO_2 is nitrogen dioxide) but do usually include it to distinguish between two compounds of the same pair of elements (CO is carbon monoxide; CO_2 is carbon dioxide).

Table 4.3 Prefixes That Indicate the Number of Atoms of an Element in a Covalent Compound

Prefix	Number of Atoms
Mono-	1
Di-	2
Tri-	3
Tetra-	4
Penta-	5
Hexa-	6
Hepta-	7
Octa-	8
Nona-	9
Deca-	10

EXAMPLE 4.7   Naming Covalent Compounds

What are the names of (a) SCl_2 and (b) P_4S_3?

Solution

a. With one sulfur atom and two chlorine atoms, SCl_2 is sulfur dichloride.
b. With four phosphorus atoms and three sulfur atoms, P_4S_3 is tetraphosphorus trisulfide.

■ **EXERCISE 4.7**
What are the names of (a) BrF_3, (b) BrF_5, (c) N_2O, and (d) N_2O_5?

EXAMPLE 4.8   Formulas of Covalent Compounds

Give the formula for tetraphosphorus hexoxide.

Solution

The *tetra-* indicates four phosphorus atoms, and the *hex-* specifies six oxygen atoms. The formula is P_4O_6.

■ **EXERCISE 4.8**
Give the formulas for (a) phosphorus trichloride, (b) dichlorine heptoxide, (c) nitrogen triiodide, and (d) disulfur dichloride.

Self-Assessment Questions

1. A bond formed by two atoms sharing a pair of electrons is said to be
   a. covalent   b. ionic   c. nonpolar   d. polar

2. The Lewis formula for Br_2 shows a bond that is
   a. double covalent   b. ionic   c. single covalent   d. triple covalent

3. The Lewis formula for N_2 shows a bond that is
   a. double covalent   b. ionic   c. single covalent   d. triple covalent

4. The formula for phosphorus trichloride is
   a. KCl_3 b. K_3Cl c. PCl_3 d. P_3Cl

5. The formula for disulfur difluoride is
   a. SF_2 b. S_2F_2 c. SFe_2 c. S_2Fl_2

6. The correct name for N_2S_4 is
   a. dinitrogen disulfide   b. dinitrogen tetrasulfide
   c. tetrasulfodinitrogen   d. tetrasulfur dinitride

7. The correct name for I_2O_5 is
   a. diiodine pentoxide   b. diiodopentoxide
   c. iridium pentoxide   d. pentaoxodiiodine

Answers: 1, a; 2, c; 3, d; 4, c; 5, b; 6, b; 7, a

4.7 Unequal Sharing: Polar Covalent Bonds

So far we have seen that atoms combine in two different ways. Atoms that are quite different in electron structure (from opposite sides of the periodic table) react by the complete transfer of one or more electrons from one atom to another to form an ionic bond. Atoms that are identical combine by sharing a pair of electrons to form a covalent bond. Now let's consider bond formation between atoms that are different, but not different enough to form ionic bonds.

Hydrogen Chloride

Hydrogen and chlorine react to form a colorless gas called hydrogen chloride. This reaction may be represented as

$$\text{H}\cdot + \cdot\ddot{\underset{\cdot\cdot}{\text{Cl}}}: \longrightarrow \text{H}:\ddot{\underset{\cdot\cdot}{\text{Cl}}}:$$

Ignoring the lone pair electrons and using a dash to represent the covalent bond, we can write the hydrogen chloride molecule as H—Cl. Both hydrogen and chlorine need an electron to achieve a noble gas configuration—a helium configuration for hydrogen and an argon configuration for chlorine. They achieve these configurations by sharing a pair of electrons to form a covalent bond.

Hydrogen and chlorine both actually consist of diatomic molecules; the reaction is more accurately represented by the scheme

$$\text{H}:\text{H} + :\ddot{\underset{\cdot\cdot}{\text{Cl}}}:\ddot{\underset{\cdot\cdot}{\text{Cl}}}: \longrightarrow$$

$$2\,\text{H}:\ddot{\underset{\cdot\cdot}{\text{Cl}}}:$$

We use the individual atoms in order to focus on the sharing of electrons to form a covalent bond.

EXAMPLE 4.9   Covalent Bonds from Lewis Structures

Use Lewis structures to show the formation of a covalent bond **(a)** between two fluorine atoms and **(b)** between a fluorine atom and a hydrogen atom.

Solution

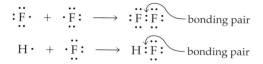

■ **EXERCISE 4.9**

Use Lewis structures to show the formation of a covalent bond between **(a)** two bromine atoms, **(b)** between a hydrogen atom and a bromine atom, and **(c)** between an iodine atom and a chlorine atom.

One might reasonably ask why a hydrogen molecule and a chlorine molecule react at all. Have we not just explained that they themselves were formed to provide more stable arrangements of electrons? Yes, indeed, we did say that. But there is stable, and there is *more* stable. The chlorine molecule represents a more stable arrangement than two separate chlorine atoms. But given the opportunity, a chlorine atom selectively forms a bond with a hydrogen atom rather than with another chlorine atom.

For convenience and simplicity, the reaction of hydrogen (molecule) and chlorine (molecule) to form hydrogen chloride often is represented as

$$\text{H}_2 + \text{Cl}_2 \longrightarrow 2\,\text{HCl}$$

The bonds between the atoms and the lone pairs on the chlorine atoms are not shown explicitly, but remember that they are there.

Each molecule of hydrogen chloride consists of one atom of hydrogen and one atom of chlorine. These unlike atoms share a pair of electrons. *Share*, however, does not necessarily mean "share equally." Chlorine atoms have a greater attraction for a shared pair of electrons than hydrogen atoms do; chlorine is said to be more *electronegative* than hydrogen.

Electronegativity

The **electronegativity** of an element is a measure of the attraction of an atom *in a molecule* for a pair of shared electrons. The atoms to the right in the periodic table are, in general, more electronegative than those to the left. The ones on the right are precisely the atoms that, in forming ions, tend to gain electrons and form negative ions. The ones on the left—metals—tend to give up electrons and become positive ions. The more electronegative an atom is, the greater its tendency to pull the electrons in the bond toward its end of the bond when it is involved in covalent bonding. The American chemist Linus Pauling (1901–1994) devised a scale of relative electronegativity values by assigning fluorine, the most electronegative element, a value of 4.0. Figure 4.5 displays the electronegativity values for some of the common elements that we will encounter in this text.

1A																	Noble gases
H 2.1	2A											3A	4A	5A	6A	7A	
Li 1.0	Be 1.5											B 2.0	C 2.5	N 3.0	O 3.5	F 4.0	
Na 0.9	Mg 1.2	3B	4B	5B	6B	7B		8B		1B	2B	Al 1.5	Si 1.8	P 2.1	S 2.5	Cl 3.0	
K 0.8	Ca 1.0													As 2.0	Se 2.4	Br 2.8	
																I 2.5	

◀ **Figure 4.5** Pauling electronegativity values of several common elements.

QUESTION: Can you estimate a value for the electronegativity of germanium (Ge)? For the electronegativity of rubidium (Rb)?

Chlorine (3.0) is more electronegative than hydrogen (2.1). In the hydrogen chloride molecule, the shared electrons are held more tightly by the chlorine atom, and this results in the chlorine end of the molecule being more negative than the hydrogen end. When the electrons in a covalent bond are not equally shared, the bond is said to be *polar*. Thus, the bond in a hydrogen chloride molecule is described as a **polar covalent bond**, whereas the bond in a hydrogen molecule or a chlorine molecule is a **nonpolar covalent bond**. A polar covalent bond is not an ionic bond. In an ionic bond, one atom completely loses an electron. In a polar covalent bond, the atom at the positive end of the bond (hydrogen in HCl) still has some share in the bonding pair of electrons (Figure 4.6). To distinguish this arrangement from that in an ionic bond, the following notation is used:

$$\overset{\delta+}{H} \overset{\delta-}{-Cl}$$

The line between the atoms represents the covalent bond, a pair of shared electrons. The $\delta+$ and $\delta-$ (read "delta plus" and "delta minus") signify which end is partially positive and which is partially negative. (The word *partially* is used to distinguish this charge from the full charge on an ion.)

We can use the electronegativity values for the atoms in a compound to predict the type of bonding. When the electronegativity difference is zero or very small (<0.5), the bond is *nonpolar covalent*, with essentially equal sharing of the electrons. When the electronegativity difference is large (>2.0), complete electron transfer occurs and an *ionic* bond is formed, as in the case of sodium and chlorine. When the electronegativity difference is between 0.5 and 2.0, *polar covalent* bonds are formed.

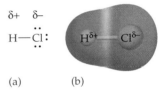

▲ **Figure 4.6** Representation of the polar hydrogen chloride molecule. (a) The electron-dot formula, with the shared electron pair shown nearer the chlorine atom. The symbols $\delta+$ and $\delta-$ indicate partial positive and partial negative charges, respectively. (b) An *electrostatic potential* diagram depicting the unequal distribution of electron density in the hydrogen chloride molecule.

EXAMPLE 4.10	Covalent Bonds from Lewis Structures

Use data from Figure 4.5 to classify bonds between each of the following pairs of atoms as nonpolar covalent, polar covalent, or ionic:

a. H, H **b.** O, H **c.** C, H

Solution

a. Two H atoms have exactly the same electronegativity; the electronegativity difference is 0; the bond is nonpolar covalent.

b. The electronegativity difference is $3.5 - 2.1 = 1.4$; the bond is polar covalent.

c. The electronegativity difference is $2.5 - 2.1 = 0.4$; the bond is nonpolar covalent.

▲ Chlorine hogs the electron blanket, leaving hydrogen partially, but positively, exposed.

2. There are three sets of electrons to consider: two C—H single bonds and one C=O double bond.

3. A triangular arrangement puts the three electron sets as far apart as possible.

4. All the sets are bonding pairs; the molecular shape is triangular, the same as the arrangement of the electrons.

b. Again, we follow the rules, starting with the Lewis formula.

1. The Lewis formula is

2. There are four sets of electrons on the sulfur atom to consider.

3. A tetrahedral arrangement around the central atom puts the four sets of electrons as far apart as possible.

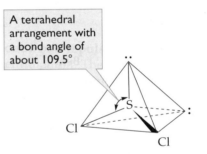

A tetrahedral arrangement with a bond angle of about 109.5°

4. Two of the sets are bonding pairs and two are lone pairs. Ignore the lone pairs; the molecular shape is *bent*, with a bond angle of about 109.5°.

■ **EXERCISE 4.14A**
What are the shapes of **(a)** the PH_3 molecule and **(b)** the nitrate ion (NO_3^-)?

■ **EXERCISE 4.14B**
What are the shapes of **(a)** the carbonate ion (CO_3^{2-}) and **(b)** the hydrogen peroxide (H_2O_2) molecule?

Self-Assessment Questions

1. What is the shape of a water molecule?
   a. bent   **b.** linear   **c.** pyramidal   **d.** tetrahedral

2. What is the shape of a carbon dioxide molecule?
   a. bent   **b.** linear   **c.** pyramidal   **d.** tetrahedral

3. What is the shape of a sulfur dioxide molecule?
   a. bent   **b.** linear   **c.** pyramidal   **d.** tetrahedral

Answers: 1, a; 2, b; 3, a

Green Chemistry and Chemical Bonds

John C. Warner, Warner Babcock Institute for Green Chemistry
Kathryn Parent, ACS Green Chemistry Institute®

You have learned in this chapter that when atoms bond to form compounds, the properties of the materials usually change considerably. In molecules, atoms join by forming covalent bonds that cause the molecules to adopt specific geometric shapes. Each atom is held somewhat rigidly in an exact position relative to other atoms in the molecule.

The specific geometry of a molecule controls how it will react with other substances. Biological molecules are recognized by enzymes and other biological receptors, rather like a key is recognized by a lock. The shape of the molecule and receptor fit together to form complexes. Complexes result from various noncovalent intermolecular forces, such as hydrogen bonds, dipole-dipole forces, and dispersion forces (Section 6.3). When the molecule binds to the receptor it triggers a response, causing a physiological event to begin or end.

Medicinal chemists design drug molecules (Chapter 18) that are similar enough in shape to biological molecules that they still bind to the enzyme or receptor, but they are different enough to behave differently once the complex is formed. The science by which molecules interact via geometric orientations of atoms is called *molecular recognition*. The 1987 Nobel Prize in Chemistry was awarded to Donald J. Cram, University of California, Los Angeles, Jean-Marie Lehn, Université Louis Pasteur, Strasbourg, and College de France, Paris, and research chemist Charles J. Pedersen, Du Pont, Wilmington, Delaware, for pioneering work in this area. Scientists are now learning to use these noncovalent mechanisms to control chemical and physical properties by mimicking nature.

Molecular recognition occurs in many biochemical reactions. For example, in protein synthesis (Chapter 16), DNA replication, the formation of messenger RNA (mRNA), the docking of an mRNA molecule with a ribosome, and the attachment of a specific amino acid molecule to a transfer RNA molecule all involve molecular recognition.

Molecular recognition is involved in the process by which our bodies fight off disease. When an antigen (a foreign substance such as a bacterial or viral protein, pollen, or certain chemicals) enters the body, it combines with an antibody (a type of protein produced and secreted by the immune system) to produce an immune complex. This antigen-antibody reaction serves as a defense against microorganisms and other foreign substances and protects us against infection. This reaction can be detrimental in that it can trigger the release of histamines and thus cause an allergic reaction (Chapter 18). It can be even more harmful if the antibody attacks the body itself, which occurs in autoimmune diseases such as rheumatoid arthritis.

What does all this have to do with green chemistry? Green Chemistry Principle #9 says that we should use catalysts when possible. Using a large excess of a reactant to drive a reaction will work, but it is inefficient and creates much waste. A catalyst is much more efficient. Chemists do not simply pick catalysts at random and hope that they work. Modern catalysts require careful design. That in turn means careful consideration of the bonding and geometry of both the catalyst and the reacting molecules. For bioproducts in particular, constructing the most effective catalyst requires much time and effort. The reward is a dramatic reduction in the environmental impact of that product.

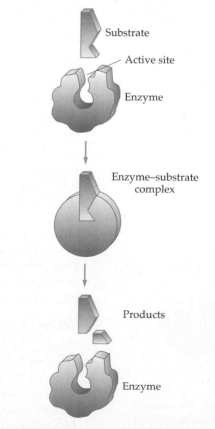

An enzyme is a biochemical catalyst. The shapes of the reacting substance or substrate and the enzyme are complementary. That is what makes the enzyme so specific in its action, and it is why enzymes can be so useful in green chemistry.

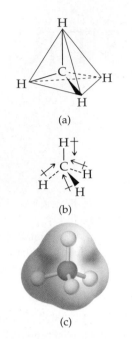

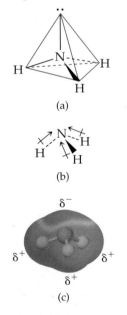

▲ Figure 4.9 The methane molecule. In (a) black lines indicate covalent bonds; the red lines outline a tetrahedron; all bond angles are 109.5°. The slightly polar C—H bonds cancel each other (b), resulting in a nonpolar, tetrahedral molecule (c).

QUESTION: How does the electrostatic potential diagram in (c) look different from the one in Figure 4.6?

▲ Figure 4.10 The ammonia molecule. In (a), black lines indicate covalent bonds; the red lines outline a tetrahedron. The polar N—H bonds do not cancel each other completely (b) resulting in a polar, pyramidal molecule (c).

QUESTION: How does the electrostatic potential diagram in (c) look similar to the one in Figure 4.6?

4.13 Shapes and Properties: Polar and Nonpolar Molecules

Section 4.7 discusses polar and nonpolar bonds. A diatomic molecule is nonpolar if its bond is nonpolar, as in H_2 or Cl_2, and polar if its bond is polar, as in HCl. Recall from Section 4.7 that the partial charges in a polar bond are indicated by $\delta+$ and $\delta-$. In the case of a **dipole**, a molecule with a positive end and a negative end, partial charges are commonly indicated with an arrow with a plus sign at the tail end ($\longmapsto$). The plus sign indicates the part of the molecule with a partial positive charge, and the head of the arrow signifies the end of the molecule with a partial negative charge.

$$H—H \text{ and } Cl—Cl \qquad \overset{\delta^+ \ \ \delta^-}{H—Cl} \ \text{ or } \ \overset{\longmapsto}{H—Cl}$$
$$\text{Nonpolar} \qquad\qquad\qquad \text{Polar}$$

For molecules with three or more atoms, we must consider the polarity of the individual bonds as well as the molecular geometry of the molecule to determine whether the molecule as a whole is polar. A **polar molecule** has separate centers of positive and negative charge, just as a magnet has north and south poles. Many properties of compounds—such as melting point, boiling point, and solubility—depend on the polarity of their molecules.

Methane: A Tetrahedral Molecule

There are four pairs of electrons on the central carbon atom in methane, CH_4. Using the VSEPR theory, we would expect a tetrahedral arrangement and bond angles of 109.5° (Figure 4.9a). The actual bond angles are 109.5°, as predicted by the theory. All four electron pairs are shared with hydrogen atoms so all four pairs occupy identical volumes. Each carbon-to-hydrogen bond is slightly polar (Figure 4.9b), but the methane molecule as a whole is symmetric. The slight bond polarities cancel out, leaving the methane molecule, as a whole, nonpolar (Figure 4.9c).

Ammonia: A Pyramidal Molecule

With ammonia, NH_3, several things change because we have only three bonds and a lone pair. The N—H bonds of ammonia are more polar than the C—H bonds of methane. More important, the molecule has a different geometry as a result of three bonding pairs and one lone pair about the nitrogen atom. The VSEPR theory predicts a tetrahedral arrangement of the four sets of electrons, giving bond angles of 109.5° (Figure 4.10a). The lone pair of electrons occupies a greater volume than a bonding pair, pushing the BPs slightly closer together. The actual bond angles are slightly less, about 107°. The pyramidal geometry can be envisioned as a tripod with a hydrogen atom at the end of each leg and the nitrogen atom with its lone pair sitting at the top. Each nitrogen-to-hydrogen bond is somewhat polar (Figure 4.10b). The asymmetric structure makes the ammonia molecule polar, with a partial negative charge on the nitrogen atom and partial positive charges on the three hydrogen atoms (Figure 4.10c).

Water: A Bent Molecule

The O—H bonds in water are even more polar than the N—H bonds in ammonia because oxygen is more electronegative than nitrogen. (Electronegativity increases from left to right in the periodic table.) Just because a molecule contains polar bonds, however, does not mean that the molecule as a whole is polar. If the atoms in the water molecule were in a straight line (that is, in a linear arrangement as in CO_2), the two polar bonds would cancel one another out and the molecule would be nonpolar.

In its physical and chemical properties, however, water acts like a polar molecule. Molecules such as water and ammonia, in which the polar bonds do not cancel out, act as dipoles; such molecules have a positive end and negative end.

We can understand the dipole in the water molecule by using VSEPR theory. The two bonds and two lone pairs should form a tetrahedral arrangement (Figure 4.11a). Ignoring the lone pairs, the molecular shape has the atoms in a bent arrangement, with a bond angle of about 104.5°. As with ammonia, the difference between the actual angle and the predicted angle is explained by the greater volume of space occupied by the two lone pairs than by the bonding pairs. The larger space occupied by the lone pairs reduces the space in which the bonding pairs reside, pushing them closer together. The two polar O—H bonds do not cancel each other (Figure 4.11b), leaving a polar, bent water molecule (Figure 4.11c).

Self-Assessment Questions

1. Which of the following molecules is polar?
   a. Br_2 **b.** CH_4 **c.** CO_2 **d.** NH_3

2. Which of the following molecules is polar?
   a. C_2H_2 (linear)   **b.** Cl_2 **c.** CCl_4 **d.** PF_3

3. Which of the following molecules is polar?
   a. CF_4 **b.** C_2H_4 (planar)   **c.** NF_3 **d.** O_2

4. Which of the following molecules has polar bonds but is not polar?
   a. CCl_4 **b.** Cl_2 **c.** NCl_3 **d.** OF_2

Answers: 1, d; 2, d; 3, c; 4, a

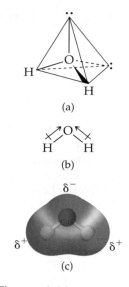

▲ **Figure 4.11** The water molecule. In (a), black lines indicate covalent bonds; the red lines outline a tetrahedron. The polar O—H bonds do not cancel each other (b), resulting in a polar, bent molecule (c).

4.14 A Chemical Vocabulary

Learning chemical symbolism is much like learning a foreign language. Once you have learned a basic "vocabulary," the rest is a lot easier. Initially, the task is complicated by different chemical species or definitions sounding a lot alike (sodium atoms versus sodium ions). Another complication is that we have several different representations of the *same* chemical species (Figure 4.12).

In Chapter 1, we introduced symbols for the chemical elements. Chapter 3 introduced the structure of the nucleus and symbolism to distinguish different isotopes from one another. In the first part of this chapter, we introduced chemical names and chemical formulas. Now you also know how to write Lewis symbols and formulas using dots to represent valence electrons, and you can predict the shapes of thousands of different molecules with the VSEPR theory. In the next chapter you will add the mathematics of chemistry to your toolbox.

More to Explore

Atkins, P. W. *Atkins' Molecules*, 2nd ed., Cambridge, UK: Cambridge University Press, 2003. A friendly and colorful discussion about some common molecules and their chemical bonds.

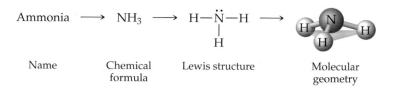

Ammonia	NH_3	H—N̈—H with H below	molecular model
Name	Chemical formula	Lewis structure	Molecular geometry

◀ **Figure 4.12** Several representations of the ammonia molecule.

Critical Thinking Exercises

Apply knowledge that you have gained in this chapter and one or more of the FLaReS principles (Chapter 1) to evaluate the following statements or claims.

4.1 Some people believe that crystals have special powers. Crystal therapists claim that they can use quartz crystals to restore balance and harmony to a person's spiritual energy. Do you think their claim is credible?

4.2 A "fuel-enhancer" device is being sold by an entrepreneur. The device contains a powerful magnet and is placed on the fuel line of an automobile. The inventor claims that the device "separates the positive and negative charges in the hydrocarbon fuel molecules, increasing their polarity and allowing them to react more readily with oxygen." Evaluate this claim for credibility.

4.3 Sodium chloride, NaCl, is a metal–nonmetal compound held together by ionic bonds. A scientist has studied mercury(II) chloride, $HgCl_2$, and says that the atoms are held together by covalent bonds. The scientist says that since a water solution of the substance does not conduct an electric current, it does not contain an appreciable amount of ions. In your opinion, is the scientist correct?

4.4 Another scientist, noting that the noble gas xenon does not contain any ions, states that xenon atoms must be held together by covalent bonds. Is this statement plausible?

■ SUMMARY

Section 4.1—Outermost shell electrons are called **valence electrons**; those in all other shells are **core electrons.** The noble gases are inert because they have an octet of valence electrons. Other elements become less reactive by gaining or losing electrons to attain an electron configuration that is **isoelectronic** with, or has the same electron configuration as, a noble gas.

Section 4.2—The number of valence electrons for most main group elements is the same as the group number. **Electron-dot symbols** or **Lewis symbols** use dots singly or in pairs to represent valence electrons of an atom or ion.

Section 4.3—The element sodium reacts violently with elemental chlorine to give sodium chloride or table salt. In this reaction, an electron is transferred from a sodium atom to a chlorine atom. The ions formed have opposite charges and are strongly attracted to one another; this attraction is called an **ionic bond.** The positive and negative ions arrange themselves in a regular array, forming a **crystal** of sodium chloride. The ions of an element have very different properties from the atoms of that element.

Section 4.4—Metal atoms tend to give up their valence electrons to become ions, while nonmetal atoms tend to accept (8 − group number) electrons to become ions. In either case the ions formed (except hydrogen and lithium ions) tend to have eight valence electrons; the ions are said to follow the **octet rule.** The symbol for an ion is given as the element symbol with a superscript to indicate the number and type (+ or −) of charge. Some elements, especially the transition metals, can form ions of different charges. The formula of an ionic compound always has the same number of positive charges as negative charges.

Section 4.5—A **binary compound** contains two different elements; a binary ionic compound contains cation(s) from a metal and anion(s) from a nonmetal. The **formula** of a binary ionic compound represents the number and type of ions in the compound. The formula is found by "crossing over" the charges of the ions.

Binary ionic compounds are named by naming the cation first and then putting an -*ide* ending on the stem of the anion. Examples include calcium chloride, potassium oxide, aluminum nitride, and so on. A metal that can form differently charged cations is named with a Roman numeral to indicate the charge: Fe^{2+} is iron(II) ion and Fe^{3+} is iron(III) ion.

Section 4.6—Nonmetal atoms can bond by sharing pairs of electrons, forming a **covalent bond.** A single bond is one pair of shared electrons; a **double bond** is two pairs; and a **triple bond** is three pairs. Shared pairs of electrons are called **bonding pairs (BPs)** whereas unshared pairs are called **nonbonding pairs** or **lone pairs (LPs).** Most binary covalent compounds are named by naming the first element in the formula, then the stem of the second element with an -*ide* ending, and using prefixes such as *mono-, di-, tri-,* and so on, to indicate the number of atoms.

Section 4.7—Bonding pairs are shared equally when the two atoms are the same but may be unequally shared when the atoms are different. The **electronegativity** of an element is the atom's attraction in a molecule for a bonding pair. Electronegativity generally increases to the right and up on the periodic table, so fluorine is the most electronegative element. When electrons are shared unequally, the more electronegative element takes on a partial negative charge ($\delta-$), the other element takes on a partial positive charge ($\delta+$), and the bond is said to be a **polar covalent bond.** The greater the difference in electronegativity, the more polar is the bond. A bonding pair shared equally is a **nonpolar covalent bond.**

Section 4.8—Carbon tends to form four bonds, nitrogen three, oxygen two, and hydrogen can only form one bond. A water molecule has two BPs (O—H bonds) and two LPs, an ammonia molecule has three BPs (N—H bonds) and one LP, and a methane molecule has four BPs (C—H bonds).

Section 4.9—A **polyatomic ion** is a charged particle containing two or more covalently bonded atoms. Compounds containing polyatomic ions are named, and formulas are written, in the same fashion as compounds containing monatomic ions, except that parentheses are placed around a polyatomic ion if there is more than one of the ion. Names of polyatomic ions often end in -*ate*, but if there are two anions containing the same elements (including oxygen), the one with one fewer oxygen atom ends in -*ite*.

Section 4.10—A **Lewis (electron-dot) formula** shows the arrangement of atoms, bonds, and lone pairs in a molecule or polyatomic ion. To draw a Lewis formula we (a) draw a reasonable skeletal structure, showing the arrangement of atoms; (b) count the valence electrons; (c) connect bonded pairs with a dash (one electron pair); (d) place electron pairs around outer atoms to give each an octet; and (e) place remaining electrons on the central atom. A multiple bond(s) is used if there are not enough electrons to give each atom (except hydrogen) an octet.

Section 4.11—There are exceptions to the octet rule. Atoms and molecules with unpaired electrons are called **free radicals.** They are highly reactive and short-lived. Examples of free radicals are NO and ClO_2.

Section 4.12—The shape of a molecule can be predicted with **VSEPR theory,** which assumes that sets of electrons around a central atom will get as far away from each other as possible. Two sets of electrons are 180° apart, three sets are about 120° apart, and four sets are about 109° apart. Once these angles have been established, we determine the shape of the molecule by examining only the bonded atoms. Simple molecules may have shapes described as linear, bent, triangular, pyramidal, or tetrahedral.

Section 4.13—Like a polar bond, a **polar molecule** has separation between its centers of positive charge and negative charge. A molecule with nonpolar bonds is nonpolar. A molecule with polar bonds is nonpolar if its shape causes the polar bonds to "cancel." If the polar bonds do not cancel, the molecule is polar or is said to be a **dipole.** The water molecule has polar O—H bonds and is bent, and so it is polar.

■ REVIEW QUESTIONS

1. How does sodium metal differ from sodium ions (in sodium chloride, for example) in properties?

2. What are the structural differences among chlorine atoms, chlorine molecules, and chloride ions? How do their properties differ?

3. Indicate charges on simple ions formed from the following elements.
   a. group 3A   **b.** group 6A
   c. group 1A   **d.** group 7A

4. In what group of the periodic table would elements that form ions with the following charges likely be found?
   a. 2+   **b.** 3−   **c.** 1−

5. How many covalent bonds do each of the following usually form? You may refer to the periodic table.
   a. H   **b.** C   **c.** O
   d. F   **e.** N   **f.** Br

6. Of the elements H, O, N, and C, which one(s) can readily form triple bonds?

■ PROBLEMS

Lewis Symbols for Elements

7. Give Lewis symbols for each of the following elements. You may use the periodic table.
   a. magnesium   **b.** oxygen   **c.** silicon

8. Give Lewis symbols for each of the following elements. You may use the periodic table.
   a. nitrogen   **b.** chlorine   **c.** boron

Lewis Structures for Ionic Compounds

9. Give Lewis structures for each of the following.
   a. calcium fluoride   **b.** sodium sulfide
   c. potassium iodide   **d.** aluminum chloride

10. Give Lewis structures for each of the following.
   a. magnesium oxide   **b.** strontium chloride
   c. sodium bromide   **d.** lithium nitride

Names and Symbols for Simple Ions

11. Without referring to Table 4.2, name the following ions.
   a. Mg^{2+} **b.** Na^+ **c.** O^{2-}
   d. Cl^- **e.** Ag^+ **f.** Cu^+

12. Without referring to Table 4.2, name the following ions.
   a. K^+ **b.** S^{2-} **c.** Br^-
   d. F^- **e.** Ca^{2+} **f.** Fe^{3+}

13. Refer to page 97. Then use that information to name the following ions.
   a. Cr^{2+} **b.** Cr^{3+} **c.** Cr^{6+}

14. Refer to page 97. Then use that information to name the following ions.
   a. Mo^{4+} **b.** Mo^{6+}

15. Refer to page 97. Then write symbols for the following.
   a. cobalt(II) ion   **b.** indium(I) ion
   c. indium(III) ion

16. Refer to page 97. Then write symbols for the following.
   a. manganese(II) ion   **b.** manganese(III) ion
   c. manganese(VII) ion

Names and Formulas for Binary Ionic Compounds

17. Name the following binary ionic compounds.
   a. NaBr   **b.** KCl   **c.** Ag_2O
   d. MgF_2 **e.** $FeCl_2$ **f.** $FeCl_3$

18. Name the following binary ionic compounds.

a. LiF   **b.** $CaCl_2$ **c.** MgS
d. AgI   **e.** CuO   **f.** Cu_2O

19. One of two binary ionic compounds is often added to toothpaste. One of the compounds contains sodium and fluorine; the other contains tin(II) ions and fluorine. Write the formulas for these two compounds and name them.

20. There are two common binary ionic compounds formed from chromium and oxygen. One of them contains chromium(III) ions; the other contains chromium(VI) ions. Write formulas for the two compounds and name them.

Names and Formulas for Ionic Compounds with Polyatomic Ions

21. Give formulas for the following.
   a. potassium hydroxide   **b.** magnesium carbonate
   c. iron(III) oxalate   **d.** iron(II) cyanide

22. Give formulas for the following.
   a. silver nitrate   **b.** lithium chromate
   c. magnesium nitrite   **d.** copper(I) phosphate

23. Name the following.
   a. $NaHSO_4$ **b.** $AgNO_2$
   c. NH_4F **d.** CuOH

24. Name the following.
   a. $KMnO_4$ **b.** $CaCO_3$
   c. $Li_2C_2O_4$ **d.** $Cu(OH)_2$

Molecules: Covalent Bonds

25. Use Lewis symbols to show the sharing of electrons between a hydrogen atom and a fluorine atom.

26. Use Lewis symbols to show the sharing of electrons between two iodine atoms to form an iodine (I_2) molecule. Label all electron pairs as bonding pairs (BPs) or lone pairs (LPs).

27. Use Lewis symbols to show the sharing of electrons between a phosphorus atom and hydrogen atoms to form a molecule in which phosphorus has an octet of electrons.

28. Use Lewis symbols to show the sharing of electrons between a silicon atom and hydrogen atoms to form a molecule in which silicon has an octet of electrons.

29. Use Lewis symbols to show the sharing of electrons between a carbon atom and fluorine atoms to form a molecule in which each atom has an octet of electrons.

30. Use Lewis symbols to show the sharing of electrons between a nitrogen atom and chlorine atoms to form a molecule in which each atom has an octet of electrons.

Names and Formulas for Covalent Compounds

31. Give formulas for the following covalent compounds.
   a. dinitrogen tetroxide   **b.** bromine trichloride
   c. nitrogen triiodide

32. Give formulas for the following covalent compounds.
   a. oxygen difluoride   **b.** chlorine trifluoride
   c. tricarbon dioxide

33. Name the following covalent compounds.
   a. CS_2 **b.** PF_5 **c.** N_2S_4

34. Name the following covalent compounds.
   a. CBr_4 **b.** Cl_2O_7 **c.** P_4S_{10}

Lewis Formulas for Covalent Compounds

35. Give Lewis formulas that follow the octet rule for the following covalent molecules.
   a. SiF_4 **b.** N_2H_4 **c.** CH_5N
   d. COF_2 **e.** NOH_3 **f.** H_3PO_3

36. Give Lewis formulas that follow the octet rule for the following covalent molecules.
   a. NCl_3 **b.** C_2H_4 **c.** H_2SO_4
   c. C_2H_2 **d.** CH_2O **f.** SCl_2

37. Give Lewis formulas that follow the octet rule for the following ions.
   a. ClO^- **b.** HPO_4^{2-} **c.** BrO_3^-

38. Give Lewis formulas that follow the octet rule for the following ions.
   a. CN^- **b.** ClO_2^- **c.** HSO_4^-

Electronegativity: Polar Covalent Bonds

39. Classify the following covalent bonds as polar or nonpolar.
   a. $H-O$ **b.** $N-Cl$ **c.** $B-F$

40. Classify the following covalent bonds as polar or nonpolar.
   a. $H-N$ **b.** $Be-F$ **c.** $P-Cl$

41. Use the symbol ($\longleftrightarrow$) to indicate the direction of the dipole in each polar bond in Problem 39.

42. Use the symbol ($\longleftrightarrow$) to indicate the direction of the dipole in each polar bond in Problem 40.

43. Use the symbols $\delta+$ and $\delta-$ to indicate partial charges, if any, on the following bonds.
   a. $Si-Cl$ **b.** $Cl-Cl$ **c.** $O-F$

44. Use the symbols $\delta+$ and $\delta-$ to indicate partial charges, if any, on the following bonds.
   a. $N-H$ **b.** $C-F$ **c.** $C-C$

Classifying Bonds

45. Classify the bonds in the following as ionic or covalent. For bonds that are covalent, indicate whether they are polar or nonpolar.
   a. KF   **b.** IBr   **c.** MgO

46. Classify the bonds in the following as ionic or covalent. For bonds that are covalent, indicate whether they are polar or nonpolar.
   a. NO   **b.** CaO   **c.** NaBr

47. Classify the bonds in the following as ionic or covalent. For bonds that are covalent, indicate whether they are polar or nonpolar.
   a. Br_2 **b.** F_2 **c.** HCl

48. Classify the bond in Problem 25 as polar or nonpolar. Label the ends of the molecule with symbols that indicate polarity.

VSEPR Theory: The Shapes of Molecules

49. Use VSEPR theory to predict the shape of each of the following molecules.
   a. silane (SiH_4)
   b. hydrogen selenide (H_2Se)
   c. arsine (AsH_3)
   d. silicon tetrachloride ($SiCl_4$)
   e. oxygen difluoride (OF_2)
   f. phosphorus trifluoride (PF_3)

50. Use VSEPR theory to predict the shape of each of the following molecules.
   a. chloroform ($CHCl_3$)
   b. boron trichloride (BCl_3)
   c. carbon tetrafluoride (CF_4)
   d. sulfur dichloride (SCl_2)
   e. nitrogen triiodide (NI_3)
   f. dichlorodifluoromethane (CCl_2F_2)

Polar and Nonpolar Molecules

51. The molecule BeF_2 is linear. Is it polar or nonpolar? Explain.

52. The molecule SF_2 is bent. Is it polar or nonpolar? Explain.

53. Look again at the molecules in Problem 49. For each, are the bonds polar? What are the approximate bond angles? Is the molecule as a whole polar?

54. Look again at the molecules in Problem 50. For each, are the bonds polar? What are the approximate bond angles? Is the molecule as a whole polar?

Molecules That Are Exceptions to the Octet Rule

55. Which of the following species (atoms or molecules) are free radicals?
   a. Br   **b.** F_2 **c.** CCl_3

56. Which of the following atoms or molecules are free radicals?
   a. S   **b.** NO_2 **c.** N_2O_4

57. Free radicals are one class of molecules in which atoms do not conform to the octet rule. Another exception involves atoms with fewer than eight electrons, as seen in elements of group 3 and in beryllium. In some covalent molecules, these atoms can have six or four electrons, respectively. Write Lewis structures for the following covalent molecules.
   a. $AlBr_3$ **b.** BeH_2 **c.** BH_3

58. Exceptions to the octet rule include molecules with atoms having more than eight valence electrons, most typically 10 or 12. Atoms heavier than Si can "expand their valence shell," meaning the "extra" electrons are accommodated in those atoms in unoccupied, higher-energy orbitals. Draw Lewis structures for the following molecules or ions.
   a. XeF_4 **b.** I_3^- **c.** SF_4 **d.** KrF_2

■ ADDITIONAL PROBLEMS

59. Why does neon tend not to form chemical bonds?

60. Draw a charge-cloud picture for the H_2S molecule. Use the symbols $\delta+$ and $\delta-$ to indicate the polarity of the molecule.

61. The gas phosphine (PH_3) is used as a fumigant to protect stored grain and other durable produce from pests. Phosphine is generated in situ by adding water to aluminum phosphide or magnesium phosphide. Give formulas for the two phosphides.

62. There are two different covalent molecules with the formula C_2H_6O. Give Lewis formulas for the two molecules.

63. Solutions of iodine chloride (ICl) are used as disinfectants. Are the molecules of ICl ionic, polar covalent, or nonpolar covalent?

64. Consider the hypothetical elements X, Y, and Z with the following Lewis formulas.

$$:\ddot{X}\cdot \qquad :\ddot{Y}\cdot \qquad :\ddot{Z}\cdot$$

a. To which group in the periodic table would each element belong?
b. Give the Lewis formula for the simplest compound of each with hydrogen.
c. Give Lewis formulas for the ions formed when X reacts with sodium and when Y reacts with sodium.

65. Potassium is a soft, silvery metal that reacts violently with water and ignites spontaneously in air. Your doctor recommends you take a potassium supplement. Would you take potassium metal? If not, what would you take?

66. Is there such a thing as a sodium chloride molecule? Explain.

67. Use subshell notations to give electron configurations for the most stable simple ion formed by each of the following elements.
a. Ba
b. K
c. Se
d. I
e. N
f. Te

68. Why is Na^+ smaller than Na? Why is Cl^- larger than Cl?

69. The halogens (F, Cl, Br, and I) tend to form only one single bond in binary molecules. Explain.

70. A science magazine for the general public contains the statement, "Some of these hydrocarbons are very light, like methane gas— just a single carbon molecule attached to three hydrogen molecules." Evaluate the statement and correct any inaccuracies.

71. Scientists estimate that the atmosphere of Titan (a moon of Saturn) consists of about 98.4% nitrogen and 1.6% methane. They have also found traces of organic molecules with the molecular formulas C_2H_4, C_3H_4, C_4H_2, HCN, HC_3N, and C_2N_2. Write possible Lewis formulas for each of the molecules.

72. Sodium tungstate is Na_2WO_4. What is the formula for a tungstate ion?

73. What is the formula for the compound formed by the imaginary ions Q^{2+} and ZX_4^{3-}?

■ COLLABORATIVE GROUP PROJECTS

Prepare a PowerPoint, poster, or other presentation (as directed by your instructor) for presentation to the class. Projects 74 and 75 are best done in a group of four students.

74. Make copies of the form in Problem 89, page 37, except replace the column headings "Word" and "Definition" with "Name" and "Lewis Structure." Student 1 should write the name of an element or compound from the following list in the first column of the form and its Lewis structure in the second column. Proceed as in Problem 89, page 37, ending by comparing the name in the last column with that in the first column. Discuss any differences in the two names. If the name in the last column differs from that in the first column, determine what went wrong in the process.
a. bromide ion
b. calcium fluoride
c. phosphorus trifluoride
d. carbon disulfide

75. Make copies of the form in Problem 89, page 37, using the column headings "Name" and "Structure." Then Student 1 should write a name from the following list in the first column of the form and its structure in the second column. Proceed as in Problem 89, page 37, ending by comparing the name in the last column with that in the first column. Discuss any differences in the two names. If the name in the last column differs from that in the first column, determine what went wrong in the process.
a. ammonium nitrate
b. potassium phosphate
c. lithium carbonate
d. copper(I) chloride

76. Starting with 20 balloons, blown up to about the same size and tied, tie two balloons together. Next, tie three balloons together. Repeat this with four, five, and six balloons. (*Suggestion*: Three sets of two balloons can be twisted together to form a six-balloon set, and so on.) Show how these balloon sets can be used to explain the VSEPR theory.

77. Prepare a brief biographical report on one of the following individuals.
a. Gilbert N. Lewis
b. Linus Pauling

78. There are several different definitions and scales for electronegativity. Search for information on some of them and write a brief compare-and-contrast essay on two of them.

The worker is reacting oxygen and acetylene to produce a hot flame for welding and cutting metal. But the proportions of oxygen and acetylene must be adjusted carefully. Too much acetylene and the flame is yellow, smoky, and not hot enough to do the job. Too much oxygen and the flame goes out. In this chapter we will investigate chemical reactions and learn how the amounts of reactants and products are determined.

CHEMICAL ACCOUNTING

CHAPTER 5

QUESTIONS YOU MAY HAVE ASKED YOURSELF

1. How can a car make more pollution than the gasoline it uses?
2. What is the difference between 14K gold and 18K gold?
3. What is the difference between whole milk and 2% milk?
4. What does a cholesterol level of 200 mean?

Mass and Volume Relationships

Much chemistry can be discussed and understood with little or no mathematics, but there are also many interesting quantitative aspects. In this chapter, we will consider some of the basic calculations used in chemistry and related fields such as biology and medicine. Chemists use many kinds of mathematics, from simple arithmetic to sophisticated calculus and complicated computer algorithms, but our calculations here will require no more than simple algebra.

5.1 Chemical Sentences: Equations

Chemistry is a study of matter and the changes it undergoes and of the energy that brings about these changes or is released when these changes occur. In Chapter 4, we discussed the symbols and formulas used to represent elements and compounds. Now that we have learned the letters (symbols) and words (formulas) of our chemical language, we are ready to write sentences (chemical equations). A **chemical equation** is a shorthand way of describing chemical change using symbols and formulas to represent the elements and compounds involved in the change.

We can describe a chemical reaction in words. For example,

Carbon reacts with oxygen to form carbon dioxide.

We can also describe the same reaction with chemical symbols and formulas.

$$C + O_2 \longrightarrow CO_2$$

The plus sign ($+$) indicates that carbon and oxygen are added together or combined in some way. The arrow ($\longrightarrow$) is read "yield(s)" or "react(s) to produce." Substances on the left of the arrow ($C + O_2$ in this case) are **reactants** or *starting materials*. Those on the right (here, CO_2) are the **products** of the reaction. Reactants and products need not be written in any particular order in a chemical equation, except that all reactants must be to the left of the arrow and all products to the right. In other words, we could also write the preceding equation as

$$O_2 + C \longrightarrow CO_2$$

At the submicroscopic (atomic or molecular) level, the chemical equation means that one atom of carbon (C) reacts with one molecule of oxygen (O_2) to produce one molecule of carbon dioxide (CO_2).

Sometimes we indicate the physical states of the reactants and products by writing the initial letter of the state immediately following the formula. Thus, (g) indicates a gaseous substance, (l) a liquid, and (s) a solid. The label (aq) indicates an aqueous solution—that is, a water solution. Using these labels, our equation becomes

$$C(s) + O_2(g) \longrightarrow CO_2(g)$$

Balancing Chemical Equations

We can represent the reaction of carbon and oxygen to form carbon dioxide quite simply, but many other chemical reactions require more thought. For example, hydrogen reacts with oxygen to form water. Using formulas, we can represent this reaction as

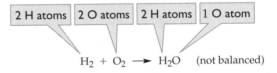

However, this representation shows two oxygen atoms in the reactants (as O_2) but only one in the product (in H_2O). Because matter is neither created nor destroyed in a chemical reaction (the law of conservation of matter, Section 2.2), the equation must be balanced to represent the chemical reaction correctly. That means the same number of each type of atom must appear on both sides. To balance the oxygen atoms, we need only place the coefficient 2 in front of the formula for water.

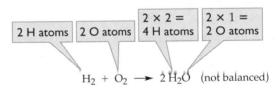

This coefficient means that two molecules of water are produced. As is the case with subscripts, a coefficient of 1 is understood when no other number appears. A coefficient preceding a formula multiplies *everything* in the formula. In the preceding equation, the coefficient 2 not only increases the number of oxygen atoms to two but also increases the number of hydrogen atoms to four on the product side.

But the equation is still not balanced. As we took care of the oxygen, we unbalanced the hydrogen. To balance the hydrogen, we place a coefficient 2 in front of H_2.

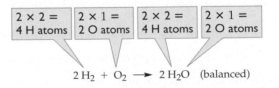

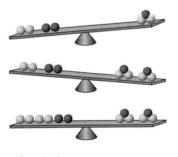

= hydrogen
= oxygen

▲ **Figure 5.1** To balance the equation for the reaction in which hydrogen and oxygen combine to form water, the same number of each kind of atom must appear on each side (atoms are conserved). When the equation is balanced, there are four H atoms and two O atoms on each side.

QUESTION: Why can't we balance the equation by removing one of the oxygen atoms from the left side of the top balance?

Now there are four hydrogen atoms and two oxygen atoms on each side of the equation. Atoms are conserved: The equation is balanced (Figure 5.1) and the law of conservation of mass is obeyed. Figure 5.2 illustrates two common pitfalls in the process of balancing equations, as well as the correct method. Remember not to add to or change chemical species on either side but use coefficients only to equate the numbers of atoms of each type.

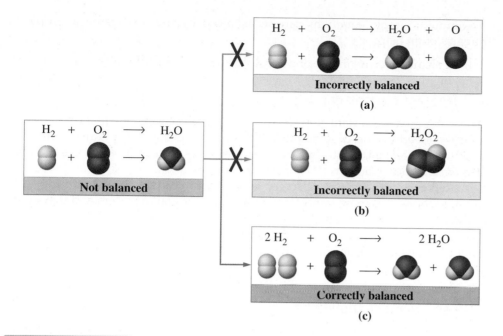

◀ **Figure 5.2** Balancing the equation for the reaction between hydrogen and oxygen to form water. (a) Incorrect. There is no atomic oxygen (O) as a product. Extraneous products cannot be introduced simply to balance an equation. (b) Incorrect. The product of the reaction is water (H_2O), not hydrogen peroxide (H_2O_2). A formula can't be changed simply to balance an equation. (c) Correct. An equation can be balanced only through the use of correct formulas and coefficients.

EXAMPLE 5.1 Balancing Equations

Balance the following representation of the chemical reaction that occurs when an airbag deploys.

$$NaN_3 \longrightarrow Na + N_2$$

Solution

The sodium (Na) atoms are balanced, but the nitrogen atoms are not. For this sort of problem, we will use the concept of the least common multiple. There are three nitrogen atoms on the left (reactants side) and two on the right (products side). The least common multiple of 2 and 3 is 6. To get *six* nitrogen atoms on each side we need *three* N_2 and *two* NaN_3:

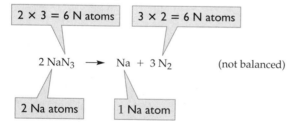

$$2\,NaN_3 \longrightarrow Na + 3\,N_2 \quad \text{(not balanced)}$$

We now have *two* sodium atoms on the left. We can get two on the right by placing the coefficient 2 in front of Na.

$$2\,NaN_3 \longrightarrow 2\,Na + 3\,N_2 \quad \text{(balanced)}$$

Checking, we count two Na atoms and six N atoms on each side. The equation is balanced.

■ EXERCISE 5.1A

The reaction between hydrogen and nitrogen to give ammonia, called the Haber process, is typically the first step in the industrial production of nitrogen fertilizers, represented as

$$H_2 + N_2 \longrightarrow NH_3$$

Balance the equation.

■ EXERCISE 5.1B

Iron ores like Fe_2O_3 are *smelted* by reaction with carbon to produce metallic iron and carbon dioxide, represented as

$$Fe_2O_3 + C \longrightarrow CO_2 + Fe$$

Balance the equation.

Although simple equations can be balanced by trial and error, a couple of strategies often help:

1. If an element occurs in just one substance on each side of the equation, try balancing that element *first*.

2. Balance any reactants or products that exist as the free element *last*.

Perhaps the most important step in any strategy is to check an equation to ensure that it is indeed balanced. Remember that for each element, the same number of atoms of the element must appear on each side of the equation; atoms are conserved in chemical reactions.

We have made the task of balancing equations appear easy by considering fairly simple reactions. It is more important at this point for you to understand the principle than to be able to balance complicated equations. You should know what is meant by a balanced equation and be able to handle simple reactions. It is essential that you understand the information that is contained in these balanced equations.

Self-Assessment Questions

1. Which of the following equations are balanced? (You need not balance the equations; just determine whether they are balanced as written.)

   I. $Ca + 2 H_2O \longrightarrow Ca(OH)_2 + H_2$

   II. $4 LiH + AlCl_3 \longrightarrow 2 LiAlH_4 + 2 LiCl$

   III. $2 LiOH + CO_2 \longrightarrow Li_2CO_3 + H_2O$

   IV. $2 Sn + 2 H_2SO_4 \longrightarrow 2 SnSO_4 + SO_2 + 2 H_2O$

   a. I and II   **b.** I and III   **c.** II and III   **d.** II and IV

2. Which of the following equations are balanced?

   I. $4 BF_3 + 3 H_2O \longrightarrow H_3BO_3 + 3 HBF_4$

   II. $12 K + 2 KNO_3 \longrightarrow 6 K_2O + N_2$

   III. $6 NaOH + 3 Cl_2 \longrightarrow 5 NaCl + NaClO_3 + 3 H_2O$

   IV. $2 NH_3 + O_2 \longrightarrow 3 H_2O + N_2$

   a. I and II   **b.** I and III   **c.** I and IV   **d.** II and III

3. Consider the equation for the reaction of phosphine with oxygen to form tetraphosphorus decoxide:

   $$PH_3 + O_2 \longrightarrow P_4O_{10} + H_2O \quad \text{(not balanced)}$$

   When the equation is balanced, how many molecules of water are produced for each molecule of P_4O_{10} formed?

   a. 1   **b.** 4   **c.** 6   **d.** 12

Items 4–6 refer to the following equation for the reaction of calcium hydroxide with phosphoric acid.

$$Ca(OH)_2 + H_3PO_4 \longrightarrow H_2O + Ca_3(PO_4)_2 \quad \text{(not balanced)}$$

4. When the equation is balanced, how many molecules of water will be produced for each molecule of phosphoric acid?

   a. 1   **b.** 2   **c.** 3   **d.** 6

5. When the equation is balanced, how many molecules of phosphoric acid are consumed for each formula unit of calcium phosphate produced?

   a. 1   **b.** 2   **c.** 3   **d.** 6

6. When the equation is balanced, how many formula units of calcium hydroxide will be consumed if six molecules of phosphoric acid react?

   a. 1   **b.** 1.5   **c.** 6   **d.** 9

Answers: 1, b; 2, b; 3, c; 4, c; 5, b; 6, d

5.2 Volume Relationships in Chemical Equations

So far we have looked at chemical reactions in terms of individual atoms and molecules. In the real world, however, chemists work with quantities of matter that contain billions of billions of atoms. John Dalton postulated that atoms of different elements had different masses. Therefore, equal masses of different elements would contain different numbers of atoms. Consider the analogous situation of golf balls and Ping-Pong balls. A kilogram of golf balls contains a smaller number of balls than a kilogram of Ping-Pong balls. One could determine the number of balls in each case simply by counting them. For atoms, however, counting is not so straightforward; the smallest visible particle of matter contains more atoms than could be counted in ten lifetimes! We can, however, determine the number of atoms in a substance, as you will learn in this chapter.

Studies of gases in the 1700s advanced our understanding of how atoms involved in chemical reactions might be accounted for. While we will explore the interesting nature and behavior of gases in Chapter 6, we note here some historical experiments with gases that revealed how chemical equations can be balanced. Experiments with gases led French scientist Joseph Louis Gay-Lussac (1778–1850) to an approach to quantifying atoms. In 1809 he announced the results of some chemical reactions that he had carried out with gases and summarized these experiments in a new law: The **law of combining volumes** states that when all measurements are made at the same temperature and pressure, the volumes of gaseous reactants and products are in a small whole-number ratio. One such experiment is illustrated in Figure 5.3. When hydrogen reacts with nitrogen to form ammonia, three volumes of hydrogen combine with one volume of nitrogen to yield two volumes of ammonia. The small whole-number ratio is 3:1:2.

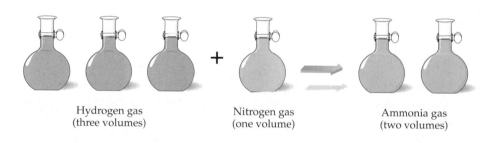

Hydrogen gas
(three volumes)

Nitrogen gas
(one volume)

Ammonia gas
(two volumes)

◀ **Figure 5.3** Gay-Lussac's law of combining volumes. Three volumes of hydrogen gas react with one volume of nitrogen gas to yield two volumes of ammonia gas.

QUESTION: Can you sketch a similar figure for the reaction $2\,SO_2(g) + O_2(g) \longrightarrow 2\,SO_3(g)$?

Gay-Lussac thought there must be some relationship between the numbers of molecules and the volumes of gaseous reactants and products. But it was Amedeo Avogadro who first explained the law of combining volumes in 1811. **Avogadro's hypothesis**, based on a shrewd interpretation of experimental facts, was that equal volumes of all gases, when measured at the same temperature and pressure, contain the same number of molecules (Figure 5.4).

The equation for the combination of hydrogen and nitrogen to form ammonia is

$$3\,H_2(g) + N_2(g) \longrightarrow 2\,NH_3(g)$$

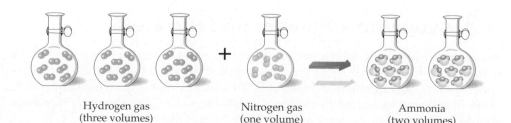

Hydrogen gas
(three volumes)

Nitrogen gas
(one volume)

Ammonia
(two volumes)

◀ **Figure 5.4** Avogadro's explanation of Gay-Lussac's law of combining volumes. Equal volumes of each of the gases contain the same number of molecules.

QUESTION: Can you sketch a similar figure for the reaction $2\,NO(g) + O_2(g) \longrightarrow 2\,NO_2(g)$?

▲ Amedeo Avogadro (1776–1856) and his hypothesis. The quotation reads, "Equal volumes of all gases at the same temperature and pressure contain the same number of molecules."

Note that the coefficients of the molecules correlate with the combining ratio of the gas volumes, 3:1:2 (Figure 5.3). The equation says that three hydrogen molecules react with one nitrogen molecule to produce two ammonia molecules. The balanced equation provides the combining ratios. If you had 3 million hydrogen molecules, you would need 1 million nitrogen molecules to produce 2 million ammonia molecules. According to the equation, three volumes of hydrogen reacts with one volume of nitrogen to produce two volumes of ammonia because each volume of hydrogen contains the same number of molecules as that same volume of nitrogen.

EXAMPLE 5.2   Volume Relationships of Gases

What volume of oxygen is required to burn 0.556 L of propane if both gases are measured at the same temperature and pressure?

$$C_3H_8(g) + 5\ O_2(g) \longrightarrow 3\ CO_2(g) + 4\ H_2O(g)$$

Solution

The coefficients in the equation indicate that each volume of $C_3H_8(g)$ requires five volumes of $O_2(g)$. Thus, we use 5 L $O_2(g)$/1 L $C_3H_8(g)$ as the ratio to find the volume of oxygen required.

$$?\ L\ O_2(g) = 0.556\ \cancel{L\ C_3H_8(g)} \times \frac{5\ L\ O_2(g)}{1\ \cancel{L\ C_3H_8(g)}} = 2.78\ L\ O_2(g)$$

■ **EXERCISE 5.2A**

Using the equation in Example 5.2, calculate the volume of $CO_2(g)$ produced when 0.492 L of propane is burned if the two gases are compared at the same temperature and pressure.

■ **EXERCISE 5.2B**

If 10.0 L each of propane and oxygen are combined at the same temperature and pressure, which gas will be left over after reaction? What volume of that gas will remain?

Self-Assessment Questions

1. In the reaction $2\ H_2(g) + O_2(g) \longrightarrow 2\ H_2O(g)$, with all substances at the same temperature and pressure, the combining ratios by volume, for H_2, O_2, and H_2O, respectively, are
   a. 1:0:1   **b.** 2:0:2   **c.** 1:1:1   **d.** 2:1:2

2. How many molecules of carbon dioxide are produced in the following reaction when 50 molecules of O_2 are consumed?

   $$2\ C_8H_{18}(l) + 25\ O_2(g) \rightarrow 16\ CO_2(g) + 18\ H_2O(l)$$

   a. 25 molecules   **b.** 50 molecules   **c.** 32 molecules   **d.** 16 molecules

3. In the reaction $N_2(g) + 3\ H_2(g) \longrightarrow 2\ NH_3(g)$, with all substances at the same temperature and pressure, what volume of ammonia would be produced when 4.50 L of nitrogen reacts with excess hydrogen?
   a. 3.00 L   **b.** 4.50 L   **c.** 6.75 L   **d.** 9.00 L

Answers: 1, d; 2, c; 3, d

5.3 Avogadro's Number and the Mole

Avogadro's hypothesis, which has been verified many times and in several ways over the years, states that equal volumes of gas at the same temperature and pressure contain equal numbers of molecules. This means that if we weigh equal volumes of different gases, the gases won't weigh the same. The ratio of their masses should be the same as the mass ratio of the molecules themselves.

Avogadro's Number: 6.02×10^{23}

Avogadro had no way of knowing how many molecules were in a given volume of gas. Scientists since his time have determined the number of atoms in various weighed samples of substances. Atoms are so incredibly small that these numbers are extremely large, even for tiny samples. Recall that the atomic mass of a carbon-12 atom is exactly 12 atomic mass units (u) (Section 2.5) because the carbon-12 atom sets the standard for atomic mass. The number of carbon-12 atoms in a 12-g sample of carbon-12 is called **Avogadro's number** and has been determined experimentally to be 6.0221367×10^{23}. For our purposes, we usually round it off to three significant figures: 6.02×10^{23}.

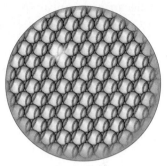

6.02×10^{23} baseballs   roughly equals   Mass of the earth

6.02×10^{23} drops of water   roughly equals   Mass of the oceans

◀ How big is Avogadro's number? Atoms are so tiny that we need a huge number to represent a visible portion of matter. If Avogadro's number is applied to common objects, the result is enormous.

Whose Number Is It, Anyway?

Avogadro died before his ideas were recognized by the scientific community. Acceptance finally came in 1860 at a scientific conference at which Stanislao Cannizzaro (1826–1910) effectively communicated Avogadro's ideas from half a century earlier. Scientists knew that the number must be enormous, but it wasn't until 1865 that reasonable estimates were made. That year Josef Loschmidt (1821–1895) measured the size of air molecules and found them to be about a "millionth of a millimeter" in diameter. This rough measurement indicated a value of 4×10^{22}, not a bad estimate for a first attempt. Later measurements, using various approaches, have shown the actual diameter of air molecules to be a bit smaller than Loschmidt had determined, and the number to be closer to 6×10^{23}. In German-speaking countries, this value is usually called the Loschmidt number, but in most places, it is called Avogadro's number in spite of the fact that Avogadro never knew how big the number was.

The Mole: "A Dozen Eggs and a Mole of Sugar, Please"

We buy socks by the pair (2 socks), eggs by the dozen (12 eggs), pencils by the gross (144 pencils), and paper by the ream (500 sheets). A dozen is the same number whether we are counting a dozen melons or a dozen oranges. But a dozen oranges and a dozen melons do not weigh the same. If a melon weighs five times as much as an orange, a dozen melons will weigh five times as much as a dozen oranges.

Chemists count atoms and molecules by the *mole*. (A single carbon atom is much too small to see or weigh, but a mole of carbon atoms fills a tablespoon and weighs 12 g.) A mole of carbon and a mole of titanium each contain the same number of atoms. But a titanium atom has a mass four times that of a carbon atom, and so a mole of titanium has a mass four times that of a mole of carbon.

A **mole (mol)** is an amount of substance that contains the same number of elementary units as there are atoms in exactly 12 g of carbon-12. That number is 6.02×10^{23}, Avogadro's number. The elementary units may be atoms (such as S or Ca), molecules (such as O_2 or CO_2), ions (such as K^+ or SO_4^{2-}), or any other kind of formula unit. A mole of NaCl, for example, contains 6.02×10^{23} NaCl formula units, which means that it contains 6.02×10^{23} Na^+ ions and 6.02×10^{23} Cl^- ions.

Formula Masses

Each element has a characteristic atomic mass. Because chemical compounds are made up of two or more elements, the masses of compounds are combinations of atomic masses. For any substance, the **formula mass** is the average[1] mass of a formula unit relative to that of a carbon-12 atom. Simply put, it is the sum of the masses of the atoms represented in a formula. If the formula represents a molecule, the term *molecular mass* is often used. For example, because the formula O_2 specifies two O atoms per molecule of oxygen, the formula (or molecular) mass of oxygen (O_2) is twice the atomic mass of oxygen.

$$2 \times \text{atomic mass of O} = 2 \times 16.0 \text{ u} = 32.0 \text{ u}$$

EXAMPLE 5.3 Calculating Molecular Masses

Calculate **(a)** the molecular mass of nitrogen dioxide (NO_2), an amber-colored gas that is a constituent of smog, and **(b)** the formula mass of ammonium sulfate [$(NH_4)_2SO_4$], a fertilizer commonly used by home gardeners.

Solution

a. We start with the molecular formula: NO_2. Then, to determine the molecular mass, we need only to add the atomic mass of nitrogen to twice the atomic mass of oxygen.

$$1 \times \text{atomic mass of N} = 1 \times 14.0 \text{ u} = 14.0 \text{ u}$$
$$2 \times \text{atomic mass of O} = 2 \times 16.0 \text{ u} = \underline{32.0 \text{ u}}$$
$$\text{Formula mass of } NO_2 = 46.0 \text{ u}$$

Using a calculator, we need only write down the final answer, 46.0 u. That is, we have no need to record the numbers 14.0 and 32.0.

b. We must make certain that all the atoms in the formula unit are accounted for, which means paying particular attention to all the subscripts and parentheses in the formula. The "$(NH_4)_2$" means that both the "N" and the "H_4" must be multiplied by 2—that is, the formula indicates a total of two N atoms and eight H atoms. Combining the atomic masses, we have

$$2 \times \text{atomic mass of N} = 2 \times 14.0 \text{ u} = 28.0 \text{ u}$$
$$8 \times \text{atomic mass of H} = 8 \times 1.01 \text{ u} = 8.08 \text{ u}$$
$$1 \times \text{atomic mass of S} = 1 \times 32.0 \text{ u} = 32.0 \text{ u}$$
$$4 \times \text{atomic mass of O} = 4 \times 16.0 \text{ u} = \underline{64.0 \text{ u}}$$
$$\text{Formula mass of } (NH_4)_2SO_4 = 132.1 \text{ u}$$

■ **EXERCISE 5.3A**

Calculate the formula mass of **(a)** sodium azide (NaN_3), used in automobile airbags, and **(b)** phosphoric acid (H_3PO_4), used to flavor some cola drinks.

■ **EXERCISE 5.3B**

Calculate the formula mass of **(a)** *para*-dichlorobenzene ($C_6H_4Cl_2$), used as a moth repellent, and **(b)** calcium dihydrogen phosphate [$Ca(H_2PO_4)_2$], used as a mineral supplement in foods.

More to Explore

Ault, Addison. "How to Say How Much: Amounts and Stoichiometry." *Journal of Chemical Education*, October 2001, pp. 1345–1347.

[1]We use the word *average* when denoting the mass of an individual molecule because molecules of a compound may have different isotopes of one or more of their constituent elements.

Self-Assessment Questions

1. How many carbon-12 atoms are there in a mass of exactly 12 g of carbon-12?
   a. 1   **b.** 12   **c.** 144   **d.** 6.02×10^{23}

2. The mass of 1 mol of carbon-12 atoms is
   a. 12 g   **b.** 12.011 g   **c.** 12 u   **d.** 12.011 u

3. How many hydrogen atoms are in one formula unit of $(NH_4)_2HPO_4$?
   a. 4   **b.** 5   **c.** 8   **d.** 9

4. Which of the following has the smallest molecular mass?
   a. CH_4 **b.** HF   **c.** H_2O **d.** NH_3

5. What is the total number of moles of sulfur atoms in 1 mol of $Fe_2(SO_4)_3$?
   a. 1   **b.** 3   **c.** 15   **d.** 17

6. How many moles of hydrogen atoms are present in 2.00 moles of ammonia (NH_3)?
   a. 2.00 mol   **b.** 3.00 mol   **c.** 6.00 mol   **d.** 8.00 mol

Answers: 1, d; 2, a; 3, d; 4, a; 5, b; 6, c

5.4 Molar Mass: Mole-to-Mass and Mass-to-Mole Conversions

The **molar mass** of a substance is just what the name implies: the mass of one mole of that substance. The molar mass is numerically equal to the atomic mass or formula mass, but it is expressed in the unit *grams per mole* (g/mol). The atomic mass of sodium is 23.0 u; its molar mass is 23.0 g/mol. The molecular mass of carbon dioxide is 44.0 u; its molar mass is 44.0 g/mol. The formula mass of ammonium sulfate is 132.1 u; its molar mass is 132.1 g/mol. We can use these facts, together with the definition of the mole, to write the following relationships.

$$1 \text{ mol Na} = 23.0 \text{ g Na}$$

$$1 \text{ mol CO}_2 = 44.0 \text{ g CO}_2$$

$$1 \text{ mol (NH}_4)_2\text{SO}_4 = 132.1 \text{ g (NH}_4)_2\text{SO}_4$$

These relationships supply the conversion factors we need to make conversions between mass in grams and amount in moles, as illustrated in the following examples.

EXAMPLE 5.4   Mole-to-Mass Conversions

What mass in grams of N_2 is 0.400 moles N_2?

Solution

The molecular mass of N_2 is $2 \times 14.0 \text{ u} = 28.0 \text{ u}$. The molar mass of N_2 is therefore 28.0 g/mol. Using the molar mass as a conversion factor (red), we have

$$? \text{ g N}_2 = 0.400 \text{ mol N}_2 \times \frac{28.0 \text{ g N}_2}{1 \text{ mol N}_2} = 11.2 \text{ g N}_2$$

▪ **EXERCISE 5.4**

Calculate the mass, in grams, of **(a)** 0.0728 mol silicon, **(b)** 55.5 mol H_2O, and **(c)** 0.0728 mol $Ca(H_2PO_4)_2$.

EXAMPLE 5.5 Mass-to-Mole Conversions

Calculate the number of moles of Na in a 62.5-g sample of sodium metal.

Solution

The molar mass of Na is 23.0 g/mol. To convert from a mass in grams to an amount in moles, we must use the *inverse* of the molar mass as a conversion factor (1 mol Na/23.0 g Na) to get the proper cancellation of units. When we start with grams, we must have grams in the denominator of our conversion factor (red).

$$? \text{ mol Na} = 62.5 \ \cancel{\text{g Na}} \times \frac{1 \text{ mol Na}}{23.0 \ \cancel{\text{g Na}}} = 2.72 \text{ mol Na}$$

■ EXERCISE 5.5

Calculate the amount, in moles, of **(a)** 3.71 g Fe, **(b)** 165 g butane, C_4H_{10}, and **(c)** 0.100 g $Mg(NO_3)_2$.

Figure 5.5 is a photograph showing one-mole samples of several different chemical substances. Each container holds Avogadro's number of formula units of the substance. That is, there are just as many copper atoms in the smallest dish as there are helium atoms in the balloon.

▶ **Figure 5.5** One mole of each of several familiar substances (from left to right on the table): sugar, salt, carbon, and copper; the balloon contains helium. Each contains Avogadro's number of formula units of the substance it contains. There are 6.02×10^{23} molecules of sugar ($C_{12}H_{22}O_{11}$), 6.02×10^{23} formula units of NaCl, 6.02×10^{23} atoms of carbon, 6.02×10^{23} atoms of copper, and 6.02×10^{23} atoms of helium in the respective samples.

QUESTION: How many Na^+ ions and how many Cl^- ions are there in the dish of salt? If the balloon was filled with oxygen gas (O_2), how many oxygen atoms would it contain?

What Is a Mole?

A mole is a particular amount
Of substance—just its formulary weight
Expressed in grams, with Avogadro's count
Of units making up the aggregate.
A mole is but a single molecule
By Avogadro's number multiplied;
One entity, extremely minuscule,
A trillion trillion times intensified.
A mole is a convenient amount,
For molecules are just too small to count.

Mole and Mass Relationships in Chemical Equations

Chemists and other scientists and engineers are often confronted with questions such as "How much iron can be obtained from one ton of Fe_2O_3?" or "How much carbon dioxide is produced by burning 1000 grams of propane?" There is no simple way to calculate mass of one substance directly from mass of another. Because chemical reactions involve atoms and molecules, quantities in reactions are calculated in moles of atoms, ions, or molecules. To do such calculations, it is necessary to convert grams of a substance to moles, as we did in Example 5.5. We also must write balanced equations to determine molar ratios.

Recall that chemical equations not only represent ratios of atoms and molecules but also give us mole ratios. For example, the equation

$$C + O_2 \longrightarrow CO_2$$

tells us that one atom of carbon reacts with one molecule (two atoms) of oxygen to form one molecule of carbon dioxide (one atom of carbon and two atoms of oxygen). The equation also indicates that 1 mol (6.02×10^{23} atoms) of carbon reacts with 1 mol (6.02×10^{23} molecules) of oxygen to yield 1 mol (6.02×10^{23} molecules) of carbon dioxide. Because the molar mass in grams of a substance is numerically equal to the formula mass of the substance in atomic mass units, the equation also tells us (indirectly) that 12.0 g (1 mol) of carbon reacts with 32.0 g (1 mol) of oxygen to yield 44.0 g (1 mol) of CO_2 (Figure 5.6).

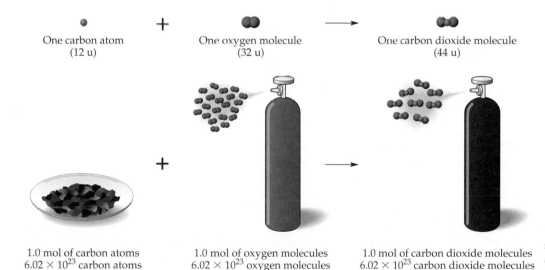

One carbon atom
(12 u)

One oxygen molecule
(32 u)

One carbon dioxide molecule
(44 u)

1.0 mol of carbon atoms
6.02×10^{23} carbon atoms
(12 g of carbon)

1.0 mol of oxygen molecules
6.02×10^{23} oxygen molecules
(32 g of oxygen)

1.0 mol of carbon dioxide molecules
6.02×10^{23} carbon dioxide molecules
(44 g of carbon dioxide)

◀ **Figure 5.6** We cannot weigh single atoms or molecules, but we can weigh equal numbers of these tiny particles.

We need not use exactly 1 mol of each reactant. The important thing is to keep the *ratio* constant. For example, in the preceding reaction, the mass ratio of oxygen to carbon is 32.0:12.0 or 8.0:3.0. We could use 8.0 g of oxygen and 3.0 g of carbon to produce 11.0 g of CO_2. In fact, to calculate the amount of oxygen needed to react with a given amount of carbon, we need only multiply the amount of carbon by the factor 32.0:12.0. The following examples illustrate these relationships.

> **CONCEPTUAL EXAMPLE 5.6**　Molecular, Molar, and Mass Relationships
>
> Nitrogen monoxide (nitric oxide), an air pollutant discharged by internal combustion engines, combines with oxygen to form nitrogen dioxide, a yellowish-brown gas that irritates the respiratory system and eyes. The equation for this reaction is
>
> $$2\,NO + O_2 \longrightarrow 2\,NO_2$$
>
> State the molecular, molar, and mass relationships indicated by the equation.

Solution

The molecular and molar relationships can be obtained directly from the equation; no calculation is necessary. The mass relationship requires a little calculation:

Molecular: Two molecules of NO react with one molecule of O_2 to form two molecules of NO_2.

Molar: 2 mol of NO reacts with 1 mol of O_2 to form 2 mol of NO_2.

Mass: 60.0 g of NO (2 mol NO × 30.0 g/mol) reacts with 32.0 g (1 mol O_2 × 32.0 g/mol) of O_2 to form 92.0 g (2 mol NO_2 × 46.0 g/mol) of NO_2.

■ EXERCISE 5.6

Hydrogen sulfide, a gas that smells like rotten eggs, burns in air to produce sulfur dioxide and water according to the equation

$$2\,H_2S + 3\,O_2 \longrightarrow 2\,SO_2 + 2\,H_2O$$

State the molecular, molar, and mass relationships indicated by this equation.

Molar Relationships in Chemical Equations

The quantitative relationship between reactants and products in a chemical reaction is called **stoichiometry**. The ratio of moles of reactants and products is given by the coefficients in a balanced chemical equation. Consider the combustion of propane, the main component of bottled gas.

$$C_3H_8 + 5\,O_2 \longrightarrow 3\,CO_2 + 4\,H_2O$$

The coefficients in the balanced equation allow us to make statements such as

*   1 mol of C_3H_8 reacts with 5 mol of O_2.
*   3 mol of CO_2 is produced for every 1 mol of C_3H_8 that reacts.
*   4 mol of H_2O is produced for every 3 mol of CO_2 produced.

We can turn these statements into conversion factors known as stoichiometric factors. (Conversion factors are explained in the Appendix.) A **stoichiometric factor** relates the amounts, in *moles*, of any two substances involved in a chemical reaction. Note that we can set up conversion factors for any two compounds involved in a reaction. Typical problems ask you to calculate how much of one compound is equivalent to a given amount of one of the other compounds. All you need to do to solve such a problem is put together a conversion factor relating the two compounds. In the examples that follow, stoichiometric factors are shown in color.

EXAMPLE 5.7   Molar Relationships

When 0.105 mol of propane is burned in a plentiful supply of oxygen, how many moles of oxygen are consumed?

$$C_3H_8 + 5\,O_2 \longrightarrow 3\,CO_2 + 4\,H_2O$$

Solution

The equation tells us that 5 mol O_2 is required to burn 1 mol C_3H_8. We can write

$$1\text{ mol }C_3H_8 \Leftrightarrow 5\text{ mol }O_2$$

where we use the symbol $\Leftrightarrow$ to mean "is stoichiometrically equivalent to." From this relationship we can construct conversion factors to relate moles of oxygen to moles of propane. The possible conversion factors are

$$\frac{1\text{ mol }C_3H_8}{5\text{ mol }O_2} \quad\text{and}\quad \frac{5\text{ mol }O_2}{1\text{ mol }C_3H_8}$$

Which one do we use? Only if we multiply the given quantity (0.105 mol C_3H_8) by the factor on the right will the answer be expressed in the units requested (moles of oxygen).

$$? \text{ mol } O_2 = 0.105 \text{ } \cancel{\text{mol } C_3H_8} \times \frac{5 \text{ mol } O_2}{1 \text{ } \cancel{\text{mol } C_3H_8}} = 0.525 \text{ mol } O_2$$

■ **EXERCISE 5.7**

For the combustion of propane in Example 5.7, **(a)** How many moles of carbon dioxide are formed when 0.529 mol of C_3H_8 is burned? **(b)** How many moles of water are produced when 76.2 mol of C_3H_8 is burned? **(c)** How many moles of carbon dioxide are produced when 1.010 mol of O_2 is consumed?

Mass Relationships in Chemical Equations

A chemical equation defines the stoichiometric relationship in terms of moles, but problems are seldom formulated in moles. Typically, you are given an amount of one substance in grams and asked to calculate how many grams of another substance can be made from it. Such calculations involve several steps.

1. Write a balanced chemical equation for the reaction.
2. Determine the molar masses of the substances involved in the calculation.
3. Write down the given quantity and use the molar mass to convert this quantity to moles.
4. Use the balanced chemical equation to convert moles of the given substance to moles of the desired substance.
5. Use the molar mass to convert moles of the desired substance to grams of the desired substance.

If we know the quantity of any substance in an equation, we can determine the quantities of all the other substances. The conversion process is diagrammed in Figure 5.7. It is best learned from examples and by working out exercises.

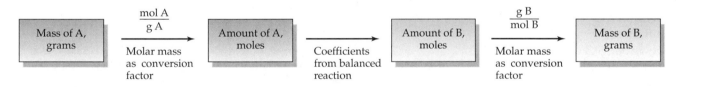

EXAMPLE 5.8 Mass Relationships

Calculate the mass of oxygen needed to react with 10.0 g of carbon in the reaction that forms carbon dioxide.

Solution

Step 1 The balanced equation is

$$C + O_2 \longrightarrow CO_2$$

Step 2 The molar masses are 12.0 g/mol for C and $2 \times 16.0 = 32.0$ g/mol for O_2.

Step 3 We convert the mass of the given substance, carbon, to an amount in moles.

$$? \text{ mol } C = 10.0 \text{ } \cancel{\text{g C}} \times \frac{1 \text{ mol C}}{12.0 \text{ } \cancel{\text{g C}}} = 0.833 \text{ mol C}$$

▲ **Figure 5.7** We can use a chemical equation to relate *moles* of any two substances represented in the equation. These substances may be a reactant and a product, two reactants, or two products. We cannot directly relate the mass of one substance to the mass of another substance. To obtain *mass* relationships, we must convert the mass of each substance to moles, relate moles of one substance to moles of the other through a stoichiometric factor, and then convert moles of the second substance to a mass.

Step 4 We use coefficients from the balanced equation to establish the stoichiometric factor (red) that relates the amount of oxygen to that of carbon.

$$0.833 \; \text{mol C} \times \frac{1 \; \text{mol O}_2}{1 \; \text{mol C}} = 0.833 \; \text{mol O}_2$$

Step 5 We convert from moles of oxygen to grams of oxygen.

$$0.833 \; \text{mol O}_2 \times \frac{32.0 \; \text{g O}_2}{1 \; \text{mol O}_2} = 26.7 \; \text{g O}_2$$

We can also combine the five steps into a single setup. Note that the units in the denominators of the conversion factors are chosen so that each cancels the unit in the numerator of the preceding term.

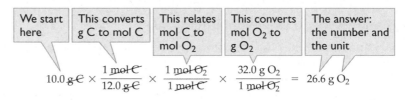

(The answers are slightly different due to rounding in the intermediate steps.)

■ **EXERCISE 5.8A**
Calculate the mass of oxygen (O_2) needed to react with 0.334 g of nitrogen (N_2) in the reaction that forms nitrogen dioxide.

■ **EXERCISE 5.8B**
Calculate the mass of carbon dioxide formed by burning 775 g of each of **(a)** methane (CH_4) and **(b)** butane (C_4H_{10}). (*Hint:* See the equation in Example 5.13.)

More to Explore
Haim, Liliana, Eduardo Cortón, Santiago Kocmur, and Lydia Galagovsky. "Learning Stoichiometry with Hamburger Sandwiches." *Journal of Chemical Education,* September 2003, pp. 1021–1022.

ANSWER

1. How can a car make more pollution than the gasoline it uses?
A mole of carbon (12 g) combines with oxygen to form a mole of carbon dioxide (44 g) as shown in Exercise 5.8B. Gasoline is mostly carbon by weight, so the mass of carbon dioxide formed is greater than the mass of gasoline burned.

Self-Assessment Questions

1. The mass of 4.75 moles of H_2SO_4 is
   a. 314 g   **b.** 456 g   **c.** 466 g   **d.** 564 g

2. A 10.0-g sample of which of the following would have the most moles?
   a. Cu   **b.** Fe   **c.** Si   **d.** S

3. How many moles of CO_2 are there in 1 lb (453.6 g) of CO_2?
   a. 0.0969 mol   **b.** 9.20 mol   **c.** 10.3 mol   **d.** 14.2 mol

4. How many moles of benzene (C_6H_6) are there in a 7.81-g sample of benzene?
   a. 0.100 mol   **b.** 1.00 mol   **c.** 7.81 mol   **d.** 610 mol

Items 5–7 refer to the equation $2 \; H_2 + O_2 \longrightarrow 2 \; H_2O$

5. At the molar level, the equation means that
   a. 2 mol H reacts with 1 mol O to form 2 mol H_2O
   b. 4 mol H reacts with 2 mol O to form 4 mol H_2O
   c. 2 mol H_2 reacts with 1 mol O_2 to form 2 mol H_2O
   d. 2 mol O_2 reacts with 1 mol H_2 to form 2 mol H_2O

6. How many moles of water can be formed from 0.500 mol H_2?
   a. 0.500 mol   **b.** 1.00 mol   **c.** 2.00 mol   **d.** 4.00 mol

7. How many moles of oxygen are required to form 0.222 mol H_2O?
   a. 0.111 mol   **b.** 0.222 mol   **c.** 0.444 mol   **d.** 1.00 mol

Answers: 1, c; 2, c; 3, c; 4, a; 5, c; 6, a; 7, a

GREEN CHEMISTRY

Atom Economy

Margaret Kerr, Worcester State College

Imagine yourself in the future. Your job related to environmental protection requires that you provide information to practicing chemists as they design new processes and reactions. What tools would you use in this job? What topics are important? How can chemists provide new products while protecting the environment?

Waste management is one of your top concerns. Almost every industry faces waste management issues. In the past waste management typically was addressed after the production of desired commodities. A greener approach considers how to minimize waste from the start of the process.

Intrinsic to greener approaches to waste management is the concept of *atom economy*, a calculation of the number of atoms that are conserved in the desired product rather than in waste. In 1998 Barry Trost of Stanford University won a Presidential Green Chemistry Challenge Award for his work in developing this concept. By calculating the number of atoms that will not become part of the desired product and thereby will enter the waste stream, chemists can precisely determine the minimum amount of waste that will be produced by chemicals used in a reaction before even running the reaction.

In this chapter, you have learned how to write and balance chemical equations. You also have learned how to calculate the molar mass of compounds, convert from grams to moles, and determine the amount of product formed from given amounts of reactants. The *theoretical yield* is the maximum amount of desired product that can be produced by the amount of reactants used. Under laboratory conditions, the theoretical yield of a reaction is not usually obtained. The actual amount of desired product that is collected is called the *experimental yield*. Chemists typically refer to their experimental results in *percent yield*, which is simply the experimental yield divided by the theoretical yield, multiplied by 100.

$$\% \text{ yield} = \frac{\text{experimental yield}}{\text{theoretical yield}} \times 100\%$$

A report of a 100% yield for a reaction seems good, but the percent yield considers only the desired product. Any other substance formed—a by-product, for instance—is not counted as part of the percent yield. By-products are often waste and become part of the chemical waste stream. Some reactions produce significantly more by-products than desired products.

Atom economy is a better way to measure reaction efficiency. Percent atom economy (% A.E.) measures the proportion of the atoms in the reaction mixture that actually become part of the desired product. Any reactant atoms that do not appear in the product are considered waste.

The percent atom economy is given by the following relationship:

$$\% \text{ A.E.} = \frac{\text{molar mass of desired product}}{\text{molar mass of all reactants}} \times 100\%$$

A reaction can have a poor atom economy even when the percent yield is near 100%.

Consider the following two ways to make ethyl bromide (CH_3CH_2Br), a compound variously used as a solvent, an anesthetic, a refrigerant, and a fumigant. (This is not the place to worry about details of these reactions; for now just focus on the idea of atom economy.) First, ethyl bromide can be made by reaction of ethylene ($CH_2{=}CH_2$) with hydrogen bromide:

$$CH_2{=}CH_2 + HBr \longrightarrow CH_3CH_2Br$$

This is an *addition reaction* (Chapter 9), in which all of the atoms in both reactants appear in the desired product CH_3CH_2Br. The percent atom economy is 100%.

$$\% \text{ A.E.} = \frac{\text{molar mass } CH_3CH_2Br}{\text{molar mass } CH_2{=}CH_2 + \text{molar mass } HBr} \times 100\%$$

$$= \frac{108.97 \text{ g}}{28.05 \text{ g} + 80.91 \text{ g}} \times 100\% = 100\%$$

All of the atoms are utilized and none go into the waste stream.

Ethyl bromide also can be made by reaction of ethyl alcohol (CH_3CH_2OH) with phosphorus tribromide (PBr_3).

$$3\ CH_3CH_2OH + PBr_3 \longrightarrow 3\ CH_3CH_2Br + H_3PO_3$$

This is a *substitution reaction*, in which the OH group of the alcohol is replaced by a Br atom. In substitution reactions, generally only one product is desired and all other products are not utilized. The wasted atoms in the preceding equation are shown in red.

We can calculate the atom economy for this reaction.

$$\% \text{ A.E.} = \frac{3 \times \text{molar mass } CH_3CH_2Br}{3 \times \text{molar mass } CH_3CH_2OH + \text{molar mass } PBr_3} \times 100\%$$

$$= \frac{3 \times 108.97 \text{ g}}{3 \times 46.07 \text{ g} + 270.69 \text{ g}} \times 100\% = 79.9\%$$

Substitution reactions never have 100% atom economy because there is always a by-product. This results in some atoms going into the waste stream.

Because the concept of atom economy offers a way to predict waste, it provides a basis for developing innovative design processes that eliminate or minimize waste. By focusing on waste from the beginning of the reaction, we reduce or eliminate cleanup at the end.

5.5 Solutions

Many chemical reactions take place in solutions. A **solution** is a homogeneous mixture of two or more substances. The substance being dissolved is the *solute*, and the substance doing the dissolving is the *solvent*. The solute is usually the component present in the lesser quantity, and the solvent is usually present in the greater quantity. There are many solvents: Hexane dissolves grease. Ethanol dissolves many drugs. Isopentyl acetate, a component of banana oil, is a solvent for the glue used in making model airplanes. Water is no doubt the most familiar solvent, dissolving as it does many common substances such as sugar, salt, and ethanol. We focus our discussion here on **aqueous solutions**, those in which water is the solvent, and we take a more quantitative look at the relationship between solute and solvent.

Solution Concentrations

We say that some substances, such as sugar and salt, are *soluble* in water. Of course, there is a limit to the quantity of sugar or salt we can dissolve in a given volume of water, but we still find it convenient to say that they are soluble in water because an appreciable quantity dissolves. Other substances, such as an iron nail or sand (silicon dioxide), we consider to be *insoluble* because the limit of solubility is near zero. Such terms as *soluble* and *insoluble* are useful, but they are imprecise and must be used with care. Two other roughly estimated but sometimes useful terms are *dilute* and *concentrated*. A **dilute solution** is one that contains a little bit of solute in lots of solvent. For example, a pinch of sugar in a cup of tea is a dilute, faintly sweet, sugar solution. A **concentrated solution** is one with much solute dissolved in a relatively small quantity of solvent. Pancake syrup, with lots of solute in a relatively small amount of water is a concentrated, very sweet solution. The lightly sweetened tea is quite "thin"—little changed in appearance from that of pure water. The concentrated pancake syrup is much thicker and pours slowly.

Scientific work generally requires more precise measurement of quantities than "a pinch of sugar in a cup of water." Further, quantitative work often requires the *mole* unit because substances enter into chemical reactions according to *molar* ratios.

Molarity

A concentration unit that chemists often use is *molarity*. For reactions involving solutions, the amount of solute is usually measured in moles and the quantity of solution in liters or milliliters. The **molarity (M)** is the amount of solute, in moles, per liter of solution.

$$\text{Molarity (M)} = \frac{\text{moles of solute}}{\text{liters of solution}}$$

EXAMPLE 5.9   Solution Concentration: Molarity and Moles

Calculate the molarity of a solution made by dissolving 3.50 mol of NaCl in enough water to produce 2.00 L of solution.

Solution

$$\text{Molarity (M)} = \frac{\text{moles of solute}}{\text{liters of solution}} = \frac{3.50 \text{ mol NaCl}}{2.00 \text{ L solution}} = 1.75 \text{ M NaCl}$$

We read 1.75 M NaCl as "1.75 molar NaCl."

■ **EXERCISE 5.9A**
Calculate the molarity of a solution that has 0.0500 mol of NH_3 in 5.75 L of solution.

■ **EXERCISE 5.9B**
Calculate the molarity of a solution made by dissolving 0.750 mol of H_3PO_4 in enough water to produce 775 mL of solution.

Usually when preparing a solution, we must *weigh* the solute. Balances do not display units of moles, so we usually work with a given mass and divide by the molar mass of the substance, as illustrated in Example 5.10.

EXAMPLE 5.10   Solution Concentration: Molarity and Mass

What is the molarity of a solution in which 333 g of potassium hydrogen carbonate is dissolved in enough water to make 10.0 L of solution?

Solution
First, we must convert grams of $KHCO_3$ to moles of $KHCO_3$.

$$333 \text{ g KHCO}_3 \times \frac{1 \text{ mol KHCO}_3}{100.1 \text{ g KHCO}_3} = 3.33 \text{ mol KHCO}_3$$

Now use this value as the numerator in the defining equation for molarity. The solution volume, 10.0 L, is the denominator.

$$\text{Molarity} = \frac{3.33 \text{ mol KHCO}_3}{10.0 \text{ L solution}} = 0.333 \text{ M KHCO}_3$$

■ **EXERCISE 5.10A**
Calculate the molarity of **(a)** 18.0 g of H_2SO_4 in 2.00 L of solution and **(b)** 3.00 g of KI in 2.39 L of solution.

■ **EXERCISE 5.10B**
Calculate the molarity of 0.206 g of HF in 752 mL of solution. (HF is used for etching glass.)

Frequently we need to know the *mass* of solute required to prepare a given volume of solution of a particular molarity. In such calculations we can use molarity as a conversion factor between moles of solute and liters of solution. Thus, in Example 5.11, the expression "0.15 M NaCl" means 0.15 mol of NaCl per liter of solution, expressed as the conversion factor

$$\frac{0.15 \text{ mol NaCl}}{1 \text{ L solution}}$$

EXAMPLE 5.11   Solution Preparation: Molarity

What mass in grams of NaCl is required to prepare 0.500 L of typical over-the-counter saline solution (about 0.15 M NaCl)?

Solution
First we use the molarity as a conversion factor to calculate moles of NaCl.

$$0.500 \text{ L solution} \times \frac{0.15 \text{ mol NaCl}}{1 \text{ L solution}} = 0.075 \text{ mol NaCl}$$

Then we use the molar mass to calculate the grams of NaCl.

$$0.75 \text{ mol NaCl} \times \frac{58.4 \text{ g NaCl}}{1 \text{ mol NaCl}} = 44 \text{ g NaCl}$$

■ **EXERCISE 5.11**
What mass in grams of potassium hydroxide is required to prepare **(a)** 2.00 L of 6.00 M KOH and **(b)** 100.0 mL of 1.00 M KOH?

Quite often, solutions of known molarity are available. For example, commercial concentrated hydrochloric acid is 12 M. How would you calculate the *volume* needed to get a certain number of moles of solute? We can again rearrange the definition of molarity to obtain

$$\text{Liters of solution} = \frac{\text{moles of solute}}{\text{molarity}}$$

More to Explore
Gorin, George. "Mole, Mole per Liter, and Molar: A Primer on SI and Related Units for Chemistry Students." *Journal of Chemical Education,* January 2003, pp. 103–104.

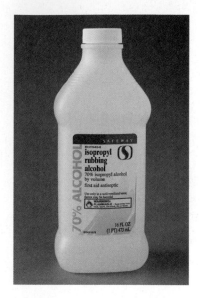

▲ Rubbing alcohol is a water–isopropyl alcohol solution that is 70% isopropyl alcohol [$(CH_3)_2CHOH$] by volume.

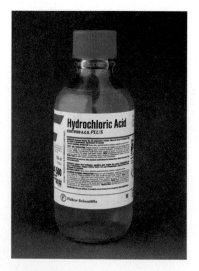

▲ "Reagent grade" concentrated hydrochloric acid is a 38% by mass solution of HCl in water.

EXAMPLE 5.12 Moles from Molarity and Volume

Concentrated hydrochloric acid has a concentration of 12.0 M HCl. What volume in milliliters of this solution would one need to get 0.425 mol of HCl?

Solution

$$\text{Liters of HCl solution} = \frac{\text{moles of solute}}{\text{molarity}} = \frac{0.425 \text{ mol HCl}}{12.0 \text{ M HCl}}$$

$$= \frac{0.425 \text{ mol HCl}}{12.0 \text{ mol HCl}/\text{L}} = 0.0354 \text{ L}$$

We would need 0.0354 L (35.4 mL) of the solution to have 0.425 mol. Remember that molarity is moles per liter of *solution,* not per liter of solvent.

■ **EXERCISE 5.12A**

What volume in milliliters of 15.0 M aqueous ammonia (NH_3) solution do you need to obtain 0.445 mol of NH_3?

■ **EXERCISE 5.12B**

What mass in grams of HNO_3 is in 500 mL of acid rain that has a concentration of 2.00×10^{-5} M HNO_3?

Percent Concentrations

For many practical applications, we often express solution concentrations in percentage composition. Then, if we require a precise quantity of solution, we simply measure out a mass or volume. There are different ways to express percentage, depending on whether mass or volume is being measured. If both the solute and solvent are liquids, **percent by volume** is often used because liquid volumes are so easily measured.

$$\text{Percent by volume} = \frac{\text{volume of solute}}{\text{volume of solution}} \times 100\%$$

In calculating percent by volume, the volume of solute and of solution may be measured in any unit of volume as long as the same unit is used for both solute and solution. For example, the volume could be measured in milliliters or in gallons. Ethanol (CH_3CH_2OH) for medicinal purposes—as per USP (an abbreviation of *United States Pharmacopeia,* the official publication of standards for pharmaceutical products)—is 95% by volume. That is, it consists of 95 mL CH_3CH_2OH per 100 mL of aqueous solution.

EXAMPLE 5.13 Percent by Volume

Two-stroke engines use a mixture of 120 mL of oil dissolved in enough gasoline to make 4.0 liters of fuel. What is the percent by volume of oil in this mixture?

Solution

$$\text{Percent by volume} = \frac{120 \text{ mL oil}}{4000 \text{ mL solution}} \times 100\% = 3.0\%$$

■ **EXERCISE 5.13A**

What is the volume percent of ethanol in a solution that has 58.0 mL water in 625 mL of an ethanol–water solution?

■ **EXERCISE 5.13B**

Assume that the volumes are additive and determine the volume percent toluene ($C_6H_5CH_3$) in a solution made by mixing 40.0 mL of toluene with 75.0 mL of benzene (C_6H_6).

Many commercial solutions are labeled with the concentration in **percent by mass**. For example, sulfuric acid is sold as a solution that is 35.7% H_2SO_4 for use in storage batteries, 77.7% H_2SO_4 for the manufacture of phosphate fertilizers, and 93.2% H_2SO_4 for pickling steel. Each of these figures is a percent by mass: 35.7 g of H_2SO_4 per 100 g of sulfuric acid solution, and so on.

$$\text{Percent by mass} = \frac{\text{mass of solute}}{\text{mass of solution}} \times 100\%$$

As with percent by volume, the mass of solute and of solution may be measured in any unit of mass, as long as both use the same unit.

EXAMPLE 5.14   Percent by Mass

What is the percent by mass of a solution of 25.5 g of NaCl dissolved in 425 g (425 mL) of water?

Solution

Use these values in the above percent-by-mass equation:

$$\text{Percent by mass} = \frac{25.5 \text{ g NaCl}}{(25.5 + 425) \text{ g solution}} \times 100\% = 5.66\% \text{ NaCl}$$

■ **EXERCISE 5.14A**

Hydrogen peroxide solutions for home use are 3.0% by mass solutions of H_2O_2 in water. What is the percent by mass of a solution of 9.40 g of H_2O_2 dissolved in 335 g (335 mL) of water?

■ **EXERCISE 5.14B**

Sodium hydroxide (NaOH, lye) is used to make soap and is very soluble in water. What is the percent by mass of a solution that contains 1.00 kg of NaOH dissolved in 950 mL of water?

EXAMPLE 5.15   Solution Preparation: Percent by Mass

Describe how to make 430 g of an aqueous solution that is 4.85% by mass $NaNO_3$.

Solution

We begin by rearranging the equation for percent by mass to solve for mass of solute.

$$\text{Mass of solute} = \frac{\text{percent by mass} \times \text{mass of solution}}{100\%}$$

Substituting, we have

$$= \frac{4.85\% \times 430 \text{ g}}{100\%} = 20.9 \text{ g}$$

Weigh 20.9 g of $NaNO_3$ and add enough water to make 430 g of solution.

■ **EXERCISE 5.15A**

Describe how you would prepare 125 g of an aqueous solution that is 4.50% glucose by mass.

■ **EXERCISE 5.15B**

Describe how you would prepare 1750 g of *isotonic saline*, a commonly used intravenous (IV) solution that is 0.89% sodium chloride by mass.

▲ **It DOES Matter!**
An "isotonic" or "normal" intravenous solution must have the proper concentration of solute to avoid damage to blood cells.

4. What does a cholesterol level of 200 mean?
Concentrations in medicine are often too low to be expressed conveniently in percentages. A common unit is "milligrams (of solute) per deciliter (of blood or other fluid)." A cholesterol level of 200 means 200 mg cholesterol per dL (100 mL) of blood.

Note that for percent concentrations, the mass or volume of the solute needed doesn't depend on what the solute is. A 10% by mass solution of NaOH contains 10 g of NaOH per 100 g of solution. Similarly, 10% HCl and 10% $(NH_4)_2SO_4$ and 10% $C_{110}H_{190}N_3O_2Br$ each contain 10 g of the specified solute per 100 g of solution. For *molar* solutions, however, the mass of solute in a solution of specified molarity is different for different solutes. A liter of a 0.10 M solution requires 4.0 g (0.10 mol) of NaOH, 3.7 g (0.10 mol) of HCl, 13.2 g (0.10 mol) of $(NH_4)_2SO_4$, or 166 g (0.10 mol) of $C_{110}H_{190}N_3O_2Br$.

Self-Assessment Questions

1. In a solution of 85.0 g of water and 5.0 g of sucrose ($C_{12}H_{22}O_{11}$), the
   a. solute is sucrose and the solution is the solvent
   b. solute is sucrose and the solvent is water
   c. solute is water and sucrose is the solvent
   d. water and sucrose are both solutes

2. What is the molarity of a sodium hydroxide solution having 0.500 mol NaOH in 250 mL of solution?
   a. 0.00200 M   b. 0.200 M
   c. 2.00 M   d. 8.00 M

3. How many moles of sugar are required to make 4.00 L of a saturated solution of sugar having a concentration of 0.600 mol/L?
   a. 1.20 mol   b. 2.40 mol
   c. 24.0 mol   d. 6.67 mol

4. What mass of NaCl in grams is needed to prepare 2.500 L of a 0.800 M NaCl solution?
   a. 2.00 g   b. 4.00 g
   c. 58.5 g   d. 117 g

5. What volume of 6.00 M HCl solution is needed to provide 1.50 mol of HCl?
   a. 250 mL
   b. 300 mL
   c. 750 mL
   d. 900 mL
   e. 9.00 L

6. What is the percent by volume of ethanol in a solution made by adding 50.0 mL of pure ethanol to enough water to make 250 mL of solution?
   a. 16.7% ethanol
   b. 20.0% ethanol
   c. 25.0% ethanol
   d. 80.0% ethanol

7. A solution contains 15.0 g sucrose and 60.0 g of water. What is the percent by mass of the sucrose solution?
   a. 15.0% sucrose
   b. 20.0% sucrose
   c. 25.0% sucrose
   d. 80.0% sucrose

8. What volume of water should be added to 23.4 g of NaCl to make 100.0 mL of 4.00 M solution?
   a. 76.6 mL
   b. 96.0 mL
   c. 100.0 mL
   d. enough to make 100.0 mL of solution

Answers: 1, b; 2, c; 3, b; 4, d; 5, a; 6, b; 7, b; 8, d

Critical Thinking Exercises

Apply knowledge that you have gained in this chapter and one or more of the FLaReS principles (Chapter 1) to evaluate the following statements or claims.

5.1 Suppose that someone has published a paper claiming a new value for Avogadro's number. The author says that he has made some very careful laboratory measurements and his calculations indicate that the true value for the Avogadro constant is 3.01875×10^{23}. Is this claim credible in your opinion? What questions would you ask the person about his claim?

5.2 A chemistry teacher asked his students, "What is the mass, in grams, of a mole of bromine?" One student said "80"; another said "160"; and several others gave answers of 79, 81, 158, and 162. The teacher stated that all of these answers were correct. Do you believe his statement?

5.3 A website on fireworks provides directions for preparing potassium nitrate, KNO_3 (molar mass 101 g/mol), using potassium carbonate (138 g/mol) and ammonium nitrate (80 g/mol), according to the following equation:

$$K_2CO_3 + 2\,NH_4NO_3 \longrightarrow 2\,KNO_3 + CO_2 + H_2O + NH_3$$

The directions claim that one should "mix one kilogram of potassium carbonate with two kilograms of ammonium nitrate. The carbon dioxide and ammonia come off as gases, and the water can be evaporated, leaving two kilograms of pure potassium nitrate." Use information from this chapter to evaluate this claim.

5.4 A battery manufacturer claims that lithium batteries deliver the same power as batteries using nickel, zinc, or lead, but the lithium batteries are much lighter because lithium has a lower atomic mass than does nickel, zinc, or lead. Evaluate this claim.

■ SUMMARY

Section 5.1—A **chemical equation** is shorthand for a chemical change, using symbols and formulas instead of words: $C + O_2 \longrightarrow CO_2$. Materials to the left of the arrow are starting materials or **reactants**, while those to the right of the arrow are the **products** or what we end with. Physical states may be indicated with (s), (l), or (g). Because matter is conserved, an equation must be balanced; it must have the same number and type of each atom on each side. Equations are balanced by placing coefficients in front of each reactant or product; we do *not* balance an equation by changing the formulas of reactants or products. The balanced equation $2\,NaN_3 \longrightarrow 2\,Na + 3\,N_2$ means that two formula units of NaN_3 react to give two atoms of Na and three molecules of N_2.

Section 5.2—Gay-Lussac's experiments led him to the **law of combining volumes**: At a given temperature and pressure, the volumes of gaseous reactants and products is in a small whole-number ratio (1:1, 2:1, 4:3, etc.). **Avogadro's hypothesis** explained this law by stating that equal volumes of all gases contain the same number of molecules at fixed temperature and pressure. **Avogadro's number** is defined as the number of ^{12}C atoms in exactly 12 grams of ^{12}C; that is, 6.02×10^{23} atoms.

Section 5.3—A **mole (mol)** is the amount of a substance that contains Avogadro's number of elementary units—atoms, molecules, or formula units, depending on the substance. The **formula mass** is the average mass of a formula unit of a substance, relative to that of a ^{12}C atom. If the formula represents a molecule, the term *molecular mass* often is used.

Section 5.4—The **molar mass** is the mass of one mole of a substance. Molar mass is the same number as formula mass, molecular mass, or atomic mass, but is expressed in units of grams per mole (g/mol). Molar mass is used to convert from grams to moles of a substance and vice versa. A balanced equation can be read in terms of atoms and molecules, or in terms of moles. **Stoichiometry** or mass relationships in chemical reactions can be found by using **stoichiometric factors**, which relate moles of one substance to moles of another in a chemical reaction. Stoichiometric factors can be obtained without calculation, directly from the balanced equation. To evaluate mass relationships, molar masses and stoichiometric factors are both used as conversion factors.

Section 5.5—A **solution** is a homogeneous mixture, consisting of a solute dissolved in a solvent. A substance is soluble in a solvent if some appreciable quantity of the substance dissolves in the solvent. Any substance that does not dissolve significantly in a solvent is insoluble in that solvent. An **aqueous solution** (aq) has water as the solvent. Concentration of a solution can be expressed in many ways. A **concentrated solution** has a relatively large amount of solute compared to solvent, and a **dilute solution** has little solute compared to the solvent. One quantitative unit of concentration is **molarity (M)**, moles of solute dissolved per liter of solution.

$$\text{Molarity (M)} = \frac{\text{moles of solute}}{\text{liters of solution}}$$

There are several types of percent concentration units. Two important ones are **percent by volume** and **percent by mass**.

$$\text{Percent by volume} = \frac{\text{volume of solute}}{\text{volume of solution}} \times 100\%$$

$$\text{Percent by mass} = \frac{\text{mass of solute}}{\text{mass of solution}} \times 100\%$$

For each concentration unit, if we know two of the three terms in the equation, we can solve for the third.

■ REVIEW QUESTIONS

1. Define or illustrate each of the following.
   a. formula unit   b. formula mass
   c. mole   d. Avogadro's number
   e. molar mass   f. molar volume

2. Explain the difference between the atomic mass of oxygen and the formula mass of oxygen (gas).

3. What is Avogadro's hypothesis? How does it explain Gay-Lussac's law of combining volumes?

4. Consider the law of conservation of mass and explain why we must work with balanced chemical equations.

5. Define or explain and illustrate the following terms.
   a. solution   b. solvent
   c. solute   d. aqueous solution

6. Define or explain and illustrate the following terms.
   a. concentrated solution   b. dilute solution
   c. soluble   d. insoluble

■ PROBLEMS

Interpreting Formula Units

7. How many oxygen atoms are indicated in one formula unit of each of the following?
   a. $Ca(H_2PO_4)_2$ b. $C_6H_5COOCH_3$
   c. $(BiO)_2SO_4$

8. How many carbon atoms are indicated in one formula unit of each of the following?
   a. $Al_2(CO_3)_3$ b. $Cu(CH_3COO)_2$
   c. $CH_3CH_2CH(CH_3)_2$

9. How many atoms of each element (N, S, H, and O) does the notation $3\,NH_4HSO_4$ indicate?

10. How many atoms of each element (Fe, C, H, and O) does the notation $4\,Fe(HOOCCOO)_3$ indicate?

Interpreting Chemical Equations

11. Consider the following equation. (a) Explain its meaning at the molecular level. (b) Interpret it in terms of moles. (c) State the mass relationships conveyed by the equation.

$$2\,Na_2O_2 + 2\,H_2O \longrightarrow 4\,NaOH + O_2$$

12. Describe each chemical equation in terms of moles.
   a. $2\,Ca + O_2 \longrightarrow 2\,CaO$
   b. $CH_4 + 2\,O_2 \longrightarrow 2\,CO_2 + 2\,H_2O$

Balancing Chemical Equations

13. Balance the following equations.
   a. $K + O_2 \longrightarrow K_2O_2$
   b. $FeCl_2 + Na_2SiO_3 \longrightarrow NaCl + FeSiO_3$
   c. $F_2 + AlCl_3 \longrightarrow AlF_3 + Cl_2$

14. Balance the following equations.
   a. $Mg + O_2 \longrightarrow MgO$
   b. $C_3H_8 + O_2 \longrightarrow CO_2 + H_2O$
   c. $H_2 + Ta_2O_3 \longrightarrow Ta + H_2O$

15. Write balanced equations for the following processes.
   a. Iron metal reacts with oxygen gas to form rust [iron(III) oxide].
   b. Calcium carbonate (the active ingredient in many antacids) reacts with stomach acid (HCl) to form calcium chloride, water, and carbon dioxide.
   c. Heptane (C_7H_{16}) burns in oxygen gas to form carbon dioxide and water.

16. Write balanced equations for the following processes.
   a. Nitrogen gas and oxygen gas react to form nitrogen oxide, NO.
   b. Ozone (O_3) decomposes into oxygen gas.
   c. Xenon hexafluoride reacts with water to form xenon trioxide and hydrogen fluoride.

Volume Relationships in Chemical Equations

17. Look again at Figure 5.3. Make a similar sketch to show that when hydrogen reacts with oxygen to form steam at 100 °C, two volumes of hydrogen unite with one volume of oxygen to yield two volumes of steam.

18. Look again at Figure 5.4. Make a similar sketch to show that when hydrogen reacts with oxygen to form steam at 100 °C, each volume of gas—hydrogen, oxygen, or steam—contains the same number of molecules.

19. Consider the following equation.

$$2\,C_4H_{10}(g) + 13\,O_2(g) \longrightarrow 8\,CO_2(g) + 10\,H_2O(g)$$

   a. What volume in liters of $CO_2(g)$ is formed when 20.6 L of $C_4H_{10}(g)$ is burned? Assume both gases are measured under the same conditions.
   b. What volume in milliliters of $C_4H_{10}(g)$ will be burned to consume 29.0 mL of $O_2(g)$? Assume both gases are measured under the same conditions.

20. Consider the following equation.

$$4\,NH_3(g) + 3\,O_2(g) \longrightarrow 2\,N_2(g) + 6\,H_2O(g)$$

   a. What volume in liters of $N_2(g)$ is formed when 125 L of $NH_3(g)$ is burned? Assume both gases are measured under the same conditions.
   b. What volume in liters of $O_2(g)$ is required to form 36 L of $H_2O(g)$? Assume both gases are measured under the same conditions.

21. Given the equation in Problem 19, the volume of $CO_2(g)$ formed will be how many times the volume of $C_4H_{10}(g)$ consumed, assuming both are measured under the same conditions?

22. Given the equation in Problem 20, the volume of $H_2O(g)$ formed will be how many times the volume of $NH_3(g)$ consumed, assuming both are measured under the same conditions?

Avogadro's Number

23. How many **(a)** phosphorus *molecules* and how many **(b)** phosphorus *atoms* are there in 1.00 mol of P_4?

24. How many **(a)** aluminum ions and how many **(b)** chloride ions are there in 1.00 mol of $AlCl_3$?

25. Choose one of the following to complete the phrase accurately: One mole of fluorine (F_2) gas
    a. has a mass of 19.0 g.
    b. contains 6.02×10^{23} F atoms.
    c. contains 12.04×10^{23} F atoms.
    d. has a mass of 6.02×10^{23} g.

26. For calcium nitrate, **(a)** how many calcium ions and how many nitrate ions are there in 1.00 mol $Ca(NO_3)_2$? **(b)** How many nitrogen atoms and how many oxygen atoms are there in 1.00 mol $Ca(NO_3)_2$?

Formula Masses and Molar Masses

You may round all atomic masses to one decimal place.

27. Calculate the molar mass of each of the following compounds.
    a. Bi_2O_3 b. $FeSO_4$
    c. $Ca(CH_3COO)_2$ d. $(NH_4)_2Cr_2O_7$

28. Calculate the molar mass of each of the following compounds.
    a. $AgNO_3$ b. $Mg(ClO)_2$
    c. $Zn(IO_4)_2$ d. $CH_3(CH_2)_3COF$

29. Calculate the mass, in grams, of each of the following.
    a. 7.57 mol $CaSO_4$ b. 0.0236 mol $CuCl_2$
    c. 2.50 mol $C_{12}H_{22}O_{11}$

30. Calculate the mass, in grams, of each of the following.
    a. 4.61 mol $AlCl_3$ b. 0.615 mol Cr_2O_3
    c. 0.158 mol IF_5

31. Calculate the amount, in moles, of each of the following.
    a. 6.63 g Sb_2S_3 b. 19.1 g MoO_3
    c. 434 g $AlPO_4$ d. 11.8 g $Be(NO_3)_2$

32. Calculate the amount, in moles, of each of the following.
    a. 16.3 g SF_6 b. 25.4 g $Pb(C_2H_3O_2)_2$
    c. 15.6 g $CoCl_3$ d. 25.3 g $(NH_4)_2C_2O_4$

Mole and Mass Relationships in Chemical Equations

33. Consider the reaction for the combustion of ethane, a minor component of natural gas.

    $$2\,C_2H_6 + 7\,O_2 \longrightarrow 4\,CO_2 + 6\,H_2O$$

    a. How many moles of CO_2 are produced when 8.12 mol of ethane is burned?
    b. How many moles of oxygen are required to burn 3.13 mol of ethane?

34. Consider the reaction for the combustion of octane.

    $$2\,C_8H_{18} + 25\,O_2 \longrightarrow 16\,CO_2 + 18\,H_2O$$

    a. How many moles of H_2O are produced when 2.81 mol of propane is burned?
    b. How many moles of CO_2 are produced when 4.06 mol of oxygen is consumed?

35. What mass in grams of **(a)** ammonia can be made from 440 g of H_2, and **(b)** hydrogen is needed to react completely with 892 g of N_2?

    $$N_2 + H_2 \longrightarrow NH_3 \quad \text{(not balanced)}$$

36. Toluene (C_7H_8) and nitric acid (HNO_3) are used in the production of trinitrotoluene (TNT, $C_7H_5N_3O_6$), an explosive.

    $$C_7H_8 + HNO_3 \longrightarrow C_7H_5N_3O_6 + H_2O \quad \text{(not balanced)}$$

    What mass in grams of **(a)** nitric acid is required to react with 454 g of C_7H_8 and **(b)** TNT can be made from 829 g of C_7H_8?

37. Phosphine (a toxic gas used as a fumigant to protect stored grain) is generated by the action of water on magnesium phosphide. The equation is

    $$Mg_3P_2 + H_2O \longrightarrow PH_3 + Mg(OH)_2 \quad \text{(not balanced)}$$

    What mass in grams of magnesium phosphide is needed to produce 134 g of PH_3?

38. In an oxyacetylene welding torch, acetylene (C_2H_2) burns in pure oxygen with a very hot flame.

    $$C_2H_2 + O_2 \longrightarrow CO_2 + H_2O \quad \text{(not balanced)}$$

    What mass in grams of oxygen is required to react with 52.0 g of C_2H_2?

Molarity of Solutions

39. Calculate the molarity of each of the following solutions.
    a. 23.4 mol of HCl in 2.50 L of solution
    b. 0.0875 mol of Li_2CO_3 in 316 mL of solution

40. Calculate the molarity of each of the following solutions.
    a. 8.82 mol of H_2SO_4 in 3.75 L of solution
    b. 0.611 mol of C_2H_5OH in 96.3 mL of solution

41. What mass in grams of solute is needed to prepare **(a)** 3.50 L of 0.400 M NaOH and **(b)** 65.0 mL of 2.90 M $C_6H_{12}O_6$?

42. What mass in grams of solute is needed to prepare **(a)** 0.500 L of 0.167 M $K_2Cr_2O_7$ and **(b)** 625 mL of 0.0100 M $KMnO_4$?

43. What volume in liters of **(a)** 6.00 M NaOH contains 1.25 mol of NaOH and **(b)** 0.0250 M KH_2AsO_4 contains 8.10 g of KH_2AsO_4?

44. What volume in liters of **(a)** 2.50 M NaOH contains 1.05 mol of NaOH and **(b)** 4.25 M $H_2C_2O_4$ contains 2.25 g of $H_2C_2O_4$?

Percent Concentrations of Solutions

45. What is the volume percent concentration of **(a)** 35.0 mL of water in 725 mL of an ethanol–water solution, and **(b)** 78.9 mL of acetone in 1550 mL of an acetone–water solution?

46. What is the volume percent concentration of **(a)** 58.0 mL of water in 625 mL of an acetic acid–water solution, and **(b)** 79.1 mL of methanol in 755 mL of a methanol–water solution?

47. Describe how you would prepare 3375 g of an aqueous solution that is 8.2% NaCl by mass.

48. Describe how you would prepare 2.44 kg of an aqueous solution that is 16.3% KOH by mass.

49. Describe how you would prepare 2.00 L of an aqueous solution that is 2.00% acetic acid by volume.

50. Describe how you would prepare 500.0 mL of an aqueous solution that is 30.0% isopropyl alcohol by volume.

■ ADDITIONAL PROBLEMS

51. Which of the following correctly represents the decomposition of potassium chlorate to produce potassium chloride and oxygen gas?

a. $KClO_3(s) \longrightarrow KClO_3(s) + O_2(g) + O(g)$

b. $2\,KClO_3(s) \longrightarrow 2\,KCl(s) + 3\,O_2(g)$

c. $KClO_3(s) \longrightarrow KClO(s) + O_2(g)$

d. $KClO_3(s) \longrightarrow KCl(s) + O_3(g)$

52. Both magnesium and aluminum react with an acidic solution to produce hydrogen. Why is it that only one of the following equations correctly describes the reaction?

$$Mg(s) + 2\,H^+(aq) \longrightarrow Mg^{2+}(aq) + H_2(g)$$

$$Al(s) + 2\,H^+(aq) \longrightarrow Al^{3+}(aq) + H_2(g)$$

53. Write a balanced chemical equation to represent **(a)** the decomposition, by heating, of solid mercury(II) nitrate to produce pure liquid mercury, nitrogen dioxide gas, and oxygen gas, and **(b)** the reaction of aqueous sodium carbonate with aqueous hydrochloric acid (hydrogen chloride) to produce water, carbon dioxide gas, and aqueous sodium chloride.

54. Joseph Priestley discovered oxygen in 1774 by heating "red calx of mercury" [mercury(II) oxide]. The calx decomposed to its elements. The equation is

$$HgO \longrightarrow Hg + O_2 \quad \text{(not balanced)}$$

What mass of oxygen is produced by the decomposition of 10.8 g of HgO?

55. Compare 0.50 mol H_2, 1.0 mol He, and 0.25 mol C_2H_2.
(a) Do the three gases have the same number of atoms?
(b) Which gas has the greatest mass?

56. What mass in grams of the magnetic oxide of iron (Fe_3O_4) can be made from 12.0 grams of pure iron and an excess of oxygen? The equation is

$$Fe + O_2 \longrightarrow Fe_3O_4 \quad \text{(not balanced)}$$

57. When heated above about 900 °C, limestone (calcium carbonate) decomposes to quicklime (calcium oxide), which is used to make cement, and carbon dioxide. What mass in grams of quicklime can be produced from 4.72×10^9 g limestone?

58. Ammonia reacts with oxygen to produce nitric acid (HNO_3) and water. What mass of nitric acid, in grams, can be made from 971 g of ammonia?

59. Hydrogen peroxide from the drugstore is 3% H_2O_2 by mass dissolved in water. How many moles of H_2O_2 are in a typical 16–fl. oz. bottle (1 fl. oz. = 29.6 mL; density = 1.00 g/mL)?

60. What is the mass percent of **(a)** NaOH in a solution of 4.12 g NaOH in 100.0 g water, and **(b)** ethanol in a solution of 5.00 mL ethanol (density = 0.789 g/mL) in 50.0 g water?

61. What is the volume percent of **(a)** ethanol in 1.32 L of an ethanol–water solution that contains 669 mL water, and **(b)** acetone in a solution of 10.00 mL acetone in 1.25 L of an acetone–water solution?

62. Laughing gas (dinitrogen monoxide, N_2O, also called nitrous oxide) can be made by heating ammonium nitrate with great care. The other product is water.
a. Write a balanced equation for the process.
b. Draw the Lewis structure for N_2O.
c. What mass in grams of N_2O can be made from 4.00 g of ammonium nitrate?

63. Mining helmets once used calcium carbide (CaC_2) as a source of light. When water is dripped on this solid, calcium hydroxide and acetylene (C_2H_2) are produced. The acetylene burns with a bright flame. What mass of CaC_2 is needed to provide light for 3.0 hours if 9.5 moles of acetylene are used every hour?

64. In what volume of solution must 31.7 g of oxalic acid ($H_2C_2O_4$) be dissolved to make a solution that is 0.0859 M in oxalic acid?

65. In tests for intoxication, blood alcohol levels are expressed as percent by volume. A blood alcohol level of 0.080% by volume means 0.080 mL ethanol per 100 mL of blood and is considered proof of intoxication. If a person's blood volume is 5.0 L, what volume of alcohol in the blood gives a blood alcohol level of 0.165% by volume?

66. Determine the molarity of **(a)** Li^+ and NO_3^- in 0.647 M $LiNO_3$ and **(b)** Ca^{2+} and I^- in 0.035 M CaI_2.

67. In Problem 34 you were asked to determine the mass of CO_2 produced from the combustion of octane. If the reaction were given in the following form, would the answer change? Explain.

$$C_8H_{18} + \frac{25}{2}O_2 \longrightarrow 8\,CO_2 + 9\,H_2O$$

68. Look again at Critical Thinking Exercise 5.3 and then calculate **(a)** the correct mass of ammonium nitrate that would react with 1.00 kg of potassium carbonate, and **(b)** the correct mass of potassium nitrate that would be obtained.

69. A truck carrying about 31,000 kg of sulfuric acid (H_2SO_4) is involved in an accident. What mass of sodium bicarbonate ($NaHCO_3$) will be needed to neutralize the sulfuric acid? The products of the reaction are sodium sulfate ($NaSO_4$), water, and carbon dioxide.

70. Look again at Problem 69 and determine the mass of sodium carbonate (Na_2CO_3) needed to neutralize the acid. (The products of the reaction are the same.)

71. How many moles of H_2O are in 1.00 L of water? ($d = 1.00$ g/mL)

■ COLLABORATIVE GROUP PROJECTS

Prepare a PowerPoint, poster, or other presentation (as directed by your instructor) for presentation to the class.

72. Prepare a brief biographical report on Amedeo Avogadro and share it with your group.

73. In the view of many scientists, hydrogen holds great promise as a source of clean energy. In 2003, the U.S. government pledged $1.2 billion for research into hydrogen-powered cars. Doing your own search, explore the proposals for using hydrogen as a fuel. Write a brief summary of what you find and use the appropriate balanced equation to explain why this is considered a clean, renewable source. More discussion of hydrogen as a fuel is found in Chapter 15.

Over 99% of all matter in the universe is in the gaseous state, in stars and interstellar matter. Planets (and moons) are unusual in that all three states of matter may be present. The upper layers of the planet Saturn shown here are mostly gaseous hydrogen and helium. The core is thought to be composed of rock and ice, topped with a layer of hydrogen in a liquid state.

GASES, LIQUIDS, SOLIDS, AND INTER-MOLECULAR FORCES

6

QUESTIONS YOU MAY HAVE ASKED YOURSELF

1. Silly Putty® can break and it can flow. Is it a liquid or a solid?
2. What is it that keeps oil and water from mixing?
3. Why doesn't "creamy" Italian dressing separate like ordinary dressing?
4. How cold can it get?
5. Why are we supposed to measure tire pressure when tires are "cold"?

We introduced the three states of matter—gas, liquid, and solid—in Chapter 1. The most familiar substance that we experience in all three states is water: as Earth's most common liquid, as a solid (ice), and as a gas (steam).

Many substances—not just water—can exist in the three different states, depending on conditions. An understanding of those conditions is important for our understanding of all matter. Which state a substance exhibits depends in part on the physical properties of that substance, and in particular on the pressure and—especially—the temperature it experiences. When it is very cold, H_2O exists as the solid known as ice. When it is warmer, water is a liquid. At still higher temperatures, water boils and becomes a gas we call steam.

Chapter 4 describes the forces called chemical bonds that bind atoms to one another within molecules. Chemical bonds are *intra*molecular forces that determine such molecular properties as geometries and polarity. Forces that exist *between* molecules—*inter*molecular forces—determine the macroscopic physical properties of liquids and solids. In fact, if there were no intermolecular attractive forces, there would be no liquids or solids—everything would be in a gaseous state. Moreover, intermolecular forces are themselves related to the kinds of chemical bonds found in substances. Figure 6.1 compares intermolecular and intramolecular forces.

The shape, size, and polarity of molecules determine how they interact with each other. The physical state of a material (whether it is a gas, liquid, or solid) depends on the strength of the intermolecular forces that hold the molecules or ions together, relative to the thermal energy (temperature) that acts to separate them.

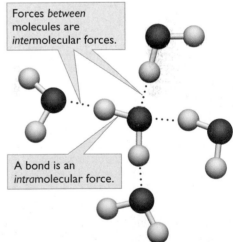

Forces *between* molecules are *inter*molecular forces.

A bond is an *intra*molecular force.

▲ **Figure 6.1** Intramolecular and intermolecular forces compared. The gray "sticks" represent covalent bonds—*intra*molecular forces or pairs of shared electrons. The dotted red lines represent the forces between molecules—*inter*molecular forces.

(a)   Solid

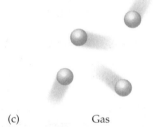

(b)   Liquid

(c)   Gas

▲ **Figure 6.2** Three states of matter. (a) Most solids have molecules (or atoms) that vibrate around fixed positions. (b) In liquids the molecules are still close together but are free to move about like people in a crowded room. (c) Molecules of gases are widely spaced apart and move freely and randomly.

6.1 Solids, Liquids, and Gases

Most solids are highly ordered assemblies of particles—atoms, molecules, or ions—in close contact with one another as suggested by Figure 6.2a. Their motion is primarily vibration. Table salt (sodium chloride) is a typical crystalline solid. In this solid, ionic bonds hold the Na^+ and Cl^- ions in position and maintain the orderly arrangement (recall Figure 4.2).

In liquids, the particles are still in close contact but are more randomly arranged and are freer to move about (Figure 6.2b). This is why liquids are able to flow to conform to the shape of their container. To get a better image of a liquid at the molecular level, think of a box of marbles being shaken continuously. The marbles move back and forth, rolling over one another. The particles of a liquid (like the marbles) are not so rigidly held in place as are particles in a solid. Even so, there must be some force attracting the particles in a liquid to one another because those particles are in contact with one another.

In gases, molecules are separated by relatively great distances and are moving quite rapidly in random directions (Figure 6.2c). In this state the atoms or molecules have essentially no interactions with one another. We shall look more closely at gases in Section 6.5.

Most solids can be changed to liquids; that is, they can be *melted*. The solid is heated, and the heat energy is absorbed by the particles of the solid. The energy causes the particles to vibrate more vigorously until, finally, the forces holding the particles in their regular arrangement are overcome. The solid becomes a liquid. The temperature at which this happens is the **melting point** of the solid. A high melting point is one indication that the forces holding a solid together are very strong.

A liquid can change to a gas or vapor in a process called **vaporization**. Again, one need only supply sufficient heat to achieve this change. Energy is absorbed by the liquid particles, which move faster and faster as a result. Finally, this increasingly violent motion overcomes the attractive forces holding the liquid particles in contact and the particles fly away from one another. The liquid becomes a gas. The temperature at which this happens is the **boiling point** of the liquid.

Removing energy from the sample and slowing down the particles can reverse the entire sequence of changes. Vapor changes to liquid in a process called

A N S W E R

1. Silly Putty® can break and it can flow. Is it a liquid or a solid?
Although it appears to be a solid, Silly Putty® is actually a liquid—but a very thick, slow-flowing (viscous) liquid.

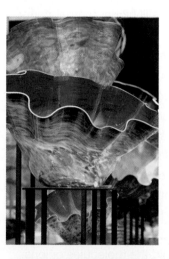

▲ Most solids are made up of highly ordered particles. Glass is an exception; the particles that make up glass (Chapter 12) are randomly arrayed.

▲ Silly Putty

GREEN CHEMISTRY

Supercritical Fluids

Doug Raynie, South Dakota State University

Many chemical reactions take place in solution (Section 5.5), and most chemical processes involve a separation of components at some stage. Water is an excellent solvent for many substances, but some substances do not dissolve in water under ordinary conditions. Chemists have traditionally used toxic solvents such as benzene (C_6H_6), a cancer-causing aromatic hydrocarbon (Section 9.5), and methylene chloride (CH_2Cl_2), a toxic chlorinated hydrocarbon (Section 9.6), to dissolve nonpolar and slightly polar substances. Wouldn't it be nice to have a nontoxic solvent that can be recycled readily?

Let's look again at the states of matter and changes of state (Section 6.1), particularly the change from liquid to gas. Have you ever used a pressure cooker or tried to cook food while backpacking in the mountains? If you have, you've probably noticed that boiling point changes with pressure. With a pressure cooker, the increase in pressure lets us cook at higher temperatures. In the mountains, where the air pressure is lower, water boils at a lower temperature, so it takes longer to prepare a meal. This behavior is shown in the following graph called a phase diagram:

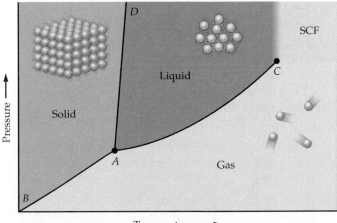

Notice that the line AC that represents the boiling point does not continue forever. Eventually continued heating will take a substance to what is called the *critical point*. Above the critical point, matter exists as neither a gas nor a liquid but kind of a hybrid called a *supercritical fluid*. We can think of supercritical fluids as having properties of both gases and liquids. Most importantly, they can dissolve things as liquids do, but they flow rapidly like gases. We can vary these properties to make the fluid more liquid-like or more gas-like by changing the temperature or pressure.

The most commonly used supercritical fluid is carbon dioxide ($scCO_2$), an inexpensive, convenient-to-use, nonflammable, and nontoxic substance. Carbon dioxide becomes a supercritical fluid at temperatures greater than 31 °C and pressures greater than 73 atm. As a solvent, carbon dioxide is a complement to water. Nonpolar molecules are soluble in $scCO_2$, giving us two environmentally friendly solvents, water and carbon dioxide. Individually or combined, these supercritical fluids can dissolve a wide range of chemical compounds. Because they can flow like gases, supercritical fluids can often make processes occur more quickly.

Supercritical fluids "generally regarded as safe" by the U.S. Food and Drug Administration are often used in the food industry. For example, coffee is sometimes decaffeinated using $scCO_2$ extraction. After the extraction, the $scCO_2$ is decompressed and returned to a gaseous state, leaving the caffeine behind. The gaseous CO_2 is cooled and compressed to return it to a liquid form to be used again. This is a much greener process that the older one using toxic methylene chloride, a liquid that had to be removed by distillation. In the Green Chemistry essay "Green Dry Cleaning" in eChapter 21, you will see that carbon dioxide can be used to replace perchloroethylene (C_2Cl_4), another toxic chlorinated hydrocarbon long used in dry cleaning.

Flavors and fragrances also are often isolated using $scCO_2$. In the cosmetic industry, oils such as palm oil or jojoba oil are isolated from their plant sources using $scCO_2$. Supercritical fluids also are used to clean or degrease precision parts in electronic instruments. In more specialized cases, chemical reactions are carried out using supercritical fluid solvents. The main drawback to the use of supercritical fluids is that special equipment is needed to control the temperatures and high pressures required.

Nearly all chemical processes require the separation of one chemical from a mixture of others. Water and $scCO_2$ give us a pair of inexpensive and safe solvents that dissolve a wide range of chemical compounds—an excellent application of Green Chemistry Principle #5, which urges the use of less toxic solvents and safer reaction conditions.

condensation; liquid changes to solid in a process called **freezing**. Figure 6.3 presents a diagram of the changes in state that occur as energy is added to or removed from a sample. Some substances go directly from the solid state to the gaseous state, a process called **sublimation**. For example, ice cubes left in a freezer for a long time "shrink," and solid mothballs slowly "disappear" in a drawer or closet. In each case the solid undergoes sublimation.

▶ **Figure 6.3** Diagram of changes in state of matter on heating or cooling.

QUESTION: What factors determine the actual temperatures at which the changes of state occur?

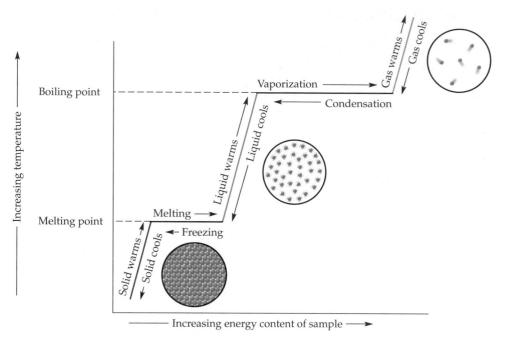

▲ **It DOES Matter!**

The reverse of sublimation is *deposition*, a process that occurs on cold mornings. Water vapor deposits directly onto the grass as solid frost.

Self-Assessment Questions

1. Molecules are farthest apart in a
   a. covalent solid   **b.** gas   **c.** ionic solid   **d.** liquid

2. Liquids are
   a. difficult to compress and have no definite shape
   b. difficult to compress and have no definite volume
   c. easy to compress and have a definite shape
   d. easy to compress and have no definite volume

3. When a gas changes into a liquid, the process is called
   a. condensation   **b.** fusion   **c.** gasification   **d.** solidification

4. When a liquid changes into a solid, the process is called
   a. condensation   **b.** freezing   **c.** fusion   **d.** melting

5. When a solid changes directly into a gas, the process is called
   a. evaporation   **b.** fusion   **c.** sublimation   **d.** vaporization

Answers: 1, b; 2, a; 3, a; 4, b; 5, c

6.2 Comparing Ionic and Molecular Compounds

Because ionic substances such as sodium chloride are not made of molecules, *intermolecular forces* is not an appropriate term. Nevertheless, we need first to consider the forces between ions in order to compare them to the forces between molecules. First, let us consider some typical differences between ionic compounds and molecular substances.

- Nearly all ionic compounds are solids at room temperature. Examples include KBr, $CaCO_3$, and NH_4NO_3. Some molecular compounds are liquids or gases at room temperature, although many are solids. Ethane (CH_3CH_3), hydrogen sulfide (H_2S), and chlorine (Cl_2) are gases; ethanol (CH_3CH_2OH), bromine (Br_2) and phosphorus trichloride (PCl_3) are liquids; and sulfur (S_8), glucose ($C_6H_{12}O_6$), and iodine (I_2) are solids.

- Ionic compounds generally have much higher melting points and boiling points than molecular compounds. *Ionic bonds* are strong forces of attraction between positive and negative ions, and high temperatures are necessary to overcome them. Less energy is required to overcome the weak attractions between molecules (Section 6.3). For example, sodium chloride ($NaCl$) must be heated to 801 °C before it melts, but the molecular compound ethane (CH_3CH_3) melts at −89 °C.

- It generally takes 10 to 100 times the energy to melt one mole of a typical ionic compound than it does to melt one mole of a typical molecular compound. To melt one mole of sodium chloride requires 29.3 kJ, while it takes only 0.583 kJ to melt one mole of ethane.

- Many familiar ionic compounds dissolve in water; those that do form solutions that conduct electricity. The dissociated ions carry the electric current through the solution. Most molecular compounds, even those that dissolve in water, do not dissociate into ions and thus do not form conducting solutions. A solution of sodium chloride conducts electricity; a solution of methanol (CH_3OH) does not.

- Most ionic compounds are crystalline, hard, and often quite brittle. Solid molecular compounds are usually much softer than ionic ones. See again Figure 1.4.

We will now consider some other interactions that hold the particles of solids and liquids together.

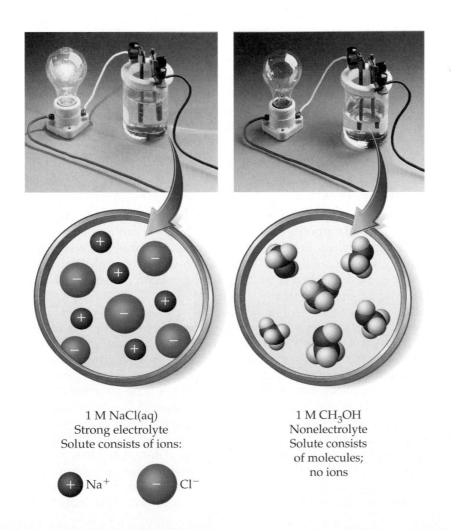

◀ (Left) The bulb lights because the solution of sodium chloride contains ions and conducts electricity. (Right) When a solution of methanol is substituted for the salt solution, the bulb does not light.

1 M NaCl(aq)
Strong electrolyte
Solute consists of ions:

+ Na$^+$ − Cl$^-$

1 M CH$_3$OH
Nonelectrolyte
Solute consists
of molecules;
no ions

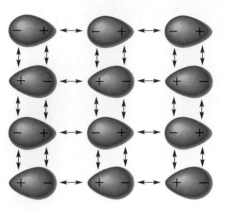

(a)

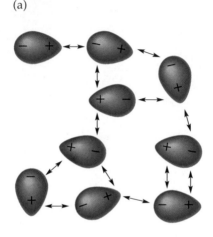

(b)

▲ **Figure 6.4** An idealized representation of dipolar forces in (a) a solid and (b) a liquid. In a real liquid or solid, interactions are more complex.

Dispersion forces are weak for small, nonpolar molecules such as H_2, N_2, and CH_4. For comparable molecular sizes and shapes, intermolecular forces range in strength from dispersion forces (the weakest) to dipolar interactions, and then hydrogen bonding (the strongest).

Self-Assessment Questions

1. Which of the following is not composed of particles that exhibit intermolecular forces?
   a. Br_2 **b.** N_2
   c. NaBr   **d.** NBr_3

2. Which of the following is likely to require the most energy input to melt it?
   a. O_2 **b.** N_2O
   c. KCl   **d.** NH_3

Answers: 1, c; 2, c

6.3 Forces between Molecules

As we have noted, many substances, such as water and carbon dioxide, do not consist of ions but are molecular compounds. Molecular compounds are characterized by co-valent bonds—*intra*molecular forces. For example, in a hydrogen chloride molecule a covalent bond holds the hydrogen and chlorine *atoms* together. But what makes one HCl molecule interact with another in the solid or liquid state? In this section we will examine several types of *inter*molecular forces—forces *between* molecules.

Dipole–Dipole Forces

Recall that the hydrogen chloride molecule is a *dipole*: a molecule with a positive end and a negative end (see Figure 4.6, page 103). When oppositely charged ends of two dipoles are brought close enough together, they attract one another. In solid HCl, the molecules line up so that the positive end of one molecule attracts the negative end of neighboring molecules, as suggested by Figure 6.4a. If we heat the solid sufficient-ly, the orderly arrangement is undone and the solid melts. In the liquid state, the op-positely charged dipoles in the liquid still attract one another, but in a more random fashion, as shown in Figure 6.4b. These **dipole–dipole** (or **dipolar**) **forces** occur be-tween any two polar molecules. Generally, dipolar forces are much weaker than ionic bonds, but they are stronger than the forces between nonpolar molecules of comparable size. The more polar the molecules, the stronger are the dipolar forces.

Dispersion Forces

It is easy enough to understand how ions or polar molecules maintain contact with one another; after all, opposite charges attract. But how can we explain the fact that *nonpolar* compounds can exist in the liquid and solid states? Even hydrogen can exist as a liquid or a solid if the temperature is low enough (its melting point is $-259\ °C$). Some force must be holding these nonpolar molecules close to one an-other in the liquid and solid states.

While depictions such as Figure 6.5a may suggest the electrons in a covalent bond are held in place and are evenly distributed between the two atoms sharing the bond, recall that electrons are *not* stationary but move continuously within their orbitals. Even within a covalent bond, electrons are moving around their respective nuclei. On *average*, the two electrons in a nonpolar bond are between and equidis-tant from the two nuclei, but at any given instant, it is possible for the electrons in one molecule to be at one end of that molecule (Figure 6.5b). The clustering of elec-trons at one end of the molecule creates a temporary negative charge at that pole, which causes electrons in the adjacent molecule to move toward the opposite end of the adjacent molecule as in Figure 6.5c. Thus, at this instant an attractive force is cre-ated between the electron-rich end of one molecule and the electron-poor end of the next. These momentary, usually weak, attractive forces between molecules are called **dispersion forces**. Because the electrons are in constant motion, dispersion

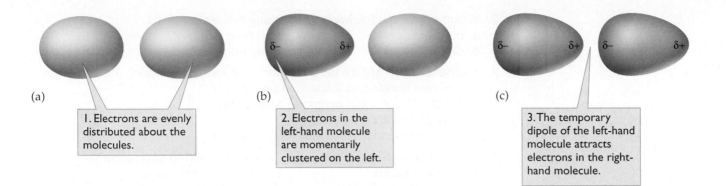

(a)

1. Electrons are evenly distributed about the molecules.

(b)

2. Electrons in the left-hand molecule are momentarily clustered on the left.

(c)

3. The temporary dipole of the left-hand molecule attracts electrons in the right-hand molecule.

forces can be thought of as dipolar forces in which the dipoles are constantly shifting about. To a large extent, dispersion forces determine the physical properties of nonpolar compounds. However, dispersion forces do exist between *any* two particles, whether polar, nonpolar, or ionic.

▲ **Figure 6.5** Mechanism of dispersion forces.

Hydrogen Bonds

Certain polar molecules exhibit stronger attractive forces than would be expected on the basis of ordinary dipolar interactions. These forces are strong enough to be given the special name *hydrogen bonds*. The name is a bit misleading because it emphasizes only one component of the interaction. Not all compounds containing hydrogen exhibit this strong attractive force. In most cases, the hydrogen *must* be attached to a small, very electronegative atom such as fluorine, oxygen, or nitrogen. These atoms permit us to offer an explanation for the great strength of hydrogen bonds as compared with other dipolar forces. Fluorine, oxygen, and nitrogen are all highly electronegative, and they are small (they are at the top right of the periodic table). A hydrogen–fluorine bond, for example, is strongly polar, with a negative fluorine end and a positive hydrogen end. Both hydrogen and fluorine are small atoms, and so the electrons on the fluorine end of one dipole can approach very closely the hydrogen end of a second molecule. The unusually strong interaction between the molecules containing hydrogen bound to small electronegative atoms is called a **hydrogen bond**. Hydrogen bonds are often explicitly represented by *dotted* lines (see Figure 6.6) to emphasize their exceptional strength as compared to other intermolecular forces.

Water has both an unusually high melting point and an unusually high boiling point for a compound with such small molecules. These abnormal values are attributed to water's ability to form hydrogen bonds. Remember, however, that during melting and boiling, only hydrogen bonds *between molecules* are overcome. The covalent bonds between the hydrogen atoms and oxygen atoms in each water molecule remain intact. There is no chemical change because the covalent bonds are many times stronger than hydrogen bonds.

The unique properties of water resulting from hydrogen bonding are discussed in Chapter 14. Hydrogen bonding also plays an important role in biological molecules. The three-dimensional structures of proteins and enzymes and the arrangement of DNA (Chapter 16) are dependent on the presence and strength of hydrogen bonds.

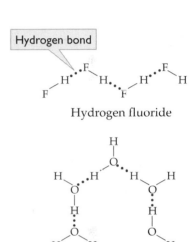

Hydrogen fluoride

Water

▲ **Figure 6.6** Hydrogen bonding in hydrogen fluoride and in water. See also Figure 6.1

Self-Assessment Questions

1. What intermolecular forces are most significant in accounting for the high boiling point of liquid water relative to other substances of similar molecular weight?
   a. dispersion forces
   b. dipolar attractions
   c. hydrogen bonding
   d. ionic bonds

2. Which one of the following interactions is the strongest?
   a. covalent bond   b. dipole force
   c. dispersion force   d. hydrogen bond

3. In ethanol (CH_3CH_2OH) the attractions between neighboring molecules are
   a. covalent bonds and dispersion forces only
   b. covalent bonds and hydrogen bonds only
   c. covalent bonds only
   d. hydrogen bonds and dispersion forces only

4. The molecules of liquid water in a beaker are held to each other by
   a. covalent bonds   b. intramolecular forces
   c. intermolecular forces   d. ionic bonds

5. Which of the compounds, CH_3OH, H_2, HF, and H_2O, can form hydrogen bonds?
   a. H_2 only   b. H_2O only
   c. CH_3OH, HF, and H_2O d. all four

6. Which one of the following molecules does *not* form hydrogen bonds?
   a. ammonia, NH_3 b. chloroform, $CHCl_3$
   c. ethyl alcohol, CH_3CH_2OH d. ethylene glycol, $HOCH_2CH_2OH$

Answers: 1, c; 2, a; 3, d; 4, c; 5, c; 6, b

6.4 Forces in Solution

To complete our look at intermolecular forces, we shall briefly examine the interactions that occur in solutions. A **solution** is an intimate, homogeneous mixture of two or more substances. By *intimate* we mean that mixing occurs down to the level of individual ions and molecules, as suggested in Figure 6.7. For example, a sugar-in-water solution has separate sugar molecules randomly distributed among water molecules. *Homogeneous* means that the mixing is thorough. All parts of the solution have the same distribution of components. A sugar solution is equally sweet at the top, bottom, and middle of the solution. The substance being dissolved, and usually present in a lesser amount (sugar, in a sugar solution), is called the **solute.** The substance doing the dissolving, and usually present in a greater amount, is the **solvent** (water in the sugar–water solution).

Ordinarily, solutions form most readily when the substances involved have *similar* strengths of forces between their molecules (or atoms or ions). An old chemical adage is, "Like dissolves like." Nonpolar solutes dissolve best in nonpolar solvents. For example, oil and gasoline, both nonpolar, mix (Figure 6.8a), but oil and water do not (Figure 6.8b). Ethyl alcohol molecules hydrogen-bond to one another. So do water molecules. Thus, ethyl alcohol readily dissolves in water because the

● = Solvent molecule
● = Solute molecule

▲ **Figure 6.7** In a solution, solute molecules (orange spheres) are randomly distributed among solvent molecules (purple spheres).

▶ **Figure 6.8** (a) Lawn mowers with two-cycle engines are fueled and lubricated with a solution of nonpolar lubricating oil in nonpolar gasoline. (b) In a salad dressing, polar vinegar and nonpolar olive oil are mixed. However, the two liquids do not form a solution and separate on standing. (c) Wine is a solution of polar ethyl alcohol in polar water.

QUESTION: What are some other examples of nonpolar–polar mixtures?

two substances can hydrogen bond with one another (Figure 6.8c). In general, a solute dissolves when attractive forces between it and the solvent overcome the attractive forces operating in the pure solute and in the pure solvent.

Why, then, does salt dissolve in water? Ionic solids are held together by strong ionic bonds. We have already indicated that high temperatures are required to melt ionic solids and break these bonds. Yet if we simply place sodium chloride in water at room temperature, the salt dissolves. And when such a solid dissolves, its bonds *are* broken. The difference between the two processes is the difference between brute force and persuasion. In the melting process, we simply put in enough energy (as heat) to break the crystal apart. In the dissolving process, we offer the ions an attractive alternative to the ionic interactions in the crystal.

Here is how a salt crystal dissolves. Water molecules surround the crystal. The molecules that approach a negative ion align themselves so that the positive ends of their dipoles point toward the ion. (With a positive ion, the process is reversed, and the negative end of the water dipole points toward the ion.) Although the attraction between a dipole and an ion is not as strong as that between two ions, there are several water molecules surrounding each ion, and in this way the many *ion–dipole* interactions overcome the ionic bonds (Figure 6.9).

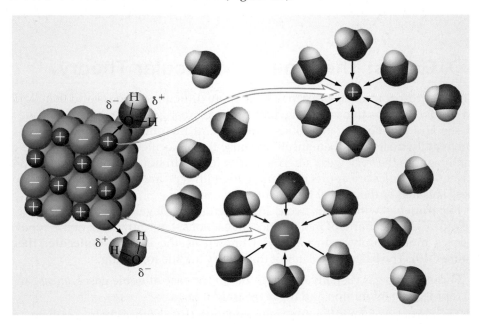

In an ionic solid, the positive and negative ions are strongly bonded together in an orderly crystalline arrangement. In solution, cations and anions move about more or less independently, each surrounded by a cage of solvent molecules. Water—including the water in our bodies—is an excellent solvent for many ionic compounds. Water also dissolves many molecules that are polar covalent like itself. These solubility principles explain how nutrients reach the cells of our bodies (dissolved in blood, which is mostly water) and how many kinds of pollutants get into our water supplies.

A N S W E R

2. What is it that keeps oil and water from mixing?
Oil is nonpolar, whereas water is very polar. The weak forces between oil and water aren't strong enough to separate the water molecules so that the oil can mix with the water.

A N S W E R

3. Why doesn't "creamy" Italian dressing separate like ordinary dressing?
Both kinds of dressings contain oil and vinegar. However, proteins (from egg yolk or other sources) are mixed into the creamy dressing. The proteins have both polar parts and nonpolar parts in their molecules and can keep tiny droplets of oil suspended in the vinegar.

◀ **Figure 6.9** An ionic solid (sodium chloride) dissolving in a polar solvent (water). The hydrogen ends of the water molecules surround the negatively charged chloride ions, while the oxygen ends of the water molecules surround the positively charged sodium ions.

Self-Assessment Questions

1. A solution is a mixture that is
   a. heterogeneous   **b.** homogeneous
   c. homologous   **d.** humectic

2. Which of the following pairs of substances is *least likely* to form a solution?
   a. an ionic compound in a nonpolar solvent
   b. an ionic compound in a polar solvent
   c. a nonpolar compound in a nonpolar solvent
   d. a polar compound in a polar solvent

3. A positive test for iodine is a purple solution of I_2 in hexane (C_6H_{14}). What type of solute–solvent interaction is most important in a solution of nonpolar I_2 in nonpolar C_6H_{14}?
   a. dipole–dipole
   b. dispersion–dispersion
   c. ion–dipole
   d. ion–dispersion

4. What type of solute–solvent interaction is most important in a solution of calcium chloride ($CaCl_2$) in water?
   a. dipole–dipole
   b. ion–dipole
   c. ion–dispersion
   d. ion–hydrogen bond

5. What type of solute–solvent interaction is most important in a solution of ethanol (CH_3CH_2OH) in water?
   a. dipole–dipole
   b. dipole–hydrogen bond
   c. hydrogen bond–hydrogen bond
   d. ion–hydrogen bond

6. Which of the following solutes will dissolve in octane, C_8H_{18}, a component of gasoline?
   a. $CH_3(CH_2)_4CH_3$
   b. $CaCl_2$
   c. KCl
   d. MgO

Answers: 1, b; 2, a; 3, b; 4, b; 5, c; 6, a

6.5 Gases: The Kinetic–Molecular Theory

At the macroscopic level, gases may seem more difficult to understand than liquids and solids. Perhaps this is in large part because most gases are invisible, while we can see and feel liquids and solids. However, at the microscopic level, gases are more readily treated mathematically because the molecules are so far apart we can usually ignore the very weak intermolecular forces. In liquids and solids the forces between molecules are appreciable and vary with type and distance between molecules. This makes theoretical treatment quite complex.

Experiments with gases were instrumental in developing the concepts of Avogadro's number and of molar ratios in reactions. Let's now look at the behavior of gases more closely, using a model known as the **kinetic–molecular theory** (Figure 6.10). The basic postulates of this theory are the following.

1. Particles of a gas (usually *molecules,* but in the case of noble gases, *atoms*) are in rapid, constant motion and move in straight lines.

2. The particles of a gas are tiny compared with the distances between them.

3. Because they are so far apart, there is very little attraction between particles of a gas.

4. Particles collide with one another. Energy is *conserved* in these collisions; energy lost by one particle is gained by the other.

5. Temperature is a measure of the average kinetic energy of the gas molecules.

Section 6.3 describes how intermolecular forces hold molecules together in solids and liquids. In gases the particles are separated by relatively great distances and are moving about randomly. Although the molecules can collide briefly, they seldom interact.

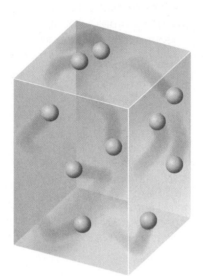

▲ **Figure 6.10** According to the kinetic–molecular theory, molecules of a gas are in constant, random motion. They move in straight lines and undergo collisions with each other and with the walls of the container.

Self-Assessment Questions

1. Which of the following is *not* a postulate of the kinetic–molecular theory of gases?
   a. Particles of a gas are in constant chaotic motion.
   b. The pressure and volume of a gas are inversely related.
   c. Temperature is a measure of the average kinetic energy of the molecules.
   d. The volume of the molecules is negligible compared to the distances between them.

2. According to the kinetic–molecular theory of gases, a gas particle
   a. collides with another particle with no loss of kinetic energy for either
   b. follows a well-defined pathway
   c. has a volume that is negligible compared to the volume occupied by the gas
   d. repels another particle when the two come close together

3. In collisions between ideal gas molecules, the total energy of the gas
   a. decreases slightly   b. decreases considerably
   c. increases slightly   d. remains the same

Answers: 1, b; 2, c; 3, d

6.6 The Simple Gas Laws

The behavior of gases can be described by mathematical relationships called gas laws. These equations use four variables to specify a sample of gas in calculations: its amount in moles (n), its volume (V), its temperature (T), and its pressure (P). These variables are related through simple laws that show how one of the variables (for example, V) changes as a second variable (for example, P) changes and the other two (for example, n and T) remain constant.

Boyle's Law: Pressure and Volume

The first gas law was discovered by Robert Boyle in 1662. It describes the relationship between the pressure and volume of a gas. **Boyle's law** states that *for a given amount of gas at a constant temperature, the volume of the gas varies inversely with its pressure.* That is, in a closed container of gas, when the pressure increases, the volume decreases; when the pressure decreases, the volume increases.

A bicycle pump illustrates Boyle's law. When you push down on the plunger, you increase the pressure of the air in the pump. The volume of the air decreases, and the higher-pressure air flows into the bicycle tire.

Think of gases as pictured in the kinetic–molecular theory. A gas exerts a particular pressure because its molecules bounce against the container walls with a certain frequency and speed (Figure 6.11). If the volume of the container is increased while the amount of gas remains fixed, the number of molecules per unit volume of gas decreases. The frequency with which molecules strike a unit area of the container walls decreases, and the gas pressure decreases. Thus, as the volume of a gas is increased, its pressure decreases.

Mathematically, for a given amount of gas at a constant temperature, Boyle's law is written

$$V \propto \frac{1}{P}$$

where the symbol $\propto$ means "is proportional to." This relationship can be changed to an equation by inserting a proportionality constant, a.

$$V = \frac{a}{P}$$

Multiplying both sides of the equation by P, we get

$$PV = a \quad \text{(at constant temperature and amount of gas)}$$

Another way to state Boyle's law, then, is that for a given amount of gas at a constant temperature, the product of the pressure and volume is a constant. This is an elegant and precise, if somewhat abstract, way of summarizing a lot of

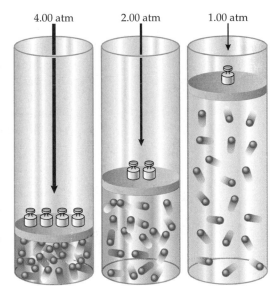

4.00 atm   2.00 atm   1.00 atm

▲ **Figure 6.11** A kinetic–molecular theory view of Boyle's law. As the pressure is reduced from 4.00 atm to 2.00 atm and then to 1.00 atm, the volume of the gas doubles and then doubles again.

QUESTION: What would happen to the volume if the pressure were changed to 0.500 atm?

▶ **Figure 6.12** A graphic representation of Boyle's law. As the pressure of the gas is increased, its volume decreases. When the pressure is doubled ($P_2 = 2 \times P_1$), the volume of the gas decreases to one-half its original value ($V_2 = \frac{1}{2} \times V_1$). The pressure–volume product is a constant ($PV = a$).

QUESTION: What would happen to the volume if the pressure were quadrupled?

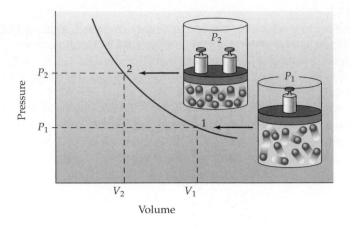

experimental data. If the product $P \times V$ is constant, then when V increases P must decrease, and vice versa. This relationship is demonstrated in Figure 6.12 by a pressure–volume graph.

Boyle's law has a number of practical applications perhaps best illustrated by some examples. In Example 6.1, we see how to estimate an answer. Sometimes this is all we need. Even when we want a quantitative answer, however, the estimate helps us to determine whether or not our answer is reasonable.

EXAMPLE 6.1 Boyle's Law: Pressure–Volume Relationships

A gas is enclosed in a cylinder fitted with a piston. The volume of the gas is 2.00 L at 0.524 atm. The piston is moved to increase the gas pressure to 5.15 atm. Which of the following is a reasonable value for the volume of the gas at the greater pressure?

0.20 L   0.40 L   1.00 L   16.0 L

Solution

The pressure increase from 0.524 atm to 5.15 atm is almost tenfold. The volume should drop to about one-tenth of the initial value. We estimate a volume of 0.20 L. (The calculated value is 0.203 L.)

■ **EXERCISE 6.1**

A gas is enclosed in a 10.2-L tank at 1208 mmHg. (The *mmHg* is a pressure unit; 760 mmHg = 1 atm.) Which of the following is a reasonable value for the pressure when the gas is transferred to a 30.0-L tank?

300 mmHg   400 mmHg   3600 mmHg   12,000 mmHg

Example 6.2 illustrates quantitative calculations using Boyle's law. Note that in these applications any units can be used for pressure and volume as long as the same units are used throughout a calculation. As long as we use the same sample of a confined gas at a constant temperature, the product of the initial pressure (P_1) times the initial volume (V_1) is equal to the product of the final pressure (P_2) times the final volume (V_2). Thus, the following useful equation representing Boyle's law can be written.

$$P_1 V_1 = P_2 V_2$$

EXAMPLE 6.2 Boyle's Law: Pressure–Volume Relationships

A cylinder of oxygen has a volume of 2.25 L. The pressure of the gas is 1470 pounds per square inch (psi) at 20 °C. What volume will the oxygen occupy at standard atmospheric pressure (14.7 psi) assuming no temperature change?

Solution

We find it helpful to first separate the initial from the final condition.

Initial	Final	Change
$P_1 = 1470$ psi	$P_2 = 14.7$ psi	↓ The pressure goes down; therefore,
$V_1 = 2.25$ L	$V_2 = ?$	↑ the volume goes up.

Then use the equation $P_1 V_1 = P_2 V_2$ and solve for the desired volume or pressure. In this case, we solve for V_2.

$$V_2 = \frac{P_1 V_1}{P_2}$$

$$V_2 = \frac{1470 \text{ psi} \times 2.25 \text{ L}}{14.7 \text{ psi}} = 225 \text{ L}$$

Because the final pressure (14.7 psi) is *less than* the initial pressure (1470 psi), we expect the final volume (225 L) to be *larger than* the original volume (2.25 L), and we see that it is.

■ **EXERCISE 6.2A**

A sample of air occupies 73.3 mL at 98.7 atm and 0 °C. What volume will the air occupy at 4.02 atm and 0 °C?

■ **EXERCISE 6.2B**

A sample of helium occupies 535 mL at 988 mmHg and 25 °C. If the sample is transferred to a 1.05-L flask at 25 °C, what will be the gas pressure in the flask?

Charles's Law: Temperature and Volume

In 1787 the French physicist Jacques Charles (1746–1823), a pioneer hot-air balloonist, studied the relationship between the volume and temperature of gases. He found that when a fixed mass of gas is cooled at constant pressure, its volume decreases. When the gas is heated, its volume increases. Temperature and volume vary directly; that is, they rise or fall together. But this law requires a bit more thought. If a quantity of gas that occupies 1.00 L is heated from 100 °C to 200 °C at constant pressure, the volume does not double but only increases to about 1.27 L. The relationship between temperature and volume is not as tidy as it may seem at first.

Zero pressure or zero volume really means *zero*; no pressure or volume can be measured. Zero degrees Celsius (0 °C), on the other hand, simply signifies the freezing point of water. This zero point is arbitrarily set, much as mean sea level is set as the arbitrary zero for measuring altitudes on Earth. Temperatures below 0 °C are often encountered, as are altitudes below sea level.

Charles noted that for each degree Celsius rise in temperature, the volume of a gas increases by $\frac{1}{273}$ of its volume at 0 °C. If we plot volume against temperature, we get a straight line (Figure 6.13). We can extrapolate the line beyond the range of measured temperatures to the temperature at which the volume of the gas would become zero. This temperature is −273.15 °C. In 1848 William Thomson (Lord Kelvin) made this temperature the zero point on an absolute temperature scale now called the Kelvin scale. As noted in Chapter 1, the unit of temperature on this scale is the kelvin (K).

A modern statement of **Charles's law** is that *the volume of a fixed amount of a gas at a constant pressure is directly proportional to its absolute temperature.* Mathematically, this relationship is expressed as

$$V = bT \quad \text{or} \quad \frac{V}{T} = b$$

4. How cold can it get?
If we could reach 0 K (absolute zero, −273.15 °C), we would find that molecular motion stops. Since molecules cannot move any slower than "standing still," no lower temperature is possible.

ANSWER

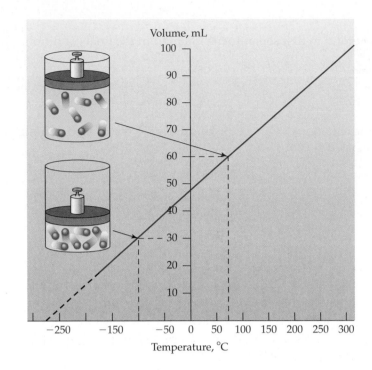

▶ **Figure 6.13** Charles's law relates gas volume to temperature at constant pressure. When the gas shown has been cooled to about 70 °C, its volume is 60 mL. In the temperature interval from about 70 °C to 100 °C, the volume drops to 30.0 mL. The volume continues to fall as the temperature is lowered. The extrapolated line intersects the temperature axis (corresponding to a volume of zero) at about −273 °C.

QUESTION: What would be the approximate volume at a temperature of 150 °C?

where b is a proportionality constant. To keep $\frac{V}{T}$ equal to a constant value, when the temperature increases, the volume must also increase. When the temperature decreases, the volume must decrease accordingly (Figure 6.14).

As long as we use the same sample of trapped gas at a constant pressure, the initial volume (V_1) divided by the initial absolute temperature (T_1) is equal to the final volume (V_2) divided by the final absolute temperature (T_2). We can use the following equation to solve problems involving Charles's law.

$$\frac{V_1}{T_1} = \frac{V_2}{T_2}$$

The kinetic–molecular model readily explains the relationship between gas volume and temperature. When we heat a gas, we supply the gas molecules with energy and they begin to move faster. These speedier molecules strike the walls of the container harder and more often. For the pressure to stay the same, the volume of the container must increase so that the increased molecular motion will be distributed over a greater space.

▶ **Figure 6.14** A dramatic illustration of Charles's law. (a) Liquid nitrogen (boiling point, −196 °C) cools the balloon and its contents to a temperature far below room temperature. (b) As the balloon warms back to room temperature, the volume of air increases proportionately (about fourfold).

(a)   (b)

EXAMPLE 6.3   Charles's Law: Temperature–Volume Relationship

A balloon indoors, where the temperature is 27 °C, has a volume of 2.00 L. What would its volume be **(a)** in a hot room where the temperature is 47 °C, and **(b)** outdoors where the temperature is −23 °C? (Assume no change in pressure in either case.)

Solution

First, and most important, convert all temperatures to the Kelvin scale:

$$T(K) = t(°C) + 273$$

The initial temperature (T_1) in each case is $(27 + 273) = 300$ K and the final temperatures are **(a)** $(47 + 273) = 320$ K and **(b)** $(-23 + 273) = 250$ K.

a. We start by separating the initial from the final condition.

Initial	Final	Change
$t_1 = 27 °C$	$t_2 = 47 °C$	⇑
$T_1 = 300$ K	$T_2 = 320$ K	⇑
$V_1 = 2.00$ L	$V_2 = ?$	⇑

Solving the equation

$$\frac{V_1}{T_1} = \frac{V_2}{T_2}$$

for V_2, we have

$$V_2 = \frac{V_1 T_2}{T_1}$$

$$V_2 = \frac{2.00 \text{ L} \times 320 \text{ K}}{300 \text{ K}} = 2.13 \text{ L}$$

As expected, because the temperature increases, the volume must also increase.

b. We have the same initial conditions as in **(a)** but different final conditions.

Initial	Final	Change
$t_1 = 27 °C$	$t_2 = -23 °C$	⇓
$T_1 = 300$ K	$T_2 = 250$ K	⇓
$V_1 = 2.00$ L	$V_2 = ?$	⇓

Again using Charles's law, we solve the equation for V_2:

$$V_2 = \frac{V_1 T_2}{T_1}$$

$$V_2 = \frac{2.00 \text{ L} \times 250 \text{ K}}{300 \text{ K}} = 1.67 \text{ L}$$

As we expected, the volume decreased because the temperature decreased.

▪ **EXERCISE 6.3A**

A sample of oxygen gas occupies a volume of 2.10 L at 25 °C. What volume will this sample occupy at 150 °C? (Assume no change in pressure.)

▪ **EXERCISE 6.3B**

At what Celsius temperature will the initial volume of oxygen in Exercise 6.3A occupy 0.750 L? (Assume no change in pressure.)

Molar Volume

As we described in Section 5.2, Amedeo Avogadro proposed a hypothesis to explain the volume ratios in which gases combine during chemical reactions. His hypothesis states that equal numbers of molecules of different gases at the same temperature and pressure occupy equal volumes. Thus, Avogadro's hypothesis relates an amount of gas (numbers of molecules, or moles) and gas volume when

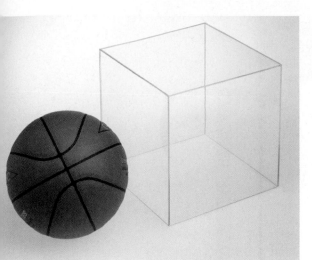

▲ **Figure 6.15** The cube holds 22.4 liters—one mole of a gas at STP. The basketball is shown for size comparison.

QUESTION: How many C_2H_6 molecules would the cube hold at STP?

temperature and pressure remain constant. We call the simple gas law implied by this relationship *Avogadro's law*:

At a fixed temperature and pressure, the volume of a gas is directly proportional to the amount of gas (that is, to the number of moles of gas, n, or to the number of molecules of gas).

If we double the number of moles of gas at a fixed temperature and pressure, the volume of the gas doubles. Mathematically, we can state Avogadro's law as

$$V \propto n \quad \text{or} \quad V = cn \quad \text{(where } c \text{ is a constant)}$$

A mole of gas contains Avogadro's number of molecules (atoms if it is a noble gas). Furthermore, Avogadro's number of molecules of gas occupies the same volume (at a given temperature and pressure) regardless of the size or mass of the individual molecules. The volume occupied by 1 mol of gas is the **molar volume** of a gas.

Because the volume of a gas is altered by changes in temperature or pressure, a particular set of conditions has been chosen for reference purposes as "standard." Standard pressure is 1 atmosphere (atm), which is the normal pressure of the air at sea level; and standard temperature is 0 °C, which is the freezing point of water. A mole of any gas at **standard temperature and pressure (STP)** occupies a volume of about 22.4 L. This is known as the standard molar volume of a gas.

A cube that measures 28.2 cm along each edge has a volume of 22.4 L (Figure 6.15). It is just a bit smaller than a cubic foot. At a temperature of 0 °C and 1 atm pressure, a 22.4-L container holds 28.0 g of N_2, 32.0 g of O_2, 44.0 g of CO_2, and so on; it holds 1 mol of any gas.

We can readily calculate the density (g/L) of a gas at STP. We begin with the molar mass of the gas (g/mol) and use the conversion factor 1 mol gas = 22.4 L. Because the conversion factor can be inverted, we also can calculate molar mass from density, using the same factor 1 mol gas = 22.4 L.

EXAMPLE 6.4 Density of a Gas at STP

Calculate the density of **(a)** nitrogen gas and **(b)** methane (CH_4) gas, both at STP.

Solution

a. The molar mass of N_2 gas is 28.0 g/mol. We multiply by the conversion factor 1 mol N_2 = 22.4 L, arranged to cancel units of *moles*.

$$\frac{28.0 \text{ g } N_2}{1 \text{ mol } N_2} \times \frac{1 \text{ mol } N_2}{22.4 \text{ L } N_2} = 1.25 \text{ g/L}$$

b. The molar mass of CH_4 gas is (1×12.0) g/mol + (4×1.01) g/mol = 16.0 g/mol. Again we use the conversion factor 1 mol CH_4 = 22.4 L.

$$\frac{16.0 \text{ g } CH_4}{1 \text{ mol } CH_4} \times \frac{1 \text{ mol } CH_4}{22.4 \text{ L } CH_4} = 0.714 \text{ g/L}$$

■ **EXERCISE 6.4A**
Calculate the density of helium at STP.

■ **EXERCISE 6.4B**
Estimate the density of air at STP (assume 78% N_2 and 22% O_2) and compare this value to the value of He you calculated in Exercise 6.4A.

Self-Assessment Questions

1. Gas pressure is caused by
   a. gas molecules colliding with each other
   b. gas molecules colliding with vessel walls
   c. gas molecules condensing to form a liquid
   d. measurement with a barometer

2. Boyle's law states that the pressure of a gas is inversely proportional to its
   a. amount   **b.** mass   **c.** temperature   **d.** volume

3. A gas occupies 4.00 L at 2.00 atm pressure. At what pressure would the volume be 6.00 L if the temperature remains constant?
   a. 1.00 atm   **b.** 1.33 atm   **c.** 2.67 atm   **d.** 3 atm

4. When the temperature is changed at constant pressure, the volume of a gas is halved. How is the new temperature related to the original temperature?
   a. The Celsius temperature is doubled.
   b. The Celsius temperature is halved.
   c. The Kelvin temperature is doubled.
   d. The Kelvin temperature is halved.

5. One 1-L flask (Flask A) contains carbon monoxide gas and another identical flask (Flask B) contains carbon dioxide gas, both at STP. Flask A has
   a. less mass and fewer particles than B
   b. less mass but the same number of particles as B
   c. less mass but more particles than B
   d. more mass and more particles than B

6. How many moles are present in 5.60 L of a gas at STP?
   a. 0.250 mol   **b.** 0.500 mol   **c.** 4.00 mol   **d.** 125 mol

Answers: 1, b; 2, d; 3, b; 4, d; 5, b; 6, a

6.7 The Ideal Gas Law

Boyle's law, Charles's law, and Avogadro's law are useful when two terms of four (P, V, n, T) can be held constant. Often, however, both the temperature and the pressure change at the same time. In such cases the simple gas laws can be formed into a single relationship called the **combined gas law**. Mathematically, this law is written

$$\frac{PV}{T} = c \quad \text{or} \quad PV = cT$$

where c is a constant. As with the simple gas laws, for comparing the same gas under two different sets of conditions, the law can be written

$$\frac{P_1V_1}{T_1} = \frac{P_2V_2}{T_2}$$

Because the amount of gas n may change as well, we can incorporate this variable to obtain an expression called the **ideal gas law**, which involves all four variables.

$$\frac{PV}{nT} = R \quad \text{or} \quad PV = nRT$$

In this equation R is a constant (called the *gas constant*), which can be calculated from the fact that 1 mol of gas occupies 22.4 L at 273 K (0 °C) and 1 atm (Section 6.6). If P is in atmospheres, V in liters, and T in kelvins, then R has a value of

$$0.0821\frac{\text{L} \cdot \text{atm}}{\text{mol} \cdot \text{K}}$$

The ideal gas equation can be used to calculate any of the four variables—P, V, n, or T—if the other three are known.

ANSWER

5. Why are we supposed to measure tire pressure when tires are "cold"?
Tire pressure is measured when the tires are cold because a tire has nearly constant volume. According to the combined gas law, if the temperature increases (by friction from the road) the pressure increases too.

EXAMPLE 6.5   Combined Gas Law

A balloon used for underwater salvage is inflated to 50.0 liters at a depth of 200 feet, where the pressure is 6.89 atm and the temperature is 3 °C. The balloon rises to the surface (22 °C and 0.988 atm). What is the new volume of the balloon?

Solution

We start by solving the combined gas law equation for V_2.

$$\frac{P_1 V_1}{T_1} = \frac{P_2 V_2}{T_2}$$

$$V_2 = \frac{P_1 V_1 T_2}{T_1 P_2}$$

$T_1 = 3 + 273 = 276$ K and $T_2 = 22.0 + 273 = 295$ K.

$$V_2 = \frac{6.89 \text{ atm} \times 50.0 \text{ L} \times 295 \text{ K}}{276 \text{ K} \times 0.988 \text{ atm}} = 373 \text{ L}$$

■ EXERCISE 6.5

What would be the final volume of a 400-mL gas sample initially at 22.0 °C and 760 mmHg pressure when the temperature is changed to 30.0 °C and the pressure to 360 mmHg?

EXAMPLE 6.6   Ideal Gas Law

Use the ideal gas law to calculate (**a**) the volume occupied by 1.00 mol of nitrogen gas at 244 K and 1.00 atm pressure, and (**b**) the pressure exerted by 0.500 mol of oxygen in a 15.0-L container at 303 K.

Solution

a. We start by solving the ideal gas equation for V.

$$V = \frac{nRT}{P}$$

$$V = \frac{1.00 \text{ mol}}{1.00 \text{ atm}} \times \frac{0.0821 \text{ L} \cdot \text{atm}}{\text{mol} \cdot \text{K}} \times 244 \text{ K} = 20.0 \text{ L}$$

b. Here we solve the ideal gas equation for P.

$$P = \frac{nRT}{V}$$

$$P = \frac{0.500 \text{ mol}}{15.0 \text{ L}} \times \frac{0.0821 \text{ L} \cdot \text{atm}}{\text{mol} \cdot \text{K}} \times 303 \text{ K} = 0.829 \text{ atm}$$

■ EXERCISE 6.6A

Determine (**a**) the pressure exerted by 0.0330 mol of oxygen in an 18.0-L container at 313 K, and (**b**) the volume occupied by 0.200 mol of nitrogen gas at 298 K and 0.980 atm.

■ EXERCISE 6.6B

Determine the volume occupied by 132 g of nitrogen gas at 25 °C and 1.03 atm.

Self-Assessment Questions

1. In the ideal gas equation $PV = nRT$, the variable n stands for
   a. a constant
   b. number of atoms
   c. number of moles
   d. the principal quantum number

2. At STP 1.00 mol of an ideal gas occupies 22.4 L. What volume does the gas occupy at 20 °C and 1.00 atm?
   a. 20.9 L
   b. 22.4 L
   c. 24.0 L
   d. 448 L

3. The molar volume of an ideal gas at 2.00 atm and 546 K is
   a. 5.60 L
   b. 22.4 L
   c. 44.8 L
   d. 89.6 L

Answers: 1, c; 2, c; 3, b

Critical Thinking Exercises

Apply knowledge that you have gained in this chapter and one or more of the FLaReS principles (Chapter 1) to evaluate the following statements or claims.

6.1 Some automobile tire stores claim that filling your car tires with pure, dry nitrogen is much better than using plain air. They make the following claims: (1) The pressure inside nitrogen-filled tires does not rise or fall with temperature changes. (2) Nitrogen leaks out of tires much more slowly than air because the nitrogen molecules are bigger. (3) Nitrogen is not very reactive, and moisture and oxygen in air cause corrosion that shortens tire life by 25 to 30%. Use information you have gained in this chapter and from other sources as necessary to evaluate these claims.

6.2 A researcher claims to have discovered the reason that the gecko can walk on walls and ceilings. His claim is that the lizard's feet have many microscopic hairlike protrusions that get close enough to the wall or ceiling surface to allow intermolecular forces to "take over" and hold the gecko to the surface. Evaluate this claim.

6.3 A microbrewery claims to have prepared "hydrogen beer." This beer is reportedly "carbonated" with hydrogen gas instead of carbon dioxide gas. A chemist at the brewery claims that the hydrogen gas undergoes hydrogen bonding with the water molecules in the beer, so that more gas can dissolve in the beer. Evaluate this claim.

■ SUMMARY

Section 6.1—When a substance is melted or vaporized, the forces that hold the particles (molecules, atoms, or ions) close to one another are overcome. Several properties are related to the strengths of those forces. The **melting point** of a solid is the temperature at which the forces holding the particles together in a regular arrangement are overcome. The reverse of the melting process, a liquid changing to a solid, is called **freezing**. **Vaporization** is the conversion of a liquid to a gas; the reverse process is called **condensation**. The temperature at which a substance vaporizes at ordinary pressure is called its **boiling point**. Some substances undergo **sublimation**, a conversion directly from the solid to the gas state.

Section 6.2— Ionic solids have very strong forces holding the ions to one another, so most ionic solids have high melting points and high boiling points compared to molecular compounds.

Section 6.3—Not all solids are held together by ionic bonds. **Dipole–dipole forces** involve the attraction of the positive end of one dipole for the negative end of another dipole. Nonpolar molecules also have forces between them. These **dispersion forces** result from electron motion of neighboring atoms, which causes tiny, short-lived dipoles. Dispersion forces are generally weak. A special, strong type of dipole force occurs when the positive end is a hydrogen atom attached to F, O, or N, and the negative end is a pair of electrons on an atom of F, O, or N. Such a force is called a **hydrogen bond**.

Section 6.4—A **solution** is a **homogeneous** mixture, a mixture in which all parts have the same composition. The substance that is dissolved is called the **solute**, and the dissolving substance is called the **solvent**. Nonpolar solutes dissolve in nonpolar solvents. Polar substances and many ionic substances dissolve in polar solvents.

Section 6.5—The behavior of gases is explained using **kinetic–molecular theory**, which describes gas molecules: (1) They move rapidly, constantly, and in straight lines; (2) they are far apart; (3) there is little attraction between them; (4) when they collide, energy is conserved; and (5) temperature is a measure of average kinetic energy of the gas molecules.

Section 6.6—**Boyle's law** says that the volume of a fixed amount of gas varies inversely with pressure at constant temperature, or $P_1V_1 = P_2V_2$. **Charles's law** says that the volume of a fixed amount of gas varies directly with absolute (Kelvin) temperature, or $V_1/T_1 = V_2/T_2$. The **molar volume** of a gas is the volume occupied by 1 mol of any gas. At 0 °C and 1 atm, known as **standard temperature and pressure (STP)**, the molar volume is 22.4 L for any gas. We can use that molar volume to convert from moles of gas to volume and vice versa.

Section 6.7—The simple gas laws can be formed into a single relationship called the **combined gas law**. The **ideal gas law** shows the relationship among pressure (P), volume (V), number of moles (n), and absolute temperature (T): $PV = nRT$. When P = atmospheres, V = liters, n = moles of gas, and T = kelvins, the value of the constant R is 0.0821 liter · atm/ (mol · K).

■ REVIEW QUESTIONS

1. In what ways are liquids and solids similar? In what ways are they different?

2. List four types of interactions between particles in the liquid and solid states. Give an example of each type.

3. Define each of the following terms.
   a. melting
   b. vaporization
   c. condensation
   d. freezing

4. In which process is energy *absorbed* by the material undergoing the change of state?
  a. melting or freezing
  b. condensation or vaporization

5. In the combined gas law, is the volume of a gas directly proportional or inversely proportional to pressure? Is volume directly or inversely proportional to absolute temperature?

6. Label each arrow with the term listed in Question 3 that correctly identifies the process presented.

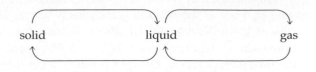

■ PROBLEMS

Intermolecular Forces

7. For which of the following would hydrogen bonding be an important intermolecular force?

8. In which of the following are dispersion forces the only type of intermolecular force: Br_2, HF, HCl?

9. In which of the following are dipole interactions an important intermolecular force: Br_2, HCl, NaCl?

10. In which of the following are ionic bonds important: Br_2, HBr, NaBr?

11. Benzene (C_6H_6) is a nonpolar solvent. Would you expect NaCl to dissolve in benzene? Explain.

12. Which of the following would you expect to dissolve in water and which in benzene (C_6H_6)?
  a. KBr   **b.** C_5H_{12} **c.** C_8H_{18} **d.** $CaCl_2$

Boyle's Law

13. A sample of helium occupies 1521 mL at 719 mmHg. Assume that the temperature is held constant and determine **(a)** the volume of the helium at 752 mmHg and **(b)** the pressure, in mmHg, if the volume is changed to 315 mL.

14. A decompression chamber used by deep-sea divers has a volume of 10.3 m^3 and operates at an internal pressure of 4.50 atm. What volume, in cubic meters, would the air in the chamber occupy if it were at 1.00 atm pressure, assuming no temperature change?

15. Oxygen used in respiratory therapy is stored at room temperature under a pressure of 150 atm in a gas cylinder with a volume of 60.0 L.
  a. What volume would the gas occupy at 0.987 atm? Assume no temperature change.
  b. If the oxygen flow to the patient is adjusted to 8.00 L/min, at room temperature and 0.987 atm, how long will the tank of gas last?

16. The pressure within a 2.25-L balloon is 1.10 atm. If the volume of the balloon increases to 7.05 L, what will be the final pressure within the balloon if the temperature does not change?

Charles's Law

17. A gas at a temperature of 100 °C occupies a volume of 154 mL. What will the volume be at a temperature of 10 °C, assuming no change in pressure?

18. A balloon is filled with helium. Its volume is 5.90 L at 26 °C. What will its volume be at 78 °C, assuming no pressure change?

19. A 567-mL sample of a gas at 305 °C and 1.20 atm is cooled at constant pressure until its volume becomes 425 mL. What is the new gas temperature?

20. A sample of gas at STP is to be heated at constant pressure until its volume triples. What is the new gas temperature?

Molar Volume and Gas Densities

21. What is the volume of each of the following gases at STP?
  a. 1.00 mole N_2 **b.** 1.00 mole H_2
  c. 2.00 moles C_2H_6

22. What is the mass of one molar volume of each of the gases in Problem 21?

23. Calculate the density of krypton (Kr) gas, in grams per liter, at STP.

24. Calculate the density of carbon monoxide (CO) gas, in grams per liter, at STP.

25. Calculate the molar mass of **(a)** a gas that has a density of 2.12 g/L at STP, and **(b)** an unknown liquid the vapor of which has a density of 2.97 g/L at STP.

26. Calculate the molar mass of **(a)** a gas that has a density of 1.98 g/L at STP, and **(b)** an unknown liquid for which 3.33 liters at STP is found to weigh 10.88 g.

The Ideal Gas Law

27. Will the volume of a fixed amount of gas increase, decrease, or remain unchanged with
  a. an increase in pressure at constant temperature?
  b. a decrease in temperature at constant pressure?
  c. a decrease in pressure coupled with an increase in temperature?

28. Will the pressure of a fixed amount of a gas increase, decrease, or remain unchanged with
  a. an increase in temperature at constant volume?
  b. a decrease in volume at constant temperature?
  c. an increase in temperature coupled with a decrease in volume?

29. According to the kinetic–molecular theory, **(a)** what change in temperature occurs if the molecules of a gas begin to move more slowly, on average, and **(b)** what change in pressure occurs when molecules of the gas strike the walls of the container less often?

30. For each of the following, indicate whether a given gas would have the same or different densities in the two containers. If the densities are different, in which container is the density greater?
   a. Containers A and B have the same volume and are at the same temperature, but the gas in A is at a higher pressure.
   b. Containers A and B are at the same pressure and temperature, but the volume of A is greater than that of B.
   c. Containers A and B are at the same pressure and volume, but the gas in A is at a higher temperature.

31. Calculate (a) the volume, in liters, of 1.12 mol $H_2S(g)$ at 62 °C and 1.38 atm, and (b) the pressure, in atmospheres, of 4.64 mol CO(g) in a 3.96-L tank at 29 °C.

32. Calculate (a) the volume, in liters, of 0.00600 mol of a gas at 31 °C and 0.870 atm, and (b) the pressure, in atmospheres, of 0.0108 mol $CH_4(g)$ in a 0.265-L flask at 37 °C.

33. How many moles of Kr(g) are there in 2.22 L of the gas at 0.918 atm and 45 °C?

34. How many grams of CO(g) are there in 745 mL of the gas at 1.03 atm and 36 °C?

■ ADDITIONAL PROBLEMS

35. What mass of He(g) will occupy the same volume at STP as 0.75 mol H_2 at STP? (*Hint*: it is not necessary to find the volume of the H_2 at STP.)

36. How many liters of hydrogen gas (at STP) are produced from the electrolysis of 1.00 L $H_2O(l)$? (*Hint*: 1.00 L of H_2O weighs 1.00 kg; how many moles is this?)

37. Which of the following laws is correctly described by the mathematical proportionality?
   a. Boyle's law: $V \propto P$
   b. Charles's law: $V \propto T$

38. Which of the following would *not* result in an increase in the volume of a gas?
   a. an increase in temperature
   b. an increase in pressure
   c. a decrease in temperature
   d. a threefold increase in pressure together with a twofold reduction in Kelvin temperature

39. Which of the following gases has the greatest density at STP? Explain how this may be answered *without* calculating the densities.
   a. Cl_2 b. SO_3 c. N_2O d. PF_3

40. Choose the answer that correctly completes the statement: At 0 °C and 0.500 atm, 4.48 L $NH_3(g)$ (a) contains 0.20 mol NH_3; (b) has a mass of 3.40 g; (c) contains 6.02×10^{22} molecules; (d) contains 0.40 mol NH_3.

41. Look again at Figure 6.14. If the balloon has a volume of a 1.50 L at 20 °C, what will the volume be in (a), assuming the air in the balloon reaches the temperature of the liquid nitrogen? You may ignore the stretching forces in the rubber.

42. Mining helmets once used calcium carbide (CaC_2) as a source of light. When water is dripped on the solid, calcium hydroxide and acetylene (C_2H_2) are produced, the latter of which when burned produces a bright flame. (a) What mass of CaC_2 is needed to provide light for 3.0 hours if 9.5 mole acetylene (at STP) is used every hour? (b) What volume, in liters, would the acetylene occupy at STP if it were not burned?

43. Calculate the density of air at STP (assume 22% oxygen and 78% nitrogen by volume). Calculate the density of steam (100% H_2O vapor) under the same conditions. Comment, without doing a calculation, on whether you think the density of moist air would be greater than or less than that of dry air.

44. A 14.4-g sample of an unknown gas has a volume of 8.00 L at 760 mmHg and 25 °C. What is the molar mass of the gas?

45. Another simple gas law, sometimes called *Amontons's law*, states that pressure of a fixed amount of gas in a constant volume is proportional to its Kelvin temperature. Use this relationship to answer the following question. A basketball is inflated to an internal pressure of 1.32 atm at 25 °C. If the ball is taken outside where the temperature is 10 °C, what will be the pressure in the ball?

46. A spacecraft has a cabin air volume of 50.0 m^3. It carries an emergency tank of air with a volume of 0.250 m^3. To what pressure must the tank be filled to in order to restore the cabin pressure to 1.00 atm if it completely loses pressure?

47. A gas at 750 mmHg pressure and 27 °C has a density of 2.32 g/L. Which of the following could it be: CO_2, Kr, H_2S, or C_4H_{10}?

48. A sample of carbon dioxide has a volume of 2.54 L at STP. What will be the volume when the pressure is 1.48 atm and the temperature is 26 °C?

■ COLLABORATIVE GROUP PROJECTS

Prepare a PowerPoint, poster, or other presentation (as directed by your instructor) for presentation to the class.

49. Prepare a brief biographical report on one of the following individuals and share it with your group.
   a. Joseph Louis Gay-Lussac   b. Robert Boyle
   c. Jacques Charles

50. Most ionic compounds are solids, but in recent years *ionic liquids* have been made and have found numerous uses. Search the Web for information on these ionic liquids. In particular, find out why they are liquids rather than solids, and report on some of their current and potential applications.

51. Search the Web for information on the hydrogen bonding (a) in DNA, or (b) in synthetic polymers. What are some of the important results of hydrogen bonding in these systems?

52. Two types of solids that we have not examined in this chapter are (a) amorphous solids and (b) covalent solids. Search more advanced textbooks for information on these types of solids and how their intermolecular forces differ from those of ionic and molecular solids.

7.1 Acids and Bases: Experimental Definitions

Acids and bases are chemical opposites, and so their properties are quite different—often opposite. Let us begin by listing a few of their properties.

An *acid* is a compound that

- causes litmus indicator dye to turn red.
- tastes sour.
- dissolves active metals such as zinc and iron, producing hydrogen gas.
- reacts with bases to form water and ionic compounds called salts.

A *base* is a compound that

- causes litmus indicator dye to turn blue.
- tastes bitter.
- feels slippery on the skin.
- reacts with acids to form water and salts.

We can identify foods that are acidic by their sour taste. Vinegar and lemon juice are examples. Vinegar is a solution of acetic acid (about 5%) in water. Lemons, limes, and other citrus fruits contain citric acid. Lactic acid gives yogurt its tart taste, and phosphoric acid is often added to carbonated drinks to impart tartness. The bitter taste of tonic water, on the other hand, is attributable to the presence of quinine, a base. Figure 7.1 shows some common acids and bases.

A litmus test is a common way to identify a substance as an acid or a base (Figure 7.2). Litmus is an **acid–base indicator**, one of many such compounds. If you dip a strip of neutral (violet-colored) litmus paper into an unknown solution and it turns pink, the solution is acidic. If it turns blue, the solution is basic. If the strip does not turn pink or blue, the solution is neither acidic nor basic. Many natural food colors, such as those in grape juice, red cabbage, and blueberries, are acid–base indicators. So are the colors in many flower petals.

<table>
<tr>
<td>
1. Are all acids corrosive?
It is clear from Table 7.1 that "acid" does not necessarily mean "corrosive." Many acids are innocuous enough to be included in foods that we eat, and some are necessary to life.
</td>
</tr>
</table>

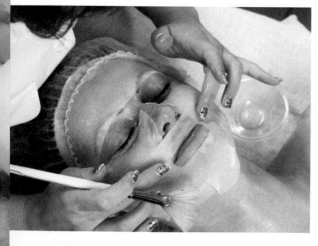

▲ Many "skin peel" preparations used by cosmetologists contain alpha-hydroxy acids such as glycolic acid or lactic acid.

▲ **Figure 7.1** Acids (left), bases (center), and salts (right) are components of many foods and familiar consumer products.

QUESTION: How would each of the three classes of compounds affect the indicator dye litmus? (See Figure 7.2.)

Self-Assessment Questions

1. Which of the following is *not* a property of acids?
   a. feel slippery on the skin
   b. react with Zn to form hydrogen gas
   c. taste sour
   d. turn litmus red

▲ **Figure 7.2** Strips of paper impregnated with litmus dye are often used to distinguish between acids and bases. The sample on the left turns blue litmus red and is therefore acidic. The sample on the right turns red litmus blue and is basic.

2. Which of the following is *not* a property of bases?
   a. feels slippery on the skin
   b. react with salts to form acids
   c. taste bitter
   d. turn litmus blue

3. In general, when water solutions of an acid and base are mixed
   a. a new acid and a salt are formed
   b. a new base and a salt are formed
   c. no reaction occurs
   d. a salt and water are formed

4. A common substance that contains lactic acid is
   a. salad oil   **b.** soap
   c. vinegar   **d.** yogurt

5. Lemons taste sour due to
   a. ammonia   **b.** acetic acid
   c. carbonic acid   **d.** citric acid

Answers: 1, a; 2, b; 3, d; 4, d; 5, d

▲ **It DOES Matter!**
Hydrangea is one of the many types of flowers that may show different colors depending on the acidity of the soil in which it is planted. Some varieties will be blue (top) when planted in slightly acidic soil, but will bloom pink (bottom) when planted in more basic soil.

7.2 Acids, Bases, and Salts

Acids and bases have certain characteristic properties. But why do they have these properties? We use several different theories to explain these properties.

The Arrhenius Theory

Svante Arrhenius developed the first successful theory of acids and bases in 1887. According to Arrhenius's concept, an **acid** is a molecular substance that breaks up in aqueous solution into hydrogen ions (H^+) and anions. (Because a hydrogen ion is

▲ Swedish chemist Svante Arrhenius (1859–1927) proposed the theory that acids, bases, and salts in water are composed of ions. He also was the first to relate carbon dioxide in the atmosphere to the greenhouse effect (Chapter 13).

a hydrogen atom from which the sole electron has been removed, H^+ ions are also called *protons*.) The acid is said to *ionize*.

In water, then, the properties of acids are those of the H^+ ion. It is the hydrogen ion that turns litmus red, tastes sour, and reacts with active metals and bases. Table 7.1 lists some common acids. Notice that each formula contains one or more hydrogen atoms. Chemists often indicate an acid by writing the formula with the H atoms first. HCl, H_2SO_4, and HNO_3 are acids; NH_3 and CH_4 are not. The formula $HC_2H_3O_2$ (acetic acid) indicates that one H atom ionizes and three do not.

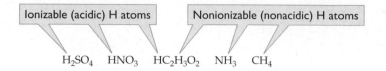

An Arrhenius **base** is defined as a substance that releases hydroxide ions (OH^-) in aqueous solution. Some bases are ionic solids that contain OH^-, such as sodium hydroxide (NaOH) and calcium hydroxide [$Ca(OH)_2$]. These compounds simply release hydroxide ions into water when the solid is dissolved in water:

$$NaOH(s) \xrightarrow{H_2O} Na^+(aq) + OH^-(aq)$$

Other bases are molecular substances such as ammonia that ionize to produce OH^- when placed in water (see page 177).

Experimental evidence indicates that the properties of bases in water are due to OH^-. Table 7.2 lists some common bases. Most of these are ionic compounds containing positive metal ions, such as Na^+ or Ca^{2+}, and negative hydroxide ions. When these compounds dissolve in water, they all provide OH^- ions, and thus they are all bases. The properties of bases are those of hydroxide ions, just as the properties of acids are those of hydrogen ions.

Arrhenius further proposed that the essential reaction between an acid and a base, **neutralization,** is the combination of H^+ and OH^- to form water. The cation that was originally associated with the OH^- combines with the anion associated with the H^+ to form an ionic compound, a **salt**.

$$\text{An acid + a base} \longrightarrow \text{a salt + water}$$

Table 7.1 Some Familiar Acids

Name	Formula	Acid Strength	Common Uses/Notes
Sulfuric acid	H_2SO_4	Strong	Battery acid; ore processing, fertilizer manufacturing, oil refining; extremely corrosive
Nitric acid	HNO_3	Strong	Manufacture of fertilizers, explosives
Hydrochloric acid	HCl	Strong	Cleaning of metals, bricks; removing scale from boilers
Phosphoric acid	H_3PO_4	Moderate	Manufacture of fertilizers; colas; rust removers
Hydrogen sulfate ion	HSO_4^-	Moderate	Toilet bowl cleaners ($NaHSO_4$)
Lactic acid	$CH_3CHOHCOOH$	Weak	Yogurt; acidulant (food additive to increase tartness); lotion additive
Acetic acid	CH_3COOH	Weak	Vinegar; acidulant
Carbonic acid	H_2CO_3	Weak	Unstable; formed in aqueous CO_2
Boric acid	H_3BO_3	Very weak	Antiseptic eye wash, roach poison
Hydrocyanic acid	HCN	Very weak	Plastics manufacture; extremely toxic

Table 7.2 Common Bases

Name	Formula	Classification	Common Uses/Notes
Sodium hydroxide	NaOH	Strong	Acid neutralization; soap making; dehorning calves
Potassium hydroxide	KOH	Strong	Making liquid soaps; absorbing CO_2
Lithium hydroxide	LiOH	Strong	Alkaline storage batteries
Calcium hydroxide	$Ca(OH)_2$	Strong*	Plaster; cement; water purification; agriculture
Magnesium hydroxide	$Mg(OH)_2$	Strong*	Antacid, laxative
Ammonia	NH_3	Weak	Fertilizer, household cleansers

*Although these bases are classified as strong, they are not very soluble. Calcium hydroxide is only slightly soluble in water, and magnesium hydroxide is practically insoluble.

CONCEPTUAL EXAMPLE 7.1 Ionization of Acids and Bases

Write equations to show **(a)** the ionization of nitric acid (HNO_3) in water, and **(b)** the ionization of solid potassium hydroxide (KOH) in water.

Solution

a. An HNO_3 molecule ionizes to form a hydrogen ion and a nitrate ion. Because this reaction occurs in water, we can use (aq) to indicate that these substances are in aqueous solution.

$$HNO_3(aq) \longrightarrow H^+(aq) + NO_3^-(aq)$$

b. Potassium hydroxide (KOH), an ionic solid, simply dissolves in the water, forming separate $K^+(aq)$ and $OH^-(aq)$ ions.

$$KOH(s) \xrightarrow{H_2O} K^+(aq) + OH^-(aq)$$

■ **EXERCISE 7.1A**

Write equations to show **(a)** the ionization of HBr (hydrobromic acid) in water, and **(b)** the ionization of solid calcium hydroxide in water.

■ **EXERCISE 7.1B**

Historically, formulas for carboxylic acids (compounds containing a —COOH group; Chapter 9) are often written with the ionizable hydrogen *last*. For example, instead of writing acetic acid as $HC_2H_3O_2$, it often is written CH_3COOH. Write an equation to show the ionization of CH_3COOH in water.

2. What is an amino acid?
An amino acid contains both a –COOH acid group and an *amino* group (–NH_2, Section 9.14), which is basic. Amino acids undergo acid–base reactions with one another and link together to form the proteins in living tissue.

ANSWER

Limitations of the Arrhenius Theory

The Arrhenius theory is limited in several ways.

- A simple free proton does not exist in water solution. The H^+ ion has such a high positive charge density that it immediately seeks out a negative charge. It finds a lone pair of electrons on the O atom of an H_2O molecule and attaches itself to form a *hydronium ion*, H_3O^+.

$$H:\overset{\cdot\cdot}{\underset{\overset{\cdot\cdot}{H}}{O}}: \ + \ H^+ \longrightarrow \left[H:\overset{\cdot\cdot}{\underset{\overset{\cdot\cdot}{H}}{O}}:H \right]^+$$

Water   Hydronium ion

- It does not explain the basicity of ammonia and related compounds. Ammonia seems out of place in Table 7.2 because it contains no hydroxide ions.
- It applies only to reactions in aqueous solution.

In water the H^+ ion is probably associated with several H_2O molecules—for example, four H_2O molecules in the ion $H(H_2O)_4^+$ or $H_9O_4^+$. For most purposes, however, we simply use H^+ and ignore the associated water molecules. However, we should understand that H^+ is a simplification of the real situation: When we mention protons in water, we really mean their sources (hydronium ions).

GREEN CHEMISTRY

Sustainability: It's Basic (and Acidic)

Irv Levy, Gordon College

Green chemistry is sometimes called *sustainable chemistry*. Sustainability is much in the news these days. Sustainability has been defined as "meeting the needs of the present generation without compromising the ability of future generations to meet their needs." Green chemistry is a critical tool in attaining sustainability.

Understanding of acids and bases has led to greener, more sustainable methods for producing consumer products such as soap and renewable fuels (*biofuels*) made from plant and animal materials. Soap traditionally has been made from fats left from cooking meats. The fats were then mixed with lye (sodium hydroxide) solution to make soap. Soaps are salts of carboxylic acids. Commercial soaps today often are made by heating vegetable oils or animal fats with superheated steam to form carboxylic acids, which are then neutralized with a base such as aqueous sodium hydroxide. In either case, glycerol (also called glycerin) is formed as a by-product. The carboxylic acids used to make soaps usually have a long chain of carbon atoms. A typical soap is

$$CH_3(CH_2)_{14}COO^-Na^+$$

The chemistry of *biodiesel* fuel production (also see Chapter 15) is quite similar to that of soap making. Both methods use fats and oils, substances composed of three long chains of carbon atoms connected to a central set of three carbon atoms, an arrangement called a *triglyceride*. To make soap, triglycerides are reacted with 3 moles of base for each mole of triglyceride, producing 3 moles of soap and 1 mole of by-product glycerol. By using a methanol (CH_3OH) solution of base instead of a water solution, the triglycerides are converted to compounds called *esters* and the by-product glycerol. A typical biodiesel molecule is

$$CH_3(CH_2)_{14}COOCH_3$$

Compare this to a typical diesel molecule from petroleum.

$$CH_3(CH_2)_{14}CH_3$$

What does this have to do with sustainability and green chemistry? The first of the Twelve Principles of Green Chemistry is the admonition to prevent waste. Wastes are often dumped, some of which are harmful to the environment. Wouldn't it be better to redirect apparently useless materials into something useful?

Consider the following scenario: Each year billions of gallons of vegetable oils and animal fats are used in the production of fried foods. Fryer oils degrade fairly quickly in commercial kitchens and, for many years, were carted off by a disposal company as waste. Today much of that used vegetable oil is being converted into biodiesel fuel, providing an alternative fuel that is not based on depleting fossil petroleum reserves, satisfying the green chemistry principle of using renewable resources while providing new use for what was once considered waste.

Both biodiesel production and soap making produce glycerol as a by-product. Glycerol has many uses. For example, it can be used as an ingredient in soap. Faculty and students at Gordon College, in Wenham, Massachusetts, have found that by adding the biodiesel glycerol by-product to the fats and oils traditionally used in soap making, the "unwanted" glycerol can be incorporated into a useful consumer product while decreasing the quantity of fats and oils needed to produce the soap. Much more glycerol by-product is produced than can be used by the soap industry, however, so the search continues for other green uses for glycerol.

Acids get involved in the "green cleaning" arena as well. For example, polylactic acid (PLA) is a type of biodegradable and renewable plastic made from chemicals derived from corn (also see "Greener Polymers" essay in Chapter 10). PLA plastic is found in many consumer materials, such as plastic cups made by NatureWorks®. Faculty and students at Simmons College, in Boston, Massachusetts, have developed a technique to convert used PLA plastic cups from the dining services at the college into an antimicrobial cleaning solution containing lactic acid. The shredded plastic is mixed with alcohol and base to break the plastic back down ("depolymerize" it) to sodium lactate. The basic sodium lactate solution is then neutralized with acid to form the lactic acid cleaner. And the story comes full circle when the lactic acid cleaner is used to wipe away soap scum.

In the quest for a sustainable future, acids and bases are an essential part of the toolbox that will continue to be used as chemists search for better ways to prevent waste and utilize renewable starting materials. In this sense one might argue that sustainability is basic—not to mention acidic.

Like many scientific theories, a better one based on newer data has supplanted Arrhenius's theory.

The Brønsted–Lowry Acid–Base Theory

The shortcomings of the Arrhenius theory were largely overcome by a theory proposed independently, in 1923, by J. N. Brønsted in Denmark and T. M. Lowry in Great Britain. In the Brønsted–Lowry theory,

- an **acid** is a *proton donor*
- a **base** is a *proton acceptor*

The theory describes the ionization of hydrogen chloride in this way:

$$HCl(aq) + H_2O \longrightarrow H_3O^+(aq) + Cl^-(aq)$$

The acid molecules donate hydrogen ions (protons) to the water molecules, and so the acid (HCl) acts as a proton donor.

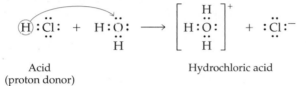

Acid
(proton donor)
Hydrochloric acid

The HCl molecule donates a proton to the water molecule, producing a hydronium ion and a chloride ion, a solution called hydrochloric acid. Other acids react similarly; they donate hydrogen ions to water to produce hydronium ions. If we let HA represent any acid, the reaction is

$$HA(aq) + H_2O \longrightarrow H_3O^+(aq) + A^-(aq)$$

Even when the solvent is something other than water, the acid acts as a proton donor, transferring H^+ ions to the solvent molecules.

CONCEPTUAL EXAMPLE 7.2 | Brønsted–Lowry Acids

Write an equation to show the reaction of HNO_3 as a Brønsted–Lowry acid with water. What is the role of water in the reaction?

Solution

As a Brønsted–Lowry acid, HNO_3 donates a proton to water, forming a hydronium ion and a nitrate ion.

$$HNO_3(aq) + H_2O \longrightarrow H_3O^+(aq) + NO_3^-(aq)$$

The water molecule accepts a proton from HNO_3; water is a Brønsted–Lowry base in this reaction.

■ **EXERCISE 7.2A**

Write an equation to show the reaction of HBr as a Brønsted–Lowry acid with water.

■ **EXERCISE 7.2B**

Write an equation to show the reaction of HBr as a Brønsted–Lowry acid with methanol, CH_3OH.

Where does the OH^- come from in the ionization of bases such as ammonia (NH_3)? The Arrhenius theory is inadequate in answering this question, but the Brønsted–Lowry theory explains how ammonia acts as a base in water. Ammonia is a gas at room temperature. When it is dissolved in water, some of the ammonia molecules react as shown by the following equation.

$$NH_3(aq) + H_2O \longrightarrow NH_4^+(aq) + OH^-(aq)$$

Acid   Base

▶ **Figure 7.3** A Brønsted–Lowry acid is a proton donor. A base is a proton acceptor.

QUESTION: Can you write an equation in which the acid is represented as HA and the base as :B$^-$?

An ammonia molecule accepts a proton from a water molecule; NH_3 acts as a Brønsted–Lowry base. (Recall that the N atom of ammonia has a lone pair of electrons to which a proton can attach.) The water molecule acts as a proton donor—an acid. The ammonia molecule becomes an ammonium ion. When a proton leaves a water molecule, it leaves behind the electron pair that joined it to the O atom. The water molecule becomes a negatively charged hydroxide ion.

In general, then, *a base is a proton acceptor* (Figure 7.3). This definition includes not only hydroxide ions but also neutral molecules such as ammonia. It also includes other negative ions such as oxide (O^{2-}), carbonate (CO_3^{2-}), and bicarbonate (HCO_3^-). The idea of an acid as a proton donor and a base as a proton acceptor greatly expands our concept of acids and bases.

Salts

Salts, formed from the neutralization reaction of acids and bases, are ionic compounds composed of cations and anions. These ions can be simple ions, such as sodium ion (Na^+) and chloride (Cl^-), or polyatomic ions, such as ammonium ion (NH_4^+), sulfate ion (SO_4^{2-}), and acetate (CH_3COO^-). Sodium chloride, ordinary table salt, is probably the most familiar salt.

Salts that conduct electricity when dissolved in water are called *electrolytes*. Various electrolytes, in certain amounts, are critical for various bodily functions including nerve conduction, heartbeat, and fluid balance. In medical practice, blood tests check electrolyte levels of Na^+, K^+, Cl^-, and HCO_3^- (eChapter 19).

There are many common salts with familiar uses. Sodium chloride and calcium chloride are used to melt ice on roads and sidewalks in winter. Copper(II) sulfate is used to kill tree roots in sewage lines. We will encounter many in following chapters as dietary minerals (Chapter 17), fertilizers (eChapter 20), and more. Table 7.3 lists some salts used in medicine.

Table 7.3 Some Salts with Present or Past Uses in Medicine

Name	Formula	Uses
Silver nitrate	$AgNO_3$	Germicide and antiseptic
Stannous fluoride [Tin(II) fluoride]	SnF_2	In toothpaste to prevent dental caries
Calcium sulfate (plaster of Paris)	$(CaSO_4)_2 \cdot H_2O$	Plaster casts
Magnesium sulfate (Epsom salts)	$MgSO_4 \cdot 7\,H_2O$	Cathartic
Potassium permanganate	$KMnO_4$	Cauterizing agent, antiseptic
Zinc oxide	ZnO	Base powder for calamine lotion
Ferrous sulfate [Iron(II) sulfate]	$FeSO_4$	Prescribed for iron deficiency anemia
Zinc sulfate	$ZnSO_4$	To treat skin conditions such as eczema
Barium sulfate	$BaSO_4$	Ingredient of a "barium cocktail": provides the contrast material for gastrointestinal radiographs
Mercurous chloride [Mercury(I) chloride; calomel]	Hg_2Cl_2	Cathartic

Self-Assessment Questions

For items 1–5, match each formula with its application.

1. CH_3COOH (a) battery acid
2. H_3BO_3 (b) drain cleaner
3. HCl   (c) mild antiseptic
4. H_2SO_4 (d) stomach acid
5. NaOH   (e) vinegar
6. When dissolved in water, an acid usually forms
     a. a covalent compound   **b.** H^+ ions and anions
     c. OH^- ions and cations   **d.** a salt

For items 7–10, match the term with the proper definition.

7. Arrhenius acid   (a) H^+ ion acceptor
8. Arrhenius base   (b) H^+ ion donor
9. Brønsted acid   (c) produces H^+ in water
10. Brønsted base   (d) produces OH^- in water

Answers: 1, e; 2, c; 3, d; 4, a; 5, b; 6, b; 7, c; 8, d; 9, b; 10, a

7.3 Acidic and Basic Anhydrides

In the Brønsted–Lowry view, many metal oxides act directly as bases because the oxide ion can accept a proton. These metal oxides also react with water to form metal hydroxides, compounds that are bases in the Arrhenius sense. Similarly, many nonmetal oxides react with water to form acids.

Nonmetal Oxides: Acidic Anhydrides

Many acids are made by the reaction of nonmetal oxides with water. For example, sulfur trioxide reacts with water to form sulfuric acid.

$$SO_3 + H_2O \longrightarrow H_2SO_4$$

Similarly, carbon dioxide reacts with water to form carbonic acid.

$$CO_2 + H_2O \longrightarrow H_2CO_3$$

In general, nonmetal oxides react with water to form acids.

$$\text{Nonmetal oxide} + H_2O \longrightarrow \text{acid}$$

Nonmetal oxides that act in this way are called **acidic anhydrides**. *Anhydride* means "without water." These reactions explain why rainwater is acidic (Section 7.8).

CONCEPTUAL EXAMPLE 7.3   Acidic Anhydrides

Give the formula for the acid formed when sulfur dioxide reacts with water.

Solution

Simply write the equation for the reaction, following the pattern in the preceding examples.

$$SO_2 + H_2O \longrightarrow H_2SO_3$$

The formula for the acid is derived by adding the two H atoms and one O atom of water to SO_2. It is H_2SO_3.

▪ **EXERCISE 7.3A**

Give the formula for the acid formed when selenium dioxide (SeO_2) reacts with water.

▪ **EXERCISE 7.3B**

Give the formula for the acid formed when dinitrogen pentoxide (N_2O_5) reacts with water. (*Hint:* Two molecules of acid are formed.)

► **Figure 7.4** Metal oxides are basic because the oxide ion reacts with water to form two hydroxide ions.

QUESTION: Can you write an equation that shows how solid sodium oxide, Na_2O, reacts with water to form sodium hydroxide?

O^{2-} H_2O OH^- OH^-

Metal Oxides: Basic Anhydrides

Just as acids can be made from nonmetal oxides, many common hydroxide bases can be made from metal oxides. For example, calcium oxide (lime) reacts with water to form calcium hydroxide (slaked lime).

$$CaO + H_2O \longrightarrow Ca(OH)_2$$

Another example is the reaction of lithium oxide with water to form lithium hydroxide.

$$Li_2O + H_2O \longrightarrow 2\ LiOH$$

In general, metal oxides react with water to form bases (Figure 7.4). These metal oxides are called **basic anhydrides**.

$$\text{Metal oxide} + H_2O \longrightarrow \text{base}$$

CONCEPTUAL EXAMPLE 7.4   Basic Anhydrides

Give the formula for the base formed by the addition of water to barium oxide (BaO).

Solution

Simply write the equation for the reaction. (Because barium ion has a 2+ charge, the O in BaO is present as an O^{2-} ion. As shown in Figure 7.4, the oxide ion reacts with water to form *two* hydroxide ions.)

$$BaO + H_2O \longrightarrow Ba(OH)_2$$

The formula for the base, barium hydroxide, is $Ba(OH)_2$.

■ **EXERCISE 7.4A**

Give the formula for the base formed by the addition of water to strontium oxide (SrO).

■ **EXERCISE 7.4B**

What base is formed by the addition of water to potassium oxide (K_2O)? (*Hint:* Two moles of base are formed for each mole of potassium oxide.)

Self-Assessment Questions

1. Selenic acid, H_2SeO_4, is an extremely corrosive acid that when heated is capable of dissolving gold. It is very soluble in water. The anhydride of selenic acid is
   a. SeO   **b.** SeO_2 **c.** SeO_3 **d.** SeO_4

2. Zinc hydroxide, $Zn(OH)_2$, is used as an absorbent in surgical dressings. The anhydride of zinc hydroxide is
   a. ZnO   **b.** ZnOH   **c.** ZnO_2 **d.** Zn_2O_3

Answers: 1, c; 2, a

7.4 Strong and Weak Acids and Bases

When gaseous hydrogen chloride (HCl) reacts with water, it reacts completely to form hydronium ions and chloride ions. Essentially no HCl molecules remain.

$$HCl + H_2O \longrightarrow H_3O^+ + Cl^-$$

For many purposes, we simply write the reaction as the ionization of HCl and use (aq) to indicate the involvement of water.

$$HCl(aq) \longrightarrow H^+(aq) + Cl^-(aq)$$

The poisonous gas hydrogen cyanide (HCN) also ionizes in water to produce hydrogen ions and cyanide ions. But HCN reacts only to a slight extent. In a solution that has 1 mol of HCN in 1 L of water, only one HCN molecule in 40,000 reacts to produce a hydrogen ion.

$$HCN(aq) \rightleftharpoons H^+(aq) + CN^-(aq)$$

We indicate this slight ionization by using a double arrow. The arrows indicate that the reaction is reversible; the ions can combine to form HCN molecules. A short arrow points to the right with a longer arrow pointing left to indicate that most of the HCN remains intact as HCN molecules.

- An acid such as HCl that reacts completely with water is called a **strong acid**.
- An acid such as HCN that reacts only slightly with water is a **weak acid**.

There are not many strong acids. The first three acids listed in Table 7.1 (sulfuric, nitric, and hydrochloric) are the common ones. Most acids are weak acids.

Bases are also classified as strong or weak.

- A **strong base** is completely ionized in water.
- A **weak base** is only slightly ionized in water.

Perhaps the most familiar strong base is sodium hydroxide (NaOH), commonly called lye. It exists as sodium ions and hydroxide ions even in the solid state. Other strong bases include potassium hydroxide (KOH) and the hydroxides of all the other group 1A metals. Except for $Be(OH)_2$, group 2A hydroxides are also strong bases. However, $Ca(OH)_2$ is only slightly soluble in water, and $Mg(OH)_2$ is nearly insoluble. The concentration of hydroxide ions in water from either is, therefore, not very high.

The most familiar weak base is ammonia (NH_3). It reacts with water to a slight extent to produce ammonium ions (NH_4^+) and hydroxide ions (Figure 7.5).

$$NH_3 + H_2O \rightleftharpoons NH_4^+ + OH^-$$

In its reaction with HCl (page 177), water acts as a base (proton acceptor). In its reaction with NH_3, water acts as an acid (proton donor). A substance, such as water, that can either donate a proton or accept a proton is said to be *amphiprotic* (see also Additional Problem 64).

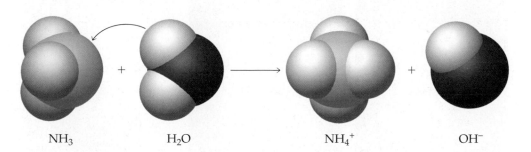

NH₃   H₂O   NH₄⁺   OH⁻

A strong acid or base is one that is almost completely ionized in solution. The word *strong* does not refer to the amount of acid or base in the solution. A solution that contains a relatively large amount of acid or base, whether strong or weak, in a given volume of solution is called a *concentrated* solution. A solution with only a little solute in that same volume of solution is a *dilute* solution.

▼ **Figure 7.5** Ammonia is a base because it accepts a proton from water. A solution of ammonia in water contains ammonium ions and hydroxide ions. Only a small fraction of the ammonia molecules react, however; most remain unchanged. Ammonia is therefore a *weak* base.

QUESTION: Amines are related to ammonia in that one or more of the H atoms of NH_3 is replaced by a carbon-containing group that we represent as R. Amines act in the same way as ammonia. Can you write an equation that shows how RNH_2 reacts with water to form R-substituted ammonium ions and hydroxide ions?

Self-Assessment Questions

For items 1–7, select the correct classification (more than one substance may fit a given classification)

1. $Ca(OH)_2$ (a) strong acid
2. HCN   (b) strong base
3. HF   (c) weak acid
4. HNO_3 (d) weak base
5. H_3PO_4
6. KOH
7. NH_3

8. Acetic acid reacts with water to form
   a. $CH_3COO^- + H_2O$ b. $CH_3COOH + OH^-$
   c. $CH_3COO^- + H_3O^+$ d. $CH_3COO^- + OH^-$

9. Ammonia reacts with water to form
   a. $NH_3 + H_2O$ b. $NH_4^+ + OH^-$
   c. $NH_2^- + H_3O^+$ d. $NH_3 + OH^-$

Answers: 1, b; 2, c; 3, c; 4, a; 5, c; 6, b; 7, d; 8, c; 9, b

7.5 Neutralization

When an acid reacts with a base, the products are water and a salt. If a solution containing hydrogen ions (an acid) is mixed with another solution containing exactly the same amount of hydroxide ions (a base), the resulting solution no longer affects litmus, and it no longer dissolves zinc or iron. Nor does it feel slippery on the skin. It is no longer either acidic or basic; it is neutral. As mentioned, the reaction of an acid with a base is called **neutralization** (Figure 7.6). In water it is simply the reaction of hydrogen ions with hydroxide ions to form water molecules.

$$H^+ + OH^- \longrightarrow H_2O$$

If sodium hydroxide is neutralized by hydrochloric acid, the products are water and sodium chloride (ordinary table salt).

▶ **Figure 7.6** The amount of acid (or base) in a solution is determined by careful neutralization. Here a 5.00-mL sample of vinegar, some water, and a few drops of phenolphthalein (an acid–base indicator) are added to a flask (a). A solution of 0.1000 M NaOH is added slowly from a buret (a device for precise measurement of volumes of solutions) (b). As long as the acid is in excess, the solution is colorless. When the acid has been neutralized and a tiny excess of base is present, the phenolphthalein indicator turns pink (c).

QUESTION: Can you write an equation for the reaction of acetic acid ($HC_2H_3O_2$), the acid in vinegar, with aqueous NaOH?

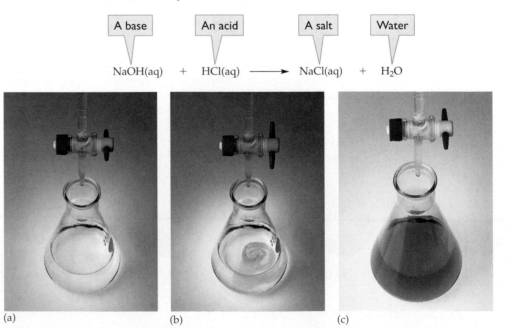

A base	An acid	A salt	Water
NaOH(aq) +	HCl(aq) ⟶	NaCl(aq) +	H_2O

(a)   (b)   (c)

EXAMPLE 7.5 Neutralization Reactions

Potassium nitrate, a component of black powder gunpowder and some fertilizers, was obtained from the late Middle Ages through the nineteenth century by precipitation from urine. Commonly called *saltpeter*, it can be prepared by the reaction of nitric acid with potassium hydroxide. Write the equation for this neutralization reaction.

Solution

The OH^- of the base and the H^+ of the acid combine to form water. The cation of the base (K^+) and the anion of the acid (NO_3^-) form a solution of the salt potassium nitrate (KNO_3).

$$KOH(aq) + HNO_3(aq) \longrightarrow KNO_3(aq) + H_2O(l)$$

■ **EXERCISE 7.5A**

Countertop spills of lye solutions (aqueous sodium hydroxide) can be neutralized with vinegar (aqueous acetic acid). Write the equation for the neutralization reaction.

■ **EXERCISE 7.5B**

A toilet bowl cleaner contains hydrochloric acid. An emergency first-aid treatment is to drink a teaspoon of milk of magnesia (magnesium hydroxide). Write the equation for the neutralization reaction between magnesium hydroxide and hydrochloric acid. (*Hint:* Be sure to write the correct formulas for reactants and the salt before attempting to balance the equation.)

ANSWER

3. What is the difference between baking soda and baking powder?
Both generate CO_2 gas by neutralization reactions; this gas makes cakes and muffins light and fluffy. Baking soda is sodium bicarbonate, $NaHCO_3$. An acid, such as lactic acid from buttermilk, must also be added to generate CO_2. Baking powder contains both baking soda and a solid acid. It needs only water to cause the reaction to occur.

Self-Assessment Questions

1. When acids and bases are mixed,
   a. the acid becomes stronger   **b.** the base becomes stronger
   c. no reaction occurs   **d.** they neutralize each other

2. What amount in moles of hydrochloric acid is needed to neutralize 1.5 mol of sodium hydroxide?
   a. 1.0 mol   **b.** 1.5 mol   **c.** 3.0 mol   **d.** 4.5 mol

3. What amount in moles of hydrochloric acid is needed to neutralize 2.4 mol of calcium hydroxide?
   a. 1.2 mol   **b.** 2.4 mol   **c.** 3.0 mol   **d.** 4.8 mol

4. What amount in moles of sodium hydroxide is needed to neutralize 1.5 mol of phosphoric acid?
   a. 0.5 mol   **b.** 3.0 mol   **c.** 4.5 mol   **d.** 6.0 mol

Answers: 1, d; 2, b; 3, d; 4, c

7.6 The pH Scale

In solutions, the concentrations of ions are measured in moles per liter. Because hydrogen chloride is completely ionized in water, a solution of 1 molar hydrochloric acid (1 M HCl), for example, contains 1 mol of H^+ ions per liter of solution. In other words, 1 L of 1 M HCl contains 6.02×10^{23} hydrogen ions, and 0.500 L of 0.00100 M HCl contains $0.500 \text{ L} \times 0.00100 \text{ mol/L} = 0.000500 \text{ mol } H^+$ or $3.01 \times 10^{20} H^+$ ions.

We can describe the acidity of a particular solution in moles per liter: The hydrogen ion concentration of a 0.00100 M HCl solution is 1×10^{-3} mol/L. However, exponential notation isn't terribly convenient. More often we see the acidity of this solution reported simply as pH 3.

We usually use the **pH** scale, first proposed in 1909 by the Danish biochemist S. P. L. Sorensen, to describe the degree of acidity or basicity. Most solutions have a pH that lies in the range 0 to 14. The neutral point on the scale is 7, with values

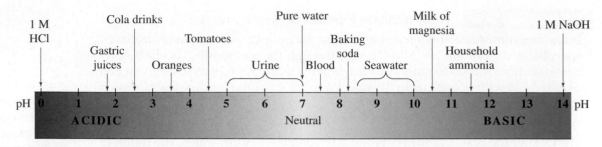

▲ **Figure 7.7** The pH scale. A change in pH of one unit means a tenfold change in the hydronium ion concentration.
QUESTION: How does the acidity of tomatoes compare to that of oranges?

Table 7.4	Relationship between pH and Concentration of Hydronium Ions

Concentration of H_3O^+(mol/L)	pH
1×10^{-0}	0
1×10^{-1}	1
1×10^{-2}	2
1×10^{-3}	3
1×10^{-4}	4
1×10^{-5}	5
1×10^{-6}	6
1×10^{-7}	7
1×10^{-8}	8
1×10^{-9}	9
1×10^{-10}	10
1×10^{-11}	11
1×10^{-12}	12
1×10^{-13}	13
1×10^{-14}	14

The relationship between pH and [H^+] is perhaps easier to see when the equation is written in the following form.
$$[H^+] = 10^{-pH}$$

More to Explore
Kolb, Doris. "The pH Concept." *Journal of Chemical Education*, January 1979, pp. 49–53, and Jensen, William B. "The Symbol for pH." *Journal of Chemical Education*, January 2004, p. 21.

below 7 indicating increasing acidity and those above 7 increasing basicity. Thus, pH 6 is slightly acidic, whereas pH 12 is strongly basic (Figure 7.7).

The numbers on the pH scale are directly related to the hydrogen ion concentration. We might expect that pure water would be completely in the form of H_2O molecules, but it turns out that about 1 out of every 500 million molecules is split into H^+ and OH^- ions. This gives a concentration of hydrogen ions and of hydroxide ions in pure water of 0.0000001 mol/L, or 1×10^{-7} M. Can you see why 7 is the pH of pure water? It is simply the power of 10 for the molar concentration of H^+, with the negative sign removed. (The H in pH stands for "hydrogen," and the p represents "power.") Thus, we define pH as the negative logarithm of the molar concentration of hydrogen ions (Table 7.4):

$$pH = -\log[H^+]$$

The brackets about the H^+ indicate molar concentration.

Although pH is an acidity scale, note that its value goes down when acidity goes up. Not only is the relationship an inverse one, but it is also logarithmic. A decrease of 1 pH unit represents a tenfold increase in acidity, and when pH goes down by 2 units, acidity increases by a factor of 100. This relationship may seem strange at first, but once you understand the pH scale, you will appreciate its convenience.

Table 7.4 summarizes the relationship between hydrogen ion concentration and pH. A pH of 4 means a hydrogen ion concentration of 1×10^{-4} mol/L, or 0.0001 M. If the concentration of hydrogen ions is 0.01 M, or 1×10^{-2} M, the pH is 2. The pH values for various common solutions are listed in Table 7.5.

Table 7.5	The Approximate pH Values of Some Common Solutions

Solution	pH
Hydrochloric acid (4%)	0
Gastric juice	1.6–1.8
Lemon juice	2.1
Vinegar (4%)	2.5
Soda pop	2.0–4.0
Milk	6.3–6.6
Urine	5.5–7.0
Rainwater*	5.6
Saliva	6.2–7.4
Pure water	7.0
Blood	7.4
Fresh egg white	7.6–8.0
Bile	7.8–8.6
Milk of magnesia	10.5
Washing soda	12.0
Sodium hydroxide (4%)	13.0

*Rainwater saturated with carbon dioxide from the atmosphere but unpolluted.

EXAMPLE 7.6 pH from Hydrogen Ion Concentration

What is the pH of a solution that has a hydrogen ion concentration of 1×10^{-5} M?

Solution

The hydrogen ion concentration is 1×10^{-5} M. The exponent is -5; the pH is therefore the negative of this exponent, or 5.

■ **EXERCISE 7.6A**

What is the pH of a solution that has a hydrogen ion concentration of 1×10^{-11} M?

■ **EXERCISE 7.6B**

What is the pH of a solution that is 0.0010 M HCl? (*Hint:* HCl is a strong acid.)

EXAMPLE 7.7 Hydrogen Ion Concentration from pH

What is the hydrogen ion concentration of a solution that has a pH of 4?

Solution

The pH value is 4. This means the exponent of 10 is -4. The hydrogen ion concentration is therefore 1×10^{-4} M.

■ **EXERCISE 7.7A**

What is the hydrogen ion concentration of a solution that has a pH of 2?

■ **EXERCISE 7.7B**

What is the concentration of a HNO_3 solution that has a pH of 3? (*Hint:* HNO_3 is a strong acid.)

EXAMPLE 7.8 pH from Hydrogen Ion Concentration

Which of the following is a reasonable pH for a solution that is 8×10^{-4} M in H^+?

a. 2.9 **b.** 3.1 **c.** 4.2 **d.** 4.8

Solution

The $[H^+]$ is greater than 1×10^{-4} M, and so the pH must be less than 4. That rules out (c) and (d). The $[H^+]$ is less than 10×10^{-4} (or 1×10^{-3} M), and so the pH must be greater than 3. That rules out (a). The only reasonable answer is (b), a value between 3 and 4.

(With a scientific calculator, you can get the actual pH by the following keystrokes: [8] [exp] [4][±] [log]. This gives a value of -3.1 for the log of 8×10^{-4}. Because the pH is the negative log of the H^+ concentration, the pH is 3.10.)

■ **EXERCISE 7.8**

Which of the following is a reasonable pH for a solution that is 2×10^{-10} M in H^+?

a. 2.0 **b.** 8.7 **c.** 9.7 **d.** 10.2

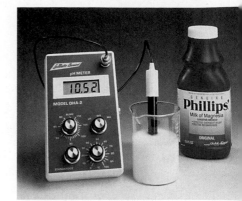

▲ A pH meter is a simple, rapid, accurate device for determining pH.

QUESTION: Is the solution in the beaker more basic or less basic than household ammonia? (Refer to Figure 7.7.)

4. What is a "pH balanced shampoo?"
Shampoo on either end of the pH scale would damage hair (and probably skin as well!). Most shampoos, therefore, are formulated to be neutral (pH 7) or slightly basic.

A N S W E R

Self-Assessment Questions

1. The hydrogen ion concentration, $[H^+]$, of a 0.0010 M HNO_3 solution is
   a. 1.0×10^{-4} M **b.** 1.0×10^{-3} M
   c. 1.0×10^{-2} M **d.** 10 M

2. What is the pH of a solution that has a hydrogen ion concentration of 1.0×10^{-11} M?
   a. 1 **b.** 3 **c.** 10 **d.** 11

3. Pool water with a pH of 8 has a hydrogen ion concentration of
   a. 8.0 M **b.** 8.0×10^{-8} M
   c. 1.0×10^{-8} M **d.** 1.0×10^{8} M

4. The pH of pure water is
   a. 0   **b.** 1   **c.** 7   **d.** 10   **e.** 14
5. Which of the following is a reasonable pH for 0.15 M HCl?
   a. 0.15   **b.** 0.82   **c.** 8.24   **d.** 13.18
6. Which of the following is a reasonable pH for 0.15 M NaOH?
   a. 0.15   **b.** 0.82   **c.** 8.24   **d.** 13.18
7. Physiological pH (7.4) is the average pH of blood. Which of the following is a reasonable hydrogen ion concentration of a solution at physiological pH?
   a. −7.4 M   **b.** 0.6 M   **c.** 6×10^{-7} M
   d. 1×10^{-8} M   **e.** 4×10^{-8}

Answers: 1, b; 2, d; 3, c; 4, c; 5, b; 6, d; 7, e

$$pH = -\log[H^+]$$

For coffee it's 5; for tomatoes it's 4;
While household ammonia's 11 or more.
It's 7 for water, if in a pure state,
But rainwater's 6 and seawater's 8.
It's basic at 10, quite acidic at 2,
And well above 7 when litmus turns blue.
Some find it a puzzlement. Doubtless their fog
Has something to do with that negative log!

7.7 Buffers and Conjugate Acid–Base Pairs

In the Brønsted–Lowry theory, a pair of compounds or ions that differ by one proton (H^+) is called a **conjugate acid–base pair**. HF and F^- are a conjugate acid–base pair, as are NH_3 and NH_4^+ ion. When a base (for example, NH_3) accepts a proton, it becomes an acid because it now has a proton that it can donate. (NH_4^+ is an acid; it can donate its "extra" proton.) Similarly, when an acid (for example, HF) donates a proton it becomes a base, because it can now accept a proton. (F^- is a base; it can accept a proton.)

CONCEPTUAL EXAMPLE 7.9   Conjugate Acid–Base Pairs

What is the conjugate base of **(a)** HBr and **(b)** of HNO_3 and what is the conjugate acid of **(c)** OH^- and **(d)** of HSO_4^-?

Solution

a. Removing a proton from HBr leaves Br^-; the conjugate base of HBr is Br^-.
b. Removing a proton from HNO_3 leaves NO_3^-; the conjugate base of HNO_3 is NO_3^-.
c. Adding a proton to OH^- gives H_2O; the conjugate acid of OH^- is H_2O.
d. Adding a proton to HSO_4^- gives H_2SO_4; the conjugate acid of HSO_4^- is H_2SO_4.

■ **EXERCISE 7.9**

What is the conjugate base of **(a)** HCN and **(b)** of H_3O^+ and what is the conjugate acid of **(c)** SO_4^{2-} and **(d)** of HCO_3^-?

Buffer Solutions

A **buffer solution** maintains an essentially constant pH when small amounts of a strong acid or strong base are added. These solutions have many important applications in industry, in the laboratory, and in living organisms because some chemical reactions consume acids, others produce acids, and many are catalyzed by acids. A buffer solution consists of a weak acid and its conjugate base (for example, $HC_2H_3O_2$ and $C_2H_3O_2^-$) or a weak base and its conjugate acid (for example, NH_3 and NH_4^+).

Consider the equation for the ionization of acetic acid:

$$HC_2H_3O_2 \, (aq) \; \rightleftharpoons \; H^+(aq) + C_2H_3O_2^-(aq)$$

where the slight ionization is indicated by a double arrow. If we add sodium acetate to a solution of acetic acid, we are adding the conjugate base of acetic acid—that is, acetate ion—thus forming a buffer solution. If we add a little strong base to this solution, it will react with the weak acid:

$$OH^- + HC_2H_3O_2 \longrightarrow H_2O + C_2H_3O_2^-$$

Do you see that the solution no longer contains the strong base? Instead, it has a little more weak base and a little less weak acid than it did before the reaction. The pH remains very nearly constant. Likewise, if a little strong acid is added, it will react with the weak base:

$$H^+ + C_2H_3O_2^- \longrightarrow HC_2H_3O_2$$

The strong acid is consumed, a weak acid takes its place, and the solution pH increases slightly. This action contrasts sharply with unbuffered solutions, in which any added strong acid or base changes the pH greatly.

A dramatic and essential example of the action of buffers is found in our blood. Blood must maintain a pH very close to 7.4 or it cannot carry oxygen from the lungs to cells. The most important buffer for maintaining acid–base balance in the blood is the carbonic acid–bicarbonate ion (H_2CO_3/HCO_3^-) buffer.

Self-Assessment Questions

1. Which of the following pairs is a conjugate acid–base pair?
   a. CH_3COOH and OH^- **b.** HCN and CN^-
   c. HCN and OH^- **d.** HCl and OH^-

2. Which of the following is *not* paired with its Brønsted–Lowry conjugate base or conjugate acid?
   a. CH_3COO^-/CH_3COOH **b.** F^-/HF
   c. H_2O/H_3O^+ **d.** NH_3/H_3O^+

3. A buffer solution is made from formic acid ($HCOOH$) and sodium formate ($HCOONa$). Added acid will react with
   a. $HCOO^-$ **b.** $HCOOH$ **c.** Na^+ **d.** OH^-

4. Which of the following pairs could be combined to make a buffer?
   a. C_6H_5COOH and C_6H_5COONa **b.** HCl and $NaCl$
   c. HCl and $NaOH$ **d.** NH_3 and NO_3^-

Answers: 1, b; 2, d; 3, a; 4, a

7.8 Acid Rain

Carbon dioxide is the anhydride of carbonic acid (Section 7.3). Raindrops falling through the air absorb CO_2, which is converted to H_2CO_3. Rainwater is therefore a dilute solution of carbonic acid, a weak acid. Rain saturated with carbon dioxide has a pH of 5.6. In many areas of the world, particularly those downwind from industrial centers, rainwater is much more acidic, with a pH as low as 3 or less. Rain with a pH below 5.6 is called **acid rain**.

Acid rain is due to acidic pollutants in the air. As we shall see in Chapter 13, several air pollutants are acid anhydrides. These include sulfur dioxide (SO_2), mainly from burning high-sulfur coal in power plants and metal smelters, and nitrogen dioxide (NO_2) and nitric oxide (NO), largely from automobile exhaust fumes.

Some acid rain is due to natural pollutants, such as those resulting from volcano eruptions and lightning. Volcanoes give off sulfur oxides and sulfuric acid, and lightning produces nitrogen oxides and nitric acid.

Acid rain is an important environmental problem that involves both air pollution (Chapter 13) and water pollution (Chapter 14). It can have serious effects on plant and animal life.

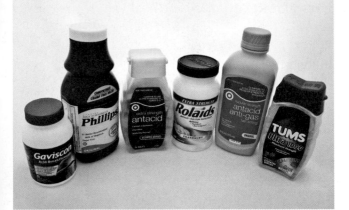

▲ **Figure 7.8** A great variety of antacids is available to consumers. All antacids are basic compounds that act by neutralizing hydrogen ions in stomach acid. Claims of "fast action" are almost meaningless because all acid–base reactions are almost instantaneous. Some tablets may dissolve a little more slowly than others. You can speed their action by chewing them.

You can make your own aspirin-free "Alka-Seltzer®." Simply place half a teaspoon of baking soda in a glass of orange juice. What is the acid and what is the base in this reaction?

7.9 Antacids: A Basic Remedy

The stomach secretes hydrochloric acid to aid in the digestion of food. Sometimes overindulgence or emotional stress leads to *hyperacidity* (too much acid secreted). Hundreds of brands of antacids (Figure 7.8) are sold in the United States to treat this condition. Despite the many brand names, there are only a few different antacid ingredients, all of which are bases. Common ingredients are sodium bicarbonate, calcium carbonate, aluminum hydroxide, magnesium carbonate, and magnesium hydroxide. Even for a given brand name, there is often a variety of products. For example, there are more than a dozen varieties of Alka-Seltzer®.

Sodium bicarbonate ($NaHCO_3$), commonly called *baking soda*, is one of the earliest antacids and is still sometimes used. The bicarbonate ions react with the acid to form carbonic acid, which then breaks down to carbon dioxide and water.

$$HCO_3^-(aq) + H^+(aq) \longrightarrow H_2CO_3(aq)$$
$$H_2CO_3(aq) \longrightarrow CO_2(g) + H_2O(l)$$

The CO_2 (g) is largely responsible for the burps that result from the use of bicarbonate-containing antacids. Sodium bicarbonate is the principal antacid in most forms of Alka-Seltzer for heartburn relief. Overuse of sodium bicarbonate can make the blood too alkaline, a condition called **alkalosis**. Antacids that contain sodium ion are not recommended for people with hypertension (high blood pressure).

Calcium carbonate ($CaCO_3$) is safe in small amounts, but regular use can cause constipation. It also appears that large amounts of calcium carbonate can actually result in increased acid secretion after a few hours. Tums® and many store-brand antacids have calcium carbonate as the only active ingredient.

Aluminum hydroxide [$Al(OH)_3$], like calcium carbonate, can cause constipation. There is also some concern that antacids containing aluminum ions can deplete the body of essential phosphate ions. Aluminum hydroxide is the active ingredient in Amphojel®.

Why Doesn't "Stomach Acid" Dissolve the Stomach?

We know that strong acids are corrosive to skin. Look back at Table 7.1 and you will see that the gastric juice in your stomach is extremely acidic. Gastric juice is a solution containing about 0.5% hydrochloric acid. Why doesn't the acid in your stomach destroy your stomach lining? The cells that line the stomach are protected by a layer of mucus, a viscous solution of a sugar–protein complex called *mucin*, and other substances in water. The mucus serves as a physical barrier, but its role is not simply passive. Rather, the mucin acts like a sponge that soaks up bicarbonate ions from the cellular side and hydrochloric acid from within the stomach. The bicarbonate ions neutralize the acid within the mucus. It used to be thought that stomach acids were the cause of ulcers, but it is now known that stomach acids play only a minor role in ulcer formation. Research now shows that most ulcers develop as a result of infection with a bacterium called *Helicobacter pylori* (*H. pylori*) that damages the mucus, exposing cells in the stomach lining to harsh stomach acids. Other agents, such as aspirin and alcohol, may be contributing factors in the development of ulcers. Treatment usually involves antibiotics that kill off *H. pylori*.

A suspension of magnesium hydroxide [Mg(OH)$_2$] in water is sold as "milk of magnesia." Magnesium carbonate (MgCO$_3$) is also used as an antacid. In small doses, these magnesium compounds act as antacids, but in large doses they act as laxatives.

Many antacid products have a mixture of antacids. Rolaids® and Mylanta® contain calcium carbonate and magnesium hydroxide. Maalox® liquid has aluminum hydroxide and magnesium hydroxide. These products balance the tendency of magnesium compounds to cause diarrhea with that of aluminum and calcium compounds to cause constipation.

Although antacids are generally safe for occasional use, they can interact with other medications. Anyone who has severe or repeated attacks of indigestion should consult a physician. Self-medication can sometimes be dangerous.

Self-Assessment Questions

1. Which of the following is a common antacid ingredient?
   a. CaCO$_3$ **b.** Ca(OH)$_2$ **c.** HCl   **d.** KOH
2. When a person with excess stomach acid takes an antacid, the pH of the person's stomach changes
   a. from a low value to a value nearer 7.
   b. from 7 to a much higher value.
   c. from a low value to an even lower value.
   d. from a high value to a lower value.

Answers: 1, a; 2, a

More to Explore
"Heartburn: Picking the Right Remedy." *Consumer Reports*, September 2002, pp. 42–45.

▲ Drugs such as ranitidine (Zantac®), famotidine (Pepcid AC®), and cimetidine (Tagamet HB®) are not antacids. Rather than neutralizing stomach acid, these drugs act on cells in the lining of the stomach, reducing the amount of acid that is produced.

7.10 Acids and Bases in Industry and in Us

Acids and bases play an important role in industry, both as products for use in many areas and as by-products that can damage the environment. Their use requires caution, and their misuse can be dangerous to human health. Acids and bases are also important participants in the biochemistry of every living thing.

Acids and Bases in Industry and at Home

Sulfuric acid is by far the leading chemical product in both the United States (about 40 billion kg produced each year) and the world (about 190 billion kg annually). Most of it is used for making fertilizers and other industrial chemicals. Around the home we use sulfuric acid in automobile batteries and in some special kinds of drain cleaners.

Hydrochloric acid is used in industry to remove rust from metal, in construction to remove excess mortar from bricks and etch concrete for painting, and in the home to remove lime deposits from fixtures and toilet bowls. The product used in the home is often called *muriatic acid,* an old name for hydrochloric acid. Concentrated solutions (about 38% HCl) cause severe burns, but dilute solutions can be used safely in the home if handled carefully. Annual U.S. production of hydrochloric acid is more than 4 billion kg.

Lime (CaO) is the cheapest and most widely used commercial base. It is made by heating limestone (CaCO$_3$) to drive off CO$_2$.

$$CaCO_3(s) + heat \longrightarrow CaO(s) + CO_2(g)$$

Annual U.S. production of calcium oxide is about 22 billion kg. Much of the lime is "slaked" by adding water, forming calcium hydroxide [Ca(OH)$_2$], which is generally safer to handle than lime. Slaked lime is used to make mortar and cement.

Sodium hydroxide (commonly known as *lye*) is the strong base most often used in the home. It is employed as an oven cleaner in products such as Easy Off®, to open clogged drains in products such as Drano®, and in both commercial and homemade soaps. Annual U.S. production of sodium hydroxide is about 9 billion kg.

More to Explore
O'Neill, Maryadele J., Patricia E. Heckelman, Cherie B. Koch, Kristin J. Roman, and Catherine M. Kenny. *The Merck Index: An Encyclopedia of Chemicals, Drugs, and Biologicals*, 14th edition. Whitehouse Station, NJ: Merck & Co., 2006. *The Merck Index* is also available online.

▲ **It DOES Matter!**
The proper pH is as important for plant growth as is fertilizer. Soil that is "sour" or too acidic (often the case) is "sweetened" by adding slaked lime (calcium hydroxide).

Ammonia is produced in huge volume, mainly for use as fertilizer. Annual U.S. production is nearly 11 billion kg. Ammonia is used around the home in a variety of cleaning products (eChapter 21).

Acids and Bases in Health and Disease

When they are misused, acids and bases can be damaging to human health. Concentrated strong acids and bases are corrosive poisons (eChapter 22) that can cause serious chemical burns. Once the chemical agents are removed, the injuries are similar to burns caused by heat, and they are often treated in the same way. Besides being a strong acid, sulfuric acid is also a powerful dehydrating agent that can react with water in the cells.

Strong acids and bases, even in dilute solutions, break down or *denature* the protein molecules in living cells, much as cooking does. Generally, the fragments are not able to carry out the functions of the original proteins. In cases of severe exposure, this fragmentation continues until the tissue has been completely destroyed.

Acids and bases affect human health in more subtle ways. A delicate balance must be maintained between acids and bases in the blood, body fluids, and cells. If the acidity of the blood changes too much, the blood loses its capacity to carry oxygen. In living cells, proteins function properly only at an optimum pH. If the pH changes too much in either direction, the proteins can't carry out their usual functions. Fortunately, the body has a complex but efficient mechanism for maintaining a proper acid–base balance.

Safety Alert
Concentrated acids and bases can cause severe burns. Use great care in working with them. Always follow directions carefully. Wear safety goggles to protect your eyes and protective clothing to protect your skin and clothes.

Self-Assessment Questions

1. The leading chemical product of U.S. industry is
   a. ammonia   **b.** lime   **c.** pesticides   **d.** sulfuric acid

2. The acid used in the automobile storage battery is
   a. citric acid   **b.** hydrochloric acid
   c. nitric acid   **d.** sulfuric acid

3. The base often used in soap making is
   a. $Ca(OH)_2$ **b.** $Mg(OH)_2$ **c.** NaOH   **d.** NH_3

Answers: 1, d; 2, d; 3, c

Critical Thinking Exercises

Apply knowledge that you have gained in this chapter and one or more of the FLaReS principles (Chapter 1) to evaluate the following statements or claims.

7.1 A television advertisement claimed that the antacid Maalox neutralizes stomach acid faster and, therefore, relieves heartburn faster than Pepcid AC®, a drug that inhibits the release of stomach acid. To illustrate this claim, two flasks of acid were shown. In one, Maalox rapidly neutralized the acid. In the other, Pepcid AC did not neutralize the acid. Did this visual demonstration validate the claim made in the advertisement?

7.2 Canadians in the province of Ontario claim that industrial plants that burn coal in the United States are polluting the air in Canada and causing damage to their buildings, trees, fish, and other wildlife. Is their claim reasonable?

7.3 In arguing a case, a witness makes the following statement: "Although runoff from our plant did appear to contaminate a stream, the pH of the stream before the contamination was 6.4, and after contamination it was 5.4. So the stream is now only slightly more acidic than before." Evaluate the statement.

7.4 A Jamaican recipe for fish uses the juice of several limes, but no heat is used to cook the fish. The directions state that the lime juice in effect cooks the fish. Is this claim reasonable?

7.5 An advertisement claims that vinegar in a glass-cleaning product will remove the spots left on glass by tap water. The spots are largely calcium carbonate deposits. Is the claim reasonable?

■ SUMMARY

Section 7.1—Acids taste sour, turn litmus red, and react with active metals to form hydrogen. Bases taste bitter, turn litmus blue, and feel slippery to the skin. Acids and bases react to form salts and water. An **acid–base indicator** such as litmus has different colors in acid and in base and is used to determine whether something is acidic or basic.

Section 7.2—According to Arrhenius's definition, an **acid** produces hydrogen ions (H⁺, protons) in water, and a **base** produces hydroxide ions (OH⁻). **Neutralization** is the combination of H⁺ and OH⁻ to form water. The remaining cations and anions make up an ionic **salt**. In the more general Brønsted–Lowry theory, an acid is a proton donor and a base is a proton acceptor. When a Brønsted–Lowry acid dissolves in water, the H_2O molecules pick up H⁺ to form hydronium ions (H_3O^+). A Brønsted–Lowry base accepts a proton from water, forming OH⁻ in water.

Section 7.3—Some nonmetal oxides (such as CO_2 and SO_3) are **acidic anhydrides** in that they react with water to form acids. Some metal oxides (such as Li_2O and CaO) are **basic anhydrides**; they react with water to form bases.

Section 7.4—A **strong acid** is one that reacts completely with water to form H⁺ and an anion. A **weak acid** reacts only slightly with water, and most of the acid exists as intact molecules. Common strong acids are sulfuric, hydrochloric, and nitric acids. Likewise, a **strong base** is completely ionized in water, and a **weak base** is only slightly ionized. Sodium hydroxide and potassium hydroxide are common strong bases.

Section 7.5—The reaction between an acid and base is called neutralization. In aqueous solution, it is the combination of H⁺ and OH⁻ to form water. The remaining cations and anions give rise to an ionic salt.

Section 7.6—The **pH** scale is an acidity and basicity scale, defined as pH = $-\log[H^+]$, where [H⁺] is "molar concentration of H⁺." A pH of 7 ([H⁺] = 1×10^{-7} M) is neutral; pH values lower than 7 represent increasing acidity, and pH values greater than 7 represent increasing basicity. A change in pH of one unit represents a tenfold change in [H⁺].

Section 7.7—A pair of compounds or ions that differ by one proton (H⁺) is called a **conjugate acid–base pair**. A **buffer solution** is a mixture of a weak acid and its conjugate base or a weak base and its conjugate acid. A buffer maintains an essentially constant pH when small amounts of a strong acid or strong base are added.

Section 7.8—**Acid rain** is rainwater with a pH less than 5.6. Acid rain arises from sulfur oxides and nitrogen oxides from natural sources as well as from industry and automobile exhaust fumes. Acid rain can have serious effects on plant and animal life.

Section 7.9—An antacid is a base such as sodium bicarbonate, magnesium hydroxide, aluminum hydroxide, or calcium carbonate that is taken to relieve hyperacidity. Overuse of some antacids can make the blood too alkaline (basic), a condition called **alkalosis**.

Section 7.10—Sulfuric acid is the number-one chemical product in the United States, used for making fertilizers and other industrial chemicals. Hydrochloric acid is used for rust removal and etching mortar and concrete. Lime (calcium oxide) is made from limestone and is the cheapest and most widely used base. Lime is an ingredient in plaster and cement and is used in agriculture. Sodium hydroxide is used to make many industrial products as well as soap. Ammonia is a weak base produced mostly for use as a fertilizer. Concentrated strong acids and bases are corrosive poisons that can cause serious burns. Living organisms have an optimum pH for their continued good health.

■ REVIEW QUESTIONS

1. Define and illustrate the following terms.
   a. acid   **b.** base   **c.** salt
2. Describe the effect on litmus and the action on iron or zinc of a solution that has been neutralized.
3. List four general properties each of **(a)** acidic solutions and **(b)** basic solutions.
4. Can a substance be a Brønsted–Lowry acid if it does not contain H atoms? Are there any characteristic atoms that must be present in a Brønsted–Lowry base?
5. Strong acids and weak acids both have properties characteristic of hydrogen ions. How do strong acids and weak acids differ?
6. What is meant by the proton as used in acid–base chemistry? How does it differ from a nuclear proton (Chapter 3)?
7. What is an acidic anhydride? A basic anhydride?
8. Describe the neutralization of an acid or base.
9. Magnesium hydroxide is completely ionic, even in the solid state, yet it can be taken internally as an antacid. Explain why it does not cause injury as sodium hydroxide would.
10. What is alkalosis? What antacid ingredient might cause alkalosis if taken in excess?
11. What are the effects of strong acids and strong bases on the skin?
12. According to the Arrhenius theory, all acids have one element in common. What is that element? Are all compounds containing that element acids? Explain.

■ PROBLEMS

Acids and Bases: The Arrhenius Theory

13. Write an equation that represents the action of iodic acid (HIO_3) as an Arrhenius acid.
14. Write an equation that represents the action of hydrogen phosphate ion (HPO_4^{2-}) as an Arrhenius acid. (Be sure to include the correct charges for ions.)
15. Write an equation showing that lithium hydroxide (LiOH) acts as an Arrhenius base in water.
16. Write an equation showing that calcium hydroxide acts as an Arrhenius base in water.

Acids and Bases: The Brønsted–Lowry Acid–Base Theory

17. Use the definitions of *acid* and *base* to identify the first compound in each equation as an acid or a base. (*Hint:* What is produced by the reaction?)
    a. $CH_3NH_2 + H_2O \longrightarrow CH_3NH_3^+ + OH^-$
    b. $H_2O_2 + H_2O \longrightarrow H_3O^+ + HO_2^-$
    c. $NH_2Cl + H_2O \longrightarrow NH_3Cl^+ + OH^-$

18. Use the definitions of *acid* and *base* to identify the first compound in each equation as an acid or a base.
    a. $(CH_3)_2NH + H_2O \longrightarrow (CH_3)_2NH_2^+ + OH^-$
    b. $C_6H_5NH_2 + H_2O \longrightarrow C_6H_5NH_3^+ + OH^-$
    c. $C_6H_5SO_2NH_2 + H_2O \longrightarrow C_6H_5SO_2NH^- + H_3O^+$

19. Write the equation that shows hydrogen chloride gas reacting as a Brønsted–Lowry acid in water. What is the name of the acid formed?

20. Write the equation that shows hydrogen bromide gas reacting as a Brønsted–Lowry acid in water. Based on the answer to Problem 19, suggest a name for the acid formed.

21. Write the equation that shows how ammonia acts as a Brønsted–Lowry base in water.

22. Hydroxylamine ($HONH_2$) is not an Arrhenius base even though it has OH, but it *is* a Brønsted–Lowry base. Write an equation that shows how hydroxylamine acts as a Brønsted–Lowry base in water.

Acids and Bases: Names and Formulas

23. Give formulas for the following acids and bases.
    a. hydrochloric acid   b. strontium hydroxide
    c. potassium hydroxide   d. boric acid

24. Give formulas for the following acids and bases.
    a. rubidium hydroxide   b. aluminum hydroxide
    c. hydrocyanic acid   d. nitric acid

25. Name the following and classify each as an acid or a base.
    a. HNO_3 b. CsOH   c. H_2CO_3

26. Name the following and classify each as an acid or a base.
    a. $Mg(OH)_2$ b. NH_3 c. CH_3COOH

27. Often, an acid ending in *-ic acid* will have a corresponding acid ending with *-ous acid*. The *-ous* acid has one less oxygen atom than the *-ic* acid. With this information, write formulas for (a) nitrous acid and (b) phosphorous acid.

28. Refer to Problem 27 and Table 7.1. Selenium is chemically similar to sulfur. Use this information to write the formulas for (a) selenic acid and (b) selenous acid.

Acidic and Basic Anhydrides

29. Give the formula for the compound formed when (a) sulfur trioxide reacts with water and (b) magnesium oxide reacts with water. In each case, is the product an acid or a base?

30. Give the formula for the compound formed when (a) potassium oxide reacts with water and (b) carbon dioxide reacts with water. In each case, is the product an acid or a base?

Strong and Weak Acids and Bases

31. Thallium hydroxide (TlOH) is ionic in the solid state and is quite soluble in water. Classify TlOH as a strong acid, weak acid, weak base, or strong base.

32. Hydrogen bromide (HBr) gas reacts completely with water to form hydronium ions and iodide ions. Classify HBr as a strong acid, weak acid, weak base, or strong base.

33. Hydrogen sulfide (H_2S) gas reacts slightly with water to form relatively few hydronium ions and hydrogen sulfide ions (HS^-) Classify H_2S as a strong acid, weak acid, weak base, or strong base.

34. Methylamine (CH_3NH_2) gas reacts slightly with water to form relatively few hydroxide ions and methylammonium ions ($CH_3NH_3^+$). Classify CH_3NH_2 as a strong acid, weak acid, weak base, or strong base.

35. Identify each of the following substances as either a strong acid, a weak acid, a strong base, a weak base, or a salt.
    a. H_3PO_4 b. LiOH   c. NH_4NO_3 d. HCl

36. Identify each of the following substances as either a strong acid, a weak acid, a strong base, a weak base, or a salt.
    a. Na_2SO_4 b. KOH   c. NH_4I d. $BaCl_2$

37. Which of the following aqueous solutions has the highest concentration of H^+ ion? Which has the lowest?
    a. 0.10 M HCl   b. 0.10 M NH_3 c. 0.10 M CH_3COOH

38. Place the following aqueous solutions in order from highest to lowest concentration of H^+ ion.
    a. 0.10 M HNO_3 b. 0.10 M CH_3NH_2
    c. 0.08 M KOH   d. 0.10 M HNO_2

Ionization of Acids and Bases

39. Write equations showing the ionization of the following as Arrhenius acids or bases.
    a. HI   b. LiOH   c. $HClO_2$

40. Write equations showing the ionization of the following as Arrhenius acids or bases.
    a. HNO_2 b. $Ba(OH)_2$ c. HBr

41. Write equations showing the ionization of the following as Brønsted–Lowry acids.
    a. $HClO_2$(aq)   b. HNO_2(aq)   c. HCN(aq)

42. Write equations to show the ionization of the following as Brønsted–Lowry acids in water.
    a. HBr   b. HIO_3 c. $CH_3CHOHCOOH$

Neutralization

43. Write equations for the reaction of (a) potassium hydroxide with hydrochloric acid and (b) lithium hydroxide with nitric acid.

44. Write equations for the reaction of (a) 1 mol calcium hydroxide with 2 mol hydrochloric acid and (b) 1 mol sulfuric acid with 2 mol potassium hydroxide.

45. Write the equation for the reaction of 1 mol phosphoric acid with 3 mol sodium hydroxide.

46. Write the equation for the reaction of 1 mol sulfuric acid with 1 mol calcium hydroxide.

The pH Scale

47. Indicate whether each of the following pH values represents an acidic, basic, or neutral solution.
    a. 4   b. 7   c. 3.5   d. 9

48. Lime juice is quite sour. Which of the following is a reasonable pH for lime juice?
    a. 2   b. 7   c. 9   d. 12

49. What is the pH of a solution that has a hydrogen ion concentration of 1.0×10^{-11} M?

50. What is the pH of a solution that has a hydrogen ion concentration of 1.0×10^{-3} M?

51. What is the hydrogen ion concentration of a solution that has a pH of 12?

52. What is the hydrogen ion concentration of a solution that has a pH of 4?

53. Milk of magnesia has a hydrogen ion concentration between 1.0×10^{-10} M and 1.0×10^{-11} M. The pH of milk of magnesia is between what two whole-number values?

54. Oven cleaner has a pH between 13 and 14. Its hydrogen ion concentration is between what two whole-number values of x in the expression 1.0×10^{-x} M?

Buffers and Conjugate Acid–Base Pairs

55. In the following reaction, identify **(a)** the conjugate base of HCl and **(b)** the conjugate acid of NH_3.

$$HCl(aq) + NH_3(aq) \longrightarrow Cl^-(aq) + NH_4^+(aq)$$

56. In the following reaction, identify **(a)** the conjugate base of propanoic acid (CH_3CH_2COOH) and **(b)** the conjugate acid of H_2O.

$$CH_3CH_2COOH + H_2O \longrightarrow CH_3CH_2COO^- + H_3O^+$$

Antacids

57. Mylanta liquid has 200 mg $Al(OH)_3$ and 200 mg $Mg(OH)_2$ per teaspoonful. Write the equations for the neutralization of stomach acid [HCl(aq)] by each of these substances.

58. What is the Brønsted–Lowry base in each of the following compounds used as ingredients in antacids?
    a. $NaHCO_3$ **b.** $Mg(OH)_2$ **c.** $MgCO_3$ **d.** $CaCO_3$

ADDITIONAL PROBLEMS

59. According to the Arrhenius theory, is every compound containing an OH group a base? Explain.

60. Lime deposits on brass faucets are mostly $CaCO_3$. The deposits can be removed by soaking the faucet in hydrochloric acid. Write an equation for the reaction that occurs.

61. Strontium iodide can be made by the reaction of solid strontium carbonate ($SrCO_3$) with hydroiodic acid (HI). Write the equation for this reaction.

62. The conjugate base of a very weak acid is a strong base, and the conjugate base of a strong acid is a weak base.
    a. Is Cl^- a strong or a weak base?
    b. Is CN^- a strong or a weak base?

63. A term akin to pH, called pOH, is related to $[OH^-]$ just as pH is related to $[H^+]$. What is the pOH **(a)** of a solution that has a hydroxide ion concentration of 1.0×10^{-2} M? **(b)** Of a 0.001 M KOH solution?

64. Like water, the hydrogen phosphate ion (HPO_4^{2-}) is amphiprotic. That is, it can act either as a Brønsted–Lowry acid or as a Brønsted–Lowry base. Write equations that illustrate both these reactions with water.

65. The pOH of a solution (Problem 63) is related to the pH of the solution by the relationship pH + pOH = 14. What is the pH **(a)** of a solution that has a pOH of 3? **(b)** Of a 0.01 M NaOH solution?

66. Three varieties of Tums have calcium carbonate as the only active ingredient: Regular Tums tablets have 500 mg; Tums E-X, 750 mg; and Tums ULTRA, 1000 mg. How many regular Tums would you have to take to get the same quantity of calcium carbonate as you would get with two Tums E-X? With two Tums ULTRA tablets?

67. Milk of magnesia has 400 mg of $Mg(OH)_2$ per teaspoon. Calculate the mass of stomach acid that can be neutralized by 1.00 teaspoon of milk of magnesia, assuming the stomach acid is 0.50% HCl by mass.

68. The active ingredient in the antacid Basajel is a gel of solid aluminum carbonate. Write the equation for the neutralization of aluminum carbonate by stomach acid (aqueous HCl).

69. When a well is drilled into rock that contains sulfide minerals, some of the minerals often dissolve in the well water and may give "sulfur water" that smells like rotten eggs. The water from these wells feels slightly slippery and contains HS^- ions. From this information, write an equation representing the reaction of HS^- ions with water.

70. Sulfuric acid is produced from elemental sulfur by a three-step process. (1) Sulfur is burned to produce sulfur dioxide. (2) Sulfur dioxide is then oxidized to sulfur trioxide using oxygen over a vanadium(V) oxide catalyst. (3) Finally the sulfur trioxide is reacted with water to produce 98% sulfuric acid. Write equations for the three reactions.

▲ Sulfur being mined

COLLABORATIVE GROUP PROJECTS

Prepare a PowerPoint, poster, or other presentation (as directed by your instructor) for presentation to the class.

71. Prepare a brief report on one of the following acids or bases. List sources (including local sources for the acid or base, if available) and commercial uses.
    a. ammonia   **b.** hydrochloric acid
    c. phosphoric acid   **d.** nitric acid
    e. sodium hydroxide **f.** sulfuric acid

72. Examine the labels of at least five antacid preparations. Make a list of the ingredients in each. Look up the properties (medical use, side effects, toxicity, and so on) of each ingredient on the Web or in a reference book such as *The Merck Index*.

73. Examine the labels of at least five toilet bowl cleaners and five drain cleaners. Make a list of the ingredients in each. Look up the formulas and properties of each ingredient on the Web or in a reference book such as *The Merck Index*. Which ingredients are acids? Which are bases?

74. Examine the labels of at least three different substances or mixtures sold to adjust the pH of swimming pool water, and list the acidic or basic ingredients of each and their properties (from the Web or *The Merck Index* or a similar reference). Also explain how pool water is tested for pH and the desired pH for pool water.

▲ **Figure 8.1** Oxidation and reduction always occur together. On the left we see orange ammonium dichromate burning. In this reaction the ammonium ion (NH_4^+) is oxidized and the dichromate ion ($Cr_2O_7^{2-}$) is reduced. Considerable heat and light are evolved. The equation for the reaction is

$$(NH_4)_2Cr_2O_7 \longrightarrow Cr_2O_3 + N_2 + 4\,H_2O$$

The water is driven off as vapor, and the nitrogen gas escapes, leaving pure Cr_2O_3 as the visible product (right).

8.1 Oxidation and Reduction: Three Views

The term *oxidation* stems from the early recognition of oxygen's involvement in oxide formation; *reduction* then meant removal of oxygen from an oxide. When oxygen combines with other elements or compounds, the process is called **oxidation**. The substances that combine with oxygen are said to have been *oxidized*. Originally the term *oxidation* was limited to reactions involving combination with oxygen. As chemists came to realize that combination with chlorine (or bromine or other active nonmetals) was not all that different from reaction with oxygen, they broadened the definition of oxidation to include these and other related reactions, as we shall shortly see.

Reduction is the opposite of oxidation. When hydrogen burns, it combines with oxygen to form water.

$$2\,H_2 + O_2 \longrightarrow 2\,H_2O$$

The hydrogen is oxidized in this reaction, but at the same time the oxygen is reduced. Whenever oxidation occurs, reduction must occur also. Oxidation and reduction always happen at the same time and in exactly equivalent amounts.

Because oxidation and reduction are chemical opposites and constant companions, their definitions are linked together. We can view oxidation and reduction in at least three different ways (Figure 8.2).

1. *Oxidation* is a gain of oxygen atoms.
   Reduction is a loss of oxygen atoms.

At high temperatures (such as those in automobile engines), nitrogen, which is normally quite unreactive, combines with oxygen to form nitric oxide.

$$N_2 + O_2 \longrightarrow 2\,NO$$

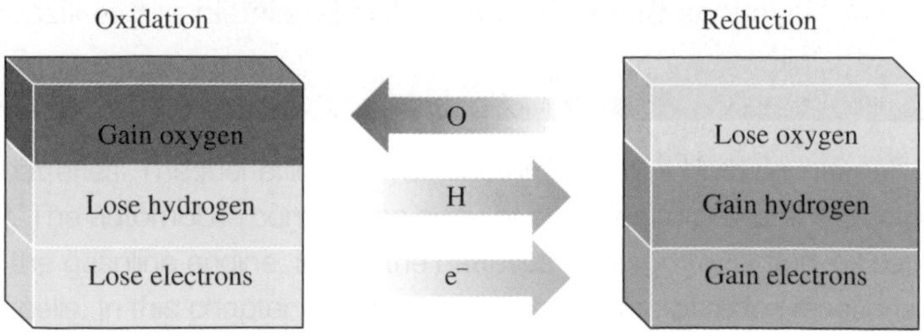

▶ **Figure 8.2** Three different views of oxidation and reduction.

Oxidation		Reduction
Gain oxygen	O	Lose oxygen
Lose hydrogen	H	Gain hydrogen
Lose electrons	e^-	Gain electrons

Nitrogen gains oxygen atoms; there are no O atoms in the N_2 molecule and one O atom in each of the NO molecules. Therefore, nitrogen is oxidized.

Now consider what happens when methane is burned to form carbon dioxide and water. Both carbon and hydrogen gain oxygen atoms, and so both elements are oxidized.

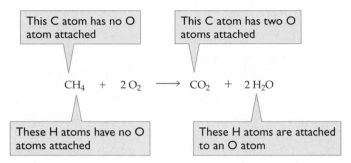

When lead dioxide is heated at high temperatures, it decomposes as follows.

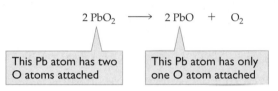

The lead dioxide loses oxygen, and so it is reduced.

CONCEPTUAL EXAMPLE 8.1) Redox—Gain or Loss of Oxygen Atoms

In each of the following reactions, is the reactant undergoing oxidation or reduction? (These are not complete chemical equations.)

a. $Pb \longrightarrow PbO_2$ b. $SnO_2 \longrightarrow SnO$
c. $KClO_3 \longrightarrow KCl$ d. $Cu_2O \longrightarrow 2\,CuO$

Solution
a. Lead gains oxygen atoms (it has none on the left and two on the right); it is oxidized.
b. Tin loses an oxygen atom (it has two on the left and only one on the right); it is reduced.
c. There are three O atoms on the left and none on the right. The compound loses oxygen; it is reduced.
d. The two copper atoms on the left share a single oxygen atom; half an oxygen atom each. On the right, each Cu atom has an O atom all its own. Cu has gained oxygen; it is oxidized.

■ EXERCISE 8.1
In each of the following reactions, is the reactant undergoing oxidation or reduction? (These are not complete chemical equations.)

a. $3\,Fe \longrightarrow Fe_3O_4$ b. $NO \longrightarrow NO_2$
c. $Cr_2O_3 \longrightarrow CrO_3$ d. $C_3H_6O \rightarrow C_3H_6O_2$

A second view of oxidation and reduction involves hydrogen atoms.

2. *Oxidation* is a loss of hydrogen atoms.
   Reduction is a gain of hydrogen atoms.

Look once more at the burning of methane:

$$CH_4 + 2\,O_2 \longrightarrow CO_2 + 2\,H_2O$$

The oxygen gains hydrogen to form water; the oxygen is reduced. (From our first definition we see that the carbon and hydrogen of CH_4 gain oxygen; CH_4 is oxidized.)

Methyl alcohol (CH_3OH), when passed over hot copper gauze, forms formaldehyde and hydrogen gas.

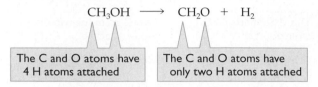

$$CH_3OH \longrightarrow CH_2O + H_2$$

| The C and O atoms have 4 H atoms attached | The C and O atoms have only two H atoms attached |

Because the methyl alcohol loses hydrogen, it is oxidized in this reaction.

Methyl alcohol can be made by reaction of carbon monoxide with hydrogen.

$$CO + 2 H_2 \longrightarrow CH_3OH$$

Because the carbon monoxide gains hydrogen atoms, it is reduced.

Biochemists often find the gain or loss of hydrogen atoms a useful way to look at oxidation–reduction processes. For example, a substance called NAD^+ is changed to NADH in a variety of biochemical redox reactions. The actual molecules are rather complex, but we can write the equation for the oxidation of ethyl alcohol to acetaldehyde, one step in the metabolism of alcohol, as

$$CH_3CH_2OH + NAD^+ \longrightarrow CH_3CHO + NADH + M + H^+$$

We can see that ethyl alcohol is oxidized (loses hydrogen) and NAD^+ is reduced (gains hydrogen).

CONCEPTUAL EXAMPLE 8.2 | Redox—Gain or Loss of Hydrogen Atoms

In each of the following reactions, is the reactant undergoing oxidation or reduction? (These are not complete chemical equations.)

a. $C_2H_6O \longrightarrow C_2H_4O$ **b.** $C_2H_2 \longrightarrow C_2H_6$

Solution

a. There are six H atoms in the reactant on the left and only four in the product on the right. The reactant loses H atoms; it is oxidized.

b. There are two H atoms in the reactant on the left and six in the product on the right. The reactant gains H atoms; it is reduced.

■ EXERCISE 8.2

In each of the following reactions, is the reactant undergoing oxidation or reduction? (These are not complete chemical equations.)

a. $C_6H_6 \longrightarrow C_6H_{12}$ **b.** $C_3H_6O \longrightarrow C_3H_4O$

A third view of oxidation and reduction involves gain or loss of electrons.

3. *Oxidation* is a loss of electrons.
Reduction is a gain of electrons.

When magnesium metal reacts with chlorine, magnesium ions and chloride ions are formed.

$$Mg + Cl_2 \longrightarrow Mg^{2+} + 2 Cl^-$$

Because the magnesium atom loses electrons, it is oxidized; because the chlorine atoms gain electrons, they are reduced.

It is easy to see that when magnesium atoms become Mg^{2+} ions they lose electrons, and that when chlorine atoms become Cl^- ions they must gain electrons. But which is oxidation and which is reduction? Perhaps Figure 8.3 can help. The charge on a simple ion is often referred to as its *oxidation number*. An increase in oxidation number (increase in positive charge) is oxidation; a decrease in oxidation number is reduction. The charge on a Mg atom is zero; thus, conversion to Mg^{2+} is an increase in oxidation number (oxidation). For Cl atoms the charge is also zero, and a change to Cl^- is therefore a decrease in oxidation number (reduction).

This view of oxidation and reduction as a gain or loss of electrons is especially useful in electrochemistry (Section 8.3).

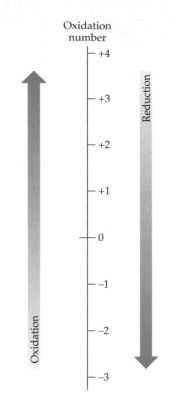

CONCEPTUAL EXAMPLE 8.3   Redox—Gain or Loss of Electrons

In each of the following reactions, is the reactant undergoing oxidation or reduction? (These are not complete chemical equations.)

a. $Zn \longrightarrow Zn^{2+}$ **b.** $Fe^{3+} \longrightarrow Fe^{2+}$
c. $S^{2-} \longrightarrow S$ **d.** $AgNO_3 \longrightarrow Ag$

Solution

a. In forming a 2+ ion, a Zn atom loses two electrons. Its oxidation number increases from 0 to +2. Zinc is oxidized.

b. To go from a 3+ ion to a 2+ ion, Fe gains an electron. Its oxidation number decreases from +3 to +2. Iron(III) ion is reduced.

c. To go from a 2− ion to an atom with no charge, S loses two electrons. Its oxidation number increases. Sulfide ion is oxidized.

d. To answer this question, you must recognize that $AgNO_3$ is an ionic compound with Ag^+ and NO_3^- ions. In going from Ag^+ to Ag, silver gains an electron. Its oxidation number decreases. Silver ion is reduced.

■ **EXERCISE 8.3**

In each of the following reactions, is the reactant undergoing oxidation or reduction? (These are not complete chemical equations.)

a. $Cu^{2+} \longrightarrow Cu$ **b.** $MnO_4^{2-} \longrightarrow MnO_4^-$
c. $Sn^{2+} \longrightarrow Sn^{4+}$ **d.** $Cu \longrightarrow CuSO_4$

▲ **Figure 8.3** An increase in oxidation number means a loss of electrons and is therefore oxidation. A decrease in oxidation number means a gain of electrons and is therefore reduction.

QUESTION: Manganese goes from oxidation number +4 in MnO_2 to +7 in MnO_4^-. Is it oxidized or reduced? Sulfur goes from oxidation number 0 in S_8 to −2 in S^{2-}. Is it oxidized or reduced?

Why do we have different ways to look at oxidation and reduction? Oxidation as a gain of oxygen is historical and specific, but the definition in terms of electrons applies more broadly. Which one should we use? Usually we use the clearest or most convenient definition. For the combustion of carbon

$$C + O_2 \longrightarrow CO_2$$

it is most convenient to see that carbon is oxidized by gaining oxygen atoms. Similarly, for the reaction

$$CH_2O + H_2 \longrightarrow CH_4O$$

it is easy to see that the reactant gains hydrogen atoms and is thereby reduced. Finally, in the case of the reaction

$$3\,Sn^{2+} + 2\,Bi^{3+} \longrightarrow 3\,Sn^{4+} + 2\,Bi$$

it is clear that tin is oxidized because it loses electrons, increasing in oxidation number from +2 to +4. Similarly, we see that bismuth is reduced because it gains electrons, decreasing in oxidation number from +3 to 0.

More to Explore
Cox, Amy L., and James R. Cox. "Determining Oxidation–Reduction on a Simple Number Line." *Journal of Chemical Education*, August 2002, pp. 965–967.

Self-Assessment Questions

1. Which of the following is *not* an example of how we view oxidation?
a. electrons gained   **b.** electrons lost
c. hydrogen atoms lost   **d.** oxygen atoms gained

2. Which of the following (partial) equations represents an oxidation of nitrogen?
a. $NH_3 \longrightarrow NH_4^+$ **b.** $N_2O_4 \longrightarrow NI_3$
c. $NO_3^- \longrightarrow NO$ **d.** $NO_2 \longrightarrow N_2O_5$

Two mnemonics:

Leo the lion.

         LEO says GER

Loss of

  Electrons is

    Oxidation.

Gain of

  Electrons is

    Reduction.

        OIL RIG

Oxidation

  Is

    Loss of electrons.

Reduction

  Is

    Gain of electrons.

3. When MnO_4^- reacts to form Mn^{2+}, the manganese in MnO_4^- is
   a. reduced; it loses electrons
   b. reduced; it loses oxygen
   c. oxidized; its oxidation number increases
   d. oxidized; it gains positive charge

4. Which of the following partial equations represents a reduction of molybdenum?
   a. $MoO_2 + H_2O \longrightarrow H_2MoO_3$ b. $MoO_2 + 4 H_2 \longrightarrow Mo + 2 H_2O$
   c. $2 MoO_2 + O_2 \longrightarrow 2 MoO_3$ d. $MoO_2 + 4 H^+ \longrightarrow Mo^{4+} + 2 H_2O$

5. In the following reaction, the Cl_2 is

   $$3 Cl_2 + 2 Al \longrightarrow 2 AlCl_3$$

   a. both oxidized and reduced   b. neither oxidized nor reduced
   c. oxidized only   d. reduced only

6. In the following partial reaction, the magnesium

   $$Mg \longrightarrow Mg^{2+} + 2 e^-$$

   a. gains electrons and is oxidized   b. gains electrons and is reduced
   c. loses electrons and is oxidized   d. loses electrons and is reduced

7. When an element is oxidized, its oxidation number
   a. decreases as electrons are gained
   b. decreases as electrons are lost
   c. increases as electrons are gained
   d. increases as electrons are lost

Answers: 1, a; 2, d; 3, b; 4, b; 5, d; 6, c; 7, d

8.2 Oxidizing and Reducing Agents

Oxidation and reduction occur together. However, in looking at oxidation–reduction reactions, we often find it convenient to focus on the role played by a particular reactant. For example, in the reaction

$$CuO + H_2 \longrightarrow Cu + H_2O$$

we see that copper oxide is reduced and hydrogen is oxidized. We can also see that if one substance is oxidized, the other must *cause* it to be oxidized. In the preceding example, CuO causes H_2 to be oxidized. Therefore, CuO is called the **oxidizing agent**. Conversely, H_2 causes CuO to be reduced, and so H_2 is the **reducing agent**. Each oxidation–reduction reaction has an oxidizing agent and a reducing agent among the reactants. The reducing agent is the substance being oxidized; the oxidizing agent is the substance being reduced.

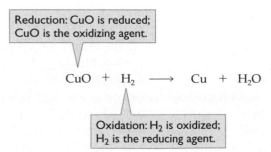

| CONCEPTUAL EXAMPLE 8.4 | Oxidizing and Reducing Agents

Identify the oxidizing agents and reducing agents in the following reactions.

a. $2 C + O_2 \longrightarrow 2 CO$ **b.** $N_2 + 3 H_2 \longrightarrow 2 NH_3$

c. $SnO + H_2 \longrightarrow Sn + H_2O$ **d.** $Mg + Cl_2 \longrightarrow Mg^{2+} + 2 Cl^-$

Solution

We can determine the answers by one or more of the preceding methods.

a. C gains oxygen and is oxidized, and so it must be the reducing agent. O_2 is therefore the oxidizing agent.

b. N_2 gains hydrogen and is reduced, and so it is the oxidizing agent. H_2 therefore is the reducing agent.

c. SnO loses oxygen and is reduced, and so it is the oxidizing agent. H_2 is therefore the reducing agent.

d. Mg loses electrons and is oxidized, and so it is the reducing agent. Cl_2 is therefore the oxidizing agent.

■ **EXERCISE 8.4**

Identify the oxidizing agents and reducing agents in the following reactions.

a. $Se + O_2 \longrightarrow SeO_2$ **b.** $CH_3CN + 2 H_2 \longrightarrow CH_3CH_2NH_2$

c. $V_2O_5 + 2 H_2 \longrightarrow V_2O_3 + 2 H_2O$ **d.** $2 K + Br_2 \longrightarrow 2 K^+ + 2 Br^-$

Self-Assessment Questions

1. A reducing agent

  a. gains electrons   **b.** gains protons

  c. loses electrons   **d.** loses protons

2. Which substance is the oxidizing agent in the following reaction?

$$Cu(s) + 2 Ag^+(aq) \longrightarrow Cu^{2+}(aq) + 2 Ag(s)$$

  a. Ag   **b.** Ag^+ **c.** Cu   **d.** Cu^{2+}

3. In the following reaction, the reducing agent is

$$Al(s) + Cr^{3+}(aq) \longrightarrow Al^{3+}(aq) + Cr(s)$$

  a. Al(s)   **b.** $Al^{3+}(aq)$ **c.** $Cr^{3+}(aq)$ **d.** Cr(s)

4. What is the oxidizing agent in the following reaction?

$$Zn(s) + 2 Tl^+(aq) \longrightarrow Zn^{2+}(aq) + 2 Tl(s)$$

  a. Tl(s)   **b.** $Tl^+(aq)$ **c.** Zn(s)   **d.** $Zn^{2+}(aq)$

5. What is the reducing agent in following equation?

$$12 H^+(aq) + 2 IO_3^-(aq) + 10 Fe^{2+}(aq) \longrightarrow 10 Fe^{3+}(aq) + I_2(s) + 9 H_2O(l)$$

  a. $Fe^{2+}(aq)$ **b.** $H^+(aq)$ **c.** $I_2(s)$ **d.** $IO_3^-(aq)$

Answers: 1, c; 2, b; 3, a; 4, b; 5, a

8.3 Electrochemistry: Cells and Batteries

An electric current in a wire is simply a flow of electrons. Oxidation–reduction reactions in which electrons are transferred from one substance to another can be used to produce electricity. This is what happens in dry cell and storage batteries.

When a coil of copper wire is placed in a solution of silver nitrate (Figure 8.4a), the copper atoms give up their outer electrons to the silver ions. (We can omit the nitrate ions from the equation because they do not change.) The copper metal dissolves, going into solution as copper(II) ions that color the solution blue. The silver

▶ **Figure 8.4** The left photograph shows a coil of copper wire immersed in a colorless solution of Ag^+ ions. As the reaction progresses, the more active copper displaces the less active silver from solution, producing (right) needle-like crystals of silver metal and a blue solution of Cu^{2+} ions. The equation for the reaction is
$$2\,Ag^+(aq) + Cu(s) \longrightarrow 2\,Ag(s) + Cu^{2+}(aq).$$

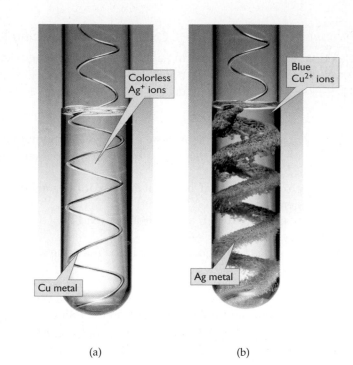

(a)　　　　　　　(b)

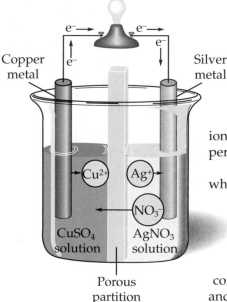

▲ **Figure 8.5** A simple electrochemical cell. The half-reactions are

Oxidation (anode):

$$Cu(s) \longrightarrow Cu^{2+}(aq) + 2\,e^-$$

Reduction (cathode):

$$Ag^+(aq) + e^- \longrightarrow Ag(s)$$

QUESTION: To balance the equation for the overall reaction we need two Ag^+ and two $Ag(s)$. Why?

We saw in Chapter 3 that electricity can produce chemical change, a process called *electrolysis*. For example, an electric current through molten sodium chloride produces sodium metal and chlorine gas. Here we see the reverse process in which chemical change produces electricity.

ions come out of solution as beautiful needles of silver metal (Figure 8.4b). The copper is oxidized; the silver ions are reduced.

Because the reaction simply involves transfer of electrons, it can occur even when the silver ions are separated from the copper metal. By placing the reactants in separate compartments and connecting them with a wire, the electrons will flow through the wire to get from the copper metal to the silver ions. This flow of electrons constitutes an electric current, and it can be used to run a motor or light a lamp.

In the **electrochemical cell** pictured in Figure 8.5, there are two separate compartments. One contains copper metal in a blue solution of copper(II) sulfate, and the other contains silver metal in a colorless solution of silver nitrate. Copper atoms give up electrons much more readily than silver atoms, and so electrons flow away from the copper and toward the silver. The copper metal slowly dissolves as copper atoms give up electrons to form copper(II) ions. The electrons flow through the wire to the silver, where silver ions pick them up to become silver atoms.

As time goes by, the copper bar slowly disappears and the silver bar gets bigger. The blue solution becomes darker blue as Cu atoms are converted to Cu^{2+} ions. Those Cu^{2+} ions give the left compartment a positive charge, so (negative) nitrate ions move from the right compartment, through the porous partition, into the copper sulfate solution. (That is why the partition is porous. If the nitrate ions were unable to move through the barrier, the cell would not work.) For each copper atom that gives up two electrons, two silver ions each pick up one electron, and two nitrate ions move from the right compartment to the left compartment.

The two pieces of metal where electrons are transferred are called **electrodes**. The electrode where oxidation occurs is called the **anode**. The one where reduction occurs is the **cathode**. In our cell, copper gives up electrons, it is oxidized, and the copper bar is therefore the anode. Silver ions gain electrons and are reduced, so the silver bar is the cathode.

Electrochemical reactions are often represented as two *half-reactions*. The following representation shows the two half-reactions for the copper–silver cell and how they are added to give the overall cell reaction.

Oxidation:	$Cu(s) \longrightarrow Cu^{2+}(aq) + 2\,e$
Reduction:	$2\,Ag^+(aq) + 2\,e \longrightarrow 2\,Ag(s)$
Overall reaction:	$Cu(s) + 2\,Ag^+(aq) \longrightarrow Cu^{2+}(aq) + 2\,Ag(s)$

Note that the electrons cancel when the two half-reactions are added.

| CONCEPTUAL EXAMPLE 8.5 | Oxidation and Reduction Half-Reactions |

Represent the following reaction as two half-reactions and label them as an oxidation half-reaction and a reduction half-reaction.

$$Mg + Cl_2 \longrightarrow Mg^{2+} + 2\,Cl^-$$

Solution

Magnesium is oxidized from Mg to Mg^{2+}, a process that involves loss of two electrons from the Mg atom. The oxidation half-reaction is, therefore,

$$Oxidation: \quad Mg \longrightarrow Mg^{2+} + 2\,e^-$$

The reduction half-reaction involves chlorine. Each of the two Cl atoms in the Cl_2 molecule must gain an electron to form a Cl^- ion. The reduction half-reaction is, therefore,

$$Reduction: \quad Cl_2 + 2\,e^- \longrightarrow 2\,Cl^-$$

■ **EXERCISE 8.5**

Represent the following reaction as two half-reactions and label them as an oxidation half-reaction and a reduction half-reaction.

$$2\,Al + 3\,Br_2 \longrightarrow 2\,Al^{3+} + 6\,Br^-$$

| EXAMPLE 8.6 | Balancing Redox Equations |

Balance the following half-reactions and combine them to give a balanced overall reaction.

$$Sn^{2+} \longrightarrow Sn^{4+}$$
$$Bi^{3+} \longrightarrow Bi$$

Solution

Atoms are balanced in both, but electric charge is not. To balance charge in the first half-reaction, we add two electrons on the right side.

$$Sn^{2+} \longrightarrow Sn^{4+} + 2\,e^-$$

The second half-reaction requires three electrons on the left side.

$$Bi^{3+} + 3\,e^- \longrightarrow Bi$$

Before we can combine the two, however, we must set electron loss equal to electron gain. (Electrons lost by the substance being oxidized must be gained by the substance being reduced.) To do this, we multiply the first half-reaction by 3 and the second by 2.

$$3 \times (Sn^{2+} \longrightarrow Sn^{4+} + 2\,e^-) = 3\,Sn^{2+} \longrightarrow 3\,Sn^{4+} + \cancel{6\,e^-}$$
$$\underline{2 \times (Bi^{3+} + 3\,e^- \longrightarrow Bi) = 2\,Bi^{3+} + \cancel{6\,e^-} \longrightarrow 2\,Bi}$$
$$3\,Sn^{2+} + 2\,Bi^{3+} \longrightarrow 3\,Sn^{4+} + 2\,Bi$$

Note that both atoms and charges balance in the overall reaction.

■ **EXERCISE 8.6A**

Balance the following half-reactions and combine them to give a balanced overall reaction.

$$Fe \longrightarrow Fe^{3+}$$
$$Mg^{2+} \longrightarrow Mg$$

■ **EXERCISE 8.6B**

Balance the following half-reactions and combine them to give a balanced overall reaction.

$$Pb \longrightarrow Pb^{2+}$$
$$Ag(NH_3)_2^+ \longrightarrow Ag + 2\,NH_3$$

1. Why do penlight batteries and flashlight batteries have the same voltage?
Cell voltage depends on the overall cell reaction. Cells with the same cell reaction will have the same voltage. However, a larger cell ("flashlight battery") contains more reactants and will last longer than a smaller cell ("penlight battery").

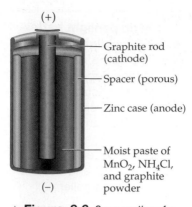

▲ Figure 8.6 Cross section of a zinc–carbon cell. The half-reactions are

Oxidation (anode):

$$Zn(s) \longrightarrow Zn^{2+}(aq) + 2\,e^-$$

Reduction (cathode):

$$2\,MnO_2(s) + H_2O + 2\,e^- \longrightarrow Mn_2O_3(s) + 2\,OH^-(aq)$$

2. Why does a battery go dead?

When one or more of the reactants (zinc and MnO_2 in a dry cell) is essentially used up, the cell can no longer provide useful energy.

3. Why are alkaline batteries better than ordinary zinc–carbon batteries?

The zinc of a dry cell reacts slowly with the ammonium chloride, so a dry cell "dies" within a year or so even if it is not used. This undesirable reaction occurs much less in the presence of hydroxide ion ("alkaline") so that alkaline cells have a much longer shelf life than ordinary zinc–carbon cells.

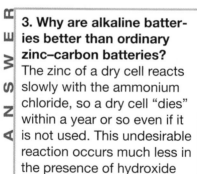

▶ Figure 8.7 One cell of a lead–acid battery has two anode plates and two cathode plates. Six such cells make up the common 12-V car battery. The half-reactions are

Oxidation (anode):

$$Pb(s) + SO_4^{2-}(aq) \longrightarrow PbSO_4(s) + 2\,e^-$$

Reduction (cathode):

$$PbO_2(s) + 4\,H^+(aq) + SO_4^{2-}(aq) + 2\,e^- \longrightarrow PbSO_4(s) + 2\,H_2O$$

Dry Cells

The familiar *dry cell* (Figure 8.6) is used in flashlights and many other small portable devices. It has a zinc anode—in this case, the container itself. A carbon rod in the center of the cell is the cathode. The space between the cathode and the anode contains a moist paste of graphite powder (carbon), manganese dioxide (MnO_2), and ammonium chloride (NH_4Cl). The anode reaction is the oxidation of the zinc cylinder to zinc ions. The cathode reaction involves reduction of manganese dioxide. A simplified version of the overall reaction is

$$Zn + 2\,MnO_2 + H_2O \longrightarrow Zn^{2+} + Mn_2O_3 + 2\,OH^-$$

Alkaline cells are similar, but most contain moist manganese dioxide and potassium hydroxide. They are more expensive than ordinary zinc–carbon cells, but they last longer both in storage and in use.

Lead Storage Batteries

Although we often refer to dry cells as batteries, a **battery** is actually a collection of electrochemical cells. The 12-volt (V) storage battery used in automobiles, for example, is a series of six 2-V cells. Each cell (Figure 8.7) contains a pair of electrodes, one lead and the other lead dioxide, in a chamber filled with sulfuric acid. Lead–acid batteries have been in existence for 150 years. Used mainly in automobiles, 300 million are made every year.

An important feature of the lead storage battery is that it can be recharged. It discharges as it supplies electricity when you turn on the ignition to start a car or when the motor is off and the lights are on. But it is recharged when the car is moving and an electric current is supplied to the battery by the mechanical action of the car. The net reaction during discharge is

$$Pb + PbO_2 + 2\,H_2SO_4 \longrightarrow 2\,PbSO_4 + 2\,H_2O$$

The reaction during recharge is just the reverse.

$$2\,PbSO_4 + 2\,H_2O \longrightarrow Pb + PbO_2 + 2\,H_2SO_4$$

Lead storage batteries are durable, but they are heavy and contain corrosive sulfuric acid.

Other Batteries

Much of current battery technology involves the use of lithium, which has an extraordinarily low density and provides a fairly high voltage. Lithium cells make today's lightweight laptop computers possible. Lithium–SO_2 cells are used in sub-

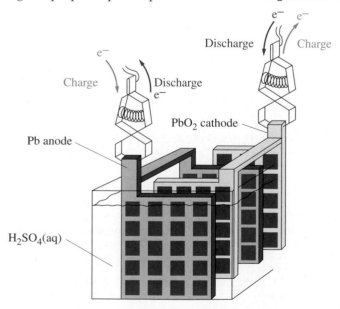

marines and rocket vehicles (such as the Jupiter probe). Lithium–iodine cells are used in pacemakers, and lithium–FeS$_2$ batteries are used in cameras, radios, and compact disc players.

Lithium cells provide so much energy in such a compact volume that there have been incidents of the cells bursting or even catching fire while being rapidly charged or discharged. Between 2005 and 2008, several computer manufacturers recalled millions of computer batteries because of the potential for such incidents.

The rechargeable Ni–Cad cell (Cd anode, NiO cathode) has long been popular for portable radios and cordless appliances. Its biggest competitor is the nickel–metal hydride cell, which replaces cadmium with a hydrogen-absorbing nickel alloy such as ZrNi$_2$ or LaNi$_5$.

The small "button" cells used in hearing aids and hand calculators formerly contained mercury (zinc anode, HgO cathode), but they are gradually being phased out and replaced with zinc–air cells. Tiny silver oxide cells (zinc anode, Ag$_2$O cathode) are used mainly in watches and cameras.

Fuel Cells

An interesting kind of battery is the fuel cell. When fossil fuels, our major energy source, are burned to generate electricity, at most 35–40% of their energy of combustion is actually harnessed. In a **fuel cell**, the fuel is oxidized at the anode and oxygen is reduced at the cathode with 70–75% efficiency. As we shall see in Chapter 15, most present-day fuel cells use hydrogen as fuel with platinum, nickel, or rhodium electrodes in a solution of potassium hydroxide.

More to Explore
"Car Batteries That Last."
Consumer Reports, October 2002, pp. 29–31, and "Picking the Right Battery." *Consumer Reports*, December 2002, p. 55.

▲ Today consumers see a wide range of cells and batteries, many of which are tailored for specific applications.

More to Explore
Smith, Michael J., and Colin A. Vincent. "Structure and Content of Some Primary Batteries." *Journal of Chemical Education*, April 2001, pp. 519–521.

Self-Assessment Questions

1. In the electrochemical cell with the reaction described by the following equation, the negative electrode is

$$Zn(s) + Cu^{2+}(aq) \longrightarrow Zn^{2+}(aq) + Cu(s)$$

   a. Cu(s) **b.** Cu^{2+}(aq) **c.** Zn(s) **d.** Zn^{2+}(aq)

2. In a dry cell, the anode is ____ and the cathode is ____.
   a. Zn, MnO$_2$ **b.** Zn, C **c.** C, Zn **d.** C, MnO$_2$

3. In an electrochemical cell composed of two half-cells, ions flow from one half-cell to another by means of
   a. electrodes **b.** an external conductor
   c. a porous partition **d.** a voltmeter

4. For the following reaction, the oxidation half-equation is
   $$Mg + CuO \longrightarrow MgO + Cu$$

   a. Cu^{2+} + 2 e$^-$ $\longrightarrow$ CuO **b.** CuO $\longrightarrow$ Cu^{2+} + 2 e$^-$
   c. Mg^{2+} + 2 e$^-$ $\longrightarrow$ MgO **d.** Mg $\longrightarrow$ Mg^{2+} + 2 e$^-$

5. In the lead storage battery, the product of both the cathode and anode reactions is
   a. H$_2$SO$_4$ **b.** Pb **c.** PbO$_2$ **d.** PbSO$_4$

6. The fluid in the chamber of a lead-acid battery is
   a. hydrochloric acid **b.** hydrofluoric acid
   c. methanol **d.** sulfuric acid

7. The overall electricity-producing reaction in the H$_2$/O$_2$ fuel cell is
   $$2 H_2(g) + O_2(g) \longrightarrow 2 H_2O(g)$$

   The reaction at the cathode is
   a. H$_2$ $\longrightarrow$ 2 H$^+$ + 2 e$^-$ **b.** 2 H$^+$ + 2 e$^-$ $\longrightarrow$ H$_2$
   c. O$_2$(g) + 4 H$^+$ + 4 e$^-$ $\longrightarrow$ 2 H$_2$O **d.** 2 H$_2$O $\longrightarrow$ O$_2$(g) + 4 H$^+$ + 4 e$^-$

Answers: 1, c; 2, b; 3, c; 4, d; 5, d; 6, d; 7, c

8.4 Corrosion

A redox process of particular economic importance is the corrosion of metals. A 2002 study estimated that in the United States alone corrosion cost $276 billion a year. Perhaps 20% of all the iron and steel production in the United States each year goes to replace corroded items. Let's look first at the corrosion of iron.

The Rusting of Iron

In moist air, iron is oxidized, particularly at a nick or scratch.

$$Fe(s) \longrightarrow Fe^{2+}(aq) + 2\,e^-$$

As iron is oxidized, oxygen is reduced.

$$O_2(g) + 2\,H_2O(l) + 4\,e^- \longrightarrow 4\,OH^-(aq)$$

The net result, initially, is the formation of insoluble iron(II) hydroxide, which appears dark green or black.

$$2\,Fe(s) + O_2(g) + 2\,H_2O(l) \longrightarrow 2\,Fe(OH)_2(s)$$

This product is usually further oxidized to iron(III) hydroxide.

$$4\,Fe(OH)_2(s) + O_2(g) + 2\,H_2O(l) \longrightarrow 4\,Fe(OH)_3(s)$$

Iron(III) hydroxide [$Fe(OH)_3$], sometimes written as $Fe_2O_3 \cdot 3\,H_2O$, is the familiar iron rust.

Oxidation and reduction often occur at separate points on the metal's surface. Electrons are transferred through the iron metal. The circuit is completed by an electrolyte in aqueous solution. In the Snow Belt, this solution is often the slush from road salt and melting snow. The metal is pitted in an anodic area, where iron is oxidized to Fe^{2+}. These ions migrate to the cathodic area, where they react with hydroxide ions formed by the reduction of oxygen.

$$Fe^{2+}(aq) + 2\,OH^-(aq) \longrightarrow Fe(OH)_2(s)$$

As indicated previously, this iron(II) hydroxide is then oxidized to $Fe(OH)_3$, or rust. This process is diagrammed in Figure 8.8. Notice that the anodic area is protected from oxygen by a water film, whereas the cathodic area is exposed to air.

Protection of Aluminum

Aluminum is more reactive than iron, and yet corrosion is not a serious problem with aluminum. We use aluminum foil, we cook with aluminum pots, and we buy beverages in aluminum cans. Even after many years, they have not corroded. How is this possible?

The aluminum surface reacts with oxygen in the air to form a thin layer of oxide. However, instead of being porous and flaky like iron oxide, aluminum oxide is hard, tough, and adheres strongly, protecting the metal from further oxidation.

Nonetheless, corrosion can sometimes be a problem with aluminum. Certain substances, such as salt, can interfere with the protective oxide coating on aluminum, allowing the metal to oxidize. This problem has caused mag wheels on automobiles to crack, and some planes with aluminum landing gear have had wheels shear off.

4. Why does steel rust, whereas aluminum does not?
Both steel (iron) and aluminum undergo oxidation. But the product from iron flakes off, leaving a fresh iron surface exposed that will rust again. The aluminum oxide formed when aluminum "rusts" adheres strongly to the aluminum, protecting it from further oxidation.

▶ **Figure 8.8** The corrosion of iron requires water, oxygen, and an electrolyte.

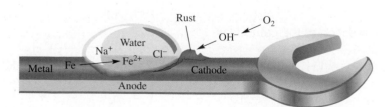

Silver Tarnish

The tarnish on silver results from oxidation of the silver surface by hydrogen sulfide (H_2S) in the air or from food. It produces a film of black silver sulfide (Ag_2S) on the metal surface. You can use silver polish to remove the tarnish, but in doing so, you also lose part of the silver. An alternative method involves the use of aluminum metal to reduce the silver ions back to silver metal.

$$3\,Ag^+(aq) + Al(s) \longrightarrow 3\,Ag(s) + Al^{3+}(aq)$$

This reaction also requires an electrolyte, and sodium bicarbonate ($NaHCO_3$) is usually used. The tarnished silver is placed in contact with aluminum metal (such as foil) and covered with a hot solution of sodium bicarbonate. A precious metal is conserved at the expense of a cheaper one.

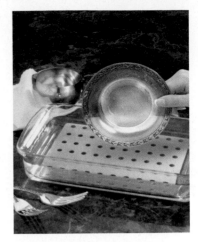

▲ Silver tarnish (Ag_2S) is readily removed from silver metal upon contact with aluminum metal in a baking soda solution. Working in a sink, add about 1 cup baking soda to 1 gallon hot water. Immerse a piece of aluminum and the tarnished silver object so that they touch.

QUESTION: To what is the Ag_2S converted? To what is the Al converted?

Self-Assessment Questions

1. What type of reaction is the corrosion of iron?
   a. acid–base
   b. addition
   c. reduction–oxidation
   d. substitution

2. Corrosion of aluminum is not as big a problem as corrosion of iron because aluminum
   a. forms an oxide layer that is firmly stuck to the surface
   b. is less reactive than Fe
   c. needs water to corrode
   d. has a higher oxidation number

3. Silver tarnish is mainly
   a. AgCl   **b.** Ag_2O **c.** Ag_2S **d.** Fe_2O_3

Answers: 1, c; 2, a; 3, c

8.5 Explosive Reactions

Chemical explosions (that is, those not based on nuclear fission or fusion) are usually the result of redox reactions. Explosions happen when a chemical reaction occurs rapidly and with a considerable increase in volume. They often involve compounds of nitrogen, such as nitroglycerine (the active ingredient in dynamite), ammonium nitrate (a common fertilizer), or trinitrotoluene (TNT). These compounds can decompose readily, yielding nitrogen gas as one of the products.

Explosive mixtures are used for mining, earthmoving projects, and the demolition of buildings, as well as in making bombs. Trucks loaded with ammonium nitrate (NH_4NO_3) mixed with fuel oil have been used in terrorist attacks around the world. This mixture, sometimes called ANFO (ammonium nitrate–fuel oil), is an explosive with enormous destructive power. The ammonium nitrate is both oxidized and reduced. Ammonium ion is the reducing agent, whereas nitrate ion is the oxidizing agent. The fuel oil provides additional oxidizable material. Using $C_{17}H_{36}$ as the formula for a typical molecule in fuel oil, we can write an equation for the explosive reaction as follows.

$$52\,NH_4NO_3(s) + C_{17}H_{36}(l) \longrightarrow 52\,N_2(g) + 17\,CO_2(g) + 122\,H_2O(g)$$

Notice that this reaction includes 53 mol of starting material, most of it solid and the rest in the liquid state, and 191 mol of products, all gases. A reaction of solid and/or liquid reactants that generates gaseous products involves a huge volume increase and a possibility of explosion. In this case, the reaction is rapid and the volume increase is enormous.

Ammonium nitrate is a widely used fertilizer, and fuel oil is used in many home furnaces. The fact that both these materials are so accessible to potential terrorists is cause for serious concern.

Self-Assessment Questions

1. In a typical explosive reaction, which of the following usually occurs?
   a. gases are converted from one form to another
   b. liquids or solids are converted to aqueous solutions
   c. liquids or solids are converted to gases
   d. nitrogen nuclei are split

2. Black powder is a mixture of charcoal (carbon), potassium nitrate, and sulfur. The oxidizing agent in black powder is
   a. charcoal
   b. nitroglycerin
   c. potassium nitrate
   d. sulfur

Answers: 1, c; 2, c

8.6 Oxygen: An Abundant and Essential Oxidizing Agent

The structure and chemistry of Earth are discussed in Chapter 12, the atmosphere in Chapter 13, and water in Chapter 14.

▲ **It DOES Matter!**
Physical activity under conditions in which plenty of oxygen is available to body tissues is called *aerobic* exercise. When the available oxygen is insufficient, the exercise is called *anaerobic*, and it often leads to weakness and pain resulting from "oxygen debt." (See eChapter 19.)

Oxygen itself is the most common oxidizing agent. Making up one-fifth of the air, it oxidizes the wood in our campfires and the gasoline in our automobiles. It even "burns" the food we eat to give us the energy to move and to think.

Certainly one of the most important elements on Earth, oxygen is one of about two dozen elements essential to life. It is found in most compounds that are important to living organisms. Foodstuffs—carbohydrates, fats, and proteins—all contain oxygen. The human body is approximately 65% water (by mass). Because water is 89% oxygen by mass and many other compounds in your body also contain oxygen, almost two-thirds of your body mass is oxygen.

Oxygen comprises about half of the accessible portion of Earth by mass. It occurs in all three subdivisions of the outermost structure of Earth. Oxygen occurs as O_2 molecules in the atmosphere (the gaseous mass surrounding Earth). In the hydrosphere (the oceans, seas, rivers, and lakes), oxygen is combined with hydrogen in water. In the lithosphere (the outer solid portion), oxygen is combined with silicon (sand is SiO_2), aluminum (in clays), and other elements.

The atmosphere is about 21% elemental oxygen (O_2) by volume. [The rest is mainly nitrogen (N_2), which is rather unreactive.] The free, uncombined oxygen in the air is taken into our lungs, passes into our bloodstreams, is carried to our body tissues, and reacts with the food we eat. This is the process that provides us with all our energy. Fuels such as natural gas, gasoline, and coal also need oxygen to burn and release their stored energy. Oxidation, or combustion, of these fossil fuels currently supplies about 86% of the energy that turns the wheels of civilization.

Pure oxygen is obtained by liquefying air and then letting the nitrogen and argon boil off. Nitrogen boils at −196 °C, argon at −186 °C, and oxygen at −183 °C. About 20 billion kg of oxygen is produced annually in the United States, but most of it is used directly by industry, much of it by steel plants. About 1% is compressed into tanks for use in welding, hospital respirators, and other purposes.

Not everything that oxygen does is desirable. In addition to causing corrosion (Section 8.4), it promotes food spoilage and wood decay.

Reactions with Other Elements

Many oxidation reactions involving atmospheric oxygen are quite complicated, but we can gain an understanding by looking at some of the simpler chemical reactions of oxygen. For example, as we have seen, oxygen combines with many metals to form metal oxides and with nonmetals to form nonmetal oxides.

| EXAMPLE 8.7 | Writing Equations: Reaction of Oxygen with Other Elements |

Magnesium combines readily with oxygen when ignited in air. Write the equation for this reaction.

Solution

Magnesium is a group 2A metal. Oxygen occurs as diatomic molecules (O_2). The two react to form MgO. The reaction is

$$Mg(s) + O_2(g) \longrightarrow MgO(s) \quad (not\ balanced)$$

To balance the equation, we need 2 MgO and 2 Mg.

$$2\,Mg(s) + O_2(g) \longrightarrow 2\,MgO(s)$$

■ **EXERCISE 8.7**

Write equations for the following reactions.
a. Zinc burns in air to form zinc oxide (ZnO).
b. Selenium (Se) burns in air to form selenium dioxide.

▲ Fuels burn more rapidly in pure oxygen than in air. Huge quantities of liquid oxygen are stored in the orange propellant tank of the space shuttle, to send the orbiter on its missions.

Reactions with Compounds

Oxygen reacts with many compounds, oxidizing one or more of the elements in the compound. As we have seen, combustion of a fuel is oxidation and produces oxides. There are many other examples. Hydrogen sulfide, a gaseous compound with a rotten-egg odor, burns, producing oxides of hydrogen (water) and of sulfur (sulfur dioxide).

$$2\,H_2S(g) + 3\,O_2(g) \longrightarrow 2\,H_2O(l) + 2\,SO_2(g)$$

| EXAMPLE 8.8 | Writing Equations: Reaction of Oxygen with Compounds |

Carbon disulfide, a highly flammable liquid, combines readily with oxygen, burning with a blue flame. What products are formed? Write the equation.

Solution

Carbon disulfide is CS_2. The products are oxides of carbon (carbon dioxide) and of sulfur (sulfur dioxide). The balanced equation is

$$CS_2(g) + 3\,O_2(g) \longrightarrow CO_2(g) + 2\,SO_2(g)$$

■ **EXERCISE 8.8A**

When heated in air, lead sulfide (PbS) combines with oxygen to form lead(II) oxide (PbO) and sulfur dioxide. Write a balanced equation for the reaction.

■ **EXERCISE 8.8B**

Write a balanced equation for the combustion (reaction with oxygen) of ethanol (C_2H_5OH).

Ozone

In addition to diatomic O_2, the normal form of oxygen, the element also has a triatomic form (O_3) called *ozone*. Ozone is a powerful oxidizing agent and a harmful air pollutant. It can be extremely irritating to both plants and animals and is especially destructive to rubber. On the other hand, a layer of ozone in the upper stratosphere serves as a shield that protects life on Earth against ultraviolet radiation from the sun (Chapter 13). Ozone clearly illustrates that the same substance can be extremely beneficial or quite harmful, depending on where it happens to be.

Self-Assessment Questions

1. When CH_4 is burned in plentiful air, the products are
   a. $C + H_2$ **b.** $CH_2 + H_2O$ **c.** $CO_2 + H_2$ **d.** $CO_2 + H_2O$
2. When sulfur burns in plentiful air, the product is
   a. CO_2 **b.** H_2S **c.** H_2SO_4 **d.** SO_2

Answers: 1, d; 2, d

Safety Alert

Strong oxidizing agents such as 30% aqueous hydrogen peroxide can cause severe burns. In general, strong oxidizing agents are corrosive and many are highly toxic. They can also initiate a fire or add to the severity of a fire when brought into contact with combustible materials.

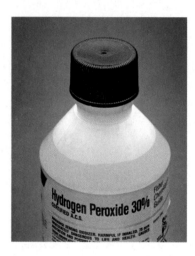

▲ Hydrogen peroxide is a powerful oxidizing agent. Dilute solutions are used as disinfectants and to bleach hair.

Solutions of potassium permanganate ($KMnO_4$), a common laboratory oxidizing agent, are good disinfectants. The solutions are a deep purplish pink. Highly diluted solutions, a pale pink (called "pinky water" in India), are used in developing countries to disinfect foods. Their use is limited in that the permanganate ion is reduced to brown manganese dioxide, leaving a brown stain on the treated surface.

8.7 Other Common Oxidizing Agents

Many oxidizing agents—sometimes called *oxidants*—are important in the laboratory, in industry, and in the home. They are used as antiseptics, disinfectants, and bleaches and play a role in many chemical syntheses.

Disinfectants and antiseptics are alike in that both are materials used to destroy microorganisms (that is, they are germicidal), but antiseptics are applied primarily to living tissue and disinfectants to nonliving things. A good example of a disinfectant is chlorine, which is used to kill disease-causing microorganisms in drinking water and some swimming pools. The actual disinfecting agent is hypochlorous acid (HOCl), formed by the reaction of chlorine with water.

$$Cl_2(g) + H_2O(l) \longrightarrow HOCl(aq) + HCl(aq)$$

Swimming pools are often chlorinated with calcium hypochlorite [$Ca(OCl)_2$]. Because calcium hypochlorite is alkaline, it also raises the pH of the water. (When a pool becomes too alkaline, the pH is lowered by adding hydrochloric acid. Swimming pools are usually maintained at pH 7.2–7.8.)

Hydrogen peroxide (H_2O_2) is a common oxidizing agent that has the advantage of being converted to water in most reactions. Pure hydrogen peroxide is a syrupy liquid. It is available (in laboratories) as a dangerous 30% solution that has powerful oxidizing power, or as a 3% solution sold in stores for various uses around the home. When combined with a specially designed catalyst, hydrogen peroxide can be used to oxidize colored compounds and water impurities.

As a dilute aqueous solution, hydrogen peroxide is used to bleach hair (thus the phrase "peroxide blonde"). Dilute H_2O_2 is used medically to clean and deodorize wounds and ulcers.

Iodine is usually used in an alcoholic solution (called tincture of iodine) or as Lugol's iodine solution as a pre- and postoperative antiseptic.

An oxidizing agent sometimes used in the laboratory is potassium dichromate ($K_2Cr_2O_7$). It is an orange material that turns green when it is reduced to chromium(III) compounds. One of the compounds that potassium dichromate oxidizes is ethyl alcohol. The old Breathalyzer test for intoxication made use of a dichromate solution. The exhaled breath of the person being tested was mixed with the acidic dichromate, and the degree of color change indicated the level of alcohol.

Ointments for treating acne often contain 5–10% benzoyl peroxide, a powerful antiseptic and also a skin irritant. It causes old skin to slough off and be replaced by new, fresher-looking skin. When used on areas exposed to sunlight, however, benzoyl peroxide may promote skin cancer.

Oxidizing agents are also used as disinfectants. A good example is chlorine, which is used to kill disease-causing microorganisms in drinking water. In recent years, concern has been raised over disinfectant by-products that form when chlorine is used to disinfect water containing certain impurities. Although chlorine kills harmful microorganisms, it can also oxidize smaller molecules to form harmful by-products, such as trichloromethane.

Bleaches are oxidizing agents, too. A **bleach** removes unwanted color from fabrics or other material. Nearly any oxidizing agent could do the job, but some might be unsafe, harmful to fabrics, or perhaps too expensive.

GREEN CHEMISTRY

Responsible Paper Production

Carol Pritchard-Martinez, Freelance Editor

Ever wonder why paper is white when the wood from which it is made is not? The high-quality paper used in the printing of this book was produced in a multistage process that includes bleaching steps that whiten and brighten wood pulp.

The bad news is that wood pulp bleaching comes at a cost to the environment. Huge quantities of water are used in the papermaking process, and toxic by-products of the chlorine bleaching process, including dioxins, are released into waters flowing from pulp mills. Dioxins, which accumulate in fatty tissues in humans and wildlife, are known teratogens and mutagens and are likely human carcinogens. Dioxins are a group of chemical compounds that have similar structures and characteristics, the most toxic being 2,3,7,8-tetrachlorodibenzo-p-dioxin (TCDD).

The good news is that green chemistry has identified ways to reduce or even eliminate the use of chlorine in the bleaching process.

Wood pulp consists of two polymers, cellulose and lignin. Chemical components of the lignin are responsible for the colors of pulp. Bleaching is a multistep process in which the first stages remove lignin and the later stages brighten the cellulose pulp.

Until recently, bleaching used elemental chlorine, capable of forming hundreds of organochlorine by-products. In the 1990s the U.S. Environmental Protection Agency (EPA) mandated the use of chlorine dioxide as the principal bleaching agent in paper manufacturing because it forms fewer such by-products. Chlorine dioxide is sometimes used in combination with chlorine, but it is used alone in *ECF* (elemental chlorine-free) bleaching sequences.

Chlorine dioxide is a strong oxidant because of chlorine's high oxidation number in the compound. Recall that oxidation and reduction reactions can be defined in terms of a gain or loss of electrons. The oxidation number of the chlorine atom in chlorine dioxide is +4. In the reduction to chloride ion (oxidation number -1), five electrons are exchanged. The oxidation of lignin substantially changes its chemical structure and, consequently, its physical properties.

Replacing elemental chlorine with chlorine dioxide in paper manufacturing has significantly reduced (though not eliminated) the generation of dioxins. Chlorine dioxide oxidatively breaks lignin away from the solid cellulose in two forms, small fragments and large fragments. Most of the lignin solubilized by bleaching is in the form of small fragments, which are digested by bacteria and other organisms in artificial oxidation ponds before the effluent is released into a natural body of water. The big lignin fragments containing color residues are too large to be digested by the microorganisms. The color is recalcitrant and ends up staining natural waterways, interfering with the absorption of light by aquatic ecosystems.

Body burdens (the total amount in a body) of a number of organochlorines have been used to measure progress in the reduction of bioaccumulative compounds from paper mills. The EPA's 2006 Toxics Release Inventory (TRI) ranks pulp and paper third among U.S. manufacturing industries in the release of dioxin and dioxin-like compounds to waterways.

Pulps that completely avoid chlorine are called *TCF pulps* (totally chlorine free). TCF processes are not yet prevalent because of cost considerations and high demand for paper. But a new bleaching process made possible by Terry Collins's group at the Institute for Green Science at Carnegie Mellon University holds the promise of completely eliminating the formation of dioxin waste.

In a triumph of green chemistry, Professor Collins has developed the TAML® oxidant activators as environmentally friendly alternatives to chlorine-based bleaching processes. Tetraamido macrocyclic ligands (TAML) used with a transition metal effectively catalyze (speed up) hydrogen peroxide oxidation. Hydrogen peroxide oxidizes by either adding oxygen or removing hydrogen atoms and produces no organochlorine-type pollutants. Made from natural biochemical elements, TAMLs are highly selective catalysts that reduce energy costs and not only prevent pollution but are also useful in pollution remediation. Collins received the Presidential Green Chemistry Challenge Award in 1999.

▲ Pure water (left) has little effect on a dried tomato sauce stain. Sodium hypochlorite bleach (right) removes the stain by oxidizing the colored tomato pigments to colorless products.

More to Explore

"Stain Removers: Which Are Best?" *Consumer Reports*, March 2000, pp. 52–53. Provides some effective home remedies.

Laundry bleaches (**eChapter 21**)* are usually sodium hypochlorite (NaOCl) as an aqueous solution (in products such as Purex® and Clorox®), often called "chlorine bleaches," or calcium hypochlorite [Ca(OCl)$_2$], known as bleaching powder. The powder is usually preferred for large industrial operations, such as the whitening of paper or fabrics. Nonchlorine bleaches contain sodium percarbonate (a combination of Na$_2$CO$_3$ and H$_2$O$_2$) or sodium perborate (a combination of NaBO$_2$ and H$_2$O$_2$). These are often called "oxygen bleaches."

Stain removal is more complicated than bleaching. A few stain removers are oxidizing agents, but some are reducing agents, some are solvents or detergents, and some have quite different action. Stains often require rather specific stain removers.

Self-Assessment Questions

1. Which of the following is a common oxidizing agent?
   a. HCl   **b.** H$_2$O$_2$ **c.** NaOH   **d.** NaCl

2. Which of the following is a common oxidizing agent?
   a. Cl$_2$ **b.** H$_2$ **c.** HCl   **d.** KOH

3. The main active ingredient in common household bleach, often called "chlorine bleach," is
   a. HCl   **b.** NaOCl   **c.** NaOH   **d.** NaCl

4. Bleach for hair usually contains
   a. ammonia   **b.** hydrochloric acid
   c. hydrogen peroxide   **d.** sodium hypochlorite

5. Many disinfectants are
   a. oxidizing agents   **b.** reducing agents
   c. strong acids   **d.** strong bases

Answers: 1, b; 2, a; 3, b; 4, c; 5, a

8.8 Some Reducing Agents of Interest

In every reaction involving oxidation the oxidizing agent is reduced, and the substance undergoing oxidation acts as a reducing agent. Let's now consider reactions in which the purpose of the reaction is reduction.

Most metals occur in nature as compounds. In order to prepare the free metals, the compounds must be reduced. Metals are often freed from their ores with coal or *coke* (elemental carbon obtained by heating coal to drive off volatile matter). Tin(IV) oxide is one of the many ores that can be reduced with coal or coke.

$$SnO_2(s) + C(s) \longrightarrow Sn(s) + CO_2(g)$$

Sometimes a metal can be obtained by heating its ore with a more active metal. Chromium oxide, for example, can be reduced by heating it with aluminum.

$$Cr_2O_3(s) + 2\,Al(s) \longrightarrow Al_2O_3(s) + 2\,Cr(s)$$

Antioxidants

In food chemistry, certain reducing agents are called **antioxidants**. Ascorbic acid (vitamin C) can prevent the browning of fruit (such as sliced apples or pears) by inhibiting air oxidation. Whereas vitamin C is water soluble, tocopherol (vitamin E) and beta-carotene (vitamin A precursor) are fat-soluble antioxidants. All these vitamins are believed to retard various oxidation reactions that are potentially damaging to vital components of living cells (Chapter 17).

▲ **It DOES Matter!**

In making a black-and-white photograph, silver bromide in the film is exposed to light. The silver ions so exposed can then be reduced to black silver metal, forming a photographic negative.

*Chapters 19–22 are found online at www.chemplace.com

Hydrogen as a Reducing Agent

Hydrogen is an excellent reducing agent that can free many metals from their ores, but it is generally used to produce more expensive metals, such as tungsten (W).

$$WO_3(s) + 3 H_2(g) \longrightarrow W(s) + 3 H_2O(l)$$

Hydrogen can be used to reduce many kinds of chemical compounds. Ethylene, for example, can be reduced to ethane.

$$C_2H_4(g) + H_2(g) \longrightarrow C_2H_6(g)$$

The reaction requires a **catalyst**, a substance that increases the rate of a chemical reaction without itself being used up. (Catalysts are further explored in Chapter 15.) Nickel is the catalyst used in this case.

Hydrogen also reduces nitrogen, from the air, in the industrial production of ammonia.

$$N_2(g) + 3 H_2(g) \longrightarrow 2 NH_3(g)$$

Ammonia is the source of most nitrogen fertilizers in modern agriculture. This reaction employs an iron catalyst.

A stream of pure hydrogen burns quietly in air with an almost colorless flame, but when a mixture of hydrogen and oxygen is ignited by a spark or a flame, an explosion results. The product in both cases is water.

$$2 H_2(g) + O_2(g) \longrightarrow 2 H_2O(l)$$

Certain metals, such as platinum and palladium, have an unusual affinity for hydrogen. They absorb large volumes of the gas. Palladium can absorb up to 900 times its own volume of hydrogen. It is interesting to note that hydrogen and oxygen can be mixed at room temperature with no perceptible reaction. But if a piece of platinum gauze is added, the gases react violently at room temperature. The platinum acts as a catalyst. It lowers the *activation energy* for the reaction. (The **activation energy** for a chemical reaction is the minimum energy needed to get the reaction started.) The heat from the initial reaction heats up the platinum, making it glow; it then ignites the hydrogen–oxygen mixture, causing an explosion.

Nickel, platinum, and palladium are often used as catalysts for reactions involving hydrogen. These metals have the greatest catalytic activity when they are finely divided and have lots of active surface area. Hydrogen adsorbed on the surface of these metals is more reactive than ordinary hydrogen gas. Catalysts are an important area of research in green chemistry. Chemists are able to develop more energy efficient and sustainable processes by designing catalysts to enable chemical reactions to occur at lower temperatures or pressures. Using catalysts can also allow a reaction to occur using milder chemicals, such as oxygen or hydrogen peroxide instead of chlorine to oxidize stains in laundry or harmful microorganisms in drinking water.

A nickel catalyst is used in converting unsaturated fats to saturated fats (Chapter 17). Unsaturated fats ordinarily react very slowly with hydrogen, but in the presence of nickel metal, the reaction proceeds readily. The nickel increases the rate of the reaction, but it does not increase the amount of product formed.

Self-Assessment Questions

1. Which of the following is a common reducing agent?
   a. Cl_2 **b.** F_2 **c.** H_2 **d.** I_2

2. Which of the following is a common reducing agent?
   a. C   **b.** Cl_2 **c.** F_2 **d.** I_2

3. All the following metals are acceptable catalysts for reactions involving hydrogen gas *except*
   a. aluminum   **b.** nickel   **c.** palladium   **d.** platinum

4. A catalyst changes the rate of a chemical reaction by
   a. changing the products of the reaction
   b. changing the reactants involved in the reaction
   c. increasing the frequency of molecular collisions
   d. lowering the activation energy

Answers: 1, c; 2, a; 3, a; 4, d

8.9 A Closer Look at Hydrogen

Hydrogen ranks low in abundance by mass on Earth. If we look beyond our home planet, however, hydrogen becomes much more significant. The sun, for example, is made up largely of hydrogen. The planet Jupiter is also mainly hydrogen. In fact, hydrogen is by far the most abundant element in the universe.

Hydrogen is an especially vital element because it and oxygen are the components of water. By mass, hydrogen makes up only about 0.9% of the outer, accessible portion of Earth. However, because of its low atomic mass, it ranks third in abundance by number of atoms. A random sample of 10,000 atoms from this portion of Earth would have 1510 atoms of hydrogen.

Unlike oxygen, hydrogen is seldom found as a free, uncombined element on Earth. Most of it is combined with oxygen in water. Some is combined with carbon in petroleum and natural gas, which are mixtures of hydrocarbons. Nearly all compounds derived from plants and animals contain combined hydrogen.

Because hydrogen has the lowest atomic number ($Z = 1$) and the smallest atoms, it is the first element in the periodic table. However, it is difficult to know exactly where to place it. Hydrogen is usually shown as the first member of group 1A because it has the same valence electronic structure (ns^1) as alkali metals. Like the atoms of elements in group 1A, a hydrogen atom can lose an electron to form a 1+ ion, but unlike the other group 1A elements, hydrogen is not a metal. Like the atoms of group 7A elements, a hydrogen atom can pick up an electron to become a 1− ion, but unlike the other group 7A elements, hydrogen is not a halogen. We place hydrogen in group 1A in the periodic table on the inside front cover, but you should keep in mind that it is a unique element, in a class by itself.

Small amounts of elemental hydrogen can be made for laboratory use by reacting zinc with hydrochloric acid (Figure 8.9).

$$Zn(s) + 2\ HCl(aq) \longrightarrow ZnCl_2(aq) + H_2(g)$$

Commercial quantities of hydrogen are obtained as by-products of petroleum refining or by the reaction of natural gas with steam. About 8 billion kg of hydrogen is produced each year in the United States. At present, hydrogen is used mainly to make ammonia and methanol. Its possible use as a fuel of the future is discussed in Chapter 15.

Hydrogen is a colorless, odorless gas and the lightest of all substances. Its density is only one-fourteenth that of air, and for this reason it was once used to fill lighter-than-air craft (Figure 8.10). Unfortunately, hydrogen can be ignited by a spark, which is what occurred in 1937 when the German airship *Hindenburg* was destroyed in a disastrous fire and explosion as it was landing in Lakehurst, New Jersey. The use of hydrogen in airships was discontinued after that, and the dirigible industry never recovered. Today the few airships that are still in service are filled with nonflammable helium, but they are used mainly in advertising.

▲ **Figure 8.9** Hydrogen gas in the laboratory can be prepared by the reaction of zinc metal with hydrochloric acid.

▲ **Figure 8.10** Hydrogen is the most buoyant gas, but it is highly flammable. The disastrous fire in the hydrogen-filled German zeppelin *Hindenburg* led to the replacement of hydrogen by nonflammable helium, which buoys the Goodyear blimp.

QUESTION: What is the equation for the reaction pictured on the left? Would a similar reaction occur if lightning struck the blimp pictured on the right? Explain.

Photochromic Glass

Eyeglasses with photochromic lenses eliminate the need for sunglasses because the lenses darken when exposed to bright light. This response to light is the result of oxidation–reduction reactions. Ordinary glass is a complex matrix of silicates (Chapter 12) that is transparent to visible light. Photochromic lenses have silver chloride (AgCl) and copper(I) chloride (CuCl) crystals uniformly embedded in the glass. Silver chloride is susceptible to oxidation and reduction by light. First the light displaces an electron from a chloride ion:

$$Cl^- \xrightarrow{\text{oxidation}} Cl + e^-$$

The electron then reduces a silver ion to a silver atom:

$$Ag^+ + e^- \xrightarrow{\text{reduction}} Ag$$

Clusters of silver atoms block the transmittance of light, causing the lenses to darken. This process occurs quite quickly. The degree of darkening depends on the intensity of the light.

To be useful in eyeglasses, the photochromic process must be reversible. The darkening process is reversed by the copper(I) chloride. When the lenses are removed from light, the chlorine atoms formed by the exposure to light are reduced by the copper(I) ions that are oxidized to copper(II) ions:

$$Cl + Cu^+ \longrightarrow Cu^{2+} + Cl^-$$

The copper(II) ions then oxidize the silver atoms.

$$Cu^{2+} + Ag \longrightarrow Cu^+ + Ag^+$$

▲ Photochromic glass darkens in the presence of light.

The net effect is that the silver and chloride atoms are converted to their original oxidized and reduced states, and the lenses become transparent once more.

---

5. Does hydrogen as a fuel have a promising future?
Gram for gram, hydrogen provides more energy than any other chemical fuel. Also, when hydrogen is oxidized with oxygen, the product is ordinary water. Unfortunately, it takes more energy to produce hydrogen than we get from burning the hydrogen. Until we can use renewable energy to generate hydrogen, its future will remain in question.

A N S W E R

Self-Assessment Questions

1. By number of atoms, what is the rank of hydrogen in abundance in the accessible portion of Earth?
   a. first
   b. second
   c. tenth
   d. third

2. Hydrogen gas can be produced by reacting an active metal such as zinc with
   a. an acid
   b. ammonia
   c. an oxidizing agent
   d. a reducing agent

3. Although helium is twice as dense as hydrogen gas, modern blimps are filled with helium rather than hydrogen gas because hydrogen is
a. highly flammable
b. more expensive
c. rarer
d. toxic

8.10 Oxidation, Reduction, and Living Things

Perhaps the most important oxidation–reduction processes are the ones that maintain life on this planet. We obtain energy for all our physical and mental activities by metabolizing food through a process called *cellular respiration*. The process has many steps, but eventually the food we eat is converted mainly into carbon dioxide, water, and energy.

Bread and many of the other foods we eat are largely made up of carbohydrates, composed of carbon, hydrogen, and oxygen (Chapter 17). If we represent carbohydrates with the simple example glucose ($C_6H_{12}O_6$), we can write the overall equation for their metabolism as follows.

$$C_6H_{12}O_6 + 6\,O_2 \longrightarrow 6\,CO_2 + 6\,H_2O + \text{energy}$$

This process constantly occurs in animals, including humans. The carbohydrate is oxidized in the process.

Meanwhile, plants need carbon dioxide, water, and energy from the sun to produce carbohydrates. The process by which plants synthesize carbohydrates is called **photosynthesis** (Figure 8.11). The chemical equation is

$$6\,CO_2 + 6\,H_2O + \text{energy} \longrightarrow C_6H_{12}O_6 + 6\,O_2$$

Notice that this process in plant cells is exactly the reverse of the process going on inside animals. In food metabolism in animals we focus on an oxidation process. In photosynthesis we focus on a reduction process.

▶ **Figure 8.11** Photosynthesis occurs in green plants. The chlorophyll pigments that catalyze the photosynthesis process give the green color to much of the land area of Earth.

The carbohydrates produced by photosynthesis are the ultimate source of all our food because fish, fowl, and other animals either eat plants or eat other animals that eat plants. Note that the photosynthesis process not only makes carbohydrates but also yields free elementary oxygen (O_2). In other words, photosynthesis provides not only all the food we eat, but it also provides all the oxygen we breathe.

There are many oxidation reactions that occur in nature (with oxygen being reduced in the process). The net photosynthesis reaction is unique in that it is a natural reduction of carbon dioxide (with oxygen being oxidized). Many reactions in nature use oxygen. Photosynthesis is the only natural process that produces it.

Self-Assessment Questions

1. The main process by which animals obtain energy from food is
   a. acidification
   b. oxidation of glucose
   c. photosynthesis
   d. reduction of CO_2

2. The only natural process that produces O_2 is
   a. acidification of carbonates
   b. photosynthesis
   c. oxidation of CO_2
   d. reduction of oxide minerals

Answers: 1, b; 2, b

Critical Thinking Exercises

Apply knowledge that you have gained in this chapter and one or more of the FLaReS principles (Chapter 1) to evaluate the following statements or claims.

8.1 Over the years some people have claimed that someone has invented an automobile engine that burns water instead of gasoline. They say that we have not heard about it because the oil companies have bought the rights to the engine from the inventor so that they can keep it from the public and people will continue to burn gasoline in their cars. Do you find this claim believable?

8.2 A friend tells you that if you take a sugar cube and hold it in a flame, it will melt but will not burn; but he claims that if you dip the sugar cube into cigarette ashes before you place it in the flame, it will burn. He explains that cigarette ashes contain something that is a catalyst for the burning of sugar. Do you think your friend's claim is valid?

8.3 A website claims that electricity can remove rust. The procedure given is as follows: Connect a wire to the object to be derusted and immerse in a bath of sodium bicarbonate (baking soda) solution. Immerse a second wire in the bath. Connect the wires to a battery so that the object to be derusted is the cathode. The site claims that hydrogen gas is produced at the cathode and oxygen at the anode, and the hydrogen somehow changes the rust back to iron metal. Evaluate this claim.

8.4 A friend claims that an automobile can be kept rust free simply by connecting the negative pole of the car's battery to the car's body. He reasons that this action makes the car the cathode, and oxidation does not occur at the cathode. He claims that for years he has kept his car from rusting by doing this. Evaluate this claim.

8.5 A salesman claims that your tap water is contaminated. As evidence, he connects a battery to a pair of "special electrodes" and dips the electrodes in a glass of your tap water. Within a few minutes, white and brown "gunk" appears in the water. The salesman sells a special purifying filter that he claims will remove the contaminants and leave the water pure and clear. Evaluate this claim.

SUMMARY

Section 8.1—Oxidation and reduction are processes that occur together. They may be defined in three ways. **Oxidation** occurs when a substance gains oxygen atoms, or loses hydrogen atoms, or loses electrons. **Reduction** is the opposite of oxidation; it occurs when a substance loses oxygen atoms, or gains hydrogen atoms, or gains electrons. We usually use the definition that is most convenient for the situation.

Section 8.2—In a redox or oxidation–reduction reaction, the substance that causes oxidation is the **oxidizing agent**. The substance that causes reduction is the **reducing agent**. In a redox reaction the oxidizing agent is reduced, and the reducing agent is oxidized.

Section 8.3—A redox reaction involves transfer of electrons. That transfer can be made to take place in an **electrochemical cell**, which has two solid **electrodes** where electrons are transferred. The electrode where oxidation occurs is the **anode**, and the one at which reduction occurs is the **cathode**. Electrons are transferred from anode to cathode through a wire, which is an electric current. When the oxidation half-reaction is added to the reduction half-reaction, the electrons in the half-reactions cancel and we get the overall reaction. Knowing this, we can balance many redox reactions.

A dry cell (flashlight battery) is an electrochemical cell with a zinc case and a central carbon rod and contains a paste of manganese dioxide. As the cell is used, the zinc is oxidized and the MnO_2 is reduced. Alkaline cells are similar but contain potassium hydroxide. A **battery** is a series of connected cells. An automotive lead storage battery contains six cells. Each cell contains lead and lead dioxide plates in sulfuric acid, and the battery can be recharged. A **fuel cell** generates electricity by oxidizing a fuel in oxygen in a special cell.

Section 8.4—Corrosion of metal is ordinarily an undesirable redox reaction. When iron corrodes, the iron is first oxidized to Fe^{2+}, and oxygen gas is reduced to hydroxide ion. The $Fe(OH)_2$ that forms is further oxidized to $Fe(OH)_3$ or common rust. Oxidation and reduction often occur in separate places on the metal surface, and an electrolyte (such as salt) on the surface can speed the corrosion reaction. Aluminum corrodes, but the oxide layer is tough and adherent and protects the underlying metal. Silver corrodes in the presence of sulfur compounds, forming black silver sulfide. The sulfide can be removed electrochemically.

Section 8.5—An explosive reaction is a redox reaction that occurs rapidly and with a considerable increase in volume (of gases).

Section 8.6—Oxygen is an oxidizing agent that is essential to life. It makes up about one-fifth of the air and about two-thirds of our body mass. Our food is "burned" in oxygen as we breathe, and fuels such as coal use oxygen to burn and release most of the energy used in civilization. Pure oxygen is obtained from liquefied air, and most of it is used in industry. Oxygen reacts with metals and nonmetals to form the corresponding oxides. It reacts with many compounds to form oxides of the elements in the compound. Ozone, O_3, is a form of oxygen that is an air pollutant in the biosphere but a shield from ultraviolet radiation in the upper atmosphere.

Section 8.7—Other oxidizing agents include hydrogen peroxide, potassium dichromate, benzoyl peroxide, and chlorine. A **bleach** is an oxidizing agent that removes unwanted color from fabric or material.

Section 8.8—An **antioxidant** is a reducing agent that retards damaging oxidation reactions in living cells. Vitamins C and E and beta-carotene are antioxidants. Common reducing agents include carbon, hydrogen, and active metals. Carbon in the form of coal or coke is often used as a reducing agent to obtain metals from their oxides. Active metals such as aluminum also are used. Hydrogen can free many metals from their ores. Some reactions of hydrogen require a **catalyst** that speeds up the reaction without itself being used up. A catalyst lowers the **activation energy**—the minimum energy needed to start the reaction. Formation of ammonia from hydrogen and nitrogen uses a catalyst.

Section 8.9—Hydrogen is a vital element because it is a component of water and of living tissue. Hydrogen's lowest atomic number and smallest atoms make it unique in its behavior among the elements. Much of the hydrogen used today is produced from petroleum refining or from reactions of natural gas and is used to make ammonia and methanol. Hydrogen was once used for lighter-than-air craft but its flammability caused it to be replaced by helium.

Section 8.10—Oxidation of glucose and other substances produces the energy needed in the animal kingdom. The reduction of carbon dioxide in **photosynthesis** is the most important reduction process on Earth. We could not exist without it; it provides the food we eat and the oxygen we breathe.

REVIEW QUESTIONS

1. What happens to the oxidation number of one of its elements when a compound is oxidized and when it is reduced?

2. What is an electrochemical cell? An electrochemical battery?

3. What is the purpose of a porous plate between the two electrode compartments in an electrochemical cell?

4. What is a half-reaction? How are half-reactions combined to give an overall redox reaction?

5. How does an alkaline cell differ from a regular carbon–zinc dry cell?

6. From what material is the case of a carbon–zinc dry cell made? What purpose does this material serve? What happens to it as the cell discharges?

7. Describe what happens when a lead storage battery discharges.

8. What happens when a lead storage battery is charged?

9. Describe what happens when iron corrodes. How does road salt speed this process?

10. Relate the chemistry of photosynthesis to the chemistry that provides energy for your heartbeat.

11. How does silver tarnish? How can tarnish be removed without the loss of silver?

12. Describe how a bleaching agent, such as hypochlorite (ClO^-), works.

▪ PROBLEMS

Recognizing Oxidation and Reduction

13. Each of the following "equations" shows only part of a chemical reaction. Indicate whether the reactant shown is being oxidized or reduced. Explain.

a. $Cr^{3+} \longrightarrow Cr$ **b.** $H_2O_2 \longrightarrow O_2$

c. $Ca \longrightarrow Ca^{2+}$ **d.** $S_8 \longrightarrow 8\,S^{2-}$

e. $C_2H_6O \longrightarrow C_2H_4O$

14. In which of the following partial reactions is the reactant undergoing oxidation? Explain.

a. $C_2H_4O \longrightarrow C_2H_6O$

b. $WO_3 \longrightarrow W$

c. $Fe^{3+} \longrightarrow Fe^{2+}$

d. $C_{27}H_{33}N_9O_{15}P_2 \longrightarrow C_{27}H_{35}N_9O_{15}P_2$

e. $2\,H^+ \longrightarrow H_2$

15. Write balanced oxidation half-reaction equations of the type $Na \longrightarrow Na^+ + e^-$ for the following metals.

a. Mg   **b.** Al   **c.** Fe (two different equations)

16. Write balanced reduction half-reaction equations of the type $Cl_2 + 2\,e^- \longrightarrow 2\,Cl^-$ for the following nonmetals.

a. F_2 **b.** P   **c.** S

Oxidizing Agents and Reducing Agents

17. Identify the oxidizing agent and the reducing agent in each reaction.

a. $CuS + H_2 \longrightarrow Cu + H_2S$

b. $4\,K + CCl_4 \longrightarrow C + 4\,KCl$

c. $C_3H_4 + 2\,H_2 \longrightarrow C_3H_8$

d. $Fe^{3+} + Ce^{3+} \longrightarrow Fe^{2+} + Ce^{4+}$

18. Identify the oxidizing agent and the reducing agent in each reaction.

a. $3\,C + Fe_2O_3 \longrightarrow 3\,CO + 2\,Fe$

b. $P_4 + 5\,O_2 \longrightarrow P_4O_{10}$

c. $C + H_2O \longrightarrow CO + H_2$

d. $H_2SO_4 + Zn \longrightarrow ZnSO_4 + H_2$

19. Look again at Figure 8.4. What is the oxidizing agent?

20. What is the reducing agent in Figure 8.4?

21. Look again at Figure 8.6. Identify the oxidizing and reducing agents.

22. Look again at Figure 8.7. **(a)** Is SO_4^{2-} the oxidizing agent, reducing agent, both, or neither? **(b)** When the cell is being charged, the discharge half-reactions are reversed. During charging, is Pb^{2+} being oxidized, reduced, both, or neither?

Half-Reactions

23. Separate the following redox reactions into half-reactions and label each half as oxidation or reduction.

a. $2\,Al(s) + 6\,H^+(aq) \longrightarrow 2\,Al^{3+}(aq) + 3\,H_2(g)$

b. $2\,Cu^+(aq) + Mg(s) \longrightarrow Mg^{2+}(aq) + 2\,Cu(s)$

24. Separate the following redox reactions into half-reactions and label each half as oxidation or reduction.

a. $Fe(s) + 2\,H^+(aq) \longrightarrow Fe^{2+}(aq) + H_2(g)$

b. $2\,Al(s) + 3\,Cr^{2+}(aq) \longrightarrow 3\,Cr(s) + 2\,Al^{3+}(aq)$

25. Label each of the following half-reactions as oxidation or reduction and then combine them to obtain a balanced overall redox reaction.

a. $2\,I^- \longrightarrow I_2 + 2\,e^-$ and $Cl_2 + 2\,e^- \longrightarrow 2\,Cl^-$

b. $HNO_3 + H^+ + e^- \longrightarrow NO_2 + H_2O$ and $SO_2 + 2\,H_2O \longrightarrow H_2SO_4 + 2\,H^+ + 2\,e^-$

26. Label each of the following half-reactions as oxidation or reduction and then combine them to obtain a balanced overall redox reaction.

a. $2\,H_2O_2 \longrightarrow 2\,O_2 + 4\,H^+ + 4\,e^-$ and $Fe^{3+} + e^- \longrightarrow Fe^{2+}$

b. $WO_3 + 6\,H^+ + 6\,e^- \longrightarrow W + 3\,H_2O$ and $C_2H_6O \longrightarrow C_2H_4O + 2\,H^+ + 2\,e^-$

Oxidation and Reduction: Chemical Reactions

27. In the following reactions, which substance is oxidized? Which is the oxidizing agent?

a. $H_2CO + H_2O_2 \longrightarrow H_2CO_2 + H_2O$

b. $5\,C_2H_6O + 4\,MnO_4^- + 12\,H^+ \longrightarrow$
$5\,C_2H_4O_2 + 4\,Mn^{2+} + 11\,H_2O$

28. In the following reactions, which element is oxidized and which is reduced?

a. $2\,HNO_3 + SO_2 \longrightarrow H_2SO_4 + 2\,NO_2$

b. $2\,CrO_{3+} + 6\,HI \longrightarrow Cr_2O_3 + 3\,I_2 + 3\,H_2O$

29. Acetylene (C_2H_2) reacts with hydrogen to form ethane (C_2H_6). Is the acetylene oxidized or reduced? Explain.

30. Unsaturated vegetable oils react with hydrogen to form saturated fats. A typical reaction is

$$C_{57}H_{104}O_6 + 3\,H_2 \longrightarrow C_{57}H_{110}O_6$$

Is the unsaturated oil oxidized or reduced? Explain.

31. Molybdenum metal, used in special kinds of steel, can be manufactured by the reaction of its oxide with hydrogen.

$$MoO_3 + 3\,H_2 \longrightarrow Mo + 3\,H_2O$$

Which substance is reduced? Which is the reducing agent?

32. To test for iodide ions (for example, in iodized salt), a solution is treated with chlorine to liberate iodine.

$$2\,I^- + Cl_2 \longrightarrow I_2 + 2\,Cl^-$$

Which substance is oxidized? Which is reduced?

33. When the water pump failed in the nuclear reactor at Three Mile Island in 1979, zirconium metal reacted with very hot water to produce hydrogen gas.

$$Zr + 2\,H_2O \longrightarrow ZrO_2 + 2\,H_2$$

What substance was oxidized in the reaction? What was the oxidizing agent?

34. Unripe grapes are exceptionally sour because of a high concentration of tartaric acid ($C_4H_6O_6$). As the grapes ripen, this compound is converted to glucose ($C_6H_{12}O_6$). Is the tartaric acid oxidized or reduced?

35. Vitamin C (ascorbic acid) is thought to protect our stomachs from the carcinogenic effect of nitrite ions (NO_2^-) by converting the ions to nitric oxide (NO). Is the nitrite ion oxidized or reduced? Is ascorbic acid an oxidizing agent or a reducing agent?

36. In the reaction in Problem 35, ascorbic acid ($C_6H_8O_6$) is converted to dehydroascorbic acid ($C_6H_6O_6$). Is ascorbic acid oxidized or reduced in this reaction?

■ ADDITIONAL PROBLEMS

39. The dye indigo (used to color blue jeans) is formed by exposure of indoxyl to air.

$$2\,C_8H_7ON + O_2 \longrightarrow C_{16}H_{10}N_2O_2 + 2\,H_2O$$
Indoxyl   Indigo

What substance is oxidized? What is the oxidizing agent?

40. Why do some mechanics lightly coat their tools with grease or oil before storing them?

41. When aluminum wire is added to a blue solution of copper(II) chloride, the blue solution turns colorless and reddish-brown copper metal comes out of solution. Write an equation for the reaction. (Chloride ion is not involved in the reaction.)

42. Hydrogen has such a strong affinity for oxygen that it can remove oxygen atoms from many metal oxides to yield the free metal. For example, when hydrogen is passed over heated copper(II) oxide, metallic copper and water are formed. Write an equation for the reaction.

43. Lead-based paints exposed to air containing hydrogen sulfide turn black because the Pb^{2+} ions react with the H_2S to form black lead sulfide (PbS). This has caused the darkening of old oil-based paintings. **(a)** Write the equation for the reaction. Hydrogen peroxide lightens the paints by oxidizing the black sulfide (S^{2-}) to white sulfates (SO_4^{2-}). **(b)** Write the equation for this reaction.

44. To oxidize 1.0 kg of fat, our bodies require about 2000 L of oxygen. A good diet contains about 80 g of fat per day. What volume (at STP) of oxygen is required to oxidize that fat?

45. The oxidizing agent we use to obtain energy from food is oxygen (from the air). If you breathe 15 times a minute (at rest), taking in and exhaling 0.5 L of air with each breath, what volume of air do you breathe each day? Air is 21% oxygen by volume. What volume of oxygen do you breathe each day?

46. The photosynthesis reaction (page 216) can be expressed as the net result of two processes, one of which requires light and is called the *light reaction*. This reaction may be written as

$$12\,H_2O \xrightarrow{\text{light}} 6\,O_2 + 24\,H^+ + 24\,e^-$$

   a. Is the light reaction an oxidation or reduction?
   b. Write the equation for the other half-reaction, called the *dark reaction*, for photosynthesis.

Combination with Oxygen

37. Give formulas for the product(s) formed when each of the following substances react with oxygen (O_2).
   a. S   **b.** CH_4 **c.** C_3H_8

38. Give formulas for the product(s) formed when each of the following substances react with oxygen (O_2). (There may be more than one correct answer.)
   a. N_2 **b.** CS_2 **c.** $C_6H_{12}O_6$

47. Indicate whether the first-named substance in each change undergoes an oxidation, a reduction, or neither. Explain your reasoning.
   a. A violet solution of V^{2+}(aq) is converted to a green solution of V^{3+}(aq).
   b. Nitrogen dioxide converts to dinitrogen tetroxide when cooled.
   c. Carbon monoxide reacts with hydrogen to form methane.

48. As a rule, metallic elements act as reducing agents, not as oxidizing agents. Explain. (*Hint:* Consider the charges on metal ions.) Do nonmetals act only as oxidizing agents and not as reducing agents? Explain your answer.

49. Consider the following reaction of calcium hydride (CaH_2) with molten sodium metal. Identify the species being oxidized and the species being reduced according to **(a)** the second definition shown in Figure 8.2 and **(b)** the third definition in Figure 8.2. What difficulty arises?

$$CaH_2(s) + 2\,Na(l) \longrightarrow 2\,NaH(s) + Ca(l)$$

50. If a stream of hydrogen gas is ignited in air, the flame can then be immersed in a jar of chlorine gas and it will still burn. Write the balanced equation for the reaction that occurs in the jar. What is the oxidizing agent in the reaction?

51. Aluminum metal can react with water, but the oxide coating normally found on aluminum prevents that reaction. In 2007 a Purdue University researcher found that adding gallium to aluminum prevented the oxide coating and allowed the aluminum to react with water to form hydrogen gas. This process might permit a compact source of hydrogen gas for fuel. Write a balanced equation for the reaction of aluminum with water (aluminum hydroxide is the other product) and identify the oxidizing and reducing agents in the reaction.

52. Refer to the Problem 51. Aluminum has a density of 2.70 g/cm^3. What volume in cubic centimeters of aluminum metal would be needed to produce 2.0×10^4 L of hydrogen gas at STP? How many times smaller is this volume of aluminum than is the volume of hydrogen gas? (Recall that at STP, 1 mole of gas occupies 22.4 liters.)

53. Refer back to Section 8.10, and write a balanced equation for the metabolism of sucrose (table sugar, $C_{12}H_{22}O_{11}$).

54. The mass in grams of an element that provides one mole of electrons is called the *equivalent weight* of the element. Write the half-reactions for the oxidation of **(a)** magnesium, **(b)** lithium, **(c)** zinc, and **(d)** potassium, and calculate their equivalent weights. Suggest a reason why lithium is such a desirable oxidizing agent in commercial electrochemical cells.

■ COLLABORATIVE GROUP PROJECTS

Prepare a PowerPoint, poster, or other presentation (as directed by your instructor) for presentation to the class.

55. Prepare a brief report on one of the following types of electrochemical cells or batteries and share it with your group. If possible, give the half-reactions, the overall reaction, and uses.

   a. lithium–SO_2 cell
   b. lithium–iodine cell
   c. lithium–FeS_2 battery
   d. Ni–Cd cell
   e. silver oxide cell
   f. nickel–metal hydride cell

56. Prepare a brief report on one of the following methods of protecting steel from corrosion and share it with your group. Give any pertinent chemical reactions involved in the process and tell how it is related to oxidation and reduction.

   a. galvanization
   b. coating with tin
   c. cathodic protection

57. Batteries contain toxic substances and should be recycled rather than thrown in the trash. Prepare a brief report on battery recycling. Are there facilities in your area for recycling batteries?

58. Prepare a brief report on the restoration of the Statue of Liberty for its centennial in 1986. What role did oxidation–reduction play in the need for restoration?

59. Go to one or more of the websites about fuel cells. Write out the reactions that occur in the different types in development. Prepare a brief report focusing on one of the following:

   a. a cost–benefit analysis for their use in automobiles versus the use of gasoline, electricity via batteries, or natural gas
   b. safety considerations of using a hydrogen–oxygen fuel cell
   c. political–economic factors that may be hindering or promoting fuel cell research

The compound capsaicin, a molecular model of which is shown here, is the principal compound responsible for the "hotness" of hot peppers. Capsaicin is organic in the original sense because it is obtained from a plant. It is also organic in the modern sense in that it is a compound of the element carbon. Numerous organic chemicals are obtained from plants and animals, but many of the millions of organic compounds known today are synthetic. See Problem 51, page 258.

ORGANIC CHEMISTRY

QUESTIONS YOU MAY HAVE ASKED YOURSELF

1. Are "organic compound" and "organic food" related?
2. What does "petroleum distillates," listed on some product labels, mean?
3. If a compound has *benzene* in the name, is the compound a carcinogen?
4. What gives some mouthwashes a medicinal odor?
5. Why does fresh fruit smell so good, and rotting fruit so bad?
6. Why do dead animals smell so bad?

The Infinite Variety of Carbon Compounds

The definition of organic chemistry has changed over the years. The changes serve as a good example of the dynamic character of science and of how scientific concepts change in response to experimental evidence. Until the nineteenth century, chemists believed that *organic* chemicals originated only in tissues of living *organisms* and required a "vital force" for their production. All chemicals not manufactured by living tissue were regarded as *inorganic*. Some chemists even believed that organic and inorganic chemicals followed different laws. This all changed in 1828, when Friedrich Wöhler synthesized the organic compound urea, which is found in urine, from ammonium cyanate, an inorganic compound. This important event led other chemists to attempt synthesis of organic chemicals from inorganic ones and changed the very definition of organic chemicals.

Organic chemistry is now defined as the chemistry of carbon-containing compounds. Most of these compounds do come from living things or from things that were once living, but this is not necessarily the case. Perhaps the most remarkable thing about organic compounds is that there are so many of them. Of the tens of millions of known chemical compounds, more than 95% are compounds of carbon.

1. Are "organic compound" and "organic food" related? The word *organic* has several different meanings. Organic fertilizer is organic in the original sense; it is derived from living organisms. Organic foods are those grown without synthetic pesticides or fertilizers. Organic chemistry is simply the chemistry of carbon compounds.

In addition to carbon, almost all organic compounds also contain hydrogen, many contain oxygen and nitrogen, and quite a few contain sulfur and halogens. Nearly all the common elements are found in at least a few organic compounds.

9.1 The Unique Carbon Atom

Carbon atoms are unique in their ability to bond to each other so strongly that they can form long chains. Silicon and a few other elements can form chains, but only short ones; carbon chains often contain thousands of carbon atoms.

Carbon chains can also have branches or form rings of various sizes. Add to this the fact that carbon atoms also bond strongly to other elements, such as hydrogen, oxygen, and nitrogen, and that these atoms can be arranged in many different ways, and it soon becomes obvious why there are so many carbon compounds.

In addition to the millions of carbon compounds already known, new ones are being discovered every day. Carbon can form an almost infinite number of molecules of various shapes, sizes, and compositions. We use thousands of carbon compounds every day without even realizing it because they are silently carrying out important chemical reactions within our bodies. Many of these carbon compounds are so vital that we literally could not live without them.

Self-Assessment Questions

1. Organic chemistry is the study of
   a. carbon compounds
   b. cyclic compounds
   c. natural substances
   d. saturated compounds
2. Which of the following is *not* a reason for the great number of carbon compounds?
   a. carbon atoms can form chains
   b. carbon can exhibit several oxidation numbers
   c. carbon atoms can form rings
   d. the same carbon atoms can be arranged different ways.

Answers: 1, a; 2, b

HYDROCARBONS

The simplest organic compounds are hydrocarbons. A **hydrocarbon** contains only hydrogen and carbon. There are several kinds of hydrocarbons, classified according to the type of bonding between carbon atoms. We will consider several of these in turn in the following sections.

9.2 Alkanes

Each carbon atom forms four bonds, and each hydrogen atom forms only one bond, and so the simplest hydrocarbon molecule that is possible is CH_4. Called methane, CH_4 is the main component of natural gas. It has the structure

$$
\begin{array}{c}
\text{H} \\
| \\
\text{H}-\text{C}-\text{H} \\
| \\
\text{H}
\end{array}
$$

Methane is the first member of a group of related compounds called **alkanes**, hydrocarbons that contain only single bonds. Alkanes can have from one to several hundred or more carbon atoms. The next member of the series is ethane,

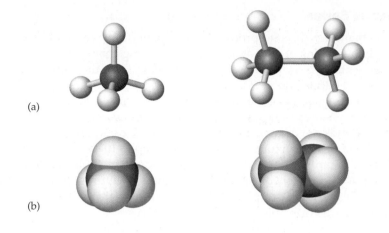

(a)

(b)

◀ **Figure 9.1** Ball-and-stick
(a) and space-filling (b) models of
methane (left) and ethane (right).

Ethane is a minor constituent of natural gas. It is seldom encountered as a pure compound in everyday life, but many compounds derived from it are common.

We saw in Chapter 4 that the methane molecule is tetrahedral, in accord with VSEPR theory. In fact, the tetrahedral shape results whenever a carbon atom is connected to four other atoms, be they hydrogen, carbon, or other elements. Figure 9.1 shows models of methane and ethane. The ball-and-stick models show the bond angles best, but the space-filling models more accurately reflect the shapes of the molecules. Ordinarily we use simple structural formulas such as the one shown previously for ethane because they are much easier to draw. A **structural formula** shows which atoms are bonded to each other, but it does not attempt to show the actual shape of the molecule.

Alkanes often are called **saturated hydrocarbons** because each carbon atom is *saturated* with hydrogen atoms, that is, bonded to the maximum number of hydrogen atoms. In constructing a formula, we connect the carbon atoms to each other through single bonds; then we add enough hydrogen atoms to give each carbon atom four bonds. All alkanes have two more than twice as many hydrogen atoms as carbon atoms. That is, alkanes can be represented by a general formula C_nH_{2n+2} in which n is the number of carbon atoms.

The three-carbon alkane is propane. Models of propane are shown in Figure 9.2. To draw its structural formula, we place three carbon atoms in a row.

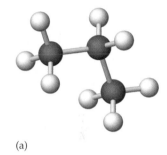

(a)

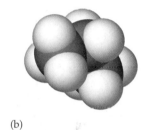

(b)

▲ **Figure 9.2** Ball-and-stick
(a) and space-filling (b) models of
propane.

$$C—C—C$$

Then we add enough hydrogen atoms (eight in this case) to give each carbon atom a total of four bonds. The structural formula of propane is, therefore,

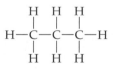

Condensed Structural Formulas

The complete structural formulas that we have used so far show all the carbon and hydrogen atoms and how they are attached to one another. But these formulas take up a lot of space, and they are quite a bit of trouble to draw or to type. For these reasons, chemists usually prefer to use **condensed structural formulas**. Condensed structures show how many hydrogen atoms are attached to each carbon atom without showing the bonds to each hydrogen atom. For example, ethane and propane are written $CH_3—CH_3$ and $CH_3—CH_2—CH_3$. These formulas can be simplified even further by omitting some (or all) of the bond lines, resulting in CH_3CH_3 and $CH_3CH_2CH_3$.

▲ A propane torch. Propane burns in air with a hot flame.

Table 9.1	Word Stems Indicating the Number of Carbon Atoms in Organic Molecules

Stem	Number
Meth-	One
Eth-	Two
Prop-	Three
But-	Four
Pent-	Five
Hex-	Six
Hept-	Seven
Oct-	Eight
Non-	Nine
Dec-	Ten

▲ Gasoline is a mixture containing dozens of hydrocarbons, many of which are alkanes.

Homologous Series

Note that methane, ethane, and propane form a pattern. We can build alkanes of any length simply by tacking carbon atoms together in long chains and adding sufficient hydrogen atoms to give each carbon atom four bonds. Even the naming of these compounds follows a pattern (Table 9.1), with a stem name indicating the number of carbon atoms. For compounds of five carbon atoms or more, each stem is derived from the Greek or Latin name for the number. The compound names end in *-ane*, signifying that the compounds are *alkanes*. Table 9.2 gives condensed structural formulas and names for continuous-chain (unbranched) alkanes up to 10 carbon atoms in length.

Notice that the molecular formula for each alkane in Table 9.2 differs from the one preceding it by precisely one carbon atom and two hydrogen atoms—that is, by a CH_2 unit. Such a series of compounds has properties that vary in a regular and predictable manner. This principle, called *homology,* gives order to organic chemistry in much the same way that the periodic table gives organization to the chemistry of the elements. Instead of studying the chemistry of a bewildering array of individual carbon compounds, organic chemists study a few members of a **homologous series** from which they can deduce the properties of other compounds in the series.

We need not have stopped at 10 carbon atoms as we did in Table 9.2. One could hook together 100 or 1000 or 1 million carbon atoms. We can make an infinite number of alkanes simply by lengthening the chain. But lengthening the chain is not the only option. Beginning with four carbon atoms, chain branching is also possible.

Isomerism

When we extend the carbon chain to four atoms and add enough hydrogen atoms to give each carbon atom four bonds, we get $CH_3CH_2CH_2CH_3$. This formula represents butane, a compound that boils at about 0 °C. A second compound, which has a boiling point of −12 °C, has the same molecular formula, C_4H_{10}, as butane. The structural formula of the second compound, however, is not the same as that of butane. Instead of having four carbon atoms connected in a continuous chain, this new compound has a continuous chain of only three carbon atoms. The fourth carbon is branched off the middle carbon of the three-carbon chain. To show the two structures more clearly, let's use complete structural formulas, showing the attachment of all the hydrogen atoms.

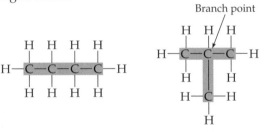

Table 9.2	The First Ten Continuous-Chain Alkanes

Name	Molecular Formula	Condensed Structural Formula	Number of Possible Isomers
Methane	CH_4	CH_4	—
Ethane	C_2H_6	CH_3CH_3	—
Propane	C_3H_8	$CH_3CH_2CH_3$	—
Butane	C_4H_{10}	$CH_3CH_2CH_2CH_3$	2
Pentane	C_5H_{12}	$CH_3CH_2CH_2CH_2CH_3$	3
Hexane	C_6H_{14}	$CH_3CH_2CH_2CH_2CH_2CH_3$	5
Heptane	C_7H_{16}	$CH_3CH_2CH_2CH_2CH_2CH_2CH_3$	9
Octane	C_8H_{18}	$CH_3CH_2CH_2CH_2CH_2CH_2CH_2CH_3$	18
Nonane	C_9H_{20}	$CH_3CH_2CH_2CH_2CH_2CH_2CH_2CH_2CH_3$	35
Decane	$C_{10}H_{22}$	$CH_3CH_2CH_2CH_2CH_2CH_2CH_2CH_2CH_2CH_3$	75

Compounds that have the same molecular formula but different structural formulas are called **isomers**. Because it is an isomer of butane, the branched four-carbon alkane is called isobutane. (Many organic compounds can be called by their common names or by the systematic names of IUPAC. As we shall see in Section 9.7, isobutane is also named 2-methylpropane because it has a methyl group attached as a substituent to the second carbon of propane.) Condensed structural formulas for the two isomeric butanes are written as follows.

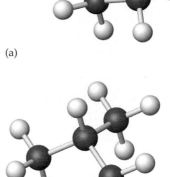

(a)

$$CH_3CH_2CH_2CH_3 \qquad \underset{\underset{\displaystyle CH_3}{|}}{CH_3CHCH_3}$$

Butane   Isobutane

Figure 9.3 shows ball-and-stick models of the two compounds.

The number of isomers increases rapidly with the number of carbon atoms (Table 9.2). There are three pentanes, five hexanes, nine heptanes, and so on. Isomerism is common in carbon compounds and provides another reason for the existence of millions of organic compounds.

Propane and the butanes are familiar fuels. Although they are gases at ordinary temperatures and under normal atmospheric pressure, they are liquefied under pressure and are usually supplied in tanks as liquefied petroleum gas (LPG). Gasoline is a mixture of hydrocarbons, mostly alkanes, with 5 to 12 carbon atoms.

Not all the possible isomers of the larger molecules have been isolated. Indeed, the task rapidly becomes more and more prohibitive as you proceed up the series. There are, for example, 4,111,846,763 possible isomers with the molecular formula $C_{30}H_{62}$.

(b)

▲ **Figure 9.3** Ball-and-stick models of (a) butane and (b) isobutane.

EXAMPLE 9.1   Hydrocarbon Formulas

Without referring to Table 9.2, give the molecular formula, the complete structural formula, and the condensed structural formula for heptane.

Solution

The stem *hept-* means seven carbon atoms, and the ending *-ane* indicates an alkane. For the complete structural formula, we can write out a string of seven carbon atoms.

$$C—C—C—C—C—C—C$$

Then we attach enough hydrogen atoms to the carbon atoms to give each carbon four bonds. This requires three hydrogen atoms on each end carbon and two each on the others.

$$\underset{\displaystyle \overset{\displaystyle |}{H}\ \overset{\displaystyle |}{H}\ \overset{\displaystyle |}{H}\ \overset{\displaystyle |}{H}\ \overset{\displaystyle |}{H}\ \overset{\displaystyle |}{H}\ \overset{\displaystyle |}{H}}{H—\overset{\overset{\displaystyle H}{|}}{C}—\overset{\overset{\displaystyle H}{|}}{C}—\overset{\overset{\displaystyle H}{|}}{C}—\overset{\overset{\displaystyle H}{|}}{C}—\overset{\overset{\displaystyle H}{|}}{C}—\overset{\overset{\displaystyle H}{|}}{C}—\overset{\overset{\displaystyle H}{|}}{C}—H}$$

For the condensed form, simply write each carbon atom's set of hydrogen atoms next to the carbon.

$$CH_3CH_2CH_2CH_2CH_2CH_2CH_3$$

For the molecular formula, we can simply count the carbon and hydrogen atoms to arrive at C_7H_{16}. Alternatively, we could use the general formula C_nH_{2n+2} with $n = 7$ to get C_7H_{16}.

■ **EXERCISE 9.1**

Give the molecular, complete structural, and condensed structural formulas for **(a)** hexane and **(b)** octane.

2. What does "petroleum distillates," listed on some product labels, mean?
Petroleum distillates are various components of crude oil, most of which are hydrocarbons. Paint thinner, lighter fluid, paraffin wax, petroleum jelly, and asphalt all are petroleum distillates.

ANSWER

Table 9.3 Physical Properties and Uses/Occurrences of Selected Alkanes

Name	Molecular Formula	Melting Point (°C)	Boiling Point (°C)	Density at 20 °C (g/mL)	Use/Occurrence
Methane	CH_4	−183	−162	(Gas)	Natural gas (main component); fuel
Ethane	C_2H_6	−172	−89	(Gas)	Natural gas (minor component); plastics
Propane	C_3H_8	−188	−42	(Gas)	LPG (bottled gas; fuel); plastics
Butane	C_4H_{10}	−138	0	(Gas)	LPG (fuel; lighter fuel)
Pentane	C_5H_{12}	−130	36	0.626	Gasoline component (fuel)
Hexane	C_6H_{14}	−95	69	0.659	Gasoline component (fuel); extraction solvent for food oils
Heptane	C_7H_{16}	−91	98	0.684	Gasoline component (fuel)
Octane	C_8H_{18}	−57	126	0.703	Gasoline component (fuel)
Decane	$C_{10}H_{22}$	−30	174	0.730	Gasoline component (fuel)
Dodecane	$C_{12}H_{26}$	−10	216	0.749	Gasoline component (fuel)
Tetradecane	$C_{14}H_{30}$	6	254	0.763	Diesel fuel component
Hexadecane	$C_{16}H_{34}$	18	280	0.775	Diesel fuel component
Octadecane	$C_{18}H_{38}$	28	316	(Solid)	Paraffin wax component
Eicosane	$C_{20}H_{42}$	37	343	(Solid)	Paraffin wax and asphalt component

Properties of Alkanes

Note in Table 9.3 that after the first few members of the alkane series, the rest show about a 20–30 °C increase in boiling point with each added CH_2 group. Note also that at room temperature alkanes with from 1 to 4 carbon atoms per molecule are gases, those with 5 to about 16 carbon atoms per molecule are liquids, and those with more than 16 carbon atoms per molecule are solids. The densities of liquid and solid alkanes are less than that of water.

Alkanes are nonpolar molecules and are essentially insoluble in water; hence, they float on top of water. They dissolve many organic substances of low polarity, such as fats, oils, and waxes.

Alkanes undergo few chemical reactions, but one important chemical property is that they burn, producing a lot of heat. Alkanes have many uses, but they mainly serve as fuels.

The physiologic properties of alkanes vary. Methane appears to be inert physiologically. We probably could breathe a mixture of 80% methane and 20% oxygen without ill effect. This mixture would be flammable, however, and no fire or spark of any kind could be permitted in such an atmosphere. Breathing an atmosphere of pure methane (the "gas" of a gas-operated stove) can lead to death—not because of the presence of methane but because of the absence of oxygen, a condition called *asphyxia*. Light liquid alkanes, such as those in gasoline, dissolve and wash away body oils when spilled on the skin. Repeated contact may cause *dermatitis*. (Other components of gasoline are less innocuous.) If swallowed, most alkanes do little

Recall that water has a density of 1.00 g/mL at room temperature. All the alkanes listed in Table 9.3 have densities less than that.

harm in the stomach; in fact, *mineral oil* is a mixture of long-chain liquid alkanes that has been used for many years as a laxative. However, in the lungs, alkanes cause chemical pneumonia by dissolving fatlike molecules from the cell membranes in the alveoli, allowing the lungs to fill with fluid. Heavier liquid alkanes, when applied to the skin, act as emollients (skin softeners). Petroleum jelly (Vaseline™ is one brand) is a semisolid mixture of hydrocarbons that can be applied as an emollient or simply as a protective film. (Skin lotions and creams are discussed in **eChapter 21.***)

Self-Assessment Questions

1. A saturated hydrocarbon is one that
   a. contains only carbon–carbon single bonds
   b. has the maximum number of carbon atoms
   c. has a vapor pressure equal to atmospheric pressure
   d. is solid at room temperature
2. Which of the following formulas represents an alkane?
   a. C_2H_2 **b.** C_2H_4 **c.** C_3H_6 **d.** C_3H_8
3. Successive members of the alkane series differ from the preceding member by one additional
   a. C atom and one H atom **b.** C atom and 2 H atoms
   c. C atom and 3 H atoms **d.** H atom and 2 C atoms
4. The members of the alkane series are similar in that each member has the same
   a. empirical formula **b.** general formula
   c. molecular formula **d.** structural formula
5. Isomers are compounds that have the same
   a. number of C atoms but a different number of H atoms
   b. number of H atoms but a different number of C atoms
   c. number and kind of atoms in a molecule but different structures
   d. kind of atoms in their molecular formulas but different numbers of these atoms
6. The formula for nonacontane, a 90-carbon alkane, is
   a. $C_{90}H_{176}$ **b.** $C_{90}H_{178}$ **c.** $C_{90}H_{180}$ **d.** $C_{90}H_{182}$
7. As the molecular mass of the members of the alkane series increases, their boiling points
   a. decrease **b.** increase **c.** remain the same **d.** vary randomly
8. For which of the following hydrocarbon molecular formulas is there more than one possible structural formula?
   a. CH_4 **b.** C_2H_6 **c.** C_3H_8 **d.** C_4H_{10}

Answers: 1, a; 2, d; 3, b; 4, b; 5, c; 6, d; 7, b; 8, d

9.3 Cyclic Hydrocarbons: Rings and Things

The hydrocarbons we have encountered so far (alkanes) have been composed of open-ended chains of carbon atoms. Carbon atoms can also connect to form closed rings. The simplest possible ring-containing hydrocarbon, or **cyclic hydrocarbon**, has the molecular formula C_3H_6.

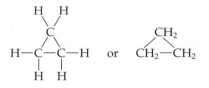

This compound is called cyclopropane (Figure 9.4).

*Chapters 19–22 are found online at www.chemplace.com

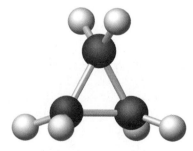

▲ **Figure 9.4** Ball-and-stick model of cyclopropane. Cyclopropane is a potent, quick-acting anesthetic with few undesirable side effects. It is no longer used in surgery, however, because it forms an explosive mixture with air at nearly all concentrations.

Names of cycloalkanes (cyclic hydrocarbons containing only single bonds) are formed by adding the prefix *cyclo-* to the name of the open-chain compound with the same number of carbon atoms as are in the ring.

Chemists often use geometric figures to represent cyclic compounds (Figure 9.5). For example, a triangle is used to represent the cyclopropane ring, and a hexagon represents cyclohexane.

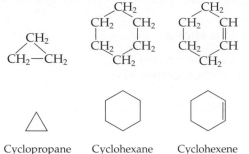

▶ **Figure 9.5** Structural formulas and symbolic representations of some cyclic hydrocarbons.

Cyclopropane   Cyclohexane   Cyclohexene

EXAMPLE 9.2   Structural Formulas of Cyclic Hydrocarbons

Give the structural formula for cyclobutane. What geometric figure is used to represent cyclobutane?

Solution

Cyclobutane has four carbon atoms arranged in cyclic fashion.

$$\begin{array}{c} C-C \\ |\ \ \ | \\ C-C \end{array}$$

Each carbon atom needs two hydrogen atoms to complete its set of four bonds.

$$\begin{array}{c} CH_2-CH_2 \\ |\ \ \ \ \ \ | \\ CH_2-CH_2 \end{array}$$

Cyclobutane can also be represented by a square.

■ **EXERCISE 9.2A**

Give the structural formula and geometric representation for cyclopentane.

■ **EXERCISE 9.2B**

What is the general formula for a cycloalkane?

Self-Assessment Questions

1. Hydrocarbons that have rings of carbon atoms are called
   a. benzenes
   b. cyclic hydrocarbons
   c. halocarbons
   d. xylenes

2. A cyclic compound with 6 carbon atoms and 12 hydrogen atoms is
   a. cyclobutane
   b. cycloheptane
   c. cyclohexane
   d. cyclosexane

Answers: 1, b; 2, c

9.4 Unsaturated Hydrocarbons: Alkenes and Alkynes

Two carbon atoms can share more than one pair of electrons. In ethylene (C_2H_4), the two carbon atoms share two pairs of electrons and are, therefore, joined by a double bond (Figure 9.6).

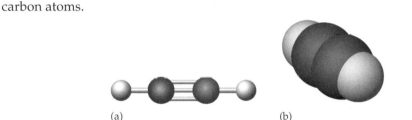

(Ethylene is the common name; the systematic or IUPAC name is ethene.) Ethylene is the simplest member of the alkene family. An **alkene** is a hydrocarbon that contains one or more carbon-to-carbon double bonds. Alkenes with one double bond have the general formula C_nH_{2n}, where n is the number of carbon atoms.

Ethylene is the most important commercial organic chemical. Annual U.S. production is over 20 billion kg. More than half goes into the manufacture of polyethylene, one of the most familiar plastics. Another 15% or so is converted to ethylene glycol, the major component of many formulations of antifreeze used in automobile radiators.

In acetylene (C_2H_2), the two carbon atoms share three pairs of electrons; the carbon atoms are joined by a triple bond (Figure 9.7).

$$H:C:::C:H \quad \text{or} \quad H-C\equiv C-H$$

Acetylene (IUPAC: ethyne) is the simplest member of the alkyne family. An **alkyne** is a hydrocarbon that contains one or more carbon-to-carbon triple bonds. Alkynes with one triple bond have the general formula C_nH_{2n-2}, where n is the number of carbon atoms.

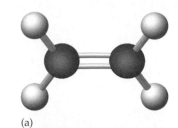

(a)

(b)

▲ **Figure 9.6** Ball-and-stick (a) and space-filling (b) models of ethylene.

◀ **Figure 9.7** Ball-and-stick (a) and space-filling (b) models of acetylene.

Acetylene is used in oxyacetylene torches for cutting and welding metals. Such torches can produce very high temperatures. Acetylene is also converted to a variety of other chemical products.

Collectively, alkenes and alkynes are called *unsaturated hydrocarbons*. An **unsaturated hydrocarbon** is so called because more hydrogen atoms can be added to it. Recall that a *saturated hydrocarbon* (alkane) has the maximum number of hydrogen atoms attached to each carbon atom; it has no double or triple bonds.

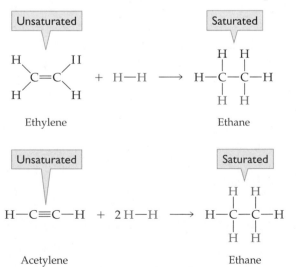

▲ Alkenes occur widely in nature. Ripening fruits and vegetables give off ethylene, which triggers further ripening. Food processors artificially introduce ethylene to hasten the normal ripening process: 1 kg of tomatoes can be ripened by exposure to as little as 0.1 mg of ethylene for 24 hours. Unfortunately, the tomatoes don't taste much like vine-ripened ones. The red color of tomatoes is due to lycopene, a hydrocarbon with many double bonds (a *polyene*). Lycopene, a potent antioxidant, is thought to aid in prevention of cancers, particularly prostate cancer.

Double bonds are common in various biochemicals important to life. Unsaturated fats are a well-known example.

EXAMPLE 9.3   Molecular Formulas of Hydrocarbons

What is the molecular formula for 1-octene? (The "1-" is a part of a systematic name that indicates the location of the double bond; it follows the first carbon atom of the chain.)

Solution

The stem *oct-* indicates eight carbon atoms, the ending *-ene* tells us that the compound is an alkene. Using the general formula C_nH_{2n} for an alkene, with $n = 8$, we see that an alkene with eight carbon atoms has the molecular formula C_8H_{16}.

■ **EXERCISE 9.3**

What are the molecular formulas for **(a)** 3-hexene and **(b)** 1-heptyne?

Properties of Alkenes and Alkynes

> An unsaturated hydrocarbon tends to burn with a smoky flame, especially in limited oxygen. Saturated hydrocarbons usually burn with a clean yellow flame. Natural gas, propane, and butane all burn almost soot free, whereas paint thinner, gasoline, and kerosene (containing unsaturated hydrocarbons) burn with sooty flames.

The physical properties of alkenes and alkynes are quite similar to those of the corresponding alkanes. Those with 2 to 4 carbon atoms per molecule are gases at room temperature, those with 5 to 18 carbon atoms are liquids, and most of those with more than 18 carbon atoms are solids. Like alkanes, alkenes and alkynes are insoluble in water and they float on water.

Like alkanes, both alkenes and alkynes burn. However, these unsaturated hydrocarbons undergo many more chemical reactions than alkanes. An alkene or alkyne can undergo an **addition reaction** across the double or triple bond. Alkenes and alkynes typically undergo addition reactions in which all the atoms of the reactants are incorporated into a single product. In the foregoing reactions, ethylene (C_2H_4) adds hydrogen (H_2) to form ethane, and acetylene (C_2H_2) adds 2 H_2 to form ethane (C_2H_6). Chlorine, bromine, water, and many other kinds of molecules also add to double and triple bonds.

One of the most unusual features of alkene (and alkyne) molecules is that they can add *to each other* to form large molecules called *polymers*. These interesting molecules are discussed in Chapter 10.

Self-Assessment Questions

1. Which of the following is an alkene?
   a. acetylene   **b.** benzene   **c.** ethylene   **d.** toluene

2. An alkene has
   a. a carbon–carbon double bond   **b.** a carbon–carbon triple bond
   c. a ring structure   **d.** only single carbon–carbon bonds

3. The hydrocarbon family that features molecules with a triple bond is the
   a. alkanes   **b.** alkenes   **c.** alkynes   **d.** aromatic hydrocarbons

4. Addition reactions occur in alkenes rather than in alkanes because alkenes have
   a. double bonds   **b.** a greater molecular mass
   c. more carbon atoms per molecule
   d. a tetrahedral arrangement of bonds

5. The general formula for the alkyne series is
   a. C_nH_n **b.** C_nH_{2n} **c.** C_nH_{2n-2} **d.** C_nH_{2n+2}

6. The following equation is a representation of what kind of reaction?

$$C_2H_4 + H_2 \longrightarrow C_2H_6$$

   a. addition   **b.** esterification   **c.** saponification   **d.** substitution

Answers: 1, c; 2, a; 3, c; 4, a; 5, c; 6, a

9.5 Aromatic Hydrocarbons: Benzene and Relatives

Another type of hydrocarbon is represented by benzene. Discovered by Michael Faraday in 1825, benzene has a molecular formula of C_6H_6. The structure of benzene puzzled chemists for decades. The formula seems to indicate an unsaturated compound, but benzene does not react as if it contains any double or triple bonds. It does not readily undergo addition reactions the way unsaturated compounds usually do.

Finally, in 1865 August Kekulé proposed a structure with a ring of six carbon atoms, each attached to one hydrogen atom.

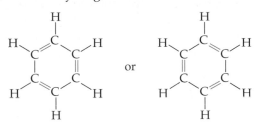

The two structures shown appear to contain double bonds, but in fact they do not. Both structures represent the same molecule, and the actual structure of this molecule is a hybrid of these two structures. The benzene molecule actually has six identical carbon-to-carbon bonds that are neither single bonds nor double bonds but something in between. In other words, the three pairs of electrons that would form the three double bonds are not tied down in one location but are spread around the ring (Figure 9.8). Today we usually represent the benzene ring with a circle inside a hexagon.

The hexagon represents the ring of six carbon atoms, and the inscribed circle the ring of six unassigned electrons. Because its ring of electrons resists being disrupted, the benzene molecule is an exceptionally stable structure.

Benzene and similar compounds are called *aromatic hydrocarbons* because quite a few of the first benzenelike substances to be discovered had strong aromas. Even though many benzene derivatives have turned out to be odorless, the name has stuck. Today an **aromatic compound** is any compound that contains a benzene ring or has certain properties similar to those of benzene.

Properties of Aromatic Hydrocarbons

Structures of some common aromatic hydrocarbons are shown in Figure 9.9. Benzene, toluene, and xylenes are all liquids that float on water. They are used mainly as solvents and fuels, but they are also used to make other benzene derivatives. Their vapors can act as narcotics when inhaled. Because benzene may cause leukemia after long exposure, its use has been restricted. Naphthalene is a volatile, white, crystalline solid used as an insecticide, especially as a moth repellant.

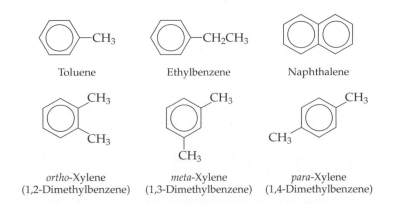

Toluene   Ethylbenzene   Naphthalene

ortho-Xylene
(1,2-Dimethylbenzene)

meta-Xylene
(1,3-Dimethylbenzene)

para-Xylene
(1,4-Dimethylbenzene)

More to Explore
Julian, Maureen M. "What Compound Was Discovered as a Result of an Insurance Claim?" *Journal of Chemical Education*, October 1981, p. 793.

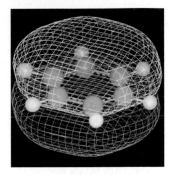

▲ **Figure 9.8** A computer-generated model of the benzene molecule. The six unassigned electrons occupy the yellow and blue areas below and above the plane of the carbon and hydrogen atoms.

More to Explore
Benfey, O. T. "August Kekulé and the Birth of the Structural Theory of Organic Chemistry in 1858." *Journal of Chemical Education*, January 1958, pp. 21–25.

◀ **Figure 9.9** Some aromatic hydrocarbons. Like benzene, toluene and the three xylenes are components of gasoline and also serve as solvents. All these compounds are intermediates in the synthesis of polymers (Chapter 10) and other organic chemicals.

A circle in a cyclic structure simply indicates an aromatic compound. It does not always mean six electrons. Naphthalene (Figure 9.9), for example, has a circle in each ring to indicate that it is an aromatic hydrocarbon. The total of unassigned electrons, however, is only 10. Some chemists still prefer to represent naphthalene (and other aromatic compounds) with alternate double and single bonds.

A N S W E R

3. If a compound has *benzene* in the name, is the compound a carcinogen?
Not necessarily; aromatic compounds often are named as derivatives of benzene simply because the structure contains a benzene ring. In fact many compounds that are useful or even necessary to life contain benzene rings, including penicillin, aspirin, and the amino acids phenylalanine and tryptophan.

Monosubstituted benzenes—in which one hydrogen is replaced by another group—are often represented in compact form, for example, toluene as $C_6H_5CH_3$ and ethylbenzene as $C_6H_5CH_2CH_3$. Similar designations are sometimes extended to disubstituted and higher aromatics by indicating the relative position of substituents, as in $1,4\text{-}C_6H_4(CH_3)_2$ for 1,4-dimethylbenzene.

Self-Assessment Questions

1. The molecular formula of benzene is
   a. C_4H_8 **b.** C_4H_{10} **c.** C_6H_6 **d.** C_6H_{10}
2. Which compound is an aromatic hydrocarbon?
   a. acetylene   **b.** butadiene   **c.** ethylene   **d.** toluene
3. Which organic molecule is represented as a hybrid of two structures?
   a. benzene   **b.** butadiene   **c.** ethylene   **d.** propyne
4. Naphthalene is an
   a. alkane   **b.** alkene   **c.** alkyne   **d.** aromatic hydrocarbon
5. The reactivity of the benzene ring reflects
   a. its aromatic odor
   b. the presence of carbon–carbon double bonds
   c. its saturated structure
   d. the spreading of six electrons over all six carbon atoms

Answers: 1, c; 2, d; 3, a; 4, d; 5, d

9.6 Chlorinated Hydrocarbons: Many Uses, Some Hazards

Many other organic compounds are derived from hydrocarbons by replacing one or more hydrogen atoms by another atom or group of atoms. For example, replacement of hydrogen atoms by chlorine atoms gives chlorinated hydrocarbons. When chlorine gas (Cl_2) is mixed with methane (CH_4) in the presence of ultraviolet (UV) light, a reaction takes place at a very rapid, even explosive, rate. The result is a mixture of products, some of which may be familiar to you (see Example 9.4 and Exercise 9.4). We can write an equation for the reaction in which one hydrogen atom of methane is replaced by a chlorine atom to give methyl chloride (chloromethane).

$$CH_4(g) + Cl_2(g) \longrightarrow CH_3Cl(g) + HCl(g)$$

EXAMPLE 9.4 Formulas of Chlorinated Hydrocarbons

What is the formula for dichloromethane (also known as methylene chloride)?

Solution

The *dichloro* indicates two chlorine atoms. The *methane* part of the name indicates that the compound is derived from methane. We conclude that two hydrogen atoms of methane (CH_4) have been replaced by chlorine atoms; the formula for dichloromethane is, therefore, CH_2Cl_2.

■ **EXERCISE 9.4**

What are the formulas for **(a)** trichloromethane (also known as chloroform) and **(b)** tetrachloromethane (carbon tetrachloride)?

Methyl chloride is used mainly in making silicone polymers (Chapter 10). Methylene chloride (dichloromethane) is a solvent used, for example, as a paint remover. Chloroform (trichloromethane), also a solvent, was used as an anesthetic in earlier times, but such use is now considered dangerous. The dosage required for effective anesthesia is too close to a lethal dose. Carbon tetrachloride (tetrachloromethane) has been used as a dry-cleaning solvent and in fire extinguishers, but it is no longer recommended for either use. Exposure to carbon tetrachloride (or most other chlorinated hydrocarbons) can cause severe damage to the liver. The use of a carbon tetrachloride fire extinguisher in conjunction with water to put out a fire can be deadly. Carbon tetrachloride reacts with water at elevated temperatures to form phosgene ($COCl_2$), an extremely poisonous gas that was used during World War I.

A variety of more complicated chlorinated hydrocarbons are of considerable interest. DDT (dichlorodiphenyltrichloroethane) and other chlorinated hydrocarbon insecticides are discussed in eChapter 20. In Chapter 10, we study polychlorinated biphenyls (PCBs). For now, let's just say that many chlorinated hydrocarbons have similar properties. Most are only slightly polar, and they do not dissolve in water, which is highly polar. Instead, they dissolve (and dissolve in) fats, oils, greases, and other substances of low polarity. This is why certain chlorinated hydrocarbons make good dry-cleaning solvents; they remove grease and oily stains from fabrics. This is also why DDT and PCBs cause problems for fish and birds and perhaps for people; the toxic substances are concentrated in fatty animal tissues rather than being easily excreted in (aqueous) urine.

Chlorofluorocarbons and Fluorocarbons

Carbon compounds containing fluorine as well as chlorine are called *chlorofluorocarbons (CFCs)*. CFCs have been used as the dispersing gases in aerosol cans, for making foamed plastics, and as refrigerants. Three common CFCs are best known by their industrial code designations:

$$CFCl_3 \qquad CF_2Cl_2 \qquad CF_2ClCF_2Cl$$
$$\text{CFC 11} \qquad \text{CFC 12} \qquad \text{CFC 114}$$

At room temperature, CFCs are gases or liquids with low boiling points. They are essentially insoluble in water and inert toward most other substances. These properties make them ideal propellants for use in aerosol cans for deodorant, hair spray, and food products. Unfortunately, the inertness of these compounds allows them to persist in the environment. They diffuse into the stratosphere, where they undergo chemical reactions that lead to depletion of the ozone layer that protects Earth from harmful UV radiation. We look at this problem and some attempts to solve it in Chapter 13.

Fluorinated compounds have found some interesting uses. Some have been used as blood extenders. Oxygen is quite soluble in certain *perfluorocarbons*. (The prefix *per-* means that all hydrogen atoms have been replaced, in this case by fluorine atoms.) These compounds can therefore serve as temporary substitutes for hemoglobin, the oxygen-carrying protein in blood. Perfluoro compounds have been used to treat premature babies, whose lungs are often underdeveloped.

Polytetrafluoroethylene (PTFE or Teflon®, Chapter 10) is a perfluorinated polymer. It has many interesting applications because of its resistance to corrosive chemicals and to high temperatures. It is especially noted for its unusual nonstick properties.

▲ Perfluorinated greases are used in highly corrosive or reactive environments that would cause most hydrocarbon greases to burst into flame.

Self-Assessment Questions

1. The compound CH_3Cl is named
   a. chloroform
   b. methyl chloride
   c. methylene chloride
   d. methyl ether
2. The compound CF_2Cl_2 is
   a. a chlorofluorocarbon
   b. chloroform
   c. methyl chloride
   d. methylene chloride
3. Which of the following is an isomer of 2-chloropropane, $CH_3CHClCH_3$?
   a. butane
   b. 2-chlorobutane
   c. 1-chloropropane
   d. propane

Answers: 1, b; 2, a; 3, c

OXYGEN-CONTAINING ORGANIC COMPOUNDS

Many organic compounds contain oxygen as well as carbon and hydrogen. In Sections 9.8 through 9.13, we introduce several important families of oxygen-containing organic compounds. First, however, we introduce another fundamental concept of organic chemistry in Section 9.7.

9.7 The Functional Group

Double and triple carbon–carbon bonds and halogen substituents are examples of **functional groups**, groups of atoms that give a family of organic compounds its characteristic chemical and physical properties. Table 9.4 lists a number of common functional groups found in organic molecules. Each of these groups is the basis for its own homologous series.

Table 9.4 Selected Organic Functional Groups

Name of Class	Functional Group*	General Formula of Class
Alkane	None	R—H
Alkene	$-C=C-$	$R_2C=CR_2$
Alkyne	$-C\equiv C-$	$RC\equiv CR$
Alcohol	$-C-OH$	R—OH
Ether	$-C-O-C-$	R—O—R′
Aldehyde	$-\overset{O}{\overset{\|}{C}}-H$	$R-\overset{O}{\overset{\|}{C}}-H$
Ketone	$-\overset{O}{\overset{\|}{C}}-$	$R-\overset{O}{\overset{\|}{C}}-R'$
Carboxylic acid	$-\overset{O}{\overset{\|}{C}}-OH$	$R-\overset{O}{\overset{\|}{C}}-O-H$
Ester	$-\overset{O}{\overset{\|}{C}}-O-C-$	$R-\overset{O}{\overset{\|}{C}}-O-R'$
Amine	$-C-N-$	$R-\overset{H}{\underset{}{N}}-H$
		$R-\overset{H}{\underset{}{N}}-R'$
		$R-\overset{R'}{\underset{}{N}}-R''$
Amide	$-\overset{O}{\overset{\|}{C}}-N-$	$R-\overset{O}{\overset{\|}{C}}-N-H$ with H
		$R-\overset{O}{\overset{\|}{C}}-N-R'$ with H
		$R-\overset{O}{\overset{\|}{C}}-N-R'$ with R″

* Neutral functional groups are shown in green, acidic groups in red, and basic groups in blue.

In many simple molecules, a functional group is attached as a substituent to a hydrocarbon stem called an *alkyl group*. An **alkyl group** is derived from an alkane by removing a hydrogen atom. The methyl group (CH_3—), for example, is derived from methane (CH_4), and the ethyl group (CH_3CH_2—) from ethane (CH_3CH_3). Propane yields two different alkyl groups, depending on whether the hydrogen atom is missing from the end or from the middle carbon atom.

Table 9.5 lists the four alkyl groups derived from the two butanes.

Often the letter R is used to stand for alkyl groups in general. Thus, ROH is a general formula for alcohols, and RCl represents any alkyl chloride.

Table 9.5 Common Alkyl Groups

Name		Structural Formula	Condensed Structural Formula
Derived from Propane	Propyl		$CH_3CH_2CH_2$—
	Isopropyl		CH_3CHCH_3
Derived from Butane	Butyl		$CH_3CH_2CH_2CH_2$—
	Secondary butyl (*sec*-butyl)		$CH_3CHCH_2CH_3$
Derived from Isobutane	Isobutyl		CH_3CHCH_2— with CH_3
	Tertiary butyl (*tert*-butyl)		CH_3—C—CH_3 with CH_3

Self-Assessment Questions

1. A specific arrangement of atoms that gives characteristic properties to an organic molecule is called a(n)
   a. alkyl group   **b.** carboxyl group
   c. functional group   **d.** hydroxyl group

2. The formula CH_3CH_2- represents
   a. an alkyl group   b. ethane
   c. ethylene   d. a functional group
3. How many carbon atoms are there in an ethyl group?
   a. 1   b. 2   c. 3   d. 4
4. The formula $CH_3CH_2CH_2-$ represents a(n)
   a. butyl group   b. isobutyl group
   c. isopropyl group   d. propyl group

Answers: 1, c; 2, a; 3, b; 4, d

9.8 The Alcohol Family

When a hydroxyl group (—OH) is substituted for any hydrogen atom in an alkane (RH), the molecule becomes an **alcohol (ROH)**. Like many organic compounds, alcohols can be called by their common names or by the systematic names of IUPAC. The IUPAC names for alcohols are based on those of alkanes, with the ending changed from *-e* to *-ol*. Methanol, ethanol, and 1-propanol are the first three members of the homologous series of alcohols. A number designates the carbon to which the —OH group is attached when more than one location is possible. For example, the 1 in 1-propanol indicates that —OH is attached to an end carbon atom of the three-carbon chain, whereas the isomeric compound, 2-propanol, has the —OH group on the second carbon atom of the chain.

$$CH_3-OH \qquad CH_3CH_2-OH \qquad CH_3CH_2CH_2-OH \qquad \begin{array}{c} CH_3CHCH_3 \\ | \\ OH \end{array}$$

Methanol   Ethanol   1–Propanol   2–Propanol

Common names for these alcohols are methyl alcohol, ethyl alcohol, and propyl alcohol. The IUPAC names are the ones used in most scientific literature.

Methyl Alcohol (Methanol)

The simplest alcohol is methanol or methyl alcohol (Figure 9.10). Methanol is sometimes called wood alcohol because it was once made from wood. The modern industrial process makes methanol from carbon monoxide and hydrogen at high temperature and pressure in the presence of a catalyst.

$$CO(g) + 2\,H_2(g) \longrightarrow CH_3OH(l)$$

Methanol is important as a solvent and as a chemical intermediate. It is also an additive and possible complete replacement for gasoline in automobiles.

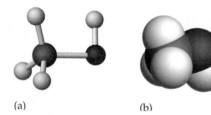

(a)   (b)

▲ **Figure 9.10** Ball-and-stick (a) and space-filling (b) models of the methanol molecule.

Ethyl Alcohol (Ethanol)

The next member of the homologous series of alcohols is ethyl alcohol (CH_3CH_2OH), also called ethanol or grain alcohol (Figure 9.11). Ethanol is made by fermentation of grain (or other starchy or sugary materials). If the sugar is glucose, the reaction is

$$C_6H_{12}O_6(aq) \xrightarrow{\text{yeast}} 2\,CH_3CH_2OH(aq) + 2\,CO_2(g)$$

Ethanol for beverages and for automotive fuel is made in this way. Ethanol for industrial use only is made by reacting ethylene with water.

$$CH_2{=}CH_2(g) + H_2O(l) \xrightarrow{H^+} CH_3CH_2OH(l)$$

This industrial alcohol is identical to that made by fermentation and is generally cheaper, but by law it cannot be used in alcoholic beverages. Since it carries no excise tax, the law requires that noxious substances be added to the alcohol to prevent people from drinking it. The resulting *denatured alcohol* is not fit to drink. This is the kind of alcohol commonly found on the shelves in chemical laboratories.

(a)   (b)

▲ **Figure 9.11** Ball-and-stick (a) and space-filling (b) models of the ethanol molecule.

Gasoline in many parts of the United States contains up to 10% ethyl alcohol from fermentation. Because the United States has a large corn (maize) surplus, there is a generous government subsidy for gasoline producers who make their ethanol this way. As a result, many factories now make ethanol by fermentation of corn.

Toxicity of Alcohols

Although ethyl alcohol is an ingredient in wine, beer, and other alcoholic beverages, alcohols in general are rather toxic. Methanol, for example, is oxidized in the body to formaldehyde (HCHO). Drinking as little as 1 oz (about 30 mL or 2 tablespoonfuls) can cause blindness and even death. Several poisonings each year result from mistaking methanol for its less toxic relative ethanol.

Ethanol is not as poisonous as methanol, but it is still toxic. About 50,000 cases of ethanol poisoning are reported each year in the United States, and dozens of people die of acute ethanol poisoning.

Generally, ethanol acts as a mild depressant; it slows down both physical and mental activity. Table 9.6 lists the effects of various doses. Although ethanol generally is a depressant, in small amounts it seems to act as a stimulant, perhaps by relaxing tensions and relieving inhibitions.

Is it poison? That seems to be a simple question, but it is a difficult one to answer. Toxicity depends on the nature of the substance, the amount, and the route by which it is taken into the body. Toxicity is discussed in detail in eChapter 22.

The *proof* of an alcoholic beverage is twice the percentage of alcohol by volume. The term originated in an old seventeenth-century English method for testing whiskey to prove that a dealer was not increasing profits by adding water to the booze. A qualitative test was to pour some of the whiskey on gunpowder and ignite it. Ignition of the gunpowder after the alcohol had burned away was considered "proof" that the whiskey did not contain too much water.

Table 9.6 Approximate Relationship among Drinks Consumed, Blood-Alcohol Level, and Behavior*

Number of Drinks[†]	Blood-Alcohol Level (percent by volume)	Behavior[‡]
2	0.05	Mild sedation; tranquility
4	0.10	Lack of coordination
6	0.15	Obvious intoxication
10	0.30	Unconsciousness
20	0.50	Possible death

* Data are for a 70-kg (154-lb) moderate drinker.
† Rapidly consumed 30-mL (1-oz) "shots" of 90-proof whiskey, 360-mL (12-oz) bottles of beer, or 150-mL (5-oz) glasses of wine.
‡ An inexperienced drinker would be affected more strongly, or more quickly, than one who is ordinarily a moderate drinker. Conversely, an experienced heavy drinker would be affected less.

Excessive ingestion of ethanol over a long period of time can alter brain cell function, cause nerve damage, and shorten life span by contributing to diseases of the liver, cardiovascular system, and practically every other organ of the body. In addition, about half of fatal automobile accidents involve at least one drinking driver. Babies born to alcoholic mothers often are small, deformed, and mentally retarded. Some investigators believe that *fetal alcohol syndrome* can occur even if mothers drink only moderately. Ethanol, by far the most abused drug in the United States, is discussed more fully in Chapter 18.

Rubbing alcohol is a 70% solution of isopropyl alcohol. Because isopropyl alcohol is also more toxic than ethyl alcohol, it is not surprising that people become ill, and sometimes die, after drinking rubbing alcohol.

▲ Warning labels on alcoholic beverages alert consumers to the hazards of excessive consumption.

Multifunctional Alcohols

Several alcohols have more than one hydroxyl group. Examples are ethylene glycol, propylene glycol, and glycerol.

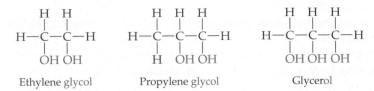

Ethylene glycol   Propylene glycol   Glycerol

Ethylene glycol is the main ingredient in many permanent antifreeze mixtures. Its high boiling point keeps it from boiling away in an automobile radiator. Ethylene glycol is a syrupy liquid with a sweet taste, but it is quite toxic. It is oxidized in the liver to oxalic acid.

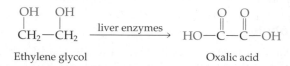

Oxalic acid forms crystals of its calcium salt, calcium oxalate (CaC_2O_4), which can damage the kidneys, leading to kidney failure and death. Propylene glycol is now marketed as a safer permanent antifreeze. It is a high-boiling antifreeze and it is not poisonous.

Glycerol (or glycerin) is a sweet, syrupy liquid made as a by-product from fats during soap manufacture (eChapter 21). It is used in lotions to keep the skin soft and as a food additive to keep cakes moist. Its reaction with nitric acid makes nitroglycerin, the explosive material in dynamite. Incidentally, nitroglycerin is also important as a vasodilator, a medication taken by heart patients to relieve angina pain.

▲ **It DOES Matter!**

Antifreeze containing ethylene glycol should always be drained into a container and disposed of properly. Pets have been known to die from licking the sweet but toxic ethylene glycol mixture from a driveway or open container.

EXAMPLE 9.5   Structural Formulas of Alcohols

Write the structural formula for *tert*-butyl alcohol, sometimes used as an octane booster in gasoline.

Solution

An alcohol can be considered an alkyl group joined to a hydroxyl group. From Table 9.5, we see that the *tert*-butyl group is

Connecting this group to an OH group gives *tert*-butyl alcohol.

$$CH_3-\underset{\underset{CH_3}{|}}{\overset{\overset{CH_3}{|}}{C}}-OH$$

■ **EXERCISE 9.5**

Write the structural formulas for **(a)** butyl alcohol and **(b)** *sec*-butyl alcohol.

Self-Assessment Questions

1. Which of the following formulas represents an alcohol?
   a. $CH_3CHOHCH_2CH_3$
   b. CH_3COCH_3
   c. CsOH
   d. HCHO
2. Which kind of reaction produces ethanol as one of the main products?
   a. esterification
   b. fermentation
   c. neutralization
   d. saponification
3. The isomers 1-propanol and 2-propanol have a different
   a. arrangement of the carbon atoms
   b. kind of functional group
   c. location of the functional group
   d. molecular formula
4. Which of the following compounds could be represented by the general formula ROH?
   a. butane
   b. 1-butanol
   c. methyl butanoate
   d. butanoic acid

5. Which of the following alcohols have molecules with more than one hydroxyl group?
a. 2-butanol   **b.** glycerol
c. 3-pentanol   **d.** 2-propanol

6. The formula for ethylene glycol is
a. CH_3CH_2OH **b.** $CH_3CH(OH)_2$
c. $C_3H_5(OH)_3$ **d.** $HOCH_2CH_2OH$

7. Which of the following formulas represents a dihydroxy alcohol?
a. $Ca(OH)_2$ **b.** $CH_3COCOOH$
c. $CH_3CHOHCH_2OH$ **d.** $HOCH(CH_2OH)_2$

Answers: 1, a; 2, b; 3, c; 4, b; 5, b; 6, d; 7, c

9.9 Phenols

When a hydroxyl group is attached to a benzene ring, the compound is called a **phenol**. Although it may appear to be an alcohol, it actually is not. The benzene ring greatly alters the properties of the hydroxyl group. In fact, phenol is a weak acid (sometimes called carbolic acid), and it is highly poisonous.

Phenol

Phenol was the first antiseptic used in an operating room, by Joseph Lister in 1867. Up until that time, surgery was not antiseptic, and many patients died from infections following surgical operations. Although phenol has a strong germicidal action, it is far from an ideal antiseptic because it causes severe skin burns and it kills healthy cells along with harmful microorganisms.

Phenol is still sometimes employed as a disinfectant for floors and furniture, but other phenolic compounds are now used as antiseptics. Hexylresorcinol, for example, is a more powerful germicide than phenol, and it is less damaging to the skin and has fewer other side effects.

Self-Assessment Questions

1. A formula that has a hydroxyl (OH) group attached to a benzene ring represents a(n)
a. alcohol   **b.** aldehyde   **c.** ketone   **d.** phenol

2. An important use of phenols is as
a. anesthetics   **b.** antiseptics   **c.** flavors   **d.** solvents

Answers: 1, d; 2, b

4. What gives some mouthwashes a medicinal odor?
Some mouthwashes and sore-throat remedies contain certain phenols, which often have what might be termed a "hospital" odor.

A
N
S
W
E
R

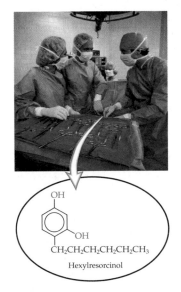

OH

OH

$CH_2CH_2CH_2CH_2CH_2CH_3$

Hexylresorcinol

▲ Phenolic compounds help to ensure antiseptic conditions in hospital operating rooms.

9.10 Ethers

Compounds with two alkyl groups attached to the same oxygen atom are called **ethers**. The general formula is ROR (or ROR' because the alkyl groups need not be alike). Best known is diethyl ether ($CH_3CH_2OCH_2CH_3$), often called simply *ether*.

Diethyl ether was once used as an anesthetic but today is used mainly as a solvent; it dissolves many organic substances that are insoluble in water. It boils at 36 °C, and so it evaporates readily, making it easy to recover dissolved materials. Although diethyl ether has little chemical reactivity, it is highly flammable, and great care must be used to avoid sparks or flames when it is in use.

Diethyl ether was introduced in 1842 as a general anesthetic for surgery and was once the most widely used anesthetic. It is rarely used for humans today because it has undesirable side effects, such as postanesthetic nausea and vomiting. We discuss modern anesthesia in Chapter 18.

As with cyclic hydrocarbons, cyclic ethers can be represented by geometric figures. Each corner of the figure (except the O) stands for a C atom with enough H atoms attached to give the carbon four bonds. Thus, ethylene oxide is represented as

Another problem with ethers is that over time they can react slowly with oxygen to form unstable peroxides, which may decompose explosively. Beware of previously opened containers of ether, especially old ones.

The ether produced in the largest amount commercially is a cyclic compound called ethylene oxide. Its two carbon atoms and an oxygen atom form a three-membered ring. Ethylene oxide is a toxic gas, much of which is used to make ethylene glycol.

$$\underset{\text{Ethylene oxide}}{\overset{\text{H}_2\text{C}-\text{CH}_2}{\underset{\text{O}}{\bigtriangleup}}} + \underset{\text{Water}}{\text{H}_2\text{O}} \xrightarrow{\text{H}^+} \underset{\text{Ethylene glycol}}{\overset{\text{CH}_2-\text{CH}_2}{\underset{\text{OH} \quad \text{OH}}{|\quad|}}}$$

Ethylene oxide is also used to sterilize medical instruments and as an intermediate in the synthesis of nonionic surfactants (**eChapter 21**). Most of the ethylene glycol produced is used in making polyester fibers (Chapter 10) and antifreeze.

CONCEPTUAL EXAMPLE 9.6 The Functional Group

Classify each of the following as an alcohol, an ether, or a phenol.

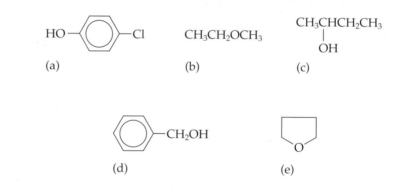

(a) (b) (c)

(d) (e)

Solution

a. The functional group, an OH, is attached directly to the benzene ring; the compound is a phenol.
b. The O atom is between two alkyl groups; the compound is an ether.
c. The OH group is attached to an alkyl group; the compound is an alcohol.
d. An alcohol; the OH is not attached directly to the benzene ring.
e. An ether; the O atom is between two C atoms.

■ **EXERCISE 9.6**

Classify the following as alcohols, ethers, or phenols.

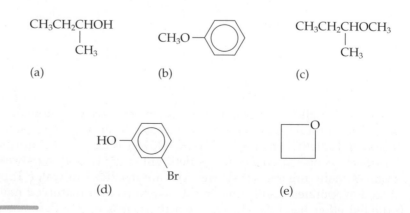

(a) (b) (c)

(d) (e)

EXAMPLE 9.7 Formulas for Ethers

Give the formula for isopropyl methyl ether.

Solution

Isopropyl methyl ether has an oxygen atom joined to an isopropyl group (three carbons joined to oxygen by the middle carbon) and a methyl group. The formula is

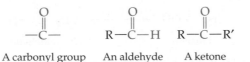

■ **EXERCISE 9.7**

Give formulas for **(a)** methyl propyl ether and **(b)** ethyl *tert*-butyl ether.

Self-Assessment Questions

1. The general formula R—O—R′ represents a(n)
   a. ester   **b.** ether   **c.** ketone   **d.** phenol

2. The formula $CH_3CH_2OCH_2CH_3$ represents
   a. 2-butanol   **b.** 2-butanone
   c. butyraldehyde   **d.** diethyl ether

3. The formula for butyl ethyl ether is
   a. $CH_3CH_2OCH_2CH_3$
   b. $CH_3OCH_2CH_2CH_3$
   c. $CH_3CH_2CH_2OCH_2CH_3$
   d. $CH_3CH_2OCH_2CH_2CH_2CH_3$

4. The compound $CH_3CHOHCH_2CH_3$ is an isomer of
   a. $CH_3COCH_2CH_2CH_3$
   b. $CH_3CH_2OCH_2CH_3$
   c. $CH_3COOCH_2CH_3$
   d. $CH_3CH_2CH_2COOH$

Answers: 1, b; 2, d; 3, d; 4, b

9.11 Aldehydes and Ketones

Two families of organic compounds that share the same functional group are **aldehydes** and **ketones**. Both families contain the **carbonyl group** ($C=O$) but aldehydes have a hydrogen atom attached to the carbonyl carbon, whereas ketones have the carbonyl carbon attached to two other carbon atoms.

$$
\begin{array}{ccc}
\overset{O}{\underset{|}{\overset{\|}{-C-}}} & R-\overset{O}{\overset{\|}{C}}-H & R-\overset{O}{\overset{\|}{C}}-R' \\
\text{A carbonyl group} & \text{An aldehyde} & \text{A ketone}
\end{array}
$$

To simplify typing, these structures are often written on one line with the $C=O$ understood:

$$
\begin{array}{ccc}
-C(=O)- \text{ or } -CO- & R-CH(=O) \text{ or } R-CHO & R-C(=O)-R' \text{ or } R-CO-R' \\
\text{A carbonyl group} & \text{An aldehyde} & \text{A ketone}
\end{array}
$$

Models of three familiar carbonyl compounds are shown in Figure 9.12.

▶ **Figure 9.12** Ball-and-stick (a) and space-filling (b) models of formaldehyde (left), acetaldehyde (center), and acetone (right).

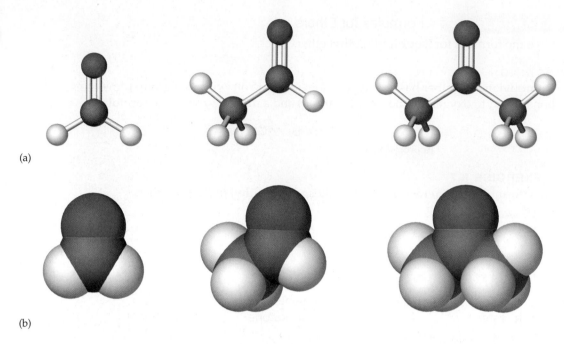

(a)

(b)

Some Common Aldehydes

The simplest aldehyde is formaldehyde (HCHO). It is a gas at room temperature but is readily soluble in water. As a 40% solution called formalin, it is used as a preservative for biological specimens and in embalming fluid. Systematic names for aldehydes are based on those of alkanes, with the ending changed from *-e* to *-al*. Thus, by the IUPAC system formaldehyde is named methanal. IUPAC names are seldom used for simple aldehydes because of possible confusion with the corresponding alcohols. For example, methanal is easily confused with methanol, either orally or (especially) when handwritten.

Formaldehyde is used in making certain plastics (Chapter 10). It also is used to disinfect homes, ships, and warehouses. Commercially, formaldehyde is made by the oxidation of methanol. This is the same net reaction that occurs in the human body when methanol is ingested and that accounts for the high toxicity of methyl alcohol.

Organic chemists often write equations that show only the organic reactants and products. Inorganic substances are omitted for the sake of simplicity.

$$CH_3OH \xrightarrow{\text{oxidation}} H-\overset{\overset{\displaystyle O}{\|}}{C}-H$$

Methanol   Formaldehyde

The next member of the homologous series of aldehydes is acetaldehyde (ethanal), formed by the oxidation of ethanol.

$$CH_3CH_2OH \xrightarrow{\text{oxidation}} CH_3-\overset{\overset{\displaystyle O}{\|}}{C}-H$$

Ethanol   Acetaldehyde

The next two members of the aldehyde series are propionaldehyde (propanal) and butyraldehyde (butanal). Both have strong, unpleasant odors. Benzaldehyde has an aldehyde group attached to a benzene ring. Also called (synthetic) oil of almond, benzaldehyde is used in perfumery and flavoring. It is the flavor ingredient in maraschino cherries.

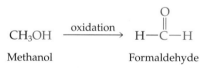

Propionaldehyde   Butyraldehyde   Benzaldehyde

Some Common Ketones

The simplest ketone is acetone, made by the oxidation of isopropyl alcohol.

Acetone is a common solvent for such organic materials as fats, rubbers, plastics, and varnishes. It also finds use in paint and varnish removers. It is a common ingredient in some fingernail polish removers. By the IUPAC system, acetone is named propanone. The systematic names for ketones are based on those of alkanes, with the ending changed from -e to -one. When necessary, a number is used to indicate the location of the carbonyl group. For example, $CH_3CH_2COCH_2CH_2CH_3$ is 3-hexanone.

Two other familiar ketones are ethyl methyl ketone and isobutyl methyl ketone, which, like acetone, are frequently used as solvents.

A ketone is a carbon chain
With carbonyl inside.
If carbonyl is on the end,
Then it's an aldehyde.

CONCEPTUAL EXAMPLE 9.8 Aldehyde or Ketone?

Identify each of the following compounds as an aldehyde or a ketone.

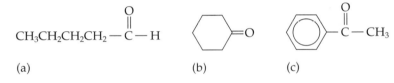

(a)   (b)   (c)

Solution

a. A hydrogen atom is attached to the carbonyl carbon atom; the compound is an aldehyde.
b. The carbonyl group is between two other (ring) carbon atoms; the compound is a ketone.
c. A ketone. (Remember that the corner of the hexagon stands for a carbon atom.)

■ **EXERCISE 9.8**
Identify each of the following compounds as an aldehyde or a ketone.

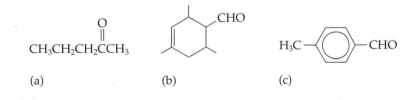

(a)   (b)   (c)

Self-Assessment Questions

1. The aldehyde functional group has a carbonyl group attached to
   a. an H atom   **b.** an N atom   **c.** an OH group   **d.** two other C atoms
2. The ketone functional group has a carbonyl group attached to
   a. an OH group   **b.** an OR group
   c. two other C atoms   **d.** two OR groups

3. The general formula R—CO—R′ represents a(n)
   a. ester   b. ether   c. ketone   d. phenol
4. A name for $CH_3COCH_2CH_3$ is
   a. butanal   b. ethyl methyl ether
   c. ethyl methyl ketone   d. 2-pentanone
5. A name for $CH_3CH_2CH_2CHO$ is
   a. butanone   b. butyraldehyde
   c. propanone   d. propylaldehyde
6. Oxidation of what alcohol produces formaldehyde?
   a. CH_3OH b. CH_3CH_2OH
   c. $CH_3CH_2CH_2OH$ d. $CH_3CHOHCH_3$
7. Oxidation of what alcohol produces acetone?
   a. CH_3OH b. CH_3CH_2OH
   c. $CH_3CH_2CH_2OH$ d. $CH_3CHOHCH_3$

Answers: 1, a; 2, c; 3, c; 4, c; 5, b; 6, a; 7, d

9.12 Carboxylic Acids

The functional group of organic acids is called the **carboxyl group**, and the acids are called **carboxylic acids**.

A carboxyl group   A carboxylic acid

Like aldehydes and ketones, these structures are often written on one line:

$$—C(=O)OH \quad \text{or} \quad —COOH \qquad R—C(=O)OH \quad \text{or} \quad R—COOH$$

A carboxyl group   A carboxylic acid

The simplest carboxylic acid is formic acid (HCOOH). The systematic (IUPAC) names for carboxylic acids are based on those of alkanes, with the ending changed from *-e* to *-oic acid*. Thus, the IUPAC name for formic acid is methanoic acid, and $CH_3CH_2CH_2CH_2COOH$ is pentanoic acid.

Formic acid was first obtained by the destructive distillation of ants (the Latin word *formica* means "ant"). The sting of an ant smarts because the ant injects formic acid as it stings. The stings of wasps and bees also contain formic acid (as well as other poisonous compounds).

Acetic acid (ethanoic acid, CH_3COOH) can be made by the aerobic fermentation of a mixture of cider and honey. This produces a solution (vinegar) containing about 4–10% acetic acid plus a number of other compounds that give vinegar its flavor. Acetic acid is probably the most familiar weak acid used in academic and industrial chemistry laboratories.

The third member of the homologous series of acids, propionic acid (propanoic acid, CH_3CH_2COOH), is seldom encountered in everyday life. The fourth member is more familiar, at least by its odor. If you've ever smelled rancid butter, you know the odor of butyric acid (butanoic acid, $CH_3CH_2CH_2COOH$). It is one of the most foul-smelling substances imaginable. Butyric acid can be isolated from butterfat or synthesized in the laboratory. It is one of the ingredients of body odor, and extremely small quantities of this and other chemicals enable bloodhounds to track fugitives.

The acid with a carboxyl group attached directly to a benzene ring is called benzoic acid.

▲ Acetic acid is a familiar weak acid in chemistry laboratories. It is also the principal active ingredient in vinegar.

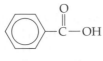

Benzoic acid

Carboxylic acid salts—calcium propionate, sodium benzoate, and others—are widely used as food additives to prevent mold (Chapter 17).

EXAMPLE 9.9 Structural Formulas of Oxygen-Containing Organic Compounds

Give the structural formula for each of the following compounds.

a. propionaldehyde
b. ethanoic acid
c. ethyl methyl ketone

Solution

a. Propionaldehyde has three carbon atoms with an aldehyde function.

$$C-C-\overset{\overset{\displaystyle O}{\|}}{C}-H$$

Adding the proper number of hydrogen atoms to the other two carbon atoms gives the structure

$$H-\overset{\overset{\displaystyle H}{|}}{\underset{\underset{\displaystyle H}{|}}{C}}-\overset{\overset{\displaystyle H}{|}}{\underset{\underset{\displaystyle H}{|}}{C}}-\overset{\overset{\displaystyle O}{\|}}{C}-H \quad \text{or} \quad CH_3CH_2CHO$$

b. Ethanoic acid has two carbon atoms with a carboxylic acid function.

$$H-\overset{\overset{\displaystyle H}{|}}{\underset{\underset{\displaystyle H}{|}}{C}}-\overset{\overset{\displaystyle O}{\|}}{C}-OH \quad \text{or} \quad CH_3COOH$$

c. Ethyl methyl ketone has a ketone function between an ethyl and a methyl group.

$$H-\overset{\overset{\displaystyle H}{|}}{\underset{\underset{\displaystyle H}{|}}{C}}-\overset{\overset{\displaystyle H}{|}}{\underset{\underset{\displaystyle H}{|}}{C}}-\overset{\overset{\displaystyle O}{\|}}{C}-\overset{\overset{\displaystyle H}{|}}{\underset{\underset{\displaystyle H}{|}}{C}}-H \quad \text{or} \quad CH_3CH_2COCH_3$$

■ **EXERCISE 9.9A**
Give the structural formula for each of the following compounds.
(a) butyric acid   **(b)** acetaldehyde   **(c)** diethyl ketone

■ **EXERCISE 9.9B**
Give the structural formula for each of the following compounds.
(a) hexanoic acid   **(b)** 3-octanone   **(c)** heptanal

Self-Assessment Questions

1. Which of the following structures represents a carboxylic acid?
   a. $CH_3CH_2CH_2OH$ **b.** CH_3CH_2COOH
   c. $CH_3CH_2COOCH_3$ **d.** CH_3CH_2CHO

2. Which of the following represents the functional group of a carboxylic acid?
   a. $-CH_2OH$ **b.** $-CHO$ **c.** $-CH_2CO-$ **d.** $-COOH$

6. Why do dead animals smell so bad?
Most amines have unpleasant odors, and many of the compounds formed when animal matter breaks down—such as putrescine and cadaverine—are amines.

CONCEPTUAL EXAMPLE 9.10   Amine Isomers: Structures and Names

Give structures and names for the other three-carbon amines.

Solution
Three carbon atoms can be in one alkyl group, and there are two such propyl groups.

$$CH_3CH_2CH_2NH_2 \qquad CH_3CHCH_3$$
$$\underset{\text{Propylamine}}{} \qquad \underset{\underset{\text{Isopropylamine}}{NH_2}}{}$$

Three carbon atoms can also be split into one methyl group and one ethyl group.

$$CH_3CH_2NHCH_3$$

Ethylmethylamine

■ **EXERCISE 9.10**
Give structures for the following amines.

a. butylamine
b. diethylamine
c. methylpropylamine
d. isopropylmethylamine

The amine with an NH_2 group attached directly to a benzene ring has the special name *aniline*. Like many other aromatic amines, aniline is used in making dyes. Aromatic amines tend to be toxic, and some are strongly carcinogenic.

Simple amines are similar to ammonia in odor, basicity, and other properties. It is the higher amines that are the most interesting, though. Figure 9.14 shows a variety of these. Notice that each structure contains one or more **amino (NH_2) groups**.

Among the most important kinds of organic molecules are amino acids. As the name implies, these compounds have carboxylic acid and amine functional groups. Amino acids are the building blocks from which proteins are constructed. The simplest amino acid, H_2NCH_2COOH, is glycine. Amino acids and proteins are considered in detail in Chapter 16.

Aniline

▶ **Figure 9.14** Some amines of interest. Amphetamine (a) is a stimulant drug (Chapter 18). Cadaverine (b) has the odor of decaying flesh. 1,6-Hexanediamine (c) is used in the synthesis of nylon (Chapter 10). Pyridoxamine (d) is a B vitamin (Chapter 17).

QUESTION: 1,6-Hexanediamine is the IUPAC name for a compound also known by the common name hexamethylenediamine. Cadaverine is a common name. What is the IUPAC name for cadaverine?

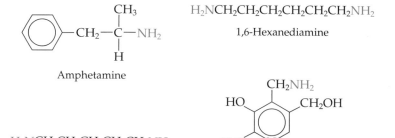

Amphetamine

$$H_2NCH_2CH_2CH_2CH_2CH_2CH_2NH_2$$
1,6-Hexanediamine

$$H_2NCH_2CH_2CH_2CH_2CH_2NH_2$$
Cadaverine

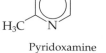
Pyridoxamine

Amides

Another important group of nitrogen-containing compounds is *amides*, which contain oxygen bonded to the same carbon as nitrogen. Thus, an **amide** has the nitrogen atom attached directly to a carbonyl group.

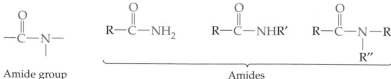

Amide group   Amides

Like those of other compounds that have $C=O$ groups, the formulas of amides are often written on one line as

$$RCONH_2 \qquad RCONHR' \qquad RCONR'R''$$

Note that urea (H_2NCONH_2), the compound that helped change the understanding of organic chemistry (page 223), is an amide.

Complex amides are of much greater interest than the simple ones considered here. Your body contains many kinds of proteins, all held together by amide linkages (Chapter 16). Nylon, silk, and wool molecules also contain hundreds of amide functional groups.

Names for simple amides are derived from those of the corresponding carboxylic acids. For example, $HCONH_2$ is formamide (IUPAC: methanamide) and CH_3CONH_2 is acetamide (IUPAC: ethanamide). Alternatively, we can say amide names are based on those of alkanes, with the ending changed from -e to -amide. For example, $CH_3CH_2CH_2CH_2CONH_2$ is pentanamide.

CONCEPTUAL EXAMPLE 9.11 Amine or Amide?

Which of the following are amides and which are amines? Identify the functional groups.

a. $CH_3CH_2CH_2NH_2$ b. CH_3CONH_2
c. $CH_3CH_2NHCH_3$ d. $CH_3COCH_2CH_2NH_2$

Solution

a. An amine; the NH_2 is an amine function; there is no $C=O$ group.
b. An amide; the $CONH_2$ is the amide function.
c. An amine; the NH is an amine function.
d. An amine; the NH_2 is an amine function; there is a $C=O$ group, but the NH_2 is not attached to it.

■ **EXERCISE 9.11**

Which of the following are amides and which are amines? Identify the functional groups.

a. CH_3NHCH_3
b. $CH_3CONHCH_3$
c. $CH_3CH_2N(CH_3)_2$
d. $CH_3NHCH_2CONH_2$

Self-Assessment Questions

1. A name for $CH_3CH_2CH_2CH_2NH_2$ is
   a. butanamide   b. butylamine
   c. propanamide   d. propylamide

2. A name for $CH_3CH_2NHCH_3$ is
   a. butylamine   b. ethylmethylamine
   c. propanamide   d. propylamine

3. A name for CH_3CONH_2 is
   a. acetamide   b. ethylamine
   c. propanamide   d. propylamine

4. The amide functional group has a carbonyl group attached to
   a. an H atom   b. an N atom
   c. an OH group   d. two other C atoms

Answers: 1, b; 2, b; 3, a; 4, b

9.15 Heterocyclic Compounds: Alkaloids and Others

Cyclic hydrocarbons feature rings of carbon atoms. Now let's look at some ring compounds that have atoms other than carbon within the ring. These **heterocyclic compounds** usually have one or more nitrogen, oxygen, or sulfur atoms.

CONCEPTUAL EXAMPLE 9.12 Heterocyclic Compounds

Which of the following structures represent heterocyclic compounds?

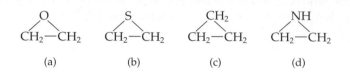

(a)   (b)   (c)   (d)

Solution

Compounds **a, b,** and **d** have oxygen, sulfur, and nitrogen atoms, respectively, in a ring structure; these represent heterocyclic compounds.

■ **EXERCISE 9.12**

Which of the following structures represent heterocyclic compounds?

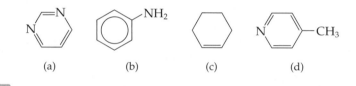

(a)   (b)   (c)   (d)

More to Explore

O'Neill, Maryadele J., Patricia E. Heckelman, Cherie B. Koch, Kristin J. Roman, and Catherine M. Kenny. *The Merck Index: An Encyclopedia of Chemicals, Drugs, and Biologicals,* 14th edition. Whitehouse Station, NJ: Merck & Co., 2006, provides formulas and properties for thousands of compounds.

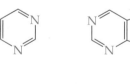

Pyrimidine   Purine

Many amines, particularly heterocyclic ones, occur naturally in plants. Like other amines, these compounds are basic. They are called **alkaloids**, which means "like alkalis." Among the familiar alkaloids are morphine, caffeine, nicotine, and cocaine. The actions of these compounds as drugs are considered in Chapter 18. Of more immediate interest are pyrimidine, which has two nitrogen atoms in a six-membered ring, and purine, which has four nitrogen atoms in two rings that share a common side. Compounds related to pyrimidine and purine are constituents of nucleic acids (Chapter 16).

Self-Assessment Questions

1. A cyclic organic compound that contains at least one atom that is not a carbon atom is called a(n)
   a. aromatic compound
   b. cyclic amide
   c. heterocyclic compound
   d. saturated compound

2. A basic organic cyclic compound that contains at least one nitrogen atom and that occurs in many plants is called a(n)
   a. alkaloid
   b. amide
   c. cyclic ester
   d. phenol

Answers: 1, c; 2, a

Critical Thinking Exercises

Apply knowledge that you have gained in this chapter and one or more of the FLaReS principles (Chapter 1) to evaluate the following statements or claims.

9.1 A television advertisement claims that gasoline with added ethanol burns cleaner than gasoline without added ethanol.

9.2 A news feature states that scientists have found that the odor of a certain ester improves workplace performance.

9.3 Alcohols, including ethanol, are toxic. An environmental activist states that all toxic chemicals should be banned from the home.

9.4 An advertisement refers to organic calcium carbonate as being superior to other forms of calcium carbonate because it contains a life force.

9.5 A Web page claims that acetylsalicylic acid (aspirin), polyester plastics, and polystyrene plastics all are carcinogens (cancer-causing agents). The reasoning given is that all three contain benzene rings in their structure, and benzene is a carcinogen.

9.6 A method is suggested for treating methanol poisoning. The method involves giving the patient a great deal of ethanol, though not enough to cause ethanol poisoning. The reasoning used is that the high concentration of ethanol will cause more ethanol to be oxidized to acetaldehyde and less methanol to be oxidized to formaldehyde in the body. With less formaldehyde produced, the toxic effects of the methanol should be reduced.

■ SUMMARY

Section 9.1—**Organic chemistry** is the study of carbon compounds. More than 95% of all known compounds contain carbon. Carbon is unique in its ability to form long chains, rings, and branches. **Hydrocarbons** contain only carbon and hydrogen.

Section 9.2—**Alkanes** are hydrocarbons that contain only single bonds; they have names ending in -*ane*. Alkanes are **saturated hydrocarbons** because each carbon atom is bonded to the maximum number of hydrogen atoms. A **structural formula** shows which atoms are bonded to one another. A **condensed structural formula** omits the C—H bond lines and is easier to write. Condensed structural formulas for the first three alkanes are methane (CH_4), ethane (CH_3CH_3), and propane ($CH_3CH_2CH_3$). The rest of the straight-chain alkanes can be generated by inserting a CH_2 unit into the previous molecule. Such a **homologous series** of compounds has properties that vary in a regular and predictable manner.

With more than three carbon atoms it is possible to have **isomers**, compounds with the same molecular formula but different structures. Alkanes are nonpolar, insoluble in water, and undergo few chemical reactions. They are used primarily as fuels.

Section 9.3—**Cyclic hydrocarbons** have one or more closed rings. Geometric figures often are used to represent cyclic compounds: a triangle for cyclopropane, a hexagon for cyclohexane, and so on.

Section 9.4—An **alkene** is a hydrocarbon with at least one carbon–carbon double bond. Ethylene ($CH_2\!=\!CH_2$) is the simplest alkene and the most important one commercially. It is used to make polyethylene plastic and ethylene glycol. An **alkyne** is a hydrocarbon with at least one carbon–carbon triple bond. Acetylene ($HC\!\equiv\!CH$) is the simplest alkyne, used in welding torches and as a starting material for other chemical products. Alkenes and alkynes are **unsaturated hydrocarbons** to which more hydrogen atoms can be added, and they undergo **addition reactions** in which a small molecule adds to the double or triple bond. They also can add to one another to form polymers.

Section 9.5—Benzene (C_6H_6) is drawn as though it has double bonds, but its ring of carbon atoms has six pairs of bonding electrons between carbon atoms, plus six unassigned electrons in a ring. Benzene and similar compounds are called **aromatic compounds** because of this structure, which is exceptionally stable. Aromatic hydrocarbons are used as solvents and fuels and to make other benzene derivatives.

Section 9.6—Chlorinated hydrocarbons are derived from hydrocarbons by replacing a hydrogen atom(s) with a chlorine atom(s). Chloroform (former anesthetic), carbon tetrachloride (former cleaning solvent), and DDT (former insecticide) are chlorinated compounds. Chlorofluorocarbons contain both chlorine and fluorine. They are used as aerosol propellants and as refrigerants, but their use has been curtailed because of their contribution to ozone layer destruction. Perfluorinated compounds, where all hydrogens have been replaced with fluorine, have important uses in plastics and in medicine.

Section 9.7—A **functional group** is a group of atoms that confers characteristic properties on a family of organic compounds. Compounds with the same functional group undergo similar reactions and have similar properties. The functional group is often attached to an **alkyl group (R—)**, an alkane with a hydrogen atom removed.

Section 9.8—A hydroxyl group (—OH) on an alkyl group produces an **alcohol (ROH)**. Methanol (CH_3OH), ethanol (CH_3CH_2OH), and 2-propanol [($CH_3)_2CHOH$] are widely used and well-known alcohols. Alcohols with more than one —OH group include ethylene glycol, used in antifreeze, and glycerol, a food additive and lotion additive.

Section 9.9—A **phenol** is a compound with an —OH group on a benzene ring. Phenols are slightly acidic, and some are used as antiseptics.

Section 9.10—An **ether** has two alkyl groups attached to the same oxygen atom (ROR′). Ethers are used as solvents, and they can react slowly with oxygen to form explosive peroxides. Diethyl ether ($CH_3CH_2OCH_2CH_3$) is a commonly used ether. Ethylene oxide is a cyclic ether used to make ethylene glycol and to sterilize instruments.

Section 9.11—An **aldehyde** contains a **carbonyl group** (C=O) with a hydrogen atom attached to the carbonyl carbon. A **ketone** has two other carbon atoms attached to the carbonyl carbon. Formaldehyde is used in making plastics and as a disinfectant. Benzaldehyde is a flavoring ingredient. Acetone, the simplest ketone, is a widely used solvent.

Section 9.12—A **carboxylic acid** has a **carboxyl group** (—COOH) as its functional group. Formic acid (HCOOH) is the acid in ant, bee, and wasp stings. Acetic acid (CH₃COOH) is the acid in vinegar. Butyric acid is the ingredient that gives rancid butter its odor. Carboxylic acid salts are used as preservatives.

Section 9.13—The structure of an **ester** (RCOOR′) is similar to a carboxylic acid, with an alkyl group replacing the hydrogen atom of the carboxyl group. Esters are made from a carboxylic acid and an alcohol or phenol. They often have fruity or flowery odors and are used as flavorings and in perfumes.

Section 9.14—An **amine** is a hydrocarbon derivative of ammonia in which one or more hydrogen atoms are replaced by alkyl or aromatic groups. Many amines simply contain an alkyl group and an **amino group** (—NH₂). Like ammonia, amines are basic and often have strong odors. Amino acids, the building blocks of proteins, contain both amine and carboxylic acid functional groups. An **amide** contains a carbonyl group whose carbon atom is attached to a nitrogen atom. Proteins, nylon, silk, and wool contain amide groups. A **heterocyclic compound** has a cyclic structure with one or more nitrogen, sulfur, or oxygen atoms in the ring. **Alkaloids** are amines, especially heterocyclic amines, that occur naturally in plants and include morphine, caffeine, nicotine, and cocaine. Heterocyclic structures are constituents of DNA and RNA.

■ REVIEW QUESTIONS

1. List three characteristics of the carbon atom that make possible the existence of millions of organic compounds.

2. Define, illustrate, or give an example of each of the following terms.
   a. hydrocarbon   b. alkyne
   c. alkane   d. alkene

3. What are isomers? How can you tell whether or not two compounds are isomers?

4. What is an unsaturated hydrocarbon? Give examples of two types of unsaturated hydrocarbons.

5. What is the meaning of the circle inside the hexagon in the modern representation of the structure of benzene?

6. What is an aromatic hydrocarbon? How can you recognize an aromatic compound from its structure?

7. Which alkanes are gases at room temperature? Which are liquids? Which are solids? State your answers in terms of the number of carbon atoms per molecule.

8. Compare the densities of liquid alkanes with that of water. When you add hexane to water in a beaker, what do you expect to observe?

9. What are the chemical names for the alcohols known by the following familiar names?
   a. grain alcohol   b. rubbing alcohol
   c. wood alcohol

10. What are some of the long-term effects of excessive ethanol consumption?

11. Give an important historical use for diethyl ether. What is its main use today?

12. How do carboxylic acids and esters differ in odor? In chemical structure?

13. For what family is each of the following the general formula?
    a. ROH   b. RCOR′   c. RCOOR′
    d. ROR′   e. RCOOH   f. RCHO

14. What is an alkaloid? Name three common alkaloids.

■ PROBLEMS

Organic and Inorganic

15. Which of the following compounds are organic?
    a. CH₃CH₂CH₂SH   b. HONH₂
    c. Cl₃CCH₂CH₃   d. HSiCl₃

16. Which of the following compounds are organic?
    a. CsO₂   b. H₂CS
    c. C₈H₁₈O   d. Co(NH₃)₆Cl₂

Names and Formulas of Hydrocarbons

17. How many carbon atoms are there in each of the following?
    a. cyclobutane   b. octane
    c. heptane   d. 2-pentene

18. How many carbon atoms are there in each of the following?
    a. propane   b. 1-pentyne
    c. ethylene   d. cyclononane

19. The general formula for a simple alkane is given on page 225. Give the formulas for the alkanes with **(a)** nine carbon atoms and **(b)** thirteen carbon atoms.

20. A particular alkyne has one triple bond and five carbon atoms. How many hydrogen atoms does it have?

21. Name the following hydrocarbons.
    a. CH₃CH₂CH₃
    b. H—C≡C—H
    c. CH₂=CH₂

22. Name the following hydrocarbons.

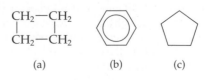

(a)   (b)   (c)

23. Give the molecular formulas and condensed structural formulas for (a) hexane and (b) octane.

24. Give the structural formulas of four-carbon alkanes (C_4H_{10}). Identify butane and isobutane.

25. Name the following alkyl groups.

    a. CH_3-

    b. $\begin{array}{c} CH_3CH_2CH- \\ | \\ CH_3 \end{array}$

26. Name the following alkyl groups.

    a. $CH_3CH_2CH_2CH_2-$

    b. $\begin{array}{c} CH_3CHCH_2- \\ | \\ CH_3 \end{array}$

Names and Formulas: Alcohols and Phenols

27. Give a structure to match the name or a name to match the structure for each of the following compounds.

    a. CH_3CH_2OH
    b. $CH_3CH_2CH_2CH_2OH$
    c. methanol
    d. 1-propanol

28. Give a structure to match the name or a name to match the structure for each of the following compounds.

    a. $\begin{array}{c} CH_3CH_2CHCH_3 \\ | \\ OH \end{array}$
    b. $\begin{array}{c} CH_3CHCH_2OH \\ | \\ CH_3 \end{array}$

    c. 1-hexanol
    d. *tert*-butyl alcohol

29. Give the structure for phenol.

30. How do phenols differ from alcohols? How are they similar?

Names and Formulas: Ethers

31. Give the structure for each of the following.
    a. diethyl ether
    b. butyl methyl ether

32. Give the structure for (a) methyl propyl ether, an anesthetic known as Neothyl, and (b) dimethyl ether, used as a compressed gas to "freeze" warts from the skin. It has the same formula as which alcohol?

Names and Formulas: Aldehydes and Ketones

33. Give a structure to match the name or a name to match the formula of each of the following compounds.
    a. acetaldehyde
    b. formaldehyde

    c. $\begin{array}{c} O \\ || \\ CH_3CH_2C-H \end{array}$
    d. $\begin{array}{c} O \\ || \\ CH_3CCH_2CH_2CH_3 \end{array}$

34. Give a structure to match the name or a name to match the formula of each of the following compounds.
    a. butyraldehyde
    b. propionaldehyde

    c. $\begin{array}{c} O \\ || \\ CH_3CH_2CCH_2CH_2CH_3 \end{array}$
    d. benzaldehyde structure

Names and Formulas: Carboxylic Acids

35. Give a structure to match the name or a name to match the structure for each of the following compounds.
    a. HCOOH
    b. CH_3CH_2COOH
    c. butanoic acid
    d. heptanoic acid

36. Give a structure to match the name or a name to match the structure for each of the following compounds.
    a. $CH_3CH_2CH_2CH_2COOH$
    b. $CH_3CH_2CH_2CH_2CH_2CH_2CH_2COOH$
    c. hexanoic acid
    d. benzoic acid

Names and Formulas: Esters

37. Give the structural formula for each of the following.
    a. ethyl acetate
    b. methyl butyrate

38. Give the structural formula for each of the following.
    a. ethyl butyrate
    b. methyl acetate

Names and Formulas: Nitrogen-Containing Compounds

39. Give a structural formula to match the name or a name to match the structure for each of the following.
    a. ethylamine
    b. dimethylamine
    c. $CH_3CH_2CH_2NH_2$
    d. $CH_3CH_2NHCH_3$

40. Give a structural formula to match the name or a name to match the structure for each of the following.
    a. methylamine
    b. isopropylamine
    c. $CH_3CH_2NHCH_2CH_3$
    d. $C_6H_5-NH_2$ (aniline structure)

Isomers and Homologs

41. Indicate whether the structures in each set represent the same compound or isomers.

    a. CH_3CH_3 and $\begin{array}{c} CH_3 \\ | \\ CH_3 \end{array}$
    b. $\begin{array}{c} CH_3CH_2 \\ | \\ CH_3 \end{array}$ and $CH_3CH_2CH_3$

    c. $\begin{array}{c} CH_3CHCH_2CH_2CH_3 \\ | \\ CH_3 \end{array}$ and $\begin{array}{c} CH_3CH_2CHCH_2CH_3 \\ | \\ CH_3 \end{array}$

42. Indicate whether the structures in each set represent the same compound or isomers.

    a. $\begin{array}{c} CH_3CHCH_2OH \\ | \\ CH_3 \end{array}$ and $\begin{array}{c} CH_3CHCH_2CH_3 \\ | \\ OH \end{array}$

    b. $\begin{array}{c} CH_3CHCH_2CH_3 \\ | \\ NH_2 \end{array}$ and $\begin{array}{c} CH_3CH_2CHNH_2 \\ | \\ CH_3 \end{array}$

43. Classify the following pairs as homologs, identical, isomers, or none of these.

    a. $CH_3CH_2CH_3$ and $CH_3CH_2CH_2CH_3$

    b. $\begin{array}{c} CH_2 \\ CH_2 \quad CH_2 \\ | \quad\quad | \\ CH_2-CH_2 \end{array}$ and $CH_3CH_2CH_2CH_2CH_3$

44. Classify the following pairs as homologs, identical, isomers, or none of these.

    a. $\begin{array}{c} CH_3CHCH_2CH_3 \\ | \\ CH_3 \end{array}$ and $\begin{array}{c} CH_3 \\ | \\ CH_3CHCH_2CH_3 \end{array}$

    b. $\begin{array}{c} CH_3-CH-CH=CH_2 \\ | \\ CH_3 \end{array}$

    and

    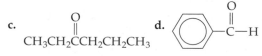

Classification of Hydrocarbons

45. Indicate whether each of the following compounds is saturated or unsaturated. Classify each as an alkane, alkene, or alkyne.

a. $CH_3C{=}CH_2$ with CH_3 below

b. $CH_3{-}\underset{\underset{CH_3}{|}}{\overset{\overset{CH_3}{|}}{C}}{-}CH_3$

46. Indicate whether each of the following compounds is saturated or unsaturated. Classify each as an alkane, alkene, or alkyne.

a. $CH_3C{\equiv}CCH_3$ b.

Functional Groups

47. Classify each of the following as an alcohol, amine, amide, ketone, aldehyde, ester, carboxylic acid, or ether. Identify the functional groups in each.

a. $CH_3CH_2\overset{\overset{O}{\|}}{C}OCH_3$ b. $CH_3CH_2\overset{\overset{O}{\|}}{C}H$

c. $CH_3CH_2CH_2NH_2$ **d.** $CH_3CH_2OCH_3$

e. $CH_3CH_2\overset{\overset{O}{\|}}{C}CH_2CH_3$ f. $CH_3CH_2\overset{\overset{O}{\|}}{C}OH$

48. Classify each of the following as an alcohol, amine, amide, ketone, aldehyde, ester, carboxylic acid, or ether. Identify the functional groups in each.
a. HCOOH   **b.** $CH_3CH_2COOCH_3$
c. $CH_3CH_2CH_2CH_2OH$ **d.** $CH_3CH_2CONHCH_2CH_2CH_3$
e. $CH_3CH_2CH_2COOH$ **f.** $HCOOCH_2CH_2CH_3$

49. Which of the following represent heterocyclic compounds? Classify each also as an amine, ether, or (cyclo)alkane.

a. $\underset{CH_2{-}CH_2}{CH_2{-}NH}$ **b.** $\underset{CH_2{-}CH_2}{CH_2{-}CH_2}$

50. Which of the following represent heterocyclic compounds? Classify each also as an amine, ether, or (cyclo)alkane.

a. $\underset{CH_2{-}CH_2}{CH_2{-}O}$ **b.** $\underset{CH_2{-}CH_2}{CH_2{-}CH{-}NH_2}$

■ ADDITIONAL PROBLEMS

51. A molecular model of capsaicin, the compound primarily responsible for the "hotness" of jalapenos and other hot peppers, is shown on page 222. The color code for the atoms is carbon (black), hydrogen (white), nitrogen (blue), and oxygen (red). What is the molecular formula of capsaicin? Name three functional groups on capsaicin.

52. Use the following color code for atoms: carbon (black), hydrogen (white), oxygen (red), nitrogen (blue), and chlorine (green). Give the structural formula and molecular formula for the compounds whose models are shown here.

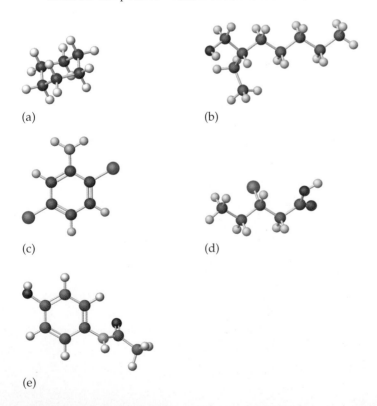

(a)   (b)

(c)   (d)

(e)

53. On page 231 we noted that alkenes and alkynes can add hydrogen to form alkanes. The reactions, called hydrogenations, are carried out using H_2 gas and a catalyst such as nickel or platinum. Write equations, using condensed structural formulas, for the complete hydrogenation of each of the following.

a. $CH_3C{\equiv}CCHCH_2CH_2CH_2CH_3$ with CH_3 below

b. $CH_2{=}CCH_2CH_2CH_3$ with CH_3 below

54. On page 238 we noted that alkenes can add water to form alcohols. The reactions, called hydrations, are carried out using H_2O gas and an acid (H^+) catalyst. Write equations, using structural formulas, for the hydration of each of the following.

a. (pentagon structure)   **b.** $CH_3CH{=}CHCH_3$

55. On page 244 we noted that alcohols can be oxidized to aldehydes and ketones. Give the structure of the alcohol that can be oxidized to each of the following.
a. $CH_3CH_2CH_2CHO$ **b.** $CH_3CH_2COCH_2CH_3$
c. $CH_3COCH(CH_3)_2$ **d.** C_6H_5CHO

56. Consider the following set of compounds. What principle does the series illustrate?
$$CH_3OH \qquad CH_3CH_2OH$$
$$CH_3CH_2CH_2OH \quad CH_3CH_2CH_2CH_2OH$$

57. Consider the following set of compounds. What principle does the series illustrate?
$$CH_3CH_2CH_2CH_2OH \qquad CH_3CH_2OCH_2CH_3$$
$$CH_3CHCH_2OH \qquad CH_3CHOCH_3$$
$$\qquad | \qquad\qquad\qquad\qquad |$$
$$\qquad CH_3 \qquad\qquad\qquad\qquad CH_3$$

58. Methanol is a possible replacement for gasoline. The complete combustion of methanol forms carbon dioxide and water.

$$CH_3OH + O_2 \longrightarrow CO_2 + H_2O$$

Balance the equation. What mass of carbon dioxide is formed by the complete combustion of 775 g of methanol?

59. Water adds to ethylene to form ethyl alcohol.

$$CH_2{=}CH_2 + H_2O \xrightarrow{H^+} CH_3CH_2OH$$

Balance the equation. What mass of ethyl alcohol is formed by the addition of water to 445 g of ethylene?

60. Refer to Table 9.7 and give the name and condensed structural formula for **(a)** the alcohol and carboxylic acid that would be needed to make rum flavoring and **(b)** the alcohol and carboxylic acid that would be needed to make pear flavoring.

61. Refer back to Section 6.3. Describe the type of intermolecular forces that exist **(a)** between dimethyl ether molecules and **(b)** between ethanol molecules. Use this information to explain why ethanol is a liquid at room temperature while dimethyl ether is a gas.

62. Diacetyl ($CH_3COCOCH_3$) is an additive used to impart buttery flavor to popcorn. Popcorn-factory workers exposed to high concentrations of diacetyl may develop a characteristic lung disease. What functional group is present in diacetyl?

63. Hydroquinone (HO—C_6H_4—OH), a phenol with the two hydroxyl groups in the 1,4- or *para* positions of the ring, is commonly used as a developer in black-and-white photography. **(a)** Give the structural formula for hydroquinone (Hint: see Figure 9.9). **(b)** In the process, hydroquinone is oxidized to *para*-benzoquinone ($C_6H_4O_2$), a compound that is no longer aromatic and that has two ketone functions. Give the structural formula for *para*-benzoquinone.

■ COLLABORATIVE GROUP PROJECTS

Prepare a PowerPoint, poster, or other presentation (as directed by your instructor) for presentation to the class.

64. The theory that organic chemistry was the chemistry of living organisms, which was overturned by Wöhler's discovery in 1828, was known as *vitalism*. Search the Internet for information about this theory. What was the effect of this philosophy on areas outside chemistry?

65. Many organic molecules contain more than one functional group. Look up structural formulas for each of the following in this text (use the index) or by using an online search or in a reference work such as *The Merck Index*. Identify and name the functional groups in each.
   a. butesin   **b.** estrone   **c.** tyrosine
   d. morphine   **e.** eugenol   **f.** methyl anthranilate

66. Prepare a brief report on one of the alcohols with three or more carbon atoms per molecule and share it with your group. List sources and commercial uses.

67. Prepare a brief report on one of the carboxylic acids with three or more carbon atoms per molecule and share it with your group. List sources and commercial uses.

The materials we call plastics are *polymers*. Polymers are made by joining small molecules (monomers) together to form giant molecules. In the form of films, fibers, and molded objects as well as starches, cellulose, and proteins, polymers pervade nearly every aspect of our lives. Thousands of consumer goods, from everyday items like juice containers (seen above) to exotic bicycle frames, are made of polymers.

POLYMERS

10

QUESTIONS YOU MAY HAVE ASKED YOURSELF

1. Why is it important to recycle plastics?
2. Why should we not burn plastic plumbing pipe?
3. Are all plastics synthetic?
4. Are silicon and silicone the same?

Giants among Molecules

Look around you. Polymers are everywhere. Around the home, carpets, curtains, upholstery, towels, sheets, floor tile, books, furniture, and most toys and containers (not to mention such things as telephones, toothbrushes, and piano keys) are made of polymers. Clothes are made of polymers. In cars, the dashboard, seats, tires, steering wheel, floor mats, ceiling, and many parts that you cannot see are made of polymers. Much of the food you eat contains polymers, and many important molecules in your body are polymers. You couldn't live without them. Some of the polymers that pervade our lives come from nature, but many are synthetic, made in chemical plants in an attempt to improve on nature in some way.

10.1 Polymerization: Making Big Ones Out of Little Ones

Polymers are composed of *macromolecules* (from the Greek *makros,* meaning "large" or "long"). Macromolecules may not seem large to the human eye (in fact, many of these giant molecules are invisible), but when compared with other molecules, they are enormous.

A **polymer** (from the Greek *poly,* meaning "many," and *meros,* meaning "parts") is made from much smaller molecules called *monomers* (from the Greek *monos,* meaning "one"). Sometimes thousands of monomer units combine to make one polymer molecule. A **monomer** is a small-molecule building block from which a polymer is made. The process by which monomers are converted to polymers is called *polymerization.* A polymer is as different from its monomer as a long strand of spaghetti is from tiny specks of flour. For example, polyethylene, the familiar waxy material used to make plastic bags, is made from the monomer ethylene, a gas.

▲ Cotton is nearly pure cellulose.

10.2 Natural Polymers

Polymers have served humanity for centuries in starches and proteins used for food; in wood used for shelter; and in wool, cotton, and silk used for clothing. Starch is a polymer made up of glucose ($C_6H_{12}O_6$) units. (Glucose is a simple sugar.) Cotton is made of cellulose, also a glucose polymer, and wood is largely cellulose as well. Proteins are polymers made up of amino acid monomers. Wool and silk are two of the thousands of different kinds of proteins found in nature.

Living things could not exist without polymers. Each plant and animal requires many different specific types of polymers. Probably the most amazing natural polymers are nucleic acids, which carry the coded genetic information that makes each individual unique. The polymers found in nature are discussed in Chapter 16. In this chapter we focus mainly on macromolecules made in the laboratory.

Self-Assessment Questions

1. A long-chain protein molecule is an example of a(n)
   a. fat   **b.** monomer   **c.** isomer   **d.** polymer
2. Which is *not* a polymer?
   a. wool   **b.** silk
   c. cellulose   **d.** glucose

Answers: 1, d; 2, d

10.3 Celluloid: Billiard Balls and Collars

The oldest attempts to improve on nature simply involved chemical modification of natural macromolecules. The synthetic material **celluloid**, as its name implies, was derived from natural cellulose (from cotton and wood, for example). When cellulose is treated with nitric acid, a derivative called cellulose nitrate is formed. In response to a contest to find a substitute for ivory for use in billiard balls, American inventor John Wesley Hyatt (1837–1920) found a way to soften cellulose nitrate by treating it with ethyl alcohol and camphor. The softened material could be molded into smooth, hard balls. Thus, Hyatt brought the game of billiards within the economic reach of more people—and saved a few elephants.

Celluloid was also used in movie film and for stiff collars (so they didn't require laundering and repeated starching). Because of its dangerous flammability (cellulose nitrate is also used as smokeless gunpowder), celluloid was removed from the market when safer substitutes became available. Today, movie film is made mainly from polyethylene terephthalate, a polyester (Section 10.7). And high, stiff collars on men's shirts are out of fashion.

It didn't take long for the chemical industry to recognize the potential of synthetics. Scientists found ways to make macromolecules from small molecules rather than simply modifying large ones. The first such truly synthetic polymers were phenol–formaldehyde resins (such as Bakelite), first made in 1909. These complex polymers are discussed in Section 10.7. Let's look at some simpler ones first.

A century ago, celluloid was widely used as a substitute for more expensive substances such as ivory, amber, and tortoiseshell. The movie industry was once known as the "celluloid industry." The 1988 oscar-winning Italian movie *Cinema Paraiso* tells the story of the evolution of film technology and includes a tragic scene in which the heat from a bulb in the movie projector ignites a fire in a film that failed to wind properly through the machine.

Self-Assessment Questions

1. The first semisynthetic polymer was actually made from a natural polymer. It was
   a. Bakelite, based on bacon grease
   b. celluloid, based on cellulose from cotton
   c. polypropylene, based on isopropyl alcohol
   d. polypantene, based on human hair

2. The first truly synthetic polymer was
   a. Bakelite, made from phenol and formaldehyde
   b. cellulose, made from glucose
   c. polyethylene, made from ethanol
   d. rubber, made from polyisoprene

Answers: 1, b; 2, a

10.4 Polyethylene: From the Battle of Britain to Bread Bags

The prevalent plastic polyethylene is the simplest and least expensive synthetic polymer. It is familiar today in the plastic bags used for packaging fruit and vegetables, in garment bags for dry-cleaned clothing, in garbage-can liners, and in many other items. Polyethylene is made from ethylene (CH_2=CH_2), an unsaturated hydrocarbon (Chapter 9) produced in large quantities from the cracking of petroleum, a process by which large hydrocarbon molecules are broken down into simpler hydrocarbons.

With pressure and heat and in the presence of a catalyst, ethylene monomers join together in long chains.

$$\cdots + \begin{matrix} H & H \\ | & | \\ C = C \\ | & | \\ H & H \end{matrix} + \begin{matrix} H & H \\ | & | \\ C = C \\ | & | \\ H & H \end{matrix} + \begin{matrix} H & H \\ | & | \\ C = C \\ | & | \\ H & H \end{matrix} + \begin{matrix} H & H \\ | & | \\ C = C \\ | & | \\ H & H \end{matrix} + \cdots \longrightarrow \sim \begin{matrix} H & H & H & H & H & H & H & H \\ | & | & | & | & | & | & | & | \\ C - C - C - C - C - C - C - C \\ | & | & | & | & | & | & | & | \\ H & H & H & H & H & H & H & H \end{matrix} \sim$$

> The ellipses ($\cdots$) and tildes ($\sim$) are like et ceteras; they indicate that the number of monomers and the polymer structure are extended for many units in each direction.

Using condensed formulas, this becomes

$$\cdots + CH_2\!=\!CH_2 + CH_2\!=\!CH_2 + CH_2\!=\!CH_2 + CH_2\!=\!CH_2 + \cdots \longrightarrow \sim CH_2CH_2CH_2CH_2CH_2CH_2CH_2CH_2 \sim$$

These equations can be tedious to draw, and so we often use abbreviated forms like these:

$$n\,\begin{matrix} H & H \\ | & | \\ C = C \\ | & | \\ H & H \end{matrix} \longrightarrow \left[\begin{matrix} H & H \\ | & | \\ C - C \\ | & | \\ H & H \end{matrix}\right]_n \qquad \text{or} \qquad n\,CH_2\!=\!CH_2 \longrightarrow \left[CH_2CH_2\right]_n$$

The molecular fragment enclosed within the brackets is called the *repeat unit* of the polymer. In the formula for the polymer product, the repeat unit is placed within brackets with bonds extending to both sides. The subscript n indicates that this unit is repeated many times in the full polymer structure.

The simplicity of the abbreviated formula facilitates certain comparisons between the monomer and the polymer. Note that the monomer ethylene contains a double bond and polyethylene does not. The double bond of the reactant contains two pairs of electrons. One of these pairs is used to connect one monomer unit to the next in the polymer (indicated by the lines sticking out to the sides in the repeat unit). This leaves only a single pair of electrons—a single bond—between the two carbon atoms of the repeat unit. Note also that each repeat unit in the polymer has the same composition (C_2H_4) as the monomer.

Molecular models provide three-dimensional representations. Figure 10.1 presents models of a tiny part of a very long molecule, which can vary in number of carbon atoms from a few hundred to several thousand.

Polyethylene was invented shortly before the start of World War II. It proved to be tough and flexible, an excellent electric insulator, and able to withstand both high and low temperatures. Before long, it was used for insulating cables in radar, a top-secret invention that helped British pilots detect enemy aircraft before the

▶ **Figure 10.1** Ball-and-stick (a) and space-filling (b) models of a short segment of a polyethylene molecule.

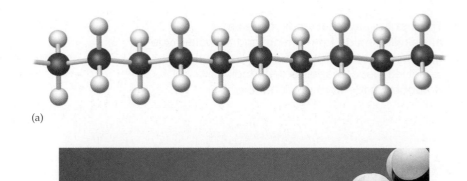

(a)

(b)

1. Why is it important to recycle plastics?

Polyethylene and most other polymers are made of petroleum. Since our supply of petroleum is limited, we can best make use of that supply by reusing those polymers. Most plastic items are imprinted with a code number to aid in proper sorting for recycling.

LDPE has densities ranging from 0.910 to 0.940 g/cm³ and HDPE from 0.941 to 0.960 g/cm³.

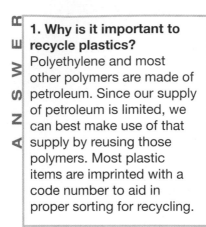

▲ **Figure 10.2** Two bottles, both made of polyethylene, were heated in the same oven for the same length of time.

QUESTION: Which of these bottles is made of HDPE and which of LDPE? Explain.

More to Explore

LaJeunesse, Sara. "Plastic Bags." *Chemical and Engineering News*, September 20, 2004, p. 54.

aircraft could be spotted visually. Without polyethylene, the British could not have had effective radar, and without radar, the Battle of Britain might have been lost. The invention of this simple plastic helped to change the course of history.

Today, there are three principal kinds of polyethylene. *High-density polyethylenes* (*HDPEs*) have mostly linear molecules that pack closely together and can assume a fairly ordered, crystalline structure. HDPEs, therefore, are rather rigid and have good tensile strength. They are used for such items as threaded bottle caps, toys, bottles, and milk jugs.

Low-density polyethylenes (*LDPEs*), on the other hand, have a lot of side chains branching off the polymer molecules. The branches prevent the molecules from packing closely together and assuming a crystalline structure. LDPEs are waxy, bendable plastics that are lower melting than high-density polyethylenes. Objects made of HDPE hold their shape in boiling water, whereas those made of LDPE are severely deformed (Figure 10.2). LDPEs are used to make plastic bags, plastic film, squeeze bottles, electric wire insulation, and many common household products where flexibility is important.

The third type of polyethylene, called *linear low-density polyethylenes* (*LLDPEs*), is actually a **copolymer**, a polymer formed from two or more different monomers. LLDPEs are made by polymerizing ethylene with a branched-chain alkene such as 4-methyl-1-pentene.

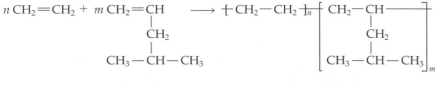

Ethylene   4-Methyl-1-pentene   An LLDPE

LLDPEs are used to make such things as plastic films for use as landfill liners, trash cans, tubing, and automotive parts.

Fullerenes: Buckyballs and Nanotubes

We saw in Chapter 9 that carbon atoms can form long chains, branched chains, and rings. Here we see in polymer molecules just how long some of these carbon chains can be. Carbon atoms, with their four bonds, can form still other structures. In 1985 scientists discovered a variety of molecules formed exclusively of carbon atoms. A particularly prominent one had a molecular mass of 720 u, corresponding to the formula C_{60}. The molecule is a roughly spherical collection of hexagons and pentagons very much like a soccer ball. Because C_{60} resembles the geodesic-domed structures that architect R. Buckminster Fuller pioneered, the scientists named it "buckminsterfullerene" in Fuller's honor. The general name *fullerenes* is now used for C_{60} and similar molecules with formulas such as C_{70}, C_{74}, and C_{82}. These substances are often colloquially called "buckyballs."

Later, scientists discovered tube-shaped carbon molecules called *nanotubes*. We can visualize a nanotube as a fullerene that has been stretched out into a hollow cylinder by the insertion of many, many more C atoms. We can also picture them as a two-dimensional array of hexagonal rings of carbon atoms, rather like ordinary "chicken wire." The "wire" is then rolled into a cylinder and capped at each end by half a C_{60} molecule. Nanotubes have unusual mechanical and electrical properties and are good conductors of heat. Carbon nanotubes are the strongest and stiffest materials known. Depending on size and atomic arrangement, the electrical conductivities of nanotubes vary from values like those of metals to those of semiconductors. The unusual properties of carbon nanotubes makes it very likely that they are "materials of the future." Their utility has only begun to be explored.

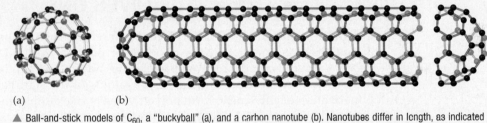

(a)   (b)

▲ Ball-and-stick models of C_{60}, a "buckyball" (a), and a carbon nanotube (b). Nanotubes differ in length, as indicated here by a break in the structure. They can be either single walled (shown) or multiwalled (tubes inside tubes).

Thermoplastic and Thermosetting Polymers

Polyethylene is one of a variety of thermoplastic polymers. A **thermoplastic polymer** can be softened by heat and pressure and then reshaped. It can be repeatedly melted down and remolded. Thermoplastics can be reshaped because their molecules can slide past one another when heat and pressure are applied. Total production of thermoplastic polymers in the United States in 2007 was about 43 billion kg, of which 18 billion kg was polyethylene.

Not all polymers can be readily melted. About 7% of U.S. production is made up of **thermosetting resins**, which harden permanently when formed. They cannot be softened by heat and remolded; instead, strong heating causes them to discolor and decompose. The permanence of thermosetting plastics is due to cross-linking (connecting side-to-side) of the polymer chains. We look at some thermosetting polymers later in this chapter.

More to Explore
Geim, Andre K. and Philip Kim. "Carbon Wonderland," *Scientific American*, April 2008, pp. 90–97. Discusses new forms of carbon.

More than 205 billion kg of plastics and rubber was manufactured globally in 2006. About 21.4 billion kg of rubber was made into tires and other rubber goods.

Self-Assessment Questions

1. When addition polymers form, the monomer units are joined by
   a. delocalized electrons that flow along the polymer chain
   b. dispersion forces
   c. double bonds
   d. new bonds formed by electrons from the double bond in the monomer
2. Compared to LDPE, HDPE has a higher melting point because it
   a. is made from more polar monomers
   b. is composed of molecules of much higher molecular mass
   c. has more cross-links
   d. has fewer branches in the chains allowing the molecules to pack more closely

3. Which of the following is *not* a common use of polyethylene?
   a. electric wire insulation   **b.** nonstick coatings for pans
   c. plastic grocery bags   **d.** squeeze bottles
4. A copolymer is made
   a. by blending two simple polymers
   b. from carbon monoxide and ethylene
   c. from two identical monomers
   d. from two different monomers
5. A polymer that can be melted and reshaped before hardening again when cooled is said to be
   a. highly crystalline   **b.** rubbery
   c. thermoplastic   **d.** thermosetting

Answers: 1, d; 2, d; 3, b; 4, d; 5, c

10.5 Addition Polymerization: One + One + One + ... GIVES ONE

There are two general types of polymerization reactions: addition polymerization and condensation polymerization. In **addition polymerization** (also called *chain-reaction polymerization*), the monomer molecules add to one another in such a way that the polymeric product contains all the atoms of the starting monomers. The polymerization of ethylene to form polyethylene is an example. In polyethylene, as we noted in Section 10.4, the two carbon atoms and the four hydrogen atoms of each monomer molecule are incorporated into the polymer structure. In *condensation polymerization* (Section 10.7), a portion of the monomer molecule is not incorporated in the final polymer but is split out as the polymer is formed.

Polypropylene

Most of the many familiar addition polymers are made from derivatives of ethylene in which one or more of the hydrogen atoms are replaced by another atom or group. Replacing one of the hydrogen atoms with a methyl group gives the monomer propylene (propene). The polypropylene molecule looks like polyethylene, except that there are methyl groups (CH_3) attached to every other carbon atom.

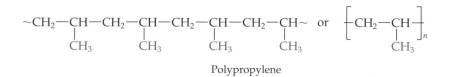

Polypropylene

The chain of carbon atoms is called the polymer *backbone*. Groups attached to the backbone, such as the CH_3 of polypropylene are called *pendant groups*.

Polypropylene is a tough plastic material that resists moisture, oils, and solvents. It is molded into hard-shell luggage, battery cases, and various kinds of appliance parts. It is also used to make packaging material, fibers for textiles such as upholstery fabrics and carpets, and ropes that float. Because of its high melting point (121 °C), polypropylene objects can be sterilized with steam.

Polystyrene

Replacing one of the hydrogen atoms in ethylene with a benzene ring gives a monomer called styrene, with the formula $C_6H_5CH\!=\!CH_2$, where C_6H_5 represents the benzene ring. Polymerization of styrene produces polystyrene, which has benzene rings as pendant groups.

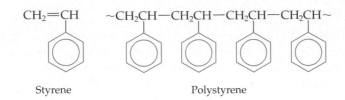

Styrene   Polystyrene

▲ Polystyrene foam insulation saves energy by reducing the transfer of heat from a warm house to the outside in winter or from the hot outside to the cooled inside in summer.

Polystyrene is the plastic used to make transparent disposable drinking cups. With color and filler added, it is the material of thousands of inexpensive toys and household items. When a gas is blown into polystyrene liquid, it foams and hardens into the familiar material of ice chests and disposable coffee cups. The polymer can easily be formed into shapes as packing material for shipping instruments and appliances, and it is widely used for home insulation.

Vinyl Polymers

Would you like a tough synthetic material that looks like leather at a fraction of the cost? Perhaps a clear, rigid material from which unbreakable bottles could be made? Do you need an attractive, long-lasting floor covering? Or lightweight, rustproof, easy-to-construct plumbing? Polyvinyl chloride (PVC) has all these properties—and more.

Replacing one of the hydrogen atoms of ethylene with a chlorine atom gives vinyl chloride (CH_2=CHCl), a compound that is a gas at room temperature. Polymerization of vinyl chloride yields the tough thermoplastic material PVC. A segment of the PVC molecule is illustrated below.

More to Explore
Cook, Perry A., Sue Hall, and Jill Donahue. "Pondering Packing Peanut Polymers." *Journal of Chemical Education*, November 2003, pp. 1288A–1288B.

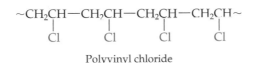

Polyvinyl chloride

PVC is readily formed into various shapes. The clear, transparent polymer is used in plastic wrap and clear plastic bottles. Adding color and other ingredients to vinyl plastics yields artificial leather. Most floor tile and shower curtains are also made from vinyl plastics, and they are widely used to simulate wood in home siding panels and window frames. About 40% of the PVC produced is molded into pipes.

2. Why should we not burn plastic plumbing pipe?
When PVC is burned, it can release the carcinogenic vinyl chloride monomer of which it is made.

A
N
S
W
E
R

(a)   (b)

◀ Polyvinyl chloride polymers can be coated onto copper wire for insulation, made into colorful resilient flooring, or formed into many other familiar consumer products.

The monomer from which vinyl plastics are made is a carcinogen. Several people who worked closely with vinyl chloride gas later developed a kind of cancer known as angiosarcoma. (Carcinogens are discussed in **eChapter 22**.)

PTFE: The Nonstick Coating

In 1938, young Roy Plunkett, a chemist at DuPont, was working with the gas tetra-fluoroethylene ($CF_2{=}CF_2$). He opened the valve on a tank of the gas—and nothing came out. Rather than discarding the tank, he decided to investigate. The tank was found to be filled with a waxy, white solid. He attempted to analyze the solid but ran into a problem: It simply wouldn't dissolve, even in hot concentrated acids. Plunkett had discovered the polymer of tetrafluoroethylene, called polytetrafluoroethylene (PTFE), and best known by its trade name, Teflon®.

$$\sim CF_2{-}CF_2{-}CF_2{-}CF_2{-}CF_2{-}CF_2{-}CF_2{-}CF_2\sim$$

Teflon

Because its C—F bonds are exceptionally strong and resistant to heat and chemicals, PTFE is a tough, unreactive, nonflammable material. It is used to make electric insulation, bearings, and gaskets. It is also widely used to coat surfaces of cookware to give them nonsticking properties.

▲ The plumber's tape, the coating on the muffin pan, and the insulation on the wire are all made of Teflon.

▲ Granular polymer resins are the basic stock for many molded polymer goods.

▲ A giant bubble of tough, transparent plastic film emerges from a die of an extruding machine. The film is used in packaging, consumer products, and food services.

CONCEPTUAL EXAMPLE 10.1 Repeat Units in Polymers

What is the repeat unit in polyvinylidene chloride? A segment of the polymer is represented as

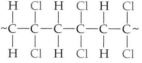

Solution

Inspection of the foregoing segment reveals that it consists of three repeat units. Placing a double bond between the carbon atoms of one unit we get

Joining hundreds of these units would form a molecule of the polymer.

■ **EXERCISE 10.1**

What is the repeat unit in polyacrylonitrile? A segment of the polymer is represented as

$$\sim CH_2CHCH_2CHCH_2CHCH_2CHCH_2CHCH_2CH\sim$$
$$\quad | \qquad | \qquad | \qquad | \qquad | \qquad |$$
$$\quad CN \quad CN \quad CN \quad CN \quad CN \quad CN$$

EXAMPLE 10.2 Structure of Polymers

Give the structure of the polymer made from vinyl fluoride ($CH_2{=}CHF$). Show at least four repeat units.

Solution

The carbon atoms become bonded in a chain with only single bonds between the carbon atoms. The fluorine atom is a substituent on the chain. (Two of the electrons in the double bond of the monomer are used to join the units.) The polymer is

$$\sim CH_2CHCH_2CHCH_2CHCH_2CH\sim$$
$$\quad | \qquad | \qquad | \qquad |$$
$$\quad F \qquad F \qquad F \qquad F$$

■ **EXERCISE 10.2**

Give the structure of the polymers made from **(a)** methyl vinyl ether ($CH_2{=}CHOCH_3$) and **(b)** vinyl acetate ($CH_2{=}CHOCOCH_3$). Show at least four repeat units of each polymer.

In everyday life, many polymers are called plastics. In chemistry, a *plastic* material is one that can be made to flow under heat and pressure. The material can then be shaped in a mold or in other ways. Plastic products often are made from powders. In *compression molding,* heat and pressure are applied directly to the polymer powder in the mold cavity. In *transfer molding,* the powder is softened by heating outside the mold and then poured into molds to harden.

There also are several methods of molding molten polymers. In *injection molding,* the plastic is melted in a heating chamber and then forced by a plunger into cold molds to set. In *extrusion molding* the melted polymer is extruded through a die in continuous form to be cut into lengths or coiled. Bottles and similar hollow objects often are *blow-molded;* a "bubble" of molten polymer is blown up like a balloon inside a hollow mold.

Table 10.1 lists some of the more important addition polymers, along with a few of their uses.

Table 10.1 Some Addition Polymers

Monomer	Polymer	Polymer Name	Some Uses
$CH_2=CH_2$		Polyethylene	Bags, bottles, toys, electrical insulation
$CH_2=CH-CH_3$		Polypropylene	Carpeting, bottles, luggage
$CH_2=CH-$⟨⟩		Polystyrene	Simulated wood furniture, insulation, cups, toys, packing materials
$CH_2=CH-Cl$		Polyvinyl chloride (PVC)	Food wrap, simulated leather, plumbing, garden hoses, floor tile
$CH_2=CCl_2$		Polyvinylidene chloride (Saran)	Food wrap, seatcovers
$CF_2=CF_2$		Polytetrafluoroethylene (Teflon)	Nonstick coating for cooking utensils, electrical insulation
$CH_2=CH-C\equiv N$		Polyacrylonitrile (Acrilan, Creslan, Dynel)	Yarns, wigs, paints
$CH_2=CH-OCOCH_3$		Polyvinyl acetate	Adhesives, textile coatings, chewing gum resin, paints
$CH_2=C(CH_3)COOCH_3$		Polymethyl methacrylate (Lucite, Plexiglas)	Glass substitute, bowling balls

Conducting Polymers: Polyacetylene

Acetylene (H—C≡C—H) has a triple bond instead of a double bond, but it can still undergo addition polymerization, forming polyacetylene (Figure 10.3). Notice that, unlike polyethylene, which has a carbon chain containing only single bonds, every other bond in polyacetylene is a double bond.

~CH=CH—CH=CH—CH=CH—CH=CH—CH=CH—CH=CH~

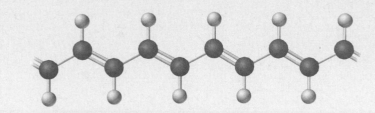

▲ **Figure 10.3** A ball-and-stick model of polyacetylene.

QUESTION: Write an equation for the formation of polyacetylene similar to that we wrote for polyethylene on page 263, in which the repeat polymer unit is placed within brackets.

The alternating double and single bonds form a *conjugated* system. They make it easy for electrons to travel along the chain, and so this polymer is able to conduct electricity. (Most plastics are electric insulators!) Polyacetylene and similar conjugated polymers can be used as lightweight substitutes for metal. In fact, the plastic even looks like metal, having a silvery luster.

Polyacetylene, the first conducting polymer, was discovered in 1970. Since then, a number of other polymers that conduct electricity have been made.

▲ New conducting polymers have special conjugated systems that appear red, green, or blue when electricity is passed through them.

Self-Assessment Questions

1. Which class of organic compounds serves as the starting materials for many polymers?
   a. alkanes   **b.** alkenes   **c.** carboxylic acids
   d. esters   **e.** ethers

2. The monomer styrene ($C_6H_5CH=CH_2$, where C_6H_5 represents the benzene ring) forms a polymer that has chains with repeating
   a. eight-carbon units in the backbone.
   b. six-carbon units in the backbone and two-carbon atom groups as pendants.
   c. CH_2CH units in the backbone and C_6H_5 groups as pendants.
   d. $CH_2=CH$ units in the backbone and C_6H_5 groups as pendants.

3. Which of the following molecules could *not* be used as the monomer for an addition polymer?
   a. $C_2H_2F_2$ **b.** C_2F_4 **c.** C_2H_5Cl **d.** $C_2H_3C_6H_5$

4. A buckyball is
   a. a carbon molecule (C_{60})
   b. carbon nanotubes wound up like a ball of string
   c. a long polymer molecule wound up like a ball of string
   d. three-dimensional cyclobutane molecules

Answers: 1, b; 2, c; 3, c; 4, a

10.6 Rubber and Other Elastomers

Although rubber is a natural polymer, it was the basis for much of the development of the synthetic polymer industry. During World War II, Japanese occupation of Malaysia and Indonesia cut off most of the Allies' supply of natural rubber. The search for synthetic substitutes resulted in much more than just a replacement for natural rubber. The plastics industry, to a large extent, developed out of the search for synthetic rubber.

Natural rubber can be broken down into a simple hydrocarbon called isoprene. Isoprene is a volatile liquid, whereas rubber is a semisolid, elastic material. Chemists can make polyisoprene, a substance identical to natural rubber, except that the isoprene comes from petroleum refineries rather than from the cells of rubber trees.

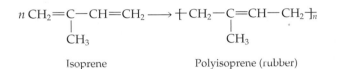

Isoprene   Polyisoprene (rubber)

Chemists have also developed several synthetic rubbers and devised ways to modify these various polymers to change their properties.

Vulcanization: Cross-Linking

The long-chain molecules that make up rubber can be coiled and twisted and intertwined with one another. When rubber is stretched, its coiled molecules are straightened. Natural rubber is soft and tacky when hot. It can be made harder by reaction with sulfur. This process, called **vulcanization**, cross-links the hydrocarbon chains with sulfur atoms (Figure 10.4). Charles Goodyear discovered vulcanization and was issued U.S. Patent 3633 in 1844.

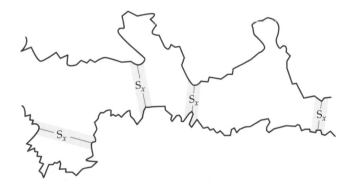

Its three-dimensional cross-linked structure makes vulcanized rubber a harder, stronger substance that is suitable for automobile tires. Surprisingly, cross-linking also improves the elasticity of rubber. With just the right degree of cross-linking, the individual chains are still free to uncoil and stretch somewhat. When stretched vulcanized rubber is released, the cross-links pull the chains back to their original arrangement (Figure 10.5). Rubber bands owe their snap to this sort of molecular structure. Materials that act in this stretchable way are called **elastomers**.

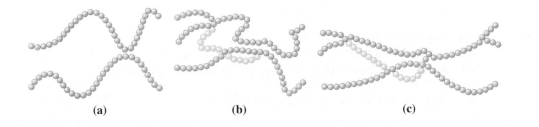

(a)   (b)   (c)

3. Are plastics all synthetic?
Latex is just one example of a natural polymer. All of the proteins, cellulose, and starches in living organisms are also natural polymers.

ANSWER

◀ **Figure 10.4** Vulcanized rubber has hydrocarbon chains (represented here by red lines) cross-linked by sulfur atoms. The subscript *x* indicates a small, indefinite number, usually not more than 4.

▼ **Figure 10.5** Vulcanization of rubber cross-links the molecular chains. (a) In unvulcanized rubber, the chains slip past one another when the rubber is stretched. (b) Vulcanization involves the addition of sulfur cross-linkages between the chains. (c) When vulcanized rubber is stretched, the sulfur cross-linkages prevent the chains from slipping past one another. Vulcanized rubber is stronger than unvulcanized rubber.

● = Carbon

○ = Sulfur

Synthetic Rubber

Natural rubber is a polymer of isoprene, and some synthetic elastomers are closely related. For example, polybutadiene is made from the monomer butadiene ($CH_2{=}CH{-}CH{=}CH_2$), which differs from isoprene only in that it lacks a methyl group on the second carbon atom. Polybutadiene is made rather easily from the monomer.

$$n\ CH_2{=}CH{-}CH{=}CH_2 \longrightarrow \ {+}CH_2{-}CH{=}CH{-}CH_2{+}_n$$

However, it has only fair tensile strength and poor resistance to gasoline and oils. These properties limit its value for automobile tires, the main use of elastomers.

Another synthetic elastomer, polychloroprene (Neoprene), is made from a monomer similar to isoprene but with a chlorine in place of the methyl group on isoprene.

$$n\ CH_2{=}\underset{\underset{Cl}{|}}{C}{-}CH{=}CH_2 \longrightarrow \ {+}CH_2{-}\underset{\underset{Cl}{|}}{C}{=}CH{-}CH_2{+}_n$$

Neoprene is more resistant to oil and gasoline than other elastomers are. It is used to make gasoline hoses and similar items used at automobile service stations.

Styrene-butadiene rubber (SBR) is a copolymer of styrene (about 25%) and butadiene (about 75%). A segment of an SBR molecule might look something like this.

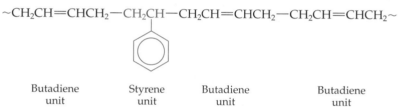

$${\sim}CH_2CH{=}CHCH_2{-}CH_2CH{-}CH_2CH{=}CHCH_2{-}CH_2CH{=}CHCH_2{\sim}$$

| Butadiene unit | Styrene unit | Butadiene unit | Butadiene unit |

SBR is more resistant to oxidation and abrasion than natural rubber, but its mechanical properties are less satisfactory.

Like those of natural rubber, SBR molecules contain double bonds and can be cross-linked by vulcanization. SBR accounts for about a third of the total U.S. elastomer production and is used mainly for making tires.

Polymers in Paints

A surprising use for elastomers is in paints and other coatings. The substance in a paint that hardens to form a continuous surface coating, often called the *binder* or resin, is a polymer, usually an elastomer. Paint made with elastomers is more resistant to cracking over time. Various kinds of polymers can be used as binders, depending on the specific qualities desired in the paint. Over the last few decades, the popularity of latex paint has soared. Organic solvents are not needed; the latex binder is dispersed in water, so that the brushes and rollers are easily cleaned in soap and water. This is a good example of green chemistry in that the hazardous organic solvents historically used in paints are replaced with water.

▲ **It DOES Matter!**
A polymer related to SBR is poly(styrene-butadiene-styrene), or SBS. Called a *block copolymer*, SBS has molecules made up of three segments. One end has a chain of polystyrene repeat units; the middle, a long chain of polybutadiene repeat units; and the other end another chain of polystyrene repeat units. SBS is a hard rubber used for such things as shoe soles and tire treads where durability is important.

▲ Synthetic polymers serve as binders in paints. Pigments in paint provide color or opacity. Titanium dioxide (TiO_2), a white solid, is the pigment most widely used.

Self-Assessment Questions

1. Natural rubber is a polymer of
   a. butadiene   **b.** ethylene   **c.** isoprene   **d.** propylene

2. Which of the following is the structure of isoprene?
   a. $CH_2{=}CH{-}C{\equiv}N$ **b.** $CH_2{=}CH{-}CH{=}CH_2$
   c. $CH_2{=}CCl{-}CH{=}CH_2$ **d.** $CH_2{=}C(CH_3){-}CH{=}CH_2$

3. Vulcanization of natural rubber
   a. cross-links the rubber, making it harder and tougher
   b. makes the rubber more crystalline and thus more glassy
   c. makes the rubber totally amorphous
   d. makes the rubber tacky, making rubber cement

4. Charles Goodyear discovered that natural rubber could be cross-linked by heating it with
   a. a binder   **b.** isoprene   **c.** neoprene   **d.** sulfur

5. Styrene-butadiene rubber is an example of a
   a. copolymer   **b.** natural polyester
   c. natural elastomer   **d.** polyamide

6. Polymers, usually elastomers, are used in paints as the
   a. binder   **b.** initiator   **c.** pigment   **d.** solvent

Answers: 1, c; 2, d; 3, a; 4, d; 5, a; 6, a

10.7 Condensation Polymers: Splitting Out Water

The polymers considered so far are all addition polymers. All the atoms of the monomer molecules are incorporated into the polymer molecules. In a *condensation polymer*, part of the monomer molecule is not incorporated in the final polymer. During **condensation polymerization**, also called *step-reaction polymerization*, small molecules, such as water, ammonia, or HCl, are split out as by-products.

Nylon and Other Polyamides

As an example, let's consider the formation of nylon. (There are several different nylons, each prepared from a different monomer or set of monomers, but all share certain common structural features.) The monomer in one type of nylon, called nylon 6, is a six-carbon carboxylic acid with an amino group on the sixth carbon atom, 6-aminohexanoic acid ($HOOCCH_2CH_2CH_2CH_2CH_2NH_2$).

In the polymerization reaction, a carboxyl group of one monomer molecule forms an amide bond with the amine group of another.

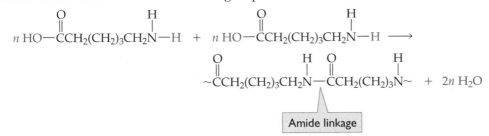

Water molecules are formed as a by-product. This formation of a nonpolymeric by-product distinguishes condensation polymerization from addition polymerization. Note that the formula of a repeat unit is not the same as that of the monomer.

Because the linkages holding the polymer together are amide bonds, nylon 6 is a **polyamide**. Another nylon is made by the condensation of two different monomers, 1,6-hexanediamine ($H_2NCH_2CH_2CH_2CH_2CH_2CH_2NH_2$) and adipic acid ($HOOCCH_2CH_2CH_2CH_2COOH$). Each monomer has six carbon atoms; the polymer is called nylon 66.

$$n \; H{-}\overset{H}{\underset{|}{N}}CH_2CH_2CH_2CH_2CH_2CH_2\overset{H}{\underset{|}{N}}{-}H \;+\; n \; HO{-}\overset{O}{\overset{\|}{C}}(CH_2)_4\overset{O}{\overset{\|}{C}}{-}OH \longrightarrow$$

1,6-Hexanediamine   Adipic acid

$$\underset{}{{+}}\overset{H}{\underset{|}{N}}CH_2CH_2CH_2CH_2CH_2CH_2\overset{H}{\underset{|}{N}}{-}\overset{O}{\overset{\|}{C}}CH_2CH_2CH_2CH_2\overset{O}{\overset{\|}{C}}\underset{n}{{+}} \;+\; 2n \; H_2O$$

Amide linkage

This was the original nylon polymer discovered in 1937 by DuPont chemist Wallace Carothers. Note that one monomer has two amino groups and the other has two carboxyl groups, but the product is still a polyamide, quite similar to nylon 6. Silk and wool, which are protein fibers, are natural polyamides.

Although nylon can be molded into various shapes, most nylon is made into fibers. Some is spun into fine thread to be woven into silklike fabrics, and some is made into yarn that is much like wool. Carpeting, which was once made primarily from wool, is now made largely from nylon.

Polyethylene Terephthalate and Other Polyesters

A **polyester** is a condensation polymer made from molecules with alcohol and carboxylic acid functional groups. The most common polyester is made from ethylene glycol and terephthalic acid. It is called polyethylene terephthalate (PET).

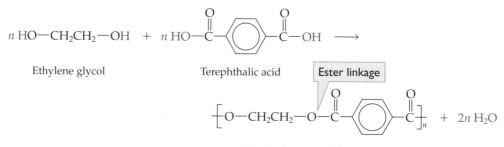

Ethylene glycol   Terephthalic acid

Polyethylene terephthalate

The hydroxyl groups in ethylene glycol react with the carboxylic acid groups in terephthalic acid to produce long chains held together by many ester linkages.

PET can be molded into bottles for beverages and other liquids. It can also be formed as a film, used in laminating documents and to make tough packaging tape. Polyester finishes are used on premium wood products such as guitars, pianos, and the interiors of vehicles and boats. Polyester fibers are strong, quick drying, wrinkle and mildew resistant, and resistant to stretching and shrinking. They are used in home furnishings such as carpets, curtains, sheets and pillow cases, and upholstery. Because polyester fibers do not absorb water, they are ideal for outdoor clothing to be used in wet and damp environments, and for insulation in boots and sleeping bags. For other clothing, they are often blended with cotton for a more natural feel. A familiar use of polyester film (Mylar®) is in party balloons that are filled with helium to celebrate special occasions.

Phenol–Formaldehyde and Related Resins

Let us now go back to Bakelite, the original synthetic polymer. Bakelite, a phenol–formaldehyde resin, was first synthesized by Leo Baekeland, who received U.S. Patent 942,699 for the process in 1909.

Phenol–formaldehyde resins are formed by splitting out water molecules, the hydrogen atoms coming from the benzene ring and the oxygen atoms from the aldehyde. The reaction proceeds stepwise, with formaldehyde adding first to the 2- or 4-position of the phenol molecule.

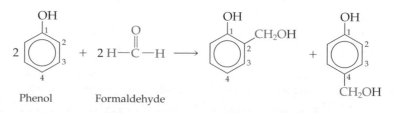

Phenol   Formaldehyde

The substituted molecules then react by splitting out water. (Remember that there are hydrogen atoms at all the unsubstituted corners of a benzene ring.) The hookup of molecules continues until an extensive network is achieved.

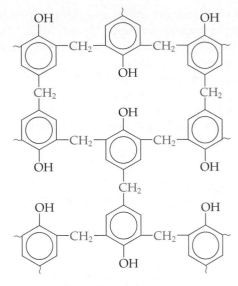

Phenol–formaldehyde resin

Water is driven off by heat as the polymer sets. The structure of the polymer is extremely complex, a three-dimensional network somewhat like the framework of a giant building. Note that the phenolic rings are joined together by CH_2 units from the formaldehyde. These polymers are thermosetting resins; they cannot be melted and remolded. Instead they decompose when heated to high temperatures.

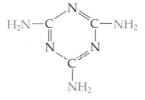

Formaldehyde is also condensed with urea $[H_2N(C{=}O)NH_2]$ to make urea–formaldehyde resins and with melamine to form melamine–formaldehyde resins. (Melamine is formed by condensation of three molecules of urea.)

Melamine

Both resins, like phenol–formaldehyde polymers, are thermosetting. The polymers are complex three-dimensional networks formed by the splitting out of H_2O from formaldehyde ($H_2C{=}O$) molecules and amino ($-NH_2$) groups. Urea–formaldehyde resins are used to bind wood chips together in panels of particle board. Melamine–formaldehyde resins are used in plastic (Melmac) dinnerware and laminate countertops.

Other Condensation Polymers

There are many other kinds of condensation polymers, but we will mention only a few.

Polycarbonates are tough, "clear as glass" polymers strong enough to be used in bulletproof windows. They are also used in protective helmets, safety glasses, clear plastic water bottles, baby bottles, and even in dental crowns. One type of polycarbonate is made from bisphenol A (BPA) and phosgene ($COCl_2$).

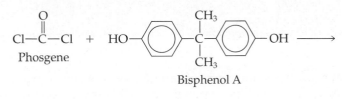

Phosgene   Bisphenol A

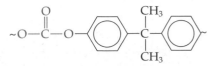

Polyurethanes are similar to nylon in structure. The repeat unit in a common polyurethane is

$$\sim\overset{O}{\overset{\|}{C}}-NH-CH_2CH_2CH_2CH_2CH_2CH_2-NH-\overset{O}{\overset{\|}{C}}-O-CH_2CH_2CH_2CH_2-O\sim$$

▲ **It DOES Matter!**
Bisphenol A (BPA) acts as an endocrine disruptor in laboratory animals; that is, it can act like an estrogenic hormone. BPA leaches out of polycarbonate bottles, especially when heated, and tests find 5–8 ng/mL BPA in the water or milk. Concern that long-term, low-dose exposure to BPA could induce chronic toxicity in humans, especially babies, led Canada to ban use of BPA in baby bottles in 2008. Threats of legislation in the United States and elsewhere caused many manufacturers of polycarbonate bottles to switch to production of BPA-free bottles.

Polyurethanes may be elastomers or tough and rigid, depending on the monomers used. They are popular in foamed padding ("foam rubber") in cushions, mattresses, and padded furniture. They are also used for skate wheels, in running shoes, and in protective gear for sports activities.

Epoxy resins make excellent surface paints and coatings. They are used to protect steel pipes and fittings from corrosion. The insides of metal cans are often coated with epoxy to prevent rusting, especially for use with acidic foods like tomatoes. Epoxies also make powerful adhesives. A common unmodified epoxy is made from epichlorohydrin and BPA.

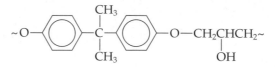

Epoxy adhesives usually have two components that are mixed just before they are used. The polymer chains become cross-linked, and the bonding is extremely strong.

Composite Materials

Composite materials are made up of high-strength fibers (of glass, graphite, synthetic polymers, or ceramics) held together by a polymeric matrix, usually a thermosetting condensation polymer. The fiber reinforcement provides the support, and the surrounding plastic protects the fibers from breaking.

The most commonly used composite materials thus far have been polyester resins reinforced with glass fibers. They are widely used in boat hulls, molded chairs, automobile panels, and in sports gear, such as tennis rackets. Some composite materials have the strength and rigidity of steel at a fraction of the weight.

Silicones

Not all polymers are based on chains of carbon atoms. A silicone (polysiloxane) is a good example of a different type of polymer. A **silicone** is a polymer based on a series of alternating silicon and oxygen atoms.

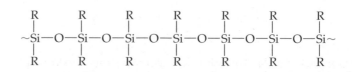

(In simple silicones, R represents a hydrocarbon group, such as methyl, ethyl, or butyl.)

More to Explore

Jacoby, Mitch. "Composite Materials." *Chemical and Engineering News*, August 30, 2004, pp. 34–39.

▶ **It DOES Matter!**

Because composite structures are strong and light, they save energy—and can even make old technology new again. Here, test driver Don Wales tries out the cockpit of the British Steam Car as engineer Mike Horne fits the composite carbon fiber bodywork. Although steam power for automobiles was abandoned in the early 20th century, lightweight composite construction of the body and engine components may make steam power both practical and environmentally sound.

Silicon is in the same group (4A) as carbon and, like carbon, is tetravalent and able to form chains. However, carbon can form chains of just carbon atoms (as in polyethylene), whereas silicone polymers have chains of alternating silicon and oxygen atoms.

Silicones can be linear, cyclic, or cross-linked networks. They are heat-stable and resistant to most chemicals, and they are excellent waterproofing materials. Depending on chain length and amount of cross-linking, silicones can be oils or greases, rubbery compounds, or solid resins. Silicone oils are used as hydraulic fluids and lubricants, whereas other silicones are used in making such products as sealants, auto polish, shoe polish, and waterproof sheeting. Fabrics for raincoats and umbrellas are frequently treated with silicone.

An interesting silicone toy is Silly Putty. It can be molded like clay or rolled up and bounced like a ball. But it is actually a liquid (see page 150) and flows very slowly on standing.

Perhaps the most remarkable silicones of all are the ones used for synthetic human body parts. Kinds of silicone replacements range from finger joints to eye sockets. Artificial ears and noses are also made from silicone polymers. They can even be specially colored to match the surrounding skin.

For many years, silicone gel was used for breast implants. In most cases, these implants have remained perfectly stable over the years. However, because of a manufacturing error, some batches of gel were not sufficiently cured. As a result, the gel implants later disintegrated, causing leakage into body tissues and leading to health problems in some cases. The FDA banned silicone implants for breast augmentation in 1992, but they remained available for breast reconstruction after a mastectomy. In 2006 two manufacturers were again allowed to offer silicone implants for augmentation. Saline-filled implants remained available all the while.

▲ Cookware made of silicone is colorful, flexible, and nonstick.

4. Are silicon and silicone the same?
Although the names are very similar, the materials are quite different. *Silicon* is a hard, brittle element. A *silicone* is a polymer that contains silicon as well as carbon, hydrogen, and oxygen. Most silicones are flexible.

A N S W E R

EXAMPLE 10.3   Condensed Structural Formulas for Polymers

Write the condensed structural formula for the polymer formed from dimethylsilanol, $(CH_3)_2Si_2(OH)_2$.

Solution

Let's start by writing out the structural formula of the monomer.

Because there are no double bonds in this molecule, we do not expect addition polymerization. Rather, we expect a condensation reaction in which an OH group of one molecule and an H atom of another combine to form a molecule of water. Moreover, because there are two OH groups per molecule, each monomer can form bonds with the neighbors on both sides. This is a key requirement for polymerization. We can represent the reaction as follows.

$$\text{HO}-\underset{\underset{CH_3}{|}}{\overset{\overset{CH_3}{|}}{Si}}-\text{OH} + \text{HO}-\underset{\underset{CH_3}{|}}{\overset{\overset{CH_3}{|}}{Si}}-\text{OH} + \text{HO}-\underset{\underset{CH_3}{|}}{\overset{\overset{CH_3}{|}}{Si}}-\text{OH} + \cdots \longrightarrow \left[\begin{array}{c} CH_3 \\ | \\ Si-O \\ | \\ CH_3 \end{array}\right]_n + n\,H_2O$$

■ **EXERCISE 10.3**

(a) Write the structural formula for the polymer formed from 3-hydroxypropanoic acid ($HOCH_2CH_2COOH$), showing at least four repeat units. **(b)** Write a condensed structural formula in which the repeat unit is shown in brackets.

Self-Assessment Questions

1. In a condensation polymerization, the two products formed are a polymer and (usually)
   a. an acid   **b.** a base   **c.** carbon dioxide   **d.** water

2. Which of the following pairs of molecules can form a polyester?
   a. $HOCH_2CH_2OH$ and $HOCH_2CH_2CH_2CH_2CH_2OH$
   b. $H_2NCH_2CH_2CH_2CH_2CH_2NH_2$ and $HOOCCH_2CH_2CH_2CH_2CH_2COOH$
   c. $H_2NCH_2CH_2CH_2CH_2CH_2NH_2$ and $HOCH_2CH_2CH_2CH_2CH_2OH$
   d. $HOOCCH_2CH_2CH_2CH_2CH_2COOH$ and $HOCH_2CH_2CH_2CH_2CH_2OH$

3. Which of the following molecules can serve as the sole monomer for a polyamide?
   a. $H_2NCH_2CH_2CH_2CH_2OH$ **b.** $H_2NCH_2CH_2CH_2CH_2CH_2NH_2$
   c. $HOOCCH_2CH_2CH_2CH_2CH_2NH_2$ **d.** $HOOCCH_2CH_2CH_2CONH_2$

4. Which of the following molecules can serve as the sole monomer for a polyester?
   a. $H_2NCH_2CH_2CH_2COOH$ **b.** $H_2NCH_2CH_2CH_2CH_2CH_2OH$
   c. $HOCH_2CH_2COOH$ **d.** $HOOCCH_2CH_2CH_2COOH$

5. Thermosetting polymers
   a. are easily recycled
   b. are usually addition polymers
   c. cannot be melted and remolded
   d. usually melt at a lower temperature than do thermoplastic polymers

6. The backbone of a silicone polymer chain is composed of repeating units of
   a. $\sim OCH_2CH_2OSi\sim$ **b.** $\sim OCH_2CH_2Si\sim$ **c.** $\sim OSi\sim$ **d.** $\sim OSiSi\sim$

Answers: 1, d; 2, d; 3, c; 4, c; 5, c; 6, c

10.8 Properties of Polymers

Polymers differ from substances consisting of small molecules in three main ways. First, the long chains can be entangled with one another; the polymer molecules form a tangled mass much like a dish of spaghetti. It is difficult to untangle the polymer molecules, especially at low temperatures. This lends strength to polymers in plastics, elastomers, and other materials.

Second, although intermolecular forces affect polymers just as they do small molecules, these forces are greatly multiplied in large molecules. The larger the molecules are, the greater the intermolecular forces between them. Even when ordinarily weak dispersion forces are the only operative intermolecular forces, they can strongly bind polymer chains together. This, too, makes for strong polymeric materials. For example, polyethylene is nonpolar with only dispersion forces between molecules, but (as we indicated in the marginal note on page 266) ultra-high-molecular-weight polyethylene forms fibers so strong they can be used in bulletproof vests.

Third, large polymer molecules move more slowly than small molecules do. A group of small molecules (monomers) can move around more rapidly and more randomly when independent than they can when joined together in a long chain (polymer). The slower speed of molecules makes a polymeric material different from one made of small molecules. For example, a polymer dissolved in a solvent will form a solution that is a lot more viscous than the pure solvent.

Crystalline and Amorphous Polymers

Some polymers are highly crystalline, and their molecules line up neatly to form long fibers of great strength. Other polymers are largely amorphous, composed of randomly oriented molecules that get tangled up with one another (Figure 10.6). Crystalline polymers tend to make good synthetic fibers, whereas amorphous polymers often make good elastomers.

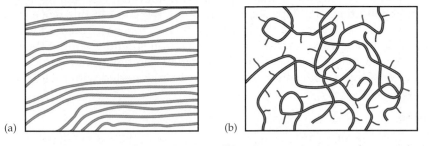

(a)   (b)

◀ **Figure 10.6** Organization of polymer molecules. (a) Crystalline arrangement. (b) Amorphous arrangement.

Sometimes the same polymer is crystalline in one region and amorphous in another. For example, scientists have designed spandex fibers (used for stretch fabrics [Lycra] in ski pants, exercise clothing, and swimsuits) so as to combine the tensile strength of crystalline fibers with the elasticity of amorphous rubber. Two molecular structures are combined in one polymer chain, with blocks of crystalline character alternating with amorphous blocks. The amorphous block is soft and rubbery; the crystalline part is quite rigid. The resulting polymer exhibits both sets of properties—flexibility and rigidity.

The Glass Transition Temperature

An important property of most thermoplastic polymers is the **glass transition temperature (T_g)**. Above this temperature, the polymer is rubbery and tough; below it, the polymer is like glass: hard, stiff, and brittle. Each polymer has a characteristic T_g. We want automobile tires to be tough and elastic, and so we use materials with low T_g values. On the other hand, we want plastic substitutes for glass to be glassy; thus, they have T_g values well above room temperature.

Fiber-Forming Properties

Not all synthetic polymers can be converted to useful fibers, but those that can often have properties superior to those of natural fibers. More than half of the 60 billion kg of textile fibers produced annually in the United States are synthetic.

Silk fabrics are beautiful and have a luxurious "feel," but nylon fabrics are also beautiful and feel much like silk. Moreover, nylon fabrics wear longer, are easier to care for, and are less expensive than silk.

Polyesters such as PET can substitute for either cotton or silk, and they outperform the natural fibers in many ways. Polyesters are not subject to mildew as cotton is, and many polyester fabrics do not need ironing. Fabric made by blending polyester fibers with 35–50% cotton combines the comfort of cotton with the no-iron easy care of polyester fabrics.

Acrylic fibers, made from polyacrylonitrile, can be spun into yarns that look like wool. Acrylic sweaters have the beauty and warmth of wool, but they do not shrink in hot water, are not attacked by moths, and do not cause the allergic skin reactions caused by wool in some people.

We can make use of the T_g concept in everyday life. For example, we can remove chewing gum from clothing by applying ice to lower the temperature of the gum below the T_g of the polyvinyl acetate resin that makes up the bulk of the gum. The cold, brittle resin then crumbles readily and can be removed.

▲ Formation of fibers by extrusion through a spinneret. A melted polymer is forced through the tiny holes to make fibers that solidify as they cool.

Self-Assessment Questions

1. The glass transition of a polymer is the temperature
   a. above which the polymer becomes highly crystalline
   b. at which the polymer changes to a silicate mineral
   c. at which the polymer melts to a clear liquid
   d. below which the polymer becomes glassy and brittle

2. Crystalline polymers are
   a. elastic   b. flexible   c. highly cross-linked   d. used as fibers

Answers: 1, d; 2, d

10.9 Disposal of Plastics

An advantage of plastics is that they are durable and resistant to many environmental conditions. Perhaps some of them are too good in this respect; they last almost forever. Once they are dumped, they do not go away. You see them littering our parks, our sidewalks, and our highways; if you should go out to the middle of the ocean, you would see them there, too. Small fish have been found dead with their digestive tracts clogged by bits of plastic foam ingested with their food.

Landfills

Plastics make up about 12% by mass of solid waste in the United States, but by volume they make about 25%. This has created a problem because 55% of all solid waste goes into landfills, and it is increasingly difficult to find suitable landfill space. (Solid wastes in general are discussed in more detail in Chapter 12.)

Incineration

One way to dispose of discarded plastics is to burn them. Most plastics have a high fuel value. For example, a pound of polyethylene has about the same energy content as a pound of fuel oil. Some communities actually generate electricity with the heat from garbage incinerators. Some utility companies burn powdered coal mixed with a few percent of ground-up rubber tires. They not only obtain extra energy from the tires but also help solve the problem of tire disposal.

On the other hand, the burning of plastics and rubber can lead to some new problems. For example, PVC produces toxic hydrogen chloride and vinyl chloride gases when it burns, and burning automobile tires give off soot and a stinking smoke. Incinerators are corroded by acidic fumes and clogged by materials that are not readily burned.

Degradable Plastics

About half of our waste plastic is from packaging. One approach to the plastics disposal problem is to make plastic packages that are biodegradable or photodegradable (broken down in the presence of bacteria or light). Of course, it is important that the package remain intact and not start to decompose while it is still being used. For now, many people seem reluctant to pay extra for garbage bags that are designed to fall apart. However, recently there are more and more examples of degradable, more environmentally friendly plastics for sale, including plastic tableware, cups, bottles, and delicatessen food containers.

Recycling

Recycling is perhaps the best way to handle waste plastics. The plastics must be collected, sorted, chopped, melted, and then remolded. Collection works well when there is strong community cooperation.

The separation step is simplified by code numbers stamped on plastic containers. Once the plastics have been separated, they can be chopped into flakes, melted, and remolded or spun into fiber.

At present, the only plastic items being recycled on a large scale are those bearing recycling codes 1 (PET) and 2 (HDPE). For example, PET bottles are made into fiber, mainly for carpets, and HDPE containers are remolded into detergent bottles. Recycling is good news for green chemistry, because recycling keeps plastics out of landfills and prevents the use of petroleum-derived monomers or other hazardous chemicals. We have a long way to go: As of 2006, only about 7% of the 27 billion kg per year of plastic waste in the United States is recycled.

▲ Recycled plastic can be used for many things. This kitchen ware is made from recycled plastic, which can be recycled again.

10.10 Plastics and Fire Hazards

The accidental ignition of fabrics, synthetic or otherwise, has caused untold human misery. The U.S. Department of Health and Human Services estimates that fires involving flammable fabrics kill several thousand people annually and injure 150,000 to 200,000 each year.

Research has led to a variety of flame-retardant fabrics. Many incorporate chlorine and bromine atoms within the polymeric fiber. In particular, federal regulations require that children's sleepwear be made of such flame-retardant materials.

Another synthetic fabric, meta-aramid, or Nomex (from DuPont), has such heat resistance that it is used for protective clothing for firefighters and race car drivers. The fibers don't ignite or melt when exposed to flames or high heat. Nomex is also used in electric insulation and for machine parts exposed to high heat.

Burning plastics often produce toxic gases. Hydrogen cyanide is formed in large quantities when polyacrylonitrile and other nitrogen-containing polymers burn. Lethal amounts of cyanide found in the bodies of plane crash victims have been traced to burned plastics. Firefighters often refuse to enter burning buildings without gas masks for fear of being overcome by fumes from burning plastics. Smoldering fires also produce lethal quantities of carbon monoxide. To make plastics safer when burned, green chemists are making new kinds of polymers that don't burn or that don't generate toxic chemicals when burned.

10.11 Plasticizers and Pollution

Chemicals used in plastics manufacture can also present problems. Plasticizers are an important example. Some plastics, particularly vinyl polymers, are hard and brittle and thus difficult to process. A **plasticizer** can make them more flexible and less brittle by lowering the glass transition temperature, T_g. Unplasticized PVC is rigid and is used for water pipes. Thin sheets of pure PVC crack and break easily, but plasticizers make them soft and pliable. Plastic raincoats, garden hoses, and seat covers for automobiles can be made from plasticized PVC. Plasticizers are liquids of low volatility and are generally lost by diffusion and evaporation as a plastic article ages. The plastic becomes brittle and then cracks and breaks.

Once used widely as plasticizers, but now banned, polychlorinated biphenyls (PCBs) are derived from biphenyl ($C_{12}H_{10}$), a hydrocarbon that has two benzene rings joined at a corner. In PCBs, some of the hydrogen atoms of biphenyl are replaced with chlorine atoms (Figure 10.7). Note that PCBs are structurally similar to the insecticide DDT.

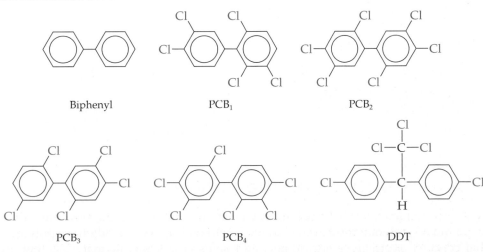

◄ **Figure 10.7** Biphenyl and some of the PCBs derived from it. These are but a few of the hundreds of possible PCBs. DDT is shown for comparison.

PCBs were also widely used as insulating materials in electric equipment (transformers and condensers) because their stability and low polarity gave them high electric resistance and ability to absorb heat. The same properties that made

PCBs so desirable as industrial chemicals cause them to be an environmental hazard. They degrade slowly in nature, and their solubility in nonpolar media—animal fat as well as vinyl plastics—leads to their concentration in the food chain. PCB residues have been found in fish, birds, water, and sediments. The physiologic effect of PCBs is similar to that of DDT. Monsanto Corporation, the only company in the United States that produced PCBs, discontinued production in 1977, but the compounds still remain in the environment.

Today the most widely used plasticizers for vinyl plastics are phthalate esters, a group of diesters derived from phthalic acid (1,2-benzenedicarboxylic acid). Phthalate plasticizers (Figure 10.8) have low acute toxicity, and although some studies have suggested possible harm to young children, the FDA has recognized these plasticizers as being generally safe. They pose little threat to the environment because they degrade fairly rapidly. Another approach that green chemists adopt is to make plastics that have the right amount of flexibility and do not require any plasticizers to be used.

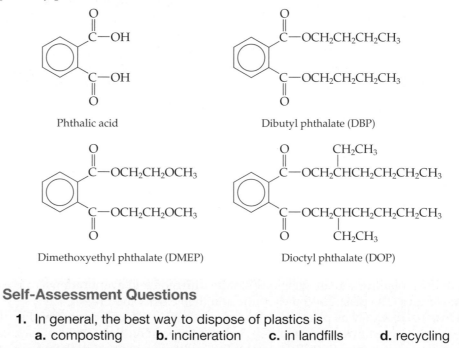

▶ **Figure 10.8** Phthalic acid and some esters derived from it. Dioctyl phthalate is also called di-2-ethylhexyl phthalate.

Phthalic acid

Dibutyl phthalate (DBP)

Dimethoxyethyl phthalate (DMEP)

Dioctyl phthalate (DOP)

Self-Assessment Questions

1. In general, the best way to dispose of plastics is
   a. composting   b. incineration   c. in landfills   d. recycling

2. Many plastics are fire hazards because they burn and produce
   a. CFCs   b. PCBs   c. solid wastes   d. toxic gases

3. A plasticizer acts by
   a. initiating addition polymerization
   b. lowering the glass transition temperature
   c. participating in condensation polymerization
   d. wetting polymer powders

4. Many flame-retardant fabrics incorporate
   a. Br and Cl atoms   b. CFCs   c. PCBs   d. plasticizers

Answers: 1, d; 2, d; 3, b; 4, a

10.12 Plastics and the Future

Widely used today, synthetic polymers are the materials of the future. There will be new kinds of polymers and even wider use in the future. We already have polymers that conduct electricity, amazing new adhesives, and synthetic materials that are stronger than steel but much lighter in weight. Plastics present problems, but they have become such an important part of our daily lives that we would find it difficult to live without them.

In medicine, body replacement parts made from polymers have become common. There were about 250,000 hip replacements in the United States in 2007. Artificial lungs and artificial hearts are available today but are enormously expensive and used for the most part only during recovery from injury or illness or until donor organs are available for transplantation. It is likely that costs of these replacements will drop and their efficacy will improve in the future.

Home construction today uses PVC water pipes, siding, and window frames, plastic foam insulation, and polymeric surface coatings. Some homes also contain lumber and wall panels of artificial wood, made from recycled plastics.

Synthetic polymers are used extensively in airplane interiors, and the bodies and wings of some planes are made of lightweight composite materials. Many automobiles now have bodies made from plastic composites. Electrically conducting polymers will aid in making lightweight batteries for electric automobiles. The electronics industry will use increasing amounts of electrically conducting thermoplastics in its miniaturized circuits.

But here is something to think about. Most synthetic polymers are made from petroleum or natural gas. Both these natural resources are nonrenewable, and our supplies are limited. We are likely to run out of petroleum during this century. You might suppose that we would be actively conserving this valuable resource, but unfortunately this is not the case. We are taking petroleum out of the ground at a rapid rate, converting most of it to gasoline and other fuels, and then simply burning it. There are other sources of energy, but is there anything that can replace petroleum as the raw material for making plastics? Yes! Today, several new types of plastics, such as polylactic acid and polyhydroxybutyrates, are made from renewable resources such as corn, soybeans, and sugarcane. Polymers are an active area of green chemistry research.

More to Explore
Bren, Linda. "Joint Replacement: An Inside Look." *FDA Consumer*, March–April 2004, pp. 12–19, and Ashley, Steven. "Artificial Muscles." *Scientific American*, October 2003, pp. 52–59.

Self-Assessment Questions

1. Most synthetic polymers are made from
   a. coal
   b. cotton
   c. petroleum and natural gas
   d. wood chips

2. A polymer made from a renewable resource is
   a. LDPE
   b. polyacrylonitrile
   c. polylactic acid
   d. styrene-butadiene rubber

Answers: 1, c; 2, c

Petroleum

That viscous, tarry liquid nature laid beneath the ground

A hundred million years before man came upon the scene

Can be transformed to marvelous new products we have found—

Like nylon, orlon, polyesters, polyethylene,

Synthetic rubber, plastics, films, adhesives, drugs, and dyes,

And other things, some that we now can only dream about.

Who knows what wondrous products man might some day synthesize

From oil! Except, alas, that our supplies are running out.

The time is near when Earth's prodigious flow of oil may stop

(We've taken so much from the ground, with no way to return it.)

Meanwhile, we strive to find and draw out every precious drop,

And then . . . incredibly . . . we take the bulk of it and burn it.

GREEN CHEMISTRY

Greener Polymers

Jennifer L. Young, ACS Green Chemistry Institute®

Many products we use today are made of plastic. How can plastics be made sustainably and with green chemistry? The Twelve Principles of Green Chemistry can be applied to plastics and polymers. We focus here on the recycling and biodegradation of plastic and its production from renewable resources.

You probably use polyethylene bags, polystyrene cups, polyethylene terephthalate (PET) soft drink bottles, nylon clothing and carpeting, and styrene-butadiene rubber tires. These polymers and most of the polymers on the market do not biodegrade, and they are made from nonrenewable resources. After use, the polymers often end up discarded into the environment or buried in landfills. However, many kinds of polymers you use every day are recyclable. By recycling, you are reducing the amount of plastic that ends up in landfills and also reducing the amount of new plastic that must be made from fossil fuels.

To facilitate recycling, many plastic objects are stamped with a triangular recycle symbol and a number code. These numbers, called *resin identification codes,* identify types of plastics as follows: #1 polyethylene terephthalate (PET), #2 high density polyethylene (HDPE), #3 polyvinyl chloride (PVC), #4 low density polyethylene (LDPE), #5 polypropylene (PP), #6 polystyrene (PS), and #7 for all other kinds of polymers.

PET (#1) and HDPE (#2) plastics are the plastics most commonly recycled. Other types of plastics may also be accepted for recycling in some locales. In the recycling process, the plastics first need to be separated by resin identification code. The plastic is then chopped up, softened by raising the temperature above the polymer's glass transition temperature, molded into a new shape, and cooled back to room temperature to harden. Because the polymer needs to be softened by heat to be recycled, only thermoplastics (not thermosetting polymers) can be recycled. Usually, recycled plastic is made into different kinds of objects than the original polymer because the properties are usually different. For example, park benches and plastic outdoor decking

material are made from recycled plastic. Just because a polymer falls into the "#7 other" category does not mean that the polymer cannot be recycled. For instance, nylon-based carpet is commonly recycled.

Some polymers including PET and polylactic acid (see below) can also be depolymerized back to a monomer. This means the polymer can undergo reactions that reverse the polymerization, breaking the polymer down into the original monomer units. The monomer can then be polymerized again to make polymer that is as pure as the original polymer, with the same properties as the original polymer, and used to make the same objects as the original polymer. Some companies that make these types of plastic will accept their plastic for recycling, to depolymerize and to reuse the monomer.

Some types of plastics can be degraded into carbon dioxide and water. For a polymer to biodegrade, it needs to be exposed to the right conditions of temperature, moisture, and microorganisms over some period of time. Different official bodies apply different definitions; for instance, the USA ASTM D6400-99 Standards, the European Standard EN13432, and the Japanese GreenPla Standard each define different conditions and periods of time. Biodegradable polymers should not be thrown in the trash because landfills do not provide the right conditions for degradation.

New kinds of polymers are being made from renewable resources, such as corn, soy, grass, and other kinds of biomass. One example is polylactic acid (PLA), made from corn. PLA is used to make plastic bottles, salad containers, coffee mugs, and even fibers for clothing. PLA can be depolymerized back to its monomer and can also degrade by composting. Another example is polyhydroxyalkanoates (PHAs) made from plant sugars and oils. PHAs are biodegradable and can even be made using bacteria. For many plastics the most sustainable solution is using renewable resources—especially nonfood crops— as the source of the monomer.

◄ Biodegradable plastic products such as those made from polylactic acid (PLA) are now widely available. Estimates see demand for degradable plastics growing rapidly as prices and properties become more competitive with conventional polymers.

■ SUMMARY

Section 10.1—A **polymer** is a giant molecule made from smaller molecules. The small "building block" molecules that make a polymer are called **monomers**.

Section 10.2—There are many natural polymers, including starch, cellulose, nucleic acids, and proteins.

Section 10.3—**Celluloid** was the first semisynthetic polymer, made from natural cellulose modified with nitric acid.

Section 10.4—The simplest, least expensive, and highest-volume synthetic polymer is polyethylene, made from the monomer ethylene. High-density polyethylene molecules can pack closely together for a rigid, strong structure. Low-density polyethylene molecules have many side chains and the product is more flexible. Linear low-density polyethylene is a **copolymer**, formed from two (or more) different monomers.

Polyethylene is one of many **thermoplastic polymers** that can be softened and reshaped with heat and pressure. **Thermosetting resins** are polymers that decompose rather than softening when heated and cannot be reshaped.

Section 10.5—In **addition polymerization** the monomer molecules add to one another; the polymer contains all atoms of the monomer(s). Addition polymers, such as polyethylene, polypropylene, polystyrene, and PVC, are made from monomers containing double or triple bonds. Polypropylene is tough and resists moisture, oils, and solvents. Polystyrene is used primarily to make Styrofoam. PVC or polyvinyl chloride is used for plumbing, artificial leather, and flexible tubing. PTFE or Teflon has great nonstick properties and is chemically inert. Thermoplastic polymers can be molded in many different ways.

Section 10.6—Natural rubber is an **elastomer**, a polymer that can be stretched yet returns to its shape. The monomer of natural rubber, isoprene, can be made artificially and polymerized. Polyisoprene is soft and tacky when hot. In a process called **vulcanization**, polyisoprene is reacted with sulfur. The sulfur cross-links the hydrocarbon chains together, making the rubber harder, stronger, and more elastic.

Other elastomers are similar in structure to polyisoprene, including polybutadiene, polychloroprene, and styrene-butadiene rubber (SBR). All of these have molecules that can coil and uncoil.

Section 10.7—In **condensation polymerization**, parts of the monomer molecules are not incorporated in the product. Small molecules such as water are split out as by-products. Nylon is a condensation polymer that is a **polyamide**, with amide [—(CO)NH—] linkages joining the monomers. A **polyester** is a condensation polymer made from monomers with alcohol and carboxylic acid functional groups. Polyethylene terephthalate is the most common polyester. Bakelite, the first fully synthetic polymer, condenses from phenol and formaldehyde. Other similar resins can be prepared using urea or melamine instead of phenol. Other types of condensation polymers include polycarbonates (bulletproof windows), polyurethanes (tough rubber), epoxies (glues), and **silicones**, which have silicon and oxygen atoms instead of carbon atoms. Combining a polymer with high-strength fibers gives a composite material that exploits the best properties of both materials.

Section 10.8—Properties of polymers are very different from those of their monomers. Polymer strength arises partly because the molecules entangle one another and partly because the large molecules have relatively strong dispersion forces. Polymers may be crystalline or amorphous. Above the **glass transition temperature (T_g)** a polymer is rubbery and tough; below it, the polymer is brittle. Strong fibers are made from crystalline polymers that have molecules neatly aligned with one another. Polymer fibers often have advantages over their natural counterparts.

Section 10.9—The longevity of plastics can be a problem. They make up a large fraction of the content of landfills. Proper incineration is one way of disposing of discarded plastics, but it has problems. Some polymers are engineered to be degradable. Recycling of plastics requires that they be collected and sorted according to the code numbers. Recycled plastics are seeing increasing use.

Section 10.10—Polymers used for clothing often are made of flame-retardant materials that incorporate chlorine or bromine atoms. The chemicals used in manufacture can present problems.

Section 10.11—A **plasticizer** makes a polymer more flexible. Polychlorinated biphenyls (PCBs) were once used as plasticizers but are now banned. Phthalates are the most common plasticizers today.

Section 10.12—Synthetic polymers are the materials of the future. They will be used even more for body replacement parts, construction, automobiles, batteries, and electronics. However, most synthetic polymers come from petroleum, which is a limited and nonrenewable natural resource. This is another important reason for recycling plastics.

■ REVIEW QUESTIONS

1. Define the following terms.
   a. macromolecule   **b.** elastomer
   c. copolymer   **d.** plasticizer

2. How does the structure of PVC differ from that of polyethylene? List several uses of PVC.

3. What is addition polymerization? What structural feature usually characterizes molecules used as monomers in addition polymerization?

4. What is Teflon? What unique property does it have? What are some of its uses?

5. What polymer is used to make clear, brittle, disposable drinking cups? From what monomer are disposable foamed plastic coffee cups made?

6. What plastic is used to make **(a)** gallon milk jugs and **(b)** 2-L soft drink bottles?

7. What is Bakelite? From what monomers is it made?

8. What is a thermosetting polymer? Give an example.

9. What type of fibers, natural or synthetic, is used most in the United States? Why?

10. What are PCBs? Why are they no longer used as plasticizers?

11. What problems arise when plastics are
    a. discarded into the environment?
    b. disposed of in landfills?
    c. disposed of by incineration?

12. What steps must be taken in order to recycle plastics?

■ PROBLEMS

Polyethylene

13. Describe the structure of high-density polyethylene (HDPE). How does this structure explain the properties of this polymer?

14. Describe the structure of low-density polyethylene (LDPE). How does this structure explain the properties of this polymer?

Addition Polymerization

15. Give the structure of the monomer from which each of the following polymers is made.
    a. polyvinyl chloride
    b. PTFE (Teflon)

16. Give the structure of the monomer from which each of the following polymers is made.
    a. polystyrene
    b. polyacrylonitrile

17. Give the structure of a segment of polymer chain that is at least eight carbon atoms long for each of the following.
    a. polyethylene
    b. vinylidene fluoride (CH_2=CF_2)

18. Give the structure of a segment of polymer chain that is at least four monomer units long for polymers formed from the following monomers.
    a. polypropylene
    b. methyl methacrylate

$$CH_2 = \overset{\overset{\displaystyle CH_3}{|}}{C}COOCH_3$$

19. Give the structure of the polymer made from each of the following. Show at least four repeat units.
    a. vinyl acetate (CH_2=CH—O—CO—CH_3)
    b. methyl acrylate (CH_2=CHCOOCH$_3$)

20. Give the structure of the polymer made from each of the following. Show at least four repeat units.
    a. 1-pentene (CH_2=CHCH$_2$CH$_2$CH$_3$)
    b. methyl cyanoacrylate

$$CH_2 = \overset{\overset{\displaystyle C\equiv N}{|}}{C}COOCH_3$$

Rubber and Other Elastomers

21. Give the structure of the monomer from which polybutadiene is made.

22. Describe the process of vulcanization. How does vulcanization change the properties of rubber?

23. Explain the elasticity of rubber.

24. How does polybutadiene differ from natural rubber in properties? In structure?

25. Name three synthetic elastomers and the monomer(s) from which they are made.

26. How is SBR made? From what monomer(s) is it made?

Condensation Polymerization

27. Nylon 44 is made from the monomers $H_2NCH_2CH_2CH_2CH_2NH_2$ and $HOOCCH_2CH_2COOH$. Give the structure of nylon 44 showing at least two units from each monomer.

28. Give the structure of a polymer made from glycolic acid (hydroxyacetic acid, $HOCH_2COOH$). Show at least four repeating units. (*Hint:* Compare with nylon 6 in Section 10.7.)

29. Kodel is a polyester fiber. The monomers are terephthalic acid (page 274) and 1,4-cyclohexanedimethanol (in the following figure). Write a condensed (bracketed) repeating structure of the Kodel polyester molecule.

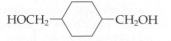

1,4-Cyclohexanedimethanol

30. Kevlar, a polyamide used to make bulletproof vests, is made from terephthalic acid (page 266) and *para*-phenylenediamine (1,4-benzenediamine). Write a condensed (bracketed) repeating structure of the Kevlar molecule.

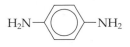

para-Phenylenediamine

Properties of Polymers

31. What three factors give polymers properties that are different from materials made up of small molecules?

32. What does the word *plastic* mean **(a)** in everyday life and **(b)** in chemistry?

33. What is the glass transition temperature (T_g) of a polymer? For what uses do we want polymers with a low T_g? With a high T_g?

34. How do plasticizers make polymers less brittle?

■ ADDITIONAL PROBLEMS

35. In the following equation, identify the parts labeled a, b, and c as monomer, polymer, and repeat unit. What type of polymerization (addition or condensation) is represented?

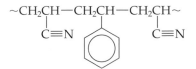

36. Is nylon 44 (Problem 27) an addition polymer or a condensation polymer? Explain.

37. One type of Saran has the structure shown below. Give the structures of the two monomers from which it is made.

$$\sim CH_2CCl_2-CH_2CHCl-CH_2CCl_2-CH_2CHCl\sim$$

38. From what monomers could the following copolymer, called poly(styrene-co-acrylonitrile) (SAN), be made?

~CH₂CH—CH₂CH—CH₂CH~
    | | |
   C≡N   C≡N

39. Cyanoacrylates, such as those made from methyl cyanoacrylate (Problem 20b), are used in instant-setting "super" glues. However, poly(methyl cyanoacrylate) can irritate tissues. Cyanoacrylates with longer alkyl ester groups are less harsh and can be used in surgery in place of sutures. A good example is poly(octyl cyanoacrylate). Write a polymerization reaction for the formation of poly(octyl cyanoacrylate) from octyl cyanoacrylate, using a condensed (bracketed) repeat structure for the polymer.

40. Polyethylene terephthalate has a high T_g. How can the T_g be lowered so that a manufacturer can permanently crease a pair of polyethylene terephthalate slacks?

41. Isobutylene [$CH_2{=}C(CH_3)_2$] polymerizes to form polyisobutylene, a sticky polymer used as an adhesive. Give the structure of polyisobutylene. Show at least four repeat units.

42. Copolymerized with isoprene, isobutylene (Problem 41) forms butyl rubber. Give the structure of butyl rubber. Show at least three isobutylene repeat units and one isoprene repeat unit.

43. The bacteria *Alcalgenes eutrophus* produce a polymer called polyhydroxybutyrate with the structure shown. Give the structure of the hydroxy acid from which this polymer could be made.

$$\begin{array}{cc} CH_3 & O \\ | & \| \\ {+}O-CH-CH_2-C{+}_n \end{array}$$

44. DSM Engineering Plastics makes two specialty polymers, Stanyl® nylon 46 and polytetramethylene terephthalamide (PA₄T). (A methylene group is CH_2; tetramethylene means four methylene groups: $CH_2CH_2CH_2CH_2$.) Both are made from 1,4-diaminobutane as the diamine. **(a)** What diacid is used for each? Write a condensed (bracketed) repeat structure of **(b)** Stanyl® nylon 46 and of **(c)** PA₄T.

45. Based on the condensed repeating structural formula of poly(ethylene naphthalate) (PEN) given below, give **(a)** the structures of the monomer(s) and **(b)** state the polymer is a polyester or a polyamide.

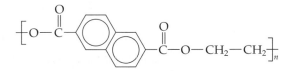

46. A student writes the formula for polypropylene as shown. Describe the error(s) in the formula.

$$+CH_2-CH-CH_3+_n$$

47. What kind of polymer is involved in the following phenomena? A rubber ball dropped from 100 cm will bounce back up to about 60 cm. Balls made from polybutadiene, called Superballs, will bounce back up to about 85 cm. Golf balls, made from cross-linked polybutadiene, bounce back up to 89 cm. Which kind of ball exhibits the phenomenon to the greatest degree?

48. Draw a structure of the likely product(s) if ethanol is substituted for ethylene glycol in the reaction with terephthalic acid (page 266).

49. Shell Chemical developed Carilon® polymers, which are highly versatile and have been used to make machine parts, fibers, laboratory equipment, specialty medical supplies, electrical connectors, hoses, and bearings (below). What functional group is present in the polymer chain? Carilon polymers are copolymers of carbon monoxide and an alkene. What alkene serves as a monomer in the following molecule?

$$\sim CH_2-CH_2-\overset{\overset{\displaystyle O}{\|}}{C}-CH_2-CH_2-\overset{\overset{\displaystyle O}{\|}}{C}-CH_2-CH_2-\overset{\overset{\displaystyle O}{\|}}{C}\sim$$

50. On page 276 we showed nylon 6 being formed from 6-aminohexanoic acid by condensation polymerization. In practice, however, nylon 6 is usually made from a cyclic compound called ∈ −caprolactam (see equation). Is this an addition polymerization or a condensation polymerization? Can you see why some chemists prefer the labels *chain-reaction polymerization* for those processes involving monomers with double bonds and *step-reaction polymerization* for those that do not?

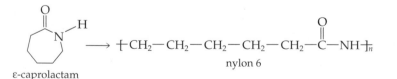

ε-caprolactam   nylon 6

51. Ethylene propylene diene monomer (EPDM) is an extremely durable elastomer that is used for pond liners, weatherstripping, and as a waterproof membrane for flat roofing. One structure for EPDM is shown below. **(a)** Review the discussion of HDPE and LDPE on page 264, and suggest whether EPDM is closer to HDPE or LDPE in its structure. **(b)** Recall Figure 10.6; do you think EPDM can form a crystalline arrangement? Justify your answer.

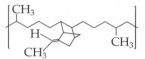

52. Consider the two alkenes below. They are different compounds but when each undergoes addition polymerization, the same polymer is formed. Explain.

$$CH_3-\underset{\underset{\displaystyle Cl}{|}}{C}=\underset{\underset{\displaystyle Br}{|}}{C}-CH_2CH_3 \qquad CH_3-\underset{\underset{\displaystyle Cl}{|}}{C}=\underset{\underset{\displaystyle CH_2CH_3}{|}}{C}-Br$$

53. Consult Table 16.3. Each *amino acid* shown there has an amino group, represented as $-NH_3^+$, and a carboxylic acid group, $-COO^-$. (In each case the proton has been donated from the carboxylic group to the amino group.) Proteins are polymers of amino acids. Draw the structure of the protein that would result from polymerization of glycine, showing four monomer units.

54. Epoxies have one or more heterocyclic rings in their structures, similar to those shown in Example 9.12. Some heterocyclic rings can polymerize by breaking one of the bonds between a carbon atom and the oxygen, nitrogen, or sulfur atom. These broken rings can undergo an addition reaction. Draw the structure of the polymer that could result from polymerization of the molecules shown in Example 9.12, parts (a) and (b). Show four repeat units for each.

▪ COLLABORATIVE GROUP PROJECTS

Prepare a PowerPoint, poster, or other presentation (as directed by your instructor) for presentation to the class.

55. Prepare a brief report on possible sources of plastics when Earth's supplies of coal and petroleum become too expensive.

56. Prepare a brief biographical report on one of the following and share it with your group.
   a. Charles Goodyear   **b.** Wallace Carothers
   c. Stephanie Kwolek   **d.** Roy Plunkett
   e. Leo Baekeland

57. Make a list of plastic objects (or parts of objects) that you encounter in your daily life. Try to identify a few of the kinds of polymers used in making these items. Compare your list with those of some of your classmates.

58. Do a risk–benefit analysis (Section 1.6) for the use of synthetic polymers as one or more of the following.
   a. grocery bags   **b.** building materials
   c. clothing   **d.** carpets
   e. food packaging   **f.** picnic coolers
   g. automobile tires   **h.** artificial hip sockets

59. To what extent are plastics a litter problem in your community? Survey one city block (or other area as directed by your instructor) and inventory the litter found. (You might as well pick it up while you are at it.) What proportion of the trash is plastics? What proportion of it is fast-food containers? To what extent are plastics recycled in your community? What factors limit further recycling?

60. Prepare a brief report on one of the following polymers and share it with your group. List sources and commercial uses.
   a. polybutadiene   **b.** ethylene–propylene rubber
   c. polyurethanes   **d.** epoxy resins

61. Do some research on Kevlar. Where is it used in addition to bullet-resistant vests? What makes it so strong? What is the Kevlar Survivors' Club, jointly sponsored by the International Chiefs of Police and DuPont?

62. Compare the treatment of environmental questions about plastics and other polymers at a website such as that of the American Plastics Council, an industry group, and at an environmental site such as Greenpeace.

In 1006 C.E. a tremendous stellar explosion—a supernova—lit Earth's sky. The remains of that event, shown here, can be found in the constellation Lupus. The lives and deaths of all stars, including our sun, are a series of nuclear reactions, and many of the elements found on Earth are thought to be the products of stellar nuclear processes.

NUCLEAR CHEMISTRY

QUESTIONS YOU MAY HAVE ASKED YOURSELF

1. Is radiation entirely a man-made problem?
2. Do cell phones and microwave ovens give off radiation?
3. Why can't we just burn or dissolve radioactive wastes to get rid of them?
4. Do irradiated foods contain radiation?
5. Do doctors and dentists expose us to dangerous radiation during X-rays and other medical procedures?
6. Can you get radiation sickness from someone who has been exposed to radiation?

The Heart of Matter

Many people associate the term *nuclear energy* with fearsome images of a mighty force, with giant mushroom clouds from nuclear explosions over devastated cities, with the nuclear power plant accident at Three Mile Island, and with the catastrophic legacy of the meltdown at Chernobyl. But some amazing, life-giving stories can be told about nuclear energy, too. Our sun is one huge nuclear power plant, supplying the energy that warms our planet and the light necessary for plant growth. The twinkling light from every star we see in the night sky is produced by unimaginably powerful nuclear reactions. Here on Earth, scientists have put nuclear energy to use in many useful applications other than weapons of war and power plants for electricity. Medical diagnoses and treatments of diseases like cancer, accurate dating of archaeological artifacts, and even tools of fire safety have been derived from nuclear reactions.

The discussion in Chapter 3 of the structure of an atom focused mainly on the electrons in the atom, as the electrons are the particles that determine its chemistry. In this chapter we take a closer look at that tiny speck in the center of the atom—the atomic nucleus.

If an atom is incomprehensibly small, the infinitesimal size of the atomic nucleus is completely beyond our imagination. The diameter of an atom is roughly 100,000 times greater than the diameter of its nucleus. If an atom could be blown up in size until it was as large as your classroom, the nucleus would be about as big as the period at the end of this sentence.

Yet this tiny nucleus contains almost all the atom's mass. How incredibly dense the atomic nucleus must be! A cubic centimeter of water weighs 1 g and 1 cm³ of gold about 19 g. A cubic centimeter of pure atomic nuclei would weigh more than 100 million metric tons!

Even more amazing than its density is the enormous amount of energy contained within each atomic nucleus. Some atomic nuclei undergo reactions that can fuel the most powerful bombs ever built or provide electricity for millions of people.

Much of what we hear today about nuclear technology is negative: nuclear accidents, the problem of nuclear waste disposal, and the global crisis of nuclear weapons proliferation. However, there is a positive side: The use of radioactive isotopes in medicine saves lives every day, and many applications of nuclear chemistry in science and industry have improved the human condition significantly. In this chapter you will learn about both the destructive and healing power of that infinitesimally small and incredibly dense heart of every atom—the nucleus.

11.1 Natural Radioactivity

Recall from Chapter 3 that most elements occur in nature in several isotopic forms, with nuclei differing in their number of neutrons. Many of these nuclei are unstable and undergo **radioactive decay**. The nuclei that undergo such decay are called **radioisotopes**, and the process produces one or more types of radiation.

Background Radiation

Humans have always been exposed to radiation. Even as you read this sentence you are being bombarded by *cosmic rays,* which originate from the sun and outer space. Other radiation reaches us from natural radioactive isotopes in air, water, soil, and rocks. We cannot escape radiation; it occurs in many natural processes, including those in our bodies. A naturally occurring radioactive isotope of potassium, ^{40}K, exists in all our cells. This ever-present radiation is called **background radiation**. Figure 11.1 shows that over three-fourths of the average radiation exposure comes from background radiation. Most of the remaining one-fourth comes from medical irradiation such as X-rays. Other sources, such as fallout from testing of nuclear bombs, releases from the nuclear industry, and occupational exposure, account for only a minute fraction of our total exposure.

A N S W E R

1. Is radiation entirely a man-made problem?
Although many radioactive isotopes are man-made, Figure 11.1 shows that the majority of the ionizing radiation to which we are exposed is entirely natural.

▶ **Figure 11.1** Most of our exposure to ionizing radiation comes from natural sources (blue shades). About 18% of our exposure comes from human activities (other colors).

QUESTION: What natural source of ionizing radiation contributes most to our exposure? What makes up the largest proportion of our exposure from human activities?

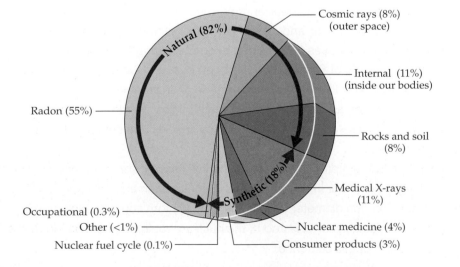

Accidents at the nuclear power plants at Three Mile Island near Harrisburg, Pennsylvania, in 1979 and at Chernobyl in Ukraine in 1986 did much to increase public apprehension about nuclear technology. No one was hurt at Three Mile Island, but at Chernobyl dozens of people were killed outright, and hundreds of others died later of severe radiation poisoning. The long-term impact on thousands more throughout Europe exposed to increased levels of radioactivity is still uncertain.

Harmful effects arise from the interaction of radiation with living tissue. Radiation with enough energy to knock electrons from atoms and molecules, converting them into ions (electrically charged atoms or groups of atoms), is called **ionizing radiation**. Nuclear radiation and X-rays are examples. Because radiation is invisible and because it has such great potential for harm, we are very much concerned with our exposure to it.

Radiation Damage to Cells

Radiation-caused chemical changes in living cells can be highly disruptive. Ionizing radiation can devastate living cells by interfering with their normal chemical processes. Molecules can be splintered into reactive fragments called *free radicals* (Section 4.11), which can disrupt vital cellular processes. White blood cells, the body's first line of defense against bacterial infection, are particularly vulnerable. Radiation also affects bone marrow, causing a drop in the production of red blood cells, which results in anemia. Radiation also has been shown to induce leukemia, a cancerlike disease of the blood-forming organs.

Ionizing radiation also can cause changes in the molecules of heredity (DNA) in reproductive cells. Such changes show up as mutations in the offspring of exposed parents. Little is known of the effects of such exposure on humans. However, many of the mutations that occurred during the evolution of present species may have been caused by background radiation.

Because of the potentially devastating effects on living things, a knowledge of radiation, radioactive decay, and nuclear chemistry in general is crucial. We first describe how to write balanced nuclear equations and how to identify the various types of radiation that are emitted.

In 1993 a Ukrainian Academy of Science study group led by Vladimir Chernousenko investigated the aftereffects of the Chernobyl accident. Chernousenko claims that 15,000 of those who helped clean up the accident site have died and that another 250,000 have been left as invalids. He also says that 200,000 children have experienced radiation-induced illnesses and that half of the children in Ukraine and Belarus have symptoms.

A N S W E R

2. Do cell phones and microwave ovens give off radiation?
Cell phones, microwave ovens, and the like *do* give off radiation, but it is not ionizing radiation and does not cause cell damage like ionizing radiation does.

Cell Phones and Microwaves and Power Lines, Oh My!

The word *radiation* has several meanings. *Ionizing* radiation that comes from decaying atomic nuclei is highly energetic and can damage tissue, as described previously. *Electromagnetic* radiation includes light in its many forms—visible light, radio waves, television broadcasts, microwaves, ultraviolet light, military ULF (ultra low frequency), and others. A few types of electromagnetic radiation—X-rays and gamma rays—have enough energy to ionize tissue, and ultraviolet light has been strongly implicated in melanoma and other skin cancers. But what about microwaves and radio waves, which don't have nearly as much energy? Although very large amounts of microwaves can cause burns (by heating), there is no conclusive evidence that low levels of microwaves pose any threat to human health. Likewise, three separate studies reported in 2000, 2003, and 2005 were unable to demonstrate a connection between cell phone usage and cancer. (A fourth study reported in 2003 did show that cell phone users were twice as likely to be involved in rear-end collisions in automobiles, so clearly there is some hazard associated with

their use.) And according to the American Cancer Society, the largest study on the subject showed no link between cancer and electromagnetic fields from electric blankets and computer monitors. Studies continue, but it appears that the effects of most nonionizing radiation on humans are, at the very most, quite small and difficult to measure accurately.

Self-Assessment Questions

1. What percentage of background radiation comes from natural sources?
   a. 5%   **b.** 18%   **c.** 45%   **d.** 82%

2. The largest artificial source of background radiation is from
   a. consumer products
   b. fallout from nuclear tests
   c. leaks from nuclear power plants
   d. medical X-rays

3. Ionizing radiation
   a. can give one superpowers
   b. comes from synthetic isotopes only
   c. is energetic enough to ionize atoms and molecules
   d. causes radioactive acids to form

4. The largest source of natural background radiation is
   a. cosmic rays
   b. isotopes in the body
   c. radon
   d. isotopes in rocks and soil

Answers: 1, d; 2, d; 3, c; 4, c

11.2 Nuclear Equations

Writing balanced equations for nuclear processes is relatively simple. Nuclear equations differ in two ways from the chemical equations discussed in Chapter 5. First, whereas chemical equations must have the same elements on both sides of the arrow, nuclear equations rarely do. Second, while we balance atoms in ordinary chemical equations, in nuclear equations we balance the *nucleons* (protons and neutrons). What this really means is that we must balance the atomic numbers (number of protons) and nucleon numbers (number of nucleons) of the starting materials and products. For this reason, we must always specify the *isotope* of each element appearing in a nuclear equation. We use nuclear symbols (Section 3.5) in writing nuclear equations because this makes the equations easier to balance.

As an example of a nuclear reaction, radon-222 atoms break down spontaneously in a process called **alpha decay,** giving off alpha (α) particles as shown in Figure 11.2a. Because alpha particles are identical to helium nuclei, this reaction can be summarized by the equation

▶ **Figure 11.2** Nuclear emission of (a) an alpha particle and (b) a beta particle.

QUESTION: What changes occur in a nucleus when it emits an alpha particle? A beta particle?

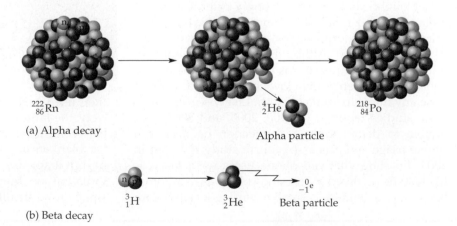

(a) Alpha decay

$^{222}_{86}\text{Rn}$ $^{4}_{2}\text{He}$ $^{218}_{84}\text{Po}$

Alpha particle

(b) Beta decay

$^{3}_{1}\text{H}$ $^{3}_{2}\text{He}$ Beta particle

$^{0}_{-1}\text{e}$

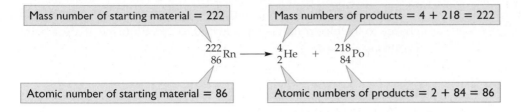

We use the symbol $^4_2$He for the alpha particle (rather than α) because it allows us to check the balance of mass and atomic numbers more readily. The atomic number $Z = 84$ identifies the new element as polonium (Po). In nuclear chemistry, the mass number A equates with the number of nucleons A in the starting material. The number of nucleons in the starting material must equal the total number of the nucleons in the products. The same is true for atomic numbers.

The heaviest isotope of hydrogen, hydrogen-3, often called tritium, decomposes by a process called **beta decay** (Figure 11.2b). Because a beta (β) particle is identical to an electron, this process can be written as

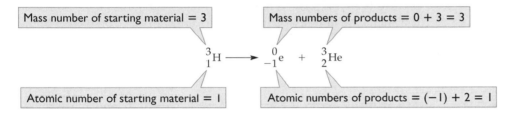

The atomic number $Z = 2$ identifies the product isotope as helium.

Beta decay is a little more complicated than indicated by the preceding reaction. In beta decay, a neutron within the nucleus is converted into a proton (which remains) and an electron (which is ejected).

$$^1_0n \longrightarrow \,^1_1p + \,^0_{-1}e$$

With beta decay, the atomic number increases, but the nucleon number remains the same.

A third kind of radiation may be emitted in a nuclear reaction: gamma (γ) rays. **Gamma decay** is different from alpha and beta decay in that gamma radiation has no charge and no mass. Neither the nucleon number nor atomic number of the emitting atoms is changed; the nuclei simply become less energetic. Table 11.1 compares the properties of alpha, beta, and gamma radiation. Note that the penetrating power of gamma rays is extremely high. While a few millimeters of aluminum are sufficient to stop most β particles, several centimeters of lead are needed to stop γ rays.

Table 11.1 Common Types of Radiation in Nuclear Reactions

Radiation	Mass (u)	Charge	Identity	Velocity*	Penetrating Power
Alpha (α)	4	2+	He^{2+}	0.1c	Very low
Beta (β)	0.00055	1−	e$^-$	<0.9c	Moderate
Gamma (γ)	0	0	High-energy photon	c	Extremely high

*c is the speed of light.

Two other types of radioactive decay are *positron emission* and *electron capture*. These two processes have the same effect on the atomic nucleus but they occur by different pathways. Both result in a decrease of 1 in atomic number with no change in nucleon number (Figure 11.3).

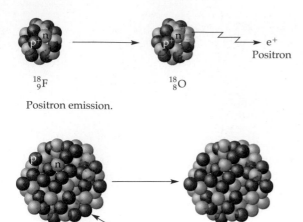

Positron emission.

Electron capture.

▲ **Figure 11.3** Nuclear change accompanying positron emission and electron capture.

QUESTION: What changes occur in a nucleus when it emits a positron? When it undergoes electron capture?

The **positron** (β^+) is a particle equal in mass but opposite in charge to the electron. It is represented as $_{+1}^{0}e$. Fluorine-18 decays by positron emission.

$$_{9}^{18}F \longrightarrow {}_{+1}^{0}e + {}_{8}^{18}O$$

We can envision positron emission as a proton in the nucleus changing into a neutron and a positron.

$$_{1}^{1}p \longrightarrow {}_{0}^{1}n + {}_{+1}^{0}e$$

After the positron is emitted, the original nucleus has one less proton and one more neutron than it had before. The nucleon number of the product nucleus is the same, but its atomic number has been reduced by 1. The emitted positron quickly encounters an electron (there are numerous electrons in all kinds of matter), and both particles are annihilated, with the production of two gamma rays.

$$_{+1}^{0}e + {}_{-1}^{0}e \longrightarrow 2\,{}_{0}^{0}\gamma$$

Electron capture (EC) is a process in which a nucleus absorbs an electron from an inner electron shell, usually the first or second. When an electron from a higher shell drops to the level vacated by the captured electron, an X-ray is released. Once inside the nucleus, the captured electron combines with a proton to form a neutron.

$$_{1}^{1}p + {}_{-1}^{0}e \longrightarrow {}_{0}^{1}n$$

Iodine-125, used in medicine to diagnose pancreatic function and intestinal fat absorption, decays by EC.

$$_{53}^{125}I + {}_{-1}^{0}e \longrightarrow {}_{52}^{125}Te$$

Notice that, unlike alpha, beta, gamma, or positron emission, the electron is a reactant (on the left side) and not a product. Conversion of a proton to a neutron (by the absorbed electron) yields a nucleus lower by 1 in atomic number but unchanged in atomic mass. Emission of a positron and absorption of an electron have the same effect on an atomic nucleus (lowering the atomic number by 1), except that positron emission is accompanied by gamma radiation and electron capture by X-radiation. Positron-emitting isotopes and those that undergo electron capture both have important medical applications (Section 11.6).

The five types of decay discussed here are summarized in Table 11.2.

Table 11.2 Radioactive Decay and Nuclear Change

Type of Decay	Decay Particle	Particle Mass (u)	Particle Charge	Change in Nucleon Number	Change in Atomic Number
Alpha decay	α	4	2+	Decreases by 4	Decreases by 2
Beta decay	β	0	1−	No change	Increases by 1
Gamma radiation	γ	0	0	No change	No change
Positron emission	β^+	0	1+	No change	Decreases by 1
Electron capture (EC)	e^- absorbed	0	1−	No change	Decreases by 1

EXAMPLE 11.1 Balancing Nuclear Equations

Write balanced nuclear equations for each of the following processes. In each case, indicate the new element that is formed.

a. Plutonium-239 emits an alpha particle when it decays.
b. Protactinium-234 undergoes beta decay.
c. Carbon-11 emits a positron when it decays.
d. Carbon-11 undergoes electron capture.

Solution

a. We start by writing the symbol for plutonium-239 and a partial equation showing that one of the products is an alpha particle (helium nucleus).

$$^{239}_{94}\text{Pu} \longrightarrow {}^{4}_{2}\text{He} + ?$$

Mass and nuclear charge are conserved. The new element must have a mass of $239 - 4 = 235$ and a charge of $94 - 2 = 92$. The nuclear charge $Z = 92$ identifies the element as uranium (U).

$$^{239}_{94}\text{Pu} \longrightarrow {}^{4}_{2}\text{He} + {}^{235}_{92}\text{U}$$

b. Write the symbol for protactinium-234 and a partial equation showing that one of the products is a beta particle (electron).

$$^{234}_{91}\text{Pa} \longrightarrow {}^{0}_{-1}\text{e} + ?$$

The new element still has a nucleon number of 234. It must have a nuclear charge $Z = 92$ in order for the total charge to be the same on each side of the equation. The nuclear charge identifies the new atom as another isotope of uranium.

$$^{234}_{91}\text{Pa} \longrightarrow {}^{0}_{-1}\text{e} + {}^{234}_{92}\text{U}$$

c. Write the symbol for carbon-11 and a partial equation showing that one of the products is a positron.

$$^{11}_{6}\text{C} \longrightarrow {}^{0}_{+1}\text{e} + ?$$

To balance the equation, a particle with $A = 11 - 0 = 11$ and $Z = 6 - 1 = 5$ (boron) is required.

$$^{11}_{6}\text{C} \longrightarrow {}^{0}_{+1}\text{e} + {}^{11}_{5}\text{B}$$

d. We write the symbol for carbon-11 and a partial equation showing it capturing an electron.

$$^{11}_{6}\text{C} + {}^{0}_{-1}\text{e} \longrightarrow ?$$

To balance the equation, the product must have $A = 11 + 0 = 11$ and $Z = 6 + (-1) = 5$ (boron).

$$^{11}_{6}\text{C} + {}^{0}_{-1}\text{e} \longrightarrow {}^{11}_{5}\text{B}$$

As we mentioned previously, positron emission and electron capture result in identical changes in atomic number and, therefore, the identical elements are formed. Also, as parts **(c)** and **(d)** illustrate, carbon-11 (and certain other nuclei) can undergo more than one type of radioactive decay.

■ EXERCISE 11.1

Write balanced nuclear equations for each of the following processes. In each case, indicate the new element that is formed.

a. Radium-226 decays by alpha emission.
b. Sodium-24 undergoes beta decay.
c. Gold-188 decays by positron emission.
d. Argon-37 undergoes electron capture.

We mentioned that nuclear equations differ greatly from ordinary chemical equations. Some important differences are summarized in Table 11.3. Some nuclear equations involve processes other than the five simple nuclear processes we have discussed here. Regardless, all nuclear equations must be balanced according to

Table 11.3 Some Differences between Chemical Reactions and Nuclear Reactions

Chemical Reactions	Nuclear Reactions
Atoms retain their identity.	Atoms usually change from one element to another.
Reactions involve only electrons and usually only the outermost electrons.	Reactions involve mainly protons and neutrons. It does not matter what the valence electrons are doing.
Reaction rates can be speeded up by raising the temperature.	Reaction rates are unaffected by changes in temperature.
Energy absorbed or given off in reactions is comparatively small.	Reactions sometimes involve enormous changes in energy.
Mass is conserved. The mass of products equals the mass of starting materials.	Huge changes in energy are accompanied by measurable changes in mass ($E = mc^2$).

nucleon numbers and atomic numbers. When an unknown particle has an atomic number that does not correspond to an atom, that particle may be a subatomic particle. A list of nuclear symbols for subatomic particles is given in Table 11.4.

Table 11.4 Nuclear Symbols for Subatomic Particles

Particles	Symbols	Nuclear Symbols
Proton	p	^1_1p or ^1_1H
Neutron	n	^1_0n
Electron	e^- or β	$^0_{-1}e$ or $^0_{-1}\beta$
Positron	e^+ or β^+	$^0_{+1}e$ or $^0_{+1}\beta$
Alpha particle	α	^4_2He or $^4_2\alpha$
Beta particle	β or β^-	$^0_{-1}e$ or $^0_{-1}\beta$
Gamma ray	γ	$^0_0\gamma$

Self-Assessment Questions

Identify items 1–4 as one of the following:
- **a.** alpha particle
- **b.** beta particle
- **c.** nucleon number
- **d.** radioactivity

1. an electron
2. atomic mass 4 u
3. emission of particles and energy from a nucleus
4. number of protons and neutrons in an atom
5. What is the mass and charge, respectively, of gamma radiation?
   - **a.** 0, −1
   - **b.** 0, 0
   - **c.** 0, +1
   - **d.** 4, +2
6. What particle is needed to complete the following nuclear reaction?

$$^9_4Be + ? \rightarrow {}^{12}_6C + {}^1_0n$$

- **a.** alpha particle
- **b.** beta particle
- **c.** neutron
- **d.** proton

Answers: 1, b; 2, a; 3, d; 4, c; 5, b; 6, a

11.3 Half-Life

Thus far we have discussed radioactivity as applied to single atoms. In the laboratory, we generally deal with great numbers of atoms—numbers far larger than the number of all the people on Earth. If we could see the nucleus of an individual atom, we could tell whether it could undergo radioactive decay by noting its composition. Certain combinations of protons and neutrons are unstable. However, we could not determine *when* the atom would undergo a change. Radioactivity is a random process, generally independent of outside influences.

With large numbers of atoms, the process of radioactive decay becomes more predictable. We can measure the *half-life*, a property characteristic of each radioisotope. The **half-life** of a radioactive isotope is the time it takes for one-half of the original number of atoms to undergo radioactive decay. Suppose, for example, we had 16.00 mg of the radioactive isotope iodine-131. The half-life of iodine-131 is 8.0 days. This means that in 8.0 days, half the iodine-131, or 8.00 mg, will have decayed, and there will be 8.00 mg left. In another 8.0 days, half of the remaining 8.00 mg will have decayed. After two half-lives or 16.0 days, one-quarter of the original iodine-131, or 4.00 mg, will remain. Two half-lives, then, do not make a whole. The concept of half-life is illustrated by the graph in Figure 11.4. Half-lives of different isotopes can differ enormously. The half-life of uranium-238 is 4.5 billion years, whereas that of boron-9 is 8×10^{-19} seconds.

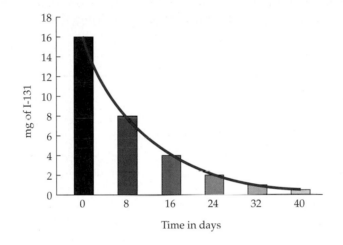

◀ **Figure 11.4** The radioactive decay of iodine-131, which has a half-life of eight days.

QUESTION: What quantity of an original 32-mg sample of iodine-131 remains after five half-lives?

We can calculate the fraction of the original isotope that remains after a given number of half-lives from the relationship

$$\text{Fraction remaining} = \frac{1}{2^n}$$

where *n* is the number of half-lives.

We cannot say when *all* the atoms of a radioactive isotope will have decayed. For most samples, we can assume that the activity is essentially gone after about 10 half-lives. (After 10 half-lives, the activity is $\left[\frac{1}{2}\right]^{10} = \frac{1}{1024}$ of the original value.) Generally, we can say that about $\frac{1}{1000}$ of the original activity remains.

EXAMPLE 11.2 Half-Lives

A 4.00-mg sample of cobalt-60, half-life 5.271 years, is to be used for radiation treatment. How much cobalt-60 remains after 15.813 years (three half-lives)?

Solution

The fraction remaining after three half-lives is

$$\frac{1}{2^n} = \frac{1}{2^3} = \frac{1}{2 \times 2 \times 2} = \frac{1}{8}$$

The amount of cobalt-60 remaining is $\left(\frac{1}{8}\right)(4.00 \text{ mg}) = 0.50$ mg.

■ **EXERCISE 11.2A**

You have 1.224 mg of freshly prepared gold-189, with a half-life of 29 min. How much of the gold-189 sample remains after five half-lives?

■ **EXERCISE 11.2B**

The half-life of phosphorus-32 is 14.3 days. What percentage of an original sample's radioactivity remains after 57.2 days?

A N S W E R

3. Why can't we just burn or dissolve radioactive wastes to get rid of them? The half-life of an isotope is a nuclear property that isn't changed by chemical reactions. For example, if we burn radioactive carbon, we get radioactive carbon dioxide as the product. Dissolve radioactive plutonium in nitric acid and we get radioactive plutonium nitrate. The isotope will be radioactive for a length of time determined entirely by its half-life—and we can't change the half-life.

EXAMPLE 11.3 Half-Lives

You obtain a 2.60-mg sample of mercury-190, half-life 20 min. How much of the mercury-190 sample remains after 2.0 h?

Solution

There are 120 min and thus $\left(\frac{120}{20}\right) = 6$ half-lives in 2.0 h. The fraction remaining after six half-lives is

$$\frac{1}{2^n} = \frac{1}{2^6} = \frac{1}{2 \times 2 \times 2 \times 2 \times 2 \times 2} = \frac{1}{64}$$

The amount of mercury-190 remaining is $\left(\frac{1}{64}\right)(2.60)$ mg $= 0.0406$ mg.

■ **EXERCISE 11.3A**

A sample of 16.0 mg of nickel-57, half-life 36.0 h, is produced in a nuclear reactor. How much of the nickel-57 sample remains after 7.5 days?

■ **EXERCISE 11.3B**

Technetium-99 decays to ruthenium-99 with a half-life of 210,000 y. Starting with 1.00 mg of technetium-99, how long will it take for 0.75 mg of ruthenium-99 to form?

Why Are Isotopes Stable—or Unstable?

Most isotopes are radioactive, but some are stable. What factors tend to make an atomic nucleus stable? Stable isotopes tend to have:

1. *Even* numbers of either protons or neutrons or, especially, of both. Of the 264 stable isotopes, 157 have *even* numbers of both protons and neutrons, and only four have odd numbers of both protons and neutrons. Elements of even Z have more stable isotopes than those of odd Z.

2. So-called "magic numbers" of either protons or neutrons. (Magic numbers are 2, 8, 20, 50, 82, and 126.)

3. An atomic number of 83 or less. All isotopes with Z > 83 are radioactive.

4. Fewer or the same number of protons as neutrons in the nucleus, and the ratio of neutrons to protons should be close to 1 if the atomic number is 20 or below. As atomic number increases, the stable n/p ratio also increases, up to about 1.5. For example, oxygen-16 is stable with an n/p ratio of 1, while the only stable isotope of manganese is Mn-55 (n/p = 1.2). There is a zone of stability within which the n/p ratio should lie for an atom of a given atomic number.

Self-Assessment Questions

1. A patient is given a 48-mg dose of technetium-99m (half-life 6.0 h) to treat a brain cancer. How much of it will remain in his body after 36 h?
   a. 0.75 mg   **b.** 1.5 mg   **c.** 3.0 mg   **d.** 12 mg

2. What is the half-life of silver-112 if a 10.0 g sample decays to 2.5 g in 12.4 h?
   a. 1.6 h   **b.** 3.1 h   **c.** 6.2 h   **d.** 12.4 h

3. Of a 64.0-g sample of strontium-90 produced by a nuclear explosion (half-life 28.5 y), what mass in grams would remain unchanged after 285 years?
   a. 0.000 g   **b.** 0.000625 g   **c.** 0.00625 g   **d.** 0.0625 g

Answers: 1, a; 2, c; 3, d

11.4 Radioisotopic Dating

The half-lives of certain isotopes can be used to estimate the ages of rocks and archaeological artifacts. Uranium-238 decays with a half-life of 4.5 billion years. The initial products of this decay are also radioactive, and breakdown continues until an isotope of lead (lead-206) is formed. By measuring the relative amounts of uranium-238 and lead-206, chemists can estimate the age of a rock. Some of the older rocks on Earth have been found to be 3.0–4.5 billion years old. Moon rocks and meteorites have been dated at a maximum age of about 4.5 billion years. Thus, the age of Earth (and the solar system itself) is generally estimated to be about 4.5 billion years.

Carbon-14 Dating

The dating of artifacts derived from plants or animals usually involves radioactive carbon-14. Of the carbon on Earth, about 99% is carbon-12 and 1% is carbon-13, both of which are stable isotopes. However, in the upper atmosphere carbon-14 is formed by the bombardment of ordinary nitrogen by neutrons from cosmic rays.

$$^{14}_{7}N + ^{1}_{0}n \longrightarrow ^{14}_{6}C + ^{1}_{1}H$$

This process leads to a tiny but steady-state concentration of carbon-14 in Earth's CO_2. Plants use CO_2 and animals consume plants and other animals, so living things constantly incorporate this isotope into their own cells. When they die, however, the incorporation of carbon-14 ceases, and the carbon-14 in the organisms decays—with a half-life of 5730 years—back to nitrogen-14. Thus, we need merely to measure the carbon-14 activity remaining in an artifact of plant or animal origin to determine the time elapsed since its death. For instance, a sample that has half the carbon-14 activity of new plant material is 5730 years old; it has been dead for one half-life. Similarly, an artifact with 25% of the carbon-14 activity of new plant material is 11,460 years old; it has been dead for two half-lives.

We have thus far described **carbon-14 dating** as though formation of the isotope was constant over the years, but this is not quite the case. For the last 7000 years or so, carbon-14 dates do correlate with those obtained from the annual growth rings of trees and with documents of known age. Calibration curves have been constructed from which accurate dates can be determined. Generally, carbon-14 is reasonably accurate for dating objects up to about 50,000 years old. Objects older than 50,000 years have too little of the isotope left for accurate measurement.

Charcoal from the fires of an ancient people, dated by determining the carbon-14 activity, is used to estimate the age of other artifacts found at the same archaeological site. Recently, carbon dating was used to confirm that ancient Hebrew writing found on a stone tablet unearthed in Israel dates from the ninth century B.C.E.

The Shroud of Turin

The Shroud of Turin is a very old piece of linen cloth, about 4 m long, bearing a faint human likeness. Since about 1350 C.E. it had been alleged to be part of the burial shroud of Christ. However, carbon-14 dating studies in 1988 by three different nuclear laboratories indicated that the flax used in making the cloth was not grown until sometime between 1260 and 1390 C.E., and that the cloth therefore could not possibly have existed at the time of Christ. Unlike the Dead Sea Scrolls, which were shown by carbon-14 dating to be authentic records from a civilization that existed about 2000 years ago, the Shroud of Turin has been shown to be less than 800 years old.

Tritium Dating

Tritium, the radioactive isotope of hydrogen, can also be used for dating. Its half-life of 12.26 years makes it useful for dating items up to about 100 years old. An interesting application is the dating of brandies. These alcoholic beverages are quite expensive when aged from 10 to 50 years. Tritium dating can be used to check the veracity of advertising claims about the age of the most expensive kinds.

Many other isotopes are useful for estimating the ages of objects and materials. Several of the more important ones are listed in Table 11.5.

EXAMPLE 11.4   Radioisotopic Dating

An old wooden implement has carbon-14 activity one-eighth that of new wood. How old is the artifact? The half-life of carbon-14 is 5730 years.

Solution

Using the relationship

$$\text{Fraction remaining} = \frac{1}{2^n}$$

we see that one-eighth is $\frac{1}{2^n}$, where $n = 3$; that is, the fraction $\frac{1}{8}$ is $\frac{1}{2^3}$. The carbon-14 has gone through three half-lives. The wood is, therefore, about $3 \times 5730 = 17,190$ years old.

■ **EXERCISE 11.4A**

How old is a piece of a fur garment that has carbon-14 activity $\frac{1}{16}$ that of living tissue? The half-life of carbon-14 is 5730 years.

■ **EXERCISE 11.4B**

Strontium-90 has a half-life of 28.5 years. How long will it take for the strontium-90 now on Earth to be reduced to $\frac{1}{32}$ of its present amount?

Table 11.5   Several Isotopes Useful in Radioactive Dating

Isotope	Half-Life (years)	Useful Range	Dating Applications
Carbon-14	5730	500 to 50,000 years	Charcoal, organic material
Hydrogen-3 (tritium)	12.26	1 to 100 years	Aged wines
Lead-210	22	1 to 75 years	Skeletal remains
Potassium-40	1.25×10^9	10,000 years to the oldest Earth samples	Rocks, the Earth's crust, the moon's crust
Rhenium-187	4.3×10^{10}	4×10^7 years to the oldest samples in the universe	Meteorites
Uranium-238	4.51×10^9	10^7 years to the oldest Earth samples	Rocks, the Earth's crust

---

Self-Assessment Questions

1. A bone had 30.0 μg of carbon-14 (half-life 5730 y) when the animal died and now has 3.75 μg of C-14. How old is the bone?
   a. 11,460 y   **b.** 17,190 y   **c.** 22,920 y   **d.** 34,380 y

2. About how old is a piece of wood from an aboriginal burial site that has a carbon-14 (half-life 5730 y) activity of 3.9 counts/min per gram of carbon? Assume that the activity of carbon-14 was 15.6 counts per minute per gram of carbon at the time of the burial.
   a. 4298 y   **b.** 5730 y   **c.** 11,460 y   **d.** 22,920 y

Answers: 1, b; 2, c

11.5 Artificial Transmutation

During the Middle Ages, alchemists tried to turn base metals, such as lead, into gold. However, they were trying to do it chemically and were, therefore, doomed to failure because chemical reactions involve only atoms' outer electrons. **Transmutation** (changing one element into another) requires altering the *nucleus*.

Thus far we have considered only natural forms of radioactivity. Other nuclear reactions can be brought about by bombardment of stable nuclei with alpha particles, neutrons, or other subatomic particles. These particles, given sufficient energy, penetrate the formerly stable nucleus and result in some form of radioactive emission. Just as in natural radioactive processes, one element is changed into another. Because the change would not have occurred naturally, the process is called *artificial transmutation*.

In 1919, a few years after his famous gold-foil experiment (Chapter 3), Ernest Rutherford reported on the bombardment of a variety of light elements with alpha particles. One such experiment, in which he bombarded nitrogen, resulted in the production of protons, as shown in this balanced nuclear equation.

$$^{14}_{7}N + ^{4}_{2}He \longrightarrow ^{17}_{8}O + ^{1}_{1}H$$

(The hydrogen nucleus is simply a proton; hence, the alternative symbol $^{1}_{1}H$ for the proton.) This provided the first empirical verification of the existence of protons in atomic nuclei, which Rutherford had first postulated in 1914.

Recall that Eugen Goldstein had produced protons in his gas discharge tube experiments in 1886 (Section 3.1). The significance of Rutherford's experiment lay in the fact that he obtained protons from the *nucleus* of an atom other than hydrogen, thus establishing their nature as constituents of nuclei. Rutherford's experiment was the first induced nuclear reaction and the first example of artificial transmutation.

▲ Ernest Rutherford (1871–1937) carried out the first nuclear bombardment experiment. Element 104 (Rf) is named rutherfordium in his honor.

EXAMPLE 11.5 Artificial Transmutation Equations

When potassium-39 is bombarded with neutrons, chlorine-36 is produced. What other particle is emitted?

$$^{39}_{19}K + ^{1}_{0}n \longrightarrow ^{36}_{17}Cl + ?$$

Solution

Write a balanced nuclear equation. To balance the equation, we need four mass units and two charge units (that is, a particle with $A = 4$ and $Z = 2$—an alpha particle).

$$^{39}_{19}K + ^{1}_{0}n \longrightarrow ^{36}_{17}Cl + ^{4}_{2}He$$

■ **EXERCISE 11.5A**

Technetium-97 is produced by bombarding molybdenum-96 with a deuteron (hydrogen-2 nucleus). What other particle is emitted?

$$^{96}_{42}Mo + ^{2}_{1}H \longrightarrow ^{97}_{43}Tc + ?$$

■ **EXERCISE 11.5B**

Some scientists want to form uranium-235 by bombarding another nucleus with an alpha particle. They expect a neutron to be produced along with the uranium-235 nucleus. What nucleus must be bombarded with the alpha particle?

Self-Assessment Questions

1. What new isotope is formed in the following artificial transmutation?
$$^{14}_{7}N + ^{1}_{0}n \longrightarrow ^{1}_{1}H + ?$$
   a. carbon-12   **b.** carbon-14   **c.** nitrogen-15   **d.** oxygen-15
2. What new isotope is formed in the following artificial transmutation?
$$^{9}_{4}Be + ^{1}_{1}H \longrightarrow ^{4}_{2}He + ?$$
   a. boron-9   **b.** boron-10   **c.** lithium-6   **d.** lithium-7
3. What new isotope is formed in the following artificial transmutation?
$$^{239}_{94}Pu + ^{4}_{2}He \longrightarrow ^{1}_{0}n + ?$$
   a. americium-242   **b.** berkelium-242
   c. curium-242   **d.** curium-234

Answers: 1, b; 2, c; 3, c

11.6 Uses of Radioisotopes

Most of the 3000 known radioisotopes are produced by artificial transmutation from stable isotopes. The value of both naturally occurring and artificial radioisotopes goes far beyond their contributions to our knowledge of chemistry.

Radioisotopes in Industry and Agriculture

Scientists in a wide variety of fields use radioisotopes as **tracers** in physical, chemical, and biological systems. Isotopes of a given element, whether radioactive or not, behave nearly identically in chemical and physical processes. Because radioactive isotopes are easily detected through their decay products, it is relatively easy to trace their movement, even through a complicated system. For example, we can use radioisotopes to

- **Detect leaks in underground pipes.** Suppose there is a leak in the pipe, which is buried beneath a concrete floor. We could locate the leak by digging up extensive areas of the floor, or we could add a small amount of radioactive material to liquid poured into the drain and trace the flow of the liquid with a Geiger counter (an instrument that detects radioactivity). Once we locate the leak, only a small area of the floor would have to be dug up to repair the leak. A compound containing a short-lived nuclide (for example, $^{131}_{53}$I, half-life 8.04 days) is usually employed.
- **Determine frictional wear in piston rings.** The ring is subjected to neutron bombardment, which converts some of the carbon in the steel to carbon-14. Wear in the piston ring is assessed by the rate at which the carbon-14 appears in the engine oil.
- **Determine the uptake of phosphorus and its distribution in plants.** This can be done by incorporating phosphorus-32, a β^- emitter with a 14.3-day half-life, into phosphate fertilizers fed to the plants. Radioisotopes are also used to study the effectiveness of weed killers, compare the nutritional value of various feeds, determine optimal insect control methods, and monitor the fate and persistence of pesticides in soil and groundwater.

One of the most successful agricultural techniques has involved inducing heritable genetic alterations known as mutations. Exposing seeds or other parts of plants to neutrons or gamma rays increases the likelihood of genetic mutations. At first glance, this procedure may not seem the most promising of endeavors. However, genetic variability is vital, not only in an effort to improve varieties but also to protect species from extinction. Emile Frison, a Belgian plant pathologist, announced in early 2003 that bananas might go extinct within 10 years due to their lack of disease resistance. The lack of genetic variability places the entire population at risk.

Radioisotopes are also used as sources to irradiate foodstuffs as a method of preservation (Figure 11.5). The radiation destroys microorganisms that cause food spoilage. Irradiated food shows little change in taste or appearance. Some people are concerned about possible harmful effects of chemical substances produced by the radiation, but there is no good evidence of harm to laboratory animals fed irradiated food, nor are there any known adverse effects in humans in countries where irradiation has been used for years. There is no residual radiation in the food after the sterilization process because gamma rays do not have nearly enough energy to change nuclei.

▲ Scientists can trace the uptake of phosphorus by a green plant by adding a compound containing some phosphorus-32 to the applied fertilizer. When the plant is later placed on a photographic film, radiation from the phosphorus isotopes exposes the film, much as light does. This type of exposure, called a *radiograph*, shows the distribution of phosphorus in the plant.

4. Do irradiated foods contain radiation?
Irradiated foods have absorbed energy in the form of gamma rays. This may change their appearance or texture slightly, but the foods are *not* radioactive.

▶ **Figure 11.5** Gamma radiation delays the decay of mushrooms. Those on the left were irradiated; the ones on the right were not.

QUESTION: Are the mushrooms on the left now radioactive?

Radioisotopes in Medicine

Nuclear medicine involves two distinct uses of radioisotopes: therapeutic and diagnostic. In radiation therapy, an attempt is made to treat or cure disease with radiation. The diagnostic use of radioisotopes is aimed at obtaining information about the state of a patient's health.

Cancer is not one disease but many. Some forms are particularly susceptible to radiation therapy. The aim of **radiation therapy** is to destroy cancerous cells before too much damage is done to healthy tissue. Radiation is most lethal to rapidly reproducing cells, and this is precisely the characteristic of cancer cells that allows radiation therapy to be successful. Radiation is carefully aimed at cancerous tissue while minimizing the exposure of normal cells. If the cancer cells are killed by the destructive effects of the radiation, the malignancy is halted.

Patients undergoing radiation therapy often get sick from the high doses of radiation needed to kill cancer cells. Nausea and vomiting are the usual early symptoms of radiation sickness. Radiation therapy can also interfere with white blood cell replenishment and increase susceptibility to infection.

Diagnostic Uses of Radioisotopes

Radioisotopes are used for diagnostic purposes to provide information about the type or extent of an illness. Table 11.6 lists some radioisotopes in common use in medicine. The list is necessarily incomplete, but even this abbreviated discussion should give you an idea of their importance. The claim that nuclear science has saved many more lives than nuclear bombs have taken is not an idle one.

Radioactive iodine-131 is used to determine the size, shape, and activity of the thyroid gland as well as to treat cancer located in this gland and to control a hyperactive thyroid. Small doses are used for diagnostic purposes and large doses for treatment of thyroid cancer. After the patient drinks a solution of potassium iodide incorporating iodine-131, the body concentrates iodide in the thyroid. A detector

Table 11.6 Some Radioisotopes and Their Medical Applications

Isotope	Name	Half-Life*	Use
^{11}C	Carbon-11	20.39 m	Brain scans
^{51}Cr	Chromium-51	27.8 d	Blood volume determination
^{57}Co	Cobalt-57	270 d	Measuring vitamin B_{12} uptake
^{60}Co	Cobalt-60	5.271 y	Radiation cancer therapy
^{153}Gd	Gadolinium-153	242 d	Determining bone density
^{67}Ga	Gallium-67	78.1 h	Scan for lung tumors
^{131}I	Iodine-131	8.040 d	Thyroid therapy
^{192}Ir	Iridium-192	74 d	Breast cancer therapy
^{59}Fe	Iron-59	44.496 d	Detection of anemia
^{32}P	Phosphorus-32	14.3 d	Detection of skin cancer or eye tumors
^{238}Pu	Plutonium-238	86 y	Provides power for pacemakers
^{226}Ra	Radium-226	1600 y	Radiation therapy for cancer
^{75}Se	Selenium-75	120 d	Pancreas scans
^{24}Na	Sodium-24	14.659 h	Locating obstructions in blood flow
^{99m}Tc	Technetium-99m	6.0 h	Imaging of brain, liver, bone marrow, kidney, lung, or other organs
^{201}Tl	Thallium-201	73 h	Detecting heart problems with treadmill stress test
^3H	Tritium	12.26 y	Determining total body water
^{133}Xe	Xenon-133	5.27 d	Lung imaging

* Abbreviations: y, years; d, days; h, hours; m, minutes.

ANSWER

5. Do doctors and dentists expose us to dangerous radiation during X-rays and other medical procedures? A number of medical procedures do use ionizing radiation or radioisotopes, but the amount of radiation is kept to a minimum. The risk from the ionizing radiation is ordinarily much less than the risk from a lack of diagnosis or treatment!

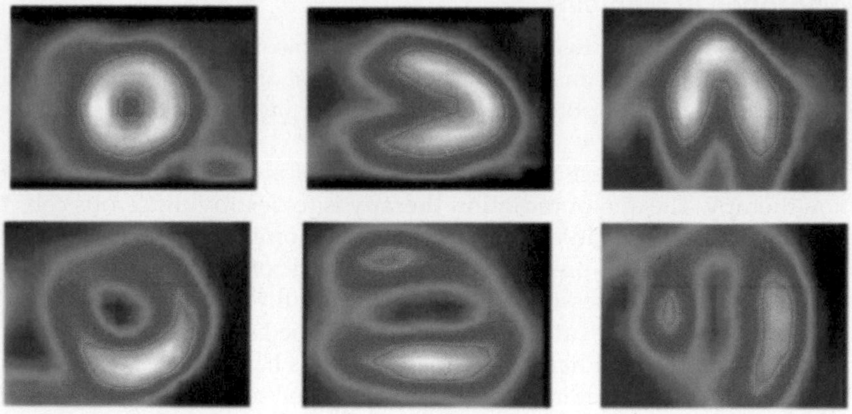

► **Figure 11.6** Gamma ray imaging using technetium-99m. Shown at the top are directional slices through a healthy human heart, and at the bottom, through a diseased heart. The lighter areas of the images indicate regions receiving adequate blood flow.

showing the differential uptake of the isotope is used in diagnosis. The resulting *photoscan* can pinpoint the location of tumors or other abnormalities in the thyroid. In cancer treatment, radiation from therapeutic (large) doses of iodine-131 kills the thyroid cells in which the radioisotope has concentrated. This occurs even in thyroid-type cells that have spread to other parts of the body.

A radioisotope widely used in medicine is gadolinium-153. This isotope is used to determine bone mineralization. Its widespread use is an indication of the large number of people, mostly women, who suffer from osteoporosis (reduction in the quantity of bone) as they grow older. Gadolinium-153 gives off two characteristic radiations, a gamma ray and an X-ray. A scanning device compares these radiations after they pass through bone. Bone densities are then determined by differences in absorption of the rays.

Technetium-99m is used in a variety of diagnostic tests (Figure 11.6). The *m* stands for "metastable," which means that this isotope gives up some energy to become a more stable version of the same isotope (same atomic number, same atomic mass). The energy it gives up is the gamma ray needed to detect the isotope.

$$^{99m}_{43}\text{Tc} \longrightarrow ^{99}_{43}\text{Tc} + \gamma$$

Notice that the decay of technetium-99m produces no alpha or beta particles which could cause unnecessary damage to the body. Technetium-99m also has a short half-life (6.0 h), which means that the radioactivity does not linger in the body long after the scan has been completed. With such a short half-life, use of the isotope must be carefully planned. In fact, the isotope itself is not what is purchased. Technetium-99m is formed by the decay of molybdenum-99.

$$^{99}_{42}\text{Mo} \longrightarrow ^{99m}_{43}\text{Tc} + ^{0}_{-1}\text{e} + \gamma$$

A container of this molybdenum isotope is obtained, and the decay product, technetium-99m, is "milked" from the container as needed.

Using modern computer technology, positron emission tomography (PET) can measure dynamic processes occurring in the body, such as blood flow or the rate at which oxygen or glucose is being metabolized. For example, PET scans have shown that the brain of a schizophrenic metabolizes only about one-fifth as much glucose as a normal brain. These scans can also reveal metabolic changes that occur in the brain during tactile learning (learning by the sense of touch). PET scans can also pinpoint the area of brain damage that triggers severe epileptic seizures. Compounds incorporating positron-emitting isotopes, such as carbon-11 or oxygen-15, are inhaled or injected prior to the scan. Before the emitted positron can travel very far in the body, it encounters an electron (numerous in any ordinary matter), and two gamma rays are produced, exiting from the body in exactly opposite directions.

$$^{11}_{6}\text{C} \longrightarrow ^{11}_{5}\text{B} + ^{0}_{+1}\text{e}$$

$$^{0}_{+1}\text{e} + ^{0}_{-1}\text{e} \longrightarrow 2\gamma$$

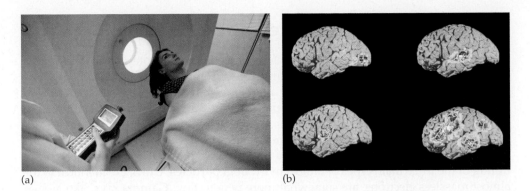

◀ **Figure 11.7** Modern computer technology used for medical diagnosis. (a) Patient in position for positron emission tomography (PET), a technique that uses radioisotopes to scan internal organs. (b) Images created by PET scanning, showing different parts of the brain involved in different functions.

Detectors, positioned on opposite sides of the patient, record the gamma rays. An image of an area in the body is formed using computerized calculations of the points at which annihilation of the positrons and electrons occurs (Figure 11.7).

EXAMPLE 11.6   Positron Emission Equations

One of the isotopes used for PET scans is oxygen-15, a positron emitter. What new element is formed when oxygen-15 decays?

Solution

First write the nuclear equation

$$^{15}_{8}O \longrightarrow \, ^{0}_{+1}e \, + \, ?$$

The nucleon number A does not change, but the atomic number Z becomes $8 - 1 = 7$; so the new product is nitrogen-15.

$$^{15}_{8}O \longrightarrow \, ^{0}_{+1}e \, + \, ^{15}_{7}N$$

■ **EXERCISE 11.6**

Phosphorus-30 is a positron-emitting radioisotope suitable for use in PET scans. What new element is formed when phosphorus-30 decays?

A N S W E R

6. Can you get radiation sickness from someone who has been exposed to radiation?
Alpha, beta, gamma rays, X-rays, and positrons absorbed by the body do cause cell damage, but by and large they are *not* energetic enough to create radioactive isotopes. One can no more "catch radiation sickness" from a person exposed to radiation than one can "catch" a sunburn.

Self-Assessment Questions

1. Iodine-131 is sometimes used for detecting leaks in underground pipes because
   a. it has a short half-life   **b.** it is highly penetrating
   c. it is nontoxic   **d.** its radiation is harmless

2. What type of radiation is used to irradiate foodstuffs to extend shelf life?
   a. alpha   **b.** beta   **c.** gamma   **d.** microwave

For items 3–7, match the radioisotope with a major use.

3. cobalt-60   **(a)** bone density

4. gadolinium-153   **(b)** brain images

5. iodine-131   **(c)** heart problems

6. technetium-99m   **(d)** radiation therapy

7. thallium-201   **(e)** thyroid disorders

8. The isotope used in a PET scan must emit what type of radiation?
   a. beta   **b.** gamma   **c.** positron   **d.** X-rays

Answers: 1, a; 2, c; 3, d; 4, a; 5, e; 6, b; 7, c; 8, c

11.7 Penetrating Power of Radiation

The danger of radiation to living organisms comes from its potential for damaging cells and tissue. The ability to inflict injury comes from the penetrating power of the radiation. The two aspects of nuclear medicine just discussed (therapeutic and diagnostic) also are dependent on the penetrating powers of the various types of radiation.

All other things being equal, the more massive the particle, the less its penetrating power. **Alpha particles**, which are helium nuclei with a mass of 4 u, are the least penetrating of the three main types of radioactivity. **Beta particles**, which are identical to the almost massless electrons, are somewhat more penetrating. **Gamma rays**, like X-rays, truly have no mass; they are considerably more penetrating than the other two types.

But all other things are not always equal. The faster a particle moves or the more energetic the radiation, the more penetrating power it has.

It may seem contrary to common sense that the biggest particles make the least headway. Consider that penetrating power reflects the ability of radiation to make its way through a sample of matter. It is as if you were trying to roll some rocks through a field of boulders. The alpha particle acts as if it were a boulder itself. Because of its size, it cannot get very far before it bumps into and is stopped by other boulders. The beta particle acts as if it were a small stone. It can sneak between and perhaps ricochet off boulders until it makes its way farther into the field. The gamma ray can be compared with a grain of sand that can get through the smallest openings.

The danger of a specific type of radiation to human tissue depends on the location of the source as well as on the penetrating power. If a radioactive substance is outside the body, alpha particles of low penetrating power are the least dangerous; they are stopped by the outer layer of skin. Beta particles also usually are stopped before they reach vital organs. Gamma rays readily pass through tissues, and so an external gamma source can be quite dangerous. People working with radioactive materials can protect themselves with one or both of the following actions:

- Move away from the source; intensity of radiation decreases with distance from the source.
- Use shielding. A sheet of paper can stop most alpha particles, a block of wood or a thin sheet of aluminum can stop most beta particles, but it takes a meter of concrete or several centimeters of lead to stop most gamma rays (Figure 11.8).

When the radioactive source is *inside* the body, as in the case of medicinal applications, the situation is reversed. The nonpenetrating alpha particles can do great damage. All such particles are trapped within the body, which must then absorb all the energy released by the particle. Alpha particles inflict all their damage in a tiny

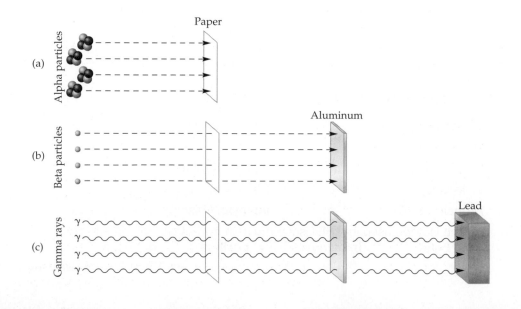

▶ **Figure 11.8** The relative penetrating powers of alpha, beta, and gamma radiation. Alpha particles are stopped by a sheet of paper (a). Beta particles will not penetrate a sheet of aluminum (b). It takes several centimeters of lead to block gamma rays (c).

area because they do not travel far. Therefore, getting the therapeutic radioisotope close to the targeted cells is vital. Beta particles distribute their damage over a somewhat larger area because they travel farther. Tissue may recover from limited damage spread over a large area; it is less likely to survive concentrated damage. Many diagnostic applications rely on the highly penetrating power of gamma rays. In most cases, the radiation created inside the body must be detected by instruments outside the body, so a minimum of absorption is desirable.

CONCEPTUAL EXAMPLE 11.7   Radiation Hazard

Most modern fire detectors contain a tiny amount of americium-241, a solid element that is an alpha emitter. The americium is in a chamber that includes thin aluminum shielding, and the device is usually mounted on a ceiling. Rate the radiation hazard of this device in normal use as either (a) high, (b) moderate, or (c) very low, and explain your choice.

Analysis and Conclusions

Alpha particles have very little penetrating ability—they are stopped by a sheet of paper or a layer of skin. The emitted alpha particles cannot exit the chamber, because even thin metal is more than sufficient to absorb the alpha particles. Even if alpha particles could escape, they would likely be stopped by the dead cells of your hair or by the air after a short distance. Therefore, the radiation hazard is probably best rated as (c)—very low.

■ **EXERCISE 11.7**

Radon-222 is a gas that can diffuse from the ground into homes. Like americium-241, it is an alpha emitter. Is the radiation hazard from radon-222 likely to be higher or lower than that from a smoke detector? Explain.

Self-Assessment Questions

1. Which of the following lists the *most* and *least* penetrating types of radiation? (most/least)
   - **a.** alpha/gamma
   - **b.** beta/gamma
   - **c.** gamma/alpha
   - **d.** X-rays/beta

2. Which form of radioactivity could readily penetrate a centimeter of lead?
   - **a.** alpha
   - **b.** beta
   - **c.** gamma
   - **d.** all of these

3. A Geiger counter registered 1850 counts/min of an unknown type of radiation. A piece of paper inserted between the source and the counter caused the reading to drop to about 45 counts/min. The radiation was mostly
   - **a.** alpha particles
   - **b.** beta particles
   - **c.** gamma rays
   - **d.** X-rays

Answers: 1, c; 2, c; 3, a

11.8 Energy from the Nucleus

We have seen that radioactivity—quiet and invisible—can be beneficial or dangerous. A much more dramatic—and equally paradoxical—aspect of nuclear chemistry is the release of nuclear energy by either fission (splitting of heavy nuclei into smaller nuclei) or fusion (combining of light nuclei to form heavier ones).

Einstein and the Equivalence of Mass and Energy

The potential power in the nucleus was worked out by Albert Einstein, a famous and most unusual scientist. Whereas most scientists work with glassware and

▲ Albert Einstein (1879–1955). Element 99 (Es) is named einsteinium in his honor.

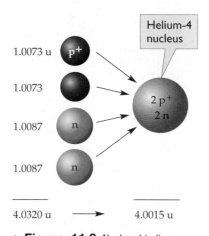

▲ **Figure 11.9** Nuclear binding energy in $^4_2$He. The mass of a helium-4 nucleus is 4.0015 u, which is 0.0305 u less than the masses of two protons and two neutrons. The missing mass is equivalent to the binding energy of the helium-4 nucleus.

instruments in laboratories, Einstein worked with a pencil and a note pad. By 1905, at the age of 26, he had already worked out his special theory of relativity and had developed his famous **mass–energy equation** in which mass (m) is multiplied by the square of the speed of light (c), which is 3.00×10^8 m/s.

$$E = mc^2$$

The equation suggests that mass and energy are just two different aspects of the same thing, and a little bit of mass can yield enormous energy. The atomic bombs that destroyed Hiroshima and Nagasaki (Section 11.9) converted less than an ounce of matter into energy.

A chemical reaction that gives off heat must lose mass in the process, but the change in mass is far too small to measure. Reaction energy must be enormous—such as the energy given off by nuclear explosions—in order for the mass loss to be measurable. (If every atom in a 1-kg lump of coal became energy, it could produce 25 billion kWh of electricity—enough to keep a 100-W lightbulb going for 29 million years. Burning 1 kg of coal in a conventional power plant produces only enough energy to keep the bulb shining for 67 hours.)

Binding Energy

As we see in the atomic bomb and in nuclear power plants, nuclear fission involves a tremendous release of energy. Where does all this energy come from? It is locked inside the atomic nucleus. When protons and neutrons combine to form atomic nuclei, a small amount of mass is converted to energy. This is the **binding energy** that holds the nucleons together in the nucleus. For example, the helium nucleus contains two protons and two neutrons. The mass of these four particles is 2×1.0073 u + 2×1.0087 u = 4.0320 u (Figure 11.9). However, the actual mass of the helium nucleus is only 4.0015 u, and the missing mass—called the *mass defect*—amounts to 0.0305 u. Using Einstein's equation $E = mc^2$, we can calculate (see Problem 55) a value of 28.3 million electron volts (MeV) for the binding energy of the helium nucleus (1 MeV = 1.6022×10^{-13} J). This is the amount of energy it would take to separate one helium nucleus into two protons and two neutrons.

When binding energy per nucleon is calculated for all the elements and plotted against nucleon number, a graph such as that in Figure 11.10 is obtained. The elements with the highest binding energies per nucleon have the most stable nuclei. They include iron and nearby elements. When uranium atoms undergo nuclear

▶ **Figure 11.10** Nuclear stability is greatest near iron in the periodic table. Fission of very large atoms or fusion of very small ones results in greater nuclear stability.

QUESTION: Which process, fission of uranium nuclei or fusion of hydrogen nuclei, releases more energy?

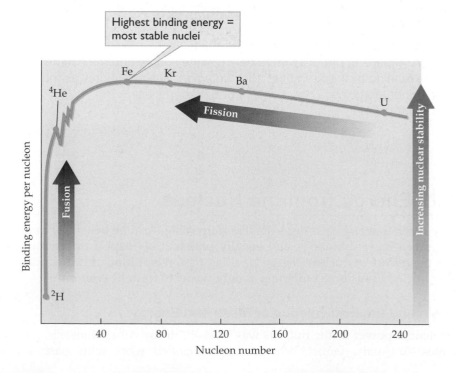

fission, they split into fragments with higher binding energies; in other words, the fission reaction converts large atoms into smaller ones with greater nuclear stability.

We can also see from Figure 11.10 that even more energy can be obtained by combining small atoms, such as hydrogen or deuterium, to form larger atoms with more stable nuclei. This kind of reaction is called **nuclear fusion**. It is what happens when a hydrogen bomb explodes, and it is also the source of the sun's energy.

Nuclear Fission

In 1934 the Italian scientists Enrico Fermi and Emilio Segrè (1905–1989) bombarded uranium atoms with neutrons. They were trying to make elements higher in atomic number than uranium, which then had the highest known number. To their surprise, they found four radioactive species among the products. One presumably was element 93, formed by the initial conversion of uranium-238 to uranium-239,

$$\ce{^{238}_{92}U + ^{1}_{0}n \longrightarrow ^{239}_{92}U}$$

which then underwent beta decay.

$$\ce{^{239}_{92}U \longrightarrow ^{0}_{-1}e + ^{239}_{93}Np}$$

They were unable to explain the remaining radioactivity.

In repeating the Fermi–Segrè experiment in 1938, German chemists Otto Hahn (1879–1968) and Fritz Strassman (1902–1980) were perplexed to find isotopes of barium among the many reaction products. Hahn wrote to Lise Meitner, his former longtime colleague, to ask what she thought about these strange results.

Lise Meitner was an Austrian physicist who had worked with Hahn in Berlin. Because she was Jewish, she had recently fled to Sweden when the Nazis took over Austria in 1938. On hearing about Hahn's work, she noted that barium atoms were only about half the size of uranium atoms. Was it possible that the uranium nucleus might be splitting into fragments? She made some calculations that convinced her that the uranium nuclei had indeed been split apart. Her nephew, Otto Frisch (1904–1979), was visiting for the winter holidays, and they discussed this new discovery with great excitement. It was Frisch who later coined the term **nuclear fission** (Figure 11.11).

Frisch was working with Niels Bohr at the University of Copenhagen, and when he returned to Denmark he took the news about the fission reaction to Bohr, who happened to be going to the United States to attend a physics conference. The discussions in the corridors about this new reaction would be the most important talks to take place at that meeting.

Meanwhile, Enrico Fermi had just received the 1938 Nobel Prize in physics. Because Fermi's wife Laura was Jewish, and the fascist Italian dictator Mussolini was an ally of Hitler, Fermi accepted the award in Stockholm and then immediately fled with his wife and children to the United States. Thus, by 1939 the United States had received news about the German discovery of nuclear fission and had also acquired from Italy one of the world's foremost nuclear scientists.

▲ Enrico Fermi (1901–1954). Element 100 (Fm) is named fermium in his honor.

▲ Lise Meitner (1878–1968). Element 109 (Mt) is named meitnerium in her honor.

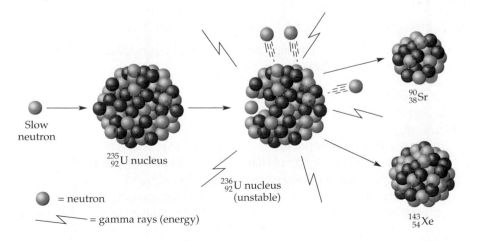

Slow neutron

$\ce{^{235}_{92}U}$ nucleus

$\ce{^{236}_{92}U}$ nucleus (unstable)

$\ce{^{90}_{38}Sr}$

$\ce{^{143}_{54}Xe}$

⬤ = neutron

⚡ = gamma rays (energy)

◀ **Figure 11.11** One possible way a uranium atom can undergo fission. The neutrons produced in the fission can split other uranium atoms, thus sustaining a chain reaction.

QUESTION: If each of the neutrons emitted by the $\ce{^{236}U}$ nucleus caused another $\ce{^{235}U}$ nucleus to split in the manner depicted, how many new neutrons would be released?

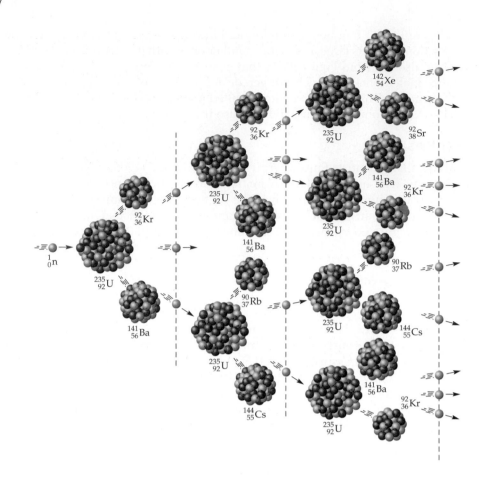

▶ **Figure 11.12** Schematic representation of a nuclear chain reaction. Neutrons released in the fission of one uranium-235 nucleus can strike other nuclei, causing them to split and release more neutrons, as well as a variety of new nuclei.

Nuclear Chain Reaction

Leo Szilard (1898–1964) was one of the first scientists to realize that nuclear fission could be a practical chain reaction. Szilard had been born in Hungary and educated in Germany, but he came to the United States in 1937 as another Jewish refugee. He saw that neutrons released in the fission of one atom could trigger the fission of other uranium atoms, thus setting off a **chain reaction** (Figure 11.12). Because massive amounts of energy could be obtained from the fission of uranium, he saw that the fission process might produce a bomb with tremendous explosive force.

Aware of the destructive forces that could be produced and concerned that Germany might develop such a bomb, Szilard prevailed on Einstein to sign a letter to President Franklin D. Roosevelt indicating the importance of the discovery. It was critical that the U.S. government act quickly.

Self-Assessment Questions

Look at the graph in Figure 11.10 to answer questions 1–3.

1. Helium has a particularly high binding energy per nucleon, which means it
   - **a.** is especially unstable compared to H and Li
   - **b.** more stable than H and Li
   - **c.** readily fused into larger nuclei
   - **d.** readily split into hydrogen-2
2. Iron nuclei are
   - **a.** less stable than Kr nuclei
   - **b.** the most stable nuclei
   - **c.** readily fused into larger nuclei
   - **d.** readily split into smaller nuclei
3. Nuclei with nucleon numbers above that of iron are
   - **a.** all radioactive
   - **b.** increasingly stable
   - **c.** less stable than iron
   - **d.** readily split into smaller nuclei

4. Which equation explains why the products of fission have less mass than the original substances?

a. $E = h\nu\lambda$ **b.** $E = h\nu$ **c.** $E = mc^2$ **d.** $PV = nRT$

5. Which subatomic particles are responsible for carrying on the chain reactions characteristic of nuclear fission?

a. beta particles   **b.** neutrons   **c.** positrons   d. protons

6. When uranium-235 absorbs a neutron and undergoes fission, barium-142 and krypton-91 are two possible fission fragments. How many neutrons are released?

$$^{235}_{92}\text{U} + ^{1}_{0}\text{n} \longrightarrow ^{142}_{56}\text{Ba} + ^{91}_{36}\text{Kr} + ?^{1}_{0}\text{n}$$

a. 0   **b.** 1   **c.** 2   **d.** 3

Answers: 1, b; 2, b; 3, c; 4, c; 5, b; 6, d

11.9 **The Building of the Bomb**

President Roosevelt launched a highly secret research project for the study of atomic energy. Called the *Manhattan Project*, it eventually became a massive research effort involving more scientific brainpower than has ever been devoted to a single project. Amazingly, it was conducted under such extreme secrecy that even Vice President Harry Truman did not know about its existence until after Roosevelt's death.

Begun in 1939, the Manhattan Project included four separate research teams trying to learn how to

- sustain the nuclear fission chain reaction
- enrich uranium to about 90% with the fissile isotope ^{235}U
- make plutonium-239 (another fissile isotope)
- construct a bomb based on nuclear fission

By that time it had been established that neutron bombardment could initiate the fission reaction, but many attributes of fission were unknown. Enrico Fermi and his group, working in a lab under the bleachers at Stagg Field on the campus of the University of Chicago, worked on the fission reaction itself and how to sustain it.

They found that the neutrons used to trigger the reaction had to be slowed down in order to increase the probability that they would hit a uranium nucleus. Because graphite slows down neutrons, a large "pile" of graphite was built to house the reaction. Then the amount of uranium "fuel" was gradually increased. The major issue was to determine the **critical mass**[1] or the amount of uranium-235 needed to sustain the fission reaction. There had to be enough fissile nuclei for the neutrons released in one fission process to have a good chance of being captured by another fissile nucleus before escaping from the pile.

On December 2, 1942, Fermi and his group achieved the first sustained nuclear fission reaction. The critical mass for uranium (enriched to about 94% ^{235}U) for this reactor turned out to be about 16 kg.

Isotopic Enrichment

Natural uranium is 99.27% uranium-238, which does not undergo fission. Uranium-235, the fissile isotope, makes up only 0.72% of natural uranium. Because making a bomb required that the uranium be enriched to about 90% uranium-235,

---

[1]The critical mass required for a nuclear explosion is not a fixed quantity. It depends on the concentration of the fissile material, its configuration, and the nature of the material surrounding the fissile isotope.

it was necessary to find a way to separate the uranium isotopes. Large-scale isotope separation became the job of a top-secret research team in Oak Ridge, Tennessee.

Chemical separation was almost impossible; ^{235}U and ^{238}U are chemically nearly identical. The separation method eventually used involved the conversion of uranium to gaseous uranium hexafluoride, UF_6. Molecules containing uranium-235 are slightly lighter and therefore move slightly faster than molecules containing the uranium-238 isotope. Uranium hexafluoride was allowed to pass through a series of thousands of pinholes, and the molecules containing uranium-235 gradually outdistanced the others. Enough enriched uranium-235 was finally obtained to make a small explosive device.

The Synthesis of Plutonium

While the tedious work of separating uranium isotopes was under way at Oak Ridge, other workers, led by Glenn T. Seaborg, approached the problem of obtaining fissionable material by another route. Although uranium-238 would not fission when bombarded by neutrons, it was found that this more common isotope of uranium *would* form a new element, named neptunium (Np). This product quickly decayed to another new element, plutonium (Pu).

$$^{238}_{92}U + {}^{1}_{0}n \longrightarrow {}^{239}_{92}U$$

$$^{239}_{92}U \longrightarrow {}^{0}_{-1}e + {}^{239}_{93}Np$$

$$^{239}_{93}Np \longrightarrow {}^{239}_{94}Pu + {}^{0}_{-1}e$$

Plutonium-239 was found to be fissile and thus was suitable material for the making of a bomb. A series of large reactors was built near Hanford, Washington, to produce plutonium.

Bomb Construction

The actual building of the nuclear bombs was carried out at Los Alamos, New Mexico, under the direction of J. Robert Oppenheimer (1904–1967). In a top-secret laboratory at a remote site, a group of scientists planned and then constructed what would become known as atomic bombs. Two different models were pursued; one based on ^{235}U, the other on ^{239}Pu.

The critical mass of uranium-235 must not be exceeded prematurely, so it was important that no single piece of fissionable material in the bomb be that large. The bomb was designed to contain a number of rings of subcritical mass, plus a neutron source to initiate the fission reaction. Then, at the chosen time, all the pieces would be forced together using an ordinary high explosive, thus triggering a runaway nuclear chain reaction.

The synthesis of plutonium turned out to be easier than the isotopic separation, and by July 1945, enough fissile material had been made for three bombs to be assembled: two using plutonium and one using uranium. The first atomic bomb (one of the plutonium devices) was tested in the desert near Alamogordo, New Mexico, on July 16, 1945. The heat from the explosion vaporized the 30-m steel tower on which the bomb was placed and melted the sand for several hectares around the site. The light produced was the brightest anyone had ever seen.

Some of the scientists were so awed by the force of the blast that they argued against its use against Japan. A few, led by Leo Szilard, suggested a demonstration of its power at an uninhabited site. But fear of a well-publicized "dud" and the desire to avoid millions of casualties in an invasion of Japan led President Harry S. Truman to order the dropping of bombs on Japanese cities. The lone uranium bomb called "Little Boy" (Figure 11.13) was dropped on Hiroshima on August 6, 1945, and caused over 100,000 casualties. Three days later, a plutonium bomb called "Fat Man" was dropped on Nagasaki with comparable results (Figure 11.14). World War II ended with the surrender of Japan on August 14, 1945.

▲ Glenn T. Seaborg (1912–1999). Element 106 (Sg) is named seaborgium in his honor. Seaborg is the only person to have an element named for him while still alive. He is shown here pointing to his namesake element on the periodic table of elements.

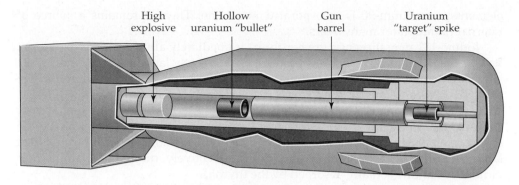

High explosive | Hollow uranium "bullet" | Gun barrel | Uranium "target" spike

▲ **Figure 11.13** An internal schematic of the "Little Boy" atomic bomb dropped on Hiroshima. Each of the rings of uranium-235 was less than a critical mass. The high explosive shot the rings down a "gun barrel" around another set of uranium-235 rings. The result was about two critical masses, and an atomic explosion occurred about a millisecond later.

Self-Assessment Questions

1. The isotope used in uranium fission reactions is
   a. all uranium isotopes   **b.** uranium-232
   c. uranium-235   **d.** uranium-238

2. What fraction of natural uranium is the fissile isotope?
   a. <1%   **b.** 10%
   c. 50%   **d.** >99%

3. Plutonium for fission reactions is
   a. made from uranium-238
   b. made from neptunium ores
   c. obtained from the minor planet Pluto
   d. obtained from plutonium ores

Answers: 1, c; 2, a; 3, a

▲ **Figure 11.14** The mushroom cloud over Nagasaki, following the detonation of "Fat Man," August 9, 1945.

11.10 Radioactive Fallout

When a nuclear explosion occurs in the open atmosphere, radioactive materials can rain down on parts of Earth thousands of miles away, days and weeks later, in what is called **radioactive fallout**. The uranium atom can split in many different ways as shown:

$$^{235}_{92}\text{U} + ^{1}_{0}\text{n} \longrightarrow ^{90}_{38}\text{Sr} + ^{143}_{54}\text{Xe} + 3\,^{1}_{0}\text{n}$$

$$\longrightarrow ^{102}_{39}\text{Y} + ^{131}_{53}\text{I} + 3\,^{1}_{0}\text{n}$$

$$\longrightarrow ^{95}_{37}\text{Rb} + ^{137}_{55}\text{Cs} + 4\,^{1}_{0}\text{n}$$

The primary fission products are radioactive. They decay to daughter isotopes, many of which are also radioactive. In all, over 200 different fission products are produced, with half-lives varying from less than a second to more than a billion years. In addition, the neutrons produced in the explosion act on molecules in the atmosphere to produce carbon-14, tritium, and other radioisotopes. Fallout is therefore exceedingly complex. We consider only two of the more worrisome isotopes here.

Of all the isotopes, strontium-90 presents the greatest hazard to people. The isotope has a half-life of 28.5 years. Strontium-90 reaches us primarily through dairy products and vegetables. Because of its similarity to calcium (both are group 2A

334 CHAPTER 11 ■ Nuclear Chemistry

elements), strontium-90 is incorporated into bone. There it remains a source of internal radiation for many years.

Iodine-131 may present a greater threat immediately after a nuclear explosion. Its half-life is only 8 days, but it is produced in relatively large amounts. Iodine-131 is efficiently carried through the food chain. In the body it is concentrated in the thyroid gland, and it is precisely this characteristic that makes a trace of iodine-131 so useful for diagnostic scanning. However, for a healthy individual, larger amounts of radioactive iodine offer only damaging side effects. To minimize the absorption of radioactive iodine, many people in the area of Chernobyl were given large amounts of potassium iodide, which effectively diluted the amount of radioactive iodine actually absorbed by the thyroid.

By the late 1950s, radioactive isotopes from atmospheric testing of nuclear weapons were detected in the environment. Concern over radiation damage from nuclear fallout led to a movement to ban atmospheric testing. Many scientists were leaders in the movement. Linus Pauling, who won the Nobel Prize in chemistry in 1954 for his bonding theories and for his work in determining the structure of proteins, was a particularly articulate advocate of banning atmospheric tests. In 1963 a nuclear test ban treaty was signed by the major powers—with the exception of France and the People's Republic of China, which continued aboveground tests. Since the signing of the treaty, other countries have joined the nuclear club. Pauling, who had endured being called a Communist and a traitor because of his outspoken position, was awarded the Nobel Prize for peace in 1962.

Self-Assessment Questions

1. Strontium-90 substitutes for calcium in bones because Sr and Ca
   a. are in the same group of the periodic table
   b. are in the same period of the periodic table
   c. have identical electron configurations
   d. have identical half-lives

2. Which of the following radioisotopes from fallout is dangerous mainly because it is concentrated in the thyroid gland?
   a. cesium-137   b. iodine-131   c. strontium-90   d. xenon-143

Answers: 1, a; 2, b

11.11 Nuclear Power Plants

A significant portion of today's electric power is generated by nuclear power plants. In the United States, one-fifth of all the electricity produced comes from nuclear power plants. Europeans rely even more on nuclear energy. France, for example, obtains more than 70% of its electric power from nuclear plants, while Belgium, Spain, Switzerland, and Sweden each generate about one-third of their power from nuclear reactors.

Ironically, the same nuclear reactions that occurred in the detonation of the bomb dropped on Hiroshima are used extensively today under the familiar concrete containment tower of a nuclear plant. The key difference is that the power plant employs a *slow, controlled release of energy* from the nuclear chain reaction, rather than an explosion. The slower process results from using uranium fuel that is less enriched (2.5–3.5% ^{235}U rather than the 90% or so for weapons-grade uranium).

One of the main problems with the production of nuclear power comes from the products of the nuclear reactions. As in nuclear fallout, most of the daughter nuclei produced by the fission of ^{235}U are themselves radioactive, some with very long half-lives. We present a further discussion of problems associated with nuclear waste in Chapter 14. Perhaps a more serious problem is the potential transformation of spent nuclear fuel into weapons-grade material (see the box titled "Nuclear Proliferation and Dirty Bombs").

GREEN CHEMISTRY

Can Nuclear Power Be Green?

Galen Suppes, University of Missouri-Columbia

The idea of nuclear power as a greener source of energy has caused renewed interest in the development of nuclear power in the United States. Nuclear power provides nearly 20% of the electricity in the United States with carbon dioxide emissions that are a small fraction of the emissions from coal and natural gas. A green approach to the nuclear power industry could have both economic and environmental benefits and solve some of our most urgent societal needs.

Like all technologies, nuclear power has advantages and disadvantages. Among the advantages are reduced CO_2 emissions and low volume of waste generation relative to coal and natural gas. Nuclear power also has lower fuel costs at ~0.7 cents/kWh of electricity versus ~1.5 to 4 cents/kWh for coal and natural gas, respectively. Nuclear power's safety history shows about one-fifth the annual occupational health deaths per kWh of electrical power versus coal. Tens of millions of years of uranium and thorium reserves offer an abundant source of nuclear power that is also reliable. Nuclear power plants operate at the industry's highest percent of nameplate capacities, i.e., maximum rated outputs. Finally, nuclear energy content greater than the world's coal and petroleum reserves is domestically available in spent fuel rods stored at approximately 100 nuclear plants in the United States.

Disadvantages of nuclear power include spent fuel handling and waste storage problems and the potential for land contamination and broad-scale lethal exposure. The potential for plutonium weapons proliferation is of significant concern. In addition, the constant base load for nuclear-generated electricity does not match peak demand, and building a nuclear power plant requires very high capital investment.

The ability of green chemistry to overcome these disadvantages is perhaps the single greatest potential advantage of nuclear power. We examine each of these disadvantages in turn.

Radioactive Waste—Nuclear waste typically contains only a small amount of radioactive material, yet the entire mixture must be handled as radioactive waste. After 30 years of storage, only 0.4% of spent nuclear fuel is radioactive. This nuclear waste can be removed using known chemistry. France and Japan partially concentrate waste today, producing a reprocessed fuel (0.9 cents/kWh). If all of the nuclear fuel used to date in the United States was fully reprocessed, that radioactive waste would fit in a building about one-third the size of a small house. It is possible to convert that waste into harmless isotopes, and future technology may make that affordable.

Contamination and Exposure—The unfortunate, large-scale Chernobyl nuclear incident (Section 15.10) resulted from unsafe design, improper operation, and a large amount of spent fuel in the reactor at the time of the release. Experience over half a century indicates that the U.S. reactor designs and operation are safe. The principles of green chemistry dictate that we should build upon this experience by designing reactors that are safer and that have less spent fuel in the reactor at any one time.

Plutonium Proliferation—It is difficult to make a nuclear bomb from reactor-grade uranium because the process involves separating uranium isotopes, and isotopes have nearly identical properties. It is easier to separate plutonium from the spent fuel (mainly uranium-238) because plutonium is a different element with different chemical properties. Plutonium-239 is purified from spent nuclear fuel in some reprocessing operations. Concerns that this ^{239}Pu could be used to make nuclear bombs might be addressed by using reprocessing technologies that do not isolate ^{239}Pu. Better yet, we might operate reactors so that plutonium-240 is produced in adequate quantities to "poison" the ^{239}Pu for weapons applications. When it absorbs a neutron, ^{240}Pu nearly always becomes plutonium-241 rather than fissioning. Weapons development is hindered when more than about 7% of the total plutonium is ^{241}Pu.

Peak Demand Power—For both safety and economy, nuclear reactors are designed for continuous operation that does not match daily variations in electrical demand. We can solve this problem while reducing oil imports. Plug-in hybrid electric vehicles can use nighttime electricity, allowing electricity demand to match nuclear base load capabilities more closely.

High Power Plants Costs—The high capital costs of nuclear power plants are caused in part by the delays between capital investment and start-up due to permitting and licensing processes. The costs can be minimized by the use of preapproved standardized designs.

Application of green chemistry principles to minimize the disadvantages of nuclear power provides greater opportunity for U.S. energy independence and reduced carbon emissions.

▲ The core of a nuclear reactor contains hollow rods filled with uranium pellets. The heat generated by fission is used to boil water, which drives turbines to generate electricity in the same fashion as a coal-fired or gas-fired generating plant. Because the pellets are only about 3% uranium-235, a nuclear explosion cannot occur, though loss of coolant may cause a meltdown.

11.12 Thermonuclear Reactions

This chapter opened with a picture of a supernova, a violent nuclear explosion. All stars, including our sun, are giant nuclear reactors. The reactions that take place in the sun are somewhat different from the ones previously discussed. They are called **thermonuclear reactions** because they require enormously high temperatures (millions of degrees) to initiate them. The intense temperatures and pressures on the sun cause nuclei to fuse and release unimaginable amounts of energy. Instead of splitting large nuclei into smaller fragments (fission), small nuclei are fused into larger ones (fusion). The principal reaction in the sun is thought to be the fusion of four hydrogen nuclei to produce one helium nucleus and two positrons.

$$4\,{}^{1}_{1}\text{H} \longrightarrow {}^{4}_{2}\text{He} + 2\,{}^{0}_{+1}\text{e}$$

The fusion of 1 g of hydrogen releases an amount of energy equivalent to the burning of nearly 20 tons of coal. Every second the sun fuses 600 tons of hydrogen, producing millions of times more energy than has been produced in the entire history of humankind. Much current research is aimed at reproducing such a reaction in the laboratory, by using ultrapowerful magnets to contain the intense heat required for ignition. Fusion technology is discussed further in Chapter 15. To date, however, fusion reactions on Earth have been limited to the uncontrolled reactions in explosion tests of hydrogen (thermonuclear) bombs and to small amounts of energy produced by very expensive experimental fusion reactors.

Nuclear Proliferation and Dirty Bombs

As power is generated in nuclear plants, important changes occur in the fuel. Neutron bombardment converts fissile ${}^{235}\text{U}$ to radioactive daughter products. Eventually, the concentration of ${}^{235}\text{U}$ becomes too low to sustain the nuclear chain reaction. The fuel rods must therefore be replaced about every three years. The rods then become high-level nuclear waste.

As ${}^{235}\text{U}$ undergoes fission, the nonfissile (and very concentrated) ${}^{238}\text{U}$ nuclei absorb neutrons and are converted to ${}^{239}\text{Pu}$. (This transmutation is described in Section 11.9.) If the rods are used long enough, ${}^{240}\text{Pu}$ is also formed. However, if the fuel rods are removed after only about three months, the fissile plutonium-239 can be easily separated from the other fission products by chemical means. Therefore, less sophisticated technology is needed to produce a nuclear weapon from plutonium than from uranium. Plutonium is thus a greater source of concern for weapons proliferation, and operation of a nuclear plant can be a method of producing materials suitable for use in a nuclear weapon. In recent years, North Korea has restarted both a nuclear power plant and a plutonium separation facility with the capability of producing enough ${}^{239}\text{Pu}$ for several weapons each year (Figure 11.15). Iran also is constructing a facility to enrich uranium, which could lead to the production of nuclear bombs. These and other developing countries raise the fear of a dangerous proliferation of nuclear weapons.

Another security concern is a "dirty bomb," a device that uses a conventional explosive, such as dynamite, to disperse radioactive material that might be stolen from a hospital or other facility that uses radioactive isotopes. In most cases, the conventional bomb would do more immediate harm than the radioactive substances. At the levels most likely to be used, the dirty bomb would not contain enough radioactive material to kill people or cause severe illness.

◀ Figure 11.15
Satellite photograph of the plutonium processing plant at Yongbyon, North Korea.

Self-Assessment Questions

1. The process represented by the equation $^2_1H + ^2_1H \longrightarrow ^4_2He +$ energy is called
   a. alpha decay   b. artificial transmutation
   c. fission   d. fusion

2. Hydrogen-2 and hydrogen-3 fuse to form helium-4. The other product is a(n)
   a. electron   b. neutron   c. positron   d. proton

Answers: 1, d; 2, b

11.13 The Nuclear Age

With the splitting of the atom, the Chinese curse, "May you live in interesting times" seems quite appropriate. The goal of the alchemists, to change one element into another, has been achieved through the application of scientific principles. New elements have been formed, and the periodic chart has been extended beyond uranium ($Z = 92$) to element 116. This modern alchemy produces plutonium by the ton; neptunium ($Z = 93$), americium ($Z = 95$), and curium ($Z = 96$) by the kilogram; and berkelium ($Z = 97$) and einsteinium ($Z = 99$) by the milligram.

As we have seen, radioactive isotopes have many uses, from killing tiny but deadly cancer cells and harmful microorganisms in food, to serving as tracers in a variety of biological experiments and imaging technologies, to generating fearsome weapons. Figure 11.16 suggests a number of other constructive uses of nuclear energy. We live in an age in which the extraordinary forces present in the atom have been unleashed as a true double-edged sword. The threat of nuclear war—and nuclear terrorism—has been a constant specter for the last six decades, and yet it is hard to believe that the world would be a better place if we had not discovered the secrets of the atomic nucleus.

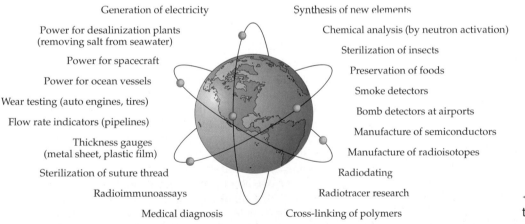

Generation of electricity
Power for desalinization plants (removing salt from seawater)
Power for spacecraft
Power for ocean vessels
Wear testing (auto engines, tires)
Flow rate indicators (pipelines)
Thickness gauges (metal sheet, plastic film)
Sterilization of suture thread
Radioimmunoassays
Medical diagnosis

Synthesis of new elements
Chemical analysis (by neutron activation)
Sterilization of insects
Preservation of foods
Smoke detectors
Bomb detectors at airports
Manufacture of semiconductors
Manufacture of radioisotopes
Radiodating
Radiotracer research
Cross-linking of polymers

◀ **Figure 11.16** Some constructive uses of nuclear energy.

Critical Thinking Exercises

Apply knowledge that you have gained in this chapter and one or more of the FLaReS principles (Chapter 1) to evaluate the following statements or claims.

11.1 A cave in Mexico is claimed to have wonderful healing powers. People who are ill sometimes sit for hours on the benches inside this cave. Studies have shown that the cave really is unusual. Its walls contain radioactive substances that constantly give off a low level of radiation. Do you think the cave really has healing power?

11.2 For more than 600 years it has been alleged that the Shroud of Turin was the burial shroud of Jesus Christ. In 1988 several laboratories carried out carbon-14 analyses indicating that the flax from which the shroud was made was grown during the period between 1260 and 1390 C.E. Recently there have been new claims that the 1988 analyses are unreliable because the shroud is coated with pollen from plants grown in the fourteenth century, and it is the age of this pollen that the scientists were actually measuring in 1988. At least one of the scientists has offered to repeat the analysis on a very carefully cleaned sample of the cloth, but further access to the shroud has been refused.

11.3 Most modern smoke detectors contain a small amount of radioactive americium, an alpha emitter. A rookie firefighter wearing a tank of breath-

ing air refuses to enter a burning home because he claims the radioactive material in the smoke detectors escapes during a fire.

11.4 A young girl refuses to visit her grandmother after the woman has undergone radiation treatment for breast cancer. The teenager is afraid that she will catch radiation sickness from her grandmother.

■ SUMMARY

Section 11.1—Some isotopes of elements are unstable and their nuclei undergo changes in nucleon number, atomic number, or energy; this process is called **radioactive decay**. The nuclei that undergo such changes are called **radioisotopes**. We are exposed to naturally occurring radioisotopes and other natural ever-present radiation, called **background radiation**. Radiation that causes harm by dislodging electrons from living tissue and forming ions is called **ionizing radiation** and includes nuclear radiation and X-rays. Radiation can disrupt normal chemical processes in cells and can damage DNA, causing mutations in some cases.

Section 11.2—Nuclear equations are used to represent nuclear processes. Equations are written and balanced so that the sum of nucleon numbers on each side is the same, and the sum of atomic numbers on each side is the same. Four types of radioactive decay are **alpha** ($^4_2$He) **decay, beta** ($^0_{-1}$e) **decay, gamma** ($^0_0\gamma$) **decay,** and **positron** ($^0_{+1}$e) **decay. Electron capture** is a fifth type of decay in which a nucleus absorbs one of the atom's electrons. There are enormous differences between nuclear reactions and chemical reactions.

Sections 11.3–11.4—The **half-life** of a radioactive isotope is the time it takes for half of a sample to undergo radioactive decay. The fraction of a radioisotope remaining after n half-lives is given by the expression

$$\text{Fraction remaining} = \frac{1}{2^n}$$

Half-lives of certain isotopes can be used to estimate the ages of various objects. **Carbon-14 dating** is the best known of the dating methods. With a half-life of 5730 years, carbon-14 dating can be used to estimate the age of once-living items up to 50,000 years old. The decay of tritium can be used to date items up to 100 years old. Other isotopes can be used to date rocks, the Earth's crust, and meteorites.

Section 11.5—**Transmutation**, the conversion of one element into another, cannot be carried out by chemical means but can be accomplished by nuclear processes. Bombarding one nucleus with other energetic particles can cause transmutation. These processes can be represented with nuclear equations.

Section 11.6—Radioisotopes have many uses. A radioisotope and a stable isotope of an element behave nearly the same, so radioisotopes can be used as **tracers** to follow processes in physical and biological systems. A radioactive atom in a molecule labels it so that it can be followed by a radiation detector. Radioisotopes have many uses in agriculture, including the production of useful mutations. Radiation can be used to irradiate foodstuffs as a method of preservation. **Radiation therapy** to destroy cancer depends on the fact that radiation is more damaging to cancer cells than to

healthy cells. Radioisotopes are used in diagnosis of various disorders. Iodine-131 is used for thyroid diagnoses, gadolinium-153 for bone mineralization examinations, and technetium-99m for a variety of diagnostic tests. Positron emission tomography (PET) involves radioisotopes that emit positrons, which are then annihilated in the body, producing gamma rays that can be used to build an image.

Section 11.7—Different types of radiation have different penetrating abilities. **Alpha particles** (helium nuclei) are relatively slow and low in penetrating power. **Beta particles** (electrons) are much faster and more penetrating. **Gamma rays** (high-energy photons) travel at the speed of light and have great penetrating power. Radiation hazard depends on the location of the source; alpha particles from a source inside the body are highly damaging. Radiation hazard can be decreased by moving away from the source or with shielding.

Sections 11.8–11.9—Einstein's **mass–energy equation**, $E = mc^2$, shows that mass and energy are different aspects of the same thing. The total mass of the nucleons in an atom is greater than the actual mass of the nucleus. The missing mass is present as **binding energy** holding the nucleons together. Binding energy can be released either by breaking down heavy nuclei into smaller ones, a process called **nuclear fission**, or by joining small nuclei to form larger ones, called **nuclear fusion**. Fermi and Segrè bombarded uranium atoms with neutrons and found radioactive species among the products, while Hahn and Strassman found light nuclei among the reaction products. Meitner, aided by Frisch, hypothesized that the uranium was undergoing fission. Szilard saw that neutrons released in the fission of one atom could trigger the fission of other atoms, setting off a **chain reaction**. The nuclear fission reaction became the center of the Manhattan Project. Its goals were (a) to achieve sustained nuclear fission and determine the **critical mass** or minimum amount of fissile material required; (b) to enrich the amount of fissile uranium-235 in ordinary uranium; (c) to synthesize plutonium-239, which is also fissile; and (d) to construct a nuclear fission bomb before the Germans were able to do so. World War II ended shortly after the dropping of atomic bombs on Hiroshima and Nagasaki.

Sections 11.10–11.13—In addition to the devastation at the site of a nuclear explosion, much radioactive debris or **radioactive fallout** is produced. Strontium-90 and iodine-131 that are produced are particularly problematic. Partly because of fallout, a nuclear test ban treaty was signed by most of the major nations; only underground testing is not banned. Nuclear power plants use the same reaction as atomic bombs, but the reaction is much slower and is controlled because of the low concentration of fissile material. Disposal of the products of nuclear power plants is an important problem facing us, as is the potential conversion of nuclear fuel into weapons. **Thermonuclear reactions**, also known as fusion reactions, combine small nuclei to form larger ones. Nuclear

fusion produces even more energy than nuclear fission. Fusion is the basis of the hydrogen bomb and the source of the sun's energy. Much research is aimed at producing controlled fusion commercially. The forces within the nucleus truly constitute a double-edged sword for civilization.

■ REVIEW QUESTIONS

1. Define or identify each of the following.
   a. half-life   **b.** positron
   c. background radiation   **d.** radioisotope

2. Match the description with the type of change.
   i. a new compound is formed   **(a)** nuclear
   ii. a new element is formed   **(b)** chemical
   iii. size, shape, appearance or   **(c)** physical
        volume is changed without
        changing the composition

3. Why are isotopes important in nuclear reactions but not particularly so in most chemical reactions?

4. Give the nuclear symbols for protium, deuterium, and tritium (which are hydrogen-1, hydrogen-2, and hydrogen-3, respectively).

5. Give the nuclear symbols for the following isotopes. You may refer to the periodic table.
   a. cobalt-59   **b.** iodine-131
   c. carbon-13   **d.** technetium-99

6. Indicate the number of protons and the number of neutrons in atoms of the following isotopes:
   a. $^{62}_{30}Zn$ **b.** $^{241}_{94}Pu$ **c.** $^{99m}_{43}Tc$ **d.** $^{81m}_{36}Kr$

7. Which of the following pairs represent isotopes? (X is a general symbol for an element.)
   a. $^{70}_{34}X$ and $^{70}_{33}X$ **b.** $^{57}_{28}X$ and $^{66}_{28}X$
   c. $^{186}_{74}X$ and $^{189}_{74}X$ **d.** $^{8}_{2}X$ and $^{6}_{4}X$

8. In which of the following atoms are there more protons than neutrons?
   a. ^{58}Fe **b.** ^{1}H **c.** ^{3}H **d.** ^{12}C

9. The longest-lived isotope of fermium (Fm) has a nucleon number of 257. How many neutrons are in the nucleus of this isotope?

10. The longest-lived isotope of technetium (Tc) has 54 neutrons. What is the nucleon number of this isotope?

11. What changes occur in the nucleon number and atomic number of the nucleus during emission of each of the following?
    a. beta particle   **b.** neutron   **c.** proton

12. What changes occur in the nucleon number and atomic number of the nucleus during emission of each of the following?
    a. alpha particle   **b.** gamma ray   **c.** positron

13. What are some of the characteristics that make technetium-99m such a useful radioisotope for diagnostic purposes?

14. From which type of radiation would **(a)** a pair of gloves be sufficient to shield the hands: the heavy alpha particles or the massless gamma rays? **(b)** heavy lead shielding be necessary to protect a worker: alpha, beta, or gamma?

15. Plutonium is especially hazardous when inhaled or ingested because it emits alpha particles. Why would alpha particles cause more damage to tissue than beta particles in this case?

16. List two ways in which workers can protect themselves from the radioactive materials with which they work.

17. Compare nuclear fission and nuclear fusion. Why is energy liberated in each case?

18. The compounds $^{235}UF_6$ and $^{238}UF_6$ are nearly chemically identical. How are they separated?

■ PROBLEMS

Nuclear Equations

19. Write a balanced equation for emission of **(a)** an alpha particle by plutonium-240, **(b)** a positron by sodium-22, and **(c)** a beta particle by copper-67.

20. Write a balanced equation for the **(a)** alpha decay of lead-210, **(b)** beta decay of chromium-55, and **(c)** capture of an electron by bromine-76.

21. Complete the following equations.
    a. $^{179}_{79}Au \longrightarrow ^{175}_{77}Ir + ?$ **b.** $^{23}_{10}Ne \longrightarrow ^{23}_{11}Na + ?$
    c. $^{121}_{51}Sb + ? \longrightarrow ^{121}_{52}Te + ^{1}_{0}n$

22. Complete the following equations.
    a. $^{10}_{5}B + ^{1}_{0}n \longrightarrow ^{4}_{2}He + ?$
    b. $^{12}_{6}C + ^{2}_{1}H \longrightarrow ^{13}_{6}C + ?$
    c. $^{154}_{62}Sm + ^{1}_{0}n \longrightarrow 2\,^{1}_{0}n + ?$

23. Radiological laboratories often have a container of molybdenum-99, which decays to form technetium-99m. What other particle is formed? Write the equation.
    $$^{99}_{42}Mo \longrightarrow ^{99m}_{43}Tc + ?$$

24. Complete the following equation for the decay of technetium-99m to technetium-99.
    $$^{99m}_{43}Tc \longrightarrow ^{99}_{43}Tc + ?$$

25. When magnesium-24 is bombarded with a neutron, a proton is ejected. What new element is formed? (*Hint:* Write a balanced nuclear equation.)

26. When silver-107 is bombarded with a neutron, a new isotope of silver forms, which then undergoes beta decay. What is the final product? (*Hint:* Write two separate nuclear equations.)

27. A radioactive isotope decays to give an alpha particle and bismuth-211. What was the original nucleus?

28. A radioisotope decays to give an alpha particle and protactinium-233. What was the original nucleus?

29. A nucleus of bismuth-210 decays by beta emission, forming a new nucleus A. Nucleus A also decays by beta emission to nucleus B. Write the complete symbol for B.

30. A proposed method of making more fissile nuclear fuel is to bombard the relatively abundant isotope thorium-232

with a neutron. The product A of this bombardment decays quickly by beta emission to nucleus B, and nucleus B decays quickly to fissile nucleus C. Write the complete symbol for C.

Half-Life

31. How long will it take for a 12.0-g sample of iodine-131 to decay to leave a total of 1.5 g of the isotope? The half-life of iodine-131 is 8.04 days.

32. Gallium-67 is used in nuclear medicine (Table 11.6). After treatment, a patient's blood gave a radioactivity of 20,000 counts per minute (counts/min). How long would it be before the activity decreased to about 5000 counts/min?

33. A lab worker reports an activity of 80,000 counts/s on a sample of magnesium-21, half-life 122 ms. Exactly 5.00 min later, another worker recorded an activity of 10 counts/min on the same sample. What best accounts for this difference is that magnesium-21
    a. gives off only neutrinos
    b. has a very short half-life and essentially every radioactive atom in the sample has decayed
    c. is an alpha emitter
    d. is a gamma emitter

34. Krypton-81m is used for lung ventilation studies. Its half-life is 13 s. How long does it take the activity of this isotope to reach one-quarter of its original value?

35. A patient is injected with a radiopharmaceutical labeled with technetium-99m, half-life 6.0 h, in preparation for a gamma ray scan in order to evaluate kidney function. If the original activity of the sample was 48 μCi, what activity remained after (a) 24 h and (b) after a total of 48 h?

36. Radium-223 has a half-life of 11.4 days. Approximately how long would it take for the radioactivity associated with a sample of ^{223}Ra to decrease to 1% of its initial value?

37. In an experiment with dysprosium-197, half-life 8.1 h, an activity of 500 counts/min was recorded at 4:00 p.m. Friday. What would be the approximate activity of this

sample at 8:30 a.m. the following Monday morning when the experiment was resumed?

38. In determining the half-life of sulfur-35, students collected the following data, with the time in days and the activity in counts/s:

time	activity	time	activity
0	1000	80	525
20	851	100	446
40	725	120	380
60	616	140	323

Without doing detailed calculations, estimate a value for the half-life.

Radioisotopic Dating

39. Living matter has a carbon-14 activity of about 16 counts/min per gram of carbon. What is the age of an artifact for which the carbon-14 activity is 8 counts/min per gram of carbon?

40. A piece of wood from an Egyptian tomb has carbon-14 activity of 980 counts per hour. A piece of new wood of the same size gave 3920 counts/h. What is the age of the wood from the tomb?

41. The ratio of carbon-14 to carbon-12 in a piece of charcoal from an archaeological excavation is found to be one-half the ratio in a sample of modern wood. Approximately how old is the site? How old would it be if the ratio were 25% of the ratio in a sample of modern wood?

42. You are offered a case of brandy supposedly bottled in the time of Napoleon (1769–1821) for a really great price. Before buying it, you insist on testing a sample of the brandy and find that it has a tritium content 12.5% of that of newly produced brandy. How long ago was the brandy bottled? Is it likely to be authentic Napoleon-era brandy?

■ ADDITIONAL PROBLEMS

43. Write a balanced nuclear equation for (a) the bombardment of $^{121}_{51}$Sb by alpha particles to produce $^{124}_{53}$I, followed by (b) the radioactive decay of the $^{124}_{53}$I by positron emission.

44. To make seaborgium (Z = 106), a 0.25-mg sample of californium-249 was used as the target. Four neutrons were emitted to yield a nucleus with 106 protons and a mass of 263 u. What was the bombarding particle?

45. One atom of meitnerium (Z = 109) with a nucleon number of 266 was produced in 1982 by bombarding a target of bismuth-209 with iron-58 nuclei for 1 week. How many neutrons were released in the process?

46. Meitnerium undergoes alpha emission to form element 107, which in turn also emits an alpha particle. What are the atomic number and nucleon number of the isotope formed by these two steps? Write balanced nuclear equations for the two reactions.

47. Radium-223 nuclei usually decay by alpha emission. For every billion alpha decays, one atom emits a carbon-14 nucleus. Write a balanced nuclear equation for each type of emission.

48. A particular uranium alloy has a density of 18.75 g/cm^3. What volume is occupied by a critical mass of 49 kg of this alloy? The critical mass can be decreased to 16 kg if the alloy is surrounded by a layer of natural uranium (which acts as a neutron reflector). What is the volume of the smaller mass? Compare your answers to the volume of a baseball, a volleyball, and a basketball.

49. Plutonium has a density of 19.1 g/cm^3. What volume is occupied by a mass of 16.3 kg of plutonium? With the proper neutron reflecting coating (made of beryllium), the critical mass can be lowered to 2.5 kg! Compare this in size to the volume of a baseball, a volleyball, and a basketball.

50. What is the new nucleus formed in each of the following processes?
    a. Lead-196 goes through two successive EC processes
    b. Bismuth-215 decays through two successive beta emissions.
    c. Protactinium-231 decays through four successive alpha emissions.
    d. Neptunium-237 undergoes a series of seven alpha and four beta decays to form a stable isotope.

51. There are few technological applications for the transuranium elements ($Z > 92$). One important one is in smoke detectors, which may use the decay of a tiny amount of americium-241 to neptunium-237. What particle is emitted from that decay process?

52. In the first step of the chain reaction of a nuclear explosion, a ^{235}U nucleus absorbs a neutron. The resulting ^{236}U nucleus is unstable and can fission into a ^{92}Kr and a ^{141}Ba nucleus as shown in the figure below. What are the other products of this reaction? In light of this reaction, explain how the chain reaction can continue.

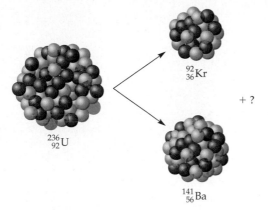

53. In 1932, James Chadwick discovered a new subatomic particle when he bombarded beryllium-9 with alpha particles. One of the products was carbon-12. What particle did Chadwick discover?

54. There are many different radioisotopes that can be used in a manner similar to carbon-14 for determining the age of rocks and other matter. In the geosciences, several different radioisotopes often are measured to confirm the age of rocks. For example, potassium-40 decays to argon-40 with a half-life of 1.2 billion years. (a) Some rocks brought back from the moon were dated at 3.6 billion years. What percentage of the potassium-40 has decayed after that time? (b) Give a reason why uranium-238 ($t_{1/2}$ = 4.6 billion years) is more useful for confirming this age than is carbon-14.

55. Einstein's mass–energy equation is $E = mc^2$, where mass is in kilograms and the speed of light is 3.00×10^8 m/s. The units of energy are joules (4.184 J = 1 cal; 1000 cal = 1 kcal; 1 J = 1 kg m^2/s^2).
    a. Calculate the energy released, in calories and kilocalories, when 1 gram of matter is converted to energy.
    b. A bowl of cornflakes supplies 110 kcal (110 food calories). How many bowls of cornflakes are equivalent to the energy in part **(a)** from one gram of matter?

56. Radioactive copper, $^{64}_{29}$Cu, is found in quantities exceeding pollution standards in the sediments of a reservoir in a routine check on Monday. The standard allows up to 14 ppm/m^3 of sediment. On Monday, 56 ppm/m^3 was measured. The half-life of $^{64}_{29}$Cu is 12.7 hours. How long will it take for the pollution level to return to 14 ppm/m^3?

57. An unidentified corpse was discovered on April 21 at 7:00 a.m. The pathologist discovered that there were 1.24×10^{17} atoms of $^{32}_{15}$P remaining in the victim's bones and placed the time of death sometime on 9 March. The half life of $^{32}_{15}$P is 14.28 days. How much $^{32}_{15}$P was present in the bones at the time of death?

■ COLLABORATIVE GROUP PROJECTS

Prepare a PowerPoint, poster, or other presentation (as directed by your instructor) for presentation to the class.

58. Write a brief report on the impact of nuclear science on one of the following.
    a. war and peace
    b. industrial progress
    c. medicine
    d. agriculture
    e. human, animal, and plant genetics

59. Write a brief report on one of the following topics.
    a. radiodating of archaeological objects
    b. use of radioisotopes in medicine
    c. the many and varied uses of nuclear energy

60. Write a brief biography of one of the following scientists.
    a. Otto Hahn   b. Enrico Fermi
    c. Glenn T. Seaborg   d. J. Robert Oppenheimer
    e. Lise Meitner   f. Albert Einstein

61. Write a brief essay on the discovery of radioactivity. (You might want to take a look at the *Journal of Chemical Education*, January 1992, p. 10.)

62. The positron is a particle of antimatter. Search the Web for information about antimatter, and the positron in particular.

63. Find a website that is strongly in favor of nuclear power plants and one that is strongly opposed. Note their sponsors and analyze their viewpoints. Try to find a website with a balanced approach.

64. Write a short research report describing the operation of smoke detectors.

65. In July 1999 researchers at Lawrence Berkeley Laboratory reported the creation of the heaviest element to date, element 118. Two years later, that report was found to be fraudulent. Report on the scientific ethical questions raised by this episode.

66. Find the location of the nuclear plant closest to where you live. Try to determine risks and benefits of the plant. To how many houses does it provide power? What are the environmental impacts of the plant under normal operating conditions?

67. Assemble two groups of two to four people each to debate the following resolution: The use of the atomic bomb on Hiroshima and Nagasaki was justified. Decide beforehand which team will take the affirmative and which the negative. Each team member is allowed a three- to six-minute speech followed by a cross-examination by the opposing team members. Have the rest of the class judge and give written or oral comments.

68. Prepare a presentation on the subject of "cold fusion." Other useful search-engine keywords are "Pons" and "Fleischmann." The presentation should discuss the differences between the hypothesized cold fusion and conventionally understood fusion reactions and should provide a balanced and impartial view of the subject.

Earth is a storehouse of chemicals—minerals, metals, and much more. Materials we extract from Earth are the basis of much of our modern civilization. People use chemistry to modify many of these materials to make them more useful.

CHEMISTRY OF EARTH

12

A
N
S
W
E
R

QUESTIONS YOU MAY HAVE ASKED YOURSELF

1. How do we know what other planets and stars are made of?
2. Why is asbestos dangerous?
3. Can you get lead poisoning from pottery?
4. If atoms are conserved, why should we need to recycle aluminum and iron?
5. Why are scrap copper and other scrap metals bringing such high prices?

Metals and Minerals

This wondrous world of ours is a fertile sphere blanketed in air, with about three-fourths of its surface covered by water. Although it is but a tiny, blue-green jewel in the vastness of space, our Spaceship Earth is about 40,000 km in circumference with a surface area of 500 million km^2.

Human astronauts have walked on the barren surface of Earth's airless moon. Our space probes have explored the desolation of Mars, the crushing pressure and hellish heat of Venus with its hurricane clouds of sulfuric acid, and Saturn's moon Titan with its liquid hydrocarbon seas and water-ice rocks. Probes have also given us close-up portraits of dry, pockmarked Mercury, and of Jupiter and Saturn with their horrendous lightning storms and their turbulent atmospheres of hydrogen and helium. Earth is but a small island in the inhospitable immensity of space, a tiny oasis uniquely suited to the life that inhabits it.

From earliest times, people have used Earth's resources. Primitive people used ordinary stones for hunting and as tools; modern people excavate ores and coal and drill for oil and gas. Spaceship Earth carries 6.7 billion passengers, and their numbers are increasing rapidly. What kinds of materials do we have aboard this spaceship? Are they sufficient for this enormous load of passengers? Let us begin by looking at the composition of Earth.

12.1 Spaceship Earth: The Materials Manifest

Earth is divided into three main regions: the core, the mantle, and the crust (Figure 12.1). The *core* is thought to consist largely of iron with some nickel. Because the core and mantle are not accessible and do not seem likely to become so, we

1. How do we know what other planets and stars are made of?

Recall Figures 3.9 and 3.11. We can view the spectra of celestial objects to identify the elements and compounds that are present. We can even get an idea of the relative amounts of the elements and compounds from the brightness of the emitted light.

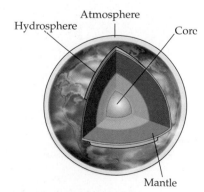

▲ **Figure 12.1** Diagram showing the structural regions of Earth. The drawing is not to scale.

won't consider them as sources of materials. Nor shall we consider the Moon, Mars, or asteroids as a source; access to any such resources is decades, if not centuries, in the future.

The *crust* is the outer solid shell of Earth, often called the *lithosphere*. The watery part, made up of the oceans, seas, lakes, rivers, and so on, is the *hydrosphere*. The *atmosphere* is the air that surrounds the planet. We discuss the atmosphere in Chapter 13 and the hydrosphere in Chapter 14. In this chapter we focus only on the lithosphere, which is about 35 km thick under the continents and about 10 km thick under the oceans. Through extensive sampling of the atmosphere, hydrosphere, and lithosphere, scientists have been able to estimate the elemental composition of the outer portion of Earth. A random sample of 10,000 atoms would have the composition summarized in Table 12.1, which features the nine most abundant elements.

More than half the atoms in our 10,000-atom sample are O atoms, which occur in the atmosphere as molecular oxygen (O_2), in the hydrosphere in combination with hydrogen as water, and in the lithosphere in combination with silicon (pure sand is largely SiO_2) and various other elements. Silicon is the second most abundant element in the sample with 1590 Si atoms, and hydrogen is third with 1510 H atoms, most of them in combination with oxygen in water. Because hydrogen is the lightest of all elements, by mass it makes up only 0.9% of Earth's crust. Just three elements—oxygen, silicon, and hydrogen—make up 8430 of the 10,000-atom sample. The top nine elements account for 9630 of the atoms, leaving only 370 atoms of all the other elements. However, these minor constituents include some elements very important to life, such as carbon, nitrogen, and phosphorus.

Table 12.1 Elemental Composition of Earth's Surface

Element	Number of Atoms in a Sample of 10,000 Atoms	Atom Percent	Percent by Mass
Oxygen	5,330	53.3	49.5
Silicon	1,590	15.9	25.7
Hydrogen	1,510	15.1	0.9
Aluminum	480	4.8	7.5
Sodium	180	1.8	2.6
Iron	150	1.5	4.7
Calcium	150	1.5	3.4
Magnesium	140	1.4	1.9
Potassium	100	1.0	2.4
All others	370	3.7	1.4
Total	10,000	100.0	100.0

Self-Assessment Questions

1. The regions of the Earth, listed in order from the center out, are
   a. core, crust, mantle, hydrosphere
   b. core, mantle, crust
   c. hydrosphere, crust, mantle, core
   d. mantle, crust, core

2. Earth's core is composed mainly of
   a. lead   b. iron   c. radioactive elements   d. sulfur

3. Three elements account for more than 80% of the Earth's crust by atom percent. In decreasing order of abundance, they are
   a. H, O, Si   b. H, Si, O   c. O, H, Si   d. O, Si, H

Answers: 1, b; 2, b; 3, d

12.2 The Lithosphere: Organic and Inorganic

The lithosphere is mainly rocks and minerals. Prominent among these are

- *silicate minerals* (compounds of metals with silicon and oxygen)
- *carbonate minerals* (metals combined with carbon and oxygen)
- *oxide minerals* (metals combined with oxygen only)
- *sulfide minerals* (metals combined with sulfur only)

Thousands of these mineral compounds make up the *inorganic* portion of the solid crust. We discuss some silicate minerals in Section 12.4. Some typical nonsilicate minerals are listed in Table 12.2.

Although much, much smaller in quantity, the *organic* portion of Earth's outer layers includes all living creatures, their waste and decomposition products, and fossilized materials (such as coal, natural gas, petroleum, and oil shale) that once were living organisms. This organic material always contains the element carbon, nearly always has combined hydrogen, and often contains oxygen, nitrogen, and other elements.

▲ **It DOES Matter!**
One of the more common rocks on Earth is limestone, which is calcium carbonate ($CaCO_3$). In nature, calcium carbonate is found in many different physical forms; marble, seashells, eggshells, coral, pearls, and chalk all are calcium carbonate. Striking examples of calcium carbonate also are found in the stalagmite and stalactite formations in caves, such as those in the Reed Flute cavern, Guilin, China.

Table 12.2 Some Nonsilicate Minerals of Economic Importance

Mineral Type	Name	Chemical Formula	Use
Oxide	Hematite	Fe_2O_3	Ore of iron; pigment
	Magnetite	Fe_3O_4	Ore of iron
	Corundum	Al_2O_3	Gemstone; abrasive
Sulfide	Galena	PbS	Ore of lead
	Chalcopyrite	$CuFeS_2$	Ore of copper
	Cinnabar	HgS	Ore of mercury
	Sphalerite	ZnS	Ore of zinc
Carbonate	Calcite	$CaCO_3$	Cement; lime

12.3 Meeting Our Needs: From Sticks to Bricks

People long have modified nature's materials to help satisfy their needs and wants, but for centuries, technological developments were based largely on trial and error. By developing an understanding of the structures of materials, modern science has greatly increased the human ability to modify natural materials.

Early people obtained food by hunting and gathering. The skins of animals provided clothing when it was needed, and shelter was found in a convenient cave or constructed from available sticks and stones and mud. After the agricultural revolution (about 10,000 years ago), people no longer were forced to search for food. Domesticated animals and plants supplied their needs for food and clothing. A reasonably assured food supply enabled them to live in villages and created a demand for more sophisticated building materials. People learned to convert natural materials into other products with superior properties. Adobe bricks, which serve well in arid areas, could be made simply by drying a mixture of clay and straw in the sun.

Fire was one of the earliest agents of chemical change. When people discovered fire, they learned that cooking improved the flavor and digestibility of meat and grains. People learned that heating adobe bricks made stronger ceramic bricks.

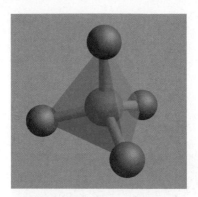

▲ **Figure 12.2** The silicate tetrahedron has an Si atom at the center and an O atom at each of the four corners. The shaded area (green) shows how the SiO₄ tetrahedron is typically represented in a mineral structure.

Similarly, firing clay at high temperatures produced ceramic pots, making cooking and storage of food much easier. They later learned to produce higher temperatures with which they could make glass and extract metals from ores. They were able to make tools from bronze and later from iron.

12.4 Silicates and the Shapes of Things

We will consider only a few representative minerals of the thousands of different ones in the lithosphere. First, let's look at some silicates. The basic unit of silicate structure is the SiO_4 tetrahedron (Figure 12.2). As shown in Table 12.3, silicate tetrahedra can exist singly as silicate anions or can be joined in a variety of ways.

Quartz is pure silicon dioxide (SiO_2). The ratio of silicon to oxygen atoms is 1:2. However, because each silicon atom is surrounded by *four* oxygen atoms, the basic unit of quartz is the SiO_4 tetrahedron. These tetrahedra are arranged in a complex three-dimensional structure. Crystals of pure quartz (rock crystal) are colorless, but various impurities produce a variety of quartz crystals sometimes used as gems (Figure 12.3).

Table 12.3 SiO_4 Tetrahedra in Some Silicate Minerals

Mineral(s)	SiO_4 Arrangement	Formula	Uses
Zircon	Simple anion (SiO_4^{4-})	$ZrSiO_4$	Ceramics; gemstones
Spudomene	Long chains of tetrahedra	$LiAl(SiO_3)_2$	Source of lithium and its compounds
Chrysotile asbestos	Double chains of tetrahedra	$Mg_3(Si_2O_5)(OH)_4$	Fireproofing (now banned)
Muscovite mica	Sheets of tetrahedra	$KAl_2(AlSi_3O_{10})(OH)_4$	Insulation; fancy paints; packing (vermiculite)
Quartz	Three-dimensional array of tetrahedra	SiO_2	Making glass (sand); gemstones (amethyst, agate, citrine)

▲ Vermiculite is a lightweight mica that has been "puffed" by heat, like popcorn. It is a useful soil additive, holding moisture much better than soil alone. It is also used as insulation and as lightweight, absorbent packing material.

● = Oxygen atom
○ = Silicon atom

▲ **Figure 12.3** A variety of quartz crystals. Counterclockwise from upper right: citrine (yellowish quartz), colorless quartz, amethyst (purple quartz), and smoky quartz. The chemical structure of quartz shows that each silicon atom is bonded to four oxygen atoms, and each oxygen atom is bonded to two silicon atoms.

Micas are composed of SiO_4 tetrahedra arranged in two-dimensional sheet-like arrays. Micas are easily cleaved into thin, transparent sheets (Figure 12.4). Pieces of mica once were used as panels for lanterns and as windows in the doors of stoves. Today, micas are used as insulators for electronics and in high-temperature furnaces.

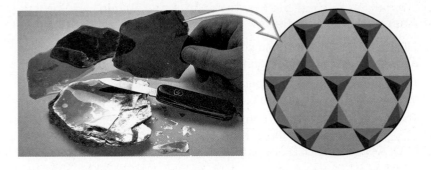

◀ **Figure 12.4** A sample of mica, showing cleavage into thin, transparent sheets. The chemical structure of mica shows sheets of SiO_4 tetrahedra. The sheets are bound together by cations, principally Al^{3+} (not shown). Mica is used as transparent "window" material in industrial furnaces.

Asbestos is a generic term for a variety of fibrous silicates. Perhaps the best known of these is *chrysotile*, a magnesium silicate (Figure 12.5). Note that chrysotile is a double chain of SiO_4 tetrahedra. The oxygen atoms that have only one covalent bond also bear a negative charge. Magnesium ions (Mg^{2+}) are associated with these negative charges.

▲ **Figure 12.5** A sample of chrysotile asbestos. The chemical structure of chrysotile shows double chains of SiO_4 tetrahedra. The two chains are joined to each other through oxygen atoms. The double chains in turn are bound to each other by cations, principally Mg^{2+} (not shown).

▲ Clothing made of asbestos fibers, shown in this old photo of a firefighter on the USS *Ranger*, was once widely used to protect workers from high temperatures and hot materials. Asbestos for this purpose has been replaced by synthetic polymers such as Nomex (Chapter 10).

Asbestos: Risks and Benefits

Asbestos is an excellent thermal insulator. In the past, it was been used widely to insulate furnaces, heating ducts, and steam pipes and to make protective clothing for firefighters and others who are exposed to flames and high temperatures. Asbestos was also used in brake linings for automobiles.

The health hazards to those who work with asbestos are well known. Inhalation of fibers 5–50 μm long over a period of 10–20 years causes asbestosis, and after 30–45 years, some asbestos workers contract lung cancer. Others get mesothelioma, a rare and incurable cancer of the linings of body cavities.

Long-term occupational exposure to asbestos increases the risk of lung cancer by a factor of two. Cigarette smoking causes a tenfold increase in the risk of lung cancer. We might expect asbestos workers who smoke to have a risk 20 times that of nonsmokers who do not work with asbestos, but instead their risk is increased by a factor of 90. Cigarette smoke and asbestos fibers act in such a way that each enhances the action of the other. Such a joint result is called a **synergistic effect**.

The harmful effects of asbestos are due mainly to a relatively rare form called crocidolite. Chrysotile, which makes up 95% of the asbestos used in the United States, appears not to be nearly as dangerous. Government regulations do not distinguish between the two types.

Public fears of developing cancer from asbestos have led to regulations that require its removal from schools and other public buildings. Costs of removal will total billions of dollars, and benefits are uncertain.

More to Explore
Bjornerud, Marcia. *Reading the Rocks: The Autobiography of the Earth.* Cambridge, MA: Westview, 2005.

2. Why is asbestos dangerous?
Asbestos fibers are extremely tiny and are very sharp. When inhaled, they are not easily expelled from the lungs and can cause long-term damage. A 2008 report suggests that carbon nanotubes may cause similar problems when inhaled.

ANSWER

3. Can you get lead poisoning from pottery?
Some types of clay contain lead compounds. If pottery made from lead-containing clay is not thoroughly glazed to make it nonporous, the lead compounds can leach into the vessel contents. Happily, pottery made in the United States is not permitted to contain toxic levels of lead.

Self-Assessment Questions

1. All silicate minerals contain silicon and
   a. carbon   **b.** hydrogen   **c.** iron   **d.** oxygen

2. The mineral composed of pure silicon dioxide (SiO_2) is
   a. asbestos   **b.** calcite   **c.** quartz   **d.** zircon

3. The silicate tetrahedron consists of one _____ atom at the center and _____ atoms at the corners
   a. O; 4 Si   **b.** O; 2 Si and 2 Al   **c.** Si; 4 O   **d.** Si; 2 O and 2 Al

For items 4–7, match the substance with the structure.
4. asbestos   **(a)** double chains
5. mica   **(b)** sheets
6. quartz   **(c)** simple anions
7. zircon   **(d)** three-dimensional array
8. In causing cancer, asbestos acts synergistically with
   a. aerogels   **b.** cigarette smoke   **c.** glass fibers   **d.** lead

Answers: 1, d; 2, c; 3, c; 4, a; 5, b; 6, d; 7, c; 8, b

12.5 Modified Silicates: Ceramics, Glass, and Cement

Pottery work, glassmaking, and cement production were among the earliest technologies developed. People learned to modify natural materials such as sand, clay, and limestone, mainly by mixing and heating, to make much more useful products.

Ceramics

Clays are complex and have widely varying compositions, but they are basically aluminum silicates. Mixed with water, clay can be molded into any shape. Firing leaves a hard, durable (but porous) product. Bricks, flowerpots, and tile are made in this manner. When porosity is not desirable—as in a cooking pot or a water jug—the pottery can be glazed by adding various salts to the surface. Heat then converts the entire surface to a glasslike matrix. Bricks and pottery are examples of **ceramics**, inorganic materials made by heating clay or other mineral matter to a high temperature at which the particles partially melt and fuse together.

Glass

Glass, a noncrystalline solid, is another technological development of ancient times. The first glass probably was made in ancient Egypt about 5000 years ago by

▲ **It DOES Matter!**
Research has led to the development of amazing new ceramic materials. Ceramics have been used to make rocket nose cones, heat-resistant tiles on the space shuttle, automobile engines, jet engine turbine blades, and memory elements in computers. *Superconducting* ceramics exhibit virtually no resistance to the flow of electricity; they make high-speed magnetically levitated ("maglev") trains possible.

▲ Aerogels are specially prepared ceramic materials that are up to 99.8% air. They are the best solid insulators in the world.

▶ Heat-softened glass can be shaped by blowing or molding. Compare the irregular arrangement of atoms to the regular arrangement of those in quartz (Figure 12.3).

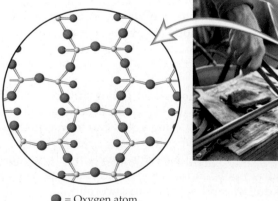

● = Oxygen atom
◐ = Silicon atom

Table 12.4 Compositions and Properties of Various Glasses

Type	Composition	Special Properties and Uses
Soda–lime glass	Sodium and calcium silicates	Ordinary glass (for windows, bottles, and so on)
Borosilicate glass	Boron oxide (instead of lime)	Heat resistant (for laboratory ware and ovenware) (Pyrex, Kimax)
Aluminosilicate glass	Aluminum oxide (instead of soda)	More highly heat resistant (for top-of-stove cookware and fiberglass)
Lead glass	Lead oxide (instead of lime)	Highly refractive (for optical glass, art glass, table crystal)
Colored glass	Selenium compounds added	Red color (ruby glass)
	Cobalt compounds added	Blue color (cobalt glass)
	Chromium compounds added	Green color
	Manganese compounds added	Violet color
	Carbon and iron oxide added	Brown color (amber glass)
Photochromic glass	Silver chloride or bromide added	Light sensitive; darkens when exposed to light (for sunglasses, hospital windows)
Laser glass	Contains neodymium	Lasers
Frosted glass	Etched with hydrofluoric acid (HF)	Satiny frosted surface

heating a mixture of sand, "soda" (sodium carbonate, Na_2CO_3), and limestone (calcium carbonate, $CaCO_3$). As the mixture melts, it becomes a homogeneous liquid. When the liquid cools, it becomes hard and transparent.

Crystalline materials have a regular, repeated arrangement of atoms, ions, or molecules. The forces among these unit particles are the same throughout. When heated, crystals melt over a narrow temperature range. For example, sodium hydroxide, an ionic crystal, melts sharply at 318 °C, and crystalline aspirin, a molecular substance, melts at 135 °C. Glass is different; when heated, it gradually softens. While soft, it can be blown, rolled, pressed, or molded into almost any shape. The properties of glass result from an irregular arrangement, in three dimensions, of SiO_4 tetrahedra. The chemical bonds in this arrangement are not all equivalent. Thus, when glass is heated, the weaker bonds break first and the glass softens gradually.

The basic ingredients in glass can be used in different proportions. Ordinary window and container glass has a typical composition of about 75% SiO_2, 15% Na_2O, and 10% CaO. Oxides of various metals can be substituted in whole or in part for the limestone, soda, or sand. Thus, many special types of glass can be made. Some examples are given in Table 12.4.

Manufacture of ordinary glass uses no vital raw materials, but the furnaces used to melt and shape glass require energy. Disposal is a potential problem because glass is one of the most permanent materials known. However, glass is easily recycled. It can be melted and formed into new objects at considerable energy savings compared with the manufacture of new glass. In fact, *cullet* (broken glass from previous melts) is usually added to each batch of sand, soda, and limestone to be melted.

Cement and Concrete

Cement, a complex mixture of calcium and aluminum silicates, is another ancient technological development. The Romans used a type of cement to construct roads, aqueducts, and the famous Roman baths. The raw materials for the production of cement are limestone and clay. The materials are finely ground, mixed, and roasted at about 1500 °C in a rotary kiln heated by burning natural gas or powdered coal. The finished product is mixed with sand, gravel, and water to form *concrete*.

Our understanding of the complex chemistry of cement is still imperfect. Nevertheless, extensive research has made various special cements available, including a fast-setting cement with high strength quickly established, a white cement, a waterproof cement, and a cement that sets at high temperatures.

More to Explore
Kolb, Kenneth E., and Doris K. Kolb. "Glass—Sand + Imagination." *Journal of Chemical Education*, July 2000, pp. 812–816.

▲ **It DOES Matter!**
Optical fibers are hair-thin threads of glass that carry messages as intermittent bursts of light. Sounds are translated into electric signals, which are converted into laser pulses that are transmitted by the glass fibers. At the end of the line, the light pulses are converted back to sound waves. A bundle of optical fibers can carry several hundred times as many messages as a copper cable of the same size, and they have replaced many old telephone lines and computer connections. They are also used in medicine to view inside the body and to guide the use of surgical instruments.

Concrete is used widely in the construction of buildings and roads because it is inexpensive, strong, chemically inert (and thus nonpolluting in itself), durable, and tolerant of a wide range of temperatures. However, its production involves extensive mining, with entire mountains being torn down for limestone rocks. The rotary kiln process consumes fossil fuels. Particulate matter from the crushing operations and smoke and sulfur dioxide from the burning of fossil fuels make air pollution from cement plants especially serious. For disposal, concrete can be broken up and used as rock fill.

▲ Portland cement is made by roasting a mixture of limestone, clay, and sand. This cement is the "glue" that holds concrete together. In 2005, 121 million tons of portland cement were produced in the United States.

Self-Assessment Questions

1. A material that is hard and durable but brittle is a
   a. ceramic   **b.** metal   **c.** polymer   **d.** semiconductor

2. The main ingredients of ordinary glass are
   a. boron oxide, sand, and limestone
   b. clay and limestone
   c. coke, sand, and limestone
   d. sand, sodium carbonate, and limestone

3. Cullet is
   a. broken glass used in the manufacture of new glass
   b. glass culled for defects
   c. glass separated by brand
   d. glass sorted by color

4. The main ingredients of ordinary cement are
   a. clay and limestone   **b.** coke, sand, and limestone
   c. sand and limestone   **d.** sand, sodium carbonate, and limestone

Answers: 1, a; 2, d; 3, a; 4, a

12.6 Metals and Ores

Human progress through the ages is often described in terms of the materials used for making tools. We speak of the Stone Age, the Bronze Age, and the Iron Age. Ancient peoples knew about gold and silver because they are often found in the *native* or uncombined state in nature, but these metals are too soft to use for tools. Metal tools were not possible until artisans learned to extract certain metals from ores by smelting.

Copper and Bronze

Copper is sometimes found in the native state, and it was probably also the first metal to be freed from its ore by early smelting techniques. Ancient Egyptian records show that copper was known 5500 years ago. They probably isolated copper by heating copper sulfide (Cu_2S) ore.

$$Cu_2S(s) + O_2(g) \longrightarrow 2\,Cu(s) + SO_2(g)$$

Many copper ores are blue to green in color and easy to identify in the ground. The ancient Egyptians produced thousands of tons of copper metal. Today copper is important mainly because of its excellent electrical conductivity. It is used primarily for electric wiring and for plumbing pipes.

More important than copper was **bronze**, a copper alloy containing about 10% tin. An **alloy** is a mixture of two or more elements, at least one of which is a metal. Bronze is harder than copper, and it could therefore be made into many useful tools. In some societies, bronze remained the most widely used metal for about 2000 years.

▲ Soaring prices for scrap metal, largely due to increased demand in China and India, have led to theft of copper and aluminum from construction sites, empty houses, and other places. Thefts have included aluminum siding from houses, bleachers from athletic fields, and wheel rims from cars. Thefts of copper pipes from vacant houses—many in foreclosure—have led to dangerous gas leaks.

Iron and Steel

The technology for iron and steel greatly lagged behind that for copper and bronze for two reasons. Iron reacts so readily with oxygen and sulfur that it is not found uncombined in nature, and it melts at a much higher temperature than copper or bronze. To produce iron from ore, carbon (coke) is employed as the reducing agent (Chapter 8). First, the carbon is converted to carbon monoxide.

$$2\,C(s) + O_2(g) \longrightarrow 2\,CO(g)$$

Then the carbon monoxide reduces the iron oxide to iron metal.

$$Fe_2O_3(s) + 3\,CO(g) \longrightarrow 2\,Fe(l) + 3\,CO_2(g)$$

Iron can be made from ore in a huge chimney-like vessel called a *blast furnace* (Figure 12.6). The raw materials fed into the furnace are iron ore, coke, and limestone. The coke (which is mainly carbon) is made by heating coal in the absence of air to drive off coal oil and tar. The limestone is added to combine with silicate impurities to form a molten **slag** that floats on top of the iron and is drawn off. Molten iron is drawn off at the bottom of the furnace. The product of the blast furnace, called *pig iron,* has a fairly low melting point, and so it is easily cast into molds. Hence, it is also known as cast iron.

Because cast iron is brittle and has many impurities, most iron from blast furnaces is alloyed. The principal alloy of iron is **steel**, an alloy with carbon. The properties of steel can be varied over a wide range by adjusting the amount of carbon in it. High-carbon steel is hard and strong and is used for knives and metalworking tools. Low-carbon steel is more ductile and malleable and is used to make sheet steel, pipes, and structural beams.

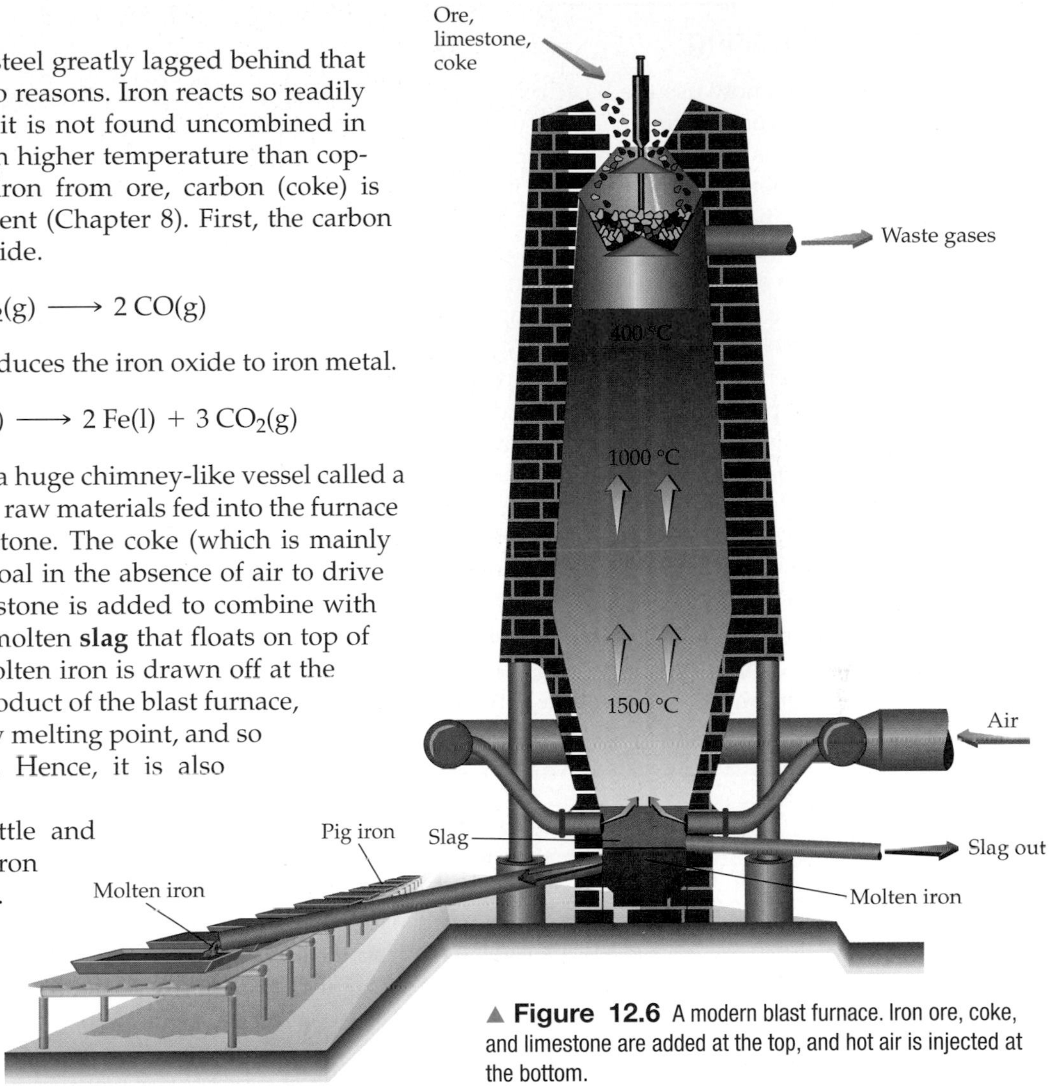

Steel is commonly alloyed with other metals to give it special properties. For example, manganese imparts hardness, tungsten gives high-temperature strength, while chromium and nickel reduce corrosion. Because it is so abundant and can be made into so many different alloys, iron is the most useful of the metals, accounting for about 95% of all metal production worldwide.

▲ Figure 12.6 A modern blast furnace. Iron ore, coke, and limestone are added at the top, and hot air is injected at the bottom.

QUESTION: What is the role of each of the three substances added at the top of the furnace?

Aluminum: Abundant and Light

Aluminum, the most abundant metal in Earth's crust, is tightly bound in compounds, and considerable energy is required to extract the metal from its ores. Nevertheless, aluminum has replaced iron for many purposes. Worldwide production of aluminum is second to iron among the metals at more than 30 billion kg per year. (Iron production is about 840 billion kg annually.) The principal ore of aluminum is *bauxite,* impure aluminum oxide. The Al_2O_3 is extracted from the impurities with a strong base, and then the melted oxide is reduced by passing electricity through it.

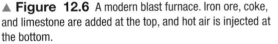

It takes 2 metric tons (t) of aluminum oxide and 17,000 kilowatt hours (kWh) of electricity to produce 1 t of aluminum.

The method used to produce aluminum today was discovered independently in 1886 by two college students, Charles Martin Hall in the United States and Paul Héroult in France. Both men were born in 1863 and both died in 1914 at the age of 51. Before the Hall-Héroult method was developed, aluminum was considered a precious metal. French emperor Napoleon III reserved his aluminum utensils for his most honored guests; others had to make do with goldware.

Modern Steelmaking

Most steel producers now use the basic oxygen process (Figure 12.7). The furnace is first charged with pig iron. Powdered limestone is added just above the molten iron, and oxygen gas at about 10 atm pressure is injected there also. Metallic impurities in the pig iron are converted to oxides, which react with SiO_2 to form a slag that floats above the liquid iron and is poured off. Then any desired alloying elements are added, often by adding scrap steel of a particular composition.

Chemistry is employed extensively during the process. As the melt progresses, samples are drawn from the furnace and analyzed on-site for the alloying metals and nonmetals. If, for example, the steel contains too much carbon, more oxygen is used to burn off the excess, or low-carbon scrap may be added. In this way the impurities are removed and the desired composition of the steel is achieved at the same time.

Some companies now make iron directly from iron ore in a single-step continuous process. They avoid the great energy requirements of the blast furnace by using temperatures below the melting points of any of the raw materials. The ore is reduced by H_2 and CO which in turn are made from natural gas. This *direct reduction* method is used mainly in developing countries that do not already have large iron and steel industries in place.

Today in the United States almost all automobiles are recycled, and about half of all steel comes from "mini-mills," which make steel from scrap iron instead of ore.

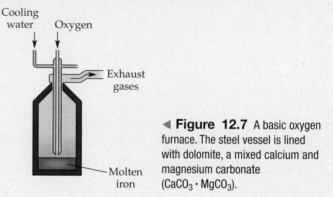

◄ **Figure 12.7** A basic oxygen furnace. The steel vessel is lined with dolomite, a mixed calcium and magnesium carbonate $(CaCO_3 \cdot MgCO_3)$.

Aluminum is light and strong. An aluminum object weighs only one-third as much as a steel article the same size. Although it is considerably more reactive than iron, aluminum corrodes much more slowly. Freshly prepared aluminum metal reacts rapidly with oxygen to form a hard, transparent film of aluminum oxide (Al_2O_3) over its surface. The film then protects the metal from further oxidation. Iron, on the other hand, forms an oxide coating that is porous and flaky. Instead of protecting the metal, the coating flakes off, allowing further oxidation.

The Environmental Costs of Iron and Aluminum

Steel and aluminum play vital roles in the modern industrial world, and it would be hard to overemphasize their economic value. But shouldn't we include the environmental cost of producing, using, and discarding these materials?

Both steel mills and aluminum plants produce considerable waste. In days gone by, steel mills discharged lime, acids, grease, oil, and iron salts into waterways. They released carbon monoxide, nitrogen oxides, particulate matter, and other pollutants into the air. Aluminum plants discharged iron, aluminum, and other metal oxides into waterways, and particulate matter, fluorides, and other pollutants into the air. Modern treatment methods have reduced the quantity of pollutants released in developed countries, but pollution from metal production in developing nations such as China and India is still a major problem.

It takes about 15 times as much fuel energy to produce aluminum as it does to produce a comparable weight of steel. Aluminum is less dense than steel, but making an aluminum can requires 6.3 times as much energy as making a steel can. On the other hand, when aluminum is recycled, the energy needed is only a fraction of that required to make the metal from ore. Further, the low density of aluminum means energy savings down the road; an airplane made largely from aluminum is so much lighter than a similar craft made entirely from steel that it takes a good deal less energy to operate it.

Other Important Metals

There are many other important metals. Some of them, with typical ores, properties, and uses, are listed in Table 12.5.

Metals and minerals are vital to a vigorous economy, but no industrial nation is 100% self-sufficient in all the vital metals and minerals. The United States has depleted much of its high-grade ores and is now dependent on imports of some 100 essential minerals, including chromium, molybdenum, manganese, and platinum. The United States is more than 90% dependent on imports for chromium—vital for steelmaking and other industries—and for platinum and palladium—elements used in making catalysts, including catalytic converters for automobiles. These minerals are found largely in South Africa and the former Soviet Union.

How can we ever run short of a metal? Aren't atoms conserved? Yes, there is as much iron on Earth as there was 100 years ago, but by using it we scatter the metal throughout the environment. Gathering it back to a factory to make new objects requires energy, just as obtaining metals from low-grade ores requires more energy than extracting them from high-grade ores.

A N S W E R

4. If atoms are conserved, why should we need to recycle aluminum and iron? The cost of melting scrap aluminum or iron—a simple physical change—is much lower than the cost of extracting the metals from their ores, which involves several chemical changes.

Table 12.5 Some Other Important Metals

Metal	Symbol	Important Ore	Selected Properties	Typical Uses
Chromium	Cr	$FeCr_2O_4$	Shiny, resists corrosion	Chrome plating, stainless steel
Gold	Au	Au	Yellow metal, soft, dense	Coinage, jewelry, dentistry, electrical contacts
Lead	Pb	PbS	Low melting, dense, soft	Plumbing, batteries
Magnesium	Mg	$MgCl_2$	Light, strong	Auto wheels, luggage, computer cases
Mercury	Hg	HgS	Dense liquid	Thermometers, barometers
Nickel	Ni	NiS	Resists corrosion	Coinage, alloy for stainless steel
Platinum	Pt	Pt	Inert, high melting	Catalyst, instruments
Silver	Ag	Ag_2S	Excellent electric conductor	Electric contacts, mirrors, jewelry, coins
Sodium	Na	NaCl	Reactive, soft	Heat transfer medium, reducing agent
Tin	Sn	SnO_2	Resists corrosion	Coating for steel cans
Tungsten	W	$CaWO_4$	Very high melting	Lightbulb filaments
Uranium	U	U_3O_8	Fissionable	Energy source
Zinc	Zn	ZnS	Forms protective coating	Galvanizing coating, die casting

Self-Assessment Questions

1. The metal most likely to be found as the free element is
   a. aluminum   **b.** copper   **c.** iron   **d.** tin

2. When copper is produced from Cu_2S, the by-product is
   a. CuO   **b.** CO_2 **c.** elemental S   **d.** SO_2

3. Bronze is an alloy of Cu and
   a. Ni   **b.** Pb   **c.** Sn   **d.** Zn

4. The raw materials for the blast furnace production of iron are iron ore,
   a. coke and limestone
   b. coke, sand, and limestone
   c. coke and sandstone
   d. sand, sodium carbonate, and limestone

5. What metal typically is alloyed with iron to make stainless steel?
   a. aluminum   **b.** chromium   **c.** copper   **d.** strontium

6. Aluminum is obtained from its ore by reduction using
   a. carbon   **b.** electrolysis   **c.** magnesium   **d.** scrap iron

7. Bauxite is impure
   a. Al_2O_3 **b.** Cu_2S **c.** Fe_2O_3 **d.** gold

Answers: 1, b; 2, d; 3, c; 4, a; 5, b; 6, b; 7, a

GREEN CHEMISTRY

We're Going to Need a Bigger Earth . . .

Denyce K. Wicht, Suffolk University

If you have seen the movie *Jaws*, you may recall the famous line "You're gonna need a bigger boat." It's uttered when the police chief, the marine scientist, and the seasoned fisherman get a look at the size of the great white shark that has been tormenting the people of fictional Amity Island, and realize that the boat they are on might be ill-equipped for the hunt. The phrase is now used colloquially to describe a situation that seems increasingly more daunting and complex as facts emerge.

We have a challenge similar to the one the *Jaws* characters faced. Finding ways to use our natural resources in a sustainable manner seems a daunting task. The principles of green chemistry are designed to tackle such challenges.

Consider the direct reduction of magnetite, Fe_3O_4, to iron, the feedstock for the steelmaking processes:

$$(1)\ Fe_3O_4 + 4\,CO \longrightarrow 3\,Fe + 4\,CO_2$$
$$(2)\ Fe_3O_4 + 4\,H_2 \longrightarrow 3\,Fe + 4\,H_2O$$

We can easily balance these chemical reactions on paper, but in practice these processes are complex and energy intensive. They require carbon monoxide and hydrogen as the reducing agents and produce carbon dioxide and water as the by-products. And iron is just one example of resource utilization. To understand the complex nature and long-term implications of our use of the Earth's materials, we must critically consider *all* of our resources.

Consider, for example, the many raw materials that are used to construct the typical modern home. The foundation is concrete, made of sand, clay, and limestone. The frame is constructed from wood boards cut from a pine forest. The main support beam is usually steel, made from iron extracted from mined ore. The exterior of the house may require clay bricks, wood, or vinyl siding. The windows, roof, plumbing, and appliances all use other types of materials, taken from the earth.

What elements make up these raw materials? Clay is mostly aluminum, silicon, and oxygen; sand is primarily silicon and oxygen; and limestone is calcium, carbon, and oxygen. Iron ore is iron and oxygen, but carbon, oxygen, and hydrogen are required to reduce the iron oxides to iron metal. Wood contains carbon, oxygen, and hydrogen, while vinyl siding is carbon, hydrogen, and chlorine. These elements may seem plentiful, but as the population and its need for adequate shelter increases, the demand for construction material also grows. In other words: We're going to need a bigger Earth.

So what should we do? We can build greener homes. A green building is a broad concept that includes the use of recycled materials such as plastics (Section 10.9) as structural components and takes into consideration the amount of energy needed to transport the materials to the site as well as how long they will last. For example, although bamboo is a green product in the sense that is a rapid-growing renewable grass (compared to wood from a tree), we must also consider the energy required to transport it over long distances.

Most experts agree that we can no longer take from Earth at the rate we have in the past. Indeed, recycling efforts have increased significantly in the recent years, but *recycle* is only one of the three R's; we also need to *reduce* consumption and find more ways to *reuse* the materials we've already mined and modified (Section 12.7).

The Twelve Principles of Green Chemistry guide science in the effort to discover and invent more sustainable chemical processes and materials. Returning to the example of the reduction of magnetite to iron, consider the preceding two balanced chemical equations. The atom economy (see page 137) of reaction (1) is 49% while that of reaction (2) is 70%. From an atom economy point of view, reaction (2) seems more desirable, but the percent yield is also important. As for by-products, water seems more benign than carbon dioxide, which contributes to global warming. As this example shows, green scientists must consider multiple factors in developing new and improved chemical processes.

Green scientists find intellectual stimulation in complex problems, such as those presented here, and the opportunity to work with an interdisciplinary community of experts is a driving force in the pursuit of their careers. These same challenges keep them engaged in working toward the goal of a better Earth.

12.7 Earth's Dwindling Resources

Dramatic fuel shortages during the 1970s brought important changes in government policies and in industrial and individual practices. But now many of the conservation efforts begun at that time have largely been forgotten, and prices have soared in recent years.

We also face potential shortages of metals and minerals. The United States has exhausted its high-grade ores of metals such as copper and iron. Australia has vast deposits of high-grade iron ore, but importing this ore is costly. Miners once found nuggets of pure gold and silver, but the "glory holes" of western North America are long gone. Now we mine ores with a fraction of an ounce of gold in a ton of gold ore (see Figure 12.9 and Problem 43). Few high-grade deposits of any metal ores remain that are readily accessible to the industrialized nations.

What's wrong with low-grade ores? Because more material must be mined for the same final amount of metal, there is more environmental disruption (Figure 12.8), and more energy is required to concentrate the ores. Both environmental cleanup and energy cost money. Consider an analogy. A bag of popcorn is useful. You can pop it and eat it. The same popcorn, scattered all over your room, would be less useful. Of course, you could gather it all up and then pop it, but that would take a lot of energy—perhaps more energy than you would care to expend and maybe more energy than you would get back when you popped the corn and ate it.

We also must import quite a few metals because their ores are not found in the United States. For example, we get tin from Bolivia, chromium from Kazakhstan and South Africa, and platinum from South Africa and Russia. As you might expect, though, the metal reserves in other countries are also being depleted.

Land Pollution: Solid Wastes

Productivity and creativity have enabled people in industrialized nations to have a seemingly endless variety of consumer goods in enormous quantities. These goods often are packaged in paper or plastic. Their wrappers litter the environment and fill our disposal sites. As the goods break, wear out, or merely become obsolete, they too add to our disposal problems. The approximate composition of municipal solid wastes (MSWs)—commonly called trash or garbage—in the United States is given in Figure 12.9.

The U.S. Environmental Protection Agency (EPA) ranks strategies for MSW as follows: (1) Source reduction (including reuse) is best, (2) recycling and composting is next best, and (3) disposal in combustion facilities and landfills is a last resort. In 2006 in the United States, 32.5% of MSW was recovered and recycled or composted, 12.5% was burned at combustion facilities, and 55% was disposed in *sanitary landfills*.

▲ When the Europeans first came to North America, native copper (shown here) was readily available and was used by Native Americans. Now we mine ore with less than 0.5% copper.

▲ Where will we get metals in the future? The sea is one possible source. Nodules rich in manganese cover vast areas of the ocean floor. These nodules also contain copper, nickel, and cobalt. Questions of who owns them and how to mine them without major environmental disruption remain to be resolved.

◀ **Figure 12.8** For the 1 ounce of gold in a necklace, miners must move 30 tons of rock. The crushed rock is leached with cyanide solution to dissove the gold. The waste cyanide solution is stored in a large containment pond. The facility shown here is part of a gold mining operation in the Mojave Desert of California.

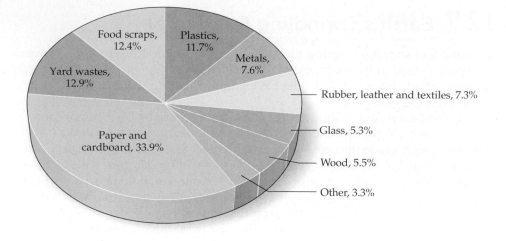

▶ **Figure 12.9** Municipal wastes in the United States are largely paper, cardboard, and yard wastes. Paper and cardboard can be recycled. Yard wastes and food wastes can be composted.

Some cities not only burn their combustible solid wastes but also use the heat from the incinerators to warm buildings, to generate power, or both. A pound of polyethylene produces about 20,000 British thermal units (Btu) of energy, about the same as a pound of no. 2 fuel oil.

In the past, most solid wastes were simply discarded in open dumps. This led to infestation by rats, flies, and other pests that often spread to nearby areas. Open burning led to offensive and unhealthful air pollution. These open dumps have now been largely phased out.

In a sanitary landfill, garbage and trash are piled into a trench, compacted, and covered over. This eliminates the problems of rats, flies, and odors. However, some landfills leak and contribute to groundwater contamination. Furthermore, materials in landfills decompose slowly. Newspapers are still readable after being entombed for 20 years. Without water and oxygen, microorganisms are unable to carry out the normal decay processes.

Disposal of solid wastes by *incineration,* when carried out in properly designed incinerators, produces minimal air pollution. However, many incinerators are old or poorly designed, and they generate considerable smoke and odor. Building new incinerators tends to be unpopular with local citizens.

Research in solid waste disposal has led to new processes. The U.S. Bureau of Mines has developed a method for converting garbage to oil with an overall yield of 25%. Ground-up rubber tires can be mixed with powdered coal and used as fuel. Waste glass and rubber tires have both been added to road-paving mixtures for highway construction.

The Three R's of Garbage: Reduce, Reuse, Recycle

There are several ways to deal with our garbage problem. Not all are equally desirable. We should choose wisely among them in order to best save energy and protect the environment.

The best way to deal with our solid waste problem is to *reduce* the amount of throwaway materials produced. This decreases the volume of materials produced and sold, saves resources and energy, and minimizes the disposal problem.

Next best is the *reuse* of materials. When possible, items should be made durable enough to withstand repeated use rather than designed for a single use and disposal. Consider the grocery bag. At the store, we are often asked, "Paper or plastic?" Most studies show that the environmental impact of a plastic bag is somewhat less than that of a paper bag, though paper should be a better choice for several reasons.

- Plastic bags can be recycled, but only a small fraction are. Of the 500 billion or more used worldwide, most are landfilled, and about 3% become litter. Bags hang from fences and trees, clog storm drains, and harm wildlife, especially sea creatures that mistake floating bags for food.

- Plastic bags are made from petroleum, a nonrenewable resource, consuming millions of liters of oil that could be used for fuel or heating. Paper is made from trees, a renewable resource. Although replacing billions of plastic bags with paper would require an enormous number of trees, paper is easier to recycle.

- Paper bags are less of a litter problem because they break down relatively quickly in the environment. Plastic bags last for years; the bags buried in land-fills may be there for centuries.

So, which should we use, paper or plastic? Neither! The best answer in most cases is reusable shopping bags made of cloth that don't need to be discarded after each use. All considerations must be examined when determining the best overall solution to a reuse problem.

The third way to reduce the volume of wastes is to *recycle* them. Recycling requires energy, and some material is unavoidably lost, but recycling saves materials. Metals such as iron and steel, aluminum, copper, and lead are largely recycled. More than 57% of steel cans and about 55% of beer and soft drink aluminum cans are recycled. About 35% of plastic soft drink bottles and 26% of glass containers are recycled. About 45% of paper and cardboard are recycled. Lawn trimmings and food wastes can be composted, converting them to soil. About 57% of yard trimmings are composted. In the future, as raw materials become scarcer, these proportions are bound to increase.

Recycling also saves energy. It takes only 5% as much energy to make new aluminum cans from old ones as it does to make them from aluminum ore. Using a ton of scrap to make new iron and steel saves 1.5 tons of iron ore and 0.3 ton of coal. It also results in a 74% savings in energy, an 86% reduction in air pollution, and a 76% reduction in water pollution.

By following the three R's—reduce, reuse, and recycle—we can protect the environment at the same time that we save materials, energy, and money.

How Crowded is Our Spaceship?

Each day, people die and other people are born, but the number born is about 200,000 more than the number who die. Every five days our world population goes up by a million people, and it reached 6.7 billion in 2008.

Figure 12.10 is a graphic picture of population growth. For tens of thousands of years, human population was never more than a few hundred million. It did not reach the 1 billion mark until about 1800. Population growth peaked in the 1960s at about 2.2% a year. The present rate of growth is about 1.3% per year. Overall, the world's population quadrupled during the twentieth century. The Population Division of the United Nations Department of Economic and Social Affairs estimates a world population of 9.0 billion by 2045.

How many people can Spaceship Earth accommodate? Many believe that we have already gone beyond the optimum population for this planet. They predict increasing conflicts over resources and increasing pollution of the environment.

Why did Earth's population start increasing so rapidly after centuries of little change? Population growth depends on both the birthrate and the death rate.

More to Explore
LaJeunesse, Sara. "Plastic Bags." *Chemical and Engineering News*, September 20, 2004, p. 54.

5. Why are scrap copper and other scrap metals bringing such high prices? One factor that contributes to the value of scrap metal is the cost of energy needed to extract the metal from its ore. The greater the cost of energy, the more it costs to convert ore to metal. Because scrap metal does not require as much energy to convert it back to useful forms, its value is increasing.

According to the Steel Recycling Institute, about 5740 kJ of energy is conserved for each pound of steel recycled. That is enough energy to light a 60-watt bulb for more than 26 hours.

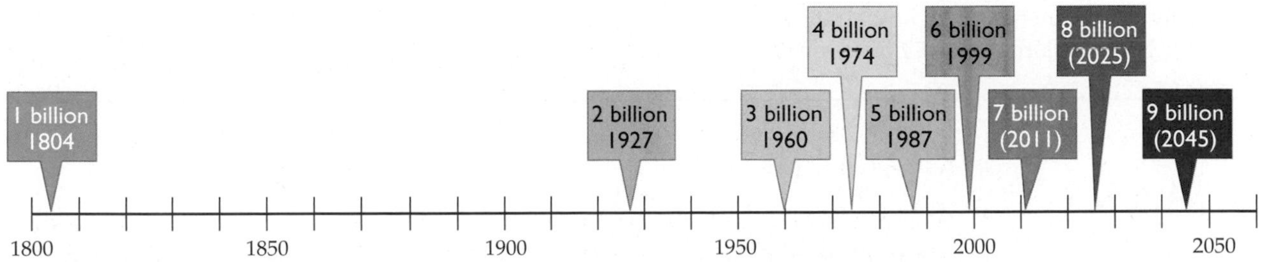

▲ **Figure 12.10** Two thousand years ago, Earth's population was about 300 million—about the same as the population of the United States today. Population changed little over the next thousand years and is estimated to have been only 310 million at the end of the first millennium. It reached 1 billion about 1804. This graph shows the intervals in which an additional billion people were added to Earth's population with projections to 2050.

Source: United Nations, *World Population Prospects: The 2006 Revision.* New York: 2007.)

Population and the food supply are discussed in **eChapter 20**.

If they are equal, population growth is zero. After being almost equal for thousands of years, scientific progress in the 1800s led to a decline in the death rate in developed countries. Yet the birthrate changed little and the population grew rapidly. These changes reached the less developed countries in the 1950s, greatly lowering the death rate. However, the birthrate remained high and the population exploded. Population growth has slowed considerably in recent years, especially in developed countries. Scientific progress continues to cause a decline in death rates, but birthrates have also declined, especially in Europe. The decrease in birthrates in the less developed countries has come more slowly. With more people comes more waste to be processed. As populations increase, we must continue to develop technologies and methods to deal with the ever-increasing waste problem.

Self-Assessment Questions

1. Which kind of material is the largest component of the garbage we put in landfills?
   a. food   **b.** metals   **c.** paper   **d.** plastic

2. Recycling an aluminum can requires what percent of the energy to make a new one from virgin aluminum?
   a. 5%   **b.** 10%   **c.** 25%   **d.** 50%

3. The three R's of garbage are
   a. reduce, recycle, and redeploy
   b. reuse, redeploy, and recycle
   c. reuse, reprocess, and recycle
   d. reuse, reduce, and recycle

Answer: 1, c; 2, a; 3, d

Critical Thinking Exercises

Apply knowledge that you have gained in this chapter and one or more of the FLaReS principles (Chapter 1) to evaluate the following statements or claims.

12.1 An economist has said that we need not worry about running out of copper because it can be made from other metals.

12.2 A citizen testifies against establishing a landfill near his home, claiming that the landfill will leak substances into the groundwater and contaminate his water well.

12.3 A citizen lobbies against establishing an incinerator near her home, claiming that plastics burned in the incinerator will release hydrogen chloride into the air.

12.4 An environmental activist claims that we could recycle all goods, leaving no need for the use of raw materials to make new ones.

12.5 A concerned parent demands that all the insulation on the steam pipes in the schools in her district be removed and replaced with new insulation. Her contention is that the pipe insulation may contain asbestos and, therefore, may cause cancer in the students.

 # SUMMARY

Section 12.1—Earth is divided into the core, the mantle, and the crust. The core is thought to be mainly iron; the mantle is mostly silicate minerals. The crust consists of the solid lithosphere, the liquid (water) hydrosphere, and the gaseous atmosphere.

Section 12.2—Nine elements make up 96% of the atoms in the crust. The lithosphere is made up largely of minerals such as silicates, carbonates, oxides, and sulfides. The organic part of the lithosphere consists of all things once and currently living.

Section 12.3—People have manipulated the lithosphere to meet their wants and needs for about 10,000 years.

Section 12.4—Silicates contain SiO_4 tetrahedra in various arrangements. **Quartz** is a crystalline mineral that consists of a three-dimensional network of these tetrahedra. Sand is mainly quartz. The fibrous mineral **asbestos** is made of chains of SiO_4 tetrahedra. Asbestos is implicated in lung cancer in asbestos workers. The incidence of lung cancer among asbestos workers who smoke is much greater than the simple combination of asbestos exposure and smoking might indicate; this is called a **synergistic effect**. The sheet-like mineral **mica** consists of sheets of SiO_4 tetrahedra.

Section 12.5—Ceramics, glass, and cement are modified silicates. **Ceramics** are made by heating clay and other inorganic minerals to fuse them partially. New kinds of ceramics have been developed that have high heat resistance, low electrical conductivity, and increased durability. **Glass** is a noncrystalline solid that does not have a fixed melting point. Most glass is made from sand with various additives to impart desired properties such as heat resistance, color, and light sensitivity. Glass does not degrade in the environment but it is easy to recycle; recycled glass aids the melting process for newly manufactured glass. **Cement** is a complex mixture of calcium and aluminum silicates made from limestone and clay. Cement with gravel, sand, water, and additives is used to make concrete.

Section 12.6—Copper was one of the first metals to be obtained from its ore but is too soft to be used for tools. **Bronze** is a mixture of copper and tin that is much harder than pure copper. Bronze implements were used for thousands of years. Iron technology came about much later. Iron is prepared in a blast furnace from iron ore, limestone, and coke. The limestone combines with impurities to produce **slag** that floats on top of the iron and is removed. Cast iron is brittle, so it is usually further processed with oxygen to burn off additional impurities. Most iron is converted to **steel**. Steel is an **alloy**, which is a mixture of a metal and at least one other element (carbon, in the case of steel). Various elements can be added to steel to give it different properties. Aluminum is less dense than steel and does not rust the way steel does. Aluminum was first isolated from its ore, bauxite, about 100 years ago, by an electrical process. There are many other useful metals and alloys.

Section 12.7—We face potential shortages of many metals and minerals because most of the high-grade ores have been used up. The United States now imports some metals and ores that it used to mine because of these shortages. It takes much less energy to recycle most metals than it does to produce them from ores. However, there are other possible sources of some metals, notably manganese nodules at the bottom of some oceans. The United States produces large amounts of waste, most of which is paper, cardboard, and yard wastes. Municipal solid waste (MSW) used to be discarded in open dumps, but now it is buried in sanitary landfills, burned, or otherwise processed to reduce the amount of MSW. Several new uses for MSW have been developed. The three R's of garbage are: Reduce the amount of garbage, reuse items such as reusable bottles when possible, and recycle when possible. Steel, aluminum, plastic, and paper are easily recycled, and yard waste can be composted. Although birthrates have declined somewhat, Spaceship Earth is crowded and is going to become more so in years to come. We must come to terms with the amounts and types of waste we produce.

■ REVIEW QUESTIONS

1. What is a synergistic effect? Give an example.
2. What environmental problems are associated with the manufacture of cement?
3. What is concrete?
4. Why was copper one of the first metals used?
5. Why did the Bronze Age come about before the Iron Age in most places?
6. What environmental problems are associated with steel mills?
7. What environmental problems are associated with aluminum production?
8. If matter is conserved, how can we ever run out of a metal?
9. List three methods of solid waste disposal. Give the advantages and disadvantages of each.
10. Explain why metals can be recycled fairly easily. What factors limit the recycling of metals?

■ PROBLEMS

Composition of Earth

11. Define *lithosphere*, *hydrosphere*, and *atmosphere*.
12. Name four kinds of minerals found in Earth's crust and give an example of each.
13. What are the third, fourth, and fifth most abundant elements by mass in Earth's crust? Why is there such a difference between the abundance by mass and abundance by atoms?
14. Consult Table 12.2 and write a balanced equation for the dissolving of galena in hydrochloric acid. The products are hydrogen sulfide gas and lead(II) chloride.
15. What materials make up the organic portion of the lithosphere?
16. How did the discovery of fire change the types of materials available to humans?

Silicate Minerals

17. In silica, four O atoms surround each Si atom, yet the formula for silica is SiO_2. Explain how this can be.

18. What is the chemical composition of quartz? How are the basic structural units of quartz arranged?

19. Why does mica occur in sheets? How are the basic structural units of mica arranged?

20. Why does asbestos occur as fibers? How are the basic structural units of chrysotile asbestos arranged?

Modified Silicates

21. What are the three principal raw materials used for making glass?

22. How does glass differ from crystalline silicates in structure and in properties?

23. How is the basic recipe for glass modified to make glasses with special properties?

24. Recycling glass is not quite as economically feasible as is the recycling of many other materials such as metals. Suggest a reason for this.

25. What are the two basic raw materials required for making cement?

26. What is the difference between cement and concrete?

Metals and Ores

27. What are the principal raw materials used in modern iron production? What is the purpose of each?

28. In iron production, what is coke? What is slag?

29. By what kind of chemical process is a metal obtained from its ore? Give an example.

30. By what chemical process does a metal corrode? Give an example.

31. What functions are served by a blast furnace in the metallurgy of iron?

32. What is the purpose of limestone in iron production?

33. What is pig iron? What are its principal impurities?

34. Which metal is the most widely used? Why?

Oxidation and Reduction

For Problems 35–38, answer the following questions. (These problems require some knowledge of material in Chapter 8.) **(a)** What substance is reduced? **(b)** What is the reducing agent? **(c)** What substance is oxidized? **(d)** What is the oxidizing agent?

35. Manganese metal may be prepared by reacting manganese(IV) oxide with silicon.

$$Si + MnO_2 \longrightarrow SiO_2 + Mn$$

36. The *Hunter* method for the production of titanium uses the reaction

$$TiCl_4 + 4\,Na \longrightarrow Ti + 4\,NaCl$$

37. The Toth Aluminum Company produces aluminum from kaolin. One step converts aluminum oxide to aluminum chloride by the following reaction.

$$Al_2O_3 + 3\,C + 3\,Cl_2 \longrightarrow 2\,AlCl_3 + 3\,CO$$

38. Copper(I) oxide, heated in a stream of hydrogen gas, is converted to copper metal.

$$Cu_2O + H_2 \longrightarrow 2\,Cu + H_2O$$

Mass Relationships

Problems 39–40 require some knowledge of material in Chapter 5.

39. Copper metal is prepared by blowing air through molten copper(I) sulfide.

$$Cu_2S + O_2 \longrightarrow 2\,Cu + SO_2$$

How much copper is obtained from 143 g of Cu_2S?

40. What mass in grams of aluminum is required to reduce 33.2 g of calcium oxide? The reaction is

$$3\,CaO + 2\,Al \longrightarrow 3\,Ca + Al_2O_3$$

■ ADDITIONAL PROBLEMS

41. An aluminum plant produces 65 million kg of aluminum per year. How much aluminum oxide is required? How much bauxite is required? (It takes 2.1 kg of crude bauxite to produce 1.0 kg of aluminum oxide.)

42. It takes 17 kWh of electricity to produce 1.0 kg of aluminum. How much electricity does the plant in Problem 41 use for aluminum production in 1 year?

43. Approximately 83,000 troy ounces of gold are mined worldwide each year. If all this gold were formed into a cube, what would be the length in meters of one edge of the cube? The density of gold is 19.3 g/cm^3 and 1 troy oz = 31.10 g.

44. The mass of gem-quality rough diamonds (on next page) mined each year is about 10 million carats. What is the mass of these diamonds in kilograms? (1 carat = 200 mg.)

45. A blast furnace produces 1.0×10^7 kg of pig iron per day. Assume the pig iron is 95% Fe by mass. How many kilograms of iron ore are consumed in this furnace per day? Assume that the ore is 82% by mass hematite (Fe_2O_3).

46. The blast furnace described in Problem 45 produces 0.50 kg of slag per kilogram of pig iron. What is the minimum daily requirement of limestone in this furnace? Assume that the limestone is 91% $CaCO_3$, that the slag is exclusively $CaSiO_3$, and that the limestone is the only source of calcium in the slag.

47. Each person in the United States produces about 4.6 pounds of trash per day. If the trash has an average density of 85 lb per cubic foot, how many cubic feet of trash are produced each year by a small town of 14,000 people? If a sanitary landfill is constructed that is 100 feet wide and 100 feet long, how deep must it be to contain this trash?

48. The equation for the reaction by which a cyanide solution dissolves gold from its ore is

$$4\,Au + 8\,NaCN + 2\,H_2O + O_2 \longrightarrow$$
$$4\,NaAu(CN)_2 + 4\,NaOH$$

What is the minimum mass of sodium cyanide required to dissolve the 10 g of Au in a ton of gold ore?

49. Refer back to Figure 12.1 and to the data on the lithosphere in Section 12.1. Earth is 12,756 km in diameter. Measure the diameter of Earth in the figure and calculate approximately how thick (in mm) the lithosphere should be on that scale. The atmosphere is about 100 km thick; can the atmosphere be shown to scale?

50. Look up the melting points of copper and iron. From this information suggest one reason why the Copper Age preceded the Iron Age.

51. Based on Figure 12.9, about what percentage of municipal waste might be burned as fuel?

52. Return to Problem 47 and examine Figure 12.9. If everyone recycled all paper and cardboard, composted yard wastes and food scraps, and recycled all plastics and metals, how might this change the answer to Problem 47?

53. One pound of plastic bags is about 67 bags. **(a)** What is the mass in kilograms of 500 billion bags? An Australian study concludes that 8.7 plastic bags equal the fuel consumed by driving a car 1 km. How far can a car travel on the equivalent of **(b)** 1 lb of bags? **(c)** 500 billion bags?

54. Compare the environmental effects of **(a)** reusable cleaning cloths, mop heads, and sponges that can be cleaned and reused to use of similar disposable products, and **(b)** cloth diapers that need to be washed to the use of disposable diapers.

■ COLLABORATIVE GROUP PROJECTS

Prepare a PowerPoint, poster, or other presentation (as directed by your instructor) for presentation to the class.

55. Prepare a brief report on one of the metals listed in Table 12.5 and share it with your group. List principal ores and describe how the metal is obtained from one of its ores. Describe some of the uses of the metal and explain how its properties make it suitable for those uses.

56. Prepare a brief report on recycling one of the following types of municipal solid wastes. List advantages and problems involved in the process.
a. glass
b. plastics
c. yard wastes
d. paper
e. lead
f. aluminum

57. Compare disposal in landfills, incineration, and recycling as methods of dealing with each of the types of municipal solid wastes listed in Problem 56. List advantages and disadvantages of each method for each type of waste.

58. Search the Web for recent mine waste disasters or long-term mine waste problems. Prepare a brief report on one of them.

59. Use the Web to find the current estimated population of the United States and of the world. What is the current rate of growth, in percent per year, in **(a)** Germany, **(b)** Kenya, **(c)** India, **(d)** United States, and **(e)** the world?

Earth's atmosphere is a thin, transparent skin of molecules that protects us from harmful ultraviolet rays, provides vital oxygen to living organisms, and warms the planet so that life can flourish. Unfortunately the human impact on the atmosphere is often clearly visible. The brown haze of air pollution over Houston, Texas casts a pall over the city, as seen here.

AIR

QUESTIONS YOU MAY HAVE ASKED YOURSELF

1. Can microwave ovens pollute the air?
2. Could we collect automobile emissions and convert them back to gasoline?
3. What are greenhouse gases?
4. Are global warming and the ozone hole the same problem?
5. What does vehicle "smog testing" actually test?

The Breath of Life

We could live about a month without food. We could even live for several days without water. But without air we cannot live more than a few minutes. Every breath we take is automatic; we seldom think about it. Under normal conditions, we cannot hold our breath for very long, even if we want to.

We may run short of food, and we may run short of fresh water, but we are not likely to run out of air. On the other hand, we might foul the air so badly in some places that it could become unfit to breathe. In some areas the air is already so bad that people become sick from breathing it, and some even die because of it. The World Health Organization estimates that a billion people live in places where the air is substandard (mostly in urban areas) and that outdoor air pollution from vehicles and industrial emissions kills 3 million people a year worldwide. Another 1.6 million die from indoor air pollution, mainly through using solid fuel. Most of the deaths occur in developing countries. We begin with a discussion of the natural chemistry of the atmosphere and then present some of the changes that human intervention is causing on that natural balance and some possible solutions.

13.1 Earth's Atmosphere: Divisions and Composition

We passengers on Spaceship Earth live under a thin blanket of air called the **atmosphere**. It is difficult to measure exactly how deep the atmosphere is. It does not end abruptly but gradually gets thinner as it fades into space. But we know that 99% of the atmosphere lies within 30 km of Earth's surface. If Earth were the diameter of a large apple, the atmosphere would be thinner than the apple's skin.

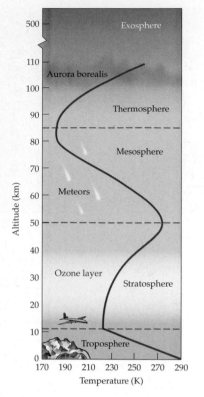

▲ **Figure 13.1** Approximate altitudes and temperature variations of the several layers of the atmosphere. The height of the troposphere varies from about 8 km at the poles to 16 km at the equator. Charged particles is the upper atmosphere interact with gases to create the glowing ring of the aurora borealis.

The atmosphere is divided into five main layers (Figure 13.1). The temperature of each layer varies with altitude. The layer nearest Earth, the *troposphere*, harbors nearly all living things and nearly all human activity. Although other planets in our solar system have atmospheres, Earth's atmosphere is unique in its ability to support higher forms of life. Many people suppose atmospheric temperature decreases as the altitude increases, but this is true primarily only for the troposphere. In the *stratosphere* and *thermosphere*, the temperature actually increases as you go higher. The *exosphere* is the boundary of outer space, and is so thin that satellites such as the International Space Station orbit there for months or even years.

Air is a mixture of gases. Dry air is (by volume) about 78% nitrogen (N_2), 21% oxygen (O_2), and 1% argon (Ar). Damp air can contain up to about 4% water vapor. There are a number of minor constituents, the most important of which is carbon dioxide (CO_2). The concentration of carbon dioxide in the atmosphere increased from about 280 parts per million (ppm) in the preindustrial world to a 2008 value of 385 ppm, and it continues to rise at 2 ppm each year. The composition of the atmosphere is summarized in Table 13.1.

Table 13.1 Composition of Clean Dry Air near Sea Level

Component	Percent by Volume
Nitrogen (N_2)	78.0842
Oxygen (O_2)	20.9463
Argon (Ar)	0.93422
Carbon dioxide (CO_2)	0.0385
Trace substances[*]	0.002

[*]Trace substances include neon (Ne), helium (He), methane (CH_4), krypton (Kr), hydrogen (H_2), dinitrogen monoxide (N_2O), xenon (Xe), ozone (O_3), sulfur dioxide (SO_2), nitrogen dioxide (NO_2), ammonia (NH_3), carbon monoxide (CO), and iodine (I_2).

Self-Assessment Questions

1. Nearly all human activity occurs in the
   a. mesosphere
   b. stratosphere
   c. thermosphere
   d. troposphere

2. The ozone layer is found in the
   a. mesosphere
   b. stratosphere
   c. thermosphere
   d. troposphere

3. The atmosphere
   a. blends into outer space
   b. ends abruptly at 30 km
   c. ends abruptly at 50 km
   d. ends abruptly at the exosphere

4. The approximate percentage of oxygen in the atmosphere is
   a. 10%   b. 20%
   c. 40%   d. 80%

5. After nitrogen and oxygen, the next most abundant gas in dry air is
   a. argon   b. carbon dioxide
   c. hydrogen   d. methane

Answers: 1, d; 2, b; 3, a; 4, b; 5, a

13.2 Chemistry of the Atmosphere

Under ordinary conditions, nitrogen and oxygen coexist in the atmosphere as a simple mixture of gases with little or no reaction between them. Under certain circumstances, however, the two elements can combine, a reaction essential to life itself.

Nitrogen makes up 78% of the atmosphere in the form of diatomic N_2 molecules. Nitrogen is essential to living things, but most animals and plants cannot assimilate N_2 directly. How then has nitrogen become a component of so many organic compounds (Chapter 9), such as the proteins and nucleic acids vital for living organisms? The answer is that organisms first must *fix* nitrogen—that is, combine nitrogen with another element. This process does not occur readily because the two nitrogen atoms are held together by a strong triple bond (recall the Lewis structure of N_2, Chapter 4). A brief look at the *nitrogen cycle* offers insight into how nitrogen is incorporated into other compounds in nature.

The Nitrogen Cycle

Several natural phenomena convert nitrogen gas to more usable forms. Lightning fixes nitrogen by causing it to combine with oxygen, forming first nitrogen monoxide (NO), commonly called nitric oxide, and then nitrogen dioxide (NO_2).

$$N_2 + O_2 + energy \xrightarrow{\text{lightning}} 2\,NO$$
$$2\,NO + O_2 \longrightarrow 2\,NO_2$$

Nitrogen dioxide reacts with water to form nitric acid (HNO_3).

$$3\,NO_2 + H_2O \longrightarrow 2\,HNO_3 + NO$$

Nitric acid falls in rainwater, adding to the supply of available nitrates in the oceans and the soil, and acidifying streams and lakes (Chapter 14).

Nitrogen is fixed industrially by combining nitrogen with hydrogen to form ammonia, a procedure called the *Haber–Bosch process* (further discussed in eChapter 20).

$$N_2 + 3\,H_2 \longrightarrow 2\,NH_3$$

This technology has greatly increased our food supply because the availability of fixed nitrogen is often the limiting factor in the production of food. Not all consequences of this intervention have been favorable, however; excessive runoff of nitrogen fertilizer has led to serious water pollution problems in some areas (which we discuss in Chapter 14).

Most nitrogen fixation is performed by bacteria in the roots of legumes (peas, beans, clover, and the like). Certain types of bacteria reduce N_2 to ammonia (NH_3). Other bacteria oxidize ammonia to nitrites (NO_2^-), and still others oxidize nitrites to nitrates (NO_3^-). Other plants are then able to take up the nitrates from the soil. Animals can get their required nitrogen compounds from the plants. The **nitrogen cycle** (Figure 13.2) is completed by the action of other types of microbes, which can use nitrate ion as their oxygen source in the decomposition of organic matter and release N_2 gas back to the atmosphere.

The Oxygen Cycle

Oxygen makes up 21% of Earth's atmosphere. As with nitrogen, a balance is set up between consumption and production of oxygen in the troposphere. Both plants and animals use oxygen in the metabolism of foods. The decay and combustion of plant and animal materials consume oxygen and produce carbon dioxide, as does the burning of fossil fuels like coal, natural gas, and oil. The rusting of metals and the weathering of rocks also consume oxygen. A simplified **oxygen cycle**, illustrated in Figure 13.3, shows the influence of green plants, including one-celled organisms (phytoplankton) in the sea, that constantly consume carbon dioxide in photosynthesis and also replenish the oxygen supply.

$$6\,CO_2 + 6\,H_2O \longrightarrow C_6H_{12}O_6 + 6\,O_2$$

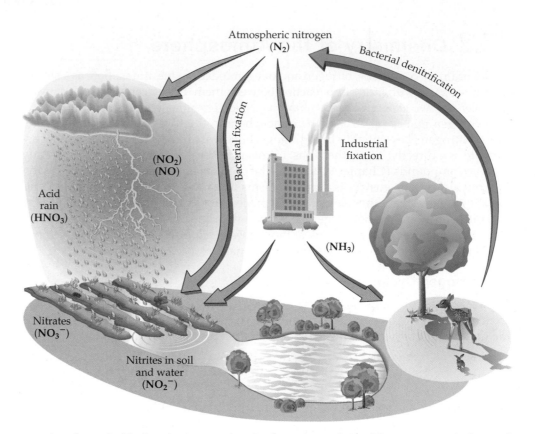

▶ **Figure 13.2** The nitrogen cycle.

Another vital balancing act occurs in the stratosphere. There oxygen is formed by the action of ultraviolet (UV) rays on water molecules. Perhaps more important, some oxygen is converted to ozone (Section 13.9).

$$3 \, O_2(g) + \text{energy (UV radiation)} \longrightarrow 2 \, O_3(g)$$

This ozone absorbs high-energy UV radiation that might otherwise make higher forms of life on Earth impossible. We discuss the atmosphere in more detail in subsequent sections, paying particular attention to changes wrought by human activities.

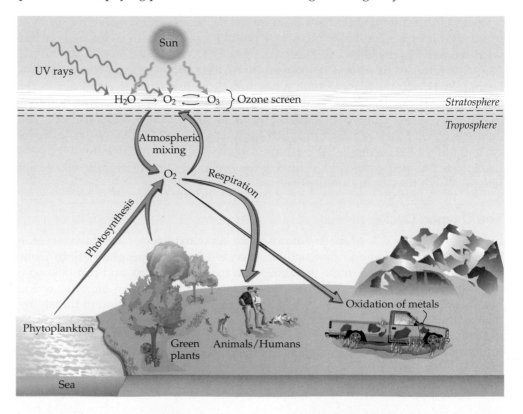

▶ **Figure 13.3** The oxygen cycle.

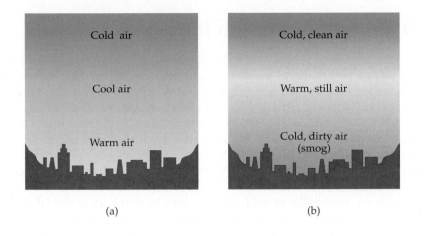

(a) (b)

◀ **Figure 13.4** Ordinarily, the air gets colder as the altitude increases (a). During an atmospheric (thermal) inversion (b), a cold layer near the surface lies beneath a warmer layer.

Temperature Inversion

Normally, air is warmest near the ground and gets cooler at moderate altitudes. On clear nights, the ground cools rapidly, but the wind usually mixes the cooler air near the ground with the warmer air above it. Occasionally the air is still, and the lower layer of cold air becomes trapped by the layer of warm air above it, a condition known as a **temperature inversion** (or a thermal inversion). Pollutants in the cooler air are trapped near the ground, and the air can become quite seriously polluted in a short time (Figure 13.4).

A temperature inversion can also occur when a warm front collides with a cold front. The less dense warm air mass slides over the cold air mass, producing a condition of atmospheric stability. Because there is no vertical air movement, the air stagnates and air pollutants accumulate.

Self-Assessment Questions

1. High-energy natural events such as lightning can convert N_2 to
   a. amino acids   **b.** ammonia   **c.** nitrogen oxides   **d.** proteins

2. Nitrogen fixation refers to conversion of atmospheric
   a. N_2 to biologically useful forms such as NH_3
   b. NH_3 to biologically useful forms such as NO_3^-
   c. NO_x into N_2
   d. NO_x into amino acids

3. The industrial process by which ammonia is made from nitrogen gas uses
   a. bacteria   **b.** electrolysis   **c.** hydrogen   **d.** ultraviolet radiation

4. Oxygen is added to the atmosphere by
   a. photosynthesis   **b.** respiration
   c. volcanic eruptions   **d.** weathering of rocks

5. Which of the following processes removes carbon dioxide from the atmosphere?
   a. photosynthesis   **b.** respiration
   c. temperature inversions   **d.** volcanic eruptions

6. Which of the following processes consumes oxygen?
   a. breakdown of ozone   **b.** corrosion of metals
   c. Haber–Bosch process   **d.** photosynthesis

7. A thermal inversion is a meteorological condition that
   a. calls for wearing thermal underwear
   b. commonly occurs over flat country
   c. concentrates air pollutants
   d. usually ends a bad air pollution episode

Answers: 1, c; 2, a; 3, c; 4, a; 5, a; 6, a; 7, c

13.3 Pollution through the Ages

▲ In a dramatic example of natural pollution, dust from an Asian storm crossed the Pacific to the United States in April 2001, reaching as far east as Maryland.

Air pollution has always been with us. Wildfires, windblown dust, and volcanic eruptions added pollutants to the atmosphere long before humans existed. Volcanoes spew ash and poisonous gases into the atmosphere. In July 2008 the Okmok Caldera on Umnak Island (one of Alaska's Aleutian Islands) erupted, sending dust and ash to 50,000 feet. The 1995 eruptions of the Soufriere Hills volcano on Montserrat covered much of the Caribbean island with ash, forcing the evacuation of more than half the country. Kilauea volcano in Hawaii emits 200–300 t of sulfur dioxide per day, leading to acid rain downwind that has created a barren region called the Kau Desert.

Dust storms, especially in arid regions, add massive amounts of particulate matter to the atmosphere. Dust from the Sahara Desert often reaches the Caribbean and South America. Dust and pollution from China blanket Japan and even reach western North America. Swamps and marshes emit noxious gases. Nature isn't always benign.

A Closed Ecosystem?

If you think for a moment about the oxygen and nitrogen cycles, you may realize that a *closed ecosystem* should be possible. In such a system, the plant and animal life would be balanced. The O_2 given off by the plants would sustain the animal life, and the CO_2 given off by the animals would be sufficient to keep the plants alive. Other animal wastes would supply the nitrogen needed by the plants, and the animals would partially consume the plants or one another for food. Trace elements would need to be present in sufficient quantity. Sunlight would provide the energy necessary to "run" the ecosystem; nothing else goes in and nothing comes out.

Earth is itself a closed ecosystem. The planet absorbs sunlight energy, but no significant amount of matter enters or leaves. But can such a system be constructed artificially? The answer is of great interest in a number of fields, including space exploration. A long space voyage would require that huge amounts of resources be carried along—at enormous expense. The required resources could be minimized by carrying along plants to provide oxygen and consume carbon dioxide and human wastes.

Tiny ecosystems are available commercially as desktop novelties. These systems have a limited lifetime of a few years. *Biosphere II* was a large-scale experiment involving humans and a semiclosed ecosystem. The experiment was cut short because of several problems, including rising levels of nitrogen oxides, water pollution, unforeseen low levels of carbon dioxide, and other difficulties that could only be examined in such an experiment. Although additional studies and experiments are slated, Earth remains the only long-term successful closed ecosystem.

▲ The Ecosphere® is a sealed, self-contained ecosystem in which plants and animals can live for a year or more with sunlight as the only "input."

The Air Our Ancestors Breathed

People have always altered their environment. The discovery of tools and the use of fire forever changed the balance between people and their environment. The problems of air pollution probably first occurred when fires were built in poorly ventilated caves or other dwellings. People cleared land, making larger dust storms possible. They built cities, and the soot from their hearths and the stench from their wastes filled the air. The Roman author Seneca wrote in 61 C.E. of the stink, soot, and "heavy air" of the imperial city. In 1257 the queen of England had to move away from the city of Nottingham because she could not endure the heavy smoke. The industrial revolution brought even worse air pollution. People burned coal to power factories and to heat homes. Soot, smoke, and sulfur dioxide filled the air. The good old days? Not in factory towns. But there were large rural areas that were relatively unaffected by air pollution.

Air pollution today is much more complex than that of our ancestors. When society changes its activities, it often changes the nature of its waste materials, including the ones it dumps into the atmosphere. Earth has gone through many changes. Change is not new, but today's rapid rate of change is unprecedented. The present pace of change is at a level that our environment may not be able to absorb.

Pollution Goes Global

Air pollution problems are no longer just local concerns. Problems are worldwide. Winds and precipitation do purify the air, but we are pouring more pollutants into the atmosphere than it can readily handle. Air pollution knows no political boundaries: Pollution originating in the United States contributes to acid rain in Canada and to strained relationships between the two countries. Contaminants from England and Germany pollute the snow in Norway. Satellites orbiting at 150 km and higher can trace plumes of pollution from China to North America. Within the United States, Los Angeles smog drifts to Colorado and beyond. Pollution from midwestern power plants leads to acid rain in the Northeast.

Air Pollution in China

As China strives to become an industrial power, its people are paying a heavy price in pollution. Coal burning supplies about three-fourths of China's commercial energy needs. China now accounts for 36% of the world's steel output, consuming huge amounts of coal and power. The coal has a high sulfur content, and emission controls are often inadequate. As a result, levels of sulfur dioxide and particulate matter (ash, soot, and dust) are among the highest in the world. According to the Worldwatch Institute, 16 of the 20 cities with the most polluted air are in China. More than 80% of 500 Chinese cities had sulfur dioxide or nitrogen oxide levels above maximum guideline levels set by the World Health Organization (WHO). Many of the cities had sulfur dioxide emissions more than double the standard. Many also exceeded WHO guidelines for particulate matter. According to World Bank estimates, some 590,000 people a year will suffer premature deaths due to urban air pollution in China in the first two decades of the twenty-first century.

China had few privately owned motor vehicles until the 1980s. However, according to the Chinese Ministry of Communications, the number on the road will grow from 20 million in 2005 to 140 million by 2020. Most vehicles are operated in large cities, and because few have effective emission controls, they contribute heavily to the smog in these cities.

China is attacking the problem by closing heavily polluting factories in some larger metropolitan areas, investing in natural gas, and using cleaner-burning briquettes as replacements for raw coal as a fuel for domestic cooking and heating.

▲ The polluted air of Beijing, China, shrouds the capital in a nasty haze on November 5, 2005. People were warned to stay indoors as the pollution index hit its highest level for the second day.

A N S W E R

1. Can microwave ovens pollute the air?

Microwaves are a form of energy, like visible light, radio waves, and ultraviolet light from tanning beds. Although high levels of microwave radiation may be harmful, microwaves are not matter and are not pollutants.

As population increases, the world is becoming more urban. Especially in developing countries, industrialization takes precedence over the environment. Cities in China, Iran, Mexico, Indonesia, and many other countries have experienced frightening episodes of air pollution. Large metropolitan areas are most afflicted by air pollution, but rural areas are affected, too. Neighborhoods around smoky factories have seen evidence of increased rates of spontaneous abortion and poor wool quality in sheep, decreased egg production and high mortality in chickens, and increased feed and care requirements for cattle. Plants are stunted, deformed, and even killed.

What is a **pollutant**? It is too much of any substance in the wrong place or at the wrong time. A chemical may be a pollutant in one place and helpful in another. For example, ozone is a natural and important constituent of the stratosphere, where it shields Earth from life-destroying UV radiation. In the troposphere, however, ozone is a dangerous pollutant (Section 13.9).

Self-Assessment Questions

1. Natural air pollution has been around
   a. for millions of years
   b. since the dawn of the human race
   c. since the last century
   d. since the start of the industrial revolution

2. Air pollution
   a. began in the twentieth century
   b. comes only from industries
   c. is a global problem
   d. consists only of synthetic chemicals

3. The air in thirteenth-century England
   a. smelled of ale
   b. smelled of roses
   c. was smoky from wood burning
   d. was pristine

Answers: 1, a; 2, c; 3, c

13.4 Coal + Fire → Industrial Smog

The word *smog* is a contraction of the words *smoke* and *fog*. There are two basic types of smog. Polluted air associated with industrial activities is often called **industrial smog**. (*Photochemical* smog is discussed in Section 13.6.) It is characterized by the presence of smoke, fog, sulfur dioxide, and particulate matter such as ash and soot. The burning of coal, especially high-sulfur coal such as that found in the eastern United States, China, and Eastern Europe, causes most industrial smog.

Chemistry of Industrial Smog

The chemistry of industrial smog is fairly simple. High-grade coal is a complex combination of organic materials and inorganic materials. The organics are mainly carbon and burn when the coal is combusted. The inorganic materials—minerals—wind up as ash. Some coal can contain 10% or more sulfur. When burned, the carbon in coal is oxidized to carbon dioxide and heat is given off.

$$C(s) + O_2(g) \longrightarrow CO_2(g) + heat$$

▲ The burning of coal in electric power plants and other industrial operations is the main source of industrial smog characterized by high levels of sulfur dioxide, particulate matter, and sometimes carbon monoxide. Industrial smog can be alleviated by the use of electrostatic precipitators, scrubbers, and other devices.

An Air Pollution Episode: London, England

So common a problem in London, industrial smog was once called *London smog*. A notorious episode of this type of smog began in London on Thursday, December 4, 1952. A large, cold air mass moved into the valley of the Thames River. A temperature inversion placed a blanket of warm air over the cold air. With nightfall, a dense fog and below-freezing temperatures caused the people of London to heap sulfur-rich coal into their stoves. These fires and coal-fired power plants poured sulfur dioxide and soot into the air. The next day (Friday), schools closed and transportation was disrupted. People continued to burn coal because the temperature remained below freezing.

Saturday was a day of darkness. For over 30 km around London, no light came through the smog. A performance of the opera *La Traviata* had to be abandoned because smog obscured the stage. The air remained cold and still, and the coal fires continued to burn throughout the weekend. On Monday, December 8, more than 100 people died of respiratory and related conditions. People tried to sleep sitting up in chairs in order to breathe a little easier. The city's hospitals overflowed with patients with pneumonia and bronchitis.

By the time a breeze cleared the air on Tuesday, December 9, more than 4000 deaths had been attributed to the smog. Many others became ill, some of whom died later. The total premature deaths before the death rate returned to normal the following summer are now estimated to be as many as 12,000 people. That is more people than were ever killed in any single tornado, mine disaster, shipwreck, or airplane crash. A second incident occurred in December 1991, in which 160 people died. Air pollution episodes may not be as dramatic as other disasters, but they can be just as deadly.

▲ The great London smog episode of 1952.

Not all the carbon is completely oxidized, and some of it winds up as carbon monoxide.

$$2\,C(s) + O_2(g) \longrightarrow 2\,CO(g)$$

Still other carbon, essentially unburned, ends up as soot.

Sulfur Oxides

The sulfur in coal also burns, forming sulfur dioxide, a choking, acrid gas.

$$S(s) + O_2(g) \longrightarrow SO_2(g)$$

Sulfur dioxide is readily absorbed in the respiratory system. It is a powerful irritant and is known to aggravate the symptoms of people who suffer from asthma, bronchitis, emphysema, and other lung diseases.

And things get worse. Some of the sulfur dioxide reacts further with oxygen in the air to form sulfur trioxide.

$$2\,SO_2(g) + O_2(g) \longrightarrow 2\,SO_3(g)$$

The two oxides of sulfur are often designated collectively as SO_x. Sulfur trioxide then reacts with water to form sulfuric acid.

$$SO_3(g) + H_2O(l) \longrightarrow H_2SO_4(l)$$

Fine droplets of this acid form an aerosol, a suspension of very small particles of a solid or droplets of a liquid dispersed in a gas. Sulfuric acid is a strong acid, and this aerosol is even more irritating to the respiratory tract than sulfur dioxide.

Particulate Matter

Usually, industrial smog also has high levels of **particulate matter (PM)**, solid and liquid particles of greater than molecular size. The largest particles are often visible

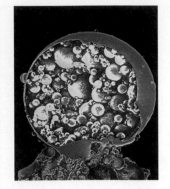

▲ False-color scanning electron micrograph of PM from a coal-burning power plant.

Although the incidence of asthma has increased dramatically in recent years it is unlikely that air pollution is the direct cause of the increase. Rather, medical studies show that pollutants at moderate levels can exert a small but measurable effect on the lung function and level of symptoms of asthmatic individuals. It seems more likely that pollutants and allergens interact to enhance the severity of asthma attacks.

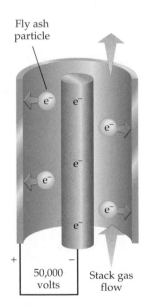

▲ **Figure 13.5** A cross section of a cylindrical electrostatic precipitator for removing particulate matter from smokestack gases. Electrons from the negatively charged discharge electrode (in the center) become attached to the particles of fly ash, giving them a negative charge. The charged particles are then attracted to and deposited on the positively charged collector plate.

in the air as dust and smoke. PM consists mainly of the mineral matter that occurs in coal and of *soot* (unburned carbon). Minerals do not burn, even in the roaring fire of a huge factory or power plant boiler. Some mineral matter is left behind as *bottom ash,* but much of it is carried aloft in the tremendous draft created by the fire. This *fly ash* settles over the surrounding area, covering everything with dust. It is also inhaled, contributing to respiratory problems in animals and humans. Small particles—those less than 10 μm in diameter (PM10 particulates)—are believed to be especially harmful. They contribute to respiratory and heart disease. The Environmental Protection Agency (EPA) also monitors fine particles, those less than 2.5 μm in diameter (PM2.5). EPA studies indicate that tens of thousands of premature deaths a year are linked to PM and that 20,000 of them are due to PM2.5. PM concentrations have been lowered significantly since nationwide monitoring began in 1999. In 2007, one in three people in the United States lived in counties with particle pollution levels higher than EPA's standards for PM2.5, PM10, or both.

Health and Environmental Effects of Industrial Smog

Minute droplets of liquid sulfuric acid and smaller particulates (PM10 and smaller) are easily trapped in the lungs. Interaction may considerably magnify the harmful effects of pollutants. A certain level of sulfur dioxide, without the presence of PM, might be reasonably safe. A particular level of PM might be fairly harmless without sulfur dioxide around, but the effect of both together might well be deadly. *Synergistic effects* such as this are quite common whenever chemicals act together. Two such synergistic interactions are those of asbestos and cigarette smoke (discussed in Chapter 12) and the synergistic interactions of certain drugs (discussed in Chapter 18).

When the pollutants in industrial smog come into contact with the alveoli of the lungs, the cells are broken down. The alveoli lose their resilience, making it difficult for them to expel carbon dioxide. Such lung damage contributes to pulmonary emphysema, a condition characterized by increasing shortness of breath.

The oxides of sulfur and the aerosol mists of sulfuric acid also damage plants. Leaves become bleached and splotchy when exposed to SO_x. The yield and quality of farm crops can be severely affected. These compounds are also major ingredients in the production of acid rain.

What to Do about Industrial Smog

Much effort goes into the prevention and alleviation of industrial smog. PM can be removed from smokestack gases in several ways:

- An **electrostatic precipitator** (Figure 13.5) induces electric charges on the particles, which are then attracted to oppositely charged plates and deposited.
- *Bag filtration* works much like the bag in a vacuum cleaner. Particle-laden gases pass through filters in a bag house. These filters can be cleaned by shaking and by periodically blowing air through them in the opposite direction.
- A *cyclone separator* is arranged so that the stack gases spiral upward with a circular motion. The particles hit the outer walls, settle out, and are collected at the bottom.
- A **wet scrubber** removes PM by passing stack gases through water. The water is usually sprayed in as a fine mist. The wastewater has to be treated to remove the particulates, which adds to the cost of this method.

The choice of device depends on the type of coal being burned, the size of the power plant, and other factors. All require energy—electrostatic precipitators use 10% of the plant's output—and the collected ash has to be put somewhere. Some of the ash is used to make concrete, as a substitute for aggregate in road base, as a soil modifier, and for backfilling mines. The rest has to be stored; most of it goes into ponds, and the rest into landfills.

It is harder to remove sulfur dioxide than PM. Sulfur can be removed by processing coal before burning, but both the flotation method and gasification or liquefaction processes (Chapter 15) are expensive. Another way to get rid of sulfur is to scrub sulfur dioxide out of stack gases after the coal has been burned. The most common scrubber uses the limestone–dolomite process. Limestone ($CaCO_3$) and dolomite (a mixed calcium–magnesium carbonate) are pulverized and heated. Heat drives off carbon dioxide to form calcium oxide (lime), a basic oxide that reacts with sulfur dioxide to form solid calcium sulfite ($CaSO_3$).

$$CaCO_3(s) + heat \longrightarrow CaO(s) + CO_2(g)$$

$$CaO(s) + SO_2(g) \longrightarrow CaSO_3(s)$$

This by-product presents a sizable disposal problem. Removal of 1 t of sulfur dioxide produces almost 2 t of solids. Some modified scrubbers oxidize the calcium sulfite to calcium sulfate, $CaSO_4$.

$$2\,CaSO_3(s) + O_2(g) \longrightarrow 2\,CaSO_4(s)$$

Calcium sulfate is much more useful than calcium sulfite because the sulfate is used to make commercial products such as plasterboard. In nature, calcium sulfate occurs as gypsum ($CaSO_4 \cdot 2\,H_2O$), which has the same composition as hardened plaster.

Self-Assessment Questions

1. Which of the following is *not* usually released from burning fossil fuels?
   a. CO_2 **b.** NO_x **c.** O_3 **d.** SO_2

2. Carbon monoxide is formed as a product of
   a. animal respiration
   b. burning coal but not burning natural gas
   c. combustion of fossil fuels with excess O_2
   d. partial combustion of fossil fuels

3. The pollutant usually described by size rather than composition is
   a. acid rain   **b.** CFCs
   c. particulates   **d.** photochemicals

4. Which of the following pairs of pollutants acts synergistically in causing lung damage?
   a. O_3 and hydrocarbons   **b.** O_3 and particulates
   c. SO_2 and NO_x **d.** SO_2 and particulates

5. A device that cleans the smoke from flue gases by producing static charges is called a(n)
   a. catalytic converter   **b.** electric generator
   c. electrostatic precipitator   **d.** particle accelerator

6. Wet scrubbers using lime remove which pollutant from flue gases?
   a. CO_2 **b.** CO   **c.** NO_x **d.** SO_2

7. A common by-product of sulfur removal from flue gases is
   a. CaO   **b.** $CaCl_2$ **c.** $CaCO_3$ **d.** $CaSO_4$

Answers: 1, c; 2, d; 3, c; 4, d; 5, c; 6, d; 7, d

13.5 Automobile Emissions

When a hydrocarbon burns in sufficient oxygen, the products are carbon dioxide and water. For example, consider the combustion of octane, one of the hundreds of hydrocarbons that make up the mixture we call gasoline:

$$2\,C_8H_{18}(l) + 25\,O_2(g) \longrightarrow 18\,H_2O(g) + 16\,CO_2(g)$$

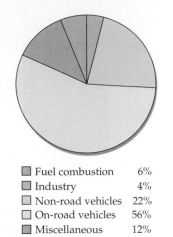

☐ Fuel combustion   6%
☐ Industry   4%
☐ Non-road vehicles   22%
☐ On-road vehicles   56%
☐ Miscellaneous   12%

▲ Sources of carbon monoxide in the United States.

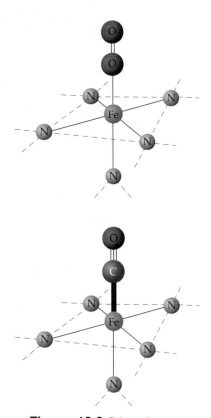

▲ **Figure 13.6** Schematic representations of a portion of the hemoglobin molecule (see Figure 16.15). Carbon monoxide bonds much more tightly than oxygen, as indicated by the heavier bond line.

Because both of the products are normal constituents of air, they are not generally considered pollutants. Unfortunately, internal combustion engines do not burn fuel with perfect efficiency, and there are compounds other than hydrocarbons in gasoline. The next sections describe several types of pollutants that result.

Carbon Monoxide: The Quiet Killer

When insufficient oxygen is present during combustion, carbon monoxide (CO) is formed. Millions of metric tons of this invisible but deadly gas are poured into the atmosphere each year. In the United States, carbon monoxide makes up more than 60% (by mass) of all air pollutants entering the atmosphere, with more than three-fourths of all CO emissions coming from transportation sources.

In urban areas, the motor vehicle contribution to carbon monoxide pollution can be 85 to 95%. Other sources of CO include industrial processes such as metals processing, residential wood burning, and natural sources such as forest fires. The U.S. EPA has set danger levels of 9 ppm carbon monoxide (average) over 8 hours and 35 ppm (average) over 1 hour. On streets, danger levels are exceeded much of the time. Even in off-street urban areas, levels often average 7–8 ppm. Such levels do not cause immediate death, but over a long period exposure can cause physical and mental impairment.

Carbon monoxide is a nonirritating, invisible, odorless, tasteless gas. We can't tell that it is around except by using a CO detector or test reagents. Drowsiness is usually the only symptom, and drowsiness is not always unpleasant. How many auto accidents are caused by drowsiness or sleep induced by carbon monoxide? No one knows for sure.

Carbon monoxide exerts its insidious effect by tying up the hemoglobin in the blood. The normal function of hemoglobin is to transport oxygen (Figure 13.6). Carbon monoxide binds to hemoglobin so strongly that the hemoglobin is hindered in carrying oxygen. The symptoms of carbon monoxide poisoning are, therefore, those of oxygen deprivation. All except the most severe cases of acute carbon monoxide poisoning are reversible, but prolonged hospital stays with oxygen therapy are sometimes necessary.

In chronic carbon monoxide poisoning, the heart has to work harder to supply oxygen to the tissues, increasing chances of a heart attack. People already suffering from heart or lung disease may be severely affected. Thousands of cases of carbon monoxide poisoning occur in the United States each year, and hundreds die. Non-lethal levels of carbon monoxide have been shown to impair time-interval discrimination. In severe cases, performance on psychomotor tests is impaired.

Nitrogen Oxides: Some Chemistry of Amber Air

Nitrogen oxides are formed when N_2 reacts with O_2 at high temperatures. Because air is 78% N_2 and 21% O_2, nitrogen oxides can be formed during the combustion of any fuel regardless of whether nitrogen is present in the fuel. Automobile exhaust and power plants that burn fossil fuels are major sources of nitrogen oxides. When nitrogen and oxygen combine, the main product is nitrogen monoxide (NO or nitric oxide).

$$N_2(g) + O_2(g) \longrightarrow 2\,NO(g)$$

The NO is then oxidized to nitrogen dioxide.

$$2\,NO(g) + O_2(g) \longrightarrow 2\,NO_2(g)$$

Nitrogen dioxide is an amber-colored gas that causes eye irritation and a brownish haze. The two nitrogen oxides are often designated collectively as NO_x; they play a vital (villain's) role in atmospheric chemistry.

At present atmospheric levels, nitrogen oxides don't seem particularly dangerous in themselves. Nitric oxide at high concentrations reacts with hemoglobin; as in carbon monoxide poisoning, this leads to oxygen deprivation. Such high levels seldom, if ever, result from ordinary air pollution, but they might be reached in areas close to industrial sources. Nitrogen dioxide is an irritant to the eyes and the respiratory system. Tests with laboratory animals indicate that chronic exposure in the range of 10–25 ppm might lead to emphysema and other degenerative diseases of the lungs.

The most serious environmental effect of nitrogen oxides is that they lead to smog formation. These gases also contribute to the fading and discoloration of fabrics and, by forming nitric acid, contribute to acid rain (Section 13.7). They also contribute to crop damage, although their specific effects are difficult to separate from those of sulfur dioxide and other pollutants.

Volatile Organic Compounds

Substances called **volatile organic compounds (VOCs)** are major contributors to smog formation. There are many sources of VOCs, including gasoline vapors, combustion products of fuels, and consumer products such as paints and aerosol sprays. Many VOCs are hydrocarbons.

Hydrocarbons are released from a variety of natural sources, such as decay in swamps. Only about 15% of all hydrocarbons found in the atmosphere are put there by people. In most urban areas, however, the processing and use of gasoline are the major sources of hydrocarbons in the environment. Gasoline can evaporate anywhere along the line, contributing substantially to the total amount of hydrocarbons in urban air. This is why it is important to have a good seal on the cap to your gas tank. The automobile's internal combustion engine also contributes by exhausting unburned and partially burned hydrocarbons.

Hydrocarbons can act as pollutants even when they aren't burned. Certain hydrocarbons, particularly alkenes (Section 9.4), combine with oxygen atoms or ozone molecules to form aldehydes. As a class, aldehydes have foul, irritating odors. Another series of reactions involving hydrocarbons, oxygen, and nitrogen dioxide leads to the formation of peroxyacetyl nitrate (PAN).

$$\overset{\overset{\displaystyle O}{\|}}{CH_3C}-O-ONO_2$$

PAN

Ozone, aldehydes, and PAN are responsible for much of the destruction wrought by smog. They make breathing difficult and cause the eyes to smart and itch. People who already have respiratory ailments may be severely affected, and the very young and the very old are particularly vulnerable.

Hydrocarbons from automobiles come largely from those with no pollution control devices or those with devices that are not operating properly. Significant amounts enter the atmosphere from spillage during automobile fueling.

2. Could we collect automobile emissions and convert them back to gasoline?
Although this might be possible, the energy needed to convert the carbon dioxide and water vapor back to gasoline and oxygen would be *greater* than the energy obtained from burning the gasoline in the first place!

ANSWER

Self-Assessment Questions

1. A colorless, odorless, and tasteless pollutant that is dangerous wherever it is found is
   a. CO   **b.** CO_2 **c.** NO   **d.** SO_2

2. Which of the following is *not* a problem caused by NO_x emissions?
   a. addiction   **b.** formation of nitric acid, leading to acid rain
   c. formation of smog   **d.** eye irritation

3. Volatile organic compounds (VOCs) are almost always composed of molecules that contain
   a. C and H   **b.** C and N   **c.** N and H   **d.** N and O

4. Volatile organic compounds (VOCs) come mainly from
   a. automobiles   **b.** dry-cleaning plants
   c. electric power plants   **d.** natural sources

5. Alkenes react with oxygen atoms to form
   a. alcohols   **b.** aldehydes   **c.** alkanes   **d.** ozone

Answers: 1, a; 2, a; 3, a; 4, d; 5, b

13.6 Photochemical Smog: Making Haze while the Sun Shines

▲ Photochemical smog results from the action of sunlight on oxides of nitrogen emitted from automobiles and other high-temperature combustion sources. An amber haze like that shown here over Paris in June 2003 characterizes this type of smog.

Individually ozone, aldehydes, and PAN are worrisome and have dramatic environmental impacts, including acid rain. Collectively, however, and with the inclusion of sunlight, these can form a complex series of reactions that produces **photochemical smog**, visible as an amber haze.

Unlike industrial smog, which accompanies cold, damp air, photochemical smog usually occurs during dry, sunny weather. The warm, sunny climate that has drawn so many people to the Los Angeles area is also the perfect setting for photochemical smog at any time of year. The principal culprits are unburned hydrocarbons (HC) and nitrogen oxides (NO_x) from automobiles.

The chemistry of photochemical smog is exceedingly complex (Figure 13.7). It starts when NO_2 absorbs a photon of sunlight and breaks apart to NO and O. The oxygen atoms are quite reactive; they combine with other components of automobile exhaust and the atmosphere to produce a variety of irritating and toxic chemicals, including ozone (O_3).

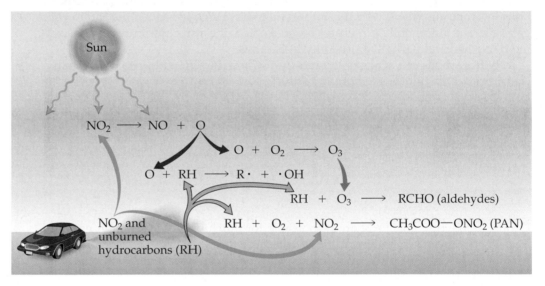

▶ **Figure 13.7** Some chemical processes in the formation of photochemical smog. Many other reactive intermediates have been omitted from this simplified scheme.

The development of air pollutants on a typical sunny summer day is shown in Figure 13.8. As traffic builds up in the early morning, NO enters the air from

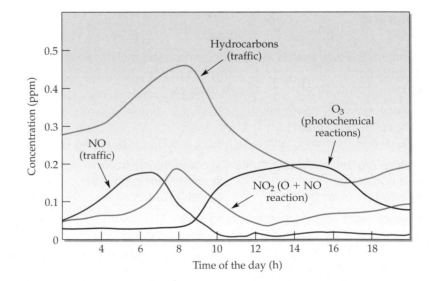

▶ **Figure 13.8** Concentrations of several urban air pollutants at different times during a typical sunny day. Hydrocarbons and NO are produced first. As they interact with sunlight, NO_2 and then ozone forms.

exhausts and begins to react with O_2 in the air to form NO_2. Hydrocarbons enter the air from exhausts as unburned components of fuel, and from fueling leaks and spills. Sunlight splits the NO_2 into NO and O atoms. The O atoms react with O_2 to form ozone (O_3). Ozone levels continue to rise all day and then decrease as the sun sets because there is less energy and fewer ingredient gases.

Solutions to Photochemical Smog

Photochemical smog requires the action of sunlight on nitrogen oxides and hydro-carbons. Reducing the quantity of any of these would diminish the amount of smog. It isn't likely that we would want to reduce the amount of sunlight, so let's focus on the other two.

Hydrocarbons have many important uses as solvents. If we could reduce the quantity of hydrocarbons entering the atmosphere, the amounts of aldehydes and PAN formed would also be reduced. Improved design of storage and dispensing systems has decreased hydrocarbon emissions from gasoline stations. Similarly, modified gas tanks and crankcase ventilation systems have reduced evaporative emissions from automobiles. The main approach, however, has been through the use of **catalytic converters** to reduce hydrocarbon and carbon monoxide emissions in automotive exhausts. The oxidation catalyst in these converters is a precious metal such as platinum (Pt) and/or palladium (Pd). Hydrocarbons and carbon monoxide react rapidly with oxygen on the surface of the metal, forming water and carbon dioxide.

Reducing the quantity of nitrogen oxides is more difficult. Whereas carbon monoxide and hydrocarbons are removed by oxidation, NO must be reduced to N_2. Lowering the operating temperature of an engine helps, but the engine then becomes less efficient. Running an engine on a richer (more fuel, less air) mixture lowers NO_x emissions but tends to raise carbon monoxide and hydrocarbon emissions. With a separate reduction catalyst, such as rhodium (Rh) or palladi-um (Pd), some of the carbon monoxide can be used to reduce the quantity of nitric oxide.

$$2\,NO(g) + 2\,CO(g) \longrightarrow N_2(g) + 2\,CO_2(g)$$

The remaining carbon monoxide and the hydrocarbons are then oxidized over the platinum catalyst. Over the past 30 years, the use of catalytic converters has pre-vented more than 12 billion tons of harmful exhaust gases from entering Earth's atmosphere.

As gasoline prices have soared, sales of *hybrid vehicles* have increased dramati-cally. A hybrid vehicle uses both an electric motor and a gasoline engine. The small gasoline engine drives the vehicle most of the time and recharges the batteries that run the electric motor. The electric motor provides additional power when it is needed (for example, when climbing a hill). Hybrid vehicles are significantly more efficient than conventional gasoline vehicles. The initial cost of a hybrid vehicle car is generally $3,000 to $6,000 higher than a conventional vehicle of a similar size but can save you money in the long run. For a typical year's driving (15,000 miles) a conventional vehicle that gets 20 mi/gal burns 750 gallons of fuel. A hybrid that gets 45 mi/gal burns just 330 gallons, a savings of 420 gallons. At $4 per gallon that would be $1680 a year or $8400 over five years. Although there is a concurrent reduction in air pollution, a complete evaluation of environmental cost would have to include the production, replacement, and recycling of the batteries. Research continues in the development of more compact, more efficient batteries that gener-ate less pollution in manufacture.

Catalytic Converters and Nanotechnology

For maximum effectiveness, the catalysts inside a catalytic converter must have a large surface area because the reactions take place on the solid surface. To achieve this, the catalyst is thinly coated on a collection of small ceramic beads or honeycomb matrix. One liter of ceramic pellets can have up to 500,000 m^2 of support surface. A honeycomb mesh support can have a surface area of about 23,000 m^2. The support is coated with 1 or 2 g of precious metals such as palladium, rhodium, and platinum. (These metals really are precious; rhodium costs about ten times as much as gold.)

Catalytic converters have used nanotechnology for several decades. A converter only uses about 4–5 g of precious metal to coat the honeycomb structure or pellets. This exposes the exhaust stream of the vehicle to a catalyst area equal to that of several football fields.

Researchers at the University of California, Davis, are studying iridium (Ir) catalysts just four atoms in size. Catalysts often are coated onto a support material (such as ceramic), but these researchers found that the catalytic nanoclusters are actually chemically bonded to the support. Previously the support material was thought to be inert; these workers found that modifying the support material under the nanoparticles actually changed the efficiency of the catalyst up to tenfold, possibly opening the way for a new class of nano-sized catalysts. This work could lead to the ability to design new catalysts for specific functions.

Experiments on cerium(IV) oxide nanoparticles carried out at Brookhaven National Laboratory may lead to catalytic converters that are better at cleaning up auto exhaust. Researchers there used a novel technique to synthesize the CeO_2 nanoparticles and then impregnated them with zirconium (Zr) in one case and with gold (Au) in another. The zirconium-doped nanoparticles store or release oxygen, depending on the engine conditions. This made the catalyst more efficient in converting CO and NO to CO_2 and N_2.

Other experiments have shown that gold-doped nanoparticles act as active catalysts in helping CO combine with O_2 to make CO_2. Gold is not catalytic in its bulk form, but when made into particles of less than 6 nm, it is a very effective catalyst.

Nanoscale catalysts open the way for many process innovations that will make various chemical processes more efficient and thus save resources.

Self-Assessment Questions

1. Nitrogen oxides and hydrocarbons combine in the presence of sunlight to cause
   a. acid rain
   b. global warming
   c. an ozone hole
   d. photochemical smog

2. An amber haze indicates the presence of
   a. HNO_3
   b. NO
   c. NO_2
   d. O_3

3. Photochemical smog occurs mainly in
   a. cold, wet weather
   b. dry, sunny weather
   c. flat, dry areas
   d. snowy, mountainous areas

4. The main source of photochemical smog in most areas is
   a. automobiles
   b. electric power plants
   c. chemical factories
   d. solid waste burning

5. Which pollutants are reduced in automobile exhaust emissions by using a catalytic converter?
   a. CO_2 and H_2O
   b. CO and hydrocarbons
   c. CO and H_2O
   d. NO_x and SO_x

6. Removing NO_x from automobile exhausts requires a(n)
   a. bag filter
   b. electrostatic precipitator
   c. reduction catalyst
   d. wet scrubber

Answers: 1, d; 2, c; 3, b; 4, a; 5, b; 6, c

13.7 Acid Rain: Air Pollution → Water Pollution

We have seen how sulfur oxides are converted to sulfuric acid (Section 13.4) and nitrogen oxides to nitric acid (Section 13.2). These acids fall on Earth as acid rain or acid snow or are deposited from acid fog or adsorbed on PM. **Acid rain** is defined as rain having a pH less than 5.6. Rain with a pH as low as 2.1 and fog with a pH of 1.8 have been reported (Figure 13.9). These values are lower than the pH of vinegar or lemon juice.

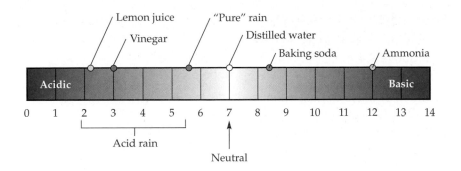

◀ **Figure 13.9** The pH of acid rain in relation to that of other familiar substances. Normal rainwater, if saturated with dissolved CO_2, has a pH of 5.6. During thunderstorms, the pH of rainwater can be much lower because of nitric acid formed by lightning. Recall (Chapter 7) that a solution with a pH value one unit lower than that of another is 10 times as acidic. A decrease in pH from 5.6 to 4.6, for example, means an increase in acidity by a factor of 10. Two pH units lower means 100 times as acidic; three pH units lower means 1000 times as acidic.

Acid rain comes mainly from sulfur oxides emitted from power plants and smelters and from nitrogen oxides discharged from power plants and automobiles. These acids are often carried far before falling as rain or snow (Figure 13.10).

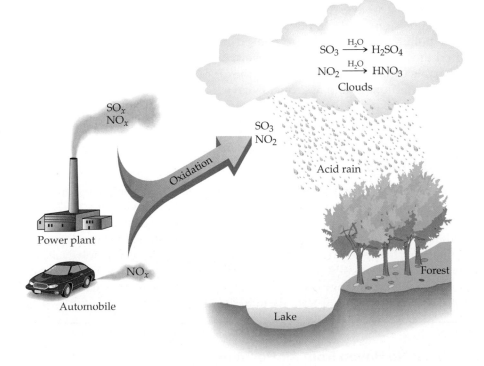

◀ **Figure 13.10** Acid rain. The two principal sources of acid rain are sulfur dioxide (SO_2) from power plants, and nitric oxide (NO) from power plants and automobiles. These oxides are converted to SO_3 and NO_2, which then react with water to form H_2SO_4 and HNO_3. The acids fall in rain, often hundreds of kilometers from their sources.

Acids corrode metals and can even erode stone buildings and statues. Sulfuric acid eats away metal to form a soluble salt and hydrogen gas.

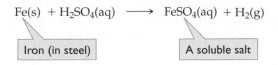

$$Fe(s) + H_2SO_4(aq) \longrightarrow FeSO_4(aq) + H_2(g)$$

▲ Marble statues are slowly eroded by the action of acid rain on marble (calcium carbonate).

The reaction shown here is oversimplified. For example, in the presence of water and oxygen (air), the iron is converted to rust (Fe_2O_3). Marble buildings and statues are disintegrated by sulfuric acid in a similar reaction that forms calcium sulfate, a slightly soluble, crumbly compound.

$$CaCO_3(s) + H_2SO_4(aq) \longrightarrow CaSO_4(aq) + H_2O + CO_2(g)$$

Marble or limestone

Self-Assessment Questions

1. Acid rain is defined as precipitation with pH
   a. above 7.4   b. above 7.0
   c. below 5.6   d. below 7.0

2. Which of these atmospheric reactions leads to acid rain?
   a. $C + O_2 \rightarrow CO_2$ b. $Cl_2 + H_2 \rightarrow 2\,HCl$
   c. $N_2 + 3\,H_2 \rightarrow 2\,NH_3$ d. $S + O_2 \rightarrow SO_2$

Answers: 1, c; 2, d

13.8 The Inside Story: Indoor Air Pollution

Indoor air pollution can pose a major health concern. It can be bad in the United States but is often much worse in developing countries where people cook and heat using wood, coal, or even dung as fuel.

EPA studies on human exposure to air pollutants indicate that indoor air levels of many pollutants may be two to five times higher, and on occasion more than a hundred times higher, than outdoor levels. Near high-traffic areas, the air indoors has the same carbon monoxide level as the air outside. In some areas, office buildings, airport terminals, and apartments have indoor CO levels that exceed government safety standards. Woodstoves, gas stoves, cigarette smoke, and unvented gas and kerosene space heaters are also sources of CO indoors.

More to Explore

The U.S. Environmental Protection Agency has a website with "Quick Finder" links to Acid Rain, Air, Clean Air Act, Climate Change, Ozone, Radon, and more.

Wood Smoke

Many people like the smell of smoke from burning wood or leaves, but wood smoke is far from benign. Each kilogram of wood burned yields 80–370 g of carbon monoxide, 7–27 g of VOCs, 0.6–5.4 g of aldehydes, 1.8–2.4 g of acetic acid, and 7–30 g of particulates. The smoke also contains hydrocarbons such as methane, benzene, and naphthalene, as well as carcinogenic polycyclic aromatic hydrocarbons (PAHs). The odor of wood smoke may evoke memories of autumns past or singing around a campfire, but it can also lead to asthma attacks and possible long-term harm to the lungs.

The Home: No Haven from Air Pollution

We try to save energy by installing better insulation and by sealing air leaks around windows and doors, but in doing so we often make indoor pollution worse by trapping pollutants inside. A kitchen with a gas range often has levels of nitrogen oxides above U.S. government standards. Freestanding kerosene stoves produce levels of NO_x up to 20 times greater than those permitted by federal regulations for outdoor air. (These regulations do not apply to indoor air.)

Cigarettes and Secondhand Smoke

Another indoor polluter is tobacco smoke. Only about 15% of cigarette smoke is inhaled by the smoker; the rest remains in the air for others to breathe. More than 40 carcinogens have been identified among the 4000 or so chemical compounds found in cigarette smoke. The EPA classifies secondhand smoke as a Class A carcinogen, a substance known to cause cancer in humans.

The health effects of smoking on the smoker, widely publicized in a series of reports released by the Surgeon General of the United States, are well known, including its role as the principal cause of emphysema. The first report, published in 1964, led to a ban on television advertisements for cigarettes. Reports have also dealt with the effects of tobacco smoke as an air pollutant. Air quality in smoke-filled rooms is poor. Smoke from just one burning cigarette can raise the level of particulate matter above government standards. Even with good room ventilation, the level of carbon monoxide is often equal to, or greater than, the legal limits for ambient air.

The risk to nonsmokers from cigarette smoke is well established. Medical research has shown that nonsmokers who regularly breathe secondhand smoke suffer many of the same diseases as active smokers, including lung cancer and heart disease. Women who have never smoked but live with a smoker have a 91% greater risk of heart disease and twice the risk of dying from lung cancer. Women who are exposed to secondhand smoke during pregnancy have a higher rate of miscarriages and stillbirths. Their children have decreased lung function and a greater risk of sudden infant death syndrome (SIDS). Children exposed to secondhand smoke are more likely to experience middle ear and sinus infections. They also have an increased frequency of asthma, colds, bronchitis, pneumonia, and other lung diseases.

These and other problems have caused many states and municipalities to ban smoking in many public places.

> The carcinogens in tobacco smoke are similar to those in wood smoke. Tobacco smoke has PAHs, aromatic amines, nitrosamines, and radioactive polonium-210.

Radon and Her Dirty Daughters

A most enigmatic indoor air pollutant is radon. A noble gas, radon is colorless, odorless, tasteless, and unreactive chemically. Radon is, however, radioactive. The decay products of radon, called **daughter isotopes**, are the problem. Radon-222 decays by alpha emission (Section 11.2) with a half-life of 3.8 days.

$$^{222}_{86}Rn \longrightarrow {}^{218}_{84}Po + {}^{4}_{2}He$$

When radon is inhaled, the polonium-218 and other daughters, including lead-214 and bismuth-214, are trapped in the lungs, where their decay damages the tissues.

Radon is released naturally from soils and rocks, particularly granite and shale, and from minerals such as phosphate ores and pitchblende. The ultimate source is uranium atoms found in these materials. Radon is only one of several radioactive materials formed during the multistep decay of uranium. However, radon is unique in that it is a gas and escapes. The others are solids and remain in the soil and rock.

Outdoors, radon dissipates and presents no problems. However, a house built on a solid concrete slab or on a basement can trap the gas inside. Radon levels build up, sometimes reaching several times the maximum safe level established by the EPA (0.15 disintegration per second per liter of air). At five times this level, the hazard is thought to equal that of smoking two packs of cigarettes a day. Scientists at the National Cancer Institute estimate that radon causes 20,000 lung cancer deaths a year; however, the exact threat isn't known. The "safe level" was set for people who work in uranium mines. These workers have high rates of lung cancer, but they also breathe radioactive dust and many smoke cigarettes. These risk factors may be synergistic and are difficult to separate.

Other Indoor Pollutants

Many homes use kerosene heaters or unvented natural gas heaters to help reduce heating costs in the winter. Heating only the occupied areas of a home seems a good idea. However, when such heaters are not properly adjusted, carbon monoxide can be generated. Carbon monoxide from gas stoves, gas furnaces, gasoline generators, even automobile exhaust from an attached garage can also contribute to the high levels seen in some houses. The EPA recommends a maximum level of CO of 9 ppm over an eight-hour period. Carbon monoxide levels from a poorly adjusted gas heater can exceed 30 ppm—continuously. Technology is now available to detect the problem. Commercial carbon monoxide detectors are reliable and easy to install. Proper ventilation is key to preventing high levels of carbon monoxide.

Mold is an often-ignored pollutant in many homes. Leaky pipes or cool areas where water vapor condenses can lead to moisture problems, and mold will grow where there is moisture. Mold spores are a form of PM that can worsen asthma, bronchitis, and other lung diseases. Generally, mold is controlled by controlling moisture.

Electronic "air cleaners" are widely advertised. It seems a great idea; just plug in the cleaner and let it remove all the bad pollutants. Unfortunately, some of these cleaners actually *produce* pollution. In particular, some cleaners generate ozone. As we shall see in the next section, ozone is itself a pollutant and is quite reactive. Ozone can indeed reduce the levels of some pollutants by reacting with them. But in doing so, it can generate additional pollutants such as formaldehyde. The EPA's recommendation is, "The public is advised to use proven methods of controlling indoor air pollution. These methods include eliminating or controlling pollutant sources, increasing outdoor air ventilation, and using proven methods of air cleaning."

Self-Assessment Questions

1. Gas ranges in kitchens are a major source of the indoor air pollutant
   a. dioxins
   b. NO_x
   c. PAHs
   d. VOCs

2. Of the following harmful indoor air contaminants, the most immediate threat is posed by
   a. carcinogens
   b. CO
   c. secondhand tobacco smoke
   d. radon

3. Radon in homes comes from
   a. gas cooking ranges
   b. fire alarms
   c. leaking heater exhausts
   d. soil and rocks

4. Which of the following is a radioactive radon daughter?
   a. cobalt-60
   b. helium-4
   c. polonium-218
   d. uranium-238

Answers: 1, b; 2, b; 3, d; 4, c

GREEN CHEMISTRY

Forty Shades of Green

Irv Levy, Gordon College

Global climate change and its impact on Earth are being discussed in classrooms, in the news, and in the political arena. Although there is some controversy about whether climate change is caused by human activity, the air around us is on our minds more than usual.

If you have ever painted a room in your home, you've noticed the odor of the paint solvents permeating the house. Perhaps you opened a window to "air out" the room. After a little fresh air, the room begins to seem normal again . . . but is it? Although the unpleasant odor is no longer noticeable, you actually have only diluted the annoying vapors by mixing them with the huge container of gases that is our atmosphere. Airing out foul or hazardous vapors simply moves the problem substances from a small space to a space so vast that we can easily forget that we ever released the solvent vapor into the air around us.

Many of the substances now found in our atmosphere did not exist before the twentieth century. Although humans may tolerate exposure to them, some of these substances may nevertheless pose serious threats. Certain substances can catalyze reactions in our upper atmosphere that upset natural balances that have existed for millions of years.

Many substances that we put into the environment are natural compounds that normally do not bother us. For example, the carbon dioxide that we exhale with every breath is essential to normal biological cycles. When we produce CO_2 more rapidly than the biosphere can absorb it, however, natural balances can be upset in alarming ways.

The Twelve Principles of Green Chemistry suggest ways for reducing the impact that human activity has on our atmosphere. One of the most obvious principles, "Prevent waste," suggests that chemicals *not* emitted into the atmosphere would have no impact on it.

Sometimes principles may present conflict. Consider, for example, the design and manufacture of computer monitors. Until recently, computer monitors and television sets contained cathode ray tubes (CRTs) that are shielded with lead. Lead must be disposed of with care because of its toxicity, and CRT displays require a lot of energy when in use. Green Chemistry Principles #4 (design safer products) and #6 (increase energy efficiency) suggest we seek a green alternative to the CRT monitor.

Liquid crystal display (LCD) monitors have become an increasingly popular replacement for CRTs in both television sets and computers. An LCD panel requires less energy than a CRT, so there is a huge increase in energy efficiency over the lifetime of the device. This leads to a decrease in emissions of carbon dioxide from coal-fired or gas-fired plants used to create that energy. And LCD production does not require the use of hazardous lead. But is LCD technology truly green? Back to the air around us. . .

Although LCD monitors are both less toxic and more energy efficient than CRT monitors, certain emissions from the manufacturing of LCDs contribute to global warming. What is worse: the dangers of acute toxicity and high energy use associated with CRT monitors, or the degree to which emissions associated with the manufacturing of LCD screens impact our atmosphere?

Manufacturing an LCD produces unwanted by-products that must be removed from the final product, often with the use of nitrogen trifluoride (NF_3). Until recently, NF_3 was a relatively uncommon material, but heavy demand by LCD manufacturers (and solar panel manufacturers) has led to a dramatic increase in use of the gas.

Scientists from the University of California (UC) at Irvine have recently reported that NF_3 contributes to global warming. Currently, the quantity of NF_3 released is much smaller than that of CO_2, the primary contributor to global warming that is released into the atmosphere mainly by the burning of fossil fuels. And it's a good thing that NF_3 emissions are smaller, because the UC Irvine studies show that NF_3 is about 17,000 times more efficient at trapping atmospheric heat than CO_2. Sadly, the companies that developed NF_3 technology for LCD panel manufacture have replaced one greenhouse gas with another.

Addressing global climate change and our impact on Earth's environment is clearly not as simple as we might wish. Dealing with the complex issue of global warming will require the cooperation of many groups of people with different perspectives. Guidance provided by the Twelve Principles will catalyze successful collaborations to meet these, and other, challenges to sustainability.

13.9 Ozone: The Double-Edged Sword

The ordinary oxygen in the air we breathe is made up of O_2 molecules. Ozone is a form of oxygen that consists of O_3 molecules. Ozone (O_3) and oxygen (O_2) are **allotropes**: two different forms of the same element. Ozone is a familiar constituent of photochemical smog. Inhaled, it is a toxic, dangerous substance. Ozone is also an important natural component of the stratosphere, where it shields Earth from life-destroying UV radiation. Ozone is a good example of just what a pollutant is: a chemical substance out of place in the environment. In the stratosphere, it helps make life possible. In the lower troposphere—the part we breathe—it makes life difficult.

Ozone as an Air Pollutant

Most U.S. urban areas have experienced ozone alerts. Table 13.2 shows the 2007 rankings of the 10 worst areas for ozone pollution. The American Lung Association estimates that one-third of the U.S. population lives in counties that exceed recommended concentrations of ozone.

Table 13.2 Metropolitan Areas with the Worst Ozone Air Pollution, 2007	
Metropolitan Area	**Rank**
Los Angeles–Long Beach–Riverside, CA	1
Bakersfield, CA	2
Visalia–Porterville, CA	3
Fresno–Madera, CA	4
Houston–Baytown–Huntsville, TX	5
Merced, CA	6
Dallas–Fort Worth, TX	7
Sacramento–Arden–Arcade–Truckee, CA, NV	8
Baton Rouge–Pierre Part, LA	9
New York–Newark–Bridgeport, NY–NJ–CT–PA	10

Source: Data from the American Lung Association www.lungusa.org

Ozone is a powerful oxidizing agent and is highly reactive. This makes it a severe irritant to the respiratory system. Episodes of high ozone levels correlate with increased hospital admissions and emergency room visits for respiratory problems. At low levels, it causes eye irritation. Repeated exposure can make people more susceptible to respiratory infection, cause lung inflammation, aggravate asthma, decrease lung function, and increase chest pain and coughing. Ozone is particularly damaging to children. The WHO estimates that 80% of all deaths attributed to air pollution are among children.

Ozone causes economic damage beyond its adverse effect on health. It is a strong oxidizing agent that causes rubber to harden and crack, shortening the life of automobile tires and other rubber items. Ozone also causes extensive damage to crops, especially tobacco and tomatoes.

The Stratospheric Ozone Shield

In the mesosphere (Figure 13.1), some ordinary oxygen molecules are split into oxygen atoms by short-wavelength, high-energy UV radiation.

▲ Ozone causes rubber to harden and crack. Tire manufacturers incorporate paraffin wax into the rubber to protect tires from ozone. The alkanes that make up the wax are among the few substances that are resistant to attack by ozone. As tires age, the paraffin wax migrates to the surface to replace that worn off.

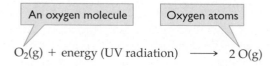

An oxygen molecule   Oxygen atoms

$$O_2(g) + \text{energy (UV radiation)} \longrightarrow 2\,O(g)$$

Some of these highly reactive atoms diffuse down to the stratosphere, where they react with O_2 molecules to form ozone.

An oxygen molecule   An ozone molecule

$$O_2(g) + O(g) \longrightarrow O_3(g)$$

The ozone in turn absorbs longer-wavelength, but still lethal, UV rays, thus shielding us from this harmful radiation. In absorbing the rays, ozone is converted back to oxygen molecules and oxygen atoms in a reversal of the previous reaction.

$$O_3(g) + \text{energy (UV radiation)} \longrightarrow O_2(g) + O(g)$$

Undisturbed, the concentration of ozone in the stratosphere is kept fairly constant; over 300 billion tons of ozone are destroyed and created every day by this cyclic process. However, because of the reduced pressure in the stratosphere, if the entire ozone layer were compressed to normal sea level pressures, it would be only 3 mm thick! This delicate barrier protects all living things from harmful UV radiation. In recent decades, human activity has threatened to upset the balance and deplete some of the protective qualities of the ozone layer.

Chlorofluorocarbons and the Ozone Hole

The less ozone there is in the stratosphere, the more harmful UV radiation reaches Earth's surface. For humans, increased exposure to UV radiation can lead to more cases of skin cancer, cataracts, and impaired immune systems. The U.S. National Research Council predicts a 2–5% increase in skin cancer for each 1% depletion of the ozone layer. Even a 1% increase in UV radiation can lead to a 5% increase in melanoma, a deadly form of skin cancer. Increased UV radiation also damages many crops resulting in lower yields. A depletion of stratospheric ozone can also cause a decreased population of phytoplankton in the ocean, disrupting the web of life in the sea and disrupting the oxygen cycle (page 347).

The thickness of the ozone layer changes with latitude and also with the season. The cycling of ozone concentration is most prevalent over Antarctica in part because the chemical reactions responsible for ozone destruction occur much more rapidly on the surfaces of ice crystals. During the Antarctic winter (June–August) more clouds containing ice crystals are formed and the levels of ozone decrease. In the early 1970s, serious concerns were raised because the ozone layer was not recovering properly during the warm months. In 1974, Mario Molina and F. Sherwood Rowland proposed a mechanism for the enhanced ozone depletion that implicated the presence of chlorofluorocarbons (CFCs) in the stratosphere. Their work was awarded the Nobel Prize in chemistry in 1995.

CFCs were used as the dispersing gases in aerosol cans, as foaming agents for plastics, and as refrigerants. At room temperature, CFCs are either gases or liquids with low boiling points. They are essentially insoluble in water and inert toward most other substances. These properties make them ideal for many uses. However, once released into the atmosphere, this lack of reactivity allows them to persist and diffuse to the higher regions of the stratosphere. There, above the protective ozone layer, energetic UV radiation cleaves one of the C—Cl bonds and initiates a series of reactions.

$$CF_2Cl_2 + \text{energy (UV light)} \longrightarrow CF_2Cl\cdot + Cl\cdot$$
$$Cl\cdot + O_3 \longrightarrow ClO\cdot + O_2$$
$$\cdot ClO + O \longrightarrow Cl\cdot + O_2$$

Both products of the initial reaction are highly reactive free radicals (Section 4.11), molecules or fragments with unpaired electrons (indicated here by a dot). The chlorine atoms then react with ozone molecules producing molecular oxygen. Note that the last step results in the formation of another chlorine atom that can break down another molecule of ozone. The second and third steps are repeated many times; the decomposition of one CFC molecule in this *chain reaction* can result in the destruction of *thousands* of molecules of ozone.

UV light sometimes is used to sterilize tools in the hairdresser's shop and safety glasses in laboratories to prevent the transfer of infectious microorganisms from one person to another. We do not want the entire planet rendered sterile, however, for we are part of the life that would be harmed.

▲ In 1986 Susan Solomon, a chemist at the Oceanic and Atmospheric Administration in Boulder, Colorado, proposed a mechanism for ozone depletion over the poles by CFCs. Later experiments confirmed the theory and led to an international ban on the use of CFCs. Solomon was awarded a 1999 National Medal of Science for her work.

International Cooperation

The United Nations has addressed the problem of ozone depletion through the 1987 Montreal Protocol, international agreements enforcing the reduction and eventual elimination of the production and use of ozone-depleting substances. As a result, CFCs have been banned in the United States and many other countries, and the search for effective substitutes has already been successful. CFCs and other chlorine- and bromine-containing compounds that might diffuse into the stratosphere have been replaced for the most part by more benign substances. Hydrofluorocarbons (HFCs) such as CH_2FCF_3, and hydrochlorofluorocarbons (HCFCs) such as $CHCl_2CF_3$ are alternatives. The C—H bonds in these compounds are more susceptible to reactive species such as hydroxyl radicals, so the molecules generally break down before reaching the stratosphere. For every solution, it seems there is a problem; CFCs, HFCs, and HCFCs all are greenhouse gases (Section 13.10).

The Montreal Protocol seems to be having the desired effects. The concentration of CFCs in the stratosphere has peaked. Chlorine concentrations leveled off in 1998 and will likely continue to decline slowly for decades. There is some indication that the ozone hole over Antarctica is *getting smaller* (Figure 13.11), but its size varies with temperature and other meteorological factors. It may be several years before we can be sure that the problem is being solved.

More to Explore
Wilson, Elizabeth. "Ozone Hole Recovery May Be Delayed." *Chemical & Engineering News,* December 12, 2005, p. 9.

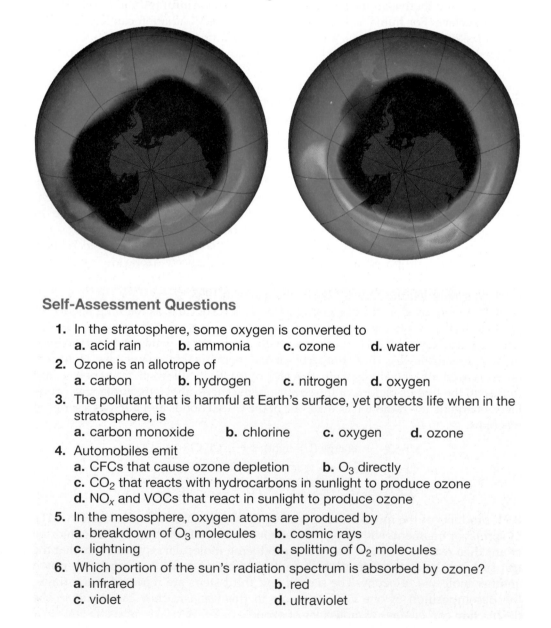

▶ **Figure 13.11** The ozone hole over Antarctica in September 2006 (left) and 2007 (right). Note the reduction in size over the course of a year.

Self-Assessment Questions

1. In the stratosphere, some oxygen is converted to
   a. acid rain **b.** ammonia **c.** ozone **d.** water

2. Ozone is an allotrope of
   a. carbon **b.** hydrogen **c.** nitrogen **d.** oxygen

3. The pollutant that is harmful at Earth's surface, yet protects life when in the stratosphere, is
   a. carbon monoxide **b.** chlorine **c.** oxygen **d.** ozone

4. Automobiles emit
   a. CFCs that cause ozone depletion **b.** O_3 directly
   c. CO_2 that reacts with hydrocarbons in sunlight to produce ozone
   d. NO_x and VOCs that react in sunlight to produce ozone

5. In the mesosphere, oxygen atoms are produced by
   a. breakdown of O_3 molecules **b.** cosmic rays
   c. lightning **d.** splitting of O_2 molecules

6. Which portion of the sun's radiation spectrum is absorbed by ozone?
   a. infrared **b.** red
   c. violet **d.** ultraviolet

7. Chlorofluorocarbons harm the ozone layer by
 a. adding more ozone   **b.** blocking UV radiation
 c. destroying oxygen   **d.** destroying ozone

Answers: 1, c; 2, d; 3, d; 4, d; 5, d; 6, d; 7, d

13.10 Carbon Dioxide and Climate Change

No matter how clean an engine or a factory is, as long as it burns coal or petroleum products it produces carbon dioxide. The concentration of carbon dioxide in the atmosphere has increased about 35% since 1832, reaching 385 ppm in 2008. Since 2000, the concentration has increased at a rate of 1.9 ppm per year. Scientists attribute the increase to global industrialization and the burning of fossil fuels. Carbon dioxide is a natural component of the environment and not considered toxic. Its immediate effect on us is slight, but what about its long-term effects?

The sun radiates many different types of radiation of which visible, infrared, and ultraviolet are most prominent. About half of this energy is either reflected or absorbed by the atmosphere; the light that gets through (mostly visible) acts to heat the surface of Earth. The three main constituents of the atmosphere—oxygen, nitrogen, and argon—are small, nonpolar molecules and do not absorb much infrared energy. However, carbon dioxide and some other gases produce a **greenhouse effect** (Figure 13.12). They let in the sun's visible light to warm the surface, but when the Earth radiates infrared energy back toward space, these greenhouse gases absorb and trap the energy. A similar thing happens in a car left in the sun with the windows closed. Visible light entering through the glass heats the interior, but the heat cannot efficiently escape back through the windows. The temperature inside the car can rise significantly above the outside temperature. Some degree of the greenhouse effect is necessary to life. Without an atmosphere, all of the radiated heat would be lost to outer space, and the Earth would be much colder than it is. In fact, just based on the distance from the sun, it is estimated that the Earth's average temperature would be a cold −18 °C.

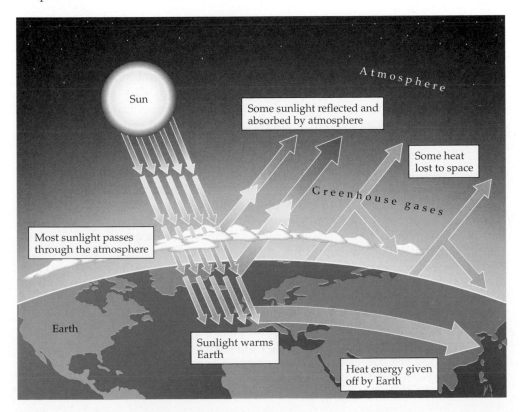

◀ **Figure 13.12** The greenhouse effect. Sunlight passing through the atmosphere is absorbed, warming Earth's surface. The warm surface re-emits energy as infrared radiation. Some of this radiation is absorbed by CO_2, H_2O, CH_4, and other gases and retained in the atmosphere as heat energy.

EXAMPLE 13.1 CO$_2$ Stoichiometry

What mass of carbon dioxide (molar mass = 44.0 g/mol) is produced by the burning of 2670 g (1 gal) of gasoline? Gasoline may be represented by the formula C$_8$H$_{18}$ (molar mass = 114.0 g/mol).

$$2 C_8H_{18}(l) + 25 O_2(g) \longrightarrow 18 H_2O(g) + 16 CO_2(g)$$

Solution

The problem gives us the mass of a reactant, and we are to find the mass of a product, so this is a stoichiometry problem, as covered in Section 5.4.

We begin by converting the mass of the given substance, C$_8$H$_{18}$, to moles.

$$2670 \text{ g C}_8\text{H}_{18} \times \frac{1 \text{ mol C}_8\text{H}_{18}}{114.0 \text{ g C}_8\text{H}_{18}} = 23.4 \text{ mol C}_8\text{H}_{18}$$

Now we use the relationship of 2 mol C$_8$H$_{18}$ ∻ 16 mol CO$_2$ to find moles of CO$_2$.

$$23.4 \text{ mol C}_8\text{H}_{18} \times \frac{16 \text{ mol CO}_2}{2 \text{ mol C}_8\text{H}_{18}} = 187 \text{ mol CO}_2$$

The final step is to use the molar mass of CO$_2$ to determine the mass of CO$_2$.

$$187 \text{ mol CO}_2 \times \frac{44.0 \text{ g CO}_2}{1 \text{ mol CO}_2} = 8240 \text{ g CO}_2$$

The mass of CO$_2$ is actually greater than the mass of gasoline consumed, because the carbon of the gasoline is combined with oxygen from the air.

■ **EXERCISE 13.1A**

Calculate the mass in grams of water vapor produced from burning 2670 g of gasoline.

■ **EXERCISE 13.1B**

Calculate the mass in kilograms of carbon dioxide produced by a class of 60 students in one week, if each student uses 10 gallons of gasoline in a week.

Greenhouse Gases and Global Warming

What many fear is an *enhanced* greenhouse effect caused by increased concentrations of carbon dioxide and other greenhouse gases in the atmosphere. This enhancement could lead to a rise in Earth's average temperature, an effect called **global warming** (Figure 13.13). Indeed, periods of high CO$_2$ concentration in Earth's past have been correlated with increased global temperatures.

Human activities add about 32 billion t (metric tons) of carbon dioxide to the atmosphere each year, of which 26 billion t comes from the burning of fossil fuels. About 55% is removed, mostly by plants and soil (30%), and the oceans (24%), leaving a net addition to the atmosphere of 14 billion t CO$_2$/year.

Methane also contributes to the greenhouse effect. The concentration of methane in the atmosphere increased from about 0.7 ppm in 1750 to about 1.70 ppm today before stabilizing over the last few years. Most atmospheric methane comes from natural wetlands, in which methanogenic (methane-producing) bacteria decompose organic material. Decomposition of organic materials in landfills is the largest human-related source of methane in the United States. Other sources include cattle and termites, which produce methane as part of microbial fermentation during their normal digestive processes. Researchers believe the stabilization of methane levels is due to reducing the release of methane from natural gas pipelines and landfill sites, and better management of rice paddies.

Molecules that are polar or are easily polarized (have moderate to strong dispersion forces) are strong greenhouse gases. Although present in much smaller amounts than carbon dioxide, trace gases such as methane, CFCs, and HCFCs are much more efficient at trapping heat. Methane is 20 to 60 times, CFCs 5000 to 14,000

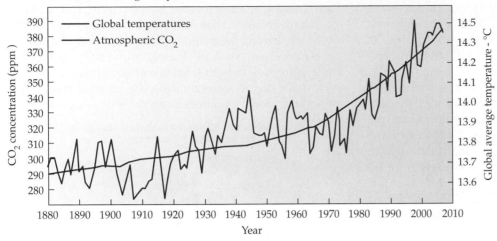

Global average temperature and carbon dioxide concentrations, 1880–2004

◀ **Figure 13.13** Carbon dioxide content of the atmosphere and surface temperature variations. The increase in carbon dioxide correlates with the beginning of the Industrial Revolution, around 1850.

times, and HCFCs 200 to 4800 times as effective as carbon dioxide at holding heat in Earth's atmosphere. Water is about 10 times less efficient at trapping the heat, though when present in significant amounts it has a noticeable effect. For instance, cloudy winter nights tend to be warmer than clear winter nights due to the presence of significant amounts of water in the clouds, which acts to trap the heat close to the Earth's surface.

Predictions and Consequences

A massive study by the United Nations' Intergovernmental Panel on Climate Change (IPCC) (*Fourth Assessment Report: Climate Change 2007*) provided a comprehensive review of past changes and predictions of future changes. The global average air temperature near the Earth's surface increased about 0.74 °C during the twentieth century. Many complex and interrelated natural variables (clouds, volcanoes, El Niño weather patterns, and such) as well as varying estimates of future greenhouse gas emissions make it difficult to predict exactly the magnitude of future warming, but the IPCC study predicts a further 1.1–6.4 °C increase by the year 2100. That temperature rise may not seem like much, but the ramifications of such warming can be severe. So far predictions made by models have underestimated the actual changes, particularly at high latitudes. Losses from the ice sheets of Greenland and Antarctica have exceeded expectations. Satellite imagery in August 2007 showed the least amount of Arctic ice ever seen. If overall trends continue, the Arctic Ocean may be ice-free in summers by 2030. The polar ice caps need not melt completely for global warming to be problematic. When water warms, it expands. The ocean rose about 3.1 mm/y from 1993 to 2003. Again, this rise may seem trivial, but even slight rises in ocean level will increase tides and result in higher, more damaging storm surges. The IPCC report predicts a rise of 28–43 cm by 2100, but in recent years sea levels have been rising at a rate that would surpass the upper end of the range predicted by the IPCC. Other studies predict rises of 100–150 cm by 2100. The IPCC shared the 2007 Nobel Peace Prize with Al Gore, former U.S. Vice President.

3. What are greenhouse gases?
Greenhouse gases are those that absorb infrared radiation. Polar molecules and those that can have a significant induced dipole (recall Section 6.3) act as greenhouse gases.

ANSWER

More to Explore
Hanson, David. "Defusing the Global Warming Time Bomb." *Scientific American*, March 2004, pp. 68–77.

Climate Change and Weather

We can look out the window and see weather, the conditions of the atmosphere over a short period of time. Weather forecasts are seldom accurate for more than a few days. Climate is an average of daily and seasonal weather events over a long period of time—decades or centuries. Even over a decade or so, short-term natural variabilty can mask long-term trends. Climate change means changes in long-term averages of daily weather—temperatures, rainfall, wind speeds, snow depths, and so on.

The IPCC panel predicted a variety of effects of global warming, some regarded as more certain than

(*continued*)

others. Climate change is expected to bring more severe droughts to many areas, as well as harsher heat waves, stronger storms, more ferocious floods, more outbreaks of agricultural pests, and increased affliction of infectious diseases.

The potential impact on food output is a particular concern: Semiarid regions, where water is already scarce and cropland overused, could experience further devastation from climate change. The humanitarian crisis in the Darfur region of the Sudan is thought to be in part caused by the effect of global warming on crops and pastureland.

The rise in sea level can bring flooding that would endanger low-lying islands and crowded river deltas such as those of Bangladesh and other southern Asian areas. In the United States, the Gulf of Mexico coastal region and Chesapeake Bay estuaries are especially susceptible.

As carbon dioxide levels in the atmosphere increase, more CO_2 will dissolve in the oceans, making the seawater more acidic. This acidification leads to the dissolving of the shells of the marine animals that make their shells out of calcium carbonate. This could damage the coral reefs, which are vital to ocean ecosystems.

Small floating animals that are a vital part of the oceanic food chain are also at risk as increased acidity erodes the shells, weakening them and making them more brittle.

Some areas, such as Russia and other northern lands, may well benefit from a warmer climate. However, many of the regions facing the greatest risks are among the world's poorest nations.

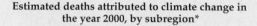

Estimated deaths attributed to climate change in the year 2000, by subregion*

■ 0 – 2
□ 2 – 4
▨ 4 – 70
■ 70 – 120
□ No data

*Change in climate compared to baseline 1961–1990 climate

Data source:
McMichael, JJ, Campbell-Lendrum D, Kovats RS, et al. Global Climate. In comparative quantification of health risks: global and regional burden of disease due to selected major risk factors. M Ezzati, Lopez, AD, Rodgers A., Murray CJL. Geneva, World Health Organization, 2004

▲ This map shows the areas most affected by global warming.

Mitigation of Global Warming

Scientists know in broad outline what we need to do to alleviate global warming: Technologically advanced countries need to make quick and dramatic cuts in emissions, and they need to transfer technology to China, India, and the rest of the developing world so that those countries can continue to develop their economies without excessive use of coal-fired power plants. The problems involved in implementing these ideas are also well known: Can we make such rapid cuts? Are governments willing to tackle these issues?

We discuss a variety of alternatives to fossil fuels for energy production in Chapter 15. However, most scientists realize that no single alternative technology, such as solar, nuclear power, biofuels, wind power, and so on, will solve the problem of increasing CO_2 emissions. We will no doubt need several strategies.

More fuel-efficient cars and appliances, better-built homes, and energy conservation will help somewhat. Large-scale plans to capture CO_2 from smokestacks and sequester it underground are more costly and less certain. **Carbon sequestration** is the removal of CO_2 from the atmosphere by collection and transport of concentrated CO_2 from large emission sources, such as power plants and factories, and subsequent storage in underground reservoirs. It also includes removal of CO_2 from the atmosphere by forestry and agricultural practices, such as planting trees and stopping deforestation. All the alternatives are more difficult than simply burning coal or petroleum.

A few nations have made progress in reducing or capping CO_2 emissions, but overall there has been little or no progress. China is building coal-fired power plants at a breakneck rate, and China surpassed the United States in CO_2 emissions in 2008. High oil prices have even led to the building of new coal-fired plants in Italy.

ANSWER

4. Are global warming and the ozone hole the same problem?
No. It should be apparent from Sections 13.9 and 13.10 that the ozone hole contributes to increased ultraviolet radiation reaching Earth, whereas global warming is the result of certain gases absorbing infrared radiation given off by Earth. However, some gases such as CFCs do act as both ozone destroyers and greenhouse gases.

Self-Assessment Questions

1. What type of radiation is trapped on Earth's surface by the greenhouse effect?
   a. gamma   **b.** infrared   **c.** ultraviolet   **d.** visible

2. During the twentieth century, the global average air temperature near the Earth's surface increased about
   a. 0.5 °C   **b.** 0.75 °C   **c.** 1.25 °C   **d.** 2.25 °C

3. The main cause of global warming is
   a. CO_2 from factories, power plants, and automobiles
   b. Earth's orbital eccentricities
   c. variations in the sun's output
   d. water vapor

4. The 2007 IPCC report on climate change suggests that over the twenty-first century the world may experience warming by as much as
   a. 1.6 °C   **b.** 4.6 °C   **c.** 6.4 °C   **d.** 9.4 °C

5. When CO_2 is dissolved in the ocean, it
   a. decreases the acidity of seawater
   b. forms an acid that dissolves the shells of sea creatures
   c. is permanently removed from the atmosphere
   d. reacts with Ca^{2+} ions to form $CaCO_3$

6. To combat global warming, a person can
   a. drain a wetland   **b.** heat with charcoal
   c. pave a dirt road   **d.** plant trees

Answers: 1, b; 2, b; 3, a; 4, c; 5, b; 6, d

The World Health Organization estimates that Earth's warming climate contributes to more than 150,000 deaths and 5 million illnesses each year. WHO predicts that this toll could double by 2030. According to data published in the journal *Nature*, a warmer climate is driving up rates of malaria, malnutrition, and diarrhea.

13.11 Who Pollutes? Who Pays?

The U.S. EPA lists six "criteria" pollutants, so called because they employ scientific criteria to measure the health effects. Since the first Earth Day in 1970, nationwide emissions of the six criteria pollutants have declined dramatically (Table 13.3). These improvements in air quality have come while the U.S. population increased 50%, and the gross domestic product and vehicle miles traveled both tripled.

Table 13.3 Ambient Air Quality Standards for Six Criteria Air Pollutants

Pollutant	Limit	% Change (1980–2006)
Carbon monoxide		−75
8-h average	9 ppm	
1-h average	35 ppm	
Nitrogen dioxide		−41
Annual average	0.053 ppm	
Ozone[a]		−21
8-h average	0.08 ppm	
1-h average	0.12 ppm	
Particulate matter		
<10 μm, 24-h average	150 μg/m^3	−30[b]
<2.5 μm, 24-h average	65 μg/m^3	−15[b]
Sulfur dioxide		−66
Annual average	0.03 ppm	
24-h average	0.14 ppm	
3-h average	0.50 ppm	
Lead		−96
3-month average	1.5 μg/m^3	

[a]Formed from precursor volatile organic compounds (VOC) and NO_x.

[b]1980–2003.

5. What does vehicle "smog testing" actually test?

Tests vary depending on the location, but unburned hydrocarbons and carbon monoxide are almost always tested. Unburned hydrocarbons may come from the tailpipe or may be evaporative (from fuel lines, gas caps, etc.) in nature. Some states and cities may also test for nitrogen oxides or particulates.

Table 13.4 lists the main sources of the seven major air pollutants along with their health and environmental effects. Note that motor vehicles are responsible for several of the major air pollutants; indeed, they are the source of nearly one-half (by mass) of all air pollutants. Our transportation system accounts for about 80% of carbon monoxide emissions, 40% of hydrocarbon emissions, and 40% of nitrogen oxide emissions. Because most transportation in the United States is by private automobile, we can conclude that cars are a major source of pollution.

On the other hand, most PM comes from power plants (about 40%) and industrial processes (about 45%). Similarly, more than 80% of sulfur oxide emissions come from power plants, with an additional 15% coming from other industries. Power plants alone contribute about 55% of nitrogen oxide emissions. Who uses electricity from power plants? We all do. An ordinary 100-W lightbulb burning for 1 year uses the electricity generated by burning 275 kg of coal.

What is the worst pollutant? Carbon monoxide is produced in huge amounts and is quite toxic. Yet it is deadly only in concentrations approaching 4000 ppm. Its contribution to cardiovascular disease, by increasing stress on the heart, is difficult to measure. The WHO rates sulfur oxides as the worst pollutants. They are powerful irritants and, according to WHO, people with respiratory illnesses are more likely to die from exposure to sulfur oxides than from exposure to any other kind of pollutant. Sulfur oxides and sulfuric acid formed from them have been linked to more than 50,000 deaths per year in the United States.

We all share the responsibility for pollution. In the words of Walt Kelly's comic-strip character Pogo, "We have met the enemy, and he is us." But we can all be part of the solution by conserving fuel and electricity. Many utility companies now offer suggestions for saving energy. Some even give free analyses of home energy use.

Table 13.4 The Seven Major Air Pollutants

Pollutant	Formula or Symbol	Major Sources	Health Effects	Environmental Effects
Carbon monoxide	CO	Motor vehicles	Interferes with oxygen transport, causing dizziness, death; contributes to heart disease	Slight
Hydrocarbons	C_nH_m	Motor vehicles, industry, solvents	Narcotic at high concentrations; some aromatics are carcinogens	Precursors of aldehydes, PAN
Sulfur oxides	SO_x	Power plants, smelters	Irritate respiratory system; aggravate lung and heart diseases	Reduce crop yields; precursors of acid rain, SO_4^{2-}
Nitrogen oxides	NO_x	Power plants, motor vehicles	Irritate respiratory system	Reduce crop yields; precursors of ozone and acid rain; produce brown haze
Particulate matter	—	Industry, power plants, dust from farms and construction sites	Irritates respiratory system; synergistic with SO_2; contains adsorbed carcinogens, toxic metals	Impairs visibility
Ozone	O_3	Secondary pollutant from NO_2	Irritates respiratory system; aggravates lung and heart diseases	Reduces crop yields; kills trees (synergistic with SO_2); destroys rubber, paint
Lead	Pb	Motor vehicles, smelters	Toxic to nervous system and blood-forming system	Toxic to all living things

Paying the Price

Air pollution costs us tens of billions of dollars each year. It wrecks our health by causing or aggravating bronchitis, asthma, emphysema, and lung cancer. It destroys crops and sickens and kills livestock. It corrodes machines and blights buildings.

Elimination of air pollution will be neither cheap nor easy. It is especially difficult to remove the last fractions of pollutants. Cost curves are exponential; costs soar toward infinity as pollutants approach zero. For example, if it costs $200 per car to reduce emissions by 50%, it usually costs about $400 to reduce them by 75%, $800 to reduce them by 87.5%, and so on. It would be extremely expensive to reduce emissions by 99%. We can have cleaner air, but how clean depends on how much we are willing to pay.

What would we gain by getting rid of air pollution? How much is it worth to see the clear blue sky or to see the stars at night? How much is it worth to breathe clean, fresh air?

More to Explore
Worldwatch Institute, 1776 Massachusetts Ave. N.W., Washington, D.C. 20036-1904, has a variety of excellent publications on environmental problems.

Self-Assessment Questions

1. The U.S. agency responsible for gathering and analyzing air pollution data is
   a. EPA
   b. FCC
   c. FDA
   d. ICC
2. Air pollutants have decreased significantly since 1970 because of
   a. decreased driving
   b. decreased population
   c. fewer electric power plants
   d. increased pollution controls
3. The largest source of carbon monoxide pollution is
   a. electric power plants
   b. industry
   c. motor vehicles
   d. smelters
4. The largest source of SO_2 pollution is
   a. electric power plants
   b. industry
   c. motor vehicles
   d. smelters
5. According to the World Health Organization, the worst pollutant in terms of health effects is
   a. CO
   b. ozone
   c. SO_2
   d. VOCs
6. The largest sources of NO_x pollution are
   a. electric power plants and motor vehicles
   b. electric power plants and industry
   c. industry and motor vehicles
   d. smelters and motor vehicles
7. The largest sources of particulate pollution are
   a. electric power plants and motor vehicles
   b. electric power plants and industry
   c. industry and motor vehicles
   d. smelters and motor vehicles

Answers: 1, a; 2, d; 3, c; 4, a; 5, c; 6, a; 7, b

Critical Thinking Exercises

Apply knowledge that you have gained in this chapter and one or more of the FLaReS principles (Chapter 1) to evaluate the following statements or claims.

13.1 A citizen claims that air pollution problems in the Los Angeles area could be solved by allowing only zero-emission vehicles to be sold and licensed in the area.

13.2 Ozone in aerosol cans was once used as a room deodorizer. An inventor proposes that such preparations be reintroduced because any ozone entering the environment would help to restore the protective ozone layer in the stratosphere.

13.3 Although Earth's surface has grown warmer during the last few decades, satellite data collected over an 18-year period show a cooling trend in the upper atmosphere. On the basis of these data, some people claim that global warming as a result of an enhanced greenhouse effect is a myth.

13.4 A notable scientist claims that we must continue to *increase* the amount of carbon dioxide in the atmosphere to prevent a coming ice age.

13.5 Some experts believe that much of global warming comes from methane produced by such natural sources as plant decay in swamps and cattle flatulence. They say that industrial activities that produce carbon dioxide raise the standard of living and should be allowed to continue without interference since they aren't the only source of global warming.

13.6 An entrepreneur promotes a new vehicle, which is an ordinary automobile adjusted to burn hydrogen. He claims that the only product of hydrogen combustion is water, and so the new vehicle is completely nonpolluting.

13.7 A scientist claims that the increase in carbon dioxide from 0.02 to 0.03% is not important in terms of global warming. His reasoning is that water vapor makes up about 2–3% of the atmosphere, and although water vapor is only about one-tenth as effective a greenhouse gas as CO_2, the 0.01% increase in CO_2 is equivalent to a trivial 0.1% increase in water vapor.

▪ SUMMARY

Section 13.1—The thin blanket of air that is our **atmosphere** is made up of layers: the troposphere (nearest Earth), the stratosphere (which includes the ozone layer), the mesosphere, and the thermosphere. The atmosphere is a mixture of gases, about 78% nitrogen, 21% oxygen, and 1% argon by volume, plus trace gases and up to 4% water vapor.

Section 13.2—Animals and most plants cannot use atmospheric nitrogen unless it has been fixed (combined with other elements). Nitrogen goes from the air into plants and animals and eventually back to the air via the **nitrogen cycle**. Oxygen from the air is involved in oxidation (of plant and animal materials) and is eventually returned to the air via the **oxygen cycle**. Oxygen is converted to ozone and then changed back to oxygen in the stratosphere. In this process, the ozone absorbs UV radiation that might otherwise make life impossible. A **temperature inversion** occurs when a layer of cold, dirty air is trapped under a layer of warmer air. The cold air can become quite polluted.

Section 13.3—A **pollutant** is a chemical that is in the wrong place and causes problems there. Air pollution has existed for millions of years. Natural air pollution includes dust storms, noxious gases from swamps, and ash and sulfur dioxide from erupting volcanoes. Today air pollution is more complex than in the past. Air pollution is a global problem because pollutants from one geographic area often migrate to another. The primary air pollutants introduced by human activity are the smoke and gases produced by the burning of fuels and the fumes and particulates emitted by factories.

Section 13.4—**Industrial smog** is the kind of smog produced in cold, damp air by excessive burning of fossil fuels. It consists of a combination of smoke and fog with sulfur dioxide, sulfuric acid, and **particulate matter (PM)**, which is solid or liquid particles of greater than molecular size. PM commonly takes the form of soot and fly ash from coal combustion. Carbon monoxide is a poisonous gas produced in most combustion that ties up hemoglobin in the blood so that it cannot transport oxygen to the cells. The pollutants that make up industrial smog can act synergistically to cause severe damage to living tissue. PM can be removed from smokestack gases using an **electrostatic precipitator**, which charges the particles, or with a **wet scrubber**, which passes the gases through water. Sulfur dioxide can be scrubbed using limestone to form calcium sulfite, which can then be oxidized to useful calcium sulfate.

Section 13.5—Complete combustion of gasoline produces carbon dioxide and water, but combustion is always incomplete. Carbon monoxide is one product of incomplete combustion, and three-fourths of the carbon monoxide produced results from transportation. Nitrogen oxides are present in automobile exhaust gases and in emissions from power plants that burn fossil fuels. The greatest problem with nitrogen oxides is their contribution to smog. The brown color of nitrogen dioxide causes the brownish haze often seen over certain large cities. **Volatile organic compounds (VOCs)** are major contributors to smog. VOCs come from gasoline vapor and other fuels and from some consumer products. Hydrocarbons can be pollutants in themselves and can react to form other pollutants such as aldehydes and peroxyacetyl nitrate (PAN), both of which are very irritating to tissues.

Sections 13.6, 13.7—Photochemical smog results when hydrocarbons, nitrogen oxides, and ozone are exposed to bright sunlight. A complex series of reactions produces a variety of pollutants. **Catalytic converters** in automobiles act to reduce photochemical smog by oxidizing unburned hydrocarbons. A separate catalyst reduces the quantity of nitrogen oxides. Hybrid vehicles are more efficient than conventional vehicles, with a reduction in emission of pollutants as well as reduced photochemical smog. Nitrogen oxides (mainly from automobiles and power plants) and sulfur oxides (mainly from power plants) can dissolve in atmospheric moisture, forming **acid rain**. Acid rain often falls far from the original sources, and it can corrode metals and erode marble buildings and statues.

Section 13.8—Indoor pollution can be as bad as, or worse than, pollution outdoors. Cigarette smoke can raise both PM and carbon monoxide above the allowed EPA levels. Because of the effects of secondhand smoke, most cities and states restrict tobacco smoking in public places. The radioactive gas radon can accumulate indoors in some cases, and inhaled radon produces **daughter isotopes** that also are radioactive and can accumulate in the lungs when breathed. The daughter isotopes can produce a significant risk of lung cancer.

Section 13.9—Ozone, an **allotrope** of oxygen, is a harmful air pollutant in the troposphere but forms a protective layer in the stratosphere that shields Earth against UV radiation from the sun. Chlorofluorocarbons (CFCs) have been linked to the hole in the ozone layer and are now widely banned. It appears that the ozone hole may be shrinking because of this ban.

Section 13.10—Carbon dioxide and other gases contribute to the **greenhouse effect**, which occurs when the infrared radiation emitted from Earth's surface cannot escape. The result is **global warming**, an increase in Earth's average temperature, which is predicted to cause many undesirable climate changes. **Carbon sequestration** refers to the capture and storage of CO_2 from large industrial emission sources.

Section 13.11—The EPA has listed six "criteria" pollutants, of which all have been reduced in recent years. Motor vehicles are a prominent source of carbon monoxide, hydrocarbons, and nitrogen oxides, whereas most PM and sulfur oxides come from industrial processes. Carbon monoxide and sulfur oxides are the most problematic pollutants. We all share responsibility for pollution—and we share the responsibility for reducing it.

■ REVIEW QUESTIONS

1. List two (former) uses of CFCs.

2. List one potential replacement for CFCs. What problems do the replacements have?

3. Describe how **(a)** a bag filter and **(b)** a cyclone separator remove particulate matter from stack gases.

4. What is bottom ash? What is fly ash? Give two uses for fly ash.

5. What is smog?

6. What are the health effects of ozone in polluted air?

7. What is a greenhouse gas? Give two examples.

8. Consider **(a)** the hole in the ozone layer and **(b)** the phenomenon of global warming. Are they the same thing? Are they related in any way?

■ PROBLEMS

The Atmosphere: Composition and Cycles

9. What is meant by nitrogen fixation? Why is it important?

10. How has industrial fixation of nitrogen to make fertilizers affected the nitrogen cycle?

11. Revisit Figure 13.1. Then consider a small airplane flying at 12,000 feet altitude. In what layer of the atmosphere is the airplane flying?

12. Revisit Figure 13.1. The X-15 was an experimental aircraft tested from 1959 to 1968. One test flew to an altitude of 67 miles. What layer of the atmosphere did the X-15 reach?

13. Revisit Figure 13.3, which shows oxidation of metals as one aspect of the oxygen cycle. Write the balanced equations for **(a)** the oxidation of iron metal by atmospheric oxygen to form iron(III) oxide and **(b)** the oxidation of chromium metal to form chromium(III) oxide.

14. Sucrose (table sugar, $C_{12}H_{22}O_{11}$) is a sugar produced by sugarcane and sugar beets. Write a balanced equation that represents the formation of sucrose from water and carbon dioxide, with oxygen gas as the other product.

Industrial Smog

15. What are the chemical components of industrial smog?

16. What weather conditions characterize industrial smog?

For Problems 17–20, give the equation for the indicated reaction.

17. Sulfur is oxidized to sulfur dioxide.

18. Sulfur dioxide is oxidized to sulfur trioxide.

19. Sulfur trioxide reacts with water to form sulfuric acid.

20. Sulfur dioxide reacts with hydrogen sulfide to form elemental sulfur and water.

21. Describe how a limestone scrubber removes sulfur dioxide from stack gases. Give the chemical reaction(s) involved.

22. What is particulate matter? What is an aerosol?

Photochemical Smog

23. What are the chemical components of photochemical smog?

24. Under what conditions do nitrogen and oxygen combine? Give the equation for the reaction.

25. Give the equation for the reaction of nitrogen monoxide with oxygen to form nitrogen dioxide.

26. What is PAN? From what is it formed? What are its health effects?

27. Describe two ways in which the level of nitrogen oxide emissions from an automobile can be reduced.

28. What are VOCs?

Carbon Monoxide

29. Is CO a free radical?

30. What are the health effects of carbon monoxide?

31. How might exposure to carbon monoxide contribute to heart disease?

32. Give the equation by which carbon monoxide reduces nitric oxide to nitrogen gas.

The Ozone Layer

For Problems 33–34, give the equation for the indicated reaction.

33. Oxygen is converted to ozone in the ozone layer.

34. CFCs destroy ozone.

35. What health effect might result from depletion of the ozone layer?

36. How are free radicals related to the ozone layer?

Acid Rain

37. What is the effect of acid rain on iron? Give an equation.

38. How might acid rain be alleviated?

39. What is the effect of acid rain on marble? Give the equation.

40. Acid rain can be formed by the reaction of one molecule of water with one molecule of sulfur dioxide to form one molecule of *sulfurous* acid. Write the equation for this reaction.

Indoor Air Pollution

41. List several indoor air pollutants. What is the source of each?

42. List some risks associated with secondhand cigarette smoke.

43. What are the health effects of radon gas? How can a radon problem in a building be alleviated?

44. How does better insulation of buildings make indoor air pollution worse?

45. When and how is carbon monoxide considered an indoor air pollutant?

46. What kind of particulate matter is found inside the home?

Carbon Dioxide and Climate

47. What is the greenhouse effect?

48. How can global warming be alleviated?

■ ADDITIONAL PROBLEMS

49. Why is zero pollution not possible?

50. The average person breathes about $20 \, \text{m}^3$ of air a day. What mass in milligrams of particulates would a person breathe in a day if the particulate level were $400 \, \mu\text{g/m}^3$?

51. The atmosphere contains $5.2 \times 10^{15} \, \text{t}$ of air. How much carbon dioxide (CO_2) is in the atmosphere if the concentration is 385 ppm?

52. Since 1978 the giant ice cap that floats in the Arctic Ocean has lost half its average thickness and has decreased in size by an area larger than the state of Texas. The ice cap atop the Antarctic continent also appears to be melting at the edges, dropping huge icebergs into the ocean. Which occurrence, if either, has caused a rise in sea level? (*Hint:* Consider whether the water level rises when an ice cube melts in a glass of water and when an ice cube is added to a glass of water.)

▲ The number and size of the icebergs that break off the continental shelf in Antarctica are increasing as a result of global warming.

53. Water is a greenhouse gas that is present in significant concentrations in the Earth's atmosphere. Water is also a product of the combustion of fossil fuels. Why is there usually little concern over increases of water vapor and its contribution to global warming?

54. Review Example 13.1. If combustion of gasoline is 99% efficient—that is, 99% of the gasoline burns to carbon dioxide—what mass of carbon *monoxide* is then produced when one gallon of gasoline is burned?

55. Elemental lead is extracted from an ore called galena (PbS). In the first step, galena is heated in air, converting it to lead(II) oxide. **(a)** Write the balanced equation for that step. **(b)** What mass of sulfur dioxide in grams is produced when 1.00×10^3 kg of galena is converted to lead oxide?

56. Pyrite (FeS_2) is an impurity in some coals. When the coal is burned, pyrite reacts with oxygen to form sulfur dioxide. **(a)** Write the balanced equation for that reaction. **(b)** How much sulfur dioxide in grams is formed when 2.00×10^4 kg of coal that is 0.0500% by mass pyrite is burned?

57. The reaction for formation of nitric acid is more complex than shown in Figure 13.10. In one step, three molecules of nitrogen dioxide react with a water molecule to form nitric acid and nitrogen oxide. In the second step, the nitrogen oxide is oxidized by atmospheric oxygen to nitrogen dioxide, and the cycle continues. Write balanced equations representing these two steps.

58. Although carbon dioxide is the actual greenhouse gas, emissions from the burning of fossil fuels are often reported as gigatons (Gt) of carbon. Emissions in China reached 1.48 Gt C in 2006. What mass in gigatons of CO_2 is that?

■ COLLABORATIVE GROUP PROJECTS

Prepare a PowerPoint, poster, or other presentation (as directed by your instructor) for presentation to the class.

59. What are the average and peak concentrations of each of the following in your community or in a nearby large city? How can you find out?
   a. carbon monoxide
   b. ozone
   c. nitrogen oxides
   d. sulfur dioxide
   e. particulate matter

60. Does your community have an air pollution problem? If so, describe it. How could it be solved?

61. Should students be allowed to smoke in school buildings? On school grounds? Defend your position.

62. Using the Internet or books, compare the amount of sulfur dioxide from Mt. Kilauea in Hawaii with that from all U.S. industry. (Remember to use the same time span for both.)

63. Substitutes for CFCs are more expensive than CFCs. What factors should you consider in a cost–benefit analysis of this substitution? Should the government subsidize the switch?

64. Air quality standards are maximum allowable amounts of various pollutants. The federal government and some state governments issue these standards. Using the Internet, find the current standards issued by the EPA. Does your state issue such standards? (*Hint*: Try the EPA site and your own state's environmental protection site.)

65. Divide your class into several teams, each taking a different view of the global warming problem. Possible positions are (1) we must act now, (2) we need more study before we act, and (3) there is no global warming. Research your position and be prepared to state your case with data to support your arguments. Hold a debate with the other groups in your class, paying close attention to the information presented by the other groups. Are their arguments persuasive? Are their sources reliable? Are yours?

Water is truly a unique substance. All three phases are present in this photograph of an Alaskan melt pond. Liquid water forms on the surface of the ice (solid water) in the Arctic summer's warmth. Water vapor from evaporation of the liquid is present but invisible. Ice floats on liquid water because when water freezes it becomes less dense. Known as the "universal solvent," water's solvent power makes it incredibly useful; it dissolves many of the substances needed for proper function of plant and animal life. It also makes it easy to contaminate and difficult to purify.

WATER

1. What prevents the creation of a filter that would remove all pollution from water?
2. Why does only one-tenth of an iceberg show above the surface?
3. Is bottled water safer than tap water?
4. About how much water do we really use each day?

Rivers of Life; Seas of Sorrows

Earth is a water world; most of its surface is covered with oceans and seas, lakes and rivers. People contain a lot of water, too: About two-thirds of our body weight is water. The water in our blood is quite similar to water in the ocean, containing a variety of dissolved ions. You might even say that we are walking sacks of seawater.

The presence of large quantities of water makes our planet unique in our solar system, the only one capable of supporting higher forms of life. Water is the only substance on Earth that commonly exists in large amounts in all three physical states. Go outside on a snowy day and you are likely to "see" all three forms at once. Gaseous water vapor is invisible (and is all around us), but tiny droplets of liquid water form clouds. Solid water falls to the ground as snow and melts into puddles of liquid water.

The Ionian philosopher Thales (ca. sixth century B.C.E.)—whom some consider the first "scientist"—held water in the highest regard, believing that it was the "primordial substance," that from which all other things are made. Little wonder from a man from an island nation. Water's unique nature makes it essential to life, and the nature of life as we know it makes it dependent on water. Our search for life on Mars is based to a large extent on evidence that the planet once had vast quantities of water and that relics of that life might still exist below the dry surface.

The properties that make water able to support life also make it easy to pollute and, therefore, potentially hazardous. Many substances are easily dispersed throughout the environment because they are soluble in water. Once dissolved, it is not easy to remove substances from water. Furthermore, many strains of bacteria and other microorganisms that are harmful to humans thrive in water.

Sol 20   Sol 24

▲ Two photographs of a trench dug on Mars by the *Phoenix* lander, taken on June 15 and June 18, 2008. Chunks of matter in the shadowed part of the left-hand photo have disappeared in the right-hand photo. The chunks are thought to be ice that underwent sublimation (Section 6.1).

1. What prevents the creation of a filter that would remove all pollution from water?
Some water pollutants are present at extremely low levels, and the removal techniques available to us just aren't specific enough to remove those pollutants. Also, water's polarity gives it a very strong attraction for many pollutants.

381

We start this chapter by discussing some of the unusual properties of water. Then we look at water pollution and water treatment. We should not underestimate the importance of clean drinking water nor take it for granted. We could live for several weeks without food, but without water we would last a few days at most.

14.1 Water: Some Unique Properties

Although water is quite familiar, it is a most unusual compound. It is the only common liquid on the surface of our planet. For most substances, the solid form is more dense than the liquid. However, the solid form of water (ice) is less dense than the liquid. This means water expands when it freezes. The consequences of this peculiar characteristic are essential for life on Earth. Ice forms on the surfaces of lakes and insulates the lower layers of water, which enables fish and other aquatic organisms to survive winter in the temperate zones. If ice were denser than liquid water, it would sink to the bottom as it formed, and even the deeper lakes of the northern United States would freeze solid in winter.

The same property—ice being less dense than liquid water—has dangerous consequences for living cells. When living tissues freeze, ice crystals are formed, and the expansion ruptures and kills cells. The slower the cooling, the larger the crystals of ice and the more damage there is to the cell. Frozen-food manufacturers make "flash-frozen" products by freezing foods so rapidly that the ice crystals are kept very small and thus do minimal damage to the cellular structure of the food.

Water also is more dense than most other familiar liquids. As a consequence, liquids less dense than water and insoluble in water float on its surface. The massive oil spills that occur when a tanker ruptures or when an offshore well gets out of control are a fairly common problem. The oil, floating on the surface of the water, washes onto beaches, where it does considerable ecological and aesthetic damage. If water were less dense than oil, the problem would certainly be different, although not necessarily less serious.

Another unusual property of water is its high specific heat. Different substances have different capacities for storing energy absorbed as heat. **Specific heat** is the quantity of heat required to change the temperature of 1 g of a substance by 1 °C. Table 14.1 gives the specific heats of several familiar substances in the old metric units—calories per gram per degree (cal/g °C)—and in the SI units—joules per gram per kelvin (J/g K). Note that it takes almost ten times as much heat to raise the temperature of 1 g of water 1 °C as to raise the temperature of 1 g of iron by the same amount. Imagine heating two iron kettles on the stove, one empty and one filled with water. It takes much longer (and much more energy) to heat the water-filled kettle. Conversely, to cause even a small drop in water temperature, a relatively large quan-

2. Why does only one-tenth of an iceberg show above the surface?
Although ice is less dense ($d = 0.92$ g/cm^3) than liquid water ($d = 1.00$ g/cm^3), the density difference is rather small. A floating object displaces its mass of water, which means that a cubic meter (1000 L) of ice will have about 920 L of its volume below the water. Only 80 L will show above the water. Ice floats a little higher in seawater because the salt water is denser than freshwater.

▶ Alskan crude oil floats on the surface of a 2.5 million-gallon oil spill test tank in Middletown Township, NJ. The salwater test tank, operated by the U.S. Navy, is used to measure the effectivecess of oil spill cleanups. Just one liter of oil can create a slick about 180 meters (almost 600 feet) in diameter.

Table 14.1   Specific Heats of Some Familiar Substances at 25 °C

Substance	Specific Heat	
	(cal/g °C)	(J/g K)
Aluminum (Al)	0.216	0.902
Copper (Cu)	0.0920	0.385
Ethanol (CH₃CH₂OH)	0.588	2.46
Iron (Fe)	0.107	0.449
Lead (Pb)	0.0306	0.128
Silver (Ag)	0.0562	0.235
Sulfur (S)	0.169	0.706
Water (H₂O)	1.00*	4.180

*Note that in cal/g °C, the specific heat of water is 1.00. As the metric system was established, the properties of water were often taken as the standard.

tity of heat must be removed. Because water stores heat so well, the water-filled kettle mentioned earlier will stay hot a lot longer than the empty kettle when the stove is turned off. The vast amounts of water on the surface of Earth thus act as a giant heat reservoir to moderate daily temperature variations. You need only consider the extreme temperature changes on the surface of the waterless moon, which range from 100 °C at noon to −173 °C at night, to appreciate this important property of water.

Water also is unique in having a high **heat of vaporization**; that is, a large amount of heat is required to evaporate a small amount of water. This is of enormous importance to us because large amounts of body heat can be dissipated by the evaporation of small amounts of water (perspiration) from the skin. This effect also accounts in part for the climate-modifying property of lakes and oceans. A large portion of the heat that would otherwise heat up the land instead vaporizes water from the surface of lakes or seas. Thus, in summer it is cooler near a large body of water than in interior land areas.

All these fascinating properties of water depend on the unique structure of the highly polar water molecule (Sections 4.12–4.13). In the liquid state, the molecules are tumbling over one another but always are associated with one another through strong hydrogen bonds. The tumbling becomes more and more violent as the temperature increases until the boiling point is reached. Under those conditions, a large fraction of the molecules have enough energy to break all of the hydrogen bonds with the other molecules surrounding them and they fly out of the liquid into the gas phase.

Conversely, when water freezes, its molecules take on a more ordered arrangement, forming four hydrogen bonds per molecule. The solid ice contains large hexagonal holes (Figure 14.1). This three-dimensional structure extends out for billions and billions of molecules. When the ice melts, the holes collapse, explaining why ice is less dense than liquid water.

(a)

(b)

◀ **Figure 14.1** Hydrogen bonds in ice. (a) Oxygen atoms are arranged in layers of distorted hexagonal rings. Hydrogen atoms lie between pairs of oxygen atoms, closer to one (covalent bond) than to the other. The yellow dashed lines indicate the hydrogen bonds. The structure has large "holes." (b) At the macroscopic level, the hexagonal arrangement of water molecules in the crystal structure of ice is revealed in the hexagonal shapes of snowflakes.

Self-Assessment Questions

1. Which of the following is *not* a property of water?
   a. expands when it freezes
   b. has a high specific heat
   c. has a low heat of vaporization
   d. is a useful solvent for many substances

2. Most of water's unique properties arise because
   a. oxygen has a stable electron configuration but hydrogen does not
   b. water molecules associate through hydrogen bonds
   c. water molecules can form covalent bonds with ionic substances
   d. water molecules can form covalent bonds with nonpolar substances

3. In a collection of water molecules, hydrogen bonds form between
   a. an H atom in one H_2O molecule and the O atom in another H_2O molecule
   b. the O atoms in different H_2O molecules
   c. the two H atoms in a single H_2O molecule
   d. two H atoms in different H_2O molecules

4. Frozen water, called ice, features an open structure with
   a. an amorphous structure
   b. large hexagonal holes
   c. molecules that are in constant motion
   d. molecules that tend to repel each other

Answers: 1, c; 2, b; 3, a; 4, b

14.2 Water in Nature

Three-fourths of the surface of Earth is covered with water, but nearly 98% of it is salty seawater—unfit for drinking and not suitable for most industrial purposes. Due to its polar nature, water readily dissolves most ionic substances, which accounts for the saltiness of the sea. Rainwater dissolves the ions of minerals, and these ions are carried by streams and rivers to the sea. There the heat of the sun evaporates part of the water, leaving the salts behind. Are the oceans growing saltier as the years go by? Apparently not. The rates of addition appear to be balanced by precipitation of minerals and absorption of ions onto clays.

Rain falls on Earth in enormous amounts, but most of it falls into the sea or on areas otherwise inaccessible. About 2% of Earth's water is frozen in the polar ice caps, leaving less than 1% available as fresh water, most of that underground. Lakes and streams account for only 0.01% of the fresh water on the planet.

Although potable (suitable for drinking) water is generally readily available in the United States, in many nations the situation is quite different. About 1.2 billion people around the world have inadequate quantities of fresh water, and 2.6 billion people live without basic sanitation. About 5 million people—most of them children—die each year because of water contaminated with bacterial, viral, and parasitic infections. According to the Pacific Institute, Oakland, California, the number of people without an adequate supply of clean water is expected to triple by 2050. Water resources around the globe are threatened by pollution, misuse, and climate change.

The Water Cycle and Natural Contaminants

Although the percentages of water apportioned to the oceans, ice caps, and freshwater rivers, lakes, and streams remain fairly constant, water is dynamically cycled among these various repositories (Figure 14.2). It constantly evaporates from both water and land surfaces, and water vapor condenses into clouds and returns to Earth as rain, sleet, and snow. This fresh water becomes part of the ice caps, runs off in streams and rivers, and fills lakes and underground pools of water in rocks and sand called *aquifers*.

More to Explore
Water for People—Water for Life: The United Nations World Water Development Report. Paris: UNESCO Publishing, 2003.

▲ An adequate supply of water is essential to life.

Gases

Rainwater is not pure H_2O. It carries down dust particles, and it dissolves some oxygen, nitrogen, and carbon dioxide as it falls through the atmosphere. The carbon dioxide makes natural rainwater slightly acidic because it reacts with water to form carbonic acid (H_2CO_3):

$$CO_2(g) + H_2O(l) \longrightarrow H_2CO_3(aq)$$

Lightning causes nitrogen, oxygen, and water vapor to combine to form nitric acid, which dissolves in rainwater as well.

Groundwater also contains the naturally occurring gas radon, a product of the decay of radioactive uranium and thorium. Although radon emits damaging ionizing radiation and produces a variety of hazardous daughter nuclei, it is only slightly soluble in water. Thus, the water used for showering, washing dishes, and cooking generally contributes only a small proportion (about 1 to 2%) of the total radon in indoor air.

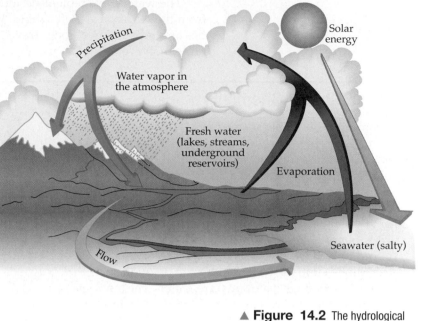

▲ **Figure 14.2** The hydrological (water) cycle.

Dissolved Minerals

As water moves along or beneath the surface of Earth, it dissolves minerals from rocks and soil. Recall that minerals (salts) are ionic and that ions are either positively charged (cations) or negatively charged (anions). Table 14.2 provides a summary of substances found in natural waters. The principal cations in natural water, shown in red in Table 14.2, are ions of sodium, potassium, calcium, magnesium, and sometimes iron (as Fe^{2+} or Fe^{3+}). The anions (shown in blue) are usually sulfate, bicarbonate, and chloride.

Water containing calcium, magnesium, or iron salts is called **hard water**. The positive ions react with the negative ions in soap to form a scum that clings to clothes and bathroom fixtures and leaves them dingy looking (eChapter 21). *Soft water* may contain ions, such as Na^+ or K^+, but these ions do not form insoluble scum with soap.

Table 14.2 Some Substances Found in Natural Waters		
Substance	**Formula**	**Source**
Carbon dioxide	CO_2	Atmosphere
Dust	—	Atmosphere
Nitrogen	N_2	Atmosphere
Oxygen	O_2	Atmosphere
Nitric acid (thunderstorms)	HNO_3	Atmosphere
Sand and soil particles	—	Soil and rocks
Sodium ions	Na^+	Soil and rocks
Potassium ions	K^+	Soil, rocks, and fertilizer
Calcium ions	Ca^{2+}	Limestone rocks
Magnesium ions	Mg^{2+}	Dolomite rocks
Iron(II) ions	Fe^{2+}	Soil and rocks
Chloride ions	Cl^-	Soil, rocks, and fertilizer
Sulfate ions	SO_4^{2-}	Soil, rocks, and fertilizer
Bicarbonate ions	HCO_3^-	Soil and rocks
Radon	Rn	Radioactive decay

The water cycle replenishes our supply of fresh water. When water evaporates from the sea, salts are left behind. When water moves through the ground, impurities are trapped in the rock, gravel, sand, and clay. This capacity to purify is not infinite, however.

Some Biblical Chemistry

According to the Biblical story of Moses, getting a dependable supply of fresh water has long been a problem. When Moses led the Israelites out of Egypt into the wilderness, he encountered a desert area where potable water was scarce. Many people know the Biblical account of how Moses struck the rock to bring forth water. An incident at Marah, where the Israelites couldn't drink the water because it was bitter, is less well known. According to Exodus, God commanded Moses to throw a certain tree into the water to sweeten (neutralize) it.

Basic (alkaline) solutions are bitter. The pool at Marah was probably basic; such *alkaline* waters are common in desert areas. A possible chemical explanation has been given for Moses's purification of the brackish water. The tree was probably dead, its cellulose bleached by the desert sun, with the alcohol groups in cellulose oxidized to carboxylic acid groups.

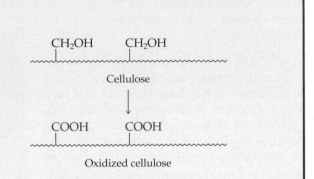

These acidic groups neutralize the alkali in the water. (Tannins, found in plant material and composed of polyphenols and other acidic compounds, also may have contributed to the neutralization.) Moses didn't have to understand the science of water purification in order to apply the appropriate technology.

Organic Matter

Rainwater dissolves matter from decaying plants and animals. In small quantities as part of the natural cycle in a forest, this enriches the soil, but in large quantities it contaminates soil and water. Organic matter can also come as traces of lubricants, fuels, some fertilizers, and pesticides. Bacteria, microorganisms, and animal wastes are all potential contaminants of natural waters.

Self-Assessment Questions

1. When water evaporates,
   a. covalent bonds are broken between H_2O molecules
   b. hydrogen bonds are broken between H_2O molecules
   c. the remaining liquid gets warmer
   d. the molecules move closer together

2. Energy to power the water cycle comes from
   a. electric power plants b. the moon c. storms d. the sun

3. Rainwater is naturally acidic due to dissolved
   a. carbon dioxide b. carbon monoxide
   c. nitrogen d. oxygen

4. The principal cations in water are
   a. Na^+, Al^{3+}, Ca^{2+}, and Mg^{2+} b. Na^+, K^+, Ca^{2+}, and Li^+
   c. Na^+, K^+, Ca^{2+}, and Mg^{2+} d. Ra^{2+}, Sr^{2+}, Ca^{2+}, and Mg^{2+}

5. Hard water may be characterized by the presence of all of the following ions *except*
   a. Ca^{2+} b. Fe^{2+} c. Mg^{2+} d. Na^+

Answers: 1, b; 2, d; 3, a; 4, c; 5, d

14.3 **Chemical and Biological Contamination**

Early people did little to pollute the water and the air, if only because their numbers were so few. Most pollutants at that time were natural: hard-water ions, alkaline contaminants, and animal wastes. The coming of the agricultural revolution and the rise of cities brought together enough *Homo sapiens* to pollute the environment seriously. Even then, pollution was mostly local and largely biological. Human wastes were dumped on the ground or into the nearest stream. Disease organisms were transmitted through food, water, and direct contact.

▲ *Cryptosporidium*, shown here in a fluo-rescent image, sickened 400,000 people and killed about 100 in the Milwaukee area in 1993. Since then, utilities have generally improved their treatment systems.

Waterborne Disease

Contamination of water supplies by *pathogenic* (disease-causing) microorganisms from human wastes was a severe problem throughout the world until about 100 years ago. During the 1830s, severe epidemics of cholera swept the Western world. Typhoid fever and dysentery were common. In 1900, for example, there were more than 35,000 deaths from typhoid in the United States. Today, as a result of chemical treatment, municipal water supplies in developed nations are generally safe. Water-borne diseases are still quite common in much of Asia, Africa, and Latin America, where even today there are occasional epidemics of cholera and typhoid. Dysentery is rampant much of the time. An estimated 80% of all the world's sickness is caused by contaminated water.

The threat of biological contamination has not been totally eliminated from de-veloped nations. The EPA estimates that 30 million people in the United States are at risk because of bacterial contamination of drinking water. One ongoing threat is *cryptosporidium*, a protozoa excreted in human and animal feces that resists stan-dard chemical disinfection. Utilities are constantly improving their treatment sys-tems to deal with this and other problems.

Biological contamination also lessens the recreational value of water, when swimming, fishing, and other recreational activities become hazardous.

▲ Contaminated beaches are closed to swimming and other recreational activities.

Acid Rain

Acids formed from sulfur oxides (SO_x) and nitrogen oxides (NO_x) come down from the sky as acid rain, fog, and snow. These acids corrode metals, dissolve limestone and marble, and even ruin the finishes on our automobiles. Acids also flow into streams from abandoned mines.

Acid rain has been somewhat alleviated over the last few decades by the reduc-tion of SO_x emissions (Section 13.7), but thousands of bodies of water in eastern North America are acidified, and thousands more have only a limited ability to neutralize the acids that enter them. The acids are presumed to originate mainly in the Ohio River Valley and Great Lakes regions. The SO_x and NO_x—mainly from coal-fired power plants, but also from other industries and automobiles—travel hundreds of kilometers downwind and fall as sulfuric and nitric acids.

Acid water is detrimental to life in lakes and streams. It has been linked to de-clining crop and forest yields. Acid waters probably affect living organisms in many ways. For example, aluminum ions, which are tightly bound in clays and other minerals, are released by acid. Aluminum ions have low toxicity for humans but they can be deadly to young fish. Many dying lakes have only old fish because none of the young survive. Ironically, lakes destroyed by excess acidity are often quite beautiful—clear and sparkling.

Acids are no threat to lakes and streams where limestone (calcium carbonate) is plentiful, as limestone can neutralize excess acid.

▲ Acid water draining from an old mine.

$$CaCO_3(s) + 2 H^+(aq) \longrightarrow Ca^{2+}(aq) + CO_2(g) + H_2O$$

<u>Limestone</u> <u>Acid</u>

Where rock is principally granite, however, no such neutralization occurs.

▲ Common sewage from homes and businesses depletes the dissolved oxygen in water.

▲ The eutrophication of a lake is a natural process, but the action can be greatly accelerated by human wastes and the runoff from farms, lawns, and golf courses.

Sewage and Dying Lakes

Pathogenic microorganisms are not the only problem caused by the dumping of human sewage into our waterways. The breakdown of organic matter by bacteria depletes **dissolved oxygen** in the water and enriches the water with plant nutrients. A stream can handle a small amount of waste without difficulty, but when massive amounts of raw sewage are dumped into a waterway, undesirable changes occur.

Most organic material can be degraded (broken down) by microorganisms. Biodegradation can be either *aerobic* or *anaerobic*. **Aerobic oxidation** occurs in the presence of dissolved oxygen. A measure of the amount of oxygen needed for this degradation is the **biochemical oxygen demand (BOD)**. The greater the quantity of degradable organic wastes, the higher the BOD. If the BOD is high enough, dissolved oxygen is depleted and no life (other than odor-producing anaerobic microorganisms) can survive in the lake or stream. Flowing streams can regenerate themselves; those with rapids soon come alive again as the swirling water dissolves oxygen. Lakes with little or no flow can remain dead for years.

With adequate dissolved oxygen, aerobic bacteria (those that require oxygen) oxidize the organic matter to carbon dioxide, water, and a variety of inorganic ions (Table 14.3). The water is relatively clean, but the ions, particularly the nitrates and phosphates, may serve as nutrients for the growth of algae, which also cause problems. When the algae die, they become organic waste and increase the BOD. This process is called **eutrophication**. The eutrophication of a lake is a natural process, but the action can be greatly accelerated by human wastes, phosphates from detergents, and the runoff of fertilizers from farms and lawns that stimulates algal bloom and die-off. Streams and lakes die because nature cannot purify them nearly as quickly as we can pollute them.

When too much organic matter depletes the dissolved oxygen in a body of water—whether from sewage, dying algae, or other sources—**anaerobic decay** processes take over. Instead of oxidizing the organic matter, anaerobic bacteria reduce it. Methane (CH_4) is formed. Sulfur is converted to hydrogen sulfide (H_2S) and foul-smelling organic compounds. Nitrogen is reduced to ammonia and odorous amines (Section 9.14). The foul odors are a good indication that the water is overloaded with organic wastes. Only anaerobic microorganisms can survive in such water.

Dumping our sewage into waterways isn't the only way we can foul up an ecological cycle. Fertilizer runoff from farm fields, golf courses, and lawns, and seepage from feedlots all add inorganic nutrients to the cycle, and an algal bloom can lead to oxygen depletion and death for the fish. Perhaps the most enigmatic influence of all comes from the introduction of new substances into the ecological water cycle: pesticides, radioisotopes, detergents, drugs, toxic metals, and industrial chemicals.

Table 14.3 Some Substances Formed in Water by the Breakdown of Organic Matter	
Substance	**Formula**
Aerobic conditions	
Carbon dioxide	CO_2
Nitrate ions	NO_3^-
Phosphate ions	PO_4^{3-}
Sulfate ions	SO_4^{2-}
Bicarbonate ions	HCO_3^-
Anaerobic conditions	
Methane	CH_4
Ammonia	NH_3
Amines	RNH_2
Hydrogen sulfide	H_2S
Methanethiol	CH_3SH

The Industrial Revolution added a new dimension to our water pollution problems. Factories were often built on the banks of streams, and wastes were dumped into the water to be carried away. The rise of modern agriculture has led to increased contamination as fertilizers and pesticides have found their way into the water system. Transportation of petroleum results in oil spills in oceans, estuaries, and rivers. Acids enter waterways from mines and factories and from acid precipitation. Household chemicals also contribute to water pollution when detergents, solvents, and other chemicals are dumped down drains.

The creation of new materials has wrought a variety of new ecological problems. Chemists and chemistry are necessary for understanding pollution—how it occurs and how it can be mitigated. Monitoring by ever more sophisticated analytical methods and instruments has helped detect and define specific ecological problems. A stinking lake is obvious to everyone, but only by using advanced analytical techniques can scientists determine the level of dangerous, yet invisible, materials in the water we drink.

Self-Assessment Questions

1. An estimated 80% of all illnesses worldwide are waterborne diseases. The principal causes of these are
   a. bacteria and viruses   **b.** DDT and dioxins
   c. lead and mercury   **d.** pesticides
2. Water in a lake made acidic by acid rain or mine drainage is often
   a. clear   **b.** eutrophic   **c.** rich in game fish   **d.** rich in lime
3. Dissolved oxygen (DO) levels are important because DO levels can affect
   a. light penetration   **b.** phosphate levels
   c. salinity levels   **d.** survival of aquatic animals
4. The biochemical oxygen demand (BOD) of a water sample is a measure of
   a. dissolved oxygen levels   **b.** organic matter levels
   c. phosphate levels   **d.** salt levels or salinity
5. How are DO and BOD related?
   a. BOD has no impact on DO   **b.** high BOD means high DO levels
   c. high BOD reduces DO as the organic matter decays
   d. low BOD means low DO
6. Aerobic oxidation produces
   a. CO_2, H_2O, and inorganic ions   **b.** CO, NH_3, and H_2O
   c. CH_4, NH_3, and H_2S **d.** sewage sludge
7. Excess phosphates in a lake or river cause
   a. aerobic decay   **b.** eutrophication
   c. fixation   **d.** increased game fish

Answers: 1, a; 2, a; 3, d; 4, b; 5, c; 6, a; 7, b

14.4 Industrial Water Use

It takes several hundred kilograms of steel to produce a typical automobile. To make a metric ton (t) of steel requires about 100 t of water. About 4 t of water is lost through evaporation. The remainder is contaminated with acids, grease and oil, lime, and iron salts. This polluted water can be cleaned up, and most of it is recycled.

Chrome plating on bumpers, grills, and ornaments was once a serious source of pollution. Waste chromium, in the form of chromate ions (CrO_4^{2-}), and cyanide ions (CN^-) are products of this process. In the past, these toxic substances were dumped into waterways. Today, chemical treatment removes most of these pollutants.

Cyanide is treated with chlorine and a base to form nitrogen gas, bicarbonate ions, and chloride ions.

$$10\,OH^- + 2\,CN^- + 5\,Cl_2 \longrightarrow N_2 + 2\,HCO_3^- + 10\,Cl^- + 4\,H_2O$$

The products are much less toxic than cyanide. The chromate is removed by reduction with sulfur dioxide to Cr^{3+} ion, and the sulfur dioxide is oxidized to sulfate.

$$2\,CrO_4^{2-} + 3\,SO_2 + 2\,H_2O \longrightarrow 2\,Cr^{3+} + 3\,SO_4^{2-} + 4\,OH^-$$

▲ Chromium plating is beautiful and durable, but the pollutants that result from the process can constitute a significant waste-disposal problem.

▲ **It DOES Matter!**

Paper production from virgin wood fibers traditionally has used enormous quantities of water. Worldwide usage is about 315 billion kg each year, or about 100 lb of water for a single ream of paper. Wood is about half cellulose, but the cellulose fibers are held together by lignin, a resinous material that must be removed to make paper. According to the U.S. Environmental Protection Agency, recycling can reduce water use in paper production by nearly 60% and energy consumption by 40%. The EPA also says that recycling causes 35% less water pollution and 74% less air pollution. (An even more effective measure is to reduce paper usage; print only what you must!)

Sulfate is generally not a serious pollutant; Cr^{3+} is relatively insoluble in alkaline solution but it is soluble enough in acidic media to constitute a problem.

With the environmental and economic costs of elastomers for tires, fabrics for upholstery, glass for windows, electricity, and so on, it is easy to see that the private automobile is an ecological problem even before it hits the road. And once on the road, it is a major contributor to air pollution (Chapter 13).

Most other industries also contribute to water pollution. Table 14.4 lists the water required (per metric ton) for the production of a variety of materials. Much of this water is cleaned up and recycled, but the need for clean water is still enormous. Industries in the United States have substantially reduced their contribution to water pollution. Most are in compliance with the Water Pollution Control Act, which requires that they use the best practicable technology. We examine here only a few examples of industrial pollution and some ways to alleviate it.

Table 14.4	Typical Quantities of Water Required to Produce Various Materials		
Industrial Products	**Water Required**[a]	**Consumer Products**	**Water Required**[b]
Steel	100	Laptop computer	10,600
Paper	20	1 kg flour	77
Copper	400	1 bowl rice	525
Rayon	800	1 L red wine	720
Aluminum	1280	1 cup coffee	140
Synthetic rubber	2400	1 XL cotton tee shirt	30,300

[a]In cubic meters per metric ton. A cubic meter of water weighs 1000 kg, or 1 t.
[b]In liters.

Water Pollution in Russia

Until the Soviet Union collapsed in 1991, their leaders covered up the environmental problems of Eastern Europe. Let's consider the case of one of the most famous rivers in the world, the mighty Volga in Russia (Figure 14.3). So many dams have been built along the Volga that its flow has been slowed to a crawl. It used to take 50 days for water to travel the 3700 km from the source to the mouth of the river, but now it takes 18 months. The sluggish flow plus the pollution from all the factories along the way have created an ecological catastrophe. Tons of industrial waste (cleaning fluids, fertilizers, pesticides, heavy metals, toxic chemicals, radioactive waste, and waste from pulp and paper mills) pour into the Caspian Sea, which appears to be dying. In some places you can see yellow, red, and black streams of water carrying sulfur, iron oxide, and various oils. The Volga once teemed with caviar-producing sturgeon, but pollution and poaching now threaten the sturgeon with extinction.

▲ **Figure 14.3** The Soviet Union built so many dams and factories along the Volga River that it is no longer the mighty, vigorous waterway it once was. The river is polluted by fertilizers and pesticides from farms and by heavy metals from the factories along its banks. Many problems were hidden from the rest of the world until the Soviet Union collapsed in 1991.

Self-Assessment Questions

1. Cyanide ions can be removed from wastewater by treatment with
   a. alum **b.** chlorine **c.** filtration **d.** fluorides

2. Chromate ions can be removed from wastewater by reduction to Cr^{3+} using
   a. activated carbon **b.** CO_2 **c.** H_2 **d.** SO_2

Answers: 1, b; 2, d

GREEN CHEMISTRY

Wear Green with Pride

Andy Jorgensen, University of Toledo

How many ways do you use water on a daily basis? In addition to the water you use directly, water is critical to the production of virtually everything in your daily life, from the food on your table, to the wood in your house, and the clothes on your back.

Green Chemistry Principle #1 is to prevent waste. When you purchase an item, the waste from its processing is distributed among various locations. We can reduce the quantities of waste going into our rivers, lakes, and seas by purchasing less and buying more prudently.

Let's consider clothing as an example. Recall the large amount of water required to produce a single T-shirt (Table 14.4). How long do you make use of that shirt? After you wear your clothes for a period of time, what happens to them? Do they stay in your closet while you add new items? Do you donate the clothes to a charity? Do you toss them into the garbage where they are buried and provide no value to anyone? The total quantity of clothes that we buy is the primary factor affecting the amount of waste deposited in water during clothing manufacturing.

Some clothing manufacturing processes affect our water more than others. There is no simple answer to the question of which fabrics can be made with the least waste in their manufacture. Waste from making clothes enters natural waters via several routes, varying by fiber source and production process.

Natural fibers, such as cotton from plants and wool from animals, are usually produced with several chemical agents to increase yield and quality. Cotton has a high biological susceptibility to predators, which means that cotton crops typically require large amounts of insecticides. Sheep provide a natural habitat for lice and other living creatures, and so farmers treat these animals with pesticides to kill the insects. Where pesticides are used in producing natural fibers, some amount of the applied chemical pesticide will get into natural bodies of water. Organic cotton or wool clothing, however, is certified to have been produced without synthetic pesticides or fertilizers. Another natural fabric is made from bamboo, which is easily grown without pesticides. Hemp (*Cannabis sativa*) produces a low-environmental-impact fiber, but U.S. laws designed to prevent cultivation of *Cannabis* strains as marijuana prevent the production of hemp. Although the fabric can be imported into the United States, you may recall from Chapter 12 that the environmental impact of transportation over long distances may offset the benefits of lower-impact production.

Artificial fibers have some advantages compared to natural ones, including the utilization of what might otherwise be waste. This is the case for rayon, a silk-like fiber that can be produced from tree pulp at much lower cost than natural silk from worms. Rayon production starts with pulp left over from making other wood products, thereby putting a wood by-product to valuable use. However, rayon fiber is structurally quite different from wood cellulose. Thus, the manufacture of rayon requires chemical treatment of the cellulose with many substances that are not innocuous. Soy fiber also uses what otherwise would be waste, in this case from the making of tofu and other food products. Since the soy is grown primarily for food, making fabric from that which is not needed adds nothing to the overall waste stream.

After production, fibers from any source typically are processed to modify their color. Bleaching with chlorine compounds was discussed in Section 8.7. Other bleaching agents, such as those made with hydrogen peroxide, can reduce or eliminate chlorine-containing waste. Dyes contain a variety of agents that give color to a fabric and provide a means for keeping that color stable over time. Some dyes contain heavy metals to help achieve these qualities but use of such dyes can lead to the subsequent deposition of the metals into natural bodies of water.

Although some natural plant and animal substances make safer dyes, even some natural chemicals can have unhealthy effects, such as allergic reactions. One very creative alternative is to design the plant to produce colored fibers. Cotton of the variety termed *Colorganic* is a modern version of a plant that has been grown in Central America for many years. This type of cotton comes from specifically bred plants that have fibers of the desired color so no dye is needed. Colorganic cotton grows in natural shades of pale green, beige, and brown.

If you want to reduce waste that enters our water system during the production of your clothes you have many choices, but some are not easy or clear-cut. However, your choices as a consumer can ultimately influence what producers and suppliers bring to market.

14.5 Groundwater Contamination → Tainted Tap Water

About half the people in the United States drink surface water (from streams and lakes). The other half get their drinking water from groundwater via wells or springs. In rural areas, 97% of the population drink groundwater. Because rocks differ in porosity and permeability, water does not move around the same way in all rocks. An aquifer forms when water-bearing rocks readily transmit water to wells and springs. Wells are drilled into an aquifer, and water is pumped out. Precipitation recharges the aquifer. For many aquifers, the rate of pumping water is much greater than the rate of recharge, and the water table drops. The dropping water table in most parts of the country requires that wells be drilled deeper and deeper.

Well water can be contaminated. Toxic chemicals have been found in the groundwater in some areas. For example,

- Many wells throughout the United States are contaminated with the gasoline additive methyl *tert*-butyl ether (MTBE) (Section 15.8), making the water undrinkable due to its offensive taste and odor.

- Wells around the Rocky Mountain Arsenal, near Denver, are contaminated by wastes from the production of pesticides.

- Wells in many areas are contaminated with perfluorooctanoic acid (PFOA) and related compounds. PFOA ($CF_3CF_2CF_2CF_2CF_2CF_2CF_2COOH$) is an industrial surfactant used for processing polytetrafluoroethylene (PTFE). These compounds persist in the environment where they accumulate in living organisms.

- Some community water supplies in New Jersey have been shut down because of contamination with industrial wastes.

Groundwater contamination is particularly alarming because, once contaminated, an underground aquifer may remain unusable for decades or longer. There is no easy way to remove the contaminants. Pumping out the water and purifying it could take years and cost billions of dollars. We will examine a few important sources of groundwater contamination in a bit more detail.

Nitrates

In many agricultural areas, well water is contaminated with nitrate ions (NO_3^-). Excessive nitrates are especially dangerous to infants. In an infant's digestive tract, nitrate ion is reduced to nitrite ion. The result is *methemoglobinemia*, or blue baby syndrome. The baby turns blue and can die after drinking the water if not treated. In some farming areas, such as California's Imperial Valley, some parents have to buy bottled water for their babies.

Nitrates in the groundwater (Figure 14.4) come from fertilizers used on farms and lawns, from the decomposition of organic wastes in sewage treatment, and from runoff from animal feed lots. They are highly soluble and thus difficult to remove from water, requiring expensive treatment.

Volatile Organic Chemicals

We mentioned volatile organic compounds (VOCs) as air pollutants (Section 13.5). VOCs are also water pollutants and add an undesirable odor to water. Many VOCs are suspected carcinogens. VOCs are used as solvents, cleaners, and fuels, and they are components of gasoline, spot removers, oil-based paints, inks, thinners, and some drain cleaners. Common VOCs are hydrocarbon solvents such as benzene (C_6H_6) and toluene (C_6H_5—CH_3), and chlorinated hydrocarbons such as carbon tetrachloride (CCl_4), chloroform ($CHCl_3$), and methylene chloride (CH_2Cl_2). Especially common is trichloroethylene (CCl_2=$CHCl$), widely used as a dry-cleaning solvent and as a degreasing compound. When spilled or discarded, VOCs enter the soil and eventually get into the groundwater.

▲ Half of Chinese cities have significantly polluted groundwater and some cities are facing a water crisis. More than 400 Chinese cities are threatened with water shortage, with 136 of them experiencing severe shortages. Here a water company worker takes a sample from the Songhua River in Harbin, northeastern China's Heilongjiang Province.

▲ **It DOES Matter!**
Homes and a school were built on the site of an old chemical dump in the Love Canal section of Niagara Falls, New York. Now weeds and tall grass grow through an abandoned sidewalk near that area. The U.S. government declared emergency evacuations of the area in 1978 and 1980, and relocated more than 800 families. Over 250 different industrial contaminants were found on the site.

Probability of Nitrate Contamination of Shallow Groundwater

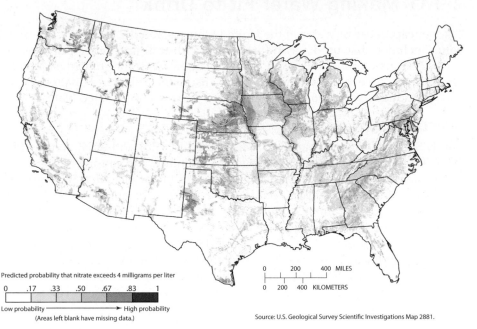

Predicted probability that nitrate exceeds 4 milligrams per liter

| 0 | .17 | .33 | .50 | .67 | .83 | 1 |

Low probability ———————→ High probability
(Areas left blank have missing data.)

0   200   400 MILES
0   200   400 KILOMETERS

Source: U.S. Geological Survey Scientific Investigations Map 2881.

◀ **Figure 14.4** Areas of the contiguous United States with groundwater likely to be contaminated by nitrates.

Chemicals buried in dumps—often years ago, before there was much awareness of environmental problems—have now infiltrated groundwater supplies. Often, as at the Love Canal site in Niagara Falls, New York, people built schools and houses on or near old dump sites. VOCs generally are only slightly soluble in water, with solubilities often in the parts-per-million or parts-per-billion range. These trace amounts are found in groundwater. Many VOCs are characterized by a lack of reactivity. They react so slowly that they are likely to be around for a long time.

Leaking Underground Storage Tanks

A major source of groundwater contamination is underground storage tanks (USTs) that contain petroleum or hazardous chemicals. These buried steel tanks last an average of about 15 years before they rust through and begin to leak. In the 1980s the U.S. EPA estimated there were about 2.5 million such tanks. More than 460,000 leaks from USTs have been detected, with petroleum products such as gasoline and certain gasoline additives such as methyl *tert*-butyl ether (MTBE) (Section 15.8) the most prevalent contaminants found in nearby wells. Laws now require replacement of old gasoline tanks and proper cleanup of any contaminated ground. About 350,000 tank sites have been cleaned up, and the frequency and severity of leaks from UST systems have been reduced greatly. However, as of 2006 there were still 113,000 remaining to be cleaned up.

More to Explore
Rogers, Peter. "Facing the Freshwater Crisis," *Scientific American*, August 2008, pp. 46–53.

Self-Assessment Questions

1. Which ion, from fertilizer runoff and sanitary wastewater discharges, can lead to methemoglobinemia in young children?
   a. Ca^{2+} **b.** Cl^- **c.** Na^+ **d.** NO_3^-

2. Which of the following is a common volatile organic compound (VOC) contaminant of groundwater?
   a. acetaldehyde from polluted air
   b. acetic acid from vinegar
   c. ethanol from alcoholic beverages
   d. methyl *tert*-butyl ether, a gasoline additive

Answers: 1, d; 2, d

14.6 Making Water Fit to Drink

The per capita use of water in the United States is about 2 million L per year. This value includes that used for industrial, agricultural, and other purposes and is twice the consumption in Europe. Although often taken for granted, significant energy and resources are spent every year to ensure the safety and quality of drinking water supplied by the 156,000 public water systems in the United States.

Safe Drinking Water Act

The U.S. Safe Drinking Water Act was passed in 1974 and amended in 1986 and 1996. The act gives the EPA power to set, monitor, and enforce national health-based standards for a variety of contaminants in municipal water supplies. As modern analytical techniques continue to improve, our ability to identify smaller and smaller concentrations of potentially harmful substances also advances. As a result, the number of regulated substances with maximum contaminant levels (MCLs) increased from 22 in 1976 to 90 in 2004. The 1996 amendments have MCLs for 7 types of bacterial or viral microorganisms; 7 standards for disinfectants, such as chlorine, or disinfectant products, such as bromate compounds; 16 inorganic compounds; 54 organic compounds; and 4 radioactive isotopes. Table 14.5 lists some of the contaminants.

Table 14.5 U.S. Environmental Protection Agency Drinking Water Standards for Selected Substances[a]

Substance	Maximum Contaminant Level (mg/L)
Primary standards: inorganic compounds	
Arsenic	0.010
Barium	2
Copper	1.3
Cyanide	0.2
Fluoride	4.0
Lead	0.015
Nitrate	10[b]
Primary standards: organic compounds	
Atrazine	0.003
Benzene	0.005
p-Dichlorobenzene	0.075
Dichloromethane	0.005
Heptachlor	0.0004
Lindane	0.0002
Toluene	1
Trichloroethylene	0.005
Secondary standards (nonenforceable)	
Chloride	250
Iron	0.3
Manganese	0.05
Silver	0.10
Sulfate	250
Total dissolved solids	500
Zinc	5

[a]A more detailed list and a more detailed explanation of the rules can be found on the EPA website.
[b]Expressed as N. Expressed as nitrate ion, the level is 45 mg/L.

Calculations of Parts per Million and Parts per Billion

We discussed solution concentrations in Chapter 5. For solutions that are extremely dilute, we often express concentrations in parts per million (ppm), parts per billion (ppb), or even parts per trillion (ppt). For example, in fluoridated drinking water, the fluoride ion concentration is maintained at about 1 ppm. A typical level of the contaminant chloroform ($CHCl_3$) in municipal drinking water taken from the lower Mississippi River is 8 ppb.

For aqueous solutions, ppm, ppb, and ppt are generally based on mass. Thus, 1 ppm of solute in a solution is the same as 1 g solute per 1×10^6 g (1 million grams) solution, and 1 ppb is 1 g solute per 10^9 g (1 billion grams) solution. Or

$$1 \text{ ppm} = \frac{1 \text{ g solute}}{10^6 \text{ g solution}} \qquad 1 \text{ ppb} = \frac{1 \text{ g solute}}{10^9 \text{ g solution}}$$

Calculations are much like those for percent by mass (Section 5.5). In Example 14.1, we introduce a useful relationship for aqueous solutions.

EXAMPLE 14.1 Contaminant Concentrations

The maximum allowable level of fluoride ion in drinking water set by the EPA is 4 mg F^- per liter. What is this level expressed in ppm?

Solution
The density of water, even if it contains traces of dissolved substances, is essentially 1.00 g/mL. One liter of water has a mass of 1000 g. So we can express the allowable fluoride level as the ratio

$$\frac{4 \text{ mg } F^-}{1000 \text{ g water}}$$

To convert this fluoride level to ppm, we need to have the numerator and denominator in the same units. By using milligrams, we make the denominator 1 million mg. The numerator then expresses the ppm of solute, that is, ppm F^-.

$$\frac{4 \text{ mg } F^-}{1000 \text{ g water}} \times \frac{1 \text{ g water}}{1000 \text{ mg water}} = \frac{4 \text{ mg } F^-}{1,000,000 \text{ mg water}} = 4 \text{ ppm } F^-$$

■ **EXERCISE 14.1A**
What is the concentration in **(a)** ppb and **(b)** ppt corresponding to a maximum allowable level in water of 0.1 μg/L of the gasoline additive MTBE (methyl *tert*-butyl ether)?

■ **EXERCISE 14.1B**
What is the molarity of the solution in Exercise 14.1A?

Some Comparisons That Put the Figures in Perspective

1 PART PER MILLION (PPM) IS
  1 inch in 16 miles
  1 minute in 2 years
  1 cent in $10,000

1 PART PER BILLION (PPB) IS
  1 inch in 16,000 miles
  1 second in 32 years
  1 cent in $10 million

1 PART PER TRILLION (PPT) IS
  1 inch in 16 million miles
  1 second in 320 centuries
  1 cent in $10 billion

But Not a Drop to Drink

Clean drinking water is scarce in Bangladesh, one of the poorest countries in the world. Ravaged frequently by floods, the surface waters that supplied drinking water for most of the country until about 40 years ago were often contaminated by disease-causing microorganisms. During the 1970s, UNICEF and the World Bank provided funds for the construction of hundreds of thousands of tubewells to provide a consistent supply of clean drinking water within 100 m of every family. The project halved the infant mortality rate during the 1980s.

(continued)

Much to everyone's surprise, the wells brought a new risk: chronic arsenic poisoning. Arsenic occurs naturally in the mineral formations of the region, and it now contaminates the water collecting in many of the wells. Arsenic poisoning threatens 85 million of Bangladesh's 140 million people. The initial stages of arsenic poisoning are characterized by skin pigmentation, warts, diarrhea, and ulcers. Arsenic also damages the lungs, kidneys, and other internal organs. In the most severe cases, arsenic poisoning causes liver and renal deficiencies or cancer that can lead to death. Deeper, more expensive wells that reach water below the arsenic-containing formations can alleviate the problem. A simple, inexpensive filter with layers of cast iron turnings, river sand, wood charcoal, and wet brick chips can remove most of the arsenic. However, nearly

▲ A village tubewell in Bangladesh.

half of Bangladesh's people live on an income less than a dollar per day, and even these cheap devices are beyond their reach. Regardless of the methods used, remediation could take decades and cost hundreds of millions of dollars.

Water Treatment Plants

About two-thirds of the population of the United States gets its drinking water from treated surface water (lakes, reservoirs) while the other third gets theirs from groundwater supplies (wells and aquifers). Contamination can threaten either of these sources. Most cities in developed nations treat their water supply before it flows into homes (Figure 14.5). The water to be purified is usually placed in a settling

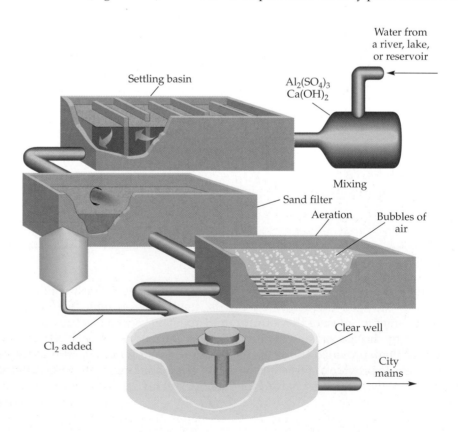

▶ **Figure 14.5** A diagram of a municipal water purification plant.

basin where it is treated with slaked lime and a flocculent such as aluminum sulfate. These materials react to form a gelatinous mass of aluminum hydroxide.

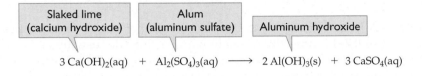

$$3\,Ca(OH)_2(aq)\ +\ Al_2(SO_4)_3(aq)\ \longrightarrow\ 2\,Al(OH)_3(s)\ +\ 3\,CaSO_4(aq)$$

The aluminum hydroxide carries down dirt particles and bacteria. The water is then filtered through sand and gravel.

Usually water is treated by *aeration*; it is sprayed into the air to remove odors and improve its taste (water without dissolved air tastes flat). Sometimes it is filtered through charcoal to remove colored and odorous compounds.

Chemical Disinfection

In the final step, chlorine is added to kill any remaining bacteria. In some communities that use river water, a lot of chlorine is needed to kill all the bacteria, and you can taste the chlorine in the water. Some people question the use of chlorine because it converts dissolved organic compounds into chlorinated hydrocarbons, including trihalomethanes such as chloroform ($CHCl_3$). Analyses of the drinking water of several cities that take their water from rivers have found chlorinated hydrocarbons, including such known carcinogens as chloroform and carbon tetrachloride. The concentration is in the parts-per-billion range, probably posing only a small threat, but it is worrisome nonetheless. (It is not nearly so worrisome, however, as the waterborne diseases that prevail in much of the world where adequate water treatment is not available.)

Chlorination is not the only way to disinfect drinking water. Ozone (O_3) is used widely in Europe and increasingly in the United States. Ozone is more expensive than chlorine, but less of it is needed. An added advantage is that ozone kills viruses on which chlorine has little, if any, effect. For example, ozone is 100 times more effective than chlorine in killing polioviruses.

Ozone acts by transferring its "extra" oxygen atom to the contaminant. Oxidized contaminants are generally less toxic than chlorinated ones. In addition, ozone imparts no chemical taste to water. Unlike chlorine, however, ozone does not provide residual protection against microorganisms. Some systems therefore use a combination of disinfectants: ozone for initial treatment and subsequent chlorine addition to provide residual protection.

Other Technologies

Water contaminated with a variety of microorganisms can be purified by irradiation with ultraviolet (UV) light. UV works rapidly, and it can be cost-effective in small-scale applications. No chemical generation, storage, or handling is required. UV is effective against *Cryptosporidium*, and there are no known by-products formed at levels that cause concern. The disadvantages include no residual protection for drinking water and no taste and odor control. UV has limited effectiveness in turbid water and generally costs more than chlorine treatment.

Fluorides

Dental caries (tooth decay) was once considered the leading chronic disease of childhood. That this is no longer true is attributed mainly to fluoride toothpastes (eChapter 21) and the addition of fluoride to municipal water supplies. The hardness of tooth enamel can be correlated with the amount of fluoride present. Tooth enamel is a complex calcium phosphate called hydroxyapatite. Fluoride ions replace some of the hydroxide ions, forming a harder mineral called fluorapatite.

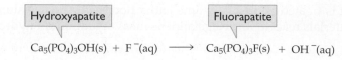

$$Ca_5(PO_4)_3OH(s) + F^-(aq) \longrightarrow Ca_5(PO_4)_3F(s) + OH^-(aq)$$

The fluoride concentration of the drinking water of many communities has been adjusted to 0.7–1.0 ppm (by mass) by adding fluoride, usually as H_2SiF_6 or Na_2SiF_6. Early studies showed reduction in the incidence of dental caries by 50 to 70%. More recent studies show a much smaller effect, perhaps because of fluoride toothpastes and the presence of fluorides in food products prepared with fluoridated water.

Some people object to water fluoridation. Fluoride salts are acute poisons in moderate to high concentrations. Indeed, sodium fluoride (NaF) is used as a poison for roaches and rats. Small amounts of fluoride ion, however, seem to contribute to our well-being through strengthening bones and teeth. There is some concern about the cumulative effects of consuming fluorides in drinking water, in the diet, in toothpaste, and from other sources. Excessive fluoride consumption during early childhood can cause mottling of tooth enamel. The enamel becomes brittle in certain areas and gradually discolors. Fluorides in high doses also interfere with calcium metabolism, with kidney action, with thyroid function, and with the actions of other glands and organs. Although there is little or no evidence that optimal fluoridation causes problems such as these, the fluoridation of public water supplies will most likely remain a subject of controversy.

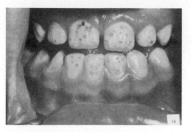

▲ Excessive fluoride consumption in early childhood can cause mottling of tooth enamel. The severe case shown here was caused by continuous use during childhood of a water supply that had an excessive natural concentration of fluoride.

Self-Assessment Questions

1. About how many substances are regulated under the U.S. EPA drinking water standards?
   a. 6   **b.** 15   **c.** 90   **d.** 250

2. A value of 3 ppm of dissolved oxygen means
   a. 3 mg O/L H_2O **b.** 3 mg O/mL H_2O
   c. 3 g O/mL H_2O **d.** 3 g O/L H_2O

3. In treatment of public water supplies, flocculation
   a. helps remove dirt and bacteria   **b.** improves taste
   c. kills bacteria   **d.** kills viruses

4. Public water supplies are aerated to
   a. improve the taste   **b.** kill bacteria
   c. remove suspended matter   **d.** remove trihalomethanes

5. Chlorine is added to public water supplies to
   a. improve the taste   **b.** kill bacteria
   c. remove suspended matter   **d.** remove trihalomethanes

6. Ozonation—treatment of water with O_3—has an advantage over chlorination in that
   a. O_3 is cheaper than Cl_2
   b. O_3 provides residual protection; Cl_2 does not
   c. oxygenated by-products are generally less toxic than chlorinated by-products
   d. sedimentation is eliminated

7. Fluoridation of drinking water strengthens tooth enamel by converting hydroxyapatite to
   a. $CaCO_3$ **b.** CaF_2 **c.** fluorapatite   **d.** fluorotartrate

Answers: 1, c; 2, a; 3, a; 4, a; 5, b; 6, c; 7, c

14.7 Wastewater Treatment

The wastes generated by nearly two-thirds of the U.S. population are gathered in sewer systems and carried to treatment plants by more than 50 billion L of water each day. The water that two-thirds of the U.S. population drinks comes from reservoirs, lakes, and rivers, much of which has been used by other municipalities upstream. Cities have to treat the used water before discharging it back into the environment, but about 10% of it passes untreated into streams or the ocean. In this section we discuss the treatment and purification of wastewater that renders it suitable for returning it to the water cycle.

Wastewater Treatment Plants

For decades, most communities treated sewage simply by holding it in settling ponds for a while before discharging it into a stream, lake, or ocean, a process now called **primary sewage treatment** (Figure 14.6). Primary treatment removes 40% to 60% of suspended solids as *sludge* and about 30% of organic matter. The effluent still has two-thirds of its original BOD and nearly all its nitrates and phosphates. All the dissolved oxygen in the pond may be used up, and anaerobic decomposition—with its resulting odors—takes over.

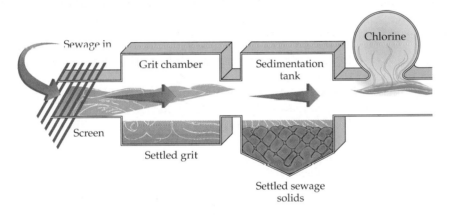

◄ **Figure 14.6** Diagram of a primary sewage treatment plant.

A **secondary sewage treatment** plant passes effluent from the primary treatment facility through sand and gravel filters. There is some aeration in this step, and aerobic bacteria convert much of the organic matter to inorganic materials. In the **activated sludge method** (Figure 14.7), a combination of primary and secondary treatment methods, the sewage is placed in tanks and aerated with large blowers.

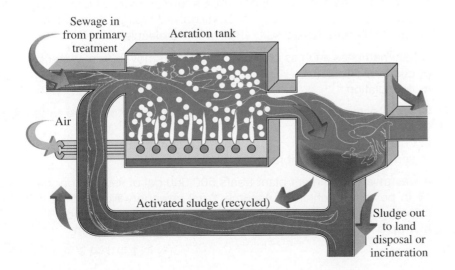

◄ **Figure 14.7** Diagram of a secondary sewage treatment plant that uses the activated sludge method of treatment.

This causes the formation of large, porous clumps called *flocs*, which filter and absorb contaminants. The aerobic bacteria further convert the organic material to sludge. Part of the sludge is recycled to keep the process going, but huge quantities must be removed for disposal. This sludge is stored on land (where it requires large areas), dumped at sea (where it pollutes the ocean), or burned in incinerators (where it requires energy—such as natural gas—and can contribute to air pollution). Some of the sludge is used as fertilizer.

Secondary treatment of wastewater removes about 90% of the BOD but often is inadequate in nitrate and phosphate removal. To an increasing extent, federal mandates require **advanced treatment** (sometimes called *tertiary treatment*). Several advanced processes are in use, and most are quite costly.

- In **charcoal filtration**, charcoal adsorbs organic molecules, including trihalomethanes such as chloroform ($CHCl_3$), that are difficult to remove by any other method. The organic molecules are adsorbed on the surface of the charcoal and thus removed from the water. After a period of time, the charcoal becomes saturated and is no longer effective. It can be regenerated by heating to drive off the adsorbed substances.

- In **reverse osmosis**, pressure forces water through a semipermeable membrane, leaving impurities behind.

- **Phytoremediation** involves passage of the effluent into large natural or constructed lagoons for storage, allowing plants such as reeds and ferns to remove metals and other contaminants.

Finding the money to finance adequate sewage treatment will be a major political problem for years to come.

The effluent from sewage plants is usually treated with chlorine to kill any remaining pathogenic microorganisms before it is returned to a waterway. Chlorination has been quite effective in preventing the spread of waterborne infectious diseases such as typhoid fever. Further, some chlorine remains in the water, providing residual protection against pathogenic bacteria. However, chlorination is not effective against some viruses such as those that cause hepatitis.

Self-Assessment Questions

1. At what stage in the treatment do bacteria consume substances in the sewage?
   - **a.** aeration
   - **b.** primary
   - **c.** secondary
   - **d.** screening and sand removal

2. Trihalomethanes such as chloroform are formed when chlorine reacts with
   - **a.** bacteria
   - **b.** dissolved organic compounds
   - **c.** ions that cause hardness
   - **d.** water molecules

3. Trihalomethanes such as chloroform can be removed from water by
   - **a.** charcoal filtration
   - **b.** chlorination
   - **c.** flocculation
   - **d.** sand filtration

4. Reverse osmosis is a way to purify water by using pressure to force the water through a
   - **a.** charcoal filter
   - **b.** sand filter
   - **c.** semipermeable membrane
   - **d.** trickling filter

5. A wastewater treatment plant treats 500,000 gal of sewage per day, using 16 lb per day of chlorine. Assume 1 gal sewage weighs about 8 lb. The chlorine dosage, in parts per million, of the wastewater is about
   - **a.** 1 ppm
   - **b.** 4 ppm
   - **c.** 8 ppm
   - **d.** 16 ppm

Answers: 1, c; 2, b; 3, a; 4, c; 5, b

14.8 The Newest Soft Drink: Bottled Water

Bottled water is the fastest growing and most profitable segment of the beverage industry. Worldwide sales of bottled water are more than 178 billion L per year, with over 33.5 billion L ($11.7 billion) sold in the United States in 2007. More than half of all Americans drink bottled water, even though it costs 240 to 10,000 times as much per liter as tap water. Per capita consumption is 112 L per year. The trend is based largely on the misconception that bottled water is safer or healthier than tap water.

The federal government considers bottled water a food, and it is therefore regulated by the Food and Drug Administration (FDA) rather than the EPA. Although the FDA has adopted EPA standards for tap water as standards for bottled water, testing of bottled waters is typically less rigorous than for municipal water.

Because municipal water supplies are usually chlorinated, tap water can have a chlorine taste. This might cause some to buy bottled water, but beware: About 25% of bottled water sold in the United States comes from municipal water supplies. Some bottled water is disinfected with ozone, just as some municipal water supplies are, but just because it comes in a bottle with a glacier and a mountain spring on the label does not necessarily mean water is "more pure." Mineral water in fact is likely to have *more dissolved ions* (typically Ca^{2+} and Mg^{2+}) than normal tap water. Fortunately, in the United States in general, both bottled water and municipal water are quite safe. However, some people will continue to buy bottled water for its convenience, taste, and falsely presumed health benefits.

It takes about 1.5 million barrels of oil to make the water bottles Americans use each year, enough to provide fuel for 100,000 cars a year instead. Only about 23 percent of those bottles are recycled. Add in fuel for transportation, and the environmental cost of bottled water is even higher.

3. Is bottled water safer than tap water?
As noted in the text, bottled water is often merely tap water that has been bottled! In any event, bottled water has never been demonstrated to be healthier or safer than tap water. Nor has it been demonstrated to be more hazardous than tap water. Bottled water is largely a convenience and not a benefit, regardless of whether it has been flavored, sweetened, fortified with vitamins, or has added minerals.

A N S W E R

14.9 Water Pollution and the Future

For most of human history, our body wastes were an integral part of Earth's natural recycling system. The wastes provided food for microorganisms that degraded them, returning nutrients to soil and water. But with the growth of cities, human waste was disconnected from the cycle. We now dump our wastes in waterways, fouling the water and wasting nutrients that could fertilize the land.

Many communities dry and sterilize sludge and then transport it to farmlands for return of nutrients to the soil. Others pump wet, suspended sludge directly to the fields. The water in the mixture irrigates the crops and the sludge provides nutrients and humus. One concern is that sludge is often contaminated with toxic metals that could be taken up by plants and eventually end up in our food.

A few communities treat their sewage by first holding it in sedimentation tanks where the solids settle out as a sludge that is removed and processed for use as fertilizer. The effluent is then diverted into a marshy area where the marsh plants filter the water and use nutrients from the sewage as fertilizer. Some plants even remove toxic metals. Effluent from the marsh is often as clean as the water in municipal reservoirs.

Other solutions are possible. Septic tanks have been used for home sewage treatment for decades, but their use is generally limited to rural and some suburban areas, and they require significant homeowner upkeep to remain effective. Toilets have been developed that compost wastes and use no energy or water. The mild heat of composting drives off water from the wastes. The system is ventilated to keep the process aerobic, and no odors enter the house from a properly installed system. Dried waste is removed about once a year. The initial cost is much higher than that of a flushed toilet, however. For each person in the United States, we flush about 35,000 L of drinking-quality water annually. Perhaps we should consider an alternative to flushing our wastes into the water we drink.

A N S W E R

4. About how much water do we really use each day?
Each person in the United States uses about 7 L per day for drinking and cooking, but we use directly a total of about 380 L per day (washing dishes, laundry, cars, flushing toilets). However, Table 14.6 shows that the vast majority of water is used indirectly for energy production.

We're the Solution to Water Pollution

We generally take our drinking water for granted. Perhaps we shouldn't. There are between 4000 and 40,000 cases of waterborne illnesses in the United States each year. Twenty million people have no running water at all, and many more obtain water from suspect sources: 10% of public water supplies do not meet one or more of the EPA standards. Another 30 million tap individual wells or springs that are often uncontrolled and of unknown quality. Much remains to be done before we can all be assured of safe drinking water.

How much water does one really need? Only about 1.5–2.0 L/day for drinking. In the United States each day, we use about 7 L per person for drinking and cooking, but we use directly a total of about 380 L. We use much more water *indirectly* in agriculture and industry (Table 14.6) to produce food and other materials; it takes 800 L of water to produce 1 kg of vegetables and 13,000 L of water to produce a steak. We also use water for recreation (for example, swimming, boating, and fishing). For most of these purposes, we need water that is free of bacteria, viruses, and parasitic organisms.

Table 14.6 Use of Water in the United States	
Use	**Percentage**
Coolant for electric power plants	48
Irrigation	34
Public water supplies	11
Industrial	5
Miscellaneous*	2

*Includes mining, livestock, aquaculture, and other uses.

The Clean Water Act has drastically reduced water pollution from industrial sources. However, we hope it is clear that complete elimination of pollution is not possible; to use water is to pollute it. All of us can do our share by conserving water and by minimizing our use of products that require vast amounts of water to make. As our population grows, it will cost a lot just to maintain the present water quality. To clean up our water, and then keep it clean, will cost even more. However, the cost of unclean water is even higher—discomfort, loss of recreation, illness, and even death.

More to Explore
The U.S. Environmental Protection Agency has a website with "Quick Finder" links to Acid Rain, Agriculture, Clean Water Act, Oil Spills, Water, Wetlands, and more.

Self-Assessment Questions

1. To produce a kilogram of vegetables requires about how much water?
   a. 8 L   **b.** 80 L   **c.** 800 L   **d.** 8000 L
2. To produce a steak requires about how much water?
   a. 0.5 L   **b.** 13 L   **c.** 130 L   **d.** 13,000 L

Answers: 1, c; 2, d

Critical Thinking Exercises

Apply knowledge that you have gained in this chapter and one or more of the FLaReS principles (Chapter 1) to evaluate the following statements or claims.

14.1 An activist group claims that all chlorine compounds should be banned because they are toxic.

14.2 A mother discovers that tests have found 1.0 ppb of trichloroethylene in the well water the family uses for drinking, bathing, cooking, and so on. The family has used the well for two years. The mother decides that the family members should be tested for cancer.

14.3 A group of citizens want to ban fluoridation of the community water supply because fluorides are toxic.

14.4 A company claims that its "super-oxygenated water can boost athletic performance." (Note that under pressure, there might be at most 0.3 g O_2/L H_2O, the amount in a 1-L breath.)

14.5 A company claims that "oxygenated and structured water has smaller molecules of water that penetrate more quickly into the cells and, therefore, hydrate your body faster and more efficiently."

■ SUMMARY

Section 14.1—Water has unusual properties: It is the only common liquid on Earth; its solid form is less dense than the liquid; and it is more dense than most other common liquids. **Specific heat** is the heat needed to raise the temperature of 1 g of a substance by 1 °C; for water it is quite high, 1 cal/(g °C). Water also has a high **heat of vaporization**, the amount of heat needed to vaporize a fixed amount of water. These properties arise because of strong hydrogen bonding between water molecules.

Section 14.2—Water covers three-fourths of Earth's surface, but only about 1% of it is available as fresh water. Water dissolves many ionic substances, which is why the sea is salty. Potable water is readily available in most parts of the United States but many other countries do not have sufficient fresh water. In the water (hydrological) cycle, water evaporates from oceans and lakes, condenses into clouds, and returns to Earth as rain or snow. During this cycle it can be contaminated by various chemicals and microorganisms, both natural and from human activities. **Hard water** is formed when minerals (salts of calcium, magnesium, and/or iron) dissolve in groundwater.

Section 14.3—Waterborne diseases such as cholera, typhoid fever, and dysentery were a severe problem until about 100 years ago and are still common in many parts of the world. Acid rain from sulfur oxides and nitrogen oxides, which in turn arise from burning coal and other fuels, has caused acidification of thousands of bodies of water in North America. In addition to its corrosive effects, acid rain is detrimental to both plant and animal life. When sewage is dumped into waterways, that sewage is biodegraded by microorganisms. **Aerobic oxidation** occurs in the presence of **dissolved oxygen**, and the **biochemical oxygen demand (BOD)** is a measure of the amount of oxygen needed. Sewage increases the BOD. If the BOD is high enough, only **anaerobic decay**, or decay in the absence of oxygen, can occur. Sewage, phosphates, and fertilizers can accelerate **eutrophication**, in which algae grow and die, thereby increasing the BOD of the water. Other new substances introduced into the water cycle may cause new problems.

Section 14.4—Building a car takes a great deal of water, which is polluted in the process and must be cleaned up and recycled.

Chromium and cyanide were once serious sources of pollution but various chemical treatments now remove them.

Section 14.5—About two-thirds the U.S. population drinks surface water and one-third drinks groundwater; the latter is especially common in rural areas. The aquifers of drilled wells are being consumed faster than they are being refilled from precipitation, and deeper wells are now required in many places. Well water may be contaminated with various chemicals including nitrates and VOCs from underground storage tanks.

Section 14.6—The U.S. Safe Water Drinking Act sets and enforces standards for water quality. In many cases, contaminants must be expressed in tiny units such as parts per million (ppm) or parts per billion (ppb). Prior to drinking, water treatment usually includes settling, filtration, aeration, and chemical disinfection with chlorine or ozone. Fluorides may also be added to prevent tooth decay.

Section 14.7—Treatment of wastewater (used water) begins with **primary sewage treatment**, allowing the wastes to settle and removing sludge before discharging into a stream, lake, or ocean. In **secondary sewage treatment** the effluent is filtered through sand and gravel filtration. Also, the **activated sludge method** may be used, in which the sewage is aerated and the sludge is removed. **Advanced** (tertiary) **treatment** may involve **charcoal filtration** to adsorb organic compounds, **reverse osmosis**, in which water is forced through a semipermeable membrane, leaving contaminants behind, or **phytoremediation**, which uses plants to remove metals and other contaminants. Increasing treatment is increasingly expensive.

Section 14.8—Bottled water is a popular but relatively expensive beverage. It is widely but incorrectly perceived that bottled water is safer or healthier than tap water. Bottled water often comes from municipal water supplies and may have more dissolved ions than normal tap water.

Section 14.9—Alternative methods for sewage treatment are being investigated. Much more water is used indirectly by consumers than directly. Maintaining water quality is expensive and will become more so, but not maintaining the quality of our water will cost much more in terms of our health and comfort.

■ REVIEW QUESTIONS

1. A barge filled with gasoline sinks and breaks open. Will the gasoline dissolve in the water? Will it float or sink?

2. Why should foods be flash-frozen rather than frozen slowly?

3. What are pathogenic microorganisms?

4. List some waterborne diseases. Why are these diseases no longer common in developed countries?

5. What problems do leaking underground storage tanks cause?

6. List some ways in which groundwater is contaminated.

7. Why do chlorinated hydrocarbons remain in groundwater for such a long time?

8. List some common industrial contaminants of groundwater.

■ PROBLEMS

Properties of Water

9. Why is ice less dense than liquid water? What consequences does this property have for life in northern lakes?

10. Define *specific heat*. Why is the high specific heat of water important to planet Earth?

For Problems 11 and 12, you may wish to consult the Appendix, part A.5, "Calculations Involving Temperature and Heat."

11. Refer to Table 14.1. How much heat, in calories, does it take to warm 540 g of water from 11.0 °C to 37.0 °C?

12. How much heat, in calories, does it take to warm 540 g of iron from 11.0 °C to 37.0 °C?

13. Why is the high heat of vaporization of water important to our bodies?

14. Why is it cooler near a lake than inland during the summer?

15. Why does the skin feel cool when one steps out of a swimming pool into a breeze?

16. Why are seas salty? Are they getting saltier?

Natural Waters

17. What impurities are present in rainwater?

18. What are the products of the breakdown of organic matter by anaerobic decay?

19. Water containing more than 3.0 grains of dissolved solids per gallon is considered to be hard. (One *grain* is equal to 64.8 mg.) What is this "hardness limit" in milligrams of dissolved solids per liter?

20. Refer to Problem 19. A pot containing 2.0 gal of tap water is allowed to evaporate. The solid material left behind is found to weigh 305 mg. Is the tap water considered to be hard?

Acidic Waters

21. List two ways in which lakes and streams have become acidic.

22. Why is acidic water especially harmful to fish?

23. Acidic water that is neutralized by natural rocks often becomes hard water. Explain.

24. Water from a pond is found to have high levels of ammonia and methane and relatively low levels of carbon dioxide and nitrate ions. What kind of decay has probably occurred in this water?

25. List several ways in which the acidity of rain can be reduced.

26. List two toxic compounds found in wastes from the chrome plating process. How is each removed?

Municipal Water Supplies

27. What is the optimal level of fluoride in drinking water?

28. Describe a nonchemical disinfecting treatment for drinking water.

29. What are some health effects of too much fluoride in the diet?

30. What is methemoglobinemia? How is it caused?

Wastewater Treatment

31. Describe a primary sewage treatment plant. What impurities does it remove?

32. Describe a secondary sewage treatment plant. What impurities does it remove?

33. What substances remain in wastewater after effective secondary treatment?

34. Describe the activated sludge method of sewage treatment.

35. Identify each of the following as a primary, secondary, or tertiary method of wastewater treatment.
    a. ion exchange   b. distillation
    c. trickling filters

36. What are the advantages and disadvantages of spreading sewage sludge on farmlands?

Chemical Equations

37. Write the balanced equation for the neutralization of acidic rain (assume HNO_3 is dissolved in water) by limestone (calcium carbonate).

38. Write the balanced equation for the reaction of slaked lime (calcium hydroxide) with alum (aluminum sulfate) to form aluminum hydroxide.

Parts per Million and Parts per Billion

39. Express the following aqueous concentrations in the unit indicated.
    a. 14 μg toluene per liter of water, as ppb toluene
    b. 0.021% $MgCO_3$, by mass, as ppm $MgCO_3$

40. Express the following aqueous concentrations in the units indicated.
    a. 25 μg trichloroethylene in 11 L water, as ppb trichloroethylene
    b. 38 g Cl_2 in 1.00×10^4 L water, as ppm of Cl_2

41. A 2.0-L sample of drinking water is found to contain 18 μg lead and 1.6 mg of barium. Do either of these exceed the standards described in Table 14.5?

42. A 0.50-L sample of drinking water is found to contain 4.0 μg benzene. Does this exceed the standards described in Table 14.5?

■ ADDITIONAL PROBLEMS

43. What types of substances are most effectively removed by each of the following methods of water purification? Which of the list of substances—$CHCl_3$, NaCl, phosphate ion, sand—is effectively removed from water by each?
    a. activated carbon bed   b. distillation
    c. filtration   d. ion exchange
    e. precipitation with Al^{3+}

44. Describe an alternative to the flush toilet.

45. Which takes more energy per gram: melting ice (solid at 0 °C to liquid at 0 °C), or boiling water (liquid at 100 °C to gas at 100 °C)? Explain briefly.

46. Express the following aqueous concentrations in the unit indicated.
    a. 2.4 ppm F^-, as molarity of fluoride ion, $[F^-]$
    b. 45 ppm NO_3^-, as molarity of nitrate ion, $[NO_3^-]$
    c. Why do some scientists prefer the ppm and ppb units over molarity with water contaminants?

47. Wastewater disinfected with chlorine must be dechlorinated before it is returned to sensitive bodies of water. The dechlorinating agent is often sulfur dioxide. The reaction is

    $$Cl_2 + SO_2 + 2\,H_2O \longrightarrow 2\,Cl^- + SO_4^{2-} + 4\,H^+$$

    Is the chlorine oxidized or reduced? Identify the oxidizing agent and reducing agent in the reaction.

48. Look back at the molecular formulas for the volatile organic compounds (VOCs) on page 392. Give structural formulas for
    a. benzene
    b. toluene
    c. chloroform
    d. methylene chloride
    e. trichloroethylene

49. Radioactive aluminum-26 is used to study the mobilization of aluminum by acidic waters. When it decays, the isotope is transmuted into magnesium-26. What kind of particle is emitted by aluminum-26?

50. The text indicates that radioactive radon gas is slightly soluble in water. What are the intermolecular forces responsible for the solubility?

51. Two identical iron teakettles are heated on the same stove. One contains 2 kg of water; the other contains 2 kg of iron pellets. After five minutes, which is hotter? Explain, using at least one key term from the chapter Summary.

52. How much calcium nitrate needs to be added to 1000 L of solution to provide 180 ppm Ca^{2+} if the water supply contains 40 ppm of Ca^{2+}? How much NO_3^-, in parts per million, does this quantity of calcium nitrate add to the solution?

■ COLLABORATIVE GROUP PROJECTS

Prepare a PowerPoint, poster, or other presentation (as directed by your instructor) for presentation to the class.

53. Consult the Web or other sources and prepare a report on the desalination of seawater by one of the following methods.
    a. distillation   b. freezing
    c. electrodialysis   d. reverse osmosis
    e. ion exchange

54. What sort of wastewater treatment is used in your community? Is it adequate?

55. Where does your drinking water come from? What steps are used in purifying it?

56. Call your water utility office or consult its website and obtain a chemical analysis of your drinking water. What substances are monitored? Are any of these substances considered problems? (If you use water from a private well, has the water been analyzed? If so, what were the results?)

57. The website of the U.S. EPA's Office of Wastewater Management (OWM) has a menu of water topics. First, notice how many water-related topics there are. Then choose one and write a few paragraphs about that aspect of water management.

58. Take a poll of your class regarding their consumption of bottled water. Why do people drink it? Do they have any concerns regarding its purity?

59. Using the EPA's OWM site or your state's environmental website, see what you can learn about the water quality where you live.

60. Find a website related to water quality and write a few paragraphs about its sponsoring organization.

61. Using the website of the Natural Resources Defense Council (NRDC) or other environmental group, investigate and write about pollution from urban stormwater runoff, which some say rivals sewage plants and factories as a source of water contamination.

The Itaipu Hydroelectric plant in Brazil is the largest in the world. Water from the Parani river spins turbines (shown here), which turn generators to produce electrical energy. Unfortunately, this renewable source makes up only a small fraction of the energy that our society produces. Most energy is obtained from the burning of fossil fuels such as oil, coal, and natural gas. Producing and using that energy forms pollutants that cause adverse health effects, economic damage, and global climate change. Cleaner alternative energy sources (such as solar and wind power) will become more and more important in the years ahead.

ENERGY

QUESTIONS YOU MAY HAVE ASKED YOURSELF

1. Is energy a form of matter?
2. Is there really an energy shortage?
3. Why aren't alternative fuels like ethanol used more?
4. How much uranium is available for power plants?
5. Is solar power the answer to the energy crisis?
6. Is hydrogen really a nonpolluting fuel?

A Fuels Paradise

We were tempted to call this chapter Fire, making the titles of Chapters 12, 13, 14, and 15 Earth, Air, Water, and Fire—the four elements of the ancient world.

Actually, fire could be an appropriate title for this chapter. We obtain most of our energy by burning fuels, which certainly involves fire. However, many sources of energy do not involve burning of anything, and we will depend more heavily on these other sources as our reserves of fossil fuels are depleted.

Energy is the ability to do work. Everything we consume or use—our food and clothes, our homes and their contents, our cars and roads—requires energy to produce, package, distribute, operate, and dispose of. The United States, with less than 5% of the world's population, uses one-fourth of all the energy currently generated on the planet. Abundant energy has enabled the United States to provide its people with tremendous benefits and conveniences and a high standard of living. But much of this energy is wasted; Canada, Australia, Japan, and the Scandinavian countries achieve higher standards with lower per capita use of energy. Consumption of energy in the United States has enormous costs to our health, to the environment, and to our national security. It affects our foreign debt and the stability of the Middle East. It pollutes the air we breathe and the water we drink.

Industry uses about 33% of all energy produced in the United States. This energy is used to convert raw materials to the many products our society seems to demand. Transportation uses about 28%, to power automobiles, trucks, trains,

---

1. Is energy a form of matter?
Energy is not matter; it has no mass and does not occupy a volume as does matter. Energy is the *ability* of matter to do work. For example, thermal energy is the energy of molecules in motion. The moving molecules can do work.

airplanes, and buses. Private homes use about 21% and commercial spaces use about 17%. Utilities use about 35% of the nation's energy production, primarily to generate electricity. (These figures add up to more than 100% because the electricity is used in industry, homes, and commercial spaces, and some of it is therefore counted twice.)

Energy lights our homes, heats and cools our living spaces, and makes us the most mobile society in the history of the human race. It powers the factories that provide us with abundant material goods. Indeed, energy is the basis of modern civilization.

15.1 Sunlight Floods Earth with Energy

All living things on Earth—us included—depend on nuclear energy for survival. Most of the energy available to us on this planet comes from the giant nuclear reactor we call the Sun. Although it is about 150 million km away, the Sun has been supplying Earth with most of its energy for billions of years, and it will likely continue to do so for billions more (Table 15.1).

The SI unit of energy is the joule (J); one watt (W), the SI unit of power, is 1 joule per second (J/s).

$$1 \text{ W} = 1\frac{\text{J}}{\text{s}}$$

A unit of more convenient size is the kilowatt: 1 kW = 1000 W. To measure energy consumption, the watt is combined with a unit of time. A familiar example is the *kilowatt-hour (kWh)*, the quantity of energy used by a 1-kW device in 1 hour; 1 kWh = 3600 kJ. (Energy and power are related in the same way that distance and speed are related. Speed is the *rate* at which distance is covered, and power is the *rate* at which energy is used.)

Recall (Chapter 11) that the Sun is a nuclear fusion reactor that steadily converts hydrogen to helium. The Sun has a power output of 4×10^{26} W. Earth receives about 1.73×10^{17} W [173,000 terawatts (TW); 1 TW = 10^{12} W] from the Sun—an amount equivalent to the output of 115 million nuclear power plants. In three days, Earth receives energy from the Sun equivalent to all our fossil fuel reserves. Yet this is only about 1 part in 50 billion of the Sun's output!

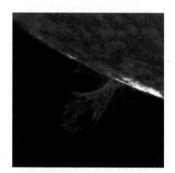

▲ The Sun fuses more than 600 million t of hydrogen into helium each second. It has enough hydrogen to last another 5 billion years. The solar flare shown here is over 20 times larger than Earth.

Table 15.1	Earth's Energy Ledger (Rough Estimates)	
Item	**Energy (terawatts, TW)**	**Approximate Percent**
Energy in		
Solar radiation	173,000	99+
Internal heat	32	0.02
Tides	3	0.002
Energy out		
Direct reflection	52,000	30
Direct heating*	81,000	47
Water cycle*	40,000	23
Winds*	370	0.2
Photosynthesis*	40	0.02

Source: M. King Hubbert, "The Energy Resources of the Earth," *Scientific American*, September 1971.
*This energy is eventually returned to space by means of long-wave radiation (heat).

EXAMPLE 15.1 Power and Energy Conversion

How much electrical energy, in joules, is consumed by a 75-W bulb burning for 1.0 hr?

Solution

One watt is 1 J/s, so we know that 75 W = 75 J/s. The bulb burns for 1 hour so

$$1\ \cancel{h} \times \frac{60\ \cancel{min}}{1\ \cancel{h}} \times \frac{60\ s}{1\ \cancel{min}} = 3600\ s$$

$$3600\ \cancel{s} \times \frac{75\ J}{1\ \cancel{s}} = 270{,}000\ J$$

■ **EXERCISE 15.1A**

How much electrical energy, in joules, does a 650-W microwave oven consume in heating a cup of coffee for 1.5 min?

■ **EXERCISE 15.1B**

In March 2006 the average cost of residential electricity in the United States was 9.86¢/kWh. A typical desktop computer uses about 150 watts. How much does it cost to run the computer for 3.75 h at that rate?

Kinetic and Potential Energy

Energy exists in two main forms. Energy due to position or arrangement is called **potential energy**. The water at the top of a dam has potential energy due to gravitational attraction. When the water is allowed to flow through a turbine to a lower level, the potential energy is converted to **kinetic energy** (the energy of motion). As the water falls, it moves faster. Its kinetic energy becomes greater as its potential energy decreases. The turbine can convert part of the kinetic energy of the water into electrical energy. The electricity thus produced can be carried by wires to homes and factories where it can be converted to light energy, to heat, or to mechanical energy. Table 15.2 gives several examples of kinetic and potential energy.

Energy and the Life-Support System

The *biosphere* is the thin (about 15 km thick) film of air, water, and soil in which all life exists. Only a small fraction of the energy the biosphere receives is used to support life. About 30% of incident radiation is immediately reflected back into space as short-wave radiation (ultraviolet and visible light). Nearly half is converted to heat, making the third planet from the Sun a warm and habitable place. About 23% of solar radiation powers the water cycle (Chapter 14), evaporating water from land and seas. The radiant energy of the Sun is converted to the potential energy of water vapor, water droplets, and ice crystals in the atmosphere. This potential energy is converted to the kinetic energy of falling rain and snow and of flowing rivers.

A tiny but most important fraction—less than 0.02%—of solar energy is absorbed by green plants, which use it to power **photosynthesis**. In the presence of green chlorophyll pigments, this energy converts carbon dioxide and water to glucose, a simple sugar rich in energy.

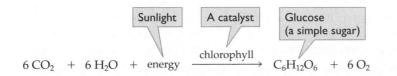

$$6\ CO_2 + 6\ H_2O + energy \xrightarrow{chlorophyll} C_6H_{12}O_6 + 6\ O_2$$

Photosynthesis also replenishes oxygen in the atmosphere. Glucose can be stored, or it can be converted to more complex foods and structural materials. All animals depend on the stored energy of green plants for survival.

▲ Boston Red Sox pitcher Diasuke Matsuzaka winding up to throw a baseball has potential energy stored in the chemicals making up his arms and legs (a). Potential energy from Matsuzaka has been converted into kinetic energy of the baseball (b).

Table 15.2 Some Examples of Potential and Kinetic Energy

Potential Energy	
Energy stored by position	Water at the top of a waterfall (hydroelectric power; Section 15.15)
	A child at the top of a slide
	A skier poised at the top of a mountain slope
	A swimmer ready to dive
	A baseball player poised to swing his bat
Energy stored in chemical bonds	Fuel (coal, gasoline, natural gas; Sections 15.6–15.8)
	Food (carbohydrates, fats, proteins; Chapter 17)
	Explosives (nitroglycerin)
Energy stored in bound nuclear particles	Nuclear energy (power plants, bombs; Section 15.10)
Energy stored by compression	A compressed spring
	A squeezed rubber ball
Kinetic Energy	
Energy of any moving object	A rolling freight train
	A spinning water turbine (hydroelectric power; Section 15.15)
	A rolling bowling ball
	A moving molecule
	A sailboat skimming across a lake
	A baseball hurtling toward home plate
	The eruption of a volcano

Self-Assessment Questions

1. The ability or capacity to do work or to produce change is a definition of
   a. force   **b.** energy   **c.** power   **d.** pressure

2. What sector of the U.S. economy uses the largest fraction of the nation's energy?
   a. residential   **b.** commercial   **c.** industrial   **d.** transportation

3. Nearly all the energy used on Earth originates in
   a. the atmosphere   **b.** the oceans
   c. radioactive rocks   **d.** the Sun

4. About how much of the world's energy is used by the United States?
   a. 0.02%   **b.** 1%   **c.** 5%   **d.** 25%

5. The kilowatt-hour (kWh) is a unit of
   a. energy   **b.** force   **c.** power   **d.** velocity

6. An apple hanging on a branch above Isaac Newton's head has
   a. force energy   **b.** kinetic energy
   c. potential energy   **d.** rotational energy

7. The largest portion of Earth's incident solar radiation
   a. powers the water cycle   **b.** is converted to heat
   c. powers photosynthesis   **d.** is reflected back into space

8. About what percentage of Earth's incident solar radiation is used in photosynthesis?
   a. 0.02%   **b.** 2%   **c.** 5%   **d.** 50%

Answers: 1, b; 2, c; 3, d; 4, d; 5, a; 6, c; 7, b; 8, a

◀ The action of a catalyst is illustrated. Hydrogen peroxide decomposes slowly to water and oxygen (a). When a lump of manganese dioxide is lowered into a solution of hydrogen peroxide (b), the reaction proceeds rapidly. The heat released during the process produces steam, and the solution froths as oxygen gas is evolved.

ENERGY: SOME SCIENTIFIC LAWS

A study of chemistry is incomplete without a discussion of energy. In calculations based on chemical equations, we can include quantities of energy. In addition, we find some scientific laws to be of considerable help in understanding chemical processes. We examine some of these ideas in the following sections.

15.2 Energy and Chemical Reactions

How fast—or how slowly—chemical reactions take place depends on several factors. One factor is *temperature*. Reactions generally proceed faster at higher temperatures. For example, coal (carbon) reacts so slowly with oxygen (from the air) at room temperature that the change is imperceptible. However, if coal is heated to several hundred degrees, it reacts rapidly. The heat evolved in the reaction keeps the coal burning smoothly.

We use the kinetic–molecular theory (Section 6.5) to explain the effect of temperature on the rates of chemical reactions. At high temperatures, molecules move more rapidly. Thus, they collide more frequently, increasing the chance for reaction. The increase in temperature also supplies more energy for the breaking of chemical bonds—a condition necessary for most reactions. We make use of our knowledge of the effect of temperature on chemical reactions in our daily lives. For example, we freeze foods to retard chemical reactions that lead to spoilage. When we want to speed up the reactions involved in cooking food, we turn up the heat.

Another factor affecting the rate of a chemical reaction is the *concentration of reactants*: The more molecules there are in a given volume of space, the more likely they are to collide. With more collisions, there are more reactions. For example, if you light a wood splint and then blow out the flame, the splint continues to glow as the wood reacts slowly with oxygen in the air. When the glowing splint is placed in pure oxygen, the splint bursts into flame, indicating a much more rapid reaction. We use the concentration of oxygen to interpret the phenomenon: Air is about 21% oxygen, and so the concentration of O_2 molecules in 100% oxygen is almost five times as great as that in air.

Catalysts also affect the rates of chemical reactions. Catalysts are of great importance in the chemical industry and in biology. Appropriate catalysts increase the rate of reactions that otherwise would be so slow as to be impractical.

▲ When liquid oxygen—much more concentrated than oxygen gas—is poured on a lit cigarette, the cigarette sparks, bursts into flame, and quickly disintegrates.

Catalysts are even more important in living organisms. Biological catalysts, called *enzymes*, mediate nearly all the chemical reactions that take place in living systems (Chapter 16).

Energy Changes and Chemical Reactions: Thermochemistry

The study of energy changes that occur during chemical reactions (and physical processes) is called **thermochemistry**. These energy changes are quantitatively related to the *amounts* of chemicals involved. For example, burning 1.00 mol (16.0 g) of methane to form carbon dioxide and water releases 803 kJ (192 kcal) of energy as heat. We can list the amount of heat evolved as a *product* of the reaction.

$$CH_4(g) + 2\,O_2(g) \longrightarrow CO_2(g) + 2\,H_2O(g) + 803\ kJ\ (192\ kcal)$$

Burning 2.00 mol (32.0 g) of methane produces twice as much heat, 1606 kJ.

To convert between the SI unit of energy, the joule (J), and the commonly used unit, the calorie (cal), we can use the equality 1 cal = 4.184 J.

EXAMPLE 15.2   Energies of Chemical Reactions

Burning 1.00 mol of propane releases 2201 kJ of energy.

$$C_3H_8(g) + 5\,O_2(g) \longrightarrow 3\,CO_2(g) + 4\,H_2O(g) + 2201\ kJ$$

How much energy in kilojoules is released when 15.0 mol of propane is burned?

Solution

We start with 15.0 mol C_3H_8 and use the balanced equation to form a conversion factor, just as we did with chemical conversions in Chapter 5.

$$15.0\ \text{mol}\ C_3H_8 \times \frac{2201\ kJ}{1\ \text{mol}\ C_3H_8} = 33{,}000\ kJ$$

■ EXERCISE 15.2A

The reaction of nitrogen and oxygen to form nitrogen monoxide (nitric oxide) requires an input of energy.

$$N_2(g) + O_2(g) + 18.07\ kJ \longrightarrow 2\,NO(g)$$

How much energy in kilojoules is absorbed when 5.05 mol of N_2 reacts with oxygen to form NO?

■ EXERCISE 15.2B

In photosynthesis carbon dioxide and water are converted to sugars in a reaction that requires an input of 15,000 kJ of energy for each 1.00 kg of glucose ($C_6H_{12}O_6$) produced. How much energy, in kilojoules, is required to make 1.00 mol of glucose?

▲ **Figure 15.1** Coal burns in a highly exothermic reaction. In a coal-fired power plant, the heat released converts water to steam that turns a turbine to generate electricity.

We usually discuss energy changes in terms of something losing energy and something else gaining it. When we warm our cold hands over a campfire, the burning wood gives off energy (as heat) and our hands gain energy, raising their temperature. In science, however, we need to describe the two "somethings" that exchange energy a little more precisely. We therefore define the **system** as the part of the universe under consideration. A real or imaginary boundary separates the system from the rest of the universe. The **surroundings** are everything else—the rest of the universe, but we usually limit our concern only to those parts of the surroundings that exchange energy or matter or both with the system. For example, a system might be a block of frozen spinach in a dish. The surroundings would be the air around the block and dish, the counter on which it sits, the rest of the kitchen, and so on.

Chemical reactions that result in the release of heat from the system to the surroundings are **exothermic** reactions. The burning of methane, gasoline, and coal (Figure 15.1) are all exothermic reactions. In each case, chemical energy is converted to heat energy that is absorbed by the surroundings.

Table 15.3 Some Exothermic and Endothermic Processes	
Exothermic Processes	**Endothermic Processes**
Freezing of water	Melting of ice
Condensation of water vapor	Evaporation of water
Metabolism in animals	Photosynthesis in plants
Making chemical bonds	Breaking chemical bonds
Discharging a battery	Charging a battery
Oxidation of Mg to MgO	Decomposition of HgO
Explosion of dynamite	Evaporation of a chlorofluorocarbon

In other reactions, such as the decomposition of water, energy must be supplied to the reactants from the surroundings. We can write the amount of heat required for this process as a reactant in the chemical equation:

$$2\ H_2O(g)\ +\ 573\ kJ\ \longrightarrow\ 2\ H_2(g)\ +\ O_2(g)$$

In an **endothermic** reaction, energy is absorbed as heat (Figure 15.2). It takes 573 kJ of energy to decompose 36.0 g (2.00 mol) of water into hydrogen and oxygen. It should be noted that exactly the same amount of energy is released when enough hydrogen is burned to form 36.0 g of water.

$$2\ H_2(g)\ +\ O_2(g)\ \longrightarrow\ 2\ H_2O(g)\ +\ 573\ kJ$$

Physical processes can also be either exothermic or endothermic. Table 15.3 lists several examples of physical and chemical processes and classifies them as exothermic or endothermic.

▲ **Figure 15.2** A striking endothermic reaction occurs when barium hydroxide octahydrate reacts with ammonium thiocyanate to produce barium thiocyanate, ammonia gas, and water.

$$Heat\ +\ Ba(OH)_2\cdot8\ H_2O(s)\ +$$
$$2\ NH_4SCN(s)\ \longrightarrow\ Ba(SCN)_2(s)\ +$$
$$2\ NH_3(g)\ +\ 10\ H_2O(l)$$

Here the reaction is carried out in a flask placed on a wet board. The temperature drops well below the freezing point of water, thus freezing the flask to the board.

EXAMPLE 15.3   Energy Changes in Chemical Reactions

How much energy in kilojoules is released when 225 g of propane (see Example 15.2) is burned?

Solution

The formula mass of propane (C_3H_8) is

$$(3\times C)\ +\ (8\times H)\ =\ (3\times12.0\ u)\ +\ (8\times1.0\ u)\ =\ 36.0\ +\ 8.0\ u\ =\ 44.0\ u$$

The molar mass is therefore 44.0 g. Next, we use the molar mass to convert grams of propane to moles of propane.

$$225\ g\ C_3H_8\ \times\ \frac{1\ mol\ C_3H_8}{44.0\ g\ C_3H_8}\ =\ 5.11\ mol\ C_3H_8$$

Now, proceeding as in Example 15.2,

$$5.11\ mol\ C_3H_8\ \times\ \frac{2201\ kJ}{1\ mol\ C_3H_8}\ =\ 11{,}200\ kJ$$

■ **EXERCISE 15.3A**

How much energy in kilojoules is released when 4.42 g of methane is burned?

$$CH_4\ +\ 2\ O_2\ \longrightarrow\ CO_2\ +\ 2\ H_2O\ +\ 803\ kJ$$

■ **EXERCISE 15.3B**

How much energy, in kilocalories, is absorbed when 0.528 g of N_2 is converted to NO? (See Exercise 15.2A.)

Self-Assessment Questions

1. The rate of a chemical reaction usually increases when
   a. a catalyst is added
   b. the chemist concentrates more intently
   c. the concentration of one of the reactants is decreased
   d. the temperature is decreased

2. Which of the following phase changes is exothermic?
   a. $H_2O(s) \longrightarrow H_2O(l)$
   b. $H_2O(l) \longrightarrow H_2O(s)$
   c. $H_2O(s) \longrightarrow H_2O(g)$
   d. $H_2O(l) \longrightarrow H_2O(g)$

3. How much heat is absorbed in the complete reaction of 0.100 mol of SiO_2 with excess carbon in the following reaction?

$$SiO_2(g) + 3\ C(s) + 624.7\ kJ \longrightarrow SiC(s) + 2\ CO(g)$$

   a. 62.5 kJ
   b. 732 kJ
   c. 1250 kJ
   d. 1870 kJ

4. When $NH_4Cl(s)$ is dissolved in water, the water gets colder, indicating that the dissolving of ammonium chloride in water is an
   a. endothermic process because it absorbs heat
   b. endothermic process because it releases heat
   c. exothermic process because it absorbs heat
   d. exothermic process because it releases heat

Answers: 1, a; 2, b; 3, a; 4, a

15.3 The Laws of Thermodynamics

We will take a look at our present energy sources and our future energy prospects shortly. To do so scientifically, we first examine some natural laws. Recall that natural laws merely summarize the results of many experiments. We won't recount all those experiments here; we will merely state the laws and some of their consequences.

Energy and the First Law: Energy Is Conserved

The **first law of thermodynamics** states that energy can be neither created nor destroyed. The first law of thermodynamics (*thermo* refers to heat; *dynamics* to motion) grew out of a variety of experiments conducted during the early 1800s. By 1840 it was clear that, although energy can be changed from one form to another, it is neither created nor destroyed. This law (also called the **law of conservation of energy**) has been restated in several ways, including, "You can't get something for nothing" and "There is no such thing as a free lunch." Energy can't be made from nothing. Neither does it just disappear, although it may go someplace else.

From the first law of thermodynamics we can conclude that we can't "win." We can't come out ahead by making a machine that produces more energy than it takes in. From the first law alone, however, we might conclude that we can't possibly run out of energy because energy is conserved. This is true enough, but it doesn't mean that we don't have problems. There is another long-armed law from which we cannot escape, the second law of thermodynamics, which we examine next.

Energy and the Second Law: You Can't Even Break Even

Despite innumerable attempts (and even several granted patents), no one has ever built a successful "perpetual-motion" machine. You can't make a machine that runs indefinitely without consuming energy. Even if an engine isn't doing any work, it loses energy (as heat) because of the friction of its moving parts. In fact, in any real engine, not only is it impossible to get as much useful energy out as you put in but you can't even break even.

Using Energy

If energy is neither created nor destroyed, why do we always need more? Won't the energy we have now last forever? The answer lies in the facts that

- energy can be changed from one form to another
- not all forms are equally useful
- more useful forms of energy are constantly being degraded into less useful forms

▲ **Figure 15.3** Energy always flows spontaneously from a hot object to a cold one, never the reverse. It flows from a hot fire to the cooler marshmallow. (A spontaneous event is one that occurs without outside influence.)

Energy flows downhill. Mechanical energy is eventually changed into heat energy. Hot objects cool off by transferring their heat to cooler objects. There is a tendency toward an even distribution of energy. Heat always flows from a hot object to a cooler one (Figure 15.3). The reverse does not occur spontaneously.

Observations of heat flow led to formulation of the **second law of thermodynamics**, one statement of which is that the energy available for work in the universe is continually decreasing. In another of many forms, the second law states that energy does not flow spontaneously from a cold object to a hot one. (A spontaneous process is one that can proceed in a specified direction without being forced by an external source of energy.) It is true that we can make energy flow from a cold region to a hot one—that's what refrigerators are all about—but we cannot do so without an input of energy and without producing changes elsewhere. We can reverse a natural process only at a price. The price, in the case of a refrigerator, is the consumption of electricity; that is, you must use energy to cause energy flow from a cold space to a warmer one.

When we change energy from one form to another, we can't concentrate all the energy in a particular source to do the job we want it to do. For example, we use energy in a fuel to push a piston in a car engine or use water rushing down from the top of a dam to run dynamos to generate electricity. In either case—indeed in all cases—some of the energy is converted to heat that is not available to do *useful* work.

Entropy

Another way to look at the second law is in terms of *entropy*. Scientists use the term **entropy** as a measure of the dispersal of energy in a system. The more the energy is spread out, the higher the entropy of the system and the less likely it is that this energy can be harnessed to do useful work. Natural processes tend toward greater entropy or are exothermic, or both.

As we mentioned in Chapter 5, molecules move about constantly. Their motion in solids is limited to rapid but tiny back-and-forth movements much like vibrations about a nearly fixed point. The molecules in liquids move more freely but still over only short distances. Those in gases move still more freely and over much greater distances.

In Chapter 3 we saw that the energy of light occurs in little packets called photons. Similarly, the energy of molecular motion is described using a concept called *microstates*. If you could take a snapshot capturing the positions of all molecules in a sample of matter at a given instant, you would have captured an image of a microstate. The number of microstates in which a given sample of matter can exist differs depending on whether it is in the solid, liquid, or gaseous state. The number of energy microstates will be lowest for a solid because the molecules in a solid are limited mainly to vibrational motion. Because molecules in the liquid state can move more freely, the number of possible energy microstates is much larger for liquids than for solids. When converted to a gas, the molecules can move still more freely and for much greater distances, so even more microstates are available to molecules in the gaseous state. As any form of matter is heated, the energy in it is spread out among more accessible microstates (see Figure 15.4). Technically, entropy is the dispersal of energy among these microstates. In practice, entropy is often thought of as being analogous to "molecular disorder." Of course, we can often reverse the tendency toward greater entropy—but only through an input of energy.

The energy to run an engine usually comes from the concentrated chemical energy inside molecules of oil or coal. In biochemistry, food molecules are the concentrated energy source. In either case, energy is spread out during the process. The spread-out energy in the product gases—CO_2 and H_2O in each case—is less useful. A car won't run on exhaust gases, nor can organisms obtain energy for life processes from respiratory products.

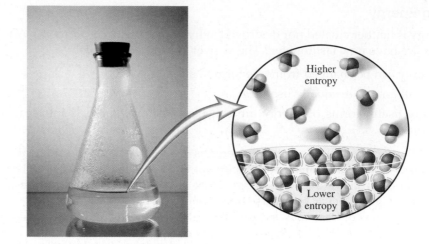

▲ **Figure 15.4** The photograph depicts the vaporization of water; a sample of liquid water [H$_2$O(l)] at room temperature spontaneously changes to H$_2$O(g) through the process of evaporation. At the macroscopic level, nothing appears to be taking place. However, in the molecular view, we see that the molecules are in motion and are much more widely spaced in the vapor state than in the liquid state. Vaporization is spontaneous because vapor has greater *entropy* than the liquid; molecules in the vapor can be "arranged" in many more ways in their spread-out spacing than can the molecules of liquid. We can make liquid water from water vapor, but only by compressing the gas and/or lowering its temperature.

Entropy and Pollution

This tendency toward the dispersal of energy helps to explain why it is relatively easy to pollute air or water and so difficult to clean it up once it is polluted. For example, it takes little energy to dump a ton of a chlorofluoro-carbon (CFC) into the air. The rapidly moving CFC molecules would quickly disperse among the nitrogen, oxygen, and other molecules in the air. Once the CFC molecules were scattered, it would take a lot of energy to concentrate them from the air. It costs a lot to clean up polluted water, soil, or air. Prevention of pollution is a far better alternative.

More to Explore

Lambert, F. L. "Entropy Is Simple, Qualitatively." *Journal of Chemical Education,* October 2002, vol. 79, pp. 1241–1246, and Jensen, William B. "Entropy and Constraint of Motion." *Journal of Chemical Education,* May 2004, pp. 639–640.

Self-Assessment Questions

1. The first law of thermodynamics is also called the law of
   a. conservation of energy
   b. conservation of mass
   c. entropy
   d. perpetual motion

2. That energy always goes from more useful forms to less useful forms is a statement of
   a. the first law of thermodynamics
   b. the second law of thermodynamics
   c. Murphy's law
   d. the standard law of energy conversion

3. According to the second law of thermodynamics, the
   a. entropy of a system always decreases
   b. entropy of a system always increases
   c. total entropy always decreases for an isolated system
   d. total entropy always increases for an isolated system

4. In the process represented by the equation H$_2$O(g) ⟶ H$_2$O(l), the entropy
   a. decreases
   b. depends on the catalyst used
   c. increases
   d. remains the same

Answers: 1, a; 2, b; 3, d; 4, a

15.4 People Power: Early Uses of Energy

Early people obtained their energy (food and fuel) by collecting wild plants and hunting wild animals. They expended this energy in hunting and gathering. Domestication of horses and oxen increased the availability of energy only slightly. The raw materials used by these work animals were natural, replaceable plant materials.

Plant materials were also the first fuels. These combustible materials kept early fires burning. As late as 1760, wood was almost the only fuel in use. Even today, wood and dried dung remain the primary fuels for about one-third of the world's people.

One of the first mechanical devices used to convert energy to useful work was the waterwheel. The Egyptians first used waterpower about 2000 years ago, primarily for grinding grain. Later on, waterpower was used for sawmills, textile mills, and other small factories. Windmills were introduced into western Europe during the Middle Ages, primarily for pumping water and grinding grain. More recently, wind power has been used to generate electricity.

Windmills and waterwheels are fairly simple devices for converting the kinetic energy of blowing wind and flowing water to mechanical energy. They were sufficient to power the early part of the Industrial Revolution, but development of the steam engine freed factories from locations along waterways. Since 1850, turbines turned by water, steam, and gas, the internal combustion engine, and a variety of other energy-conversion devices have boosted the energy available for use by an estimated factor of 10,000.

Let's turn our attention now to the fossils that fueled the Industrial Revolution and still serve as the basis for modern civilization.

▲ Waterwheels at Hama, Syria, are 2000 years old. Some are more than 20 m high. Several are still used to lift water into aqueducts. Waterwheels were rarely used in Europe, however, until the Middle Ages. The Romans used human slaves instead.

FOSSIL FUELS

The combustion of long-buried fossils (preserved remains of living organisms) initiated and sustain our modern industrial civilization. More than 90% of the energy used to support our way of life comes from **fossil fuels**—coal, petroleum, and natural gas that formed over millions of years during Earth's Carboniferous period around 300 million years ago. In the following sections, we consider the origin and chemical nature of these fuels and how they are burned to release energy that plants of ages past captured from rays of sunlight.

A **fuel** is a substance that burns readily with the release of significant amounts of energy. Fuels are *reduced* forms of matter, and the burning process is oxidation (Chapter 8). If an atom already has its maximum number of bonds to oxygen (or to other electronegative atoms such as chlorine or bromine), the atom cannot serve as a fuel. Indeed, some such substances can be used to put out fires. Figure 15.5 shows some representative fuels and nonfuels.

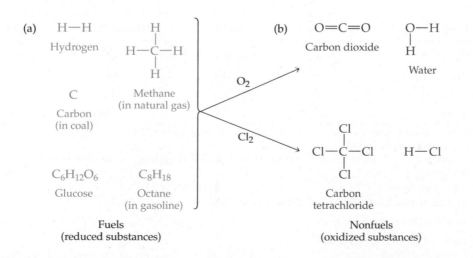

◀ **Figure 15.5** Some fuels (a). Fuels are reduced forms of matter that release relatively large quantities of heat when burned. Some nonfuels (b). These compounds are oxidized forms of matter.

2. Is there really an energy shortage?

There is no shortage of energy itself, but not all energy is equally useful to us. The supply of materials, like fossil fuels, that provide useful energy is dwindling.

By energy *production*, we mean conversion of some form of energy into a more useful form. For example, production of petroleum means pumping, transporting, and refining it. Energy *consumption* means using it in a way that changes it to a less useful form. In either case, we are neither making energy nor destroying it.

15.5 Reserves and Consumption Rates of Fossil Fuels

Earth has only a limited supply of fossil fuels. Estimated U.S. and world reserves and annual U.S. and world consumption are given in Table 15.4. Estimates of reserves vary greatly, depending on the assumptions made. Even the most optimistic estimates, however, lead to the conclusion that nonrenewable energy resources are being depleted rapidly. Indeed, in just a century, we will have used up more than half the fossil fuels that were formed over the ages. In only a few hundred more years, we will have removed from Earth and burned virtually all the remaining recoverable fossil fuels. Of all that ever existed, about 90% will have been used in a period of 300 years. Within the lifetime of today's 18-year-old, natural gas and petroleum will likely become so scarce and so expensive that they won't be used much as fuels. At the current rate of production, U.S. reserves of petroleum and natural gas will be substantially depleted sometime during this century. Coal reserves should last perhaps 300 years. However, the rate of use of all fossil fuels is increasing, especially in developing nations.

Presumably, fossil fuels are still being formed in nature, a process that is perhaps most evident in peat bogs. The rate of formation is extremely slow, estimated at only one-fifty-thousandth the rate we are using them.

Self-Assessment Questions

1. Coal, petroleum, and natural gas are called fossil fuels because they
   a. are burned to release energy
   b. are nonrenewable and will run out
   c. cause air pollution
   d. were formed over millennia from the remains of ancient plants and animals

2. Which of the following is *not* a fuel?
   a. C   b. CH_4 c. CH_3OH d. CO_2

3. Fuels are all
   a. carbon compounds   b. hydrocarbons
   c. oxidized forms of matter   d. reduced forms of matter

Answers: 1, d; 2, d; 3, d

Table 15.4 Estimated U.S. and World Reserves of Economically Recoverable Fuels and Annual Consumption of Fossil Fuels*

Fuel	Reserves		Consumption	
	United States	**World**	**United States**	**World**
Coal	121,962	501,171	545	2,342
Petroleum	4,184	156,700	904	3,507
Natural gas	4,711	158,198	517	2,122
Total	130,857	816,069	1,966	7,971

Source: World Resources Institute, Washington, DC. Website (http://www.wri.org/) includes assumptions made in making these estimates.

*Expressed in millions of metric tons of oil equivalents (Mtoe) for ready comparison of energy content.

Conversion factors:

1 metric ton oil equivalent (toe) = 7.8 barrels oil

1 toe = 1270 m^3 of natural gas

1 toe = 2.3 metric ton (t) of coal

15.6 Coal: The Carbon Rock of Ages

People probably have used small amounts of coal since prehistoric times. After the steam engine came into widespread use (by about 1850), the Industrial Revolution was powered largely by coal. By 1900 about 95% of the world's energy production came from the burning of coal.

Coal is a complex combination of organic materials that burn and inorganic materials that produce ash. Its main element is carbon, but it also contains small percentages of other elements. The quality of coal as an energy source is based on its carbon content. Complete combustion of the carbon produces carbon dioxide.

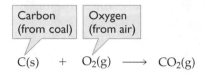

$$C(s) \ + \ O_2(g) \ \longrightarrow \ CO_2(g)$$

In limited quantities of air, however, carbon monoxide and soot are formed.

$$2\,C(s) + O_2(g) \ \longrightarrow \ 2\,CO(g)$$

Soot is mostly unburned carbon.

Coal is ranked by carbon content, from low-grade peat and lignite to high-grade anthracite (Table 15.5). The energy obtained from coal is roughly proportional to its carbon content. Soft (bituminous) coal is much more plentiful than hard coal (anthracite). Lignite and peat have become increasingly important as the supplies of higher grades of coal have been depleted.

Table 15.5 Approximate Composition (Percent by Mass) and Energy Content of Typical Grades of Coal (Dry Basis)

Grade of Coal	Carbon	Hydrogen	Oxygen	Nitrogen	Energy Content (MJ/kg)
Wood (for comparison)	50	6	43	1	—
Peat	60	6	33	2	14.7
Lignite (brown coal)	71	5	25	1	23
Bituminous (soft) coal	86–91	5	5–15	1	36
Anthracite (hard coal)	95	2–3	2–3	Trace	35.2

Source: Diessel, C. F. K. *Coal-Bearing Depositional Systems.* New York: Springer-Verlag, 1992.

◀ Coal is a fossil fuel. Giant ferns, reeds, and grasses grew during the Carboniferous period 300 million years ago. These plants were buried and through the ages were converted to the coal we burn today.

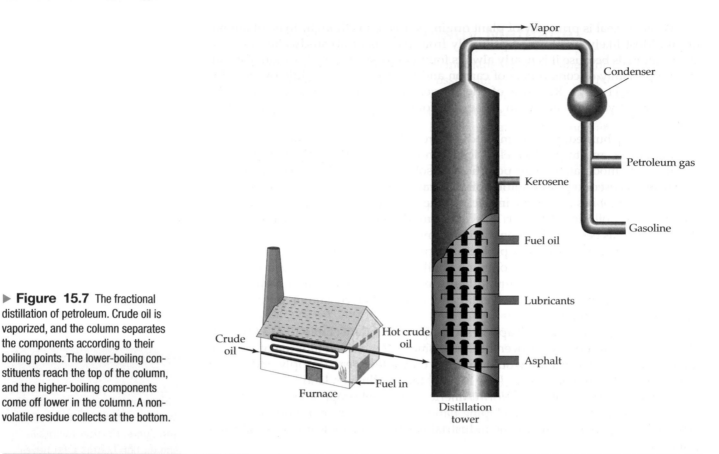

▶ **Figure 15.7** The fractional distillation of petroleum. Crude oil is vaporized, and the column separates the components according to their boiling points. The lower-boiling constituents reach the top of the column, and the higher-boiling components come off lower in the column. A non-volatile residue collects at the bottom.

Table 15.6 Typical Petroleum Fractions

Fraction	Typical Range of Hydrocarbons	Approximate Range of Boiling Points (°C)	Typical Uses
Gas	CH_4 to C_4H_{10}	Less than 40	Fuel, starting materials for plastics
Gasoline	C_5H_{12} to $C_{12}H_{26}$	40–200	Fuel, solvents
Kerosene	$C_{12}H_{26}$ to $C_{16}H_{34}$	175–275	Diesel fuel, jet fuel, home heating; cracking to gasoline
Heating oil	$C_{15}H_{32}$ to $C_{18}H_{38}$	250–400	Industrial heating, cracking to gasoline
Lubricating oil	$C_{17}H_{36}$ and up	Above 300	Lubricants
Residue	$C_{20}H_{42}$ and up	Above 350 (some decomposition)	Paraffin, asphalt

the amount of fossil fuels aboard Spaceship Earth. Scientists are seeking new sources of energy that do not depend on petroleum. Perhaps we can soon stop this profligate waste of resources. Spaceship Earth has aboard it all the supplies it will ever have. We must use them wisely.

$$CH_3CH_2CH_2CH_2CH_2CH_2CH_2CH_2CH_2CH_2CH_2CH_2CH_2CH_3 \xrightarrow[\text{catalyst}]{\text{heat}}$$

$$CH_3CH_2CH_2CH_2CH_2CH_2CH_2CH_2CH_2CH_2CH_2CH_3 \ + \ CH_2{=}CH_2$$

and

$$CH_3CH_2CH_2CH_2CH_2CH_2CH_2CH_2CH_2CH_2CH_3 \ + \ CH_3CH{=}CH_2$$

and

$$CH_3CH_2CH_2CH_2CH_2CH_2CH_2CH_3 \ + \ CH_3CH_2CH_2CH_2CH{=}CH_2$$

and so on

▶ **Figure 15.8** Formulas of a few of the possible products formed when $C_{14}H_{30}$, a typical molecule in kerosene, is cracked. In practice, a wide variety of hydrocarbons (most of which have fewer than 15 carbon atoms), hydrogen gas, and char (mostly elemental carbon) are formed. Cyclic and branched-chain hydrocarbons are also produced.

Gasoline

Gasoline, like the petroleum from which it is derived, is mainly a mixture of hydrocarbons. A commercial gasoline typically contains more than 150 different compounds, but up to 1000 have been identified in some blends. Among the hydrocarbons, a typical gasoline sample, by volume, might have 4–8% straight-chain alkanes, 25–40% branched-chain alkanes, 2–5% alkenes, 3–7% cycloalkanes, 1–4% cycloalkenes, and 20–50% aromatic hydrocarbons (0.5–2.5% benzene). The gasoline also has a variety of additives, including antiknock agents, antioxidants, antirust agents, anti-icing agents, upper-cylinder lubricants, detergents, and dyes. Typical alkanes in gasoline range from C_5H_{12} to $C_{12}H_{26}$. There are also small amounts of some sulfur- and nitrogen-containing compounds.

The gasoline fraction of petroleum as it comes from a distillation column is called *straight-run gasoline*. It doesn't burn very well in modern high-compression automobile engines, but chemists are able to modify it to make it burn more smoothly.

The Octane Ratings of Gasolines

In an internal combustion engine, the gasoline–air mixture sometimes ignites before the spark plug "fires." This is called *knocking* and can damage the engine. Early on, scientists learned that some types of hydrocarbons, especially those with branched structures, burned more evenly and were less likely to cause knocking than others. An arbitrary performance standard, called the **octane rating**, was established in 1927. Isooctane was assigned a value of 100 octane. An unbranched-chain compound, heptane, was given an octane rating of 0. A gasoline rated 90 octane was one that performed the same as a mixture that was 90% isooctane and 10% heptane.

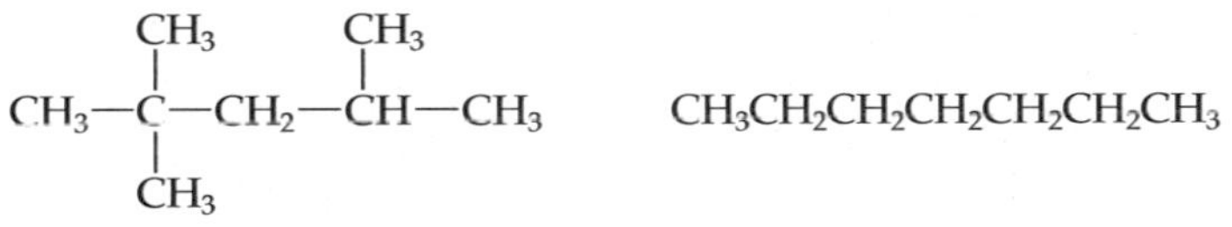

$$CH_3-\underset{\underset{CH_3}{|}}{\overset{\overset{CH_3}{|}}{C}}-CH_2-\underset{\underset{}{\overset{\overset{CH_3}{|}}{CH}}}-CH_3 \qquad CH_3CH_2CH_2CH_2CH_2CH_2CH_3$$

Isooctane   Heptane
(Octane rating 100)   (Octane rating 0)

During the 1930s, chemists discovered that the octane rating of gasoline could be improved by heating it in the presence of a catalyst such as sulfuric acid (H_2SO_4) or aluminum chloride ($AlCl_3$). This converted (*isomerized*) part of the unbranched structures to highly branched molecules. For example, heptane molecules can be isomerized to branched structures:

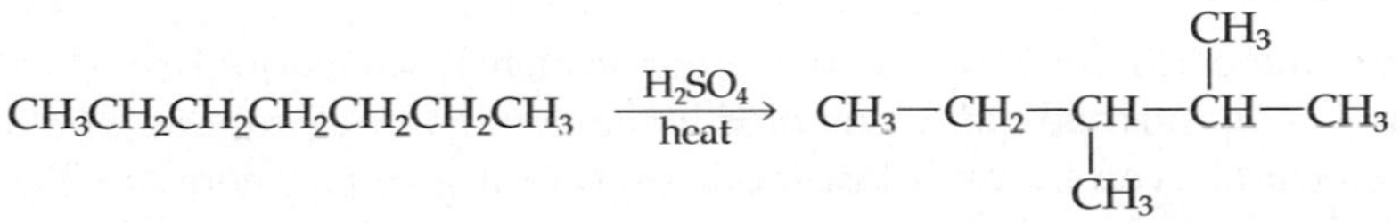

$$CH_3CH_2CH_2CH_2CH_2CH_2CH_3 \xrightarrow[\text{heat}]{H_2SO_4} CH_3-CH_2-\underset{\underset{CH_3}{|}}{CH}-\overset{\overset{CH_3}{|}}{CH}-CH_3$$

Chemists also can combine small hydrocarbon molecules (below the gasoline range) into larger ones more suitable for use as fuel. This process is called *alkylation*. In a typical alkylation reaction, shown below, isobutylene is reacted with propane.

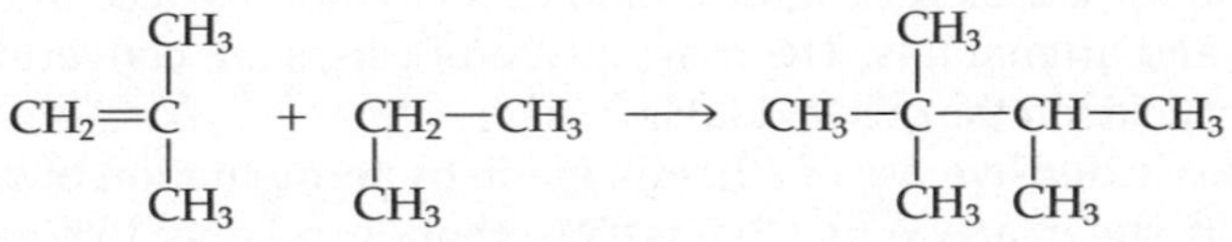

$$CH_2=\underset{\underset{CH_3}{|}}{\overset{\overset{CH_3}{|}}{C}} \;+\; \underset{\underset{CH_3}{|}}{CH_2}-CH_3 \;\longrightarrow\; CH_3-\underset{\underset{CH_3}{|}}{\overset{\overset{CH_3}{|}}{C}}-\underset{\underset{CH_3}{|}}{CH}-CH_3$$

▲ Fuel with the octane booster tetraethyllead, designated "Ethyl" gasoline, was available at many gas stations before leaded gasoline was phased out, beginning in the 1970s.

The product molecules are in the right size range for gasoline, and the highly branched molecules are high in octane number.

Certain additives substantially improve the antiknock quality of gasoline. Tetraethyllead, $Pb(CH_2CH_3)_4$, was found to be especially effective. As little as 1 mL tetraethyllead per liter of gasoline (one part per thousand) increases the octane rating by 10 or more.

Lead fouls the catalytic converters used in modern automobiles. More important, lead is especially toxic to the brain. Even small amounts can lead to learning disabilities in children. In the United States unleaded gasoline became available in 1974, and all leaded gasoline now has been phased out.

Scientists have found other ways to get high octane ratings in unleaded fuels. For example, petroleum refineries use **catalytic reforming** to convert low-octane alkanes to high-octane aromatic compounds. Hexane (with an octane number of 25) is converted to benzene (octane number, 106).

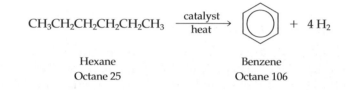

Hexane
Octane 25

Benzene
Octane 106

Octane boosters to replace tetraethyllead include ethanol, methanol, *tert*-butyl alcohol, and methyl *tert*-butyl ether (MTBE) (Chapter 9). None of these is nearly as effective as tetraethyllead in boosting the octane rating. They must therefore be used in fairly large proportions. The amount that can be used in gasoline is limited by solubility problems. For example, ethanol in excess of 10% tends to separate from the gasoline in the presence of moisture.

Unlike gasoline, which is made up of hydrocarbons, the various alcohols and their derivatives all contain oxygen and are sometimes called *oxygenates*. Not only do these additives improve the octane rating, but they also decrease the amount of carbon monoxide in auto exhaust gas. A disadvantage of oxygenates is that their energy content is lower than pure hydrocarbon fuels, and so the distance one can travel per tankful is somewhat shorter.

MTBE, like ethers in general (Chapter 9), is rather unreactive chemically, but it is soluble in water to the extent of 4.8 g per 100 g of water. When gasoline spills or leaks from storage tanks, MTBE enters the groundwater, leading to widespread contamination. MTBE is listed as a hazardous substance under the Federal Superfund law and is considered a potential human carcinogen by the EPA. The exact threat to human health is far from clear, but use of MTBE in gasoline has declined considerably in the United States.

Alternative Fuels

An automobile engine can be made to run on nearly any liquid or gaseous fuel. There are cars on the road today powered by natural gas, by propane, by diesel fuel, by fuel cells, and even by used fast-food restaurant grease. There are electric cars that can be plugged into the power grid and hybrid cars that can switch from electricity to gasoline as needed.

Diesel fuel for automobiles overlaps the kerosene fraction of petroleum (Table 15.6); it consists mainly of C_9 to C_{20} hydrocarbons with a boiling range of about 250 °C to 350 °C. Diesel fuel has a greater proportion of straight-chain alkanes than gasoline does. The standard for performance, called the *cetane number*, is based on hexadecane ($C_{16}H_{34}$), once known as cetane. A renewable fuel called *biodiesel* can be used in unmodified diesel engines. Biodiesel is made by reacting ethanol with vegetable oils and animal fats. The triacylglycerol esters are converted to ethyl esters of the fatty acids in the oils and fats.

Brazil makes extensive use of ethanol, made by fermentation of sucrose from a major local crop, sugarcane. A fuel that is 85% ethanol and only 15% gasoline, called

More to Explore
Ritter, Steve. "What's That Stuff: Gasoline." *Chemical & Engineering News*, February 21, 2005, p. 37.

E85, is available in some parts of the United States. More than 4 million American cars and trucks have the ability to run on E85 right now, but many people who own them are not aware of that. Ethanol alone is not likely to be an answer to our energy problems. Most ethanol is made from corn, and widespread use would require huge tracts of farmland to be converted to corn production. Diversion of corn to ethanol production also appears to have led to increased food prices.

▲ In 2006 only about 1200 of the 180,000 gasoline stations in the United States provided E85 fuel. Most are in corn-growing states.

Energy Return on Energy Invested

When evaluating an energy source, we need to consider the *energy return on energy invested* (EROEI) of exploiting the source. EROEIs are difficult to evaluate but can be quite important because if the EROEI of a resource is 1 or less, it becomes an energy sink and is no longer a primary source of energy.

In the early days of the petroleum industry, the EROEI for Texas crude oil may have been as much as 100 to 1; investing the equivalent of 1 barrel of oil could produce 100 barrels of petroleum. The EROEI for oil from the Middle East is still about 30 to 1. Oil from ever more remote places or ever deeper offshore wells will have ever lower EROEIs.

The EROEIs of alternative energy sources are often quite low. Canada has vast deposits of tar sands, but the EROEI is only about 1.5 to 1. Colorado has huge deposits of oil shale, but the EROEI is likely to be quite low.

The EROEI for ethanol is the subject of much debate, ranging from a low of 0.78 to 1 to the highest estimate for ethanol from corn of 1.67 to 1. Two reasons for the low EROEI for ethanol are that much energy is needed to distill or evaporate the ethanol after fermentation and that ethanol has a lower energy content than petroleum fuels.

Hydrogen has an EROEI of less than 1 to 1 because it cannot be extracted from the ground nor generated by a biosystem. Hydrogen can be generated by electricity or by heating steam to very high (2000 °C) temperatures. For now, those measures require more energy than the hydrogen produces when burned.

In addition to the EROEI, we need to consider the environmental consequences of exploiting an energy source. For example, getting oil from tar sands or oil shale is a messy process. Whatever else, our quest for plentiful energy will be costly and controversial.

3. Why aren't alternative fuels like ethanol used more?
Use of alternative fuels is increasing, but the cost (EROEI) of those fuels is very high compared to fossil fuels. Also, the current infrastructure is set up to use fossil fuels, and the cost of new infrastructure (hydrogen filling stations, more ethanol pumps and tanks, etc.) is quite high.

A N S W E R

Self-Assessment Questions

1. U.S. oil reserves
   a. peaked in 1970
   b. peaked in 1985
   c. will peak in 2010
   d. will peak in 2030

2. On what physical property is the separation of petroleum fractions based?
   a. boiling point   b. density
   c. electrostatic precipitation   d. temperature

Items 3–6 refer to petroleum processing.

3. A catalytic cracker is used to
   a. convert heavier fractions to lighter ones
   b. improve the octane rating
   c. remove sulfur
   d. separate various fractions

4. An alkylation unit
   a. converts lighter fractions to heavier ones
   b. improves the asphalt yield
   c. removes sulfur
   d. separates various fractions

5. An isomerization unit
   a. converts branched molecules to straight-chain ones
   b. converts ethers to alcohols
   c. converts ketones to aldehydes
   d. improves the octane rating

6. A catalytic reforming unit converts
   a. branched molecules to straight-chain ones
   b. branched molecules to aromatic ones
   c. straight-chain molecules to aromatic ones
   d. straight-chain molecules to branched ones

7. The octane rating for gasoline is a measure of the
   a. concentration of octane (C_8H_{18})   b. energy content in kilocalories
   c. power rating   d. tendency of fuel to knock

8. Which of the following does *not* increase the octane rating of gasoline?
   a. aromatic hydrocarbons   b. carbon chain branching
   c. straight-chain hydrocarbons   d. tetraethyllead

9. E85 fuels
   a. are 15% ethanol, 85% gasoline   b. are 85% ethanol, 15% gasoline
   c. are 85 octane   d. provide 85 horsepower

Answers: 1, a; 2, a; 3, a; 4, a; 5, a; 6, c; 7, d; 8, c; 9, b

15.9 Convenient Energy: Electricity

The convenience of a fuel depends on its physical state: Gases and liquids are convenient, while solids are much less so. Perhaps the most convenient energy form of all is electricity (Table 15.7). With electricity we can have light and hot water, and we can run motors of all sorts. We can use it to heat and cool our homes and workplaces. When looking at future energy sources, then, we look mainly at ways of generating electricity.

Any fuel can be burned to boil water, and the steam produced can turn a turbine to generate electricity. Figure 15.9 shows a coal-fired steam power plant. At present,

Table 15.7 Convenience of Fuels in Various Physical States

Physical State	Extraction	Transportation to Cities	Distribution within a City	In Use	
				Convenience	Cleanliness
Solids (coal, wood)	Shovels, borers, blasting	Trucks, trains, barges (slurry with water in pipe)	Trucks, buckets	Least	Dirtiest
Liquids (gasoline, fuel oil)	Pumps	Pipelines, tankers, barges, trucks	Trucks		
Gases (natural gas)	Pumps	Pipeline	Pipes		
Electricity* (electron flow)	—	Wires	Wires	Most	Cleanest

*Produced by burning any of the primary fuels and included for comparison.

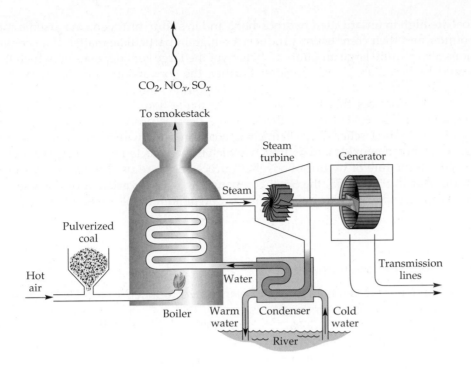

about half of U.S. electric energy comes from coal-burning plants (Figure 15.10). Such facilities are at best only about 40% efficient; 60% of the energy of the fossil fuel is wasted as heat. Some power installations use the waste heat generated to warm buildings, a technique called *cogeneration*.

Coal Gasification and Liquefaction

Coal can be converted to gas or oil. When we run short of gas and petroleum, why not make them from coal? The technology has been around for years. Gasification and liquefaction do have some advantages: Gases and liquids are easy to transport, and the process of conversion leaves some of the sulfur and minerals behind, thus reducing a serious disadvantage of coal as a fuel.

The basic process for converting coal to a synthetic gaseous fuel is a reduction of carbon by hydrogen. Passing steam over hot charcoal produces *synthesis gas*, a mixture of hydrogen and carbon monoxide.

$$C(s) + H_2O(g) \longrightarrow CO(g) + H_2(g)$$

The hydrogen can be used to reduce the carbon in coal or other substance to form methane.

$$C(s) + 2 H_2(g) \longrightarrow CH_4(g)$$

Coal can also be converted to methanol by way of synthesis gas. Part of the CO in synthesis gas is reacted with water to form more H_2.

$$CO(g) + H_2O(g) \longrightarrow CO_2(g) + H_2(g)$$

The remaining CO is combined with H_2 to form methanol.

$$CO(g) + 2 H_2(g) \longrightarrow CH_3OH(l)$$

The methanol can be used directly as fuel or converted, in turn, to gasoline-like hydrocarbons that we represent as C_nH_m.

$$n\, CH_3OH(l) \longrightarrow C_nH_m(l) + x\, H_2O(l)$$

Each step requires an appropriate catalyst.

Both gasification and liquefaction of coal require a lot of energy: Up to one-third of the energy content of the coal is lost in the conversion. Liquid fuels from

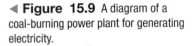

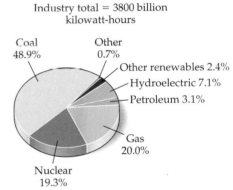

▲ **Figure 15.10** Percentages of electric power generation in the United States in 2006 from various energy sources.

coal are high in unsaturated hydrocarbons and in sulfur, nitrogen, and arsenic compounds, and their combustion products are high in particulate matter. Coal conversions also require large amounts of water, yet the large coal deposits on which they would be based are in arid regions. Further, the conversions are messy. Without stringent safeguards, plants would seriously pollute both air and water.

Other processes that have been used to make liquid fuels from coal are the Bergius and Fischer–Tropsch methods. In the Bergius process, coal is reacted with hydrogen. In the Fischer–Tropsch method, coal is reacted with steam to make a mixture of carbon monoxide and hydrogen, which is then reacted over an iron catalyst to produce a mixture of hydrocarbons. Both processes were used in Germany during World War II. Since 1955, the Sasol company has used a variation of the Fischer—Tropsch method to provide a significant portion of liquid fuels for South Africa.

Self-Assessment Questions

1. About what percentage of the electricity produced in the United States comes from coal burning plants?
   a. 25% **b.** 35% **c.** 50% **d.** 90%
2. Natural gas is transported mainly by
   a. barge **b.** pipelines **c.** railroad car **d.** truck
3. Which of the following is the main fuel produced by coal gasification?
   a. coke **b.** H_2 **c.** CH_4 **d.** $CH_3CH_2CH_3$
4. The convenience of a fuel depends on its
   a. carbon content **b.** density **c.** odor **d.** physical state

Answers: 1, c; 2, b; 3, c; 4, d

NUCLEAR ENERGY

Chapter 11 explored some nuclear reactions and noted that both nuclear fission and nuclear fusion can be used in bombs. Nuclear fission can also be controlled to generate power. So far nuclear fusion reactions cannot be controlled to provide energy; they can be employed only in bombs.

15.10 Nuclear Fission

Nuclear fission reactions can be controlled in a **nuclear reactor**. The energy released during fission can be used to generate steam, which can turn a turbine to generate electricity (Figure 15.11).

At the dawn of the nuclear age, some people envisioned nuclear power to be destined to fulfill the biblical prophecy of a fiery end to our world. Others saw nuclear power as a source of unlimited energy. During the late 1940s, some claimed that electricity from nuclear plants would become so cheap that it eventually would not have to be metered. Nuclear power has not yet brought us either paradise or perdition, but it has become a most controversial issue. Our great demand for energy indicates that the controversy will continue for years to come.

At present, nearly 20% of U.S. electricity comes from nuclear power plants. The eastern seaboard and upper midwestern states, many of which have minimal fossil fuel reserves, are heavily dependent on nuclear power for electricity.

We could rely more on nuclear power as other nations do (Table 15.8), but the public is quite fearful of nuclear power plants. This apprehension was exacerbated by the accident at Chernobyl, Ukraine. The United States has 104 operating nuclear reactors, but only one new plant has been ordered since 1978. It takes about 10 years

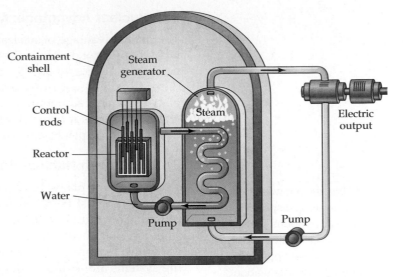

▲ **Figure 15.11** A diagram of a nuclear power plant for generating electricity. Fission of uranium heats the water in the reactor, which generates steam that drives the turbines, producing electricity. Control rods absorb neutrons to slow the fission reaction as needed. Nuclear reactors are housed in containment buildings made of steel and reinforced concrete, designed to withstand nuclear accidents without releasing radioactive substances into the environment.

Table 15.8 Electricity Generation Using Nuclear Power Plants (Selected Countries)

Country	Total Electricity from Nuclear Power (%)	Number of Operating Nuclear Power Plants	Number of Nuclear Power Plants under Construction
China	2	11	6
France	77	59	1
Germany	26	17	0
India	3	17	6
Japan	28	55	1
Russian Federation	16	31	7
Sweden	46	10	0
Ukraine	48	15	2
United States	19	104	1
World total	15	439	24

Source: International Atomic Energy Agency, *PRIS Database*. Vienna: 2007.

to build a nuclear power plant. There will have to be a dramatic change in public attitude if nuclear power is to play a big role in our future.

Types of Nuclear Power Plants

There are several types of nuclear power plants, but we won't attempt to discuss all the possible types here. The one illustrated in Figure 15.11 is a pressurized water reactor. Earlier models were mainly boiling water reactors in which the steam from the reactor was used to power the turbine directly.

Nuclear power plants use the same fission reactions employed in nuclear bombs, but nuclear power plants cannot blow up like bombs. The uranium used in power plants is enriched to only 3–4% uranium-235. A bomb requires about 90% uranium-235.

In a nuclear power plant, a *moderator* is used to slow down the fission neutrons so that they can be absorbed by U-235 atoms. Ordinary water serves as the moderator in 75% of reactors worldwide, graphite is used in 20% of reactors, and heavy water (D_2O) in 5% of reactors. The reaction is controlled by the insertion of boron steel or cadmium *control rods*. Boron and cadmium absorb neutrons readily, preventing them from participating in the chain reaction. These rods are installed when the reactor is built. Removing them partway starts the chain reaction; pushing them in all the way stops the reaction.

A N S W E R

4. How much uranium is available for power plants? The latest estimate from the USDOE indicates that U.S. readily-obtained reserves are about 890 million pounds of uranium oxide. This is equivalent to roughly 80 years of use at current levels. However, as with other forms of energy, technology is likely to make lower-grade ores accessible in the future. Use of breeder reactors could extend these reserves enormously.

The Nuclear Advantage: Minimal Air Pollution

The main advantage of nuclear power plants over those that burn fossil fuels is in what they do not do. Unlike fossil-fuel-burning plants, nuclear power plants produce no carbon dioxide to add to the greenhouse effect, and they add no sulfur oxides, nitrogen oxides, soot, or fly ash to the atmosphere. They contribute almost nothing to global warming, air pollution, or acid rain. They reduce our dependence on foreign oil and lower our trade deficit. If the 104 nuclear plants in the United States were replaced by coal-burning plants, airborne pollutants would increase by 18,000 tons/day!

Problems with Nuclear Power

Nuclear power plants have some disadvantages. Elaborate and expensive safety precautions must be taken to protect plant workers and the inhabitants of surrounding areas from radiation. The reactor must be heavily shielded and housed inside a containment building of metal and reinforced concrete. Because loss of coolant water can result in a meltdown of the reactor core, backup emergency cooling systems are required. Despite what proponents call utmost precautions, some opponents of nuclear power still fear a runaway nuclear reaction in which the containment building is breached and massive amounts of radioactivity escape into the environment. The chance of such an accident is probably exceedingly small, but if one did occur, thousands of people could be killed and large areas rendered uninhabitable for centuries. The benefit of nuclear power—abundant electric energy—is clear, but the small probability of a serious accident causes scientists and others to endlessly debate its desirability.

Another problem is that the fission products are highly radioactive and must be isolated from the environment for centuries. Again, scientists disagree about the feasibility of nuclear waste disposal. Proponents of nuclear power say that such wastes can be safely stored in old salt mines or other geologic formations. Opponents fear that the wastes may arise from their "graves" and eventually contaminate the groundwater. It is impossible to do a million-year experiment in a few years to determine who is right.

Mining and processing uranium ore produces wastes called tailings. Over 200 million tons of tailings now plague ten western states. These tailings are mildly radioactive, giving off radon gas and gamma radiation. Dust from these tailings carries problems to surrounding areas.

The problem of *thermal pollution* is unavoidable. As the energy from any material is converted to heat to generate electricity, some of the energy is released into the environment as waste heat. Nuclear power plants generate more thermal pollution than plants that burn fossil fuels, but the difference is perhaps not as important as the other problems we have mentioned.

Most of the nuclear waste currently awaiting disposal is from nuclear weapons production. High-level waste from nuclear power plant operation in the United States totals about 3000 metric tons per year.

▲ Mildly radioactive tailings are a by-product of the processing of uranium ores.

Yucca Mountain

For many, the solution to the problem of ever-accumulating radioactive spent nuclear fuel and high-level radioactive waste is situated 300 m underground about 160 km northwest of Las Vegas, Nevada. Yucca Mountain, now slated to open in 2020, will be the United States' first high-level radioactive waste depository. Currently, spent fuel rods and other radioactive materials are stored at 126 sites throughout the country.

Since 1978 scientists and engineers have been studying the suitability of Yucca Mountain for the depository site. They not only need to understand the

▶ Yucca Mountain, site of the U.S. proposed nuclear waste depository.

natural characteristics of the mountain that could affect a potential repository's safety but also must engineer repository systems that further isolate the waste. By 2008 cost estimates had reached $90 billion. In 2008 the U.S. Department of Energy (DOE) submitted an application to obtain the Nuclear Regulatory Commission license to proceed with construction of the repository.

Few argue against the need for a long-term solution to a growing problem, but there are concerns, especially over the transportation of radioactive material to Nevada. Estimates are that 19,200 rail cars and 93,000 trucks would be required to relocate the radioactive material—mostly in the form of heavy solid waste pellets—moving it through 43 states. Once on site, the pellets would be placed in casks and maneuvered by rail into tunnels deep in Yucca Mountain. Despite the protests from the citizens of Nevada, the DOE plans to begin entombing 70,000 metric tons of spent nuclear fuel in 2020 at the earliest. Most of the high-level wastes will decay away in a few decades, but the site would remain radioactive for 10,000 years or more.

Nuclear Accidents: Real and Imagined Risks

In 1979 a loss-of-coolant accident at the Three Mile Island nuclear power plant near Harrisburg, Pennsylvania, released a tiny amount of radioactivity into the environment. Although no one was killed or seriously injured, the accident whetted public fear of nuclear power. A 1986 accident at Chernobyl, Ukraine, was much more frightening. There a reactor core meltdown killed several people outright. Others died from radiation sickness in the following weeks and months, and 135,000 people were evacuated. The Ukraine Radiological Institute estimates the accident caused more than 2500 deaths overall. A large area will remain contaminated for decades. Radioactive fallout spread across much of Europe. The Ukraine thyroid cancer rate increased tenfold. Thousands of others have an increased risk of cancer because of exposure to radiation. At Three Mile Island, a containment building kept nearly all the radioactive material inside. The Chernobyl plant had no such protective structure.

There is considerable controversy over most aspects of nuclear power. Although scientists may be able to agree on the results of laboratory experiments, they don't always agree on what is best for society. The essay "Can Nuclear Power be Green?" in Chapter 11 (p. 317) provided one perspective on the advantages and disadvantages of nuclear energy.

▲ An accident at this nuclear power plant at Chernobyl, Ukraine, increased public fears of nuclear power. It released considerable amounts of radioactivity because it had no reinforced containment building. The plant used graphite as a moderator (U.S. plants use water), and the burning carbon hindered efforts to tame the runaway reaction.

Breeder Reactors: Making More Fuel Than They Burn

The supply of fissionable uranium-235 isotope is limited, making up less than 1% of naturally occurring uranium. Separation of uranium-235 leaves behind large quantities of uranium-238, which is not fissionable. However, uranium-238 can be converted to fissile plutonium-239 by bombardment with neutrons. Unstable uranium-239 is formed initially, but it rapidly decays to neptunium-239.

$$\ce{^{238}_{92}U} + \ce{^{1}_{0}n} \longrightarrow \ce{^{239}_{92}U} \longrightarrow \ce{^{239}_{93}Np} + \ce{^{0}_{-1}e}$$

Neptunium-239 then decays to plutonium:

$$\ce{^{239}_{93}Np} \longrightarrow \ce{^{239}_{94}Pu} + \ce{^{0}_{-1}e}$$

If a reactor is built with a core of fissionable plutonium surrounded by uranium-238, neutrons from the fission of plutonium convert the uranium-238 shield to more plutonium. In this way, the reactor, called a **breeder reactor**, produces more fuel than it consumes. There is enough uranium-238 to last several centuries, and so one of the disadvantages of nuclear plants could be overcome by the use of breeder reactors.

Breeder reactors have some problems of their own, however. Plutonium is fairly low melting (640 °C), and a plant is therefore limited to fairly cool, inefficient operation. Water is not adequate as a coolant; therefore, molten sodium metal is used in the primary loop. These reactors are often called *liquid-metal fast breeder reactors*.

A 20-year study of 70,000 nuclear shipyard workers found that they had 24% *lower* mortality than nonnuclear workers in the general population. Those with the highest chronic radiation exposure had the *lowest* mortality from all causes—including cancer. Could it be that small doses of radiation are beneficial?

If an accident occurred in such a breeder, the sodium could react violently with both the water and the air.

Plutonium's low melting point means that a failure of the cooling system could cause the reactor's core to melt. All reactors are required to have an emergency backup core-cooling system. Whether these systems work is a principal area of controversy.

Plutonium is highly toxic and has a half-life of about 24,000 years. It emits alpha particles, making it especially dangerous if ingested. An estimated 1 μg in a person's lungs is enough to induce lung cancer. Yet another problem is that plutonium is chemically different from uranium-238. That means it is much easier to separate it from U-238, for illicit nuclear weapons use, than is uranium-235.

An alternate breeder reaction converts thorium-232 into fissile uranium-233.

$$^{232}_{90}\text{Th} + {}^{1}_{0}\text{n} \longrightarrow {}^{233}_{90}\text{Th} \longrightarrow {}^{233}_{91}\text{Pa} + {}^{0}_{-1}\text{e}$$

$$^{233}_{91}\text{Pa} \longrightarrow {}^{233}_{90}\text{U} + {}^{0}_{-1}\text{e}$$

Uranium-233 is like plutonium in that it emits damaging alpha particles, but it is thought to be rather difficult to make bombs from reactor-grade uranium-233.

The technical and practical issues accompanying breeder reactors have made them much less prevalent than ordinary fission reactors. Only 11 breeder reactors of more than 100 MW have ever been built, none in the United States. Only four were still operating in 2004, and two of those are scheduled to be shut down.

> Thorium-232 is especially attractive for a breeder reactor because it is quite abundant. There is about as much thorium in the earth's crust as there is lead.

Self-Assessment Questions

1. In a nuclear fission reactor,
   a. carbon-14 nuclei decay to nitrogen-14
   b. hydrogen nuclei fuse to form helium
   c. uranium-235 nuclei absorb a neutron and split into two smaller nuclei
   d. uranium-238 nuclei decay to plutonium-238

2. The control rods in a nuclear reactor
   a. absorb neutrons, causing them to ignite fission
   b. absorb neutrons, preventing them from causing fission
   c. slow the neutrons down so the moderator can stop them
   d. speed the neutrons up so they will ignite fission

3. Nuclear power plants produce about what percentage of electricity in the United States?
   a. 5   b. 10
   c. 20   d. 50

4. Nuclear reactors that produce more nuclear fuel than they consume
   a. are called breeder reactors
   b. are called fusion reactors
   c. are called transuranic reactors
   d. violate the law of conservation of energy

5. Approximately how many nuclear power plants are in operation in the United States?
   a. 50   b. 100
   c. 200   d. 400

6. An advantage of nuclear power is that
   a. coal-burning plants have an appalling safety record
   b. it produces no dangerous wastes
   c. it produces no greenhouse gases
   d. the reactor core cannot melt down

Answers: 1, c; 2, b; 3, c; 4, a; 5, b; 6, c

15.11 Nuclear Fusion: The Sun in a Magnetic Bottle

Chapter 11 discusses the thermonuclear reactions that power the Sun and that occur in the explosion of hydrogen bombs. Control of these fusion reactions to produce electricity would give us nearly unlimited power. To date, fusion reactions have been useful only for making bombs, but research on the control of nuclear fusion is progressing (Figure 15.12).

Controlled fusion would have several advantages over nuclear fission reactors. The principal fuel, deuterium ($^2_1$H), is plentiful and is obtained by the fractional electrolysis (splitting apart by means of electricity) of water. (Only 1 hydrogen atom in 6500 is a deuterium atom, but we have oceans of water to work with.) The problem of radioactive wastes would be minimized. The end product, helium, is stable and biologically inert. Escape of tritium ($^3_1$H), which undergoes beta decay with a half-life of 12.3 years, might be a problem because this hydrogen isotope would be readily incorporated into organisms. Also, neutrons are emitted in most fusion reactions, and neutrons can convert stable isotopes into radioactive ones. Finally, we would still be concerned with thermal pollution: the unavoidable loss of part of the energy as heat.

Great technical difficulties have to be overcome before a controlled fusion reaction can be used to produce energy. A sustainable fusion reaction would require

- attainment of a critical ignition temperature of 100 million to 200 million °C
- confinement time of 1–2 s
- sufficient ion density of about $2-3 \times 10^{20}$ ions per cubic meter

In practice, the required conditions are measured by a product of these three values.

No molecule could hold together at the fusion temperature; no material on Earth can withstand more than a few thousand degrees. Even atoms are unstable under these conditions; atoms are stripped of their electrons, and the nuclei and free electrons form a mixture called a **plasma**. The plasma, made of charged particles (nuclei and electrons), can be contained by a strong magnetic field (Figure 15.13).

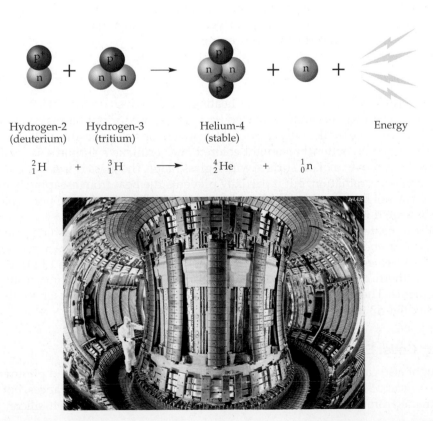

Hydrogen-2 (deuterium) Hydrogen-3 (tritium) Helium-4 (stable) Energy

$$^2_1\text{H} + ^3_1\text{H} \longrightarrow ^4_2\text{He} + ^1_0\text{n}$$

◀ **Figure 15.12** A promising fusion reaction is the deuterium–tritium reaction. A hydrogen-2 (deuterium) nucleus fuses with a hydrogen-3 (tritium) nucleus to form a helium-4 nucleus. A neutron is released along with a considerable quantity of energy.

◀ **Figure 15.13** A giant, donut-shaped electromagnet called a *tokamak* is designed to confine plasma at the extremely high temperatures and pressures required for nuclear fusion.

Nuclear fusion may well be our best hope for relatively clean, abundant energy in the future, but much work remains to be done. Even when controlled fusion is achieved in the laboratory, it will still be decades before it becomes a practical source of energy.

Self-Assessment Questions

1. The missing product in the nuclear reaction $^2_1H + ^3_1H \longrightarrow ^1_0n + ?$ is
   a. ^3_1H
   b. ^3_2He
   c. ^4_2He
   d. ^5_2He

2. Fission, not fusion, is used in nuclear power plants because fusion
   a. fuel is too expensive, requiring transuranium elements
   b. produces less heat than fission
   c. produces waste products that are radioactive for centuries
   d. requires temperatures of millions of degrees, making it difficult to contain

Answers: 1, c; 2, d

RENEWABLE ENERGY SOURCES

Burning fossil fuels leads to air pollution and to the depletion of vital resources. Using nuclear power also presents problems; nuclear fuel is not unlimited, and the waste problems remain unsolved. Are there no renewable energy resources? There are indeed, and we shall devote the remainder of this chapter to them.

15.12 Harnessing the Sun: Solar Energy

At the beginning of this chapter, we saw that nearly all the energy on Earth comes from the Sun. With all that energy from our celestial power plant, why do we need fossil fuels or nuclear power? The answer lies in the fact that this solar energy is thinly spread out and difficult to concentrate.

Solar Heating

Solar energy can be used directly for heating. As it arrives on the surface of Earth, 30% of solar energy is simply reflected back into space. About half is converted to heat. We can increase the efficiency of this conversion rather easily. A black surface absorbs radiation better than a light-colored one. To make a simple solar collector, we need only cover a black surface with a glass plate. The glass is transparent to the incoming solar radiation, but it partially prevents the heat from escaping back into space. The hot surface is used to heat water or other liquids, and the hot liquids are usually stored in an insulated reservoir.

Water heated this way can be used directly for bathing, dishwashing, and laundry, or it can be used to heat a building. Air is passed around the warm reservoir, and the warmed air is then circulated through the building (Figure 15.14). Even in cold northern climates, solar collectors can meet about 50% of home heating requirements. These installations are expensive but can pay for themselves, through fuel savings, in a few years.

Solar Cells: Electricity from Sunlight

Sunlight also can be converted directly to electricity by devices called **photovoltaic cells** or *solar cells*. These devices can be made from a variety of substances, but most are made from elemental silicon. In a crystal of pure silicon, each silicon atom

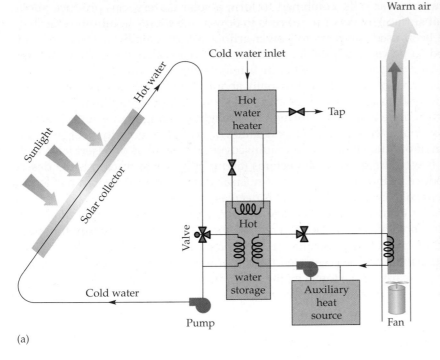

(a)

(b)

▲ **Figure 15.14** Energy in sunlight is absorbed by solar collectors and used to heat water. The hot water can be used directly, or it can be circulated to partially heat the building. This diagram shows how a solar collector furnishes hot water and warm air for heating the building (a). A variety of collectors can be used to take advantage of solar energy (b).

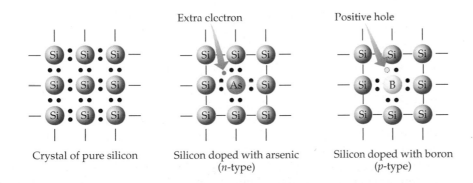

◄ **Figure 15.15** Models of silicon crystals. Crystals doped with impurities such as arsenic and boron are more conductive than pure silicon and are used in solar cells.

has four valence electrons and is covalently bonded to four other silicon atoms (Figure 15.15). To make a solar cell, extremely pure silicon is "doped" with small amounts of specific impurities and formed into single crystals.

One type of crystal has about 1 ppm of arsenic added. Arsenic atoms have five valence electrons, four of which are used to form bonds to silicon atoms. The fifth electron is relatively free to move around. Because this material has extra electrons, and electrons are negatively charged, it is called an *n-type semiconductor* (*n* = negative). Adding about 1 ppm of boron to silicon forms a different type of material. Boron has three valence electrons, producing a shortage of one electron and leaving a *positive hole* in the crystal. This boron-doped silicon is called a *p-type semiconductor*.

Joining the two types of crystals (*n*-type and *p*-type) forms a photovoltaic cell (Figure 15.16); electrons flow from the *n*-type region, which has a high concentration, to the *p*-type region. However, the holes near the junction are quickly filled by nearby mobile electrons, and the flow ceases.

When sunlight hits the photovoltaic cell, an electric current is generated. The energetic photons knock electrons out of the Si—Si bonds, creating more mobile electrons and more positive holes. Because of the barrier at the junction between the two semiconductors, electrons cannot move through the interface. When an external circuit connects the two crystals, electrons flow from the *n*-type region around the circuit to the *p*-type region and that current can be used directly.

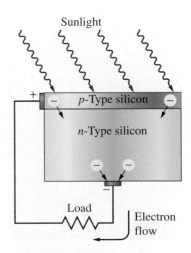

▲ **Figure 15.16** Schematic diagram of the operation of a solar cell. Electrons flow from the *n*-type (lower layer) to the *p*-type (upper layer) through the external circuit.

An array of solar cells, combined to form a solar battery, can produce about 100 W/m² of surface; it takes 1 m² of cells to power one 100-W lightbulb. Solar batteries have been used for years to power artificial Earth satellites. They are now widely used to power small devices such as electronic calculators. They are also used to provide electricity for weather instruments in remote areas.

Solar cells are not very efficient (usually about 14–19%), although experimental photovoltaic systems using lenses or mirrors to concentrate sunlight have achieved an efficiency of more than 40%. Much of the sunlight striking them is reflected back into space or converted to heat. The generation of enough energy to meet a significant portion of our demands would require covering vast areas of desert land with solar cells. It would require 2000 hectares (about 5000 acres) of cloud-free desert land to produce as much energy as one nuclear power plant. Research has led to more efficient solar cells, and their cost is decreasing.

Using solar energy requires storage of energy for use at night and on cloudy days. The technology is now widely available for the use of solar energy for space heating and for providing hot water. Several companies now sell home-sized solar electric systems that can be connected to the normal house supply and can reduce electric bills to nearly zero. However, they are quite expensive, and widespread use of solar electricity is probably still some years away.

5. Is solar power the answer to the energy crisis?
Today's solar cells are still too expensive and too inefficient to provide a complete solution to our energy needs. Also, conventional transportation requires more power than solar cells mounted on the roof can conveniently provide; some other method of powering our buses and cars will be needed.

Self-Assessment Questions

1. Solar heating is most useful for
   a. heating water
   b. powering fuel cells
   c. producing cheap electricity
   d. storing nuclear energy

2. The particles that gain enough energy to produce an electric current when sunlight falls on a photovoltaic cell are
   a. electrons
   b. photons
   c. protons
   d. positrons

3. Which of the following is an *n*-type semiconductor?
   a. arsenic doped with silicon
   b. arsenic doped with germanium
   c. silicon doped with arsenic
   d. silicon doped with boron

4. *p*-Type semiconductors feature crystals sites with
   a. photons
   b. positrons
   c. positive holes
   d. protons

Answers: 1, a; 2, a; 3, c; 4, c

▲ These willow trees are grown specifically as fuel. The biomass is harvested by cutting the trees to near ground level, which stimulates regrowth. Willow is especially suitable for biomass because of its rapid growth.

15.13 Biomass: Photosynthesis for Fuel

Why bother with solar collectors and photovoltaic cells to capture energy from the Sun when green plants do it every day? Indeed, dry plant material, called **biomass** when used as a fuel, burns quite well. It could be used to fuel a power plant for the generation of electricity. Biomass burns cleanly; the emissions are almost entirely water vapor and the carbon dioxide the plant took from the air in the first place. "Energy plantations" could grow plants for use as fuel. Biomass is a renewable resource whose production is powered by the Sun.

Unfortunately, there are several disadvantages to this scheme, too. Most available land is needed for the production of food. Even where productive land is available, plants have to be planted, harvested, and transported to the power plant. Often the land is far from where the energy is needed. However, some hybrid species of trees grow quickly and densely, reducing the needed growing area. The overall efficiency of biomass use is even less than that of solar cells—only about 3% at best. Nevertheless, there is considerable research activity in the area of biomass; nature "constructs" plants much more easily and more quickly than we can construct industrial plants.

We don't have to burn plant material directly.

- Starches and sugars from plants can be fermented to form ethanol. Wood can be distilled in the absence of air to produce methanol. Both alcohols are liquids and convenient to transport and are excellent fuels that burn quite cleanly.

- Bacterial breakdown of plant material produces methane. Under proper conditions, this process can be controlled to produce a clean-burning fuel similar to natural gas.

- Vegetable oils or animal fats can be converted to a fuel called *biodiesel*. The triacylglycerols are converted into the methyl esters of the constituent fatty acids. A typical biodiesel molecule from palm oil is methyl palmitate:

$$CH_3CH_2CH_2CH_2CH_2CH_2CH_2CH_2CH_2CH_2CH_2CH_2CH_2CH_2CH_2COOCH_3$$

Each of these conversions, however, results in the loss of a portion of the useful energy. The laws of thermodynamics tell us that we would get the most energy by burning the biomass directly rather than converting it to a more convenient liquid or gaseous fuel.

The shortage of land probably means that we will never obtain a major portion of our energy from biomass. We could, however, supplement our other sources by burning agricultural wastes, fermenting some to ethanol, producing methanol from wood where wood is plentiful, and fermenting human and animal wastes to produce methane. The technology for all these processes is readily available. Each has been used in the past, and all are now being used on a limited scale.

Hemp, the fiber crop from which rope is made, is a good energy source. Its woody stalks are 77% cellulose, and hemp can produce 10 t of biomass per acre in just 4 months. The oil from hemp seeds can be used as diesel fuel. The problem is that hemp must be imported. It is illegal to raise hemp in the United States because the leaves and flowers of the female plant constitute the drug marijuana.

Self-Assessment Questions

1. Renewable sources of energy are so called because they
   a. are clean and cost-free
   b. can be converted directly into electricity
   c. can be replenished by nature relatively quickly
   d. do not produce air pollution

2. Which of the following is a renewable energy source?
   a. biomass
   b. hydrogen
   c. nuclear power
   d. tar sands

3. Biodiesel *cannot* be made from
   a. corn starch
   b. olive oil
   c. soybean oil
   d. waste cooking fat

Answers: 1, c; 2, a; 3, a

GREEN CHEMISTRY

Biofuels

Doug Raynie, South Dakota State University

Current news reports constantly remind us of our nation's search for alternatives to fossil fuels. While some combination of solar, wind, nuclear, and geothermal sources represents the future for power generation, and progress is being made toward more efficient fuel cells (Section 15.14), our need for liquid transportation fuels will continue for some time.

The most promising source of liquid fuels is biomass, generally crops and grasses, forestry products, or agricultural wastes. The renewable biomass sources can vary with geographic region to support a local or regional energy base. Two major biofuels currently being produced are ethanol, and biodiesel.

Fermentation of biomass to make ethanol, CH_3CH_2OH, for alcoholic beverages has been common for centuries. Ethanol is useful as a fuel, usually mixed with gasoline. Henry Ford in 1925 said that ethanol was the "fuel of the future," but in fact its use as automobile fuel has been limited. As the text recounts (Section 15.8), ethanol was relatively recently promoted to replace MTBE; interest in ethanol fuel has accelerated with recent dramatic increases in gasoline prices.

Ethanol is made from starch- or sugar-based crops—corn in the United States and sugarcane in Brazil, the world's largest ethanol producer. The biomass is ground and cooked into a mash. Yeast ferments the mash to form carbon dioxide and ethanol. The ethanol is separated, distilled, and dried using a molecular sieve adsorbent, then blended with gasoline. In the United States, blends with 10% ethanol are common and can be used in any automobile. E-85, a gasoline blend containing 85% ethanol, requires a flex-fuel vehicle. The U.S. EPA is considering ethanol blends up to 20 or 30%. The IndyCar racing series uses 100% ethanol in its cars, and ethanol has proved to be safer and to provide higher track speeds than gasoline. In Brazil, ethanol accounts for about half of all liquid transportation fuels. The residue from the corn-ethanol fermentation, called distiller's grains, can be sold as cattle feed. Without the sale of by-products, ethanol as a fuel would not be as economical nor as environmentally beneficial.

The next generation of ethanol is expected to be *cellulosic ethanol,* ethanol produced from nonfood plants such as grasses or forestry and agricultural waste. Cellulose is a carbohydrate polymer that can be hydrolyzed to sugars that can then be fermented. The barrier to the production of cellulosic ethanol is the difficulty of breaking down the protective lignin sheath that gives the plant structure its solidity. Cellulosic ethanol production is expected to begin within the next few years. Other workers are researching the fermentation of biomass to larger, less polar alcohols like 1-butanol, which blend better with nonpolar gasoline.

A second type of biofuel is biodiesel, a renewable, clean-burning fuel that can be mixed with petroleum-based diesel. The development of biodiesel mirrors the development of ethanol, but in the United States diesel is much less commonly used as a transportation fuel than gasoline.

Most biodiesel manufacturing processes require high-purity virgin oils, mostly from soybeans, the price of which accounts for most of the cost of biodiesel. Biodiesel is thus not a competitive fuel without government subsidy. Research is underway on a process to make biodiesel from cheap materials such as waste frying grease from fast-food restaurants and a variety of plant oils in a flow-through reactor. The reactor uses a heterogeneous catalyst made of porous microspheres of oxides of zirconium, titanium, or aluminum, calling for less energy overall.

The first use of fuels by human beings probably occurred when a tree, struck by lightning, caught fire and provided warmth to people nearby. How ironic that we are now returning to that same type of fuel to help satisfy the energy needs of modern society!

◄ Biodiesel can be used in many vehicles that burn ordinary diesel fuel. Some users make their own biodiesel from waste food oils. And some entrepreneurs produce biodiesel on a larger scale for retail sale.

15.14 Hydrogen: Light and Powerful

Just as natural gas is sent through pipes to wherever it is needed, so could other fuel gases be sent through pipes. One such gas is hydrogen.

When hydrogen burns, it produces water and gives off energy.

$$2\,H_2(g) + O_2(g) \longrightarrow 2\,H_2O(l) + 572\ kJ$$

Gram for gram, hydrogen yields more energy than any other chemical fuel. It is also a clean fuel, yielding only water as a chemical product. Although hydrogen is the most abundant element in the universe, as elemental hydrogen (H_2) it is almost nonexistent on Earth. There is a lot of hydrogen on Earth, but it is tied up in chemical compounds, mainly water, and releasing it requires more energy than the hydrogen produces when it is burned. Because hydrogen can be made from seawater, the supply is almost inexhaustible, but finding a way to make H_2 from water economically is a complicated matter.

Fuel Cells

A **fuel cell** is a device in which fuel is oxidized in an electrochemical cell (Section 8.3) so as to produce electricity directly. Fuel cells differ from the usual electrochemical cell in two ways:

- The fuel and oxygen are fed continuously. As long as fuel is supplied, current is generated.
- The electrodes are made of an inert material such as platinum that does not react during the process.

Most of today's fuel cells use hydrogen and oxygen (Figure 15.17). At the platinum anode, H_2 is oxidized forming hydrogen ions and electrons:

$$2\,H_2(g) \longrightarrow 4\,H^+ + 4\,e^-$$

The electrons produced at the anode travel through the external circuit and arrive at the cathode, where they combine with the hydrogen ions and oxygen molecules to form water.

$$4\,e^- + O_2(g) + 4\,H^+ \longrightarrow 2\,H_2O(g)$$

The overall reaction is identical to combustion, but not all the chemical energy is converted to heat. About 40–55% of the chemical energy is converted directly to electricity, making fuel cells much more efficient than internal combustion engines.

$$2\,H_2(g) + O_2(g) \longrightarrow 2\,H_2O(l)$$

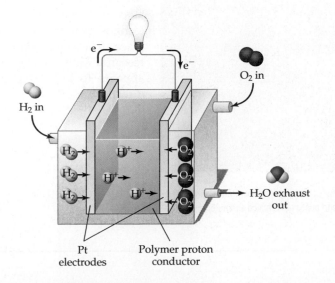

H₂ in

O₂ in

H₂O exhaust out

Pt electrodes

Polymer proton conductor

◀ **Figure 15.17** A hydrogen–oxygen fuel cell. Unlike a conventional dry cell that goes "dead" after prolonged use, the fuel cell continues to provide electricity as long as reactants are supplied.

Hydrogen in Your Future (Car)

Hydrogen can be used as a fuel for cars, either directly or in fuel cells. In his 2003 State of the Union address, President George W. Bush announced a $1.2 billion federal initiative to support hydrogen fuel cell research with the hope that it will lead to the widespread use of hydrogen-powered cars. Among the advantages of hydrogen as a fuel are that

- the exhaust is almost entirely water vapor, which cuts down on urban air pollution and the production of the greenhouse gas CO_2
- hydrogen can be produced from renewable sources
- usage could reduce our reliance on foreign oil and its associated economic and political costs

There are, however, significant obstacles to this transition to hydrogen; the roads will not be filled with hydrogen cars any time soon. Among the problems are the following:

- Hydrogen is not a *source* of energy; there are no hydrogen "mines" or "reservoirs."
- The hydrogen-powered cars available now are expensive novelties, costing $100,000 or more each.
- There is no distribution system of hydrogen fueling stations.
- Hydrogen is a low-density gas that liquefies at $-253\,°C$ (20 K). Use of the gas requires large, heavy tanks. Use as a liquid requires that it be maintained at a temperature only a few degrees above absolute zero. Therefore, storage, transportation and dispensing of the fuel are problematic.

Progress has been made in several areas. California has about two dozen pilot stations run by utilities and carmakers. Governor Arnold Schwarzenegger's "hydrogen blueprint" calls for up to 2000 hydrogen vehicles and 100 refueling stations by 2010 at an estimated cost of $54 million.

Fuel cell technology is becoming more efficient and costs are coming down, but there is still the problem of producing hydrogen gas economically. Hydrogen has to be made using another energy source: fossil fuels, nuclear power, or—preferably—some renewable source such as solar energy or wind power. Some experimental cells make hydrogen on-board from methanol or natural gas by a high-temperature process called *reforming*.

$$CH_4(g) + 2\,H_2O(g) \longrightarrow CO_2(g) + 4\,H_2(g)$$

The drawback is that reforming emits carbon dioxide, negating one of the reasons for using fuel cells in the first place. In 2007, Jerry Woodall at Purdue University developed a process for liberating hydrogen from water using an alloy of aluminum and gallium. Although electrical energy is needed to extract aluminum from its ore, the method shows promise.

Even liquid H_2, with a density of $0.07\,\text{g/cm}^3$, requires a large fuel tank, and the tank must be well insulated. It may be possible to store hydrogen in other ways: Scientists have found that various metals can absorb up to a thousand times their own volume of hydrogen gas. Special forms of carbon, such as ultrathin carbon nanotubes, may also hold large quantities.

The cost of establishing pipelines and filling stations to handle hydrogen will be immense. It takes many years for revolutionary technologies to come to the marketplace. The increased emphasis on research into hydrogen fuel cells appears to be a step in the right direction.

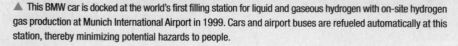

▲ This BMW car is docked at the world's first filling station for liquid and gaseous hydrogen with on-site hydrogen gas production at Munich International Airport in 1999. Cars and airport buses are refueled automatically at this station, thereby minimizing potential hazards to people.

Fuel cells are used on spacecraft to produce electricity. Fuel cells overall have a weight advantage over storage batteries—an important consideration when launching a spaceship—and the water produced can be used for drinking. Because they have no moving parts, fuel cells are tough and reliable.

On Earth, research is underway to reduce cost and design long-lasting cells. Perhaps someday fuel cells will provide electricity to meet peak needs in large power plants. Unlike huge boilers and nuclear reactors, they can be started and stopped simply by turning the fuel on or off.

Some people are afraid of using hydrogen as fuel for home heating because of the possibility of explosion, but natural gas leaks can also lead to serious explosions. Hydrogen escapes more readily than natural gas or gasoline because of its smaller molecular size, but it also dissipates more readily rather than forming a flammable low-lying vapor as gasoline sometimes does.

6. Is hydrogen really a nonpolluting fuel?
The only product from combustion of hydrogen is water. However, an internal combustion engine uses high temperatures and pressures. Under these conditions some amount of nitrogen oxides, which are pollutants, will form. If hydrogen is used in a fuel cell instead of an internal combustion engine, pollution is negligible.

Self-Assessment Questions

1. The major product from burning hydrogen gas as a fuel is
   a. CO_2 **b.** H_2 **c.** H_2O **d.** O_2

2. Which of the following fuels has the lowest EROEI?
   a. coal   **b.** hydrogen   **c.** natural gas   **d.** petroleum

3. Which of the following energy technologies does not use energy that originated in the sun?
   a. biomass   **b.** hydroelectric   **c.** nuclear   **d.** wind

4. The fuel used by almost all present day fuel-cells is
   a. electricity   **b.** hydrogen   **c.** natural gas   **d.** propane

5. Fuel cells are
   a. dependent on fuels from other sources
   b. net sources of energy
   c. producers of new fuels
   d. renewable energy sources

Answers: 1, c; 2, b; 3, c; 4, b; 5, a

15.15 Other Renewable Energy Sources

Modern civilization requires tremendous amounts of energy and will require more and more in the coming years. We are depleting our reserves of fossil fuels, and many see significant disadvantages to nuclear power. Using the Sun's energy seems to be a good idea, but it would be difficult to meet all our needs with solar energy. In this section we look at other kinds of energy sources, some of which are already in use.

Wind and Water

The Sun, by heating Earth, causes winds to blow and water to evaporate and rise into the air, later to fall as rain. The kinetic energy of blowing wind and flowing water can be used as sources of energy—as they have been for centuries.

Why not use wind power and waterpower to help solve the energy crisis? We do. Waterpower provides more than 7% of our current electricity production, most of it in the mountainous western United States. In a modern hydroelectric plant, water is held behind huge dams. Some of it is released through penstocks against the blades of water turbines. The potential energy of the stored water is converted to the kinetic energy of flowing water. The moving water imparts mechanical energy to the turbine, which drives a generator that converts mechanical energy to electrical energy.

▲ Hoover Dam, on the Colorado River, was completed in 1936. It generates 4 billion kWh a year—enough to serve 1.3 million people.

Hydroelectric plants are relatively clean, but most of the good dam sites in the United States have already been used. To obtain more hydroelectric energy, we would have to dam up scenic rivers and flood valuable cropland and recreational areas. Reservoirs silt up over the years, and sometimes dams break, causing catastrophic floods. Dams block the migration of fish upstream to spawning beds. Declining fish populations have led to closure of salmon fishing in waters off California and Oregon and to a call by many for dam removal from rivers where salmon spawn.

Worldwide, hydroelectric plants provide about 20% of the electricity, and use is still being expanded in many developing countries. As a major example, China's huge Three Gorges project on the Yangtze River to be completed in 2009 includes a large hydroelectric plant with a capacity of 18,200 Mw. It will provide more than 10% of China's electricity. The project displaced a million people. The dam may cause significant environmental damage. Water quality in the Yangtze's tributaries is already deteriorating rapidly because the dammed river disperses pollutants less effectively and algae blooms flourish. The rising water caused extensive soil erosion, riverbank collapses, and landslides. Even renewable hydroelectric power has its problems.

We could use more wind power. The kinetic energy of moving air is readily converted to mechanical energy to pump water, grind grain, or turn turbines and generate electricity. Wind power amounts to only about 1% of U.S. energy production but is now the fastest-growing energy source, increasing at an annual rate of 20 to 30%.

Wind is clean, free, and abundant. However, the wind does not always blow, and some means of energy storage or an alternative source of energy is needed. Some environmentalists oppose wind power because the rotating blades kill thousands of birds each year, but so do domestic cats and collisions with television and microwave towers. Land use might become a problem if wind power were used widely, but land under windmills can be used for farming or grazing.

What is the connection between waterpower and wind power and chemistry? Chemists produce new materials such as metal alloys, reinforced plastics, and lubricants vital to the construction and operation of dams, windmills, and turbines. Chemists also monitor water quality above and below dams, and they participate in other activities vital to protection of the environment around generating facilities.

Geothermal Energy

The interior of Earth is heated by immense gravitational forces and by natural radioactivity. This heat comes to the surface in some areas through geysers and volcanoes. **Geothermal energy** has long been used in Iceland, New Zealand, Japan, and Italy. It has some potential in the United States—it is being used now in California (Figure 15.18)—but in the near future this potential could be realized only in areas where steam or hot water is at or near the surface. One drawback of geothermal energy is that the wastewater is quite salty; its disposal could be a problem.

▶ Wind turbines, such as these in an offshore wind farm, supply about 19% of Denmark's electricity. By the end of 2007, generating capacity of over 94,100 MW was operating worldwide, producing just over 1% of the world's electricity.

(a)

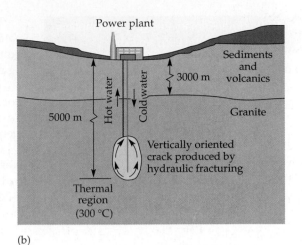

(b)

Oceans of Energy

The oceans that cover three-fourths of Earth's surface are an enormous reservoir of potential energy. The use of *ocean thermal energy* was first proposed in 1881 and was shown to be workable in the 1930s. The difference in temperature between the surface and the depths is 20 °C or more, enough to evaporate a liquid and use the vapor to drive a turbine. The liquid is condensed by the cold from the ocean depths and the cycle is repeated.

Other ways of using ocean energy have been tested. In some areas of the world, the daily rise and fall of the tide can be harnessed, using machinery similar to a hydroelectric plant. At high tide, water fills a reservoir or bay. At low tide, the water escapes through a turbine to generate electricity. Even the energy of the waves crashing to shore can be used with the appropriate technology.

▲ **Figure 15.18** Geothermal energy contributes to energy supplies in some areas. A geothermal field at Wairakei, New Zealand (a). Energy is extracted from dry, hot rock 5 km below the surface by pumping water into the rock to be heated (b).

More to Explore
Scigliano, Eric. "Wave Energy." *Discover*, December 2005, pp. 42–45.

Self-Assessment Questions

1. The renewable energy source that provides the most electrical energy in the United States today is
   a. geothermal   **b.** hydropower   **c.** solar   **d.** wind
2. The energy that comes from the internal heat of the Earth is
   a. bioenergy   **b.** geothermal energy
   c. hydroelectricity   **d.** solar energy

Answers: 1, b; 2, b

15.16 Energy: How Much Is Too Much?

Our energy future is quite uncertain. World petroleum production will peak soon—some claim it has peaked already. Developing nations will increasingly compete for fuels on the world market. Science and technology will be asked to provide an ever-growing supply of energy. The Energy Information Administration of the U.S. Department of Energy projected that world energy consumption will increase by 57% from 2002 to 2025, most of it in developing countries.

Each of the energy sources discussed in this chapter has advantages and disadvantages. How do we choose the best way to deal with energy problems? Wise choices require informed citizens who examine the process from beginning to end. We must know what is involved in the construction of power plants, the production of fuels, and the ultimate use of energy in our homes and factories. We must know that energy is wasted (as heat) at every step in the process.

Table 15.9 Total Primary Energy Consumption per Capita, 2007 (in GJ)*

Canada	469
United States	356
Russia	222
Japan	184
European Union	172
South Africa	109
Mexico	71
China	54
India	17
World	76

*1 GJ $= 10^9$ J

Will our profligate consumption of energy affect Earth's climate? Our activities have already modified the climate in and around metropolitan areas. The worldwide effects of our expanding energy consumption are harder to estimate.

What can we do as individuals? We can conserve. We can walk more and use cars less. We can reduce our wasteful use of electricity. We can buy more efficient appliances, and we can avoid purchasing energy-intensive products.

Over the past three decades we have made significant progress in energy conservation. Appliances are more energy efficient. New furnaces are 90–95% efficient, as compared with the 60% efficiency of 20-year-old furnaces. New refrigerators and air conditioners use about 35% less energy than earlier models. More people use fluorescent lamps, which are at least four times as energy efficient as incandescent lamps. Overall, American industry has reduced its energy consumption per product by almost 30%.

We could significantly reduce energy consumption in the United States by greater use of public transportation. In Europe, where there is much wider use of public transit, per capita energy use is considerably less (Table 15.9), partly because the average price for gasoline is usually more than twice that in the United States. So far, few people used to the convenience of personal cars seem eager to start using buses and trains. As the price of gasoline in the United States continues to rise, perhaps public transportation will seem more attractive.

The simple facts are that our population on Spaceship Earth is going up and our fuel resources are going down. People in developing countries want to raise their standard of living, and that will require more energy. We need to conserve energy and to look for new energy sources.

Critical Thinking Exercises

Apply knowledge that you have gained in this chapter and one or more of the FLaReS principles (Chapter 1) to evaluate the following statements or claims.

15.1 In 2002 Rusi P. Taleyarkhan and colleagues at the Oak Ridge National Laboratory claimed that experiments in a device similar to an ultrasonic cleaning bath containing deuterated acetone (CD_3COCD_3; acetone in which all the hydrogen atoms are hydrogen-2, deuterium) showed production of tritium (hydrogen-3) and neutrons consistent with nuclear fusion. (Taleyarkhan is now at Purdue University.) Bubbles are formed in such ultrasound experiments, and simulations seem to show that temperatures can reach 10,000,000 K inside the bubbles as they collapse—temperatures like those in typical fusion reactions. Other scientists have tried to repeat this work but could find no evidence of a fusion reaction. The only scientists claiming to have verified Taleyarkhan's claims are two of Taleyarkhan's students at Purdue University.

15.2 In 1996 inventor Ramar Pillai at the Indian Institute of Technology (IIT) claimed to have discovered a secret herbal formula that converted water to a hydrocarbon fuel.

15.3 An inventor claims to have discovered a catalyst that speeds the conversion of linear alkanes, such as hexane, to branched-chain isomers.

15.4 While watching TV during the summer, you hear someone proclaim that a way to cool your kitchen is to keep the refrigerator door open.

15.5 A man claims that a battery pack of car batteries will supply your house with electricity because he has a battery charger that is 200% efficient. An inverter unit hooked to the batteries provides more power output than line power from the house.

15.6 A website claims that a product called Super-Mag Fuel Extender attached to the fuel line increases gas mileage for automobiles by changing the molecular structure of gasoline.

15.7 A company claims that it can use a metal such as magnesium as a fuel for automobiles. A coil of the metal would be fed into a chamber where it would react with superheated steam to produce $H_2(g)$; the H_2 would then power a fuel cell that would run the car.

■ SUMMARY

Section 15.1—**Energy** is the ability to do work. The United States has about 5% of the world's population but uses about one-fourth of all energy generated. Most of the energy used on Earth originated from the Sun. The SI unit of energy is the joule (J), and the unit of power is the watt (W), 1 joule per second. Energy can be classified as **potential energy** (energy of position) or **kinetic energy** (energy of motion). In the process called **photosynthesis**, solar energy is absorbed by plants and stored as glucose, and oxygen is generated.

Section 15.2—Rates of reactions are affected by temperature, by concentrations of the reactants, and by the presence of catalysts. The study of energy changes that occur during chemical reactions (and physical processes) is called **thermochemistry**. We define a **system** as the part of the universe under consideration. The **surroundings** are everything else. A chemical reaction or a physical process may be either **exothermic** (giving off energy) or **endothermic** (requiring energy). When a given process is exothermic, the reverse process is endothermic to exactly the same extent. Reactions involving burning of fuels are all exothermic.

Section 15.3—The **first law of thermodynamics (law of conservation of energy)** says that energy can be neither created nor destroyed. The **second law of thermodynamics** states that in a spontaneous process, energy is degraded from more useful forms to less useful forms. **Entropy** is the dispersal of energy among the possible states of a system. There is a tendency for a system to go toward greater entropy.

Section 15.4—Ancient societies used slave labor and animal power to do work. The primary fuel was wood. The first mechanical devices for doing work were the waterwheel and the windmill. Energy from water, steam, gas, the internal combustion engine, and other sources fueled the Industrial Revolution.

Section 15.5—A **fuel** is that which will burn readily with release of energy. **Fossil fuels** are natural fuels derived from once-living plants and animals. They include coal, petroleum, and natural gas.

Section 15.6—**Coal** is a rock that is both a fuel and a source of chemicals. The main element in coal is carbon. Coal rank is determined by its carbon and energy content, with anthracite being highest in rank and peat lowest. Coal is the most plentiful fossil fuel but is inconvenient to use and produces more pollution than other fuels.

Section 15.7—**Natural gas** is mainly methane and burns relatively cleanly. It is both an important fuel and a raw material for synthesis of plastics and other products.

Section 15.8—**Petroleum** or crude oil is a thick liquid mixture of organic compounds, mainly hydrocarbons. Petroleum products can be burned fairly cleanly but inefficient combustion produces pollution. Petroleum is refined by separating it, by distillation, into different fractions. **Gasoline** is the fraction consisting of hydrocarbons from about C_5 to C_{12}. Branched-structure hydrocarbons such as isooctane burn more smoothly than straight-chain hydrocarbons. The **octane rating** of gasoline compares its antiknock performance to pure isooctane. To raise the octane rating, oil refiners use isomerization (to increase branching), **catalytic reforming** (to convert straight-chain hydrocarbons to aromatic rings), and alkylation (to convert gases to highly branched gasoline molecules). Octane boosters such as methanol, ethanol, MTBE, and *tert*-butyl alcohol are added as well. Tetraethyllead was used to improve octane rating for more than 50 years but has been discontinued.

Section 15.9—Electricity is the most convenient form of energy. About half of the electricity in the United States is generated from coal-fired steam power plants. Coal can be converted to more convenient gaseous or liquid fuels but at an energy cost.

Section 15.10—Nuclear fission reactions can be controlled in a **nuclear reactor**, which uses a low concentration (3–5%) of uranium-235. The reaction is controlled with control rods of cadmium or boron steel. The energy released generates steam that turns a turbine to generate electricity. A nuclear power plant cannot explode like a nuclear bomb, it produces minimal air pollution, and its wastes are collected rather than dispersed. But elaborate safety precautions are needed to prevent escape of radioactive material, the needed fuel is limited in availability, and ultimate disposal of radioactive waste is an ongoing problem. A **breeder reactor** converts other isotopes to useful nuclear fuel and can make more fuel than it uses.

Section 15.11—Nuclear fusion cannot yet be controlled, as there are great technical difficulties. Temperatures in the millions of degrees are needed to generate a **plasma** containing free electrons and nuclei for controlled fusion. But fusion has very important potential advantages over most other energy sources.

Section 15.12—Solar energy is rather diffuse and hard to use. However, simple solar collectors can produce hot water for heating. Solar cells or **photovoltaic cells** produce electricity directly from sunlight, using *n*-type and *p*-type semiconductor material. The cells are expensive, not very efficient, and require a storage system for night use.

Section 15.13—Plant **biomass** such as wood can be burned directly or can be converted to liquid or gaseous fuels, but biomass requires much land area and has a very low overall efficiency.

Section 15.14—Hydrogen is an energy-storage and transport method rather than an energy source. It burns cleanly but requires more energy to make than it produces. One advantage of hydrogen is that it can be used in **fuel cells**, which consume fuel and oxygen continuously to generate electricity efficiently.

Section 15.15—Other renewable energy sources include wind, which can be used to power windmills, and hydroelectric power. Windmills are expensive and not terribly efficient, and hydroelectric power can be generated only in certain areas. Heat energy from the interior of Earth is a form of **geothermal energy**, which has been used in several countries. It can only be used in areas where steam or hot water is near the surface. We can also make use of the temperature difference between surface and deep waters, and tidal energy, and the energy of the waves.

Section 15.16—All energy sources have advantages, limitations, and consequences. We can improve the energy situation locally by using more efficient appliances and simply by using less energy when practical.

■ REVIEW QUESTIONS

1. What was the principal fuel used in the United States before 1800?

2. What kind of reaction powers the Sun?

3. What was the principal fuel used in the United States from about 1850 to 1950? What has been our principal fuel since 1950?

4. List some advantages and disadvantages of hydrogen as a vehicle fuel.

■ PROBLEMS

Fuels and Combustion

5. Which of the following are fuels?
   a. sucrose ($C_{12}H_{22}O_{11}$)   **b.** acetylene (C_2H_2)
   c. H_2O_2

6. Which of the following are fuels?
   a. ethylene (C_2H_4)   **b.** C_4H_{10} **c.** iodine (I_2)

7. For each substance in Problem 5 that is a fuel, write a balanced equation for its complete combustion in oxygen gas (O_2).

8. For each substance in Problem 6 that is a fuel, write a balanced equation for its complete combustion in oxygen gas (O_2).

9. Give the equation for the complete combustion of coal, assuming that the coal is carbon.

10. Give the equation for the incomplete combustion of coal to form carbon monoxide, assuming that the coal is carbon.

11. Give the equation for the complete combustion of methane.

12. Give the equation for the incomplete combustion of methane to form carbon monoxide and water vapor.

13. Water gas is a mixture of H_2 and CO. Can water gas serve as a fuel? Explain.

14. Water gas (see Problem 13) can be made by reacting red hot charcoal with steam. Write the equation for the reaction.

Energy and Chemical Reactions

15. How does temperature affect the rate of a chemical reaction?

16. Why does wood burn more rapidly in pure oxygen than in air?

17. Burning 1.00 mol of methane releases 803 kJ of energy. How much energy is released by burning 5.25 mol of methane?

$$CH_4(g) + 2\,O_2(g) \longrightarrow CO_2(g) + 2\,H_2O(g) + 803\text{ kJ}$$

18. It takes 572 kJ of energy to decompose 2.00 mol of liquid water. How much energy does it take to decompose 55.5 mol of water?

$$2\,H_2O(l) + 572\text{ kJ} \longrightarrow 2\,H_2(g) + O_2(g)$$

19. When burned, 1.00 g of gasoline gives off 1030 cal. What is this quantity in joules?

20. How much heat, in kilojoules, is given off when 4.301 g of $H_2(g)$ reacts with $O_2(g)$ to form steam [$H_2O(g)$] according to the following equation?

$$2\,H_2(g) + O_2(g) \longrightarrow 2\,H_2O(g) + 483.6\text{ kJ}$$

21. Review Problem 17 and determine how much energy in kilojoules must be supplied to convert one mole of gaseous CO_2 and two moles of water vapor into methane gas and oxygen gas.

22. Review Problem 20 and determine how much energy in kilojoules must be supplied to convert one mole of water vapor into hydrogen gas and oxygen gas.

Energy and the Laws of Thermodynamics

23. State the first law of thermodynamics.

24. State the second law of thermodynamics in terms of energy flow and in terms of degradation of energy.

25. What is entropy? Is entropy increased or decreased when a fossil fuel is burned?

26. Energy is conserved. How can we ever run out of energy?

27. How does the statement, "You can't get something for nothing," relate to the first law of thermodynamics?

28. How does the statement, "You can't even break even," relate to the second law of thermodynamics?

Fossil Fuels

29. What are the advantages and disadvantages of coal as a fuel?

30. Why did the United States shift from coal to petroleum and natural gas when we have much larger reserves of coal than of the two hydrocarbon fuels?

Estimates in Problems 31–34 are from the U.S. Energy Information Administration.

31. The estimated proven world oil reserves in 2006 were 1293 billion barrels. The world rate of annual use was 31.0 billion barrels. How long will these reserves last if this rate of use continues? (The actual rate of use is increasing more than 2% a year.)

32. The estimated proven U.S. oil reserves in 2006 were 21.8 billion barrels. **(a)** How long will these reserves last if there are no imports or exports and if the U.S. annual rate

of use of 7.6 billion barrels continues? **(b)** Taking the most optimistic projection for oil reserves in the Arctic National Wildlife Refuge—10.4 billion barrels—how long would exploiting that resource extend our reserves at the current U.S. annual rate of use?

33. The estimated world natural gas reserves in 2006 were 6124 trillion ft^3. The world rate of annual use was 105.5 trillion ft^3. How long will these reserves last if this rate of use continues?

34. The U.S. proven natural gas reserves in 2006 were 204 trillion ft^3. How long will these reserves last if there are no imports or exports and if the U.S. annual rate of use of 22 trillion ft^3 continues?

Natural Gas

35. Write a balanced equation for the combustion of ethane (C_2H_6, a component of natural gas) in oxygen gas to produce carbon dioxide and water vapor.

36. Write a balanced equation for the combustion of butane (C_4H_{10}, a minor component of natural gas) in oxygen gas to produce carbon dioxide and water vapor.

Petroleum

37. What is thought to be the origin of petroleum?

38. What are the advantages and disadvantages of petroleum as a source of fuels?

39. How is crude petroleum modified to better meet our needs and wants? What are the advantages and disadvantages of tetraethyllead as an octane booster?

40. Use the term *molecular mass* to describe the difference between gasoline and kerosene.

41. Consult Table 15.6 and then suggest a reason why asphalt is so cheap that it is used to pave highways and driveways.

42. Which should have a higher octane rating: **(a)** hexane ($CH_3CH_2CH_2CH_2CH_2CH_3$) or 2,3-dimethylbutane [$(CH_3)_2CHCH(CH_3)_2$]? **(b)** benzene or heptane?

Nuclear Power

43. What proportion of U.S. electricity is generated by nuclear power plants?

44. What proportion of electricity in France is generated by nuclear power plants?

45. Can a nuclear power plant explode like a nuclear bomb? Explain your answer.

46. Can nuclear bombs be made from reactor-grade uranium? Explain your answer.

47. Can a nuclear bomb be made from reactor-grade plutonium?

48. How does a breeder reactor produce more fuel than it consumes?

49. What are some of the disadvantages of breeder reactors?

50. List some possible advantages of a nuclear fusion reactor over a fission reactor. What is plasma?

51. List some possible problems with nuclear fusion reactors.

52. Give nuclear equations showing how uranium-238 is converted to fissile plutonium-239.

53. Give nuclear equations showing how thorium-232 is converted to fissile uranium-233.

54. Give the nuclear equation that shows how deuterium and tritium fuse to form helium and a neutron.

55. The fission of one mole of uranium-235 can be represented by:

$$^{235}_{92}U + ^{1}_{0}n \longrightarrow 2\,^{1}_{0}n + ^{139}_{54}Xe + ^{95}_{38}Sr + 1.8 \times 10^{10}\ kJ$$

See the equation in Exercise 15.3A for the combustion of methane. How many moles of methane must be burned to generate the energy produced by fission of 1.00 mole (235 grams) of uranium-235? What is the mass of this amount of methane in metric tons (1 t = 1000 kg)?

56. Coal is mostly carbon. Burning one mole (12.0 grams) of carbon to form carbon dioxide releases 393 kJ of energy. Review Problem 55 and determine how many moles of carbon must be burned to generate the energy produced by fission of 1.00 mole (235 grams) of uranium-235. What is the mass of this amount of carbon in metric tons?

Renewable Energy Sources

57. What is a photovoltaic cell?

58. What are some problems associated with the use of solar energy?

59. The average power demand of a U.S. household with non-electric cooking is 1.2 kW. Solar cells can produce about 100 W/m^2. What area of solar cells is needed to meet this demand? What are the limits of this application?

60. Large wind turbines can produce 1.5 MW of electrical power. How many of the homes described in Problem 59 could be powered by one big wind turbine? What are the limits of this application?

61. What is plant biomass?

62. List some advantages and disadvantages of the use of biomass as a source of energy.

63. List two ways that fuel cells differ from electrochemical cells.

64. List the advantages, disadvantages, and limitations of each of the following as an energy source.
    a. wind power   **b.** geothermal power
    c. power from tides   **d.** hydroelectric power

Synthetic and Converted Fuels

65. Give the chemical equations for the basic processes by which coal (carbon) is converted to **(a)** methane and **(b)** carbon monoxide and hydrogen.

66. Give the chemical equations for **(a)** the conversion of carbon monoxide and hydrogen to methanol and **(b)** the reaction that occurs in a hydrogen–oxygen fuel cell.

ADDITIONAL PROBLEMS

67. A cold pack (photo) works by the dissolving of ammonium nitrate in water. Cold packs are carried by athletic trainers when transporting ice is not possible. Is this process endothermic or exothermic? Explain.

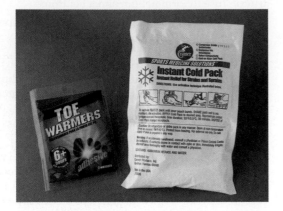

68. A hot pack foot warmer (photo) contains iron powder, water, salt, activated carbon, and vermiculite. It is activated by exposing the contents to air. Hot packs are used by hunters and other outdoor sports enthusiasts and workers where access to heaters or fires is not possible. Is this process endothermic or exothermic? Explain.

69. A woman uses 1125 kcal in running 10.0 km in 39 min 18 s. Calculate her average power output in watts (1.000 kcal = 4184 J).

70. The energy requirements of the human brain are about 20% of the total body metabolism. Calculate the power consumption (in watts) of your brain if you use 2175 kcal of energy per day.

71. What mass of carbon dioxide is formed by the combustion of 1250 kg of coal that is 51.2% carbon?

72. Heats of combustion of three gaseous fuels, in kilojoules per mole, are given below. Which yields the most energy **(a)** per kilogram and **(b)** per liter when the volume is measured under the same conditions of temperature and pressure?

Hydrogen (H_2), 286.6 kJ

Isobutane [$CH_3CH(CH_3)_2$], 2868 kJ

Neopentane [$CH_3C(CH_3)_3$], 3515 kJ

73. Indicate a natural process or processes by which carbon atoms are **(a)** removed from the atmosphere; **(b)** returned to the atmosphere; **(c)** effectively withdrawn from the carbon cycle.

74. The United States leads the major developed countries in per capita emissions of $CO_2(g)$ with 19.0 t per person per year (1 t = 1000 kg). What mass, in metric tons, of each of the following fuels would yield this quantity of CO_2?
  a. CH_4 **b.** C_8H_{18}
  c. coal that is 94.1% C by mass

75. A large coal-fired electric plant burns 2500 metric tons of coal per day. The coal contains 0.65% S by mass. Assume that all the sulfur is converted to SO_2. What mass of SO_2 is formed? If a thermal inversion traps all this SO_2 in a parcel of air that is 45 km × 60 km × 0.40 km, will the level of SO_2 in the air exceed the primary national air quality standard of 365 μg SO_2/m^3 air?

76. The contribution of the combustion of various fuels to the buildup of CO_2 in the atmosphere can be assessed in different ways. One way relates the mass of CO_2 formed to the mass of fuel burned; another relates the mass of CO_2 to the quantity of heat evolved in the combustion. Which of the three fuels C(graphite), $CH_4(g)$, or $C_4H_{10}(g)$ produces the smallest mass of CO_2 **(a)** per gram of fuel and **(b)** per kilojoule of heat evolved? The heat released per mole of the three substances is C(graphite), 393.5 kJ; $CH_4(g)$, 803 kJ; and $C_4H_{10}(g)$, 2877 kJ.

77. Estimates of recoverable energy are notoriously uncertain, but some scientists estimate that a total of only 1.05×10^7 terawatt-hours (TWh) of recoverable energy is stored in the Earth (in the form of oil, coal, natural gas, and uranium). Worldwide use is about 1.23×10^5 TWh per year. **(a)** If demand for energy remains the same as today, approximately how many years do we have before all the energy stored in the Earth is gone? **(b)** Per capita consumption in the United States (population 300,000,000) is about four times that of the rest of the world (population 6 billion). If all the countries throughout the world consumed energy at the same per capita rate as the United States, approximately how many years do we have before all the energy stored in the Earth is used up? **(c)** Evaluate your answers to parts (a) and (b).

78. The nitrogen in natural gas (see the figure on page 422) probably comes from air trapped with the natural gas during its formation. Air is about 80% nitrogen and 20% oxygen. Suggest a reason why oxygen does *not* make up any significant proportion of natural gas. Use the terms *oxidized* and *reduced* in your explanation.

■ COLLABORATIVE GROUP PROJECTS

Prepare a PowerPoint, poster, or other presentation (as directed by your instructor) for presentation to the class.

79. Which would you rather have in your neighborhood, a nuclear power plant or a coal-burning plant? Why? Explain your choice fully.

80. What characteristics should an ideal energy source have? Which of these characteristics do the different forms of energy have?

81. Social cost pricing takes into account all the costs of a product or activity that are external to market costs. For example, one of the factors in the cost of gasoline-based transportation is air pollution. Select three forms of energy and list the possible social costs that should be included.

82. Compare the quantity of electricity used for two appliances that are alike except for different efficiencies. Extend this comparison to CO_2 emissions from the appliances, assuming the electricity is generated by burning coal.

83. Compare qualitatively the efficiency of heating with natural gas versus electricity produced from burning natural gas at a power plant.

84. Discuss advantages and limitations of using solar panels on the roof of a house versus photovoltaic cells in a "solar farm."

85. Which of the following is the best fuel for heating your home? What problems with supply, use, and waste products are involved in each case?
    a. natural gas
    b. electricity
    c. coal
    d. fuel oil

86. There was an accident at a uranium processing plant in Tokai, Japan, on September 30, 1999. Using your favorite search engine, find out what happened and compare this accident with the ones at Chernobyl and Three Mile Island. How could these accidents have been avoided? Do these incidents prove that nuclear power plants should be phased out? Why or why not?

87. The use of appliances and heating equipment that employ alternative forms of energy (such as solar power) can help in individual energy conservation. Using Internet sources, find such equipment. Do companies selling these products appear to be doing so for mostly idealistic—or mainly commercial—reasons? (A good place to begin is Solar Access.com, which features solar industry directories.)

Martin Chalfie of Columbia University, Roger Y. Tsien of the University of California, San Diego, and Osamu Shimomura of the Marine Biological Laboratory and Boston University Medical School shared the 2008 Nobel Prize in Chemistry for research involving green fluorescent protein (GFP). This glowing jellyfish protein revolutionized the ability to study disease in living organisms. Researchers worldwide use GFP to track the development of brain cells, the growth of tumors and the spread of cancer cells. More whimsical uses of GFP include these genetically modified Medaka fish. The potential of biochemical manipulation to ease human suffering is enormous, but many problems have yet to be faced.

BIOCHEMISTRY

A Molecular View of Life

The human body is an incredible chemical factory, far more complex than any industrial plant. To stay in good condition and perform its varied tasks, it needs many specific chemical compounds, and it manufactures most of them in an exquisitely organized network of chemical production lines.

Every minute of every day, thousands of chemical reactions take place within each of the 100 trillion tiny cells in your body. The study of these reactions and the chemicals they produce is called *biochemistry*.

16.1 The Living Cell

Biochemistry is the chemistry of living things and life processes. The structural unit of all living things is the *cell*. Every cell is enclosed in a *cell membrane* through which it gains nutrients and gets rid of wastes. Plant cells (Figure 16.1) also have walls made of cellulose. Animal cells (Figure 16.2) do not have cell walls. Cells have a variety of interior structures that serve a multiplicity of functions. We can consider only a few of them here.

The largest interior structure is usually the *cell nucleus,* which contains the material that controls heredity. Protein synthesis takes place in the *ribosomes.* The *mitochondria* are the cell "batteries" where energy is produced. Plant cells (but not animal cells) also contain *chloroplasts* in which energy from the Sun is converted to chemical energy, which is stored in the plant in the form of carbohydrates.

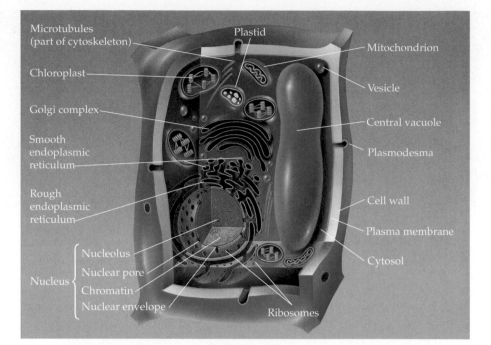

▶ **Figure 16.1** An idealized plant cell. Not all the structures shown here occur in every type of plant cell.

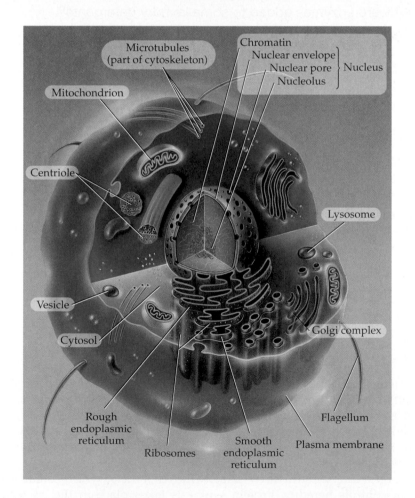

▶ **Figure 16.2** An animal cell. The entire range of structures shown here seldom occurs in a single cell, and we discuss only a few of them in this text. Each kind of plant or animal tissue has cells specific to the function of that tissue. Muscle cells differ from nerve cells, nerve cells differ from red blood cells, and so on.

Energy in Biological Systems

Life requires energy. Living cells are inherently unstable and only a continued input of energy keeps them from falling apart. Living organisms are restricted to using certain forms of energy. Supplying a plant with heat energy by holding it in a flame will do little to prolong its life. On the other hand, a green plant is uniquely able to tap sunlight, the richest source of energy on Earth. Chloroplasts in green plant cells capture the radiant energy of the Sun and convert it to chemical energy, which is then stored in carbohydrate molecules. The photosynthesis of glucose is represented by the equation

$$6\,CO_2 + 6\,H_2O \longrightarrow C_6H_{12}O_6 + 6\,O_2$$

Plant cells can also convert these carbohydrate molecules to fat molecules and, with the proper inorganic nutrients, to protein molecules.

Animals cannot directly use the energy of sunlight. They must get their energy by eating plants or by eating other animals that eat plants. Animals obtain energy from three major types of foods: carbohydrates, fats, and proteins.

Once digested and transported to a cell, a food molecule can be used as a building block to make new cell parts or to repair old ones, or it can be "burned" for energy. The entire series of coordinated chemical reactions that keeps cells alive is called **metabolism**. In general, metabolic reactions are divided into two classes. The degrading of molecules to provide energy is called **catabolism**, and the process of building up or synthesizing the molecules of living systems is termed **anabolism**.

Carbohydrates, fats, and proteins as foods are discussed in Chapter 17. In this chapter, we deal mainly with the synthesis and structure of these vital materials.

Self-Assessment Questions

1. The three major types of foods from which we obtain energy are
   a. amino acids, proteins, and carbohydrates
   b. carbohydrates, fats, and oils
   c. carbohydrates, fats, and proteins
   d. proteins, nucleic acids, and oils

2. The process of building up of molecules of living organisms is called
   a. anabolism
   b. catabolism
   c. polymerization
   d. transcription

3. The overall set of chemical reactions that keeps cells alive is called
   a. anabolism
   b. mitosis
   c. metabolism
   d. proteolysis

Answers: 1, c; 2, a; 3, c

16.2 Carbohydrates: A Storehouse of Energy

It is difficult to give a simple formal definition of *carbohydrates*. Chemically, **carbohydrates** are polyhydroxy aldehydes or ketones or compounds that can be hydrolyzed (split by water) to form such aldehydes and ketones. Composed of the elements carbon, hydrogen, and oxygen, carbohydrates include sugars, starches, and cellulose. Usually, the atoms of these elements are present in a ratio expressed by the formula $C_x(H_2O)_y$. Glucose, a simple sugar, has the formula $C_6H_{12}O_6$, which could be written as $C_6(H_2O)_6$. The term *carbohydrate* is derived from formulas such as these. We use the name even though carbohydrates are not actually hydrates of carbon.

Some Simple Sugars

Sugars are sweet-tasting carbohydrates. The simplest of these are **monosaccharides**, carbohydrates that cannot be further hydrolyzed. Three familiar dietary monosaccharides are shown in Figure 16.3. They are glucose (also called dextrose), fructose (fruit sugar), and galactose (a component of lactose, the sugar in milk). Glucose and galactose are **aldoses**, monosaccharides with an aldehyde functional group (see the functional groups in Table 9.4). Fructose is a **ketose**, a monosaccharide with a ketone functional group. Pure fructose is about 25% sweeter than table sugar (sucrose, page 457); glucose is about 25% less sweet than table sugar.

▶ **Figure 16.3** Three common dietary monosaccharides. All have hydroxyl groups. Glucose and galactose have aldehyde functional groups, and fructose has a ketone group. Glucose and galactose differ only in the arrangement of the H and OH on the fourth carbon from the top. The living cell uses glucose as a source of energy. Fructose is fruit sugar, and galactose is a component of milk sugar.

QUESTION: What is the molecular formula of each of the three monosaccharides? How are the three compounds related?

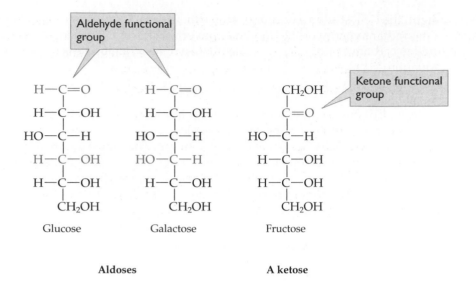

1. What is high fructose corn syrup and why is it used in so many products?
High fructose corn syrup (HFCS) has had some of its glucose enzymatically converted to fructose. This makes it significantly sweeter than ordinary corn syrup, because fructose is sweeter than glucose. Also, HFCS is quite cheap in the United States. Unfortunately, like table sugar, HFCS provides little more than empty calories.

CONCEPTUAL EXAMPLE 16.1 Classification of Monosaccharides

Shown below are structures of (left to right) erythrulose, used in some sunless tanning lotions; mannose, a sugar found in some fruits including cranberry; and ribose, a component of ribonucleic acid (RNA; Section 16.10).

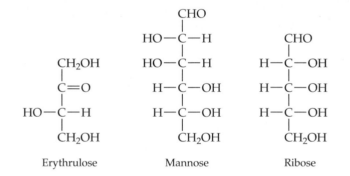

Classify erythrulose and ribose as an aldose or a ketose.

Solution
Erythrulose is a ketose. The carbonyl (C=O) group is on the second carbon atom from the top; it is between two other carbon atoms, a situation that defines a ketone. Ribose is an aldose. The carbonyl group is on an "end" carbon atom, as it is in all aldehydes.

■ **EXERCISE 16.1**
How does the structure of mannose differ from that of glucose?

These monosaccharides are represented in Figure 16.3 as open-chain compounds to show the aldehyde or ketone functional groups. However, these sugars exist in solution mainly as cyclic molecules (Figure 16.4).

▶ **Figure 16.4** Cyclic structures for glucose, galactose, and fructose. A corner with no letter represents a carbon atom. Glucose and galactose are represented as six-membered rings. They differ only in the arrangement of the H and OH (green) on the fourth carbon. Fructose is shown as a five-membered ring. Some sugars exist in more than one cyclic form. For simplicity, we show only one form of each here.

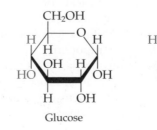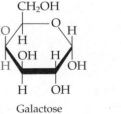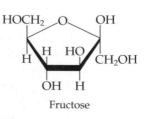

Sucrose and lactose are examples of **disaccharides**, carbohydrates consisting of molecules that can be hydrolyzed to two monosaccharide units (Figure 16.5). Sucrose is split into glucose and fructose. Hydrolysis of lactose gives glucose and galactose.

$$\text{Sucrose} + H_2O \longrightarrow \text{Glucose} + \text{Fructose}$$
$$\text{Lactose} + H_2O \longrightarrow \text{Glucose} + \text{Galactose}$$

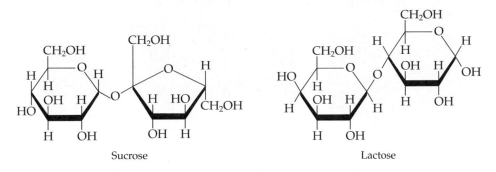

Sucrose   Lactose

◀ **Figure 16.5** Sucrose and lactose are disaccharides. On hydrolysis, sucrose produces glucose and fructose, whereas lactose yields glucose and galactose. Sucrose is cane or beet sugar and lactose is milk sugar.

QUESTION: What is the molecular formula of each of these disaccharides? How are the two compounds related? Label the two monosaccharide units in each as fructose, galactose, or glucose.

Polysaccharides: Starch and Cellulose

Polysaccharides are composed of large molecules that yield many monosaccharide units on hydrolysis. Polysaccharides include starches, which comprise the main energy storage system of many plants, and cellulose, which is the structural material of plants. Figure 16.6 shows short segments of starch and cellulose molecules. Notice that both are polymers of glucose. Starch molecules generally have from 100 to about 6000 glucose units. Cellulose molecules are composed of 1800–3000 or more glucose units.

A crucial structural difference between starch and cellulose is in the way the glucose units are hooked together. Consider starch first. With the CH_2OH at the top as a reference, the oxygen atom joining the glucose units is pointed *down*. This arrangement is called an *alpha linkage*. In cellulose, again with the CH_2OH as a point of reference, the oxygen atom connecting the glucose segments is pointed *up*, an arrangement called a *beta linkage*. This subtle but important difference in linkages determines whether the materials can be digested by humans (Chapter 17). The different linkages also result in different three-dimensional forms for cellulose and starch. For example, cellulose in the cell walls of plants is arranged in *fibrils*, bundles of parallel chains. As shown in Figure 16.7a, fibrils in turn lie parallel to each other in each layer of the cell wall. In alternate layers, the fibrils are perpendicular, an arrangement that imparts great strength to the wall.

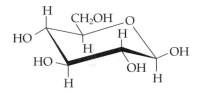

▲ Although we have represented these cyclic monosaccharides as flat hexagons, they are actually three-dimensional. Most assume a shape called a *chair conformation* because it outlines a structure that somewhat resembles a reclining chair.

(a) Segment of starch molecule

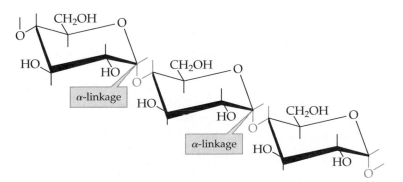

(b) Segment of cellulose molecule

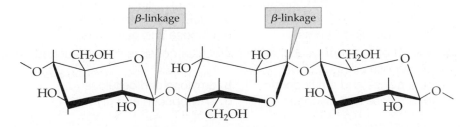

◀ **Figure 16.6** Both starch (a) and cellulose (b) are polymers of glucose. They differ in that the glucose units are joined by alpha linkages (blue) in starch and by beta linkages (red) in cellulose.

QUESTION: The formulas for most polymers can be written in condensed form (page 263). Write a condensed formula for starch. How would the condensed formula for cellulose differ from that of starch?

▶ **Figure 16.7** Cellulose molecules form fibers whereas starch (glycogen) forms granules. Electron micrographs of (a) the cell wall of an alga, made up of successive layers of cellulose fibers in parallel arrangement, and (b) glycogen granules (red) in a human liver cell.

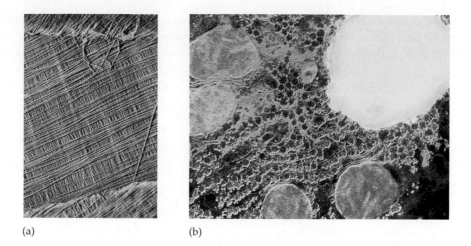

(a)   (b)

> ## 2. Why can some animals eat grass while we can't?
> Although both cellulose and starch are glucose polymers, humans do not have the enzyme to break the beta linkages in cellulose. The digestive systems of cows, horses, and deer contain bacteria that can break these linkages, so that ruminants can digest the cellulose in grass.

There are two kinds of plant starch. One, called *amylose,* has glucose units joined in a continuous chain like beads on a string. The other kind, *amylopectin,* has branched chains of glucose units. These starches are perhaps best represented schematically, as in Figure 16.8, where each glucose unit is represented by the abbreviation Glc.

Animal starch is called *glycogen.* Like amylopectin, it is composed of branched chains of glucose units. In contrast to cellulose, we see in Figure 16.7b that glycogen in muscle and liver tissue is arranged in granules, clusters of small particles. Plant starch, on the other hand, forms large granules. Granules of plant starch rupture in boiling water to form a paste, and on cooling, the paste gels. Potatoes and cereal grains form this type of starchy broth. All forms of starch are hydrolyzed to glucose during digestion.

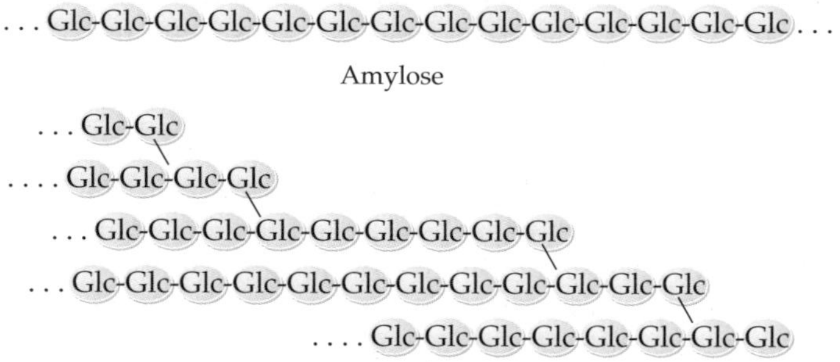

. . . Glc-Glc-Glc-Glc-Glc-Glc-Glc-Glc-Glc-Glc-Glc-Glc-Glc-Glc . . .

Amylose

. . . Glc-Glc

. . . . Glc-Glc-Glc-Glc

. . . Glc-Glc-Glc-Glc-Glc-Glc-Glc-Glc-Glc

. . . Glc-Glc-Glc-Glc-Glc-Glc-Glc-Glc-Glc-Glc-Glc-Glc

. . . . Glc-Glc-Glc-Glc-Glc-Glc-Glc-Glc

Amylopectin

▶ **Figure 16.8** Schematic representations of amylose and amylopectin. Glc stands for a glucose unit. The structure of glycogen is similar to that of amylopectin.

When cornstarch is heated at about 400 °F for an hour, the starch polymers are broken down into shorter chains to form a product called *dextrin* that has properties intermediate between sugars and starches. Mixed with water, dextrin makes a slightly thick, sticky mixture. Starch and dextrin adhesives make up about 58% of the U.S. packaging market.

Self-Assessment Questions

1. A monosaccharide with an aldehyde group is called
   a. an aldol   **b.** an aldose   **c.** aldicarb   **d.** aldosterol

2. A hexose is a
   a. chain of six monosaccharide units
   b. hydrocarbon with the sixth C atom in a CHO group
   c. monosaccharide with six C atoms
   d. peptide with six monosaccharide units

3. When most monosaccharides dissolve in water, they form
   a. disaccharides   **b.** ring structures
   c. polypeptides   **d.** polysaccharides

4. The two monosaccharide units that make up a sucrose molecule are
   a. both glucose   **b.** fructose and galactose
   c. glucose and fructose   **d.** glucose and galactose

5. Which of these polysaccharide molecules consists of unbranched chains of glucose units?
   a. amylose and amylopectin   **b.** amylose and cellulose
   c. amylose and glycogen   **d.** amylopectin and glycogen

6. Carbohydrates are stored in the liver and muscle tissue as

  a. amylose   **b.** amylopectin   **c.** cellulose   **d.** glycogen

16.3 Fats and Other Lipids

Fats are the predominant forms of a class of compounds called *lipids*. These substances are not defined by functional groups, as are most other families of organic compounds. Rather, lipids have common solubility properties. A **lipid** is a cellular constituent that is insoluble in water but soluble in organic solvents of low polarity such as hexane, diethyl ether, or chloroform. In addition to fats, the lipid family includes **fatty acids** (long-chain carboxylic acids), steroids such as cholesterol and sex hormones (Chapter 18), fat-soluble vitamins (Chapter 17), and other substances. Figure 16.9 shows three representations of palmitic acid, a typical fatty acid.

$$CH_3CH_2CH_2CH_2CH_2CH_2CH_2CH_2CH_2CH_2CH_2CH_2CH_2CH_2CH_2COOH$$
(a)

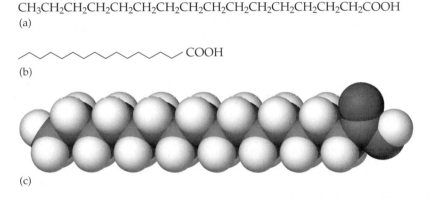

(b)

(c)

◀ **Figure 16.9** Three representations of palmitic acid: (a) Condensed structural formula. (b) Line-angle formula, in which the lines denote bonds and each bend and end of a line represents a carbon atom. (c) Space-filling model.

A **fat** is an ester of fatty acids and the trihydroxy alcohol glycerol (Figure 16.10). A fat has three fatty acid chains joined to glycerol through ester linkages (which is why fats are often called *triglycerides* or triacylglycerols). Related compounds are also classified according to the number of fatty acid chains they contain. A *monoglyceride* has one fatty acid chain joined to glycerol, and a *diglyceride* has two.

Naturally occurring fatty acids nearly always have an even number of carbon atoms. Representative ones are listed in Table 16.1. Animal fats are generally rich in

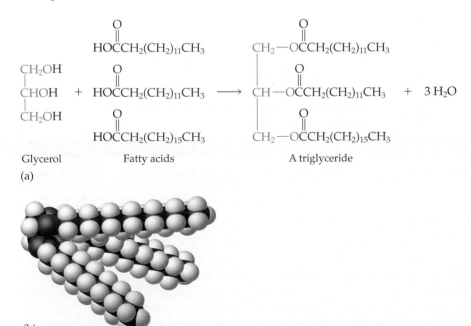

(a)

(b)

◀ **Figure 16.10** Triglycerides (triacylglycerols) are esters in which the trihydroxy (three OH groups) alcohol glycerol is esterified with three fatty acid groups. (a) The equation for the formation of a triglyceride. (b) Space-filling model of a triglyceride.

saturated fatty acids (fatty acids with no carbon-to-carbon double bonds) and have a smaller proportion of unsaturated fatty acids. At room temperature, most animal fats are solids. Oils (liquid fats) are obtained principally from vegetable sources. Oils typically have a higher proportion of unsaturated fatty acid units than do solid fats.

Table 16.1 Some Fatty Acids in Natural Fats

Number of Carbon Atoms	Condensed Structural Formula	Name	Source
4	$CH_3CH_2CH_2COOH$	Butyric acid	Butter
6	$CH_3(CH_2)_4COOH$	Caproic acid	Butter
8	$CH_3(CH_2)_6COOH$	Caprylic acid	Coconut oil
10	$CH_3(CH_2)_8COOH$	Capric acid	Coconut oil
12	$CH_3(CH_2)_{10}COOH$	Lauric acid	Palm kernel oil
14	$CH_3(CH_2)_{12}COOH$	Myristic acid	Oil of nutmeg
16	$CH_3(CH_2)_{14}COOH$	Palmitic acid	Palm oil
18	$CH_3(CH_2)_{16}COOH$	Stearic acid	Beef tallow
18	$CH_3(CH_2)_7CH{=}CH(CH_2)_7COOH$	Oleic acid	Olive oil
18	$CH_3(CH_2)_4CH{=}CHCH_2CH{=}CH(CH_2)_7COOH$	Linoleic acid	Soybean oil
18	$CH_3CH_2(CH{=}CHCH_2)_3(CH_2)_6COOH$	Linolenic acid	Fish oils
20	$CH_3(CH_2)_4(CH{=}CHCH_2)_4CH_2CH_2COOH$	Arachidonic acid	Liver

Fats are often classified according to the degree of unsaturation of the fatty acids they incorporate. A *saturated fatty acid* contains no carbon-to-carbon double bonds, a *monounsaturated fatty acid* has one carbon-to-carbon double bond per molecule, and a *polyunsaturated fatty acid* molecule has two or more carbon-to-carbon double bonds. A *saturated fat* contains a high proportion of saturated fatty acids; these fat molecules have relatively few carbon-to-carbon double bonds. A *polyunsaturated fat* (oil) incorporates mainly unsaturated fatty acids; these fat molecules have many double bonds.

The iodine number usually expresses the degree of unsaturation of a fat or oil. The **iodine number** is the mass in grams of iodine that is consumed by 100 g of fat or oil. Iodine, like other halogens, undergoes an addition reaction with carbon-to-carbon double bonds.

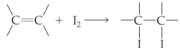

The more double bonds a fat contains, the more iodine is required for the addition reaction; thus, a high iodine number means a high degree of unsaturation. Representative iodine numbers are listed in Table 16.2. Note the generally lower values for animal fats (butter, tallow, and lard) compared with those for vegetable oils. Coconut oil, which is highly saturated, and fish oils, which are relatively unsaturated, are notable exceptions to the general rule.

▲ Many salad dressings are made of an oil and vinegar. The one shown here has olive oil floating on top of vinegar, an aqueous solution of acetic acid.

Table 16.2 Typical Iodine Numbers for Some Fats and Oils*

Fat or Oil	Iodine Number	Fat or Oil	Iodine Number
Coconut oil	8–10	Cottonseed oil	100–117
Butter	25–40	Corn oil	115–130
Beef tallow	30–45	Fish oils	120–180
Palm oil	37–54	Canola oil	125–135
Lard	45–70	Soybean oil	125–140
Olive oil	75–95	Safflower oil	130–140
Peanut oil	85–100	Sunflower oil	130–145

* Oils shown in blue are from plant sources.

Self-Assessment Questions

1. Lipids are classified based on
   a. chemical reactivity
   c. function in the body
   b. functional group
   d. solubility in nonpolar solvents

Questions 2–6 refer to the structures S, T, U, and V.

   S: $CH_3(CH_2)_{10}COOH$
   T: $CH_3(CH_2)_{16}COOH$
   U: $CH_3(CH_2)_7CH = CH(CH_2)_7COOH$
   V: $CH_3CH_2(CH = CHCH_2)_3(CH_2)_6COOH$

2. Which are saturated fatty acids?
   a. S and T   **b.** S and U   **c.** T and U   **d.** U and V

3. Which is a monounsaturated fatty acid?
   a. S   **b.** T   **c.** U   **d.** V

4. Which is a polyunsaturated fatty acid?
   a. S   **b.** T   **c.** U   **d.** V

5. Saturated fats often have a high proportion of
   a. S and T   **b.** S and U   **c.** T and U   **d.** U and V

6. Polyunsaturated fats (oils) often have a high proportion of
   a. S   **b.** T   **c.** U   **d.** V

7. The iodine number of a fat or oil is a measure of its
   a. calorie content
   c. degree of unsaturation
   b. degree of acidity
   d. digestibility

8. In general, vegetable oils differ from animal fats in that the oil molecules
   a. are polypeptides
   c. are smaller
   b. are saturated with hydrogen
   d. have more C=C double bonds

Answers: 1, d; 2, a; 3, c; 4, d; 5, a; 6, d; 7, c; 8, d

3. What makes animal fats bad for you?
The saturated fats found in most animal fats can stack closely together to form artery-clogging plaque, which contributes to heart disease. The structures of unsaturated fats do not stack neatly and do not form this plaque.

▼ **Figure 16.11** Space-filling model (a) and structural formula (b) of a short segment of a protein molecule. In the structural formula, hydrocarbon side chains (green), an acidic side chain (red), a basic side chain (blue), and a sulfur-containing side chain (amber) are highlighted. The broken lines in (b) indicate where two amino acid units join (see Section 16.5).

QUESTION: The formulas for most polymers can be written in condensed form (page 263). Those for natural proteins cannot. Explain.

PROTEINS: THE STUFF OF LIFE

Proteins are vital components of all life. No living part of the human body—or of any other organism, for that matter—is completely without protein. There is protein in blood, muscles, brain, and even tooth enamel. The smallest cellular organisms—bacteria—contain protein. Viruses, so small that they make bacteria look like giants, are little more than proteins and nucleic acids. This combination of nucleic acids and proteins is found in all cells and is the stuff of life itself.

Each type of cell makes its own kinds of proteins. Proteins serve as the structural material of animals, much as cellulose does for plants. Muscle tissue is largely protein; so are skin and hair. Silk, wool, nails, claws, feathers, horns, and hooves are proteins. All proteins contain the elements carbon, hydrogen, oxygen, and nitrogen, and most also contain sulfur. The structure of a short segment of a typical protein molecule is shown in Figure 16.11.

(a)

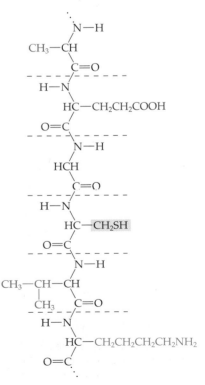
(b)

16.4 Proteins: Polymers of Amino Acids

Like starch and cellulose, *proteins* are polymers. They differ from other polymers that we have studied in that the monomer units are about 20 different amino acids (Table 16.3). The amino acids differ in their side chains. An **amino acid** has two functional groups: an amino group (—NH$_2$) and a carboxyl group (—COOH) attached to the same carbon atom (called the *alpha carbon*).

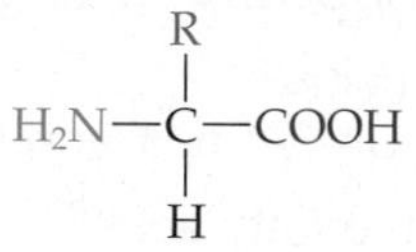

$$
\begin{array}{c}
R \\
| \\
H_2N-C-COOH \\
| \\
H
\end{array}
$$

In this formula, the R can be H (as it is in glycine, the simplest amino acid), or any of the groups shown in green in Table 16.3. The structure above shows the proper placement of these groups, but it is not really correct. Acids react with bases to form salts; the carboxyl group is acidic, and the amino group is basic. The two functional groups interact, the acid transferring a proton to the base. The resulting product is an inner salt, or **zwitterion**, a compound in which the negative charge and the positive charge are on different parts of the same molecule.

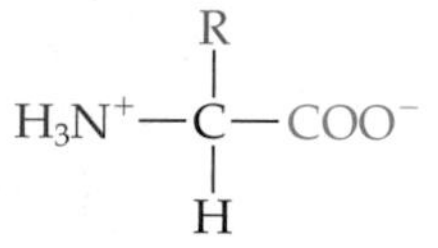

$$
\begin{array}{c}
R \\
| \\
H_3N^+-C-COO^- \\
| \\
H
\end{array}
$$

A zwitterion

Plants can synthesize proteins from carbon dioxide, water, and minerals such as nitrates (NO$_3^-$) and sulfates (SO$_4^{2-}$). Animals require proteins as one of the three major classes of foods. Humans can synthesize some of the amino acids in Table 16.3; others must be part of the proteins consumed in a normal diet. The latter are called *essential amino acids*.

Self-Assessment Questions

1. Proteins are polymers of
   a. amino acids
   b. fatty acids
   c. monoacylglycerides
   d. monosaccharides

2. The 20 amino acids that make up proteins differ mainly in their
   a. alpha carbon atoms
   b. amino groups
   c. carboxyl groups
   d. side chains

3. Which of the following is *not* an alpha amino acid?
   a. CH$_3$CH(NH$_2$)COOH
   b. (CH$_3$)$_2$CHCH(NH$_2$)COOH
   c. H$_2$NCH$_2$COOH
   d. H$_2$NCH$_2$CH$_2$COOH

4. Which of the following structures represents a zwitterion?
   a. H$_2$NCH$_2$COOH
   b. $^-$NHCH$_2$COO$^-$
   c. $^+$H$_3$NCH$_2$COOH
   d. $^+$H$_3$NCH$_2$COO$^-$

Answers: 1, a; 2, d; 3, d; 4, d

Table 16.3 The 20 Amino Acids Specified by the Genetic Code

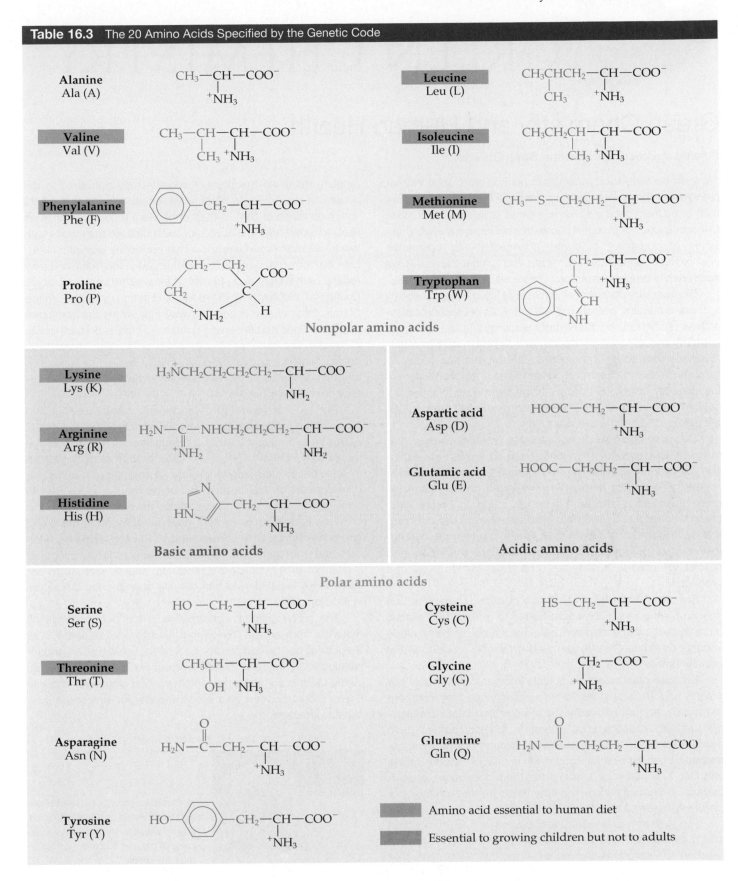

Nonpolar amino acids

Basic amino acids

Acidic amino acids

Polar amino acids

Amino acid essential to human diet

Essential to growing children but not to adults

GREEN CHEMISTRY

Green Chemistry and Human Health

Edward J. Brush, Bridgewater State College

"Why do we have toxic materials in our society?" John Warner, president of the Warner-Babcock Institute for Green Chemistry, asks in a *Chemical & Engineering News* editorial, "Unintended Consequences." Although intended to improve our lives, thousands of consumer products—pharmaceuticals, cosmetics, plastics, fuels, agricultural chemicals, and synthetic fibers—have inadvertently brought hazardous chemicals into the food chain.

Biochemistry plays a central role in all living things: in energy processes (sugars, carbohydrates), catalysis of biochemical reactions (enzymes), solubility (fatty acids and lipids), structure (proteins and other biopolymers), and heredity (nucleic acids). Given the complexity of the human body, it perhaps comes as no surprise that "unintended consequences" of chemicals like thalidomide, perchloroethylene, flame retardants, plasticizers, polychlorinated biphenyls (PCBs), and pesticides may include the disruption of metabolic processes.

Generally chemists do not intend to create hazardous materials (unless they are intentionally making a poison or explosive, for example), and many now apply the principles of green chemistry to predict and avoid unintended consequences. A green approach starts at the design stage to create safer chemical products by reducing or eliminating the use and generation of hazardous substances. Green chemistry is cost-effective and energy efficient, uses benign chemicals and processes, and reduces wastes throughout the life cycle of a product.

Chapter 10 describes synthetic plastics, some of which contain plasticizers, substances added to make a rigid plastic more flexible. Unintended consequences of plasticizers include impacts on human health, as has been widely reported in the popular press.

Because plasticizers are soluble in organic solvents of low polarity and are almost insoluble in water, when they leach out of plastic materials and enter the environment, many plasticizers are not biodegradable, persist in the environment, and accumulate in the food chain in fatty tissues. Our greatest exposure comes from eating foods high in fats, like certain meats and fish. The plasticizers then enter the bloodstream, are adsorbed, and accumulate in our fatty tissue from which they can be slowly released over time.

The phthalate plasticizers and bisphenol A (BPA) (Section 10.11) can act as *endocrine disruptors*. The endocrine system (Section 18.7) produces hormones, chemical messengers that regulate many life functions. A recent study published in the *Journal of the American Medical Association* found that adults with high levels of BPA in their urine were more than twice as likely to report having diabetes or heart disease than those with low levels. Animal studies show that endocrine disruptors confuse the hormonal system, cause changes in the brain, and may reduce birth weight and survival of fetuses. More than 2 billion pounds of bisphenol A (BPA) is used annually in the United States, most of which finds its way into numerous consumer products. Food and beverage containers are the primary routes of human exposure, especially synthetic plastics made from polycarbonate plastics (Section 10.7). Bisphenol A is also found in dental sealants, reusable water bottles and food containers, and baby bottles.

Are the low amounts of endocrine disruptors found in food and beverage containers truly harmful to human health? At this time, the Food and Drug Administration (FDA) finds that the current level of exposure to BPA through food containers is safe, but because endocrine disruptors affect growth and development of vital organs, concerns remain about BPA in baby bottles and how it may affect developing infants and children. Endocrine disruptors have been found in the blood and urine of most U.S. children and adults and have also been detected in breast milk. The U.S. Congress has investigated the use of BPA in children's products, and several states are considering legislation to ban the use of endocrine disruptors in food products.

The green chemistry alternative is the design of plastic materials from benign, renewable feedstocks (Section 10.12). Bioplastics can be designed to be flexible, avoiding the need for plasticizers. They have been incorporated in food packaging, eating utensils and drinking containers, and clothing and textile fibers. Bioplastics are also biodegradable, completing a sustainable life cycle.

◄ Plasticizers added to polyurethane skateboard wheels provide flexibility, but those plasticizers are under scrutiny because of potential effects on the human endocrine system.

16.5 The Peptide Bond: Peptides and Proteins

The human body contains tens of thousands of different proteins. Each of us has a tailor-made set. Proteins are polyamides. The amide linkage is called a **peptide bond** (shaded part of the structure below) when it joins two amino acid units.

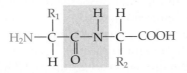

Note that there is still a reactive amino group on the left and a carboxyl group on the right. These groups can react further to join more amino acid units. This process can continue until thousands of units have joined to form a giant molecule—a polymer called a *protein*. We will examine the structure of proteins in the next section, but first let us look at some molecules called *peptides* that have only a few amino acid units.

When only two amino acids are joined, the product is a *dipeptide*.

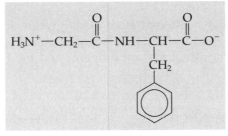

Glycylphenylalanine
(a dipeptide)

Three amino acids combine to form a *tripeptide*.

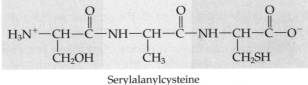

Serylalanylcysteine
(a tripeptide)

In describing peptides and proteins, scientists find it simpler to indicate the amino acids in a chain by using the abbreviations given in Table 16.3. The three-letter abbreviations are more common, but the alternate single-letter set is also used. Thus, glycylphenylalanine is written as either Gly-Phe or G-F, and serylalanylcysteine is written as Ser-Ala-Cys or S-A-C. Example 16.2 further illustrates this naming.

EXAMPLE 16.2   Names of Peptides

Give the **(a)** alternate designation using one-letter abbreviations and the **(b)** full name for the pentapeptide Met-Gly-Phe-Ala-Cys. You may use Table 16.3.

Solution
a. M-G-F-A-C
b. The endings of the names for all except the last amino acid are changed from *-ine* to *-yl*. The name is, therefore, methionylglycylphenylalanylalanylcysteine.

■ **EXERCISE 16.2A**
Give the **(a)** alternate designation using one-letter abbreviations and the **(b)** full name for the tetrapeptide His-Pro-Val-Ala.

■ **EXERCISE 16.2B**
Use **(a)** three-letter abbreviations and **(b)** one-letter abbreviations to indicate the amino acids in the peptide threonylglycylalanylalanylleucine.

Some protein molecules are enormous, with molecular weights in the tens of thousands. The molecular formula for hemoglobin, the oxygen-carrying protein in red blood cells, is $C_{3032}H_{4816}O_{780}N_{780}S_8Fe_4$, with a molar mass of 64,450 g/mol. Although they are huge compared with ordinary molecules, a billion average-sized protein molecules could still fit on the head of a pin.

A molecule with more than 10 amino acid units is often simply called a **polypeptide**. When the molecular weight of a polypeptide exceeds about 10,000, it is called a **protein**. The distinctions are arbitrary and are not always precisely applied.

The Sequence of Amino Acids

For peptides and proteins to function properly, it is not enough that they incorporate certain *amounts* of specific amino acids. The order or *sequence* in which the amino acids are connected is also of critical importance. The sequence is written starting at the end with a free amino group (—NH₂), called the *N-terminal,* and continuing to the end with a free carboxyl group (—COOH), called the *C-terminal.*

Glycylalanine is, therefore, different from alanylglycine. The sequence for glycylalanine is written Gly-Ala, and that for alanylglycine is Ala-Gly. Although the difference seems minor, the structures are different and the two substances behave differently in the body.

$$H_3N^+CH_2CO—NHCHCOO^- \qquad H_3N^+CHCO—NHCH_2COO^-$$
$$\qquad\qquad\quad CH_3 \qquad\qquad\qquad\qquad CH_3$$

Glycylalanine   Alanylglycine

As the length of a peptide chain increases, the number of possible sequential variations becomes enormous. Just as we can make millions of different words with our 26-letter English alphabet, we can make millions of different proteins with 20 amino acids. And just as we can write gibberish with the English alphabet, we can make nonfunctioning proteins by putting together the wrong sequence of amino acids.

EXAMPLE 16.3   Numbers of Peptides

How many different tripeptides can be made from the three amino acids methionine, valine, and phenylalanine, each used once? Write them, using the three-letter designations in Table 16.3.

Solution

Write out the various possibilities. Two begin with Met, two start with Val, and two with Phe:

Met-Val-Phe	Val-Met-Phe	Phe-Val-Met
Met-Phe-Val	Val-Phe-Met	Phe-Met-Val

There are six possible tripeptides.

■ **EXERCISE 16.3A**

How many different tripeptides can be made from two methionine units and one valine unit?

■ **EXERCISE 16.3B**

How many different tetrapeptides can be made from one molecule each of the four amino acids methionine, valine, tyrosine, and phenylalanine?

Although the correct sequence is ordinarily of utmost importance, it is not always absolutely required. Just as you can sometimes make sense of incorrectly spelled English words, a protein with a small percentage of "incorrect" amino acids may continue to function. It may not function as well, however. And sometimes a seemingly minor change can have a disastrous effect. Some people have hemoglobin with one incorrect amino acid unit out of about 300. That "minor" error is responsible for sickle cell anemia, an inherited condition that ordinarily proves fatal.

Self-Assessment Questions

1. In a dipeptide, two amino acids are joined through a(n)
   a. amide linkage   **b.** dipolar interaction
   c. ester linkage   **d.** hydrogen bond

2. The C-terminal amino acid of the peptide Met-Ile-Val-Glu-Cys-Tyr-Gln-Trp-Asp is
   a. Asp   **b.** Cys   **c.** Gln   **d.** Met

3. How many different tripeptides can be formed from alanine, valine, and lysine, when each appears only once in each product?
   a. 1   **b.** 3   **c.** 6   **d.** 9

4. The tripeptides represented as Ala-Val-Lys and Lys-Val-Ala are
   a. different only in secondary structure   **b.** different in molar mass
   c. exactly the same   **d.** two different peptides

Answers: 1, a; 2, a; 3, c; 4, d

16.6 Structure of Proteins

The structure of proteins has four organizational levels:

- **Primary structure:** Amino acids are linked by peptide bonds to form polypeptide chains. The primary structure of a protein molecule is simply the order of its amino acids. By convention, this order is written from the amino end (N-terminal) to the carboxyl end (C-terminal).

- **Secondary structure:** Polypeptide chains can fold into regular structures such as the alpha helix and the beta sheet (described in this section).

- **Tertiary structure:** Protein folds yield spatial relationships of amino acid units that are relatively far apart in the protein chain.

- **Quaternary structure:** With more than one polypeptide chain, the chains can assemble into multiunit structures.

Primary Structure

To specify the primary structure, we write out the sequence of amino acids. For even a small protein molecule, this sequence can be quite long. For example, it takes about half a page, using one-letter abbreviations, to give the sequence of the 451 amino acid units in a protein called hexokinase, which aids in glucose metabolism. The primary structure of angiotensin II, a peptide that causes powerful constriction of blood vessels and is produced in the kidneys, is

Asp-Arg-Val-Tyr-Ile-His-Pro-Phe

This sequence specifies an octapeptide with aspartic acid at the N-terminal, joined to arginine, valine, tyrosine, isoleucine, histidine, and proline and ending with phenylalanine at the C-terminal.

Secondary Structure

Protein chains are held together in unique configurations. The secondary structure of a protein refers to the arrangement of chains about an axis. Common arrangements are a *pleated sheet*, as in silk, and a *helix*, as in wool.

▶ **Figure 16.12** Beta pleated sheet conformation of protein chains. (a) Ball-and-stick model. (b) Model emphasizing the pleats. The side chains extend above or below the sheet and alternate along the chain. The protein chains are held together by interchain hydrogen bonds.

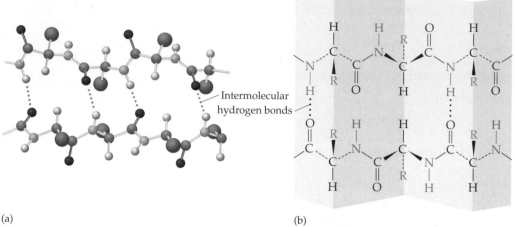

(a)

(b)

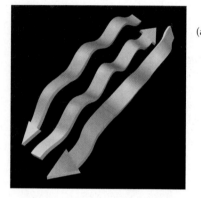

▲ The protein strands in a pleated sheet are often represented as ribbons.

In the pleated sheet conformation (Figure 16.12), protein chains exist in an extended zigzag arrangement. The molecules are stacked in extended arrays, with hydrogen bonds holding adjacent chains together. The appearance gives this type of secondary structure its name, the **beta pleated sheet** arrangement. This structure, with its multitude of hydrogen bonds, makes silk strong and flexible.

The protein molecules in wool, hair, and muscle contain large segments arranged in the form of a right-handed helix, or **alpha helix** (Figure 16.13). Each turn of the helix requires 3.6 amino acid units. The N—H groups in one turn form hydrogen bonds to carbonyl (C=O) groups in the next turn. These helices wrap around one another in threes or sevens like strands of a rope. Unlike silk, wool can be stretched, much as we can stretch a spring by pulling the coils apart.

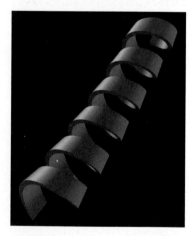

▲ The protein strands in an alpha helix are often represented as coiled ribbons.

▶ **Figure 16.13** Two representations of the alpha-helical conformation of a protein chain. (a) Intrachain hydrogen bonding between turns of the helix is shown in the ball-and-stick model. (b) The skeletal representation best shows the helix.

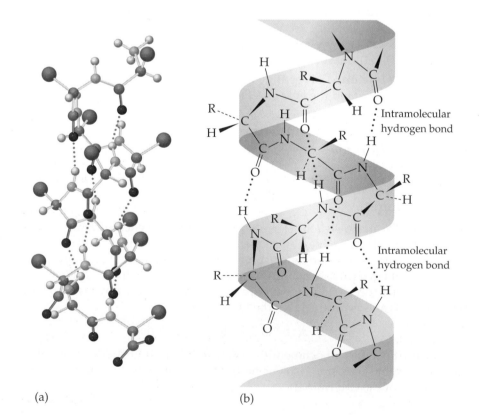

(a)

(b)

Tertiary Structure

The tertiary structure of a protein refers to the spatial relationships of amino acid units that are relatively far apart in the protein chain. In describing tertiary structure, we frequently talk about how the molecule is folded. An example is the protein chain in **globular proteins**. Figure 16.14 shows the structure of myoglobin, which is folded into a compact, spherical shape.

Quaternary Structure

Quaternary structures exist only if there is more than one polypeptide chain, in which case there can be an aggregate of subunits. Hemoglobin is the most familiar example. A single hemoglobin molecule contains four polypeptide units, and each unit is roughly comparable to a myoglobin molecule. The four units are arranged in a specific pattern (Figure 16.15). When we describe the *quaternary structure* of hemoglobin, we describe the way the four units are stacked in the hemoglobin molecule.

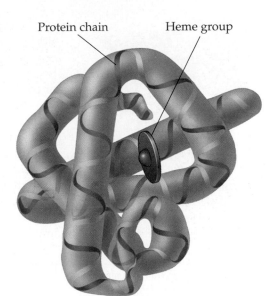

Protein chain   Heme group

▲ **Figure 16.14** The tertiary structure of myoglobin. The protein chain is folded to form a globular structure, much as a string can be folded into a ball. The disk shape represents the heme group, which binds the oxygen carried by myoglobin (or hemoglobin).

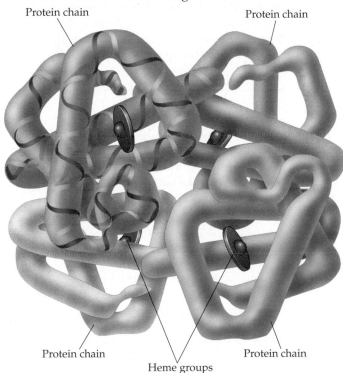

Protein chain   Protein chain

Protein chain   Protein chain

Heme groups

◀ **Figure 16.15** The quaternary structure of hemoglobin has four coiled chains, each analogous to myoglobin (Figure 16.14), stacked in a nearly tetrahedral arrangement.

Four Ways to Link Protein Chains

As we noted earlier, the primary structure of a protein is the order of the amino acids, which are held together by peptide bonds. What forces determine the secondary, tertiary, and quaternary structures? There are four kinds of forces that operate between protein chains—hydrogen bonds, ionic bonds, disulfide linkages, and dispersion forces. These forces are illustrated in Figure 16.16.

The most important hydrogen bonding in a protein involves an interaction between the atoms of one peptide bond and those of another. Thus, the amide hydrogen (N—H) of one peptide link can form a hydrogen bond to a carbonyl (C=O) oxygen located (a) some distance away on the same chain, an arrangement that occurs in the secondary structure of wool (alpha helix), or (b) on an entirely different chain, as in the secondary structure of silk (beta pleated sheet). In either case, there is usually a pattern of such interactions. Because peptide links are regularly spaced along the chain, the repetition of hydrogen bonds is also regular, both intramolecularly (in wool) and intermolecularly (in silk). Other types of hydrogen bonding can occur (such as with side chains of amino acids) but are less important.

Hydrogen bonds and dispersion forces, introduced in Chapter 6 as intermolecular forces, can be intramolecular as well, as indicated below and in Figure 16.13.

▼ **Figure 16.16** The tertiary structure of proteins is maintained by four different types of interactions.

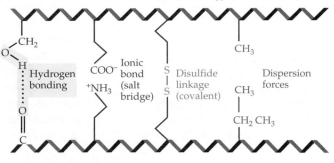

▲ **Figure 16.17** A protein chain often folds with nonpolar groups on the inside, where they are held together by dispersion forces. The outside has polar groups, visible in this space-filling model of a myoglobin molecule as oxygen atoms (red) and nitrogen atoms (blue). The carbon atoms (gray) are mainly on the inside.

Ionic bonds, sometimes called *salt bridges,* occur when an amino acid with a basic side chain appears opposite one with an acidic side chain. Proton transfer results in opposite charges, which then attract one another. These interactions can occur between relatively distant groups that happen to come in contact because of folding or coiling of a single chain. They also occur between chains.

A **disulfide linkage** is formed when two cysteine units (whether on the same chain or on two different chains) are oxidized. A disulfide linkage is a covalent bond and, thus, is much stronger than a hydrogen bond. Although far less numerous than hydrogen bonds, disulfide linkages are critically important in determining the shape of some proteins (for example, many of the proteins that act as enzymes) and the strength of others (such as the fibrous proteins in connective tissue and hair).

Dispersion forces are the only kind of forces that exist between nonpolar side chains. Recall (Section 6.3) that dispersion forces are relatively weak. These interactions can be important, however, when other types of interactions are missing or are minimized. They are made stronger by the cohesiveness of the water molecules surrounding the protein. Nonpolar side chains minimize their exposure to water by clustering together on the inside folds of the protein in close contact with one another (Figure 16.17). Dispersion forces become fairly significant in structures such as that of silk, in which a high proportion of amino acids in the protein have nonpolar side chains.

Infectious Prions: Deadly Protein

In December 2003 and June 2005 the phrase *mad cow disease* cast a shadow over the beef industry, causing uncertainty and consumer concern over beef safety. On those dates two cows in the United States were found to have the disease, properly called bovine spongiform encephalopathy (BSE).

Just what is BSE? Although the agent responsible has yet to be fully characterized and understood, it is clear that BSE is not a bacterial disease like anthrax or cholera. Nor is it a viral disease like the common cold and influenza. Instead, BSE arises from an infectious *prion,* an abnormal form of a protein. Because they are simple proteins rather than living organisms, prions can remain unchanged under conditions of considerable heat, disinfectants, and antibiotics used to destroy bacteria and viruses.

Much of the fear over BSE is unfounded. BSE is not contagious like bacterial and viral disease. It is generally spread by consumption of contaminated feed. U.S. Department of Agriculture restrictions on material that may be used in cattle feed have been quite effective. The incidence of prion-caused diseases is extremely low among people; the best-known example is *kuru* or laughing sickness, which seems to arise from cannibalistic practices in certain societies. Nonetheless, the U.S. government continues to be vigilant in avoiding any potential spread of BSE because of the near impossibility of treating the disease once contracted.

Self-Assessment Questions

1. Hydrogen bonds form between the N—H group of one amino acid and the C=O group of another, giving a protein molecule the form of an alpha helix or a beta pleated sheet. This best describes what level of protein structure?
   a. primary
   b. quaternary
   c. secondary
   d. tertiary

2. The folding of protein chains into a globular form is an example of
   a. primary structure
   b. quaternary structure
   c. secondary structure
   d. tertiary structure

3. The arrangement of protein subunits into a specific pattern such as the near tetrahedral pattern of hemoglobin is an example of
   a. primary structure
   b. quaternary structure
   c. secondary structure
   d. tertiary structure

4. The cross-links between cysteine units in protein chains of fibrous proteins such as hair are
   a. dispersion linkages
   b. disulfide linkages
   c. hydrogen bonds
   d. salt bridges

5. An acidic side chain of an amino acid unit on a protein chain reacts with the basic side chain of an amino acid unit on another chain or at a distance on the same chain to form a
   a. dispersion linkage   **b.** disulfide linkage
   c. hydrogen bond   **d.** salt bridge

6. When a nonpolar side chain of an amino acid unit in a protein chain interacts with another nonpolar side chain of an amino acid unit on another chain or at a distance on the same chain, the result is
   a. dispersion forces   **b.** a disulfide linkage
   c. a hydrogen bond   **d.** a salt bridge

7. The infectious agent of bovine spongiform encephalopathy (BSE) is
   a. herpes virus   **b.** a protein
   c. RNA molecules   **d.** viral inclusion bodies

Answers: 1, c; 2, d; 3, b; 4, b; 5, d; 6, a; 7, b

16.7 Enzymes: Exquisite Precision Machines

A highly specialized class of proteins, **enzymes**, are biological catalysts produced by cells. They have enormous catalytic power and are ordinarily highly specific, catalyzing only one reaction or a closely related group of reactions. Nearly all known enzymes are proteins.

Enzymes enable reactions to occur much more rapidly and at lower temperatures than they otherwise would. They do this by changing the reaction path. The reacting substance, called the **substrate**, attaches to an area on the enzyme called the **active site**, to form an enzyme–substrate complex. Biochemists often use the model pictured in Figure 16.18, called the *induced-fit* model, to explain enzyme action. In this model, the shapes of the substrate and the active site are not perfectly complementary, but the active site adapts to fit the substrate, much like a glove molds to fit the hand that is inserted into it. The enzyme–substrate complex decomposes to form products, and the enzyme is regenerated. We can represent the process as follows.

$$\text{Enzyme} + \text{Substrate} \longrightarrow \text{Enzyme–substrate complex} \rightleftharpoons \text{Enzyme} + \text{Products}$$

Not only must the substrate shape fit the enzyme, but the complex is also usually held together by electrical attraction. This requires that certain charged groups on the enzyme complement certain charged or partially charged groups on the substrate. The formation of new bonds to the enzyme by the substrate weakens bonds within the substrate, and these weakened bonds can then be more easily broken to form products.

Portions of the enzyme molecule other than the active site may be involved in the catalytic process. Some interaction at a position remote from the active site can change the shape of the enzyme and thus change its effectiveness as a catalyst. In this way it is possible to slow or stop the catalytic action. Figure 16.19 offers a model for the inhibition of enzyme catalysis. An inhibitor molecule attaches to the enzyme at a position remote from the active site where the substrate is bound. The enzyme changes shape as it accommodates the inhibitor, and the substrate is no longer able to bind to the enzyme. This is one of the mechanisms by which cells "turn off" enzymes when their work is done.

Some enzymes consist entirely of protein chains. In others, another chemical component, called a **cofactor**, is necessary for proper function of the enzyme. This cofactor may be a metal ion such as zinc (Zn^{2+}), manganese (Mn^{2+}), magnesium (Mg^{2+}), iron(II) (Fe^{2+}), or copper(II) (Cu^{2+}). An organic cofactor is called a **coenzyme**. By definition, coenzymes are nonprotein. The pure protein part of an enzyme is called the **apoenzyme**. Both the coenzyme and the apoenzyme must be present for enzymatic activity to take place.

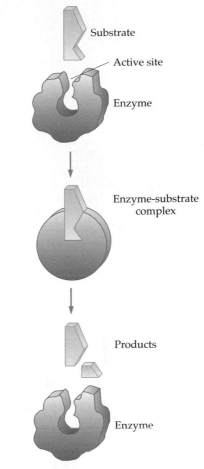

▲ **Figure 16.18** The induced-fit model of enzyme action. In this case, a single reactant molecule is broken into two product molecules.

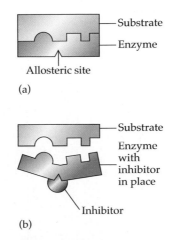

▲ **Figure 16.19** A model for the inhibition of an enzyme. (a) The enzyme and its substrate fit like a lock and key. (The allosteric site is where the inhibitor binds.) (b) With the inhibitor bound to the enzyme, the active site of the enzyme is distorted, and the enzyme cannot bind to its substrate.

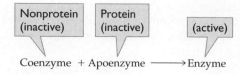

$$\text{Coenzyme} + \text{Apoenzyme} \longrightarrow \text{Enzyme}$$

Many coenzymes are vitamins or are derived from vitamin molecules. Enzymes are essential to the function of every living cell.

Enzymes in Medicine

Enzymes find many uses in medicine, industry, and everyday life. Diabetics monitor their blood glucose with meters that use test strips containing the enzyme *glucose oxidase* and a reducing agent. The glucose oxidase catalyzes the oxidation of blood glucose to gluconic acid. The gluconic acid formed then oxidizes the reducing agent, which causes a current to be generated by an electrode (Section 8.3). That current, which is proportional to the concentration of glucose in the blood sample, is measured by the meter and displayed as mg glucose per deciliter.

Clinical analysis for enzymes in body fluids or tissues is a common diagnostic technique in medicine. For example, a diseased or damaged liver can leak into the bloodstream those enzymes normally found only in the liver. Appearance of these enzymes in blood confirms liver damage. Analysis for specific enzymes in blood is a valuable diagnostic tool.

Enzymes are used to break up blood clots after a heart attack. This has increased survival rates for heart attack patients when these medicines are given as soon as possible after an attack. Other enzymes are used to do just the opposite: Blood-clotting factors are used to treat hemophilia, a disease in which the blood fails to clot normally.

Enzymes in Industry

Enzymes are widely used in a variety of industries, from baby foods to beer. Enzymes in intact microorganisms have long been used to make bread, beer, wine, yogurt, and cheese. Now the use of enzymes has been broadened to include such things as the following:

- Enzymes that act on proteins (called proteases) are used in the manufacture of baby foods to predigest complex proteins and make them easier for a baby to digest.
- Enzymes that act on starches (carbohydrases) are used to convert cornstarch to corn syrup (mainly glucose) for use as a sweetener.
- An enzyme called isomerase is used to convert the glucose in corn syrup into fructose, which is sweeter than glucose and can be used in smaller amounts to give the same sweetness in diet foods. (Glucose and fructose are isomers; both have the molecular formula $C_6H_{12}O_6$.)
- Protease enzymes are used to make beer and fruit juices clear by breaking down the proteins that cause cloudiness.
- Enzymes that degrade cellulose (cellulases) are used in stonewashing blue jeans. These enzymes break down the cellulose polymers of cotton fibers. By carefully controlling their activity, manufacturers can get the desired effect without destroying the cotton material. Varying the cellulase enzymes creates different effects to make true "designer jeans."
- Protease and carbohydrases are added to animal feed to make nutrients more easily absorbed and improve the animal's digestion.

Enzymes in Everyday Life

Enzymes are important components of some detergents. Enzymes that act on fats (called lipases) and proteins (called proteases) break down stains by attacking the protein-type stains, such as blood, meat juice, dairy products, and the fats and grease that make up many stains, breaking them down into smaller, water-soluble substances. A protease enzyme is the active ingredient in meat tenderizers, making

▲ It DOES Matter!

You can do your own protein hydrolysis experiment. Mix a gelatin dessert powder according to directions. Divide into two parts. Add fresh pineapple, kiwi, or papaya to one portion. Add cooked fruit or none at all to the other. The gelatin with the fresh fruit will not gel! These fruits contain a protease enzyme that breaks down proteins (gelatin is a protein). The enzyme is inactivated by cooking, so canned pineapple is perfectly fine in gelatin. (Why do you think cooking ham with pineapple tenderizes the ham, and eating raw pineapple can "burn" the lips?)

a tough piece of meat easier to eat. All told, the worldwide production of enzymes is worth more than $1 billion per year.

Enzymes and Green Chemistry

Scientists are hard at work trying to find enzymes to carry out all kinds of chemical conversions. Enzymes are often more efficient than the usual chemical catalysts, and they are more specific in that they produce more of the desired product and fewer by-products that have to be disposed of. Further, processes involving enzymes do not require toxic solvents, thus producing less waste.

Companies spend hundreds of millions of dollars studying the special characteristics of the enzymes inside *extremophiles,* microorganisms that live in harsh and strange environments that would kill other creatures. Scientists hope to replace ordinary enzymes with enzymes from extremophiles. They are already finding uses in making specialty chemi-

cals and new drugs. Many industrial processes can only be carried out at high temperatures and pressures. Enzymes make it possible to carry out some of these processes at fairly low temperatures and atmospheric pressure, thus reducing the energy requirements and the need for expensive equipment.

Another area of enzyme use is for cleaning up pollution. For example, polychlorinated biphenyls (PCBs), notoriously persistent pollutants, are usually destroyed by incineration at 1200 °C. Microorganisms have been isolated that can degrade PCBs and thus have the potential to destroy these toxic pollutants. However, much research is still needed to make them cost-effective and able to compete economically with their chemical processes.

Self-Assessment Questions

1. Almost all enzymes are
   a. carbohydrates   **b.** lipids   **c.** nucleic acids   **d.** proteins
2. The molecule upon which an enzyme acts is called the enzyme's
   a. catalyst   **b.** coenzyme   **c.** cofactor   **d.** substrate
3. The induced-fit model of enzyme action holds that the
   a. action of an enzyme is enhanced by an inhibitor
   b. active site is the same as the allosteric site
   c. shapes of active sites are rigid and only fit substrates with exact complementary shapes
   d. shapes of active sites can change somewhat to fit a substrate
4. An organic, nonprotein component of an enzyme molecule is called a(n)
   a. accessory enzyme   **b.** allosteric group
   c. coenzyme   **d.** stimulator
5. Enzyme cofactors are
   a. always ions such as Ca^{2+}, Mg^{2+}, and K^+
   b. always nonprotein organic molecules
   c. either ions such as Ca^{2+}, Mg^{2+}, and K^+ or nonprotein organic molecules
   d. small protein molecules
6. The pure protein part of an enzyme is called a(n)
   a. apoenzyme   **b.** coenzyme   **c.** holoenzyme   **d.** ribozyme
7. Elevated liver enzyme levels in the blood indicate
   a. an active metabolism   **b.** a poor diet
   c. a possible heart attack   **d.** possible liver damage

Answers: 1, d; 2, d; 3, d; 4, c; 5, c; 6, a; 7, d

A N S W E R

4. Why are enzymes added to some laundry detergents?
Enzymes in laundry detergents catalytically break down the proteins that used to be so difficult for detergents and bleaches to remove. Years ago, grass stains in children's clothing were almost impossible to remove because the proteins of grass have a strong attraction for cotton and other cloth. Enzymes have alleviated that problem.

NUCLEIC ACIDS: THE CHEMISTRY OF HEREDITY

Life on Earth has a fantastic range of forms, but all life arises from the same molecular ingredients: five nucleotides that serve as the building blocks for DNA and RNA (Section 16.8), and 20 amino acids (Section 16.4) that are the building blocks for proteins. These components limit the chemical reactions that can occur in cells and thus restrict what life is like.

16.8 Nucleic Acids: Parts and Structure

Nucleic acids serve as the information and control centers of the cell. There are two kinds of nucleic acids: *deoxyribonucleic acid (DNA)*, which serves as the blueprint for all the proteins of an organism, and *ribonucleic acid (RNA)*, which carries out protein assembly. DNA is a coiled threadlike molecule found primarily in the cell nucleus. RNA is found in all parts of the cell, where different forms do different jobs. Both DNA and RNA are long chains of repeating units called **nucleotides**. Each nucleotide in turn consists of three parts (Figure 16.20): a pentose (five-carbon) sugar, a phosphate unit, and a heterocyclic amine (Section 9.15) base. The sugar is either ribose (in RNA) or deoxyribose (in DNA). Looking at the sugar in the nucleotide, note that the hydroxyl group on the first carbon atom is replaced by one of five bases. The bases with two fused rings, adenine and guanine, are classified as **purines**. The **pyrimidines** cytosine, thymine, and uracil have only one ring.

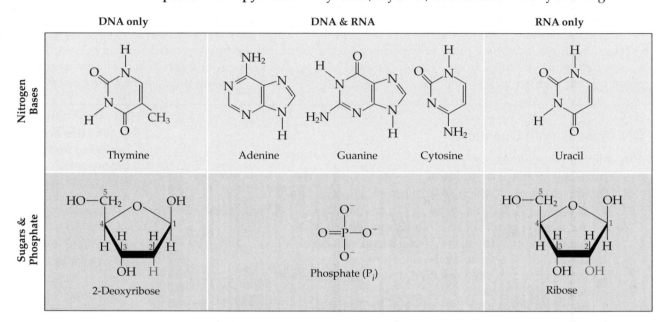

▲ **Figure 16.20** The components of nucleic acids. The sugars are 2-deoxyribose (in DNA) and ribose (in RNA). Note that deoxyribose differs from ribose in that it lacks an oxygen atom on the second carbon atom. *Deoxy* indicates that an oxygen atom is "missing." Phosphate units are often abbreviated as P_i ("inorganic phosphate"). Heterocyclic bases are found in nucleic acids. Adenine, guanine, and cytosine are found in both DNA and RNA. Thymine occurs only in DNA and uracil only in RNA. Note that thymine has a methyl group (red) that is lacking in uracil.

The hydroxyl group on the fifth carbon of the sugar unit is converted to a phosphate ester group. Adenosine monophosphate (AMP) is a representative nucleotide. In AMP, the base (blue) is adenine and the sugar (black) is ribose.

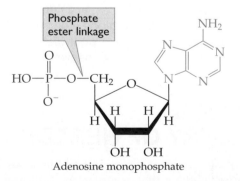

Adenosine monophosphate

Nucleotides can be represented schematically as in the following, where P_i is the biochemist's designation for "inorganic phosphate." A general representation is shown on the left and a specific schematic for AMP on the right.

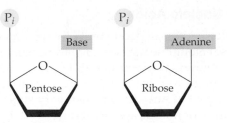

Nucleotides are joined to one another through the phosphate group to form nucleic acid chains. The phosphate unit on one nucleotide forms an ester linkage to the hydroxyl group on the third carbon atom of the sugar unit in a second nucleotide. This unit is in turn joined to another nucleotide, and the process is repeated to build up a long nucleic acid chain (Figure 16.21). The backbone of the chain consists of alternating phosphate and sugar units, and the heterocyclic bases branch off this backbone.

$$\sim sugar\text{-}P_i\text{-}sugar\text{-}P_i\text{-}sugar\text{-}P_i\text{-}sugar\text{-}P_i\text{-}sugar\text{-}P_i\text{-}sugar\text{-}P_i\sim$$
$$\quad | \qquad | \qquad | \qquad | \qquad | \qquad |$$
$$\text{base} \quad \text{base} \quad \text{base} \quad \text{base} \quad \text{base} \quad \text{base}$$

The numbers refer to positions on the pentose sugar ring (see again Figure 16.20). The 3' end has a hydroxyl group and the 5' end has a phosphate group. If the foregoing diagram represents DNA, the sugar is deoxyribose and the bases are adenine, guanine, cytosine, and thymine. In RNA the sugar is ribose and the bases are adenine, guanine, cytosine, and uracil.

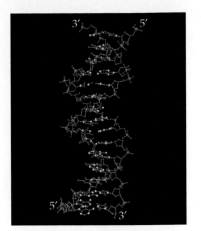

▲ **Figure 16.21** A DNA molecule. Each repeating unit is composed of a sugar, phosphate unit, and base. The sugar and phosphate units make a backbone on the outside. The bases are attached to the sugar units. Pairs of bases make "steps" on the inside of the spiral staircase.

EXAMPLE 16.4 Nucleotides

Consider the following nucleotide. Identify the sugar and the base. State whether it appears in DNA or in RNA.

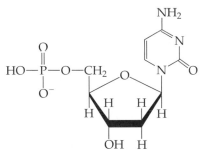

Solution

The sugar has two H atoms on the second C atom and, therefore, is deoxyribose. The base is a pyrimidine (one ring). The amino (NH_2) group helps identify it as cytosine.

■ **EXERCISE 16.4A**

Consider the following nucleotide. Identify the sugar and the base. State whether it appears in DNA or in RNA.

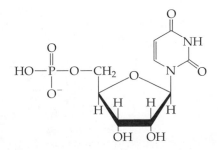

■ **EXERCISE 16.4B**

A nucleotide is composed of a phosphate unit, ribose, and thymine. Does it appear in DNA, RNA, or neither? Explain.

More to Explore

Watson, James D. *The Double Helix.* New York: New American Library, 1968. Watson's personal account of the discovery of the structure of DNA.

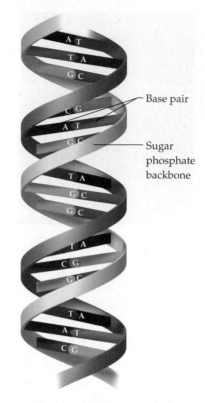

▲ **Figure 16.22** A model of a portion of a DNA double helix is given in Figure 16.21. Here we show a schematic representation of the double helix with the sugar–phosphate backbones shown as ribbons and complementary base pairs shown as steps on the spiral staircase.

Base Sequence in Nucleic Acids

All the vast genetic information needed to build living organisms is stored in the sequence of the four bases along the nucleic acid strand. Not surprisingly, these molecules are huge, with molecular masses ranging into the billions for mammalian DNA. Along these chains, the four bases can be arranged in almost infinite variations. We will examine this aspect of nucleic acid chemistry shortly, but first we consider another important feature of nucleic acid structure.

The Double Helix

By 1950 experiments designed to probe the structure of DNA showed that the molar amount of adenine (A) in DNA corresponds to the molar amount of thymine (T). Similarly, the molar amount of guanine (G) is essentially the same as that of cytosine (C). To maintain this balance, researchers concluded the bases in DNA must be paired, A to T and G to C. But how? At the midpoint of the twentieth century, it was clear that whoever answered this question would win a Nobel Prize. Many illustrious scientists worked on the problem, but two who were relatively unknown announced in 1953 that they had worked out the structure of DNA. Using data that involved quite sophisticated chemistry, physics, and mathematics, and working with models not unlike a child's construction set, James D. Watson and Francis H. C. Crick determined that DNA must be composed of two helices wound about one another. The two strands are antiparallel; they run in opposite directions. The phosphate and sugar backbone of the polymer chains form the outside of the structure, which is rather like a spiral staircase. The heterocyclic amines are paired on the inside—with guanine always opposite cytosine and adenine always opposite thymine. In our spiral staircase analogy, these base pairs are the steps (Figure 16.22).

Why do the bases pair in this precise pattern, always A to T and T to A, and always G to C and C to G? The answer is hydrogen bonding and a truly elegant molecular arrangement. Figure 16.23 shows the two sets of base pairs. You should notice two things. First, a pyrimidine is paired with a purine in each case, and the long dimensions of both pairs are identical (1.085 nm).

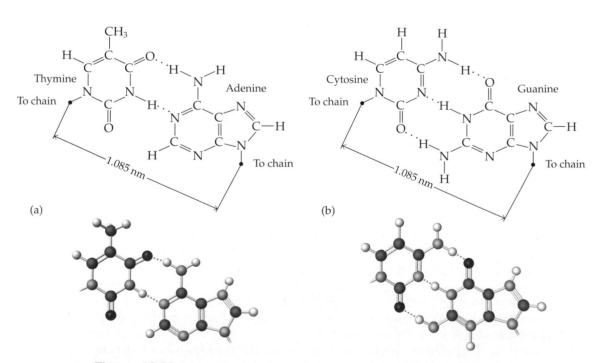

▲ **Figure 16.23** Pairing of the complementary bases thymine and adenine (a) and cytosine and guanine (b). The pairing involves hydrogen bonding, as in DNA.

The second thing you should notice in Figure 16.23 is the hydrogen bonding between the bases in each pair. When guanine is paired with cytosine, three hydrogen bonds can be formed between the bases. No other pyrimidine–purine pairing permits such extensive interaction. Indeed, in the combination shown in the figure, each pair of bases fits like a lock and key.

Other scientists around the world quickly accepted the Watson–Crick structure because it answers so many crucial questions. It explains how cells are able to divide and go on functioning, how genetic data are passed on to new generations, and even how proteins are built to required specifications. It all depends on the base pairing.

Structure of RNA

Most RNA molecules consist of single strands of nucleic acid. Some internal (intramolecular) base pairing can occur in sections where the molecule folds back on itself. Portions of some RNA molecules exist in double-helical form (Figure 16.24).

▲ Francis H. C. Crick (1916–2004) (seated) and James D. Watson (b. 1928) used data obtained by British chemist Rosalind Franklin (1920–1958) to propose a double-helix model of DNA in 1953. Franklin used a technique called X-ray diffraction, which showed DNA's helical structure. Without her permission, Franklin's colleague Maurice Wilkins shared her work with Watson and Crick. These data helped Watson and Crick decipher DNA's structure. Wilkins, Watson, and Crick were awarded the Nobel Prize for this discovery in 1962, a prize Franklin might have shared had she lived.

Self-Assessment Questions

1. The monomer units of nucleic acids are
   a. amino acids
   b. fatty acids
   c. monosaccharides
   d. nucleotides

2. A DNA nucleotide could contain
   a. A, P_i, and ribose
   b. G, P_i, and glucose
   c. T, P_i, and deoxyribose
   d. U, P_i, and deoxyribose

3. A possible base pair in DNA is
   a. A-G   **b.** A-T
   c. A-U   **d.** T-G

4. A possible base pair in RNA is
   a. A-G   **b.** A-T
   c. A-U   **d.** T-G

5. Nucleic acid–base pairs are joined by
   a. amide bonds   **b.** ester bonds
   c. glucosidic bonds   **d.** hydrogen bonds

Answers: 1, d; 2, c; 3, b; 4, c; 5, d

▲ **Figure 16.24** RNA occurs as single strands that can form double-helical portions by internal pairing of bases.

16.9 DNA: Self-Replication

Cats have kittens that grow up to be cats. Birds lay eggs that hatch and grow up to be birds. How is it that each species reproduces its own kind? How does a fertilized egg "know" that it should develop into a kangaroo and not a koala?

The physical basis of heredity has been known for a long time. Most higher organisms reproduce sexually. A sperm cell from the male unites with an egg cell from the female. The fertilized egg so formed must carry all the information needed to make the various cells, tissues, and organs necessary for the functioning of a new individual. In addition, if the species is to survive, information must be passed along in germ cells—both sperms and eggs—for the production of new individuals.

Chromosomes and Genes

The hereditary material is found in the nuclei of all cells, concentrated in elongated, threadlike bodies called *chromosomes*. Chromosomes form compressed X-shaped structures when strands of DNA coil up tightly just before a cell divides. The number of chromosomes varies with the species. In sexual reproduction, chromosomes come in pairs, with one member of the pair from each parent. Each human inherits 23 pairs; thus, our body cells have 46 chromosomes. The entire complement of chromosomes is achieved only when the egg's 23 chromosomes combine with a like number from the sperm.

Chromosomes are made of DNA and proteins. Arranged along the chromosomes are the basic units of heredity, the genes. Structurally, a **gene** is a section of the DNA molecule, although some viral genes contain only RNA. Genes control the synthesis of proteins and in this way tell cells, organs, and organisms how to function in their surroundings. The environment helps determine which genes become active at a particular time. The complete set of genes of an organism is called its *genome*. When cell division occurs, each chromosome produces an exact duplicate of itself. Transmission of genetic information, therefore, requires the **replication** (copying or duplication) of DNA molecules.

The double helix provides a ready model for replication. If the two chains of the double helix are pulled apart, each chain can direct the synthesis of a new DNA chain using nucleotides from the cellular fluid surrounding the DNA. Each of the separating chains serves as a template, or pattern, for the formation of a new complementary chain. The two DNA strands are *antiparallel*. One chain runs in a direction labeled $3' \rightarrow 5'$ and the other from $5' \rightarrow 3'$. Synthesis begins with a base on a nucleotide pairing with its complementary base on the DNA strand—adenine with thymine and guanine with cytosine (Figure 16.25). Each base unit in the separated strand can pick up only a unit identical to the one with which it was paired before. For example, a 12-base-pair strand running in the $5' \rightarrow 3'$ direction that has guanine as the third base will pair with the base cytosine as the tenth base on the complementary 12-base-pair strand running in the opposite direction.

$$3'-A-T-G-G-G-T-C-T-A-T-A-T-5'$$
$$5'-T-A-C-C-C-A-G-A-T-A-T-A-3'$$

As the nucleotides align, enzymes connect them to form the sugar–phosphate backbone of the new chain. In this way, each strand of the original DNA molecule forms a duplicate of its former partner. Whatever information was encoded in the original DNA double helix is now contained in each of the replicates. When the cell divides, each daughter cell gets one of the DNA molecules and all the information that was available to the parent cell.

How is information stored in DNA? The code for the directions for building all the proteins that comprise an organism and enable it to function resides in the *sequence* of bases along the DNA chain. Just as *cat* means one thing in English and *act* means another, the sequence of bases CGT means one amino acid, and GCT means another. Although there are only four "letters"—the four bases—in the genetic codes of DNA, their sequence along the long strands can vary so widely that essentially unlimited information storage is available. Each cell carries in its DNA all the information it needs to determine all its hereditary characteristics.

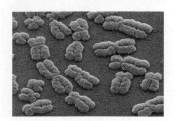

▲ Human chromosomes prior to cell division take on a distinctive X-shape.

▼ **Figure 16.25** A simplified schematic diagram of DNA replication. The original double helix is "unzipped" and new nucleotides (as triphosphate derivatives) are brought into position by the enzyme DNA polymerase. Phosphate bridges are formed, restoring the original double-helix configuration. Each newly formed double helix consists of one old strand and one new strand.

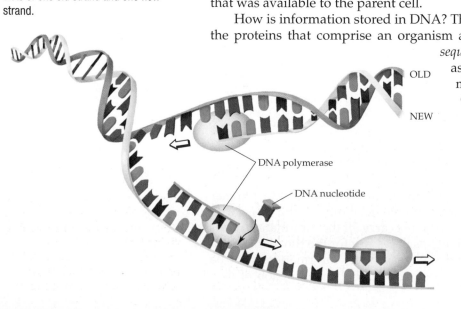

OLD

NEW

DNA polymerase

DNA nucleotide

Key

Adenine

Thymine

Guanine

Cytosine

Self-Assessment Questions

1. Replication of a DNA molecule involves breaking bonds between the
   a. base pairs **b.** pentose sugars
   c. phosphate units **d.** bases and pentoses
2. The first step of DNA replication involves
   a. bonding of free nucleotides in the correct sequence
   b. formation of a single-stranded RNA molecule
   c. linking of tRNA to an amino acid
   d. unzipping of the DNA molecule by breaking hydrogen bonds
3. If the original DNA strand had the $3' \rightarrow 5'$ base sequence T-A-G-C, the complementary $5' \rightarrow 3'$ strand would have the base sequence
   a. A-T-C-G **b.** G-C-A-T **c.** T-A-C-G **d.** U-A-C-G
4. If a portion of a gene has the base sequence T-C-G-A-A-T, what base sequence appears in the complementary DNA when the gene is replicated?
   a. A-C-G-T-T-A **b.** A-G-C-T-T-A
   c. T-C-G-A-A-T **d.** U-G-C-A-A-U

Answers: 1, a; 2, d; 3, a; 4, b

Genes have functional regions called *exons* interspersed with inactive portions called *introns*. During protein synthesis, introns are snipped out and not translated. Some genes, called housekeeping genes, are expressed in all cells at all times and are essential for the most basic cellular functions. Other genes are expressed only in specific cell types or at certain stages of development. For example, the genes that encode brain nerve cells are expressed only in brain cells, not in the liver.

16.10 RNA: Protein Synthesis and the Genetic Code

DNA carries a message that must somehow be relayed and acted on in a cell. Because DNA does not leave the cell nucleus, its information, or "blueprint," must be transported by something else. In the first step, called **transcription**, DNA transfers its information to a special RNA molecule called *messenger RNA (mRNA)*. The base sequence of DNA specifies the base sequence of mRNA. Thymine in DNA calls for adenine in mRNA, cytosine specifies guanine, guanine calls for cytosine, and adenine requires uracil (Table 16.4). Remember that in RNA molecules, uracil is used in place of DNA's thymine. Notice the similarity in the structure of these two bases (recall Figure 16.20).

Table 16.4 DNA Bases and Their Complementary RNA Bases	
DNA Base	**Complementary RNA Base**
Adenine (A)	Uracil (U)
Thymine (T)	Adenine (A)
Cytosine (C)	Guanine (G)
Guanine (G)	Cytosine (C)

The next step in creating a protein involves deciphering the code copied by mRNA and **translation** of that code into a specific protein structure. The decoding occurs when the mRNA travels from the nucleus and attaches itself to a ribosome in the cytoplasm of the cell. Ribosomes are constructed of RNA and proteins.

Another type of RNA molecule, called *transfer RNA (tRNA)*, carries amino acids from the cell fluid to the ribosomes. A tRNA molecule has the looped structure shown in Figure 16.26. At the head of the molecule is a set of three base units, a *base triplet* called the *anticodon*, that pairs with a set of three complementary bases on mRNA, called the **codon**. The codon triplet determines which amino acid is carried at the tail of the tRNA. To illustrate, the base triplet GUA on a segment of mRNA pairs with the anticodon base triplet CAU on a tRNA molecule. All tRNA molecules with the base triplet CAU always carry the amino acid valine. Once the tRNA has

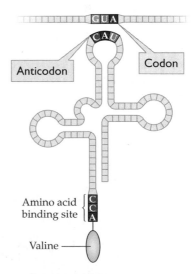

▲ **Figure 16.26** A given transfer RNA (tRNA) doubles back on itself, forming three loops with intermolecular hydrogen bonding between complementary bases. The anticodon triplet at the head of the molecule joins with a complementary codon triplet on mRNA. Here the base triplet CAU in the anticodon specifies the amino acid valine.

Table 16.5 The Genetic Code

		SECOND BASE				
		U	**C**	**A**	**G**	
FIRST BASE	**U**	UUU=Phe UUC=Phe UUA=Leu UUG=Leu	UCU=Ser UCC=Ser UCA=Ser UCG=Ser	UAU=Tyr UAC=Tyr UAA=Termination UAG=Termination	UGU=Cys UGC=Cys UGA=Termination UGG=Trp	U C A G
	C	CUU=Leu CUC=Leu CUA=Leu CUG=Leu	CCU=Pro CCC=Pro CCA=Pro CCG=Pro	CAU=His CAC=His CAA=Gln CAG=Gln	CGU=Arg CGC=Arg CGA=Arg CGG=Arg	U C A G
	A	AUU=Ile AUC=Ile AUA=Ile AUG=Met	ACU=Thr ACC=Thr ACA=Thr ACG=Thr	AAU=Asn AAC=Asn AAA=Lys AAG=Lys	AGU=Ser AGC=Ser AGA=Arg AGG=Arg	U C A G
	G	GUU=Val GUC=Val GUA=Val GUG=Val	GCU=Ala GCC=Ala GCA=Ala GCG=Ala	GAU=Asp GAC=Asp GAA=Glu GAG=Glu	GGU=Gly GGC=Gly GGA=Gly GGG=Gly	U C A G

(THIRD BASE)

Why is a genetic code that must specify 20 different amino acids based on a triplet of bases? Four "letters" can be arranged in 4 × 4 × 4 = 64 possible three-letter "words." A "doublet" code of four letters would have only 4 × 4 = 16 different words, and a "quadruplet" code of four letters can be arranged in 4 × 4 × 4 × 4 = 256 different ways. A triplet code can call for 20 different amino acids with some redundancy, whereas a doublet code is inadequate, and a quadruplet code would provide too much redundancy.

paired with the base triplet of mRNA, it releases its amino acid and returns to the cell fluid to pick up another amino acid molecule.

Each of the 61 different tRNA molecules carries a specific amino acid into place on a growing peptide chain. The protein chain gradually built up in this way is released from the tRNA as it is formed. A complete dictionary of the genetic code has now been compiled (Table 16.5). It shows which amino acids are specified by all the possible mRNA base triplets. There are 64 possible triplets and only 20 amino acids, and so there is some redundancy in the code. Three amino acids (serine, arginine, and leucine) are each specified by six different codons. Two others (tryptophan and methionine) have only one codon each. Three base triplets on mRNA are *stop* signals that call for termination of the protein chain. The codon AUG signals "start" as well as specifying methionine in the chain. Figure 16.27 provides an overall summary of protein synthesis.

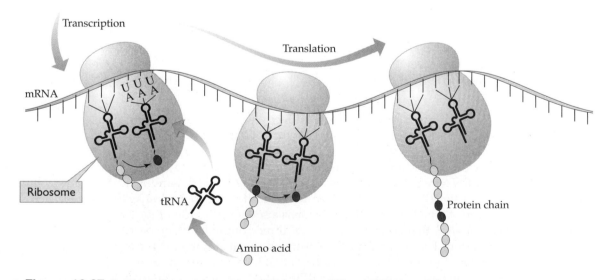

▲ **Figure 16.27** Protein synthesis visualized from DNA through mRNA and tRNA to a protein.

Self-Assessment Questions

1. The synthesis of RNA from a strand of DNA is called
   a. transaction   **b.** transcription   **c.** transition   **d.** translation

2. The base sequence of a strand of mRNA transcribed from an original strand of DNA that has the base sequence T-A-C-G is
   a. A-U-G-C   **b.** C-G-A-U   **c.** T-A-G-C   **d.** U-A-G-C

3. The anticodon triplet is found on the
   a. original DNA   **b.** ribosome   **c.** mRNA   **d.** tRNA

4. The genetic code consists of how many amino acids and how many codons?
   a. 20 amino acids, 20 codons   **b.** 20 amino acids, 64 codons
   c. 64 amino acids, 20 codons   **d.** 64 amino acids, 64 codons

5. Each amino acid except tryptophan and methionine is coded for by more than one codon, an example of
   a. evolution   **b.** redundancy
   c. sequence assurance   **d.** translation

6. Refer to Table 16.5; the codon AUG
   a. codes for Met and signals start   **b.** codes for Met and signals stop
   c. signals start only   **d.** signals stop only

Answers: 1, b; 2, a; 3, d; 4, b; 5, b; 6, a

Errors can occur at each step in the replication–transcription–translation process. In replication alone, each time a human cell divides, 4 billion bases are copied to make a new strand of DNA, and there may be up to 2000 errors. Although most such errors are corrected and others are unimportant, some have terrible consequences: Genetic disease or even death may result.

16.11 The Human Genome

Many human diseases have clear genetic components. Some are directly caused by one defective gene, whereas others have several genes involved. Some of these genes have been located on the human genome map, and the function of some has been identified. Once all the genes are identified, the ability to use this information to diagnose and cure genetic diseases will revolutionize medicine.

Humans were long thought to have 80,000 or more active genes. When scientists completed the Human Genome Project in 2000, they found only about 20,000 to 25,000 genes in the human genome.

Genetic Testing

DNA is sequenced by using enzymes to cleave it into segments of a few to several hundred nucleotides each. A technique called the *polymerase chain reaction (PCR)* is used to duplicate and amplify each DNA sequence by making millions of copies of the sequence. PCR employs enzymes called *DNA polymerases* to amplify the small amount of DNA extracted from a sample into the large quantities needed for DNA sequencing. The fragments are then separated by length from longest to shortest, and a pattern—often called a *DNA fingerprint*—is obtained. The print can be in the form of peaks on a chart or as a series of horizontal bars resembling the bar codes imprinted on packaged goods sold in supermarkets.

Patterns of DNA fragments are characteristic of certain families, and DNA fingerprints of various family members can be compared to look for inherited diseases. If the DNA pattern of a relative resembles that of a person with a genetic disease closely enough, that relative will probably develop the disease. DNA fingerprinting is used in hospitals around the world to screen newborn babies for inherited disorders such as cystic fibrosis, sickle cell anemia, Huntington's disease, hemophilia, and many others. Early detection of such disorders enables the doctors and parents to prepare for proper treatment of the child.

Because children inherit half their DNA from each parent, DNA fingerprinting also is used to establish the parentage of a child of contested origin. The odds in favor of being right in such cases are excellent—at least 100,000:1.

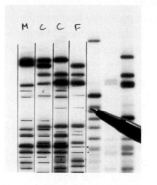

▲ The photo shows bands of DNA from a mother, father, and two children, produced by a technique called electrophoresis. Some bands are shared by related people, and the DNA fingerprints thus can be used to prove conclusively whether people are related. In this case, both children share some bands with each parent, proving that they are indeed related.

DNA Fingerprinting in Crime and Innocence

British scientist Alec Jeffreys invented DNA fingerprinting in 1985 and gave it its name. Like fingerprints, each person's DNA is unique. Any cells—skin, blood, semen, saliva, and so on—can supply the necessary DNA sample.

DNA samples from evidence found at the crime scene are amplified by PCR and compared with the DNA obtained from a suspect. The DNA fingerprints are then compared to those of any suspects and of people known to have been at the scene.

DNA fingerprinting is a major advance in criminal investigation, and thousands of criminal cases have been solved with this technology.

DNA fingerprinting has led to many criminal convictions, but perhaps more important, it can readily prove someone innocent. If the DNA does not match, the suspect could not have left the biological sample. More than 200 people have been shown to be innocent and freed, some after spending years in prison. The technique is a major advance in the search for justice.

Recombinant DNA: Using Organisms as Chemical Factories

All living organisms (except some viruses) have DNA as their hereditary material. *Recombinant DNA* is DNA that has been created artificially. DNA from two different sources is incorporated into a single recombinant DNA molecule. The first step is to cut the DNA to extract a fragment of DNA, which scientists accomplish with an enzyme obtained from bacteria called an *endonuclease*. Endonuclease is a type of *restriction enzyme,* so called because its ability to cleave DNA is restricted to certain short sequences of DNA called *restriction sites*. This enzyme cuts the two DNA molecules at the same site. There are three different methods by which recombinant DNA is made. We describe one method here.

After scientists determine the base sequence of the gene that codes for a particular protein, they can isolate it and amplify it by PCR. The gene can then be spliced into a special kind of bacterial DNA called a *plasmid*. The recombined plasmid is then inserted into the host organism (Figure 16.28). Once inside the host, the plasmid replicates, making multiple exact copies of itself. Producing many identical copies of the same recombinant molecule is called *cloning*. As the engineered bacteria multiply, they become effective factories for producing the desired protein.

What kinds of proteins might scientists wish to produce using biotechnology? Insulin, a protein coded by DNA, is required for the proper use of glucose by cells. People with diabetes, an insulin-deficiency disease, formerly had to use insulin from pigs or cattle. Now human insulin is made using recombinant DNA technology. Scientists take the human gene for insulin production and paste it into the DNA of *Escherichia coli,* a bacterium commonly found in the human digestive tract. The bacterial cells multiply rapidly, making billions of copies of themselves, and each new *E. coli* cell carries in its DNA a replica of the gene for human insulin. Diabetics are no longer dependent on insulin from cows or pigs. The hope for the future is that a functioning gene for insulin can be incorporated directly into the cells of insulin-dependent diabetics.

In addition to insulin, many other valuable materials that are used in human therapy and are difficult to obtain in any other way are now made using recombinant DNA technology. Some examples are

- human growth hormone (HGH), which replaced cadaver-harvested HGH; used to treat children who fail to grow properly
- erythropoietin (EPO) for treating anemia resulting from chronic kidney disease and from the treatment of cancer
- factor VIII (formerly harvested from blood), a clotting agent for treating males with hemophilia
- tissue plasminogen activator (TPA) for dissolving blood clots associated with diseases such as heart attack and stroke

- interferons, which assist the immune response and induce the resistance of host cells to viral infection, seen as promising anticancer agents
- bovine somatotropin (bST) used in the battle to increase milk production

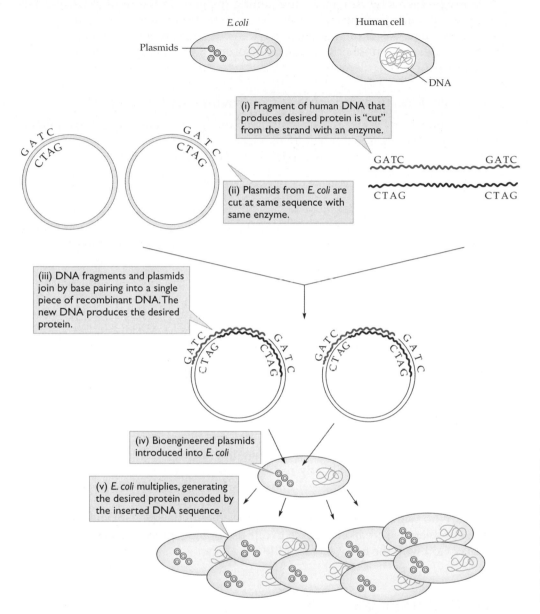

◀ **Figure 16.28** Recombinant DNA: cloning a human gene into a bacterial plasmid. The bacteria multiply and, in doing so, make multiple copies of the human gene. The gene can be one that codes for insulin, human growth hormone, or another valuable protein. Potentially, a gene from any organism—microbe, plant, or animal—can be incorporated into any other organism.

Gene Therapy

Gene therapy involves introducing a functioning gene into a person's cells to correct the action of a defective gene. Viruses are commonly used to carry DNA into cells. Current gene therapy is experimental, and human gene transplants so far have had only modest success:

- In 1999, ten children with severe combined immunodeficiency disease (SCID) were injected with working copies of a gene that helps the immune system develop. Almost all the children got better and most are healthy today, but in 2002, two of the children developed leukemia.
- Four severely blind young adults had a curative gene injected into their eyes. They can now see more light. Two of them can now read several lines of an eye chart.
- Genetically altered cells injected into the brain of patients with Alzheimer's disease appear to nourish ailing neurons and may slow the cognitive decline in these people.

More to Explore
Gibbs, W. Wayt. "The Unseen Genome: Gems Among the Junk." *Scientific American*, November 2003, pp. 46–53.

▲ **It DOES Matter!**
Genetically modified rice ("golden rice"), developed in 1999, has new genes implanted from daffodils and a bacterium. Golden rice was developed as a fortified food to contain high amounts of beta-carotene, a precursor of vitamin A. It was to be used in areas where there is a shortage of dietary vitamin A to help children who suffer from vitamin A deficiency (Chapter 17). An estimated 124 million people in Africa and South East Asia are affected by vitamin A deficiency, causing 1–2 million deaths and 500,000 cases of blindness. Opposition to and complex regulation of transgenic crops have prevented the plant from being used on farms; golden rice is still in field trials.

• In 2006 researchers at the National Cancer Institute (NCI) successfully reengineered immune cells to attack cancer cells in patients with advanced melanoma, the first successful use of gene therapy to treat cancer in humans.

Many problems remain, including getting the gene in the right place and targeting it so it only inserts itself into the cells where it is needed. The stakes in gene therapy are huge. We all carry some defective genes. About one in ten people either has or will develop an inherited genetic disorder. Gene therapy could become an efficient way of curing disease.

Genetically Modified Organisms: Controversy and Promise

Genes can now be transferred from nearly any organism to any other. New animals can be cloned from cells in adult animals. Genetically modified (GM) organisms are in widespread—and often controversial—use. For example, by 2006 in the United States, 89% of the planted area of soybeans, 83% of cotton, and 61% of maize were GM varieties.

• Genes for herbicide resistance are now incorporated into soybeans so that herbicide application will kill the weeds without killing the soybean plants. People worry that these genes will escape into wild plants (weeds), making them too resistant to herbicides.

• Genes from the bacteria *Bacillus thuringiensis* (Bt) are incorporated into corn and cotton plants. The genes cause the plants to produce a protein toxin that kills caterpillars that would otherwise feed on the plants. People worry that these toxins will hasten the time when insects will become resistant to the bacterial toxin, and that butterflies and other desirable insects will also be killed.

GM foods are foods made from GM organisms that have altered DNA. GM foods first entered the market in the early 1990s. The most common modified foods—soy beans, corn, canola oil, and cottonseed oil—are derived from GM plants. People worry that GM foods will have proteins that will cause allergic reactions in some people. To protect against such developments, strict guidelines for recombinant DNA research have been instituted. Some people think these are not strict enough.

Molecular genetics has already resulted in some impressive achievements. Its possibilities are mind-boggling—elimination of genetic defects, a cure for cancer, and who knows what else? Knowledge gives power, but it does not necessarily give wisdom. Who will decide what sort of creature the human species should be? The greatest problem we are likely to face in our use of bioengineering is choosing who is to play God with the new "secret of life."

A N S W E R

5. Are genetically modified foods safe?
There are potential problems with today's genetically modified foods, but in practice we have been using genetically modified foods for over a century—ever since Mendel discovered the concepts of genetics. Most of today's foods are hybrids, the result of genetic modification by cross-breeding. Of course, sometimes a hybrid tomato isn't tasty or is too small. Likewise, some genetically modified plants aren't exactly what we'd like.

Self-Assessment Questions

1. Recombinant DNA cloning involves
   a. combining DNA from two different organisms to make a single new organism
   b. creating a genetically identical organism from an epithelial cell
   c. isolation of a genetically pure colony of cells
   d. splitting of a fertilized egg into two cells to generate identical twins

2. In recombinant DNA cloning, the DNA is cut with
   a. DNA ligase   b. phages
   c. restriction enzymes d. vectors

3. Recombinant DNA is presently used in the biotechnology industry to
   a. create new species of animals
   b. eliminate all infectious disease in domestic animals
   c. increase the frequency of mutations
   d. synthesize insulin, interferon, and human growth hormone

4. Some geneticists suggest transferring some photosynthesis genes from an efficient crop plant to a less efficient one, thus producing a new plant variety with greater productivity. The project would most likely involve
a. amniocentesis **b.** genetic engineering
c. genetic screening **d.** transcription

Answers: 1, a; 2, c; 3, d; 4, b

Critical Thinking Exercises

Apply knowledge that you have gained in this chapter and one or more of the FLaReS principles (Chapter 1) to evaluate the following statements or claims.

16.1 In a paternity case, a woman states that a certain man is the father of her child and that DNA fingerprinting will prove her case.

16.2 DNA is an essential component of every living cell. An advertisement claims that DNA is a useful diet supplement.

16.3 A restaurant claims that "Other restaurants use saturated beef fat for frying. We use only pure vegetable oil to fry our fish and french fries, for healthier eating."

16.4 The wrapper of an energy bar claims in large print: "21,000 mg of amino acids in each bar!"

16.5 A dietician claims that a recommendation of 60 grams of protein per day for an adult is somewhat misleading. She claims that the right kinds of protein, with the proper amino acids in them, must be consumed, otherwise even a much larger amount of protein per day may be insufficient to maintain health.

16.6 A man convicted of rape claims he is innocent because the DNA fingerprinting of a semen sample taken from the rape victim shows that it contains another man's DNA.

■ SUMMARY

Section 16.1—Biochemistry is the chemistry of life processes. Life requires chemical energy. Cell metabolism is the set of chemical reactions that keeps an organism alive. Metabolism includes many processes of both **anabolism** (building up of molecules) and **catabolism** (breaking down of molecules). Food molecules include carbohydrates, fats, and proteins.

Section 16.2—A carbohydrate is a compound whose formula can be written as a hydrate of carbon. Sugars, starches, and cellulose are carbohydrates. The simplest sugars are **monosaccharides**, molecules that cannot be further hydrolyzed. A monosaccharide is either an **aldose** (aldehyde functional group) or a **ketose** (ketone functional group). Glucose and fructose are monosaccharides. A **disaccharide** can be hydrolyzed into two monosaccharide units. Sucrose and lactose are disaccharides. **Polysaccharides** contain many saccharide units linked together. Starches and cellulose are polysaccharides. Starches are polymers of glucose held together by alpha linkages; cellulose is a glucose polymer held together by beta linkages. There are two kinds of plant starches, amylose and amylopectin. Animal starch is called glycogen.

Section 16.3—A lipid is a cellular component that is insoluble in water and soluble in nonpolar solvents. Lipids include solid fats and liquid oils; both are triglycerides (esters of fatty acids with glycerol). Lipids also include steroids, hormones, and **fatty acids**, long-chain carboxylic acids. A **fat** is an ester of fatty acids and glycerol. Fats may be saturated (all carbon–carbon bonds in the fatty acids are single bonds), monounsaturated (one $C=C$), or polyunsaturated (more than one $C=C$). The **iodine number** of a fat is the number of grams of iodine that reacts with 100 grams of fat. Oils tend to have higher iodine numbers than fats because there are more double bonds in oils.

Section 16.4—An amino acid contains an amino ($-NH_2$) group and a carboxyl ($-COOH$) group. An amino acid exists as a **zwitterion** in which a proton from the carboxyl group is transferred to the amino group. Proteins are polymers of amino acids, with 20 amino acids forming thousands of different proteins. Plants can synthesize proteins from CO_2, water, and other substances, whereas animals require proteins in their diets.

Section 16.5—The bond that joins two amino acid units in a protein is called a **peptide bond**. Two amino acids joined form a dipeptide, three form a tripeptide, and more than about 10 amino acids joined form a **polypeptide**. When the molecular weight of a polypeptide exceeds about 10,000, it is called a **protein**. The sequence of the amino acids in a protein is of great importance in its function. Three-letter or single-letter abbreviations are used to designate amino acids.

Section 16.6—Protein structure has four levels. The **primary structure** of a protein is simply the sequence of its amino acids. The **secondary structure** is the arrangement of protein chains about an axis. Two such arrangements are the **beta pleated sheet**, in which the arrays of chains form zigzag sheets, and the **alpha helix**, which is a coil of proteins. The **tertiary structure** of a protein is its folding pattern; **globular proteins** such as myoglobin are folded into compact shapes. Some proteins have a **quaternary structure** in which the protein contains subunits in a specific pattern.

There are four types of forces that generate secondary, tertiary, and quaternary structures in proteins. They are (a) hydrogen bonding; (b) ionic bonds formed where acidic and basic side chains meet; (c) **disulfide linkages**, which are S—S bonds formed between cysteine groups; and (d) dispersion forces between the nonpolar regions of the molecule.

Section 16.7—Enzymes are biological catalysts. In operation, the reacting substance or **substrate** attaches to the **active site** of the enzyme to form a complex, which then decomposes to the products. An enzyme is made up of an **apoenzyme** (protein) and a **cofactor** that is necessary for proper function. An organic cofactor (such as a vitamin) is called a **coenzyme**. Sometimes a cofactor is inorganic (such as a metal ion).

Section 16.8—Nucleic acids are the information and control centers of a cell. DNA and RNA are polymers of **nucleotides**. Each nucleotide contains a pentose sugar unit, a phosphate unit, and one of four amine bases. Bases with one ring are called **pyrimidines**, and those with two fused rings are **purines**. In DNA (deoxyribonucleic acid), the sugar is deoxyribose and the bases are adenine, thymine, guanine, and cytosine. In RNA (ribonucleic acid) the sugar is ribose and the nitrogen bases are the same as those in DNA, except that uracil replaces thymine. Nucleotides are joined through their phosphate groups to form nucleic acid chains. In DNA the bases thymine and adenine are always paired, as are the bases cytosine and guanine. Pairing occurs via hydrogen bonding between the bases and gives rise to the double-helix structure of DNA. RNA consists of single strands of nucleic acids, some parts of which may hydrogen bond to form double helixes.

Section 16.9—Chromosomes are made of DNA and proteins. A **gene** is a section of DNA. Genetic information is stored in the base sequence; a three-base sequence means a particular amino acid. Transmission of genetic information requires **replication** or copying of the DNA molecules. When the chains of DNA are pulled apart, each half can replicate from nucleotides in the cellular fluid. Each base unit picks up a unit identical to that with which it was paired before. Enzymes connect the nucleotides to form the new chain. When a cell divides, one DNA molecule goes to each daughter cell, with all the accompanying genetic information.

Section 16.10—In **transcription** the DNA transfers its information to messenger RNA (mRNA). The next step is the **translation** of the code into a specific protein structure. Transfer RNA (tRNA) delivers amino acids to the growing protein chain. Each **base triplet** on mRNA, called the **codon**, codes for a specific amino acid.

Section 16.11—Many diseases have genetic components. In genetic testing, DNA is cleaved and duplicated. The patterns of DNA fragments are examined and compared to those of individuals with genetic diseases to identify and predict the occurrence of such diseases. A similar technique is used to compare DNA from different sources in DNA fingerprinting. In recombinant DNA technology, the base sequence for the gene that codes for a desired protein is determined and that new gene is spliced into a plasmid. The plasmid is inserted into a host organism, which generates the protein as it reproduces. Gene therapy—replacement of defective genes—could be used to cure diseases. Genetic engineering holds great promise and carries with it great responsibility.

■ REVIEW QUESTIONS

1. Briefly identify and state a function of each of the following parts of a cell.
   a. cell membrane   b. cell nucleus   c. chloroplasts
   d. mitochondria   e. ribosomes
2. List some of the properties of fats and oils.
3. In what parts of the body are proteins found? What tissues are largely proteins?
4. What is the chemical nature of proteins?
5. How does the elemental composition of proteins differ from those of carbohydrates and fats?
6. What is a peptide bond?
7. What is the difference between a polypeptide and a protein?
8. Of what importance is the sequence of amino acids in a protein molecule?

9. What kind of intermolecular force is involved in base pairing?
10. Describe the process of replication.
11. How do DNA and RNA differ in structure?
12. What is the relationship among the cell parts called chromosomes, the units of heredity called genes, and the nucleic acid DNA?
13. What is the polymerase chain reaction (PCR)?
14. How do DNA fragment patterns indicate whether a person will probably develop a genetic disease?
15. List the steps in recombinant DNA technology.
16. How can a virus be used to replace a defective gene in a human?

■ PROBLEMS

Carbohydrates

17. What is a monosaccharide? Name three common ones.
18. What is a disaccharide? Name three common ones.
19. What is glycogen? How does it differ from amylose? From amylopectin?
20. In what way are amylose and cellulose similar? What is the main structural difference between starch and cellulose?
21. Which of the following are monosaccharides?
    a. lactose   b. galactose
    c. cellulose   d. glucose
22. Which of the following are monosaccharides?
    a. sucrose   b. mannose
    c. fructose   d. amylopectin

23. Which of the carbohydrates in Problems 21 and 22 are disaccharides?
24. Which of the carbohydrates in Problems 21 and 22 are polysaccharides?
25. What are the hydrolysis products of each of the following?
    a. amylose   b. glycogen
26. What are the hydrolysis products of each of the following?
    a. amylopectin   b. cellulose
27. What functional groups are present in the formula for the open-chain form of glucose?
28. Give the formula for the open-chain form of fructose. What functional groups are present?
29. What functional groups are present in the formula for the open-chain form of galactose? How does the structure of galactose differ from that of glucose?

30. Mannose differs from glucose only in that the H and OH on the second carbon atom are reversed. Give the formula for the open-chain form of mannose. What functional groups are present?

Lipids: Fats and Oils

31. How do fats and oils differ in structure? In properties?
32. Define monoglyceride, diglyceride, and triglyceride.
33. Which of the following fatty acids are saturated? Which are unsaturated?
    a. linolenic acid   b. linoleic acid
    c. oleic acid   d. stearic acid
34. Which of the fatty acids in Problem 33 are monounsaturated? Which are polyunsaturated?
35. How many carbon atoms are there in a molecule of each of the following fatty acids?
    a. linolenic acid   b. palmitic acid
    c. stearic acid
36. How many carbon atoms are there in a molecule of each of the following fatty acids?
    a. linoleic acid   b. oleic acid
    c. butyric acid
37. Which of the two foodstuffs shown below would you expect to have a higher iodine number? Explain your reasoning.

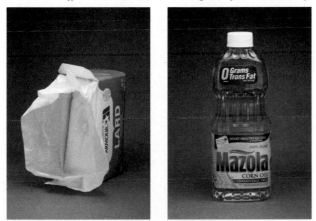

38. Which of the two fats shown below is likely to have a lower iodine number? Explain your reasoning.

Proteins

39. What functional groups are found on amino acid molecules? What is a zwitterion?
40. What is a dipeptide? A tripeptide? A polypeptide?
41. Of the amino acids glutamic acid, lysine, serine, leucine, and glycine, which are essential to the human diet?
42. Examine Table 16.3 and draw the condensed structural formula for that part of each structure that is identical for 19 of the 20 amino acids. Which amino acid does not have this particular structural feature?

43. Give structural formulas for the following amino acids.
    a. arginine   b. asparagine
44. Identify the following amino acids.

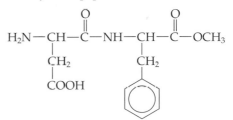

$$HOOC-CH_2CH_2-CH-COO^-$$
$$\overset{|}{{}^+NH_3}$$
(a)

$$HS-CH_2-CH-COO^-$$
$$\overset{|}{{}^+NH_3}$$
(b)

45. Give structural formulas for the following dipeptides.
    a. glycylalanine   b. alanylserine
46. Identify the following dipeptides.

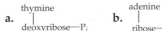

(a) $\overset{OH-CH_2}{\underset{{}^+NH_3-CH-CONH-CH-COO^-}{|}} \quad \overset{CH_2CH(CH_3)_2}{|}$

(b) $\overset{HS-CH_2}{\underset{{}^+NH_3-CH_2-CONH-CH-COO^-}{|}}$

47. The sweetener *aspartame* has the structure shown below. It consists of a CH_3O- group (at right) attached to a dipeptide. Identify the dipeptide.

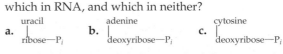

48. Refer to Problem 47 and draw the structure of the other dipeptide that contains the same amino acids as aspartame.
49. List the four different ways protein chains are bonded to one another.
50. Describe (a) the induced-fit model of enzyme action and (b) how an inhibitor deactivates an enzyme.

Nucleic Acids: Parts and Structure

51. Which of the following nucleotides would occur in DNA, which in RNA, and which in neither?
    a. thymine | deoxyribose—P_i
    b. adenine | ribose—P_i
    c. cytosine | ribose—P_i
52. Which of the following nucleotides would occur in DNA, which in RNA, and which in neither?
    a. uracil | ribose—P_i
    b. adenine | deoxyribose—P_i
    c. cytosine | deoxyribose—P_i
53. Identify the sugar and the base in the following nucleotide.

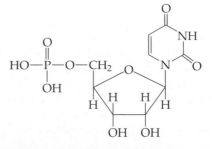

54. Identify the sugar and the base in the following nucleotide.

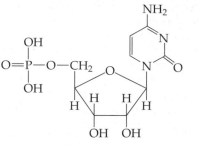

55. In DNA, which base would be paired with the base listed?
 a. cytosine
 b. adenine
 c. guanine
 d. thymine

56. In RNA, which base would pair with the base listed?
 a. adenine
 b. guanine
 c. uracil
 d. cytosine

DNA: Self-Replication

57. In replication, a parent DNA molecule produces two daughter molecules. What is the fate of each strand of the parent DNA double helix?

58. We say DNA controls protein synthesis, yet most DNA resides within the cell nucleus and protein synthesis occurs outside the nucleus. How does DNA exercise its control?

RNA: Protein Synthesis and the Genetic Code

59. Which nucleic acid or acids are involved in **(a)** the process referred to as transcription, and **(b)** the process referred to as translation?

60. Explain the roles of **(a)** mRNA and **(b)** tRNA in protein synthesis.

61. The base sequence along one strand of DNA is AATTCG. What would be the sequence of the complementary strand of DNA?

62. What sequence of bases would appear in the mRNA molecule copied from the original DNA strand shown in Problem 61?

63. If the sequence of bases along the mRNA strand is UCCGAU, what was the sequence along the DNA template?

64. What are the complementary triplets on tRNA for the following triplets on mRNA?
 a. UUU   **b.** CAU   **c.** AGC   **d.** CCG

65. What are the complementary base triplets on mRNA for the following triplets on tRNA?
 a. UUG   **b.** GAA

66. What are the complementary codons on mRNA for the following anticodons on tRNA?
 a. UCC   **b.** CAC

■ ADDITIONAL PROBLEMS

67. In the schematic at right, each circle represents a glucose unit. What substance is indicated?

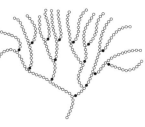

68. Which of the following is a carbohydrate? Which is a lipid? Which is a peptide?

(a)

(b)

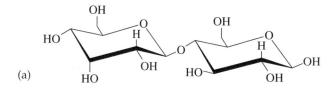

(c)

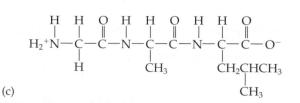

69. Human insulin molecules are made up of two chains with the following structure:

Chain A: G-I-V-E-Q-C-C-T-S-I-C-S-L-Y-Q-L-E-N-Y-C-N

Chain B: F-V-N-Q-H-L-C-G-D-H-L-V-E-A-L-Y-L-V-C-G-E-R-G-F-F-Y-T-P-K-T

The A chain has a loop formed by a disulfide linkage between the sixth and eleventh amino acids (designated A-6 to A-11) and is joined to the B chain by two disulfide linkages (A-7 to B-7 and A-20 to B-19). What level of protein structure is described by **(a)** the one-letter abbreviations describing each chain and **(b)** by the description of the loop and chain linkages?

70. Synthetic polymers can be formed from single amino acids. Write the formula for a segment of **(a)** polyglycine and **(b)** polyserine. Show at least four repeating units of each.

71. Identify the bases in the structures in **(a)** Problem 53 and **(b)** Problem 54 as purines or pyrimidines.

72. Answer the following questions for the molecule shown.
 a. Is the base a purine or a pyrimidine?
 b. Would the compound be incorporated in DNA or in RNA?

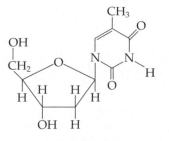

73. Answer the questions posed in Problem 72 for the following compound.

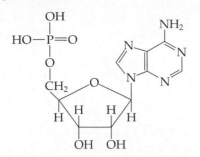

74. What amino acid is specified by the DNA triplet CTC? What amino acid is specified if a mutation changes the DNA triplet to CTT? To CTA? To CAC?

75. Many laundry detergents have enzymes to aid removal of biological stains such as food or blood. The enzyme subtilisin, produced by *Bacillus subtilis*, is a common one. The natural enzyme was not very effective because it was largely deactivated by the harsh detergent environment. A genetically modified form, in which the base sequence ~ACC-AGC-AUG-GCG~ is replaced by ~ACC-AGC-GCG-GCG~, is much more effective because it is more stable in the detergent solution. What is the amino acid sequence in this segment specified by the bases in **(a)** the original enzyme and **(b)** the genetically modified enzyme? You may consult Table 16.2.

76. In the novel *Jurassic Park* by Michael Crichton, the cloned dinosaurs were supposed to be genetically engineered so as to be unable to survive outside the island park. They were engineered so as to be unable to synthesize the amino acid lysine, which was supposed to make the dinosaurs dependent on their makers to feed them a source of lysine. Examine Table 16.3 and suggest a reason why this type of engineering would not keep the engineered creatures from surviving elsewhere.

77. With what mRNA codon would the tRNA in the diagram form a codon–anticodon base pair?

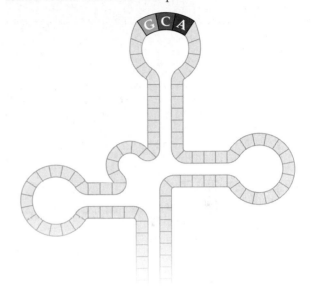

78. Write the **(a)** three-letter and **(b)** one-letter abbreviated versions of the following structural formula.

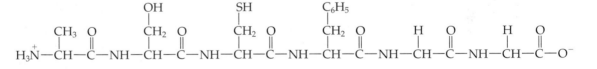

■ COLLABORATIVE GROUP PROJECTS

Prepare a PowerPoint, poster, or other presentation (as directed by your instructor) for presentation to the class.

79. Prepare a brief report on one of the following carbohydrates. List principal sources and uses of the carbohydrate.
    a. glycogen　　b. maltose　　c. fructose
    d. sucrose　　e. lactose　　f. starch

80. Prepare a brief report on one of the following lipids. List principal sources and uses of the lipid.
    a. palmitic acid　　b. oleic acid　　c. beef tallow
    d. palm oil　　e. corn oil　　f. lard

81. Prepare a brief report on one of the following applications of genetic engineering. List some advantages and disadvantages of the application.
    a. golden rice
    b. herbicide-resistant transgenic corn
    c. transgenic cotton that produces Bt toxin
    d. transgenic tobacco that produces human serum albumin
    e. transgenic tomatoes that have improved virus resistance
    f. transgenic rabbits that produce human interleukin-2, a protein that stimulates the production of T-lymphocytes that play a role in fighting selected cancers

82. Genetic testing carries both promise and peril and is an important area of bioethics today. After some online research, write a brief essay on one of the following or on a similar topic assigned by your instructor.
    a. In which cases might a person not want to be tested for a disease that his or her parent had?
    b. When is genetic testing most valuable? What privacy issues need to be addressed in this area?
    c. Does testing negative for the breast cancer genes mean that a woman doesn't need mammograms?

A balanced diet must provide many varied components. Protein from meat, fish, milk products, or plant products such as soy beans provides the building blocks of many of the body's parts. Numerous vitamins and minerals are needed for structural components and body processes. Carbohydrates, including starches and sugars, and fats provide the energy that fuels our bodies. And water, of course, is the solvent in which all the chemical reactions of the body occur.

FOOD

QUESTIONS YOU MAY HAVE ASKED YOURSELF

1. The doctor said my father has high triglycerides. What are they?
2. Is blood cholesterol primarily a problem for older people?
3. How much fat in the diet is too much?
4. Are large doses of vitamins beneficial?
5. Are some foods really "super foods" or "miracle foods"?
6. Should the government prohibit all food additives?

Molecular Gastronomy

FOOD! From holiday feasts to late-night snacks, many of life's joys involve food. Yet the prime purpose of food is not to give pleasure; it is to sustain life. Food supplies all the molecular building blocks from which our bodies are made and all the energy for our life activities. That energy comes ultimately from the Sun through photosynthesis.

In many places on Earth some people never have enough food. More than 800 million people are always hungry. According to the Hunger Project, begun in 1977, an estimated 15 million people died of hunger that year. Now that estimate is 9 million each year. Three-fourths are children under the age of five. Famine and wars make the news, but they cause just 10% of all deaths from hunger. Most result from chronic malnutrition. Worldwide, hunger and malnutrition are the greatest risk to health. And things may be getting worse. Food prices soared in 2008, triggering food riots in several countries. Perhaps as many as 100 million more people fell into poverty. The United Nations Food and Agriculture Organization (FAO) found 37 countries short of food as well as seed and fertilizer for planting new crops.

While millions starve, other locales have a surplus. Most of us in the Western world have an overabundance of food and often eat too much. Yet even in the United States, the U.S. Department of Agriculture estimates that 31 million people live in households that experience hunger or the risk of hunger. Nearly 9 million people, including more than 3.2 million children, actually experience hunger, frequently skipping meals or eating too little. These people often have lower-quality diets, or they resort to seeking food from emergency sources.

Worldwide, six children under five years old die every minute from eating food or drinking water contaminated with microorganisms. One of every eight people on Earth suffers from malnutrition severe enough to stunt physical and mental growth and to shorten life.

Many of us also frequently eat the wrong kinds of food. Our diet, often too rich in saturated fats, sugar, and alcohol, has been linked in part to at least five of the ten leading causes of death in the United States: heart disease, cancer, stroke, diabetes, and kidney disease. The 1999–2002 National Health and Nutrition Examination Survey estimated that more than 65% of the adults in the United States are overweight. Worldwide, according to the International Obesity Task Force, 1.7 billion people worldwide are overweight. This excess weight contributes to poor health, increased risk of diabetes, heart attacks, strokes, and some forms of cancer. It also leads to increased susceptibility to other diseases.

The human body is a conglomeration of chemicals. A chemical analysis of the human body would show that it is about two-thirds water. The body contains several minerals and thousands of other chemicals. Our foods are also made up of chemicals that enable children to grow, provide people with energy, and repair and replace body tissues.

In this chapter we discuss food, what it's made of, and the substances added to it—by accident or design.

THE FOOD WE EAT

The three main classes of foods are carbohydrates, fats, and proteins. These, too, are chemicals. For proper nutrition, our diet should include balanced proportions of these three foodstuffs, plus water, vitamins, minerals, and fiber.

17.1 Carbohydrates in the Diet

Some of the chemistry of carbohydrates is discussed in Section 16.3. The role of foods in health and fitness is discussed in eChapter 19.

Dietary carbohydrates include sugars and starches. Sugars are mainly monosaccharides and disaccharides; starches are polysaccharides.

Sweet Chemicals: Sugars

Sugars have been used for ages to make food sweeter. Two monosaccharides, *glucose* (or *dextrose*) and *fructose* (fruit sugar), and the disaccharide *sucrose* are the most common dietary sugars (Figure 17.1). Common table sugar is sucrose, usually obtained from sugarcane or sugar beets. Glucose is the sugar used by the cells of our bodies for energy. Because it is the sugar that circulates in the bloodstream, it is often called **blood sugar**. Fructose is found in honey and in some fruits, but much of it is made from glucose. Corn syrup, made from starch, is mainly glucose. *High-fructose corn syrup (HFCS)* is made by treating corn syrup with enzymes to convert

▶ **Figure 17.1** The principal sugars in our diet are sucrose (cane or beet sugar), glucose (corn syrup), and fructose (fruit sugar, often in the form of high-fructose corn syrup).

QUESTION: What is the molecular formula for glucose? For fructose? How are the two compounds related? How is sucrose related to glucose and fructose? *Hint:* You may refer to Chapter 16, if necessary.

much of the glucose to fructose. Fructose is sweeter than sucrose or glucose. Foods sweetened to the same degree with fructose have somewhat fewer calories than those sweetened with sucrose.

Since 1968, per-capita consumption of sugars in the United States has skyrocketed from 11 kg/year to 68 kg/year, including 28 kg of HFCS. Most are consumed in soft drinks, presweetened cereals, candy, and other highly processed foods with little or no other nutritive value. The sugars in sweetened foods provide calories and little else. They also contribute to tooth decay and obesity.

Digestion and Metabolism of Carbohydrates

Glucose and fructose are absorbed directly into the bloodstream from the digestive tract. Sucrose is hydrolyzed (split by water) during digestion to glucose and fructose.

$$\text{Sucrose} + H_2O \longrightarrow \text{Glucose} + \text{Fructose}$$

The disaccharide lactose occurs in milk. During digestion, it is hydrolyzed to two simpler sugars, glucose and galactose.

$$\text{Lactose} + H_2O \longrightarrow \text{Glucose} + \text{Galactose}$$

Nearly all human babies have the enzyme necessary to accomplish this breakdown, but many adults do not. People who lack the enzyme get digestive upsets from drinking milk, a condition called *lactose intolerance*. When milk is cooked or fermented, the lactose is at least partially hydrolyzed. People with lactose intolerance may still be able to enjoy cheese, yogurt, or cooked foods containing milk with little or no discomfort. Lactose-free milk, made by treating milk with an enzyme that hydrolyzes lactose, is available in most grocery stores.

All monosaccharides are converted to glucose during metabolism. Some babies are born with *galactosemia*, a deficiency of the enzyme that catalyzes the conversion of galactose to glucose. For proper nutrition, they require a synthetic formula in place of milk (Figure 17.2).

▲ **Figure 17.2** Infants born with galactosemia can thrive on a milk-free substitute formula.

Complex Carbohydrates: Starch and Cellulose

Starch and cellulose are both polymers of glucose, but the connecting links between the glucose units are different (Figure 16.6). Humans can digest starch but not cellulose. **Starch**, a polymer of α-glucose, is an important part of any balanced diet (Figure 17.3). **Cellulose**, a β-glucose polymer, is an important component of dietary fiber.

When digested, starch is hydrolyzed to glucose, as represented by the following equation:

$$(C_6H_{10}O_5)_n + n\,H_2O \xrightarrow{\text{Carbohydrases}} n\,C_6H_{12}O_6$$
$$\text{Starch} \qquad\qquad\qquad\qquad \text{Glucose}$$

The body then metabolizes the glucose, using it as a source of energy. Glucose is broken down through a complex set of more than 50 chemical reactions to produce carbon dioxide and water, with the release of energy.

$$C_6H_{12}O_6 + 6\,O_2 \longrightarrow 6\,CO_2 + 6\,H_2O + \text{Energy}$$

The net reaction is essentially the reverse of photosynthesis. In this way, animal organisms are able to make use of the energy from the Sun that was captured by plants in the process of photosynthesis.

Carbohydrates, which supply 4 kcal of energy per gram, are our bodies' preferred fuels. When we eat more than we can use, small amounts of carbohydrates can be stored in the liver and in muscle tissue as **glycogen** (animal starch), a highly branched polymer of α-glucose. Large excesses, however, are converted to fat for storage. Most health authorities recommend obtaining carbohydrates from a diet rich in whole grains, fruits, vegetables, and legumes (beans). We should minimize our intake of the simple sugars and refined starches found in many prepared foods.

▲ **Figure 17.3** Bread, flour, cereals, and pasta are rich in starches.

Cellulose is the most abundant carbohydrate. It is present in all plants, forming their cell walls and other structural features. Wood is about 50% cellulose, and cotton is almost pure cellulose. Unlike starch, cellulose cannot be digested by humans or other meat-eating animals. We get no caloric value from dietary cellulose because its glucose units are joined by beta linkages, and most animals lack the enzymes needed to break this beta glycoside bonding. Certain bacteria that live in the gut of termites and in the digestive tract of grazing animals such as cows do have such enzymes, however, so that these animals can convert cellulose to glucose. Cellulose does, however, play an important role in human digestion, providing *dietary fiber* that absorbs water and helps move food through the digestive tract in humans (Section 17.6).

Self-Assessment Questions

1. About _____ people in the world are hungry today.
   a. 1 in 4   **b.** 1 in 8   **c.** 1 in 12   **d.** 1 in 20

2. Digestion of sucrose produces
   a. galactose and fructose   **b.** galactose and glucose
   c. glucose and fructose   **d.** only glucose

3. People who have lactose intolerance are deficient in the
   a. enzyme that catalyzes the hydrolysis of lactose to galactose and glucose
   b. enzyme that catalyzes the conversion of galactose to glucose
   c. enzymes that catalyze the oxidation of lactose to CO_2 and H_2O
   d. taste receptors for lactose, leaving a bitter taste

4. Galactosemia is an inherited disease in which the child is deficient in the
   a. enzyme that catalyzes the hydrolysis of lactose to galactose and glucose
   b. enzyme that catalyzes the conversion of galactose to glucose
   c. enzymes that catalyze the oxidation of galactose to CO_2 and H_2O
   d. taste receptors for galactose, leaving a bitter taste

5. The monomer units of starch and cellulose, respectively, are
   a. α-amino acids and β-amino acids
   b. α-glucose and β-glucose
   c. β-glucose and α-glucose
   d. β-glucose and α-amino acids

6. Dietary carbohydrates serve us mainly as
   a. monomers for construction of cell walls
   b. precursors of hormones
   c. sources of energy
   d. substrates for fatty acid production

7. Small quantities of carbohydrates can be stored in liver and muscle tissue as
   a. amylose   **b.** cellulose   **c.** glucose   **d.** glycogen

8. Humans cannot digest cellulose because they lack enzymes for the hydrolysis of
   a. α-amino acid linkages   **b.** α-glucose linkages
   c. β-glucose linkages   **d.** sucrose

Answers: 1, b; 2, c; 3, a; 4, b; 5, b; 6, c; 7, d; 8, c

▲ **Figure 17.4** Cream, butter, margarine, cooking oils, and foods fried in fat are rich in fats. The average American diet contains too much fat; about 34% of our calories come from fat.

17.2 Fats and Cholesterol

Fats—esters of fatty acids and glycerol—are high-energy foods, yielding about 9 kcal of energy per gram. Some fats are "burned" as fuel for our activities. Others are used to build and maintain important constituents of our cells, such as cell membranes. The fat in our diet comes from many sources, some of which are shown in Figure 17.4.

Digestion and Metabolism of Fats

Dietary fats are mainly *triacylglycerols,* commonly called **triglycerides**. Fats are digested by enzymes called *lipases,* ultimately forming fatty acids and glycerol (Figure 17.5). Some fat molecules are hydrolyzed only to a monoglyceride (*monoacylglycerol;* glycerol combined with only one fatty acid) or a diglyceride (*diacylglycerol;* ester of two fatty acids and glycerol). Once absorbed, these products of fat digestion are reassembled into triglycerides, which are attached to proteins for transportation through the bloodstream.

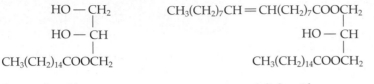

A monoglyceride   A diglyceride

Fats are stored throughout the body, principally in **adipose tissue**, in locations called **fat depots**. Storage depots around vital organs, such as the heart, kidneys, and spleen, cushion and help to prevent injury to these organs. Fat is also stored under the skin, where it helps to insulate against temperature changes.

When fat reserves are called on for energy, fat molecules are hydrolyzed back to glycerol and fatty acids. The glycerol can be burned for energy or converted to glucose. The fatty acids enter a process called the *fatty acid spiral* that removes carbon atoms two at a time. The two-carbon fragments can be used for energy or for the synthesis of new fatty acids.

$$CH_3(CH_2)_7CH{=}CH(CH_2)_7COOCH_2$$
$$CH_3(CH_2)_7CH{=}CH(CH_2)_7COOCH$$
$$CH_3(CH_2)_{14}COOCH_2$$
fat (triglyceride)

$$\xrightarrow[\text{lipase}]{\text{water}}$$

$$2\ CH_3(CH_2)_7CH{=}CH(CH_2)_7COOH$$
oleic acid
+
$$CH_3(CH_2)_{14}COOH$$
palmitic acid

+

$$HO{-}CH_2$$
$$HO{-}CH$$
$$HO{-}CH_2$$
glycerol

▲ **Figure 17.5** In the digestion of fats, triglyceride is hydrolyzed to fatty acids and glycerol in a reaction catalyzed by the enzyme lipase.

QUESTION: What diglyceride is formed by removal of the palmitic acid part of the triglyceride? What monoglyceride is formed by removal of both oleic acid parts? *Hint:* You may refer to Chapter 16, if necessary.

Fats, Cholesterol, and Human Health

Dietary *saturated fats* and cholesterol have been implicated in *arteriosclerosis* ("hardening of the arteries"). (Recall from Section 16.3 that saturated fats are those made of a large proportion of saturated fatty acids, such as palmitic and stearic.) Incidence of cardiovascular disease is strongly correlated with diets rich in saturated fats. As the disease develops, deposits form on the inner walls of arteries. Eventually these deposits harden, and the vessels lose their elasticity (Figure 17.6). Blood clots tend to lodge in the narrowed arteries, leading to a heart attack (if the blocked artery is in heart muscle) or a stroke (if the blockage occurs in an artery that supplies the brain).

1. The doctor said my father has high triglycerides. What are they? Triglycerides are a type of fat molecule in the bloodstream. High triglycerides are often the result of a diet high in fats, and they contribute to heart disease.

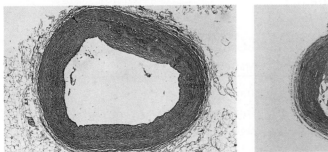

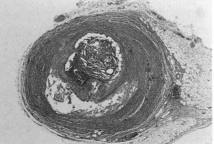

◀ **Figure 17.6** Photomicrographs of cross sections of a normal artery (a) and a hardened artery (b), showing deposits of plaque that contain cholesterol.

Emulsions

When oil and water are vigorously shaken together, the oil is broken up into tiny, microscopic droplets and dispersed throughout the water, a mixture called an *emulsion*. Unless a third substance has been added, the emulsion usually breaks down rapidly, the oil droplets recombining and floating to the surface of the water.

Emulsions can be stabilized by adding certain types of gum, a soap, or a protein that can form a protective coating around the oil droplets and prevent them from coming together. Compounds called *bile salts* keep tiny fat droplets suspended in aqueous media in the body during human digestion.

▲ Many foods are emulsions. Milk is an emulsion of butterfat in water. The stabilizing agent is a protein called *casein*. Mayonnaise is an emulsion of salad oil in water, stabilized by egg yolk.

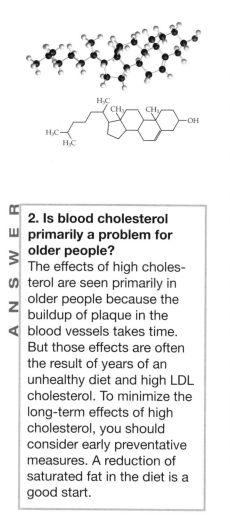

The plaque in clogged arteries is rich in cholesterol, a fatlike steroid alcohol found in animal tissues and various foods. Cholesterol is normally synthesized by the liver and is important as a constituent of cell membranes and a precursor to steroid hormones. High blood levels of cholesterol, like those of triglycerides, correlate closely with the risk of cardiovascular disease. Like fats, cholesterol is insoluble in water, as we can determine from its molecular formula ($C_{27}H_{45}OH$) and structure (margin). Cholesterol is transported in blood by water-soluble proteins. The cholesterol–protein combination is called a lipoprotein. A **lipoprotein** is any of a group of proteins combined with a lipid, such as cholesterol or a triglyceride.

Lipoproteins are usually classified according to their density (Table 17.1). Very-low-density lipoproteins (VLDLs) serve mainly to transport triglycerides, whereas low-density lipoproteins (LDLs) are the main carriers of cholesterol. LDLs carry cholesterol to the cells for use, and these lipoproteins are the ones that deposit cholesterol in arteries, leading to cardiovascular disease. High-density lipoproteins (HDLs) also carry cholesterol, but they carry it to the liver for processing and excretion. Exercise is thought to increase the levels of HDL, the lipoprotein sometimes called "good" cholesterol. High levels of LDL, called "bad" cholesterol, increase the risk of heart attack and stroke. The American Heart Association recommends a maximum of 300 mg/day of cholesterol for the general population and 200 mg/day for people with heart disease or at risk for it.

Fats differ in their effect on blood cholesterol levels. Many nutritionists advise us to use olive oil and canola oil as our major sources of dietary lipids because they contain a high percentage of monounsaturated fatty acids, which have been shown to lower LDL cholesterol. There is also statistical evidence that fish oils can prevent heart disease. For example, Greenlanders who eat a lot of fish have a low risk of heart disease despite a diet that is high in total fat and cholesterol. The probable effective agents are polyunsaturated fatty acids such as eicosapentaenoic acid (EPA) and docosahexaenoic acid (DHA):

$$CH_3(CH_2CH=CH)_5(CH_2)_3COOH$$
EPA

$$CH_3(CH_2CH=CH)_6(CH_2)_2COOH$$
DHA

2. Is blood cholesterol primarily a problem for older people?

The effects of high cholesterol are seen primarily in older people because the buildup of plaque in the blood vessels takes time. But those effects are often the result of years of an unhealthy diet and high LDL cholesterol. To minimize the long-term effects of high cholesterol, you should consider early preventative measures. A reduction of saturated fat in the diet is a good start.

Table 17.1	Lipoproteins in the Blood			
Class	**Abbreviation**	**Protein (%)**	**Density (g/mL)**	**Main Function**
Very low density	VLDL	5	1.006–1.019	Transport triglycerides
Low density	LDL	25	1.019–1.063	Transport cholesterol to the cells for use
High density	HDL	50	1.063–1.210	Transport cholesterol to the liver for processing and excretion

These fatty acids are known as *omega-3 fatty acids* because they have a carbon–carbon double bond that begins on the *third* carbon from the end opposite the COOH group—the *omega* end. Studies have also shown that diets with added omega-3 fatty acids lead to lower cholesterol and triglyceride levels in the blood.

Fats and oils containing double bonds can undergo hydrogenation. Hydrogenation of vegetable oils to produce semisolid fats is an important process in the food industry. The chemistry of this conversion process is essentially identical to the hydrogenation reaction described for alkenes in Section 9.4.

$$CH_3(CH_2)_7CH\!=\!CH(CH_2)_7COOH \xrightarrow[\text{Ni}]{H_2} CH_3(CH_2)_7CH_2CH_2(CH_2)_7COOH$$
Oleic acid (monounsaturated)   Stearic acid (saturated)

By properly controlling the reaction conditions, inexpensive vegetable oils (cottonseed, corn, soybean) can be partially hydrogenated, forming soft and pliable fats suitable for use in oleomargarine, or fully hydrogenated into harder fats like shortening. The consumer would get much greater unsaturation by using the oils directly, but most people would rather spread margarine than pour oil on their toast.

Concern about the role of saturated fats in raising blood cholesterol and clogging arteries caused many consumers to switch from butter to margarine. However, recently it was found that some of the unsaturated fats that remain after partial hydrogenation of vegetable oils have structures similar to saturated fats. These unsaturated fats raise cholesterol levels and increase the risk of coronary heart disease.

Figure 17.7 provides molecular models of three types of fatty acids. Most naturally occurring unsaturated fatty acids have a *cis* arrangement about the double bond. For example, oleic acid has the structure given by Figure 17.7a. Recall from Chapter 9 that a *cis* arrangement about double-bonded carbon atoms has both hydrogen atoms on the same side of the double bond. During hydrogenation, in some of the molecules the arrangement is changed so that the hydrogen atoms of the double-bonded carbon atoms are on opposite sides of the double bond, a *trans* arrangement Figure 17.7b). (Draw a straight line passing through the double-bonded carbon atoms. If the hydrogen atoms on those carbon atoms are on the same side of the line, the molecule represents a *cis* fatty acid. If they fall on opposite sides, it is a *trans* fatty acid.)

Note in Figures 17.7a and 17.7b that both saturated fatty acids (such as stearic acid) and *trans* fatty acids are more or less straight and can stack neatly together like logs. This maximizes the intermolecular attractive forces, making saturated and *trans* fatty acids both more likely to be solids than the *cis* fatty acids. On the other hand, *cis* fatty acids (Figure 17.7c) have a bend in their structure, fixed in position by the double bond. They have weaker intermolecular forces and are more likely to be liquids. Because *trans* fatty acids also resemble saturated fatty acids in their tendency to raise blood levels of LDL cholesterol, the FDA now requires manufacturers to label foods with the *trans* fat content.

Half of the total fat, three-fourths of the saturated fat, and all the cholesterol in a typical human diet come from animal products such as meat, milk, cheese, and eggs. Advertising that a vegetable oil (for example) contains no cholesterol is silly. No vegetable product contains cholesterol.

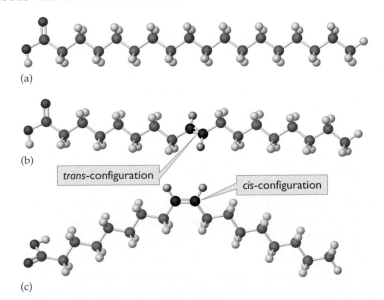

(a)

(b)

trans-configuration

cis-configuration

(c)

◄ **Figure 17.7** The structures of (a) stearic acid (a saturated fatty acid) and (b) the *trans* isomer of oleic acid (a *trans* unsaturated fatty acid) are very similar in shape and behave similarly in the body, whereas (c) oleic acid (a *cis* unsaturated fatty acid) has a very different shape.

3. How much fat in the diet is too much?
Not more than about 30% of the food energy in the diet should come from fat, and not more than 10% should come from saturated fat. In a 2000-kcal daily diet, this means less than 600 kcal from fat and less than 200 kcal from saturated fat.

Many Americans eat too much fat, especially saturated and *trans* fats. Health experts generally advise us to limit our intake of fats in general, and saturated and *trans* fats in particular. Many health professionals recommend that fat should not exceed 30% of total calories and no more than one-third of the fat should be saturated fat.

EXAMPLE 17.1 Nutrient Calculations

A single-serving pepperoni pan pizza has 38 g of fat and 780 total calories. (Recall that a food calorie is a kilocalorie.) Estimate the percentage of the total calories from fat. (Further calculations of this sort are found in eChapter 19.)

Solution

First, calculate the calories from fat. Fat furnishes about 9 kcal/g, so 38 g of fat furnishes

$$38 \text{ g fat} \times \frac{9 \text{ kcal}}{1 \text{ g fat}} = 340 \text{ kcal}$$

Now divide the calories from fat by the total calories. Then multiply by 100% to get the percentage (parts per 100).

$$\% \text{ calories from fat} = \frac{340 \text{ kcal}}{780 \text{ kcal}} \times 100\% = 44\%$$

(This answer is only an estimate for two reasons: The value of 9 kcal/g is only approximate and the "38 g of fat" is known only to the nearest gram. A proper answer is "about 44%.")

■ **EXERCISE 17.1A**

A serving of two fried chicken thighs has 51 g of fat and furnishes 720 kcal. What percentage of the total calories is from fat?

■ **EXERCISE 17.1B**

A lunch consists of a regular hamburger with 12 g of fat and 275 kcal, a serving of french fried potatoes with 12 g of fat and 240 kcal, and a chocolate milk shake with 9 g of fat and 360 kcal. What percentage of the total calories is from fat?

EXAMPLE 17.2 Fatty Acids

Following are line-angle structures of three fatty acids. **(a)** Which is a saturated fatty acid? **(b)** Which is a monounsaturated fatty acid? **(c)** Which is an omega-3 fatty acid?

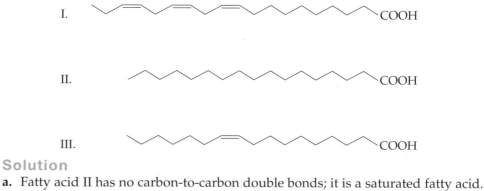

Solution

a. Fatty acid II has no carbon-to-carbon double bonds; it is a saturated fatty acid.

b. Fatty acid III has one carbon-to-carbon double bond; it is a monounsaturated fatty acid.

c. In fatty acid I, one of the carbon-to-carbon double bonds is three carbons removed from the end farthest from the carboxyl group (the omega end); it is an omega-3 fatty acid.

■ **EXERCISE 17.2A**

Which of the fatty acids in Example 17.2 is polyunsaturated?

■ **EXERCISE 17.2B**
Is fatty acid III likely to be more similar physiologically to fatty acid II or to fatty acid I? Explain.

Self-Assessment Questions

1. Fats are
   a. esters of glycerol and fatty acids
   b. esters of cholesterol and fatty acids
   c. polymeric steroids
   d. polymers of amino acids

2. Among lipoproteins, HDL is the
   a. bad form because it can block arteries
   b. form that carries cholesterol to the cells
   c. good form because it can clear arteries
   d. kind formed in the liver

Questions 3–5 refer to the following structures.
   a. $CH_3(CH_2)_{10}COOH$
   b. $CH_3(CH_2)_{16}COOH$
   c. $CH_3(CH_2)_7CH=CH(CH_2)_7COOH$
   d. $CH_3CH_2(CH=CHCH_2)_3(CH_2)_6COOH$

3. Which is a monounsaturated fatty acid?

4. Which is a polyunsaturated fatty acid?

5. Which is an omega-3 fatty acid?

6. In shape, *trans* fatty acid molecules resemble
   a. cholesterol
   b. disaccharides
   c. other unsaturated fatty acids
   d. saturated fatty acids

7. Most health professionals recommend our daily total intake of fat and of saturated fat, respectively, should be no more than what percentage of total calories?
   a. 15, 10   b. 30, 20   c. 30, 10   d. 50, 30

Answers: 1, a; 2, c; 3, c; 4, d; 5, d; 6, d; 7, c

17.3 Proteins: Muscle and Much More

As we saw in Chapter 16, proteins are polymers of amino acids. A gene carries the blueprint for a specific protein, and each protein serves a particular purpose. We require protein in our diet in order to provide the amino acids needed to make muscles, hair, enzymes, and many other cellular components vital to life.

Protein Metabolism: Essential Amino Acids

Proteins are broken down in the digestive tract into their component amino acids.

$$\text{Proteins} + n\,H_2O \xrightarrow{\text{Proteases}} \text{Amino acids}$$

From these amino acids, our bodies synthesize proteins for growth and repair of tissues. When a diet contains more protein than is needed for the body's growth and repair, the excess protein is used as a source of energy, providing about 4 kcal per gram.

The adult human body can synthesize all but nine of the amino acids needed for making proteins. These nine are called **essential amino acids**—isoleucine, lysine, phenylalanine, tryptophan, leucine, methionine, threonine, and valine—(see Table 16.3 for structures) and must be included in our diet. Each of the essential amino acids is a **limiting reactant** in protein synthesis. When the body is deficient in one of them, it can't make proper proteins.

An *adequate* (or *complete*) *protein* supplies all the essential amino acids in the quantities needed for the growth and repair of body tissues. Most proteins from

The distinction between essential and nonessential amino acids is somewhat ambiguous because some amino acids can be produced from others. For example, both methionine and cysteine contain sulfur, and although methionine cannot be synthesized from other precursors, cysteine can partially meet the need for methionine. Similarly, tyrosine can partially substitute for phenylalanine.

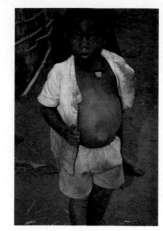

▲ **Figure 17.8** An extreme lack of proteins and vitamins causes a deficiency disease called kwashiorkor. The symptoms include retarded growth, discoloration of skin and hair, bloating, a swollen belly, and mental apathy.

animal sources contain all the essential amino acids in adequate amounts. Lean meat, milk, fish, eggs, and cheese supply adequate protein. Gelatin is one of the few inadequate animal proteins. It contains almost no tryptophan and has only small amounts of threonine, methionine, and isoleucine.

In contrast, most plant proteins are deficient in one or more amino acids. Corn protein has insufficient lysine and tryptophan, and people who subsist chiefly on corn may suffer from malnutrition even though they get adequate calories. Protein from rice is short of lysine and threonine. Wheat protein lacks enough lysine. Even soy protein, one of the best nonanimal proteins, is deficient in the essential amino acid methionine.

Protein Deficiency in Young and Old

Our requirement for protein is about 0.8 g per kilogram of body weight. Diets with inadequate protein are common in some parts of the world. A protein-deficiency disease called *kwashiorkor* (Figure 17.8) is rare in the developed countries but is common during times of famine in parts of Africa where corn is the major food. In the United States, protein deficiency occurs mainly among elderly persons in nursing homes.

Nutrition is especially important in a child's early years. This is readily apparent from the fact that the human brain reaches nearly full size by the age of two years. Early protein deficiency leads to both physical and mental retardation.

Vegetarian Diets

Green plants trap a small fraction of the energy that reaches them from the Sun. They use some of this energy to convert carbon dioxide, water, and mineral nutrients (including nitrates, phosphates, and sulfates) to proteins. Cattle eat plant protein, digest it, and convert a small portion of it to animal protein. It takes 100 g of protein feed to produce 4.7 g of edible beef or veal protein, an efficiency of only 4.7%. People eat this animal protein, digest it, and reassemble some of the amino acids into human protein. Some of the energy originally transformed by green plants is lost as heat at every step. If people ate the plant protein directly, one highly inefficient step would be skipped. A vegetarian diet conserves energy. Pork production, at a protein conversion efficiency of 12.1%, and chicken or turkey production (at 18.2%) are more efficient. Milk production (22.7%) and egg production (23.3%) are still more efficient but do not compare well with eating the protein directly.

Vegetarians generally are less likely than meat eaters to have high blood pressure. Vegetarian diets that are low in saturated fat can help us avoid or even reverse coronary artery disease. These diets also offer protection from some other diseases. However, although complete proteins can be obtained by eating a carefully selected mixture of vegetable foods, total (*vegan*) vegetarianism can be dangerous, especially for young children. Even when the diet includes a wide variety of plant materials, an all-vegetable diet is short in several nutrients, including vitamin B_{12} (a nutrient not found in plants), calcium, iron, riboflavin, and vitamin D (required by children not exposed to sunlight). A modified vegetarian (*ovolacto*) diet that includes eggs, milk, and milk products can provide excellent nutrition, with red meat totally excluded.

A variety of ethnic dishes supply relatively good protein by combining vegetable foods, usually a cereal grain with a legume (beans, peas, peanuts, and so on). The grain is deficient in tryptophan and lysine, but it has sufficient methionine. The legume is deficient in methionine, but it has enough tryptophan and lysine. A few such combinations are listed in Table 17.2. Peanut butter sandwiches are a popular American example of a legume–cereal grain combination.

Table 17.2 Ethnic Foods That Combine a Cereal Grain with a Legume

Group	Food*
Mexicans	Corn tortillas and beans
Japanese	Rice and soybean curds (tofu)
English working classes	Baked beans on toasted bread
American Indians	Corn and beans (succotash)
Western Africans	Rice and peanuts (ground nuts)
Cajuns (Louisiana)	Red beans and rice

*Cereal grains are in red, and legumes are in blue.

Self-Assessment Questions

1. Kilocalories in 1 g each of carbohydrate, fat, and protein, respectively, are
   a. 4, 7, 4   **b.** 4, 4, 7   **c.** 4, 9, 4   **d.** 4, 12, 6

2. What are the end products of the hydrolysis of a protein?
   a. amino acids   **b.** fatty acids
   c. monosaccharides   **d.** nucleotides

3. An essential amino acid is one that
   a. is necessary for the synthesis of all polypeptides
   b. is needed in greater quantities than some others
   c. is synthesized in the body   **d.** must be included in the diet

4. Which of the following foods furnishes complete protein?
   a. beans   **b.** brown rice   **c.** fish   **d.** gelatin with fruit

5. The daily protein requirement for a 60-kg male is
   a. 0.8 g   **b.** 48 g   **c.** 60 g
   d. 1.5 times that of a 60-kg female

6. One can get complete protein by combining whole wheat bread and
   a. butter   **b.** jelly   **c.** peanut butter   **d.** rice

7. Strict vegetarian diets are often deficient in
   a. B vitamins and vitamin E   **b.** calcium and folic acid
   c. vitamin B_{12} and iron   **d.** vitamin B_{12} and folic acid

Answers: 1, c; 2, a; 3, d; 4, c; 5, b; 6, c; 7, c

17.4 Minerals: Inorganic Chemicals and Life

Thirty elements, listed in Table 17.3, are known to be essential to one or more living organisms. Among these are six structural elements found in organic compounds such as carbohydrates, fats, and proteins. Several inorganic substances, called

Table 17.3 Elements Essential to Life

Element	Symbol	Form Used	Element	Symbol	Form Used
Bulk Structural Elements			**Ultratrace Elements**		
Hydrogen	H	Covalent	Manganese	Mn	Mn^{2+}
Carbon	C	Covalent	Molybdenum	Mo	Mo^{2+}
Oxygen	O	Covalent	Chromium	Cr	?
Nitrogen	N	Covalent	Cobalt	Co	Co^{2+}
Phosphorus*	P	Covalent	Vanadium	V	?
Sulfur*	S	Covalent	Nickel	Ni	Ni^{2+}
			Cadmium	Cd	Cd^{2+}
Macrominerals			Tin	Sn	Sn^{2+}
Sodium	Na	Na^+	Lead	Pb	Pb^{2+}
Potassium	K	K^+	Lithium	Li	Li^+
Calcium	Ca	Ca^{2+}	Fluorine	F	F^-
Magnesium	Mg	Mg^{2+}	Iodine	I	I^-
Chlorine	Cl	Cl^-	Selenium	Se	SeO_4^{2-}?
Phosphorus*	P	$H_2PO_4^-$	Silicon	Si	?
Sulfur*	S	SO_4^{2-}	Arsenic	As	?
			Boron	B	H_3BO_3
Trace Elements					
Iron	Fe	Fe^{2+}			
Copper	Cu	Cu^{2+}			
Zinc	Zn	Zn^{2+}			

* Note that phosphorus and sulfur each appear twice; they are structural elements and are also components of the macrominerals phosphate and sulfate.

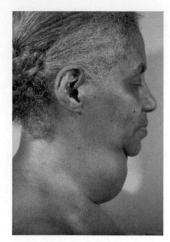

▲ **Figure 17.9** A person with goiter. The swollen thyroid gland in the neck results from a dietary deficiency of the trace element iodine.

Iron is an example of a substance that can be both essential and toxic. The National Institutes of Health recommends daily intake of 8 mg for adult males, 18 mg for adult females. The institute also warns that a 200-mg dose of iron in children can cause death.

dietary minerals, are vital to life. Minerals represent about 4% of the weight of the human body. Eleven elements that comprise the structural elements and the macrominerals make up more than 99% of all the atoms in the body. The 19 trace elements include iron, zinc, copper, and 16 others called *ultratrace elements*.

Minerals serve a variety of functions; that of iodine is quite dramatic. A small amount of iodine is necessary for proper thyroid function. An iodine deficiency has dire effects, of which goiter is perhaps the best known (Figure 17.9). Iodine is available naturally in seafood, but, to guard against iodine deficiency, a small amount of potassium iodide (KI) is often added to table salt. The use of iodized salt has greatly reduced the incidence of goiter.

Iron(II) ions (Fe^{2+}) are necessary for proper function of the oxygen-transporting compound hemoglobin. Without sufficient iron, oxygen supply to body tissues is reduced, and anemia, a general weakening of the body, results. Foods especially rich in iron compounds include red meat and liver. It appears that most adult males need very little dietary iron because iron seems to be retained by the body and lost mainly through bleeding.

Calcium and phosphorus are necessary for the proper development of bones and teeth. Growing children need about 1.5 g of each mineral per day. These elements are available from milk. The calcium and phosphorus needs of adults are less widely known but are very real just the same. For example, calcium ions are necessary for the coagulation of blood (to stop bleeding) and for maintenance of the rhythm of the heartbeat. Phosphorus occurs as phosphate units in adenosine triphosphate (ATP), the "energy currency" of the body. Phosphorus is necessary for the body to obtain, store, and use energy from foods. As phosphate units, phosphorus is an important part of the nucleic acid backbone (Section 15.9).

Sodium ions and chloride ions make up sodium chloride (salt). In moderate amounts, salt is essential to life. It is important in the exchange of fluids between cells and plasma, for example. The presence of salt increases water retention, however, and a high volume of retained fluids can cause swelling and high blood pressure (hypertension). Most physicians agree that our diets generally contain too much salt. About 65 million people in the United States suffer from hypertension, the leading risk factor for stroke, heart attack, kidney failure, and heart failure. Antihypertensives are among the most widely prescribed drugs in the United States.

Iron, copper, zinc, cobalt, manganese, molybdenum, calcium, and magnesium are essential to the proper functioning of metalloenzymes, which are life sustaining. A great deal remains to be learned about the role of inorganic chemicals in our bodies. Bioinorganic chemistry is a flourishing area of research.

Self-Assessment Questions

1. About what percentage of our body weight is minerals?
   a. 1% **b.** 4% **c.** 10% **d.** 20%
2. A deficiency of iodide results in
   a. anemia **b.** goiter **c.** osteoporosis **d.** scurvy
3. Iron in blood cells is found mainly in
   a. cytochromes **b.** hemoglobin **c.** myoglobin **d.** plasma
4. Iron is swiftly depleted from the body by
   a. blood loss **b.** energetic exercise
   c. inactivity **d.** weight loss
5. Most of the calcium and phosphorus in the body are in the
   a. blood **b.** bones and teeth
   c. liver **d.** muscle tissue
6. People with high blood pressure are usually told to restrict intake of
   a. Ca^{2+} **b.** Mg^{2+} **c.** Na^+ **d.** $SO_4{}^{2-}$

Answers: 1, b; 2, b; 3, b; 4, a; 5, b; 6, c

17.5 The Vitamins: Vital, but Not All Are Amines

Why are British sailors called "limeys"? And what does this term have to do with food? Sailors on long voyages have been plagued since early times by *scurvy*. In 1747 Scottish navy surgeon James Lind showed that this disease could be prevented by including fresh fruit and vegetables in the diet. Fresh fruits conveniently carried on long voyages (there was no refrigeration) were limes, lemons, and oranges. British ships put to sea with barrels of limes aboard, and sailors ate a lime or two every day. That is how they came to be known as "lime eaters," or simply limeys.

In 1897 the Dutch scientist Christiaan Eijkman showed that polished rice lacked something found in the hull of whole-grain rice. Lack of that substance caused the disease *beriberi*, which was a serious problem in the Dutch East Indies at that time. A British scientist, F. G. Hopkins, found that rats fed a synthetic diet of carbohydrates, fats, proteins, and minerals were unable to sustain healthy growth. Again, something was missing.

In 1912 Casimir Funk, a Polish biochemist, coined the word *vitamine* (from the Latin word *vita,* meaning "life") for these missing factors. Funk thought all these factors contained an amine group. In the United States, the final *e* was dropped after it was found that not all the factors were amines. The generic term became *vitamin.* Eijkman and Hopkins shared the 1929 Nobel Prize in physiology and medicine for their discoveries relating to vitamins.

Vitamins are specific organic compounds that are required in the diet (in addition to the usual proteins, fats, carbohydrates, and minerals) to prevent specific diseases. Some vitamins, along with their sources and deficiency symptoms, are shown in Table 17.4. The role of vitamins in the prevention of deficiency diseases, such as those shown in Figure 17.10, has been well established.

Vitamins do not share a common chemical structure. They can, however, be divided into two broad categories: *fat-soluble vitamins* (A, D, E, and K) and *water-soluble vitamins* (B vitamins and vitamin C). The fat-soluble vitamins incorporate a high proportion of hydrocarbon elements. They contain one or two oxygen atoms but are only slightly polar as a whole.

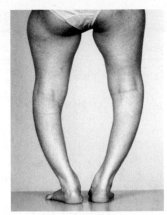

(a)

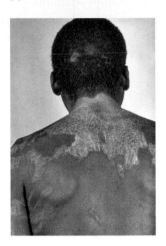

(b)

▲ **Figure 17.10** (a) This person has softened bones caused by a deficiency of vitamin D. (b) Inflammation and abnormal pigmentation characterize pellagra, caused by a deficiency of niacin.

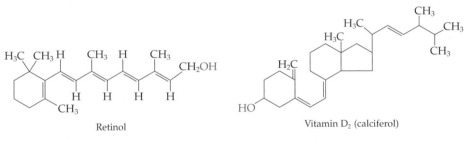

Retinol

Vitamin D₂ (calciferol)

Fat-soluble vitamins

In contrast, a water-soluble vitamin contains a high proportion of the electronegative atoms oxygen and nitrogen, which can form hydrogen bonds to water; therefore, the molecule as a whole is soluble in water.

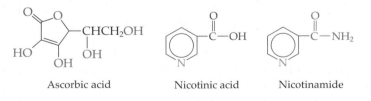

Ascorbic acid   Nicotinic acid   Nicotinamide

Water-soluble vitamins (sites for hydrogen bonding in color)

Fat-soluble vitamins dissolve in the fatty tissue of the body, where reserves can be stored for future use. An adult can store several years' supply of vitamin A. If the diet becomes deficient in vitamin A, these reserves are mobilized, and the adult remains

Vitamin*	Name	Sources	Deficiency Symptoms
Fat-Soluble Vitamins			
A	Retinol	Fish, liver, eggs, butter, cheese; also a vitamin precursor in carrots and other vegetables	Night blindness
D_2	Calciferol	Cod liver oil, irradiated ergosterol (milk supplement)	Rickets
E	α-Tocopherol	Wheat germ oil, green vegetables, egg yolks, meat	Sterility, muscular dystrophy
K_1	Phylloquinone	Spinach, other green leafy vegetables	Hemorrhage
Water-Soluble Vitamins			
B_1	Thiamine	Germ of cereal grains, legumes, nuts, milk, and brewer's yeast	Beriberi—polyneuritis resulting in muscle paralysis, enlargement of heart, and ultimately heart failure
B_2	Riboflavin	Milk, red meat, liver, egg white, green vegetables, whole wheat flour (or fortified white flour), and fish	Dermatitis, glossitis (tongue inflammation)
B_3	Niacin	Red meat, liver, collards, turnip greens, yeast, and tomato juice	Pellagra—skin lesions, swollen and discolored tongue, loss of appetite, diarrhea, various mental disorders (Figure 17.10)
B_6	Pyridoxine	Eggs, liver, yeast, peas, beans, and milk	Dermatitis, apathy, irritability, and increased susceptibility to infections; convulsions in infants
	Folic acid	Liver, kidney, mushrooms, yeast, and green leafy vegetables	Anemias (folic acid is used to treat megaloblastic anemia, a condition characterized by giant red blood cells); neural tube defects in fetuses of deficient mothers
B_{12}	Cyanocobalamine	Liver, meat, eggs and fish (not found in plants)	Pernicious anemia
C	Ascorbic acid	Citrus fruits, tomatoes, green peppers	Scurvy

Table 17.4 Some of the Vitamins

* Some vitamins exist in more than one chemical form. We name a common form of each here.

free of the deficiency disease for quite a while. However, a small child who has not built up a store of the vitamin soon exhibits deficiency symptoms. Children in developing countries often are permanently blinded as a result of vitamin A deficiency.

Because they are efficiently stored in the body, overdoses of fat-soluble vitamins can have adverse effects. Large excesses of vitamin A cause irritability, dry skin, and a feeling of pressure inside the head. Too much vitamin D can cause pain in the bones, hard deposits in the joints, nausea, diarrhea, and weight loss. Vitamins E and K are also fat soluble, but they are metabolized and excreted. They are not stored to the extent that vitamins A and D are, and excesses seldom cause problems.

The body has a limited capacity to store water-soluble vitamins. It excretes anything over the amount that can be used immediately. Water-soluble vitamins are needed frequently, every day or so. Some foods lose their vitamin content when they are cooked in water and then drained. The water-soluble vitamins go down the drain with the water.

Self-Assessment Questions

1. Vitamins
   a. are all amines
   b. are best when taken as supplements
   c. share a common functional group
   d. are specific organic compounds required in the diet
2. The fat-soluble vitamins in the body are
   a. A, B, D, E   b. A, D, E, K   c. the B complex   d. the B complex and C
3. Ascorbic acid is the chemical name for vitamin
   a. B_6 b. C   c. D   d. K

Answers: 1, d; 2, b; 3, b

17.6 Other Essentials: Fiber and Water

We need carbohydrates, proteins, fats, minerals, and vitamins, but some other items in our diets are also important. One is *fiber*; another is *water*.

Dietary Fiber

Dietary fiber may be soluble or insoluble. The insoluble fiber is usually cellulose, while the soluble fibers are generally sticky materials called gums and pectins. High-fiber diets prevent constipation and are an aid to dieters. Fiber has no calories because the body doesn't absorb it. Therefore, high-fiber foods such as fruits and vegetables are low in fat and often low in calories. Fiber takes up space in the stomach, making us feel full, so we eat less food. Soluble fiber lowers cholesterol levels, perhaps by removing bile acids that digest fat. It may also help control blood sugar by delaying stomach emptying, thus slowing sugar absorption after a meal. This may reduce the amount of insulin needed by a diabetic. These properties make dietary fiber beneficial to people with high blood pressure, diabetes, heart disease, and diverticulitis.

A N S W E R

4. Are large doses of vitamins beneficial? Although there is some evidence that large doses of antioxidants such as vitamin C may have beneficial effects, "megadoses" of most vitamins have not generally been shown to have any significant benefit. In fact, some of the fat-soluble vitamins are toxic in large amounts. There are well-documented examples of polar explorers who suffered poisoning from eating the liver of polar bears or seals, which contain extremely high levels of vitamin A.

It's a Drug! No, It's a Food! No, It's . . . a Dietary Supplement!

Almost every day we see advertisements for products that are claimed to promote rapid weight loss, improve stamina, aid memory, or provide other seemingly miraculous benefits. (There are thousands of such products; see Collaborative Group Project 72 for some examples.) Although these products may appear to act as drugs, many are not classified as such. Legally, they are *dietary supplements*. The Dietary Supplement Health and Education Act, enacted by the U.S. Congress in 1994, changed the law so that dietary supplements are now regulated as foods and not as drugs. Manufacturers are permitted to describe some specific benefits that may be attributed to use of the supplement. However, they must also include a disclaimer, such as "This statement has not been evaluated by the Food and Drug Administration. This product is not intended to diagnose, treat, cure, or prevent any disease."

Some supplements are simply combinations of various vitamins. Others contain minerals, amino acids, herbs or other plant materials, other nutrients, or ingredients and extracts of animal and plant origin.

Should you take a particular supplement? Maybe; maybe not. A supplement containing omega-3 oils might well aid in reducing LDL cholesterol. But a tablespoon a day of a special preparation of vitamins and minerals is not likely to reverse the effects of aging, no matter how brightly colored the bottle, no matter how attractive the person in the advertisement. Perhaps a judicious application of the FLaReS principles (see this chapter's Critical Thinking Exercises) to the manufacturer's claims will provide insight.

Supplement Facts

Serving size 1 Tablet

Amount Per Serving	%DV
Vitamin B-6 (as pyridoxine HCl)......5 mg	250%
Folic Acid....................................400 mcg	250%
Vitamin B-12.............................1000 mcg	16666%

DIRECTIONS: As a dietary supplement, take 1 or 2 tablets daily. Allow tablet to dissolve under tongue. Conforms to USP <2091> for weight.

This statement has not been evaluated by the Food and Drug Administration. This product is not intended to diagnose, treat, cure, or prevent any disease.

▲ Most dietary supplements do not go through the rigorous testing that is required by law for all prescription drugs and over-the-counter drugs.

In 2002, worldwide consumption of beverages included an estimated 195 billion L of milk, 139 billion L of coffee, 185 billion L of soda pop, 143 billion L of beer, and 131 billion L of bottled water.

Water

Much of the food we eat is mainly water. It should come as no surprise that tomatoes are 90% water or that melons, oranges, and grapes are largely water. But water is one of the main ingredients in practically all foods, from roast beef and seafood to potatoes and onions.

In addition to the water we get in our food, we need to drink about 1.0–1.5 L of water each day. We could satisfy this need by drinking plain water, but often we choose other beverages—milk, coffee, tea, and soft drinks. Many people also drink beverages that contain ethyl alcohol. These are made by fermenting grains or fruit juices. Beer is usually made from malted barley, and wine from grape juice. Even these alcoholic beverages are mainly water. Water as a nutrient is discussed further in eChapter 19.

Self-Assessment Questions

1. Eating foods high in fiber
   a. can cause constipation
   b. helps build muscle
   c. helps to fill you up
   d. provides quick energy

2. Soluble dietary fiber
   a. acts the same as insoluble fiber
   b. may help lower cholesterol levels
   c. is not affected by food processing
   d. is a good source of trace elements

3. In addition to the water supplied by foods, the amount of additional water we need each day is about
   a. 100 mL   b. 250 mL   c. 1.5 L   d. 4 L

4. How many kilocalories are in 1 g of water?
   a. 0   b. 1   c. 4   d. 7

Answers: 1, c; 2, b; 3, c; 4, a

Weight loss through diet and exercise is discussed in eChapter 19.

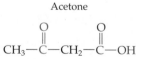

Acetone

Acetoacetic acid

B-Hydroxybutyric acid

▲ **Figure 17.11** The three ketone bodies.

QUESTION: Identify the functional groups in each of the compounds. Which ketone body is not a ketone?

17.7 Starvation and Fasting

When the human body is totally deprived of food, whether voluntarily or involuntarily, the condition is known as **starvation.** Involuntary starvation is a serious problem in much of the world. Although starvation is seldom the sole cause of death, those weakened by malnutrition succumb readily to disease. Even a seemingly minor disease, such as measles, can become life threatening.

Metabolic changes similar to those occurring in starvation take place during fasting. During total fasting, the body's glycogen stores are depleted in less than a day and the body calls on its fat reserves. Fat is first taken from around the kidneys and the heart. Then it is removed from other parts of the body, eventually even from the bone marrow.

Increased dependence on stored fats as an energy source leads to *ketosis*, a condition characterized by the appearance of compounds called *ketone bodies* in the blood and urine (Figure 17.11). Ketosis rapidly develops into *acidosis*; the blood pH drops, and oxygen transport is hindered. Oxygen deprivation leads to depression and lethargy.

Acidosis also is associated with the insulin-deficiency disease diabetes. (See the essay, "Diabetes," in Section 18.7.) Insulin enables the body's cells to take up glucose from the bloodstream. A lack of insulin causes the liver to act as though the cells are starving, and they will start burning fat for energy while blood sugar levels rise. This fat metabolism leads to the production of ketone bodies and subsequent acidosis.

In the early stages of a *total* fast, body protein is metabolized at a relatively rapid rate. After several weeks, the rate of protein breakdown slows considerably as the brain adjusts to using the breakdown products of fatty acid metabolism for its energy source. When fat reserves are substantially depleted, the body must again draw heavily on its structural proteins for its energy requirements. The emaciated appearance of a starving individual is due to the depletion of muscle proteins.

Processed Food: Less Nutrition

Malnutrition need not be due to starvation or dieting. It can be the result of eating too much highly processed food.

Whole wheat is an excellent source of vitamin B$_1$ and other vitamins. To make white flour, the wheat germ and bran are removed from the grain. This greatly increases the storage life of the flour, but the remaining material has few minerals or vitamins and little fiber. We eat the starch and use much of the germ and bran for animal food. Our cattle and hogs often get better nutrition than we do. Similarly, polished rice has had most of its protein and minerals removed, and it has almost no vitamins. The disease beriberi became prevalent when polished rice was introduced into Southeast Asia.

When many fruits and vegetables are peeled, they lose most of their vitamins, minerals, and fiber. The peels are often dumped (directly or through a garbage disposal) into a sewage system, where they contribute to water pollution (Chapter 14). The heat used to cook food also destroys some vitamins. If water is used in cooking, some of the water-soluble vitamins (C and B complex) and some of the minerals are often drained off and discarded with the water.

It is estimated that 90% of the food budget of an average family in the United States goes to buy processed foods. A diet of hamburgers, potato chips, and colas is lacking in many essential nutrients. Highly processed convenience foods threaten to leave the people of developed nations obese but poorly nourished despite their abundance of food.

Self-Assessment Questions

1. In fasting or starvation, glycogen stores are depleted in about
   a. 1 hour **b.** 1 day **c.** 1 week **d.** 1 month

2. In the intermediate stages of starvation (1–4 days), the body draws on its reserves of
   a. amino acids **b.** fat **c.** glycogen **d.** protein

3. In prolonged starvation (a few weeks), for its energy source the body uses its
   a. adipose tissue **b.** amino acid reserves
   c. glycogen stores **d.** structural proteins

4. Which of the following is *not* a ketone body?
   a. CH_3COCH_3 **b.** $CH_3CHOHCH_2COCH_3$
   c. $CH_3CHOHCH_2CHOHCH_3$ **d.** CH_3COCH_2COOH

Answers: 1, b; 2, b; 3, d; 4, c

FOOD ADDITIVES

The label reads "egg whites, vegetable oils, nonfat dry milk, lecithin, mono- and diglycerides, propylene glycol monostearate, xanthan gums, sodium citrate, aluminum sulfate, artificial flavor, iron phosphate, niacin, riboflavin, and irradiated ergosterol." Ingredients on a food label are listed in decreasing order by weight, but you recognize only a few of the ingredients. Just what is in the food we eat? There are more than 3000 substances that together comprise a list, which can be found on the FDA's website, called *Everything Added to Food in the United States* (*EAFUS*). We examine some of those in the following sections.

A N S W E R

5. Are some foods really "super foods" or "miracle foods"?
Some foods do provide special benefits. For example, flaxseed has cholesterol-reducing properties, cherries have been found to alleviate symptoms of gout in some people, and cranberries aid in preventing urinary tract infections. However, no single food or special combination of foods will magically bring perfect health, especially if the remainder of the diet is unhealthy. A few tablespoons of flaxseed with your breakfast cereal doesn't offset the adverse effects of a 500-calorie sugar-and-fat laden coffee drink.

More to Explore
Ehrenberg, Rachel. "What's Cookin'," *Science News*, March 29, 2008, pp. 202–204, and O'Driscoll, Cath. "Good Chemists Make Great Chefs." *Chemistry and Industry*, May 3, 2004, pp. 12–13. Science in the kitchen.

17.8 Additives to Enhance Nutrition and Taste

6. Should the government prohibit all food additives? Many additives provide useful improvements in food. For example, most people find that unsalted bread has little taste. Which soft drink would you prefer if they were all colorless? Canned tomatoes processed without calcium chloride may be soft and mushy.

On the label of many processed foods, most of the substances listed on the label are **food additives**, substances other than basic foodstuffs that are present in food as a result of some aspect of production, processing, packaging, or storage. Because food processing removes certain essential food substances, some additives are included in prepared food to increase its nutritional value. Other substances are added to enhance color and flavor, to retard spoilage, to provide texture, to sanitize, to bleach, to ripen (or prevent ripening), to control moisture levels, or to control foaming. There are several thousand different additives. Sugar, salt, and corn syrup are used in the greatest amounts. These three, plus citric acid, baking soda, vegetable colors, mustard, and pepper, make up more than 98% (by weight) of all additives.

Some food additives, such as salt to preserve meat and fish and spices to flavor and preserve foods, have been used since ancient times. Throughout the centuries, other additives were found to be useful. Movement of the population from farms to cities has increased the necessity for using preservatives. More widespread consumption of convenience foods has also led to greater use of additives.

In the United States, food additives are regulated by the Food and Drug Administration (FDA). The original Food, Drug, and Cosmetic Act was passed by Congress in 1938. Under this act, the FDA had to prove that an additive was unsafe before its use could be prevented. The Food Additives Amendment of 1958 shifted the burden of proof to the food industry. A company that wishes to use a food additive must first furnish proof to the FDA that the additive is safe for the intended use. The FDA can also regulate the quantity of additives that can be used.

Some people express concern about the "chemicals in our food," but food itself consists of chemicals. Table 17.5 shows the chemical composition of a typical breakfast. Many of the chemicals in this breakfast might be harmful in large amounts but are harmless in the trace amounts that occur naturally in foods. Indeed, some make important contributions to delightful flavors and aromas.

Our bodies are also collections of chemicals. If broken down into its elements (Table 17.6), your body would be worth only a few dollars. It is the unique combination and arrangement of the elements in every human body that make you different from everyone else and make each individual's worth beyond measure. Since food is chemical and we are chemical, we shouldn't have to worry about chemicals in our food—except perhaps for some specific ones in specific amounts.

Additives That Improve Nutrition

The first nutrient supplement approved by the Bureau of Chemistry of the U.S. Department of Agriculture, potassium iodide (KI), was added to table salt in 1924 to reduce the incidence of goiter (recall Figure 17.9). (The Bureau of Chemistry later became the FDA.)

Several other chemicals are added to foods specifically to prevent deficiency diseases. Addition of vitamin B_1 (thiamine) to polished rice is essential in the Far East, where beriberi still is a problem. The replacement of the B vitamins thiamine, riboflavin, folic acid, and niacin (which are removed in processing) and the addition of iron, usually as ferrous carbonate ($FeCO_3$), to flour is called **enrichment**. Enriched bread or pasta made from this flour still isn't as nutritious as bread made from whole wheat. It lacks vitamin B_6, pantothenic acid, zinc, magnesium, and fiber, nutrients usually provided by whole-grain flour. Despite these shortcomings, the enrichment of bread, corn meal, and cereals has essentially eliminated pellagra, a disease that once plagued the southern United States.

Vitamin C (ascorbic acid) is frequently added to fruit juices, flavored drinks, and beverages. Although our diets generally contain enough ascorbic acid to prevent scurvy, some scientists recommend a much larger intake than minimum daily requirements. Vitamin D is added to milk in developed countries, and this use of fortified milk has led to the almost total elimination of rickets. Similarly, vitamin A, which occurs naturally in butter, is added to margarine so that the substitute more nearly matches butter in nutritional quality.

Table 17.5 Your Breakfast—As Seen by a Chemist*

Chilled Melon

Starches	Anisyl propionate
Sugars	Amyl acetate
Cellulose	Ascorbic acid
Pectin	Vitamin A
Malic acid	Riboflavin
Citric acid	Thiamine
Succinic acid	

Scrambled Eggs

Ovalbumin	Lecithin
Conalbumin	Lipids (fats)
Ovomucoid	Fatty acids
Mucin	Butyric acid
Globulins	Acetic acid
Amino acids	Sodium chloride
Lipovitellin	Lutein
Livetin	Zeazanthine
Cholesterol	Vitamin A

Sugar-Cured Ham

Actomyosin	Adenosine triphosphate (ATP)
Myogen	Glucose
Nucleoproteins	Collagen
Peptides	Elastin
Amino acids	Creatine
Myoglobin	Pyroligneous acid
Lipids (fats)	Sodium chloride
Linoleic acid	Sodium nitrate
Oleic acid	Sodium nitrite
Lecithin	Sodium phosphate
Cholesterol	Sucrose

Coffee

Caffeine	Acetone
Essential oils	Methyl acetate
Methanol	Furan
Acetaldehyde	Diacetyl
Methyl formate	Butanol
Ethanol	Methylfuran
Dimethyl sulfide	Isoprene
Propionaldehyde	Methylbutanol

Cinnamon Apple Chips

Pectin	Propanol
Cellulose	Butanol
Starches	Pentanol
Sucrose	Hexanol
Glucose	Acetaldehyde
Fructose	Propionaldehyde
Malic acid	Acetone
Lactic acid	Methyl formate
Citric acid	Ethyl formate
Succinic acid	Ethyl acetate
Ascorbic acid	Butyl acetate
Cinnamyl alcohol	Butyl propionate
Cinnamaldehyde	Amyl acetate
Ethanol	

Toast and Coffee Cake

Gluten	Mono- and diglycerides
Amylose	Methyl ethyl ketone
Amino acids	Niacin
Starches	Pantothenic acid
Dextrins	Vitamin D
Sucrose	Acetic acid
Pentosans	Propionic acid
Hexosans	Butyric acid
Triglycerides	Valeric acid
Sodium chloride	Caproic acid
Phosphates	Acetone
Calcium	Diacetyl
Iron	Maltol
Thiamine	Ethyl acetate
Riboflavin	Ethyl lactate

Tea

Caffeine	Phenyl ethyl alcohol
Tannin	Benzyl alcohol
Essential oils	Geraniol
Butyl alcohol	Hexyl alcohol
Isoamyl alcohol	

*The chemicals listed are those found normally in the foods. The chemical listings are not necessarily complete.

Source: Manufacturing Chemists Association, Washington, DC.

Table 17.6 Approximate Elemental Analysis of the Human Body

Element	Percent by Weight in Human Body
Oxygen	65
Carbon	18
Hydrogen	10
Nitrogen	3
Calcium	1.5
Phosphorus	1
Potassium	0.35
Sulfur	0.25
Chlorine	0.15
Sodium	0.15
Magnesium	0.05
Iron	0.004
Trace elements to make 100%	

If we ate a balanced diet of fresh foods, we probably wouldn't need nutritional supplements. But many people choose to eat primarily processed foods and convenience foods. With our usual diets rich in highly processed foods, we often need the nutrients provided by vitamin and mineral food additives.

Additives That Taste Good

Spice cake, soda pop, gingerbread, sausage, and many other foods depend on spices and other additives for most of their flavor. Cloves, ginger, cinnamon, and nutmeg are examples of natural spices, and basil, marjoram, thyme, and rosemary are widely used herbs. Natural flavors are also extracted from fruits and other plant materials.

Some esters that serve as flavors are listed in Table 9.7. Other flavor compounds are shown in Figure 17.12. Chemists can analyze natural flavors and then synthesize the components to make mixtures that resemble the natural products. Major components of natural and artificial flavors are often identical. For example, both vanilla extract and imitation vanilla owe their flavor mainly to vanillin. The natural flavor is often more complex because it contains a wider variety of chemicals than the imitation. Flavor additives, whether natural or synthetic, probably present little hazard when used in moderation, and they contribute considerably to our enjoyment of food.

Artificial Sweeteners

Obesity is a major problem in most developed countries. Presumably, we could reduce the intake of calories by replacing sugars with noncaloric sweeteners, but there is little evidence that artificial sweeteners are of value in controlling obesity. People eat more and more sugar, now averaging 150 lb per capita annually in the United States.

For many years, the major artificial sweeteners were saccharin and cyclamates. Cyclamates were banned in the United States in 1970 after studies showed they caused cancer in laboratory animals. (Subsequent studies have failed to confirm these findings, but the FDA has not lifted the ban.) In 1977 saccharin was shown to cause bladder cancer in laboratory animals. However, the move by the FDA to ban it was blocked by Congress because at that time saccharin was the only approved artificial sweetener. Its ban would have meant the end of diet soft drinks and low-calorie products.

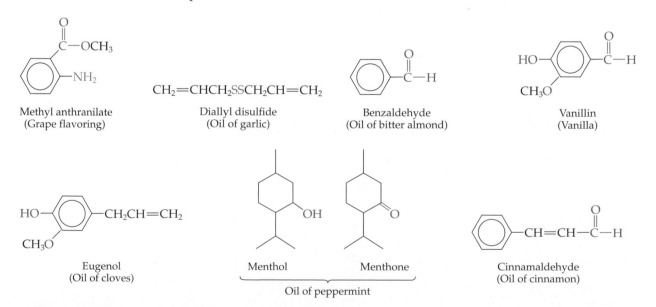

▲ **Figure 17.12** Some molecular flavorings. See Table 9.7 for others.

QUESTION: Which of those shown here have aldehyde functional groups? Which are phenols? Ethers? Alkenes? Which is an ester? A ketone? An alcohol?

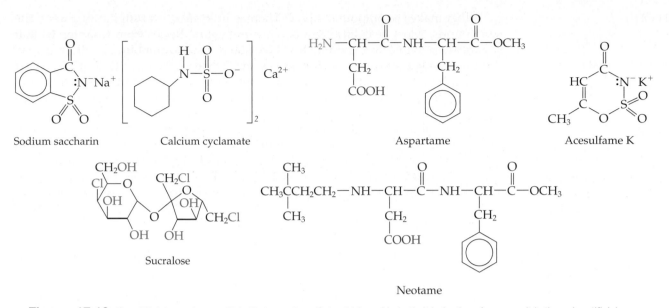

▲ **Figure 17.13** Six artificial sweeteners. Note that sucralose (Splenda) is a chlorinated derivative of sucrose. It is the only artificial sweetener actually made from sucrose.

There are now six FDA-approved sweeteners (Figure 17.13). Aspartame (the methyl ester of the dipeptide aspartylphenylalanine) was approved in 1981. A closely related compound, called neotame, has a 3,3-dimethylbutyl group $[CH_3C(CH_3)_2CH_2CH_2—]$ on the nitrogen atom of the aspartyl unit of the dipeptide. There are anecdotal reports of problems with aspartame, but repeated studies have shown it to be generally safe, except for people with phenylketonuria, an inherited condition in which phenylalanine cannot be metabolized properly. Other artificial sweeteners approved for use in the United States include acesulfame K (Sunette) and sucralose (Splenda), sweeteners that can survive the high temperatures of cooking processes, whereas aspartame is broken down by heat. Table 17.7 compares the sweetness of a variety of substances.

A natural product of some note is stevia, a South American herb long used as a sweetener by the Guarani Indians of Paraguay. Leaves of this herb (*Stevia rebaudiana*) provide a sweetness 150 to 300 times that of sucrose. Stevia, an approved food additive in Japan and South America, is available in the United States only as a dietary supplement; its approval by the FDA as a food additive is expected in 2009.

Table 17.7 Approximate Sweetness of Some Compounds	
Compound	**Relative Sweetness***
Lactose	0.16
Maltose	0.33
Glucose	0.74
Sucrose	1.00
Fructose	1.73
Cyclamate	45
Aspartame	180
Acesulfame K	200
Saccharin	300
Sucralose	600
Neotame	13,000

* Sweetness is relative to sucrose at a value of 1.

CH$_2$OH
|
CHOH   CH$_2$OH
| |
CHOH   CHOH
| |
CHOH   CHOH
| |
CHOH   CHOH
| |
CH$_2$OH   CH$_2$OH

Sorbitol   Xylitol

▲ Polyhydroxy alcohols are used to sweeten sugar-free products such as chewing gum. These products aid in control of dental caries because polyols are metabolized slowly if at all in dental plaque.

More to Explore
Selim, Jocelyn. "The Chemistry of ∙∙∙ Artificial Sweeteners." *Discover*, August 2005, pp. 18–19.

What makes a compound sweet? There is little structural similarity among the compounds. Most bear little resemblance to sugars. Recall from Chapter 16 that sugars are polyhydroxy compounds. Like sugar, many compounds with hydroxyl groups on adjacent carbon atoms are sweet. Ethylene glycol (HOCH$_2$CH$_2$OH) is sweet, although it is quite toxic. Glycerol, obtained from the hydrolysis of fats (Section 17.2), is also sweet. It is used as a food additive, principally because of its properties as a **humectant** (moistening agent), and only incidentally as a sweetener.

Other polyhydroxy alcohols (polyols) used as sweeteners are sorbitol, made by the reduction of glucose, and xylitol, which has five carbon atoms with a hydroxyl group on each. These compounds occur naturally in foods such as fruits and berries. They are used as sweeteners in "sugar-free" items such as chewing gum and candies. Unlike sugars, polyols do not cause sudden increases in blood sugar and, thus, can be used in moderation by diabetics. Large amounts—more than about 10 g—can cause gastrointestinal distress.

Several thousand sweet-tasting compounds have been discovered. They belong to more than 150 chemical classes. We now know that all the sweet substances act on a single taste receptor. (In contrast, we have more than 30 receptors for bitter substances.) Unlike any other known receptor, the sweet receptor has more than one area that can be activated by various molecules. The different areas have different affinities for the various molecules. For example, sucralose fits the receptor more tightly than sucrose does, partly because its chlorine atoms carry a stronger charge than the oxygen atoms in the OH groups they replaced. Neotame, approved by the FDA in 2002, fits the receptor so tightly it keeps the receptor firing repeatedly.

Flavor Enhancers

Some chemical substances, although not particularly flavorful themselves, are used to enhance other flavors. Common table salt (sodium chloride) is a familiar example. In addition to being a necessary nutrient, salt seems to increase sweetness and helps to mask bitterness and sourness.

Another popular flavor enhancer is monosodium glutamate (MSG). MSG is the sodium salt of glutamic acid, one of the 20 amino acids that occur naturally in proteins. It is used in many convenience foods.

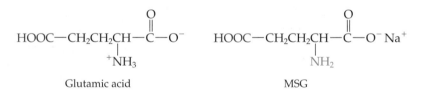

Glutamic acid   MSG

Although glutamates are found naturally in proteins, there is evidence that huge excesses can be harmful. MSG can numb portions of the brains of laboratory animals. It may also be teratogenic, causing birth defects when eaten in large amounts by women who are pregnant.

Self-Assessment Questions

1. Of the 70 kg or so of food additives we eat each year, 67 kg are
   a. artificial colors   **b.** emulsifiers
   c. natural flavors   **d.** salt and sweeteners

2. The first U.S. government–approved food additive, placed in table salt to prevent goiter, was
   a. aluminum sulfate   **b.** iron phosphate
   c. lecithin   **d.** potassium iodide

3. To market a new food additive, a manufacturer must
   a. notify the FDA of plans to use it
   b. pay the FDA to test it for safety

    c. provide the FDA with evidence that the ingredient is safe for its intended use

    d. submit an application to the FDA but does not have to prove it safe and effective

4. Processed flour that has missing vitamins and minerals added back is said to be

    a. enriched   **b.** fortified   **c.** natural   **d.** whole grain

5. Compared to synthetic flavors, natural flavorings

    a. are without chemicals   **b.** have different main ingredients

    c. come from the Spice Islands   **d.** have more chemicals

6. Which of the following is *not* an artificial sweetener?

    a. acesulfame   **b.** calcium cyclamate   **c.** maltose   **d.** sucralose

7. The flavor enhancer MSG is related to a(n)

    a. amino acid   **b.** fatty acid

    c. monosaccharide   **d.** polyhydroxy alcohol

Answers: 1, d; 2, d; 3, c; 4, a; 5, d; 6, c; 7, a

17.9 Additives to Retard Spoilage

Food spoilage can result from the growth of molds, yeasts, or bacteria. Substances that prevent this are often called *antimicrobials*. Certain carboxylic acids and their salts are commonly used. Propionic acid and its sodium and calcium salts are used to inhibit molding in bread and cheese. Sorbic acid, benzoic acid, and their salts are also used (Figure 17.14).

Some inorganic compounds are also added as spoilage inhibitors. Sodium nitrite ($NaNO_2$) is used in meat curing and to maintain the pink color of smoked hams, frankfurters, and bologna. It also contributes to the tangy flavor of processed meat products. Nitrites are particularly effective as inhibitors of *Clostridium botulinum*, the bacterium that produces botulism poisoning. However, only about 10% of the amount used to keep meat pink is needed to prevent botulism. Nitrites have been investigated as possible causes of cancer of the stomach. In the presence of the hydrochloric acid (HCl) in the stomach, nitrites are converted to nitrous acid,

$$NaNO_2 + HCl \longrightarrow HNO_2 + NaCl$$

which may then react with secondary amines (amines with two alkyl groups on nitrogen) to form *nitroso* compounds.

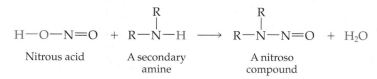

The R groups can be alkyl groups such as methyl (CH_3-) or ethyl (CH_3CH_2-), or they can be more complex. In any case, these nitroso compounds are among the most potent carcinogens known. The rate of stomach cancer is higher in countries that use prepared meats than in developing nations where people eat little or no cured meat. However, the incidence of stomach cancer is *decreasing* in the United States, perhaps because ascorbic acid (vitamin C) inhibits the reaction between nitrous acid and amines to form nitrosamines, and we often have orange juice with our breakfast bacon or sausage. Nevertheless, possible problems with nitrites have led the FDA to approve sodium hypophosphite (NaH_2PO_2) as a meat preservative.

Other inorganic food additives include sulfur dioxide and sulfite salts. A gas at room temperature, sulfur dioxide (SO_2) serves as a disinfectant and preservative,

CH_3CH_2COOH

Propionic acid

$CH_3CH_2COO^- Na^+$

Sodium propionate

$CH_3CH=CHCH=CHCOOH$

Sorbic acid

$CH_3CH=CHCH=CHCOO^- K^+$

Potassium sorbate

—COOH

Benzoic acid

—COO⁻ Na⁺

Sodium benzoate

▲ **Figure 17.14** Most spoilage inhibitors are carboxylic acids or salts of carboxylic acids.

One person in 100 has a strong allergic reaction to sulfur dioxide and related compounds called sulfiting agents. Common ones include sodium sulfite (Na_2SO_3), potassium sulfite (K_2SO_3), sodium bisulfite ($NaHSO_3$), potassium bisulfite ($KHSO_3$), sodium metabisulfite ($Na_2S_2O_5$), and potassium metabisulfite ($K_2S_2O_5$).

particularly for dried fruits such as peaches, apricots, and raisins. It is also used as a bleach to prevent the browning of wines, corn syrup, jelly, dehydrated potatoes, and other foods. Sulfur dioxide seems safe for most people when ingested with food, but it is a powerful irritant when inhaled and is a damaging ingredient of polluted air in some areas (Chapter 13). Sulfur dioxide and sulfite salts cause severe allergic reactions in some people. The FDA requires their use in foods to be indicated on the label.

Antioxidants: BHA and BHT

One class of preservatives is composed of **antioxidants** that inhibit the chemical spoilage of food that occurs in the presence of oxygen. These substances are added to foods (or their packaging) to prevent fats and oils from forming rancid products that make the food unpalatable. Antioxidants also minimize the destruction of some essential amino acids and vitamins. Packaged foods that contain fats or oils (bread, potato chips, sausage, breakfast cereal) often have antioxidants added. Compounds commonly used include butylated hydroxytoluene (BHT), butylated hydroxyanisole (BHA), propyl gallate, and *tert*-butylhydroquinone, (Figure 17.15).

Fats turn rancid, in part as a result of oxidation, a process that occurs through the formation of molecular fragments called **free radicals**, which have an unpaired electron as a distinguishing feature. (Recall from Section 4.6 that covalent bonds are shared *pairs* of electrons.) We need not concern ourselves with the details of the structures of radicals, but we can summarize the process. First, a fat molecule reacts with oxygen to form a free radical.

$$\text{Fat} + O_2 \longrightarrow \text{Free radical}$$

Then the radical reacts with another fat molecule to form a new free radical that can repeat the process. A reaction such as this, in which intermediates are formed that keep the reaction going, is called a **chain reaction**. One molecule of oxygen can lead to the decomposition of many fat molecules.

To preserve foods containing fats, processors package the products to exclude air, but it cannot be excluded completely, and so chemical antioxidants are used to stop the chain reaction by reacting with the free radicals.

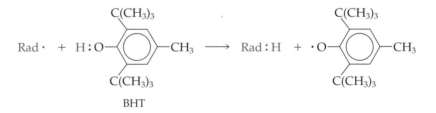

The new radical formed from BHT is rather stable. The unpaired electron doesn't have to stay on the oxygen atom but can move around in the electron cloud of the benzene ring. The BHT radical doesn't react with fat molecules, and the chain is broken.

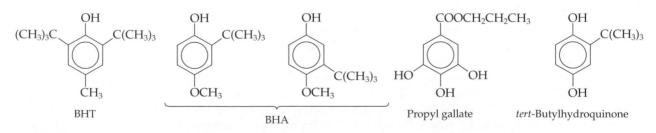

▲ **Figure 17.15** Four common antioxidants. BHA is a mixture of two isomers.

QUESTION: What functional group is common to all four compounds? What other functional group is present in BHA? In propyl gallate?

Why are the butyl groups important? Without them, the phenols would simply couple when exposed to an oxidizing agent.

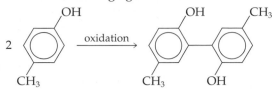

With the bulky butyl groups aboard, the rings can't get close enough together for coupling. They are free, then, to trap free radicals formed by the oxidation of fats.

Many food additives have been criticized as being harmful, and BHA and BHT are no exceptions. They have been reported to cause allergic reactions in some people. In one study, pregnant mice fed diets containing 0.5% BHA or BHT gave birth to offspring with brain abnormalities. On the other hand, when relatively large amounts of BHT were fed to rats daily, their life spans were increased by a human equivalent of 20 years. One theory about aging is that it is caused in part by the formation of free radicals. BHT retards this chemical breakdown in cells in the same way that it retards spoilage in foods. BHA and BHT are synthetic chemicals, but there are also natural antioxidants, such as vitamin E. Vitamin E (like BHT) is a phenol, with several substituents on the benzene ring. Presumably, its action as an antioxidant is quite similar to that of BHT. However, one recent study suggests that large doses of vitamin E may also be somewhat harmful.

Although too much vitamin A (from liver or supplements) can be toxic, if too much β-carotene is consumed, the body merely stores it in fat, eventually using or excreting it. Very large amounts of β-carotene have been known to cause the skin to turn orange! (The effect is temporary and harmless.)

Self-Assessment Questions

1. Calcium propionate is added to bread to
   a. act as an antioxidant
   b. enhance the flavor
   c. inhibit mold growth
   d. replace Ca^{2+} removed in processing
2. Which of the following food additives inhibits formation of botulinum toxin?
   a. BHT
   b. sodium ascorbate
   c. sodium benzoate
   d. sodium nitrite
3. Antioxidants
   a. aid in lipid transport
   b. help regulate blood sugar levels
   c. promote growth
   d. trap free radicals
4. Which of the following are antioxidants used to prevent rancidity?
   a. ascorbic acid   b. BHT   c. sodium sulfite   d. sugar
5. Which of the following food additives causes wheezing and other symptoms of asthma in some people?
   a. flavoring   b. food colors   c. sodium nitrite   d. sulfur dioxide
6. Vitamin C, vitamin E, BHA, BHT, nitrites, and sulfites are all
   a. antimicrobial
   b. antioxidants
   c. flavor enhancers
   d. spoilage inhibitors

Answers: 1, c; 2, d; 3, d; 4, b; 5, d; 6, d

17.10 Color Additives

Some foods are naturally colored. For example, the yellow compound β-carotene occurs in carrots. β-carotene (shown in the following line-angle formula) is used as a color additive in other foods such as butter and margarine. (Our bodies convert β-carotene to vitamin A by cutting the molecule in half at the central double bond; thus, it is a vitamin additive as well as a color additive.)

More to Explore
Dalton, Louisa. "What's That Stuff: Food Preservatives." *Chemical & Engineering News*, November 11, 2002, p. 40.

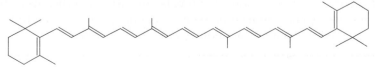

B-Carotene

Botulin toxin (Botox®) is used in minute quantities for producing long-term (months) paralysis of muscles. Originally intended for the relief of unmanageable muscle spasms, Botox® is now widely used for cosmetic purposes to paralyze facial muscles to conceal wrinkles. The treatment lasts only four to six months.

of it could kill more than a million people. The point is that a food is not inherently good simply because it is natural. Neither is it necessarily bad because a synthetic chemical substance has been added to it.

Carcinogens

There is little chance that we will suffer acute poisoning from approved food additives. But what about cancer? Could all these chemicals in our food increase our risk of cancer? The possibility exists, even though the risk is low.

Carcinogens occur naturally in food. A charcoal-broiled steak contains 3,4-benzpyrene, a carcinogen also found in cigarette smoke and automobile exhaust fumes. Cinnamon and nutmeg contain safrole, a carcinogen that has been banned as a flavoring in root beer.

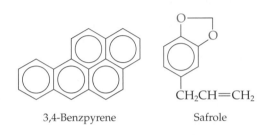

3,4-Benzpyrene   Safrole

Among the most potent carcinogens are **aflatoxins**, compounds produced by molds growing on stored peanuts and grains (Figure 17.17). Aflatoxin B_1 is estimated to be 10 million times as potent a carcinogen as saccharin, and there is no way to completely keep it out of our food. The FDA sets a tolerance of 20 ppb for aflatoxins.

Scientists estimate that our consumption of natural carcinogens is 10,000 times that of synthetic carcinogens. Should we ban steaks, spices, peanuts, and grains because they contain naturally occurring carcinogens? Probably not. The risk is slight, and life is filled with more serious risks. Should we ban additives that have been shown to be carcinogenic? Whether we do or not, we should carefully weigh the risks against any benefits the additives might provide.

Food Contaminants

Food contamination can come from many sources and at any step between producer and consumer. Pesticide residues, insect parts, hair from food workers, glass from broken containers, and antibiotics added to animal feeds are examples of contaminants or *incidental additives*. (See Collaborative Group Project 73.)

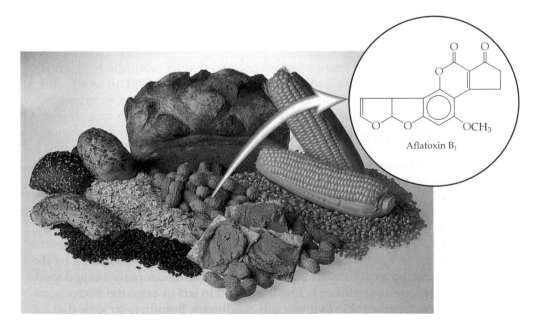

▲ **Figure 17.17** Aflatoxins, toxic and carcinogenic compounds, are produced by molds that grow on peanuts and stored grains.

Antibiotics in animal feed are a troubling example of food contamination, often showing up in the meat we eat. Antibiotics are added in low-level dosages to animal feed to promote weight gain. In fact, half or more of the 23 million kg of antibiotics produced annually in the United States goes into animal feeds. Residues of these antibiotics, sometimes found in meat, may result in the sensitization of individuals who eat the meat, thus hastening the development of allergies. Scientists fear the use of antibiotics in animal feeds will hasten the process by which bacteria become drug resistant (Section 18.4).

Present-day food contamination problems are usually biological, not chemical. For example, bacteria such as the O157:H7 strain of *Escherichia coli* (*E. coli*), a fecal contaminant found at times in raw meat or dairy products and occasionally in raw fruits and vegetables, can cause serious food poisoning. *E. coli* contamination was responsible for costly food recalls of ground beef in 2007 and 2008, as well as tomato and jalapeño pepper recalls in 2008. *Salmonella* contamination, found in poultry, eggs, unprocessed milk, and in meat and water can cause cramping, fever, dehydration, and diarrhea. *Salmonella* is also carried by pet turtles and birds. *Listeria monocytogenes* causes a food-borne illness, listeriosis, marked by fever, muscle aches, nausea, and diarrhea.

Viruses such as hepatitis A, which causes a serious contagious disease of the liver, and the Norwalk virus, which causes about 50% of all food-borne outbreaks of gastroenteritis in the United States, can be spread by way of food by infected food workers.

Parasites such as the roundworms *Trichinella spiralis*, which causes trichinosis, and *Anisakis simplex*, which causes anisakiasis, are relatively rare problems in developed countries but are fairly common in the developing world. The few cases of trichinosis in the United States result mainly from eating undercooked game meat or pork from home-raised pigs. Trichinosis is characterized by symptoms such as nausea, heartburn, and diarrhea. Anisakiasis results from eating raw or undercooked marine fish in dishes such as sushi and sashimi (Japan), cod livers (Scandinavia), fermented herrings (Netherlands), and ceviche, a citrus-marinated salad made with raw seafood, popular in Latin America.

Proper cooking and food handling procedures will reduce the chance of most food-borne diseases.

Self-Assessment Questions

1. Aflatoxins are
   a. antioxidants
   b. carcinogens
   c. on the GRAS list
   d. found only in peanuts
2. Which of the following is *not* an incidental additive?
   a. DDT
   b. metal shavings from a machine
   c. antibiotics
   d. spices such as cloves

Answers: 1, b; 2, d

GREEN CHEMISTRY

Think Global, Eat Local

Doris Lewis, Suffolk University

The *Twelve Principles of Green Chemistry* were developed as guidelines for chemists. So, if you are not a chemist, why be concerned about renewable feedstocks, atom economy, waste production, or energy efficiency? Consider that innumerable chemical reactions take place in our bodies, many of which involve food as the starting material. Many consumers are now aware that food consumption has political and ecological effects as well as personal ones, with enormous consequences for other people as well as ourselves. The principles of green chemistry can help us understand how this is so.

Fruits and vegetables from the supermarket often have traveled thousands of miles to your table. The distribution system that enables this long-distance food journey provides a large variety of "fresh" fruits and vegetables available year-round, but this bounty has hidden consequences that are becoming apparent. Increasing oil prices have made all materials that require transport more expensive. Hence, higher oil prices quickly result in higher food prices, a predictable consequence of the use of nonrenewable petroleum fuels.

Green chemistry is concerned with the use of renewable feedstocks to make things. Almost all the packaging in today's supermarket is made from petroleum products—for example, the Styrofoam tray and the clear polyethylene wrapper protecting many fresh food items. Rising petroleum prices lead to increased packaging costs and, thus, higher prices at the grocery store.

An increasingly popular way to avoid the higher cost of foods shipped long distances is to become a "locavore," a consumer who seeks out locally produced foods. The New England locavore may enjoy locally grown apples, while a Floridian would eat a fresh orange. There is loss of variety, of course, but this is partially offset by the quality and taste of truly fresh produce and the superior characteristics of varieties grown for flavor rather than the ability to survive the rigors of shipping. With this system, tomatoes and strawberries become seasonal treats, rather than year-round staples with diminished flavor and appeal. Buying directly at a farmer's market also has the advantage of supporting the local economy.

Many chefs now buy local foods for their superior quality. By avoiding food from large-scale growers who ship food all over the country, one may also lessen the chance of contaminated foods, which, although rare, add an element of risk to the food system.

Green chemistry addresses several problems in the food industry. New packaging materials are being made from corn products rather than petroleum (see the "Greener Polymers" essay in Chapter 10, page 284) and greener fertilizers are being developed. Fertilizers (see eChapter 20) enhance plant growth but at a cost to the environment, with some fertilizer ending up in waterways. For example, more than 1.6 million metric tons of nitrogen, mainly runoff from midwestern farms, end up in the Gulf of Mexico each year, leading to eutrophication (Section 14.3). Decaying plant matter has overwhelmed natural oxygen-dependent purification reactions, and a "dead zone" the size of the state of New Jersey now exists in the Gulf. All life forms are affected in such a zone; once prosperous fisheries are threatened along with the humans who depend on them. Heeding the principle of waste prevention might have prevented this ecological catastrophe.

Organic farming addresses the problem of fertilizer runoff (see eChapter 20) by using green manures and mulching in place of synthetic fertilizers and soil amendments. Because synthetic fertilizers are to some degree petroleum based, organic farming can also save energy.

Each of us can apply the principle of waste prevention in our own food consumption. A University of Arizona study concludes that 40 to 50% of the food produced in the United States is never eaten, wasting billions of dollars a year. On average, American households throw out 14% of the food purchased—$490 per household per year, or $43 billion overall. Reducing food waste by half would significantly reduce landfill use, soil depletion, and the use of fertilizers, pesticides, and herbicides. Composting food waste is another way to reduce landfill use while producing fertilizer. The next time you are at a meal, observe what happens. How much of the food is not actually consumed, and where does that food go? How might this be changed?

17.12 A World without Food Additives

Could we get along without food additives? Some of us could. But food spoilage might drastically reduce the food supply in some parts of an already hungry world, and diseases due to vitamin and mineral deficiencies might increase. Foods might cost more and be less nutritious. Food additives seem to be a necessary part of modern society. There are potential hazards associated with the use of some food additives, but the major problem with our food supply is still contamination by harmful microorganisms, rodents, and insects. Indeed, there are about 76 million illnesses, 325,000 hospitalizations, and 5000 deaths annually in the United States resulting from food poisoning caused by bacteria, viruses, and parasites. Few, if any, deaths associated with the use of intentional food additives have been documented.

What should we do about food additives? We should be sure that the FDA is staffed with qualified personnel to ensure the adequate testing of proposed food additives. People trained in chemistry are necessary for the control and monitoring of food additives and the detection of contaminants. Research on the analytical techniques necessary for the detection of trace quantities is vital to adequate consumer protection. We should demand laws adequate to prevent the unnecessary and excessive use of pesticides and other agricultural chemicals that might contaminate our food. Above all, we should be alert and informed about these problems that are so vital to our health and well-being.

More to Explore

Two fascinating books are McGee, Harold. *On Food and Cooking: The Science and Lore of the Kitchen*. New York: Scribner, 2004 (complete revision and updating of the 1984 edition), and Wolke, Robert L. *What Einstein Told His Cook: Kitchen Science Explained*. New York: W. W. Norton, 2002.

Self-Assessment Questions

1. About how many people die from food poisoning in the United States each year?
   a. 100   **b.** 500   **c.** 5000   **d.** 50,000

2. The problem with our food supply causing the most deaths and diseases is contamination with
   a. aflatoxins   **b.** chemicals   **c.** microorganisms   **d.** pesticides

Answers: 1, c; 2, c

Critical Thinking Exercises

Apply knowledge that you have gained in this chapter and one or more of the FLaReS principles (Chapter 1) to evaluate the following statements or claims.

17.1 The author of a new diet book claims that he and many other people have been able to lose a substantial amount of weight by following a diet very low in carbohydrates (such as bread, cereals, pasta) but high in proteins (such as beef and lamb).

17.2 A physician noted that the blood cholesterol of one of his patients had gone down dramatically. On the advice of a friend, the patient had been taking daily supplements of niacin, one of the B vitamins. Although he was skeptical, the doctor found six other patients who were willing to try the niacin treatment without any further medication. A year later he was surprised to find that five of these patients had lowered their cholesterol levels significantly. The doctor says he believes that niacin can be used to lower blood cholesterol.

17.3 Information on a website states that a chemical from marijuana, delta 1-tetrahydrocannabinol (delta 1-THC), is a fat-soluble vitamin they call "vitamin M." They say they can make the claim "because many of the properties of delta 1-THC are similar to those of the fat-soluble vitamins."

17.4 Information on a website states that natural vitamins may be worth the extra cost because vitamins in nature occur within a family of "things like trace elements, enzymes, cofactors, and other unknown factors that help them absorb into the human body and function to their full potential" whereas "synthetic vitamins are usually isolated into that one pure chemical, thereby missing out on some of the additional benefits that nature intended."

17.5 A friend claims that it is fine to eat all the ground beef he likes, as long as it is labeled "85% lean." He shows you the label, which says that a 100-g portion of cooked meat contains 15 g of fat and has 210 calories.

17.6   A woman's doctor orders her to follow a very-low-fat diet. She makes a bowl of fat-free popcorn and then sprays it with a "flavor spray." The label of the spray lists the ingredients as canola oil, butter, and lecithin. It also indicates serving size 1 spray (0.5 g), calories 5, fat 0 g. (The FDA allows less than 0.5 g to be rounded to zero.) She sprays the popcorn about 40 times to give it a lot of flavor, pointing out that at zero grams of fat per spray, she is still adhering to the very-low-fat regimen.

SUMMARY

Section 17.1—The three main classes of foods are carbohydrates, fats, and proteins. We also need vitamins, minerals, fiber, and water in our diets. Carbohydrates include sugars, starches, and cellulose. Glucose, also called **blood sugar**, is the sugar used directly by cells for energy. Other sugars include fructose, sucrose (table sugar), and lactose. Sucrose and lactose are broken down during digestion. **Starch** is a polymer of α-glucose, and **cellulose** is a β-glucose polymer. We can hydrolyze starch to glucose, but cellulose is not hydrolyzed, instead serving as dietary fiber. We store small amounts of carbohydrates as **glycogen**.

Section 17.2—Dietary fats are esters called **triglycerides**, digested by enzymes to form fatty acids and glycerol. Fats are stored in **adipose tissue** located in **fat depots** and are hydrolyzed back to glycerol and fatty acids when needed. A **lipoprotein** is a cholesterol– or triglyceride–protein combination transported in the blood. Low-density lipoproteins (LDLs) carry cholesterol to the cells for use and increase the risk of heart disease, whereas high-density lipoproteins (HDLs) carry it to the liver for excretion. Polyunsaturated fats (mostly from plants), *cis*-fatty acids, and omega-3 fatty acids tend to lower LDL levels. Saturated fats (mostly from animals) and *trans*-fatty acids from hydrogenation of polyunsaturated fats have been implicated in cardiovascular disease.

Section 17.3—Proteins, polymers of amino acids, are broken down to those amino acids during digestion. The **essential amino acids** must be included in the diet. Each essential amino acid is a **limiting reactant** in protein synthesis; deficiencies cause an inability to make the proper proteins. Most animal proteins have these essential amino acids; plant proteins usually are deficient and must be combined with other plant proteins for a complete diet. Protein deficiency is common in developing nations.

Section 17.4—**Dietary minerals** are inorganic substances vital to life. Bulk structural elements include carbon, hydrogen, and oxygen. Macrominerals include sodium, potassium, and calcium. We also need trace elements such as iron and ultratrace elements including manganese, chromium, and vanadium. Minerals serve a variety of functions: Iodine is needed by the thyroid; iron is part of hemoglobin; and calcium and phosphorus are important in the bones and teeth.

Section 17.5—**Vitamins** are organic compounds that our bodies need but cannot make. Most are needed as coenzymes. The B vitamins and vitamin C are water soluble and are needed frequently, whereas vitamins A, D, E, and K are all fat soluble and can be stored.

Section 17.6—Dietary fiber may be soluble (gums and pectins) or insoluble (cellulose). It prevents digestion problems such as constipation. Drinking enough water is important, but it need not be plain water. Most drinks are mainly water.

Section 17.7—**Starvation**—deprivation of food—causes metabolic changes. The body first depletes its glycogen and then metabolizes fat and muscle tissue. Malnutrition can result from starvation, dieting, or even the lack of a proper diet.

Section 17.8—**Food additives** are substances other than basic foodstuffs that are added to food to aid nutrition; inhibit spoilage; add color, flavor, or sweetness; bleach; or provide texture. There are thousands of additives. The FDA regulates which additives, and how much, can be used. In **enrichment**, chemicals are added to foods to prevent deficiency diseases: B vitamins in grains, vitamin C in drinks, vitamin D in milk. Natural herbs and spices, or their synthetic counterparts, add flavor. Artificial sweeteners reduce sugar intake but do not appear to aid in controlling obesity. Glycerol is a sweetener that is used primarily as a **humectant** or moisturizing agent. Salt and MSG are flavor enhancers.

Section 17.9—Propionate, sorbate, benzoate, and nitrite salts inhibit spoilage and bacterial growth. Fats may turn rancid from formation of **free radicals** with unpaired electrons. One free radical can cause a **chain reaction** that leads to decomposition of many fat molecules. BHT and BHA are **antioxidants** that react with free radicals and prevent rancidity.

Section 17.10—Natural food colorings, such as β-carotene, and synthetic ones improve the appearance of food. Some additives have been found to be harmful; additives are usually banned if they are shown to cause cancer in laboratory animals. The list of additives generally recognized as safe is the **GRAS list**, which is periodically reevaluated.

Section 17.11—Although toxic substances are sometimes natural ingredients of foods, occasionally they are added inadvertently to our food in the form of pesticide residues or animal feed additives. Charcoal-broiled steak, cinnamon, and nutmeg contain carcinogens. **Aflatoxins** are produced by molds on peanuts and grain and are potent carcinogens. We may consume about 10,000 times as much natural carcinogens as synthetic carcinogens. Most contaminants are not chemicals but are biologicals such as bacteria, viruses, and parasites.

Section 17.12—Although we may be able to get along without food additives, spoilage and disease would be likely to cause serious problems worldwide. Additives continue to be carefully scrutinized and regulated.

■ REVIEW QUESTIONS

1. What functions do fats serve?
2. What is the role of proteins in the diet?
3. Vitamins and minerals are discussed in this chapter. Which are organic and which are inorganic?
4. A diet high in meat products makes less efficient use of the energy originally captured by plants through photosynthesis than a vegetarian diet does. Explain why.
5. Is an excess of a water-soluble vitamin or an excess of a fat-soluble vitamin more likely to be dangerous? Why?
6. What is starvation?
7. What is a food additive?
8. What are the two major categories of food additives? Give an example of each.
9. What is MSG?
10. What is botulism? What are aflatoxins?
11. What vitamin serves as a fat-soluble antioxidant?
12. Name three artificial sweeteners. Which are approved for current use in the United States?
13. What is the chemical nature of aspartame?
14. List some incidental additives that have been found in foods.
15. What U.S. government agency regulates the use of food additives? What must be done before a food company can use a new food additive?
16. List five functions of food additives.

■ PROBLEMS

Carbohydrates

17. What are the chemical names of the following sugars?
   a. blood sugar   **b.** table sugar   **c.** fruit sugar
18. What is the principal sugar in corn syrup? How is high-fructose corn syrup made?
19. What sugar is formed when starch is digested?
20. What type of chemical reaction is involved in the digestion of starch?

Fats and Cholesterol

21. What products are formed initially when a fat is digested? What type of chemical reaction is involved?
22. Where is fat stored in the body?
23. How do animal fats differ from vegetable oils?
24. What is a *trans* fat? How are *trans* fats made?

Proteins

25. What is an adequate protein? List some foods that contain adequate proteins.
26. Which essential amino acids are likely to be lacking in corn? In beans?
27. What is a limiting reactant?
28. What is kwashiorkor? What age group in developed countries suffers a similar condition?
29. How do dietary proteins differ from most dietary carbohydrates and fats in elemental composition?
30. Ezekiel 4:9 describes a bread made from "wheat, and barley, and beans, and lentils, and millet, and spelt. . . ." Would the bread supply complete protein?

Minerals

31. Indicate a biological function for each of the following dietary minerals.
   a. iodine   **b.** iron
   c. calcium   **d.** phosphorus
32. Of the minerals Ca, Cl, Co, Mo, Na, P, and Zn, which would you expect to find in relatively large amounts in the human body? Explain.

Vitamins

33. Classify the following vitamins as water soluble or fat soluble.

Pantothenic acid

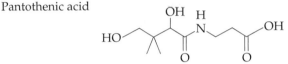

Phylloquinone

Biotin

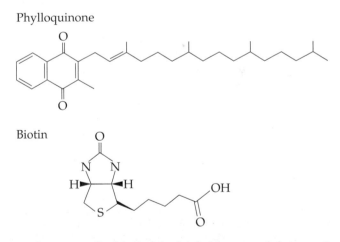

34. Coenzyme Q$_{10}$(CoQ$_{10}$) is claimed by some to have a vitaminlike activity: It prevents LDL oxidation and may act as an antihypertensive, perhaps reducing the risk of cardiovascular disease. CoQ$_{10}$ belongs to a family of substances called ubiquinones and is synthesized in the body. Is CoQ$_{10}$ (below) **(a)** a vitamin? **(b)** Fat soluble or water soluble?

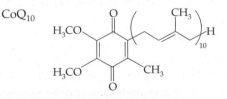

35. Which of the following are B vitamins?
   a. niacin   **b.** phylloquinone
   c. pyridoxine   **d.** calciferol
   e. ascorbic acid

36. Match the compound with its designation as a vitamin.

Compound	Designation
Ascorbic acid	Vitamin A
Calciferol	Vitamin B$_{12}$
Cyanocobalamin	Vitamin C
Retinol	Vitamin D
Tocopherol	Vitamin E

37. Identify the vitamin associated with each of the following deficiency diseases.
   a. pellagra   **b.** beriberi   **c.** pernicious anemia

38. In each case, identify the disease associated with a deficiency of the indicated vitamin.
   a. ascorbic acid   **b.** retinol
   c. α-tocopherol

39. Identify each of the following vitamins as water soluble or fat soluble.
   a. calciferol   **b.** niacin
   c. riboflavin   **d.** tocopherol

40. Identify each of the following vitamins as water soluble or fat soluble.
   a. vitamin A   **b.** vitamin B$_6$ **c.** vitamin B$_{12}$
   d. vitamin C   **e.** vitamin K

Food Additives

41. What is the function of each of the following food additives?
   a. FeCO$_3$ **b.** SO$_2$ **c.** potassium sorbate

42. What is the function of each of the following food additives?
   a. potassium iodide   **b.** vanillin
   c. MSG   **d.** sodium nitrite

43. What is the purpose of each of the following food additives?
   a. aspartame   **b.** vitamin D
   c. sodium hypophosphite

44. What is the purpose of each of the following food additives?
   a. BHA   **b.** FD & C Yellow No. 5   **c.** saccharin

45. Which of the following is true about a dietary supplement labeled "natural"? It is
   a. mild acting
   b. without risk of side effects
   c. always safe to use with other medications
   d. none of the above

46. Propyl gallate is made by esterification of gallic acid, obtained by the hydrolysis of tannins from Tara pods. Is propyl gallate a natural antioxidant?

47. What are antioxidants? Name a natural antioxidant and two synthetic antioxidants.

48. What is a chain reaction? What is the action of an antioxidant on a chain reaction?

Calorie and Nutrient Calculations

49. The label on a can of soup indicates that each half-cup portion supplies 60 kcal and has 3 g of protein, 8 g of carbohydrate, and 1 g of fat. Calculate the percentage of calories each from carbohydrate, fat, and protein.

50. The label on a can of milk substitute indicates that each half-cup serving supplies 150 kcal and has 8 g of protein, 12 g of carbohydrate, and 8 g of fat. Calculate the percentage of calories each from carbohydrate, fat, and protein.

51. A cup of low-fat cottage cheese provides 31 g protein, 4 g fat, and 8 g of carbohydrate. About how many kilocalories does the cottage cheese provide? What percentage of the calories is from fat?

52. A 6-oz portion of baked orange roughy fish provides 148 kcal, 32 g protein, and 2 g fat. How much carbohydrate does the fish provide?

53. A very active male teenager needs about 3000 kcal per day. If the guideline of 30% or less calories from fat is to be followed, how much fat, in grams, is permitted?

54. An adult female goes on a diet that provides 1200 kcal per day with no more than 25% from fat and no more than 30% of that fat being saturated fat. How much saturated fat, in grams, is permitted in this diet?

■ ADDITIONAL PROBLEMS

55. Omega-6 fatty acids are unsaturated fatty acids in which the double bond closest to the methyl (omega) end of the molecule occurs at the sixth carbon from that end. Which of the following are omega-6 fatty acids? Refer to Table 16.1 for structures not given here.
   a. palmitoleic acid, CH$_3$(CH$_2$)$_5$CH=CH(CH$_2$)$_7$CO$_2$H
   b. oleic acid
   c. linoleic acid
   d. linolenic acid
   e. arachidonic acid

56. Following are line-angle formulas for two fats. **(a)** Which is a saturated fat and which is a polyunsaturated fat? **(b)** Identify the fatty acid units in each.

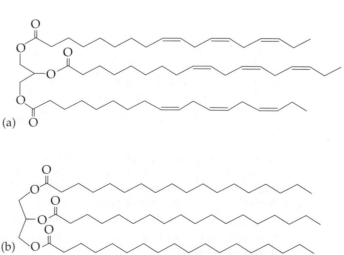
(a)

(b)

Use the following nutritional labels to answer Problems 57–61.

57. What percentage of the energy of whole milk comes from fat?

58. Based on a 2000-kcal diet, what percentage of the maximum allowed cholesterol does a cup of whole milk provide?

59. Based on a 2000-kcal diet, what percentage of the maximum allowed saturated fat does a cup of whole milk provide?

60. A student drinks four cups of skim milk in a day. If the student has the goal of consuming 60 g per day of protein, what percentage of that protein is obtained from the milk?

61. What percentage of the calories in skim milk comes **(a)** from sugars and **(b)** from protein?

62. Structurally, what differences are there between sucrose (Figure 16.5) and sucralose (Figure 17.13)?

63. Of the vitamins calciferol, riboflavin, niacin, and cyanocobolamine, which are most likely to be needed on a daily basis? Explain.

64. Sodium saccharin and sucralose (Figure 17.13) are both readily soluble in water but for somewhat different reasons. Explain.

65. Pioneers in the earlier days of the United States made heavy use of butter, beef tallow, and lard for their long treks across the country. One reason for this was the ready availability of these fats. But even if canola oil, soybean oil, and similar oils had been available, they probably would not have been used much. Consult Table 16.2 and suggest a chemical reason that is related to storage.

66. Chew an unsalted "saltine" cracker for several minutes, and you will notice a slight sweet taste. Your saliva contains enzymes that begin the digestion process. Explain the sweet taste.

67. When eating "thick" canned soups such as beef vegetable soup, you may notice that the liquid in the soup becomes thinner as you eat the soup. Suggest a reason for this phenomenon. (Hint: See Problem 66; starches are used as thickeners in some soups.)

■ COLLABORATIVE GROUP PROJECTS

Prepare a PowerPoint, poster, or other presentation (as directed by your instructor) for presentation to the class.

68. Several fat substitutes have been developed. Prepare a brief report on one of the following fake fats. List principal advantages and disadvantages of each.
   a. Simplesse, made by NutraSweet
   b. Olestra, made by Procter & Gamble
   c. Leanesse (Oatrim), a ConAgra product

69. Prepare a brief report on one of the minerals listed in Table 17.3. List principal dietary sources and uses of the mineral.

70. Prepare a brief report on one of the vitamins listed in Table 17.4. List principal dietary sources and uses of the vitamin.

71. Examine the label on a sample of each of the following.
   a. a can of soft drink   **b.** a can of beer
   c. a dried soup mix   **d.** a can of soup
   e. a can of fruit drink   **f.** a cake mix
   Make a list of the food additives in each. Try to determine the function of each additive. Do all the products offer this information?

72. Prepare a brief report on one of the following dietary supplements. What is the presumed function of each?
   a. black cohosh   **b.** coenzyme Q_{10}
   c. echinacea   **d.** fish oil
   e. glucosamine   **f.** saw palmetto
   g. St. John's wort   **f.** turmeric

73. Prepare a brief report on one of the following chemical contaminants found in food in the past. Dates are given if the incident was largely confined to a single year. Is the contaminant still found in food today?
   a. aminotriazole, 1959   **b.** daminozide (Alar), 1989
   c. DDT   **d.** DES
   e. PBBs   **f.** PCBs

Digitalis purpurea, the beautiful common foxglove, contains the heart stimulant *digoxin*. For centuries digoxin was used as a poison; ingestion of even small amounts of the leaves, flowers, or stems of the plant could be fatal. Its medical usefulness as a heart stimulant was discovered about a hundred years ago. Careful dosage of digoxin can bring patients with heart failure and irregular heartbeat an improved quality of life. Extensive research has brought about the discovery of many natural sources of drugs, as well as semisynthetic and synthetic drugs, for a wide variety of ills.

DRUGS

QUESTIONS YOU MAY HAVE ASKED YOURSELF

1. Why is aspirin not recommended for children?
2. What are the dangers of Tylenol®?
3. Why won't my doctor give me antibiotics for a cold?
4. Does eating lots of sugar cause diabetes?
5. Are "energy drinks" good for you?
6. How should I dispose of old medicine?

Chemical Cures, Comforts, and Cautions

The word *drug* originally meant a dried plant (or plant part) used as a medicine, either directly or after extracting active ingredients as a tea. Today a **drug** is defined as a chemical substance that affects the functioning of living things. Drugs are used to relieve pain, to treat an illness, and to improve one's state of health or well-being. Many people literally owe their lives to drugs. Drugs can kill bacteria, lower blood pressure, prevent seizures, relieve allergies, and do many remarkable things.

Many drugs are still obtained from natural sources, including the following obtained from plants.

- Quinine, the antimalarial drug, from the bark of *Cinchona* species
- Morphine, the analgesic, from the opium poppy
- Digoxin, for heart disorders, from *Digitalis purpurea* (page 526)
- Vinblastine and vincristine, anticancer agents isolated from the Madagascar periwinkle, *Catharanthus roseus*
- The anticancer drug paclitaxel (Taxol) initially isolated from the bark of the Pacific yew tree *Taxus brevifolia*
- Tubocurarine, the muscle relaxant, from *Chondrodendron* and *Curarea* species, used by Amazon natives in the arrow poison curare

Many other drugs are obtained from microorganisms. Some important ones are

- Antibacterial agents (penicillins) and cholesterol-lowering agents (statins) from *Penicillium* species

▲ Opium poppy.

▲ Madagascar periwinkle.

▲ Pacific yew tree *Taxus brevifolia*

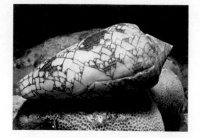

▲ Cone snail *Conus textile.*

- Immunosuppressants, such as cyclosporins, from *Streptomyces* species

Another vast resource for new drugs is marine organisms. Some drugs from the sea are

- The anticancer drug dolastatin-10 from a sea hare
- An anti-inflammatory agent, manoalide, from the sponge *Luffarriella variabilis*
- Ziconotide, a drug for treatment of severe pain, derived from cone snail venom

In addition we have a host of drugs that either have been derived from these natural sources, or have been synthesized from other substances and materials.

18.1 Scientific Drug Design

Starting with molecules from natural sources, chemists make modifications that improve the properties of these substances. These drugs are said to be *semisynthetic*. Many of the earliest scientifically designed drugs were semisynthetic. Examples include

- Salicylic acid (from willow bark) is converted to acetylsalicylic acid (aspirin)
- Morphine (from opium poppies) is converted to heroin
- Lysergic acid (from ergot fungus) is converted to LSD

Today there are many completely synthetic drugs on the market. These substances are found in nearly all categories, from antibacterial drugs to sleeping pills to narcotics such as methadone and fentanyl.

Drugs are used to treat, diagnose, and prevent diseases. People use drugs for many reasons—to relieve pain, to fight infections, to stay awake, to get to sleep, to calm anxiety, to prevent conception, and to treat various conditions from arthritis to heart disease to pneumonia to cancer. Many types of drugs can, however, do more harm than good, especially if used in the wrong amounts or for the wrong reasons.

Humans have used drugs since prehistoric times, and drug use occurs in all cultures. Most societies used alcohol, and the use of marijuana goes back to at least 3000 B.C.E. The narcotic effect of the opium poppy was known to the Greeks in the third century B.C.E, and the Indians of the Andes Mountains have long chewed leaves of the coca plant because of the stimulating effect of the cocaine.

In 1904 Paul Ehrlich (1854–1915), a German chemist, found that certain dyes used to stain bacteria to make them more visible under a microscope could also be used to kill the bacteria. He used dyes against the organism that causes African sleeping sickness. He also prepared an arsenic compound that proved somewhat effective against the organism that causes syphilis. Ehrlich realized that certain chemicals were more toxic to disease organisms than to human cells and could, therefore, be used to control or cure infectious diseases. He coined the term **chemotherapy** (from "chemical therapy"). Ehrlich was awarded the Nobel Prize in Physiology or Medicine in 1908.

The use of drugs is not new, but never before has there been such a vast array of them. There are dozens of pharmaceutical companies producing thousands of drugs. Vast research laboratories spend huge sums to develop new drugs each year. According to a 2006 study by Christopher P. Adams, a Federal Trade Commission (FTC) economist, and Van V. Brantner, an FTC research analyst [*Health Affairs*, 25, no. 2 (2006) pp. 420–428], it costs an estimated $500 million to $2 billion to develop a new drug that is not simply a slight variation of an old one. That cost means that companies must often charge high prices for the products they develop, at least for a while, making them unaffordable for many of the people who need them. After the patent rights expire, other companies can make "generic" versions of the same products at lower prices because they have incurred no research costs. Generic drugs must act identically to the original products.

▲ Some cults in ancient Greece worshiped opium. This Greek coin from that period shows an opium poppy capsule at the nape of the warrior's neck.

Drug industry sales worldwide in 2006 were more than $600 billion. Americans filled 3.4 billion prescriptions in 2006, an average of about 12 per person, and spent about $275 billion for prescription drugs.

Many drugs are sold under generic names and several different brand names, resulting in thousands of prescription medicines. Several hundred medicines are available over the counter (OTC) under thousands of brand names and in numerous combinations. Some of these drugs are vital to the lives and good health of the people who take them. Other drugs are mainly recreational and can be quite destructive. Many of these are addictive, and most are also illegal. In this chapter we look at some of the many drugs that are available today.

18.2 Pain Relievers: Nonsteroidal Anti-Inflammatory Drugs (NSAIDs)

Acetylsalicylic acid, commonly called aspirin in the United States, is widely used around the world with about 40 million kg consumed annually—enough to make 100 billion standard tablets. Its history goes back at least 2400 years to the time of Hippocrates in ancient Greece, who knew that sick people could ease their pain and lower their fever by chewing willow leaves. Even earlier, the Old Testament book of Leviticus (Chapter 23, Verse 40) tells the Israelites to "take you on the first day the boughs of goodly trees, . . . and willows of the brook." It was not until 1835 that chemists isolated the active ingredient, salicylic acid (page 249), an effective **analgesic** (pain reliever), **antipyretic** (fever reducer), and **anti-inflammatory** drug, but it also caused considerable stomach distress and bleeding. Aspirin was introduced in 1893 by the Bayer company in Germany. Acetylsalicylic acid had the desirable medicinal properties of salicylic acid, but it was gentler to the stomach.

Aspirin tablets are sold under a variety of trade names and store brands. A standard aspirin tablet contains 325 mg of acetylsalicylic acid held together with an inert binder such as starch. "Extra-strength" formulations usually contain 500 mg. "Buffered" aspirin contains added antacids to lessen the stomach irritation that sometimes occurs when aspirin is taken on an empty stomach.

Aspirin is one of several substances called **nonsteroidal anti-inflammatory drugs (NSAIDs)**, a designation that distinguishes these drugs from the more potent steroidal anti-inflammatory drugs such as cortisone and prednisone (Section 18.7). Other NSAIDs include ibuprofen (Advil®, Motrin®), ketoprofen (Orudis®), and naproxen (Aleve®) (Figure 18.1). Acetaminophen, like aspirin, relieves minor aches and reduces fever but is not anti-inflammatory and is, therefore, not an NSAID.

NSAIDs act to relieve pain and reduce inflammation by inhibiting the production of *prostaglandins,* compounds involved in sending pain messages to the brain (Section 18.7). They don't cure whatever is causing the pain; they merely "kill the messenger." Inflammation is caused by an overproduction of prostaglandin derivatives, and inhibition of their synthesis reduces the inflammatory process.

Regrettably, for those who take NSAIDs on a regular basis, prostaglandins also have an effect on the blood platelets, the kidneys, and the stomach lining. Suppression of these functions can lead to the side effects most often associated with taking NSAIDs, excessive bleeding and stomach pains. The **anticoagulant** action (inhibition of the clotting of blood) of NSAIDs often leads to complications in patients taking NSAIDs for treatment of arthritis. About 1.5% of patients with rheumatoid arthritis taking NSAIDs for one year had a serious stomach complication. This is a small proportion of patients overall, but it is a large number of people because of the vast number of people using NSAIDs.

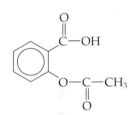

▲ Acetylsalicylic acid (aspirin).

▲ **It DOES Matter!**

Almost every single-ingredient analgesic product sold today uses one of four active ingredients: acetaminophen, aspirin, ibuprofen, and naproxen. These drugs are available in generic forms, under dozens of brand names, and in many combinations including cough/cold and allergy/sinus products. They are often combined with caffeine in OTC products and with codeine in prescription pain medicines. You can search the Web for brand names, using the generic names as keywords.

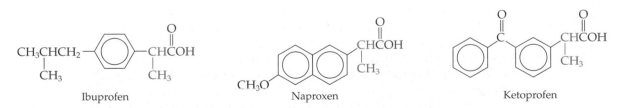

▲ **Figure 18.1** Three NSAIDs. Ibuprofen, naproxen, and ketoprofen are derivatives of propionic acid (CH_3CH_2COOH).

GREEN CHEMISTRY

Drugs

Andy Jorgensen, University of Toledo

Pharmacy shelves are filled with medicines that came directly or indirectly from nature (Section 18.1), but nearly all need some processing to yield safe and effective drugs. The manufacture of drugs is a prime area for the application of green chemistry principles.

The 2008 Presidential Green Chemistry Challenge Academic Award was presented to Robert E. Maleczka, Jr. and Milton R. Smith, III of Michigan State University for development of chemistry for the preparation of compounds with carbon–boron bonds, a common step in making pharmaceuticals. As you read the chemistry behind their work, see if you can identify which green chemistry principles they used.

Let's start with some basic chemistry. Many chemical reactions described in this or other texts involve the formation of a new bond, for example, a new carbon–oxygen bond in ester formation (Section 9.13). Often in the synthesis of new drugs a new carbon–carbon (C—C) bond is needed. This may look simple on paper, but in the lab it can be quite complicated. The new C—C bond must be formed at the correct location without destroying another part of the molecule. And you want to do this with safe reagents, at low cost, and with little waste.

A common means for making a C—C bond is to use an intermediate that has a carbon–boron bond. Let's say that you want to make a bond between two carbon-containing units such as the aromatic compounds R_1—C_6H_5 and R_2—C_6H_5 to give R_1—C_6H_4—C_6H_4—R_2. The R_1 and R_2 may represent a variety of groups. You cannot just mix R_1—C_6H_5 and R_2—C_6H_5 and expect them to bond in the proper manner. Special conditions are needed.

In general, carbon–hydrogen bonds are quite stable. An H atom on the C_6H_5 group must be "activated." One way to do this goes through an organic species R_1—C_6H_4—MgBr. This introduces bromine, which is potentially hazardous to the environment. The process also requires careful reaction conditions, such as the total absence of water.

Another way is by a catalytic process called Suzuki Coupling. One example is:

$$C_6H_5\text{—}B(OH)_2 + Br\text{—}C_6H_5 \longrightarrow$$
$$C_6H_5\text{—}C_6H_5 + \text{by-products}$$

The C_6H_5 groups represent benzene rings, either or both of which may have other atoms attached to them.

Maleczka and Smith developed a process using an iridium catalyst at low temperature for making compounds with a carbon–boron bond while leaving the other important parts of the molecules unchanged. Thus, they avoided the use of halogens and, in some cases, solvents. Their only by-product was H_2 from the H atom originally on a carbon atom and the H atom that was initially on the boron.

Another key feature of their work was the exact placement of the new C—C bond within a molecule containing several carbon atoms. Energy and space considerations often cause a preferential addition to one or only a few locations. Controlling bond location is a key requirement in making drugs because biological systems are quite sensitive to the exact geometry of the molecule. Even a slight misalignment in an interaction between a drug and a receptor can make the difference between an effective drug and a failure.

The chemistry invented by Maleczka and Smith allowed bond formation in locations that were not the most energetically favored, but in areas that led to the correct location for biological utility. This makes for a more efficient synthesis and can even lead to the creation of molecules not readily produced by any other means. This feature of their work added a valuable dimension to their accomplishments.

In writing chemical equations, it is easy to show exactly what you want in a product. Once you start the lab work behind those simple plans, complications often can make your task far more difficult than you could imagine. Maleczka and Smith have made a major contribution to laboratory synthesis of valuable substances, particularly those compounds which can improve our health and maybe even save our lives. Furthermore, they accomplished this through innovation that reduced the impact of chemical synthesis on the planet.

Aspirin for Heart Attack and Stroke Prevention

Because NSAIDs act as anticoagulants, they should not be used by people facing surgery, childbirth, or some other hazard involving the possible loss of blood (NSAIDs should be withheld for at least a week before the event). On the other hand, small daily doses seem to lower the risk of heart attack and stroke, presumably by the same anticoagulant action that causes bleeding in the stomach. Adult low-strength (usually 81 mg) aspirin tablets are now routinely used by people at risk for heart attack and stroke.

NSAIDs and Fever Reduction

Fevers are induced by substances called *pyrogens*. These compounds are produced by and released from leukocytes and other circulating cells. Pyrogens usually use prostaglandins as secondary mediators. Fevers therefore can be reduced by aspirin and other prostaglandin inhibitors.

Our bodies try to fight off infections by elevating the temperature. Mild fevers in adults (those below 39 °C or 102 °F) are usually best left untreated. High fevers, however, can cause brain damage and require immediate treatment.

How NSAIDs Work

Prostaglandins are produced in the body from a cell membrane component called arachidonic acid (Section 18.7). Scientists have identified several types of enzymes, called cyclooxygenases (COX), that catalyze the conversion. Two of the forms are called COX-1, found in stomach and kidney tissues where NSAID side effects occur, and COX-2, found in tissues where inflammation occurs. The older NSAIDs, such as aspirin, ibuprofen, ketoprofen, and naproxen, inhibit both COX-1 and COX-2, providing relief from pain and inflammation but also producing undesirable side effects. Newer drugs preferentially inhibit COX-2. Three such drugs, celecoxib (Celebrex®), rofecoxib (Vioxx®), and valdecoxib (Bextra®) are anti-inflammatory agents claimed to have fewer NSAID-type side effects than the older drugs (Figure 18.2). However, Vioxx® and Bextra® were removed from the market because of a higher risk of stroke and heart attacks. Celebrex® remains on the market but carries a new warning label, called a "Black Box" warning because it is the strongest warning that a medicine label can provide. The U.S. Food and Drug Administration (FDA) now requires labels on all prescription NSAIDs to include a boxed warning highlighting the potential for increased risk of heart problems, as well as information regarding allergic reactions and internal bleeding.

Acetaminophen

Acetaminophen, the most widely used analgesic in the United States, gives relief of pain and reduction of fever comparable to that of aspirin. Acetaminophen is chemically similar to an older analgesic, phenacetin, which was removed from the market because it is a suspected carcinogen. The total world market for acetaminophen is about 60 million kg/year. However, unlike the NSAIDS, acetaminophen is not effective against inflammation, and it is not an anticoagulant. People allergic to aspirin or susceptible to bleeding can safely take acetaminophen, but it is of limited

1. Why is aspirin not recommended for children? The use of aspirin in children suffering from the flu or chicken pox is associated with *Reye's syndrome*, a potentially fatal illness. It is for this reason that aspirin products carry a warning that they should not be used to treat children with fevers.

Pyrogens that do not work through prostaglandins are not affected by aspirin.

"Children's" NSAIDs are typically one-quarter of the normal adult dose per tablet, with fewer tablets per bottle than those for adults. This reduces significantly the possibility of an accidental overdose.

◀ **Figure 18.2** Three COX-2 inhibitors.

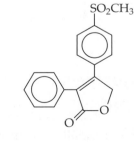

Celecoxib (Celebrex®)

Rofecoxib (Vioxx®)

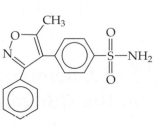

Valdecoxib (Bextra®)

use to people with arthritis. Acetaminophen is often used to relieve the pain that follows minor surgery. Regular acetaminophen tablets are 325 mg, and extra-strength forms are 500 mg. Overuse of acetaminophen, especially when combined with alcohol, has been linked to liver and kidney damage.

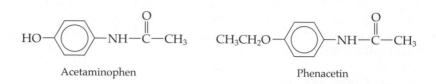

Acetaminophen   Phenacetin

Acetaminophen acts on a third variant of the cyclooxygenase enzyme, COX-3. Inhibition of COX-3 may represent a mechanism by which acetaminophen decreases pain and fever. However, COX-3 appears to have no role in inflammation.

Combination Pain Relievers

Many analgesic products are some combination of one or more NSAIDs, acetaminophen, caffeine, antihistamines (Section 18.3), or other drugs. Acetaminophen, for example, is found in more than 70 combination products for pain relief, allergy symptoms, and cold and flu treatment. These combinations are available under various brand names and as store brands. Familiar brands include Excedrin, which can be purchased in six different combinations, and Anacin, which is available in five different formulations.

Many other combination pain relievers are available. Most claim some advantage over regular aspirin. However, such advantages are marginal at best. For occasional use by most people, plain aspirin may be the cheapest, safest, and most effective product.

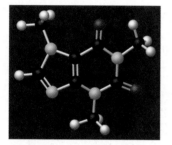

▲ Caffeine, $C_8H_{10}N_4O_2$, is a stimulant (Section 18.17) and is added to NSAIDs in some combination pain reliever formulations.

Self-Assessment Questions

1. An antipyretic is a(n)
   a. appetite suppressant   b. fever reducer
   c. fire retardant   d. gum treatment
2. Which of the following is *not* an anti-inflammatory agent?
   a. acetaminophen   b. aspirin
   c. celecoxib   d. ibuprofen
3. NSAIDs act by inhibiting the production of
   a. endorphins   b. ketone bodies
   c. norepinephrine   d. prostaglandins
4. Which of the following does *not* act as an anticoagulant?
   a. acetaminophen   b. aspirin
   c. ibuprofen   d. ketoprofen
5. Prostaglandins are formed from
   a. cell membrane components   b. cholesterol
   c. histamines   d. nucleotides
6. Newer NSAIDs inhibit
   a. COX-1   b. COX-2
   c. COX-1 and COX-2   d. COX-3

Answers: 1, b; 2, a; 3, d; 4, a; 5, a; 6, b

18.3 Chemistry, Allergies, and the Common Cold

From giant warehouse-type stores to small convenience stores, we find shelves stocked with a variety of cold and allergy medicines. They contain compounds that act as antihistamines, cough suppressants, expectorants, bronchodilators, or nasal decongestants. None of the products cure the common cold. Colds are caused by as many as 200 related viruses, and no antibiotic or other drug provides an effective cure. Most cold remedies treat just the symptoms. An FDA advisory panel reviews these substances periodically for safety and effectiveness.

Antihistamines

Many cold medicines contain an **antihistamine**, such as chlorpheniramine and diphenhydramine (Figure 18.3). Antihistamines relieve the symptoms of allergies: sneezing, itchy eyes, and runny nose. They don't cure colds, but they can temporarily relieve some of the symptoms.

When an **allergen** (a substance that triggers an allergic reaction) binds to the surfaces of certain cells, it triggers the release of *histamine*, which causes the redness, swelling, and itching associated with allergies. An antihistamine inhibits the release of histamine, but many first-generation OTC antihistamines enter the brain and act on the cells controlling sleep, thus making patients drowsy; loratadine (Claritin®, Alavert®) is an exception among the over-the-counter drugs. Prescription drugs, such as fexofenadine (Allegra®) and desloratadine (Clarinex®), inhibit the release of histamine but do not enter the brain and so do not cause drowsiness. Cetirizine (Zyrtec®), approved for OTC sales in 2007, is long-acting but is somewhat more sedating than fexofenadine and desloratadine. Many cough and cold products contain more than one ingredient (see Collaborative Group Project 68).

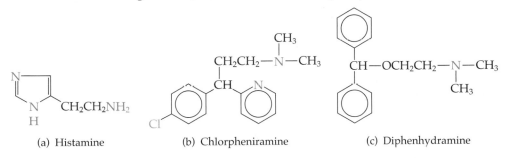

(a) Histamine   (b) Chlorpheniramine   (c) Diphenhydramine

◀ **Figure 18.3** Histamine (a) and two antihistamines: (b) chlorpheniramine (Chlor-Trimeton®) and (c) diphenhydramine (Benadryl®). Diphenhydramine is also used as a cough suppressant.

QUESTION: To what family of organic compounds do these three substances belong? How does brompheniramine differ from chlorpheniramine?

Cough Suppressants

The two *antitussives* (cough suppressants) available OTC are diphenhydramine and dextromethorphan. Codeine and hydrocodone (see Table 18.5) are effective antitussives but are available only by prescription; both are narcotics. The question arises, though: Should a cough really be suppressed? Ordinarily, a cough is functional: The respiratory tract uses the cough mechanism to rid itself of congestion. When a cough is dry or interferes with needed rest, however, temporary cough suppression may be advisable.

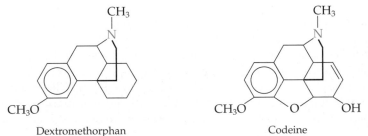

Dextromethorphan   Codeine

A 2004 study at Pennsylvania State University questioned the effectiveness of the OTC antitussives for children. The study found that diphenhydramine and dextromethorphan were no better than a placebo (page 576) for cough suppression in children ages 2 to 18.

Expectorants and Decongestants

An *expectorant* is a substance that helps bring up mucus from the bronchial passages. The only expectorant rated safe and effective is glyceryl guaiacolate (guaifenesin), and its effectiveness is not well documented.

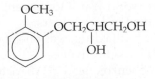

Guaifenesin

Nasal *decongestants* seem to be safe and effective for occasional use, although repeated use leads to a rebound effect in which the nasal passages swell and make congestion seem worse. Examples of long-acting nasal decongestants are oxymetazoline and xylometazoline. These are effective in a few minutes and last for 6 to 12 hours. Short-acting nasal decongestants include naphazoline and phenylephrine.

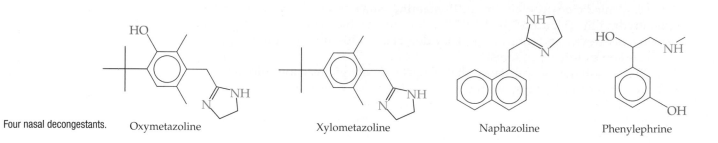

▶ Four nasal decongestants.   Oxymetazoline   Xylometazoline   Naphazoline   Phenylephrine

How should you treat a common cold? First, you should drink plenty of liquids and get lots of rest. When you are in good physical condition and have a strong immune system, colds seem to strike less often. Frequent washing of the hands, especially when colds are prevalent, can dramatically reduce transmission of the virus. Many people claim to get cold relief by using some favorite over-the-counter drug. And, of course, there are those who still swear by chicken soup!

Self-Assessment Questions

1. Allergens trigger the release of
   a. endorphins   **b.** histamines
   c. prostaglandins   **d.** pyrogens
2. Which of the following cures the common cold?
   a. antibiotics   **b.** antihistamines
   c. there is no cure   **d.** vitamin C

Answers: 1, b; 2, c

18.4 Antibacterial Drugs

A century ago, infectious diseases were the principal cause of death in the United States (Figure 18.4). At the time of the American Civil War, thousands of wounded soldiers required amputations of wounded arms or legs, and nearly 80% of the patients died after coming to a hospital, mostly from infections. Today infectious disease categories have moved toward the bottom of the top 10 list of leading causes of death. (These top 10 causes account for more than three-quarters of all causes of death in the United States.) Today many diseases have been brought under control by the use of *antibacterial drugs*.

Sulfa Drugs

The first antibacterial drugs were *sulfa drugs*, the prototype of which was discovered in 1935 by the German chemist Gerhard Domagk (1895–1964). Sulfa drugs were used extensively during World War II to prevent wound infections. Many soldiers lived who would have died in earlier wars.

Sulfanilamide, the simplest sulfa drug, was one of the first to have its action understood at the molecular level. Its effectiveness is based on a case of mistaken identity. Bacteria need *para*-aminobenzoic acid (PABA) to make folic acid, which is

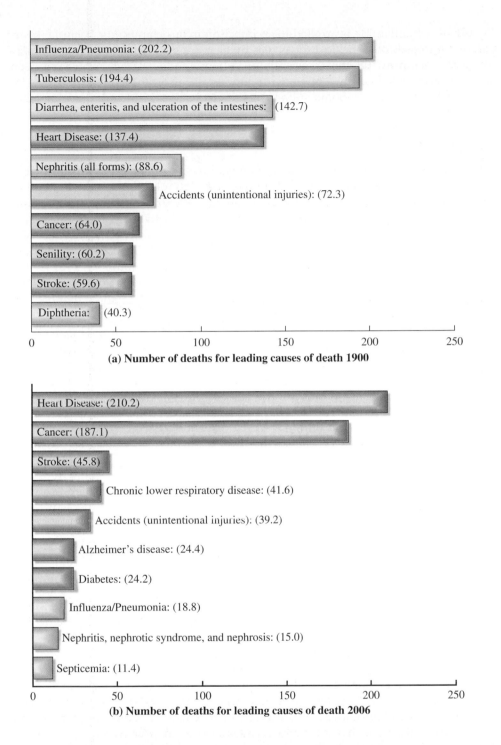

Influenza/Pneumonia: (202.2)

Tuberculosis: (194.4)

Diarrhea, enteritis, and ulceration of the intestines: (142.7)

Heart Disease: (137.4)

Nephritis (all forms): (88.6)

Accidents (unintentional injuries): (72.3)

Cancer: (64.0)

Senility: (60.2)

Stroke: (59.6)

Diphtheria: (40.3)

0   50   100   150   200   250

(a) Number of deaths for leading causes of death 1900

Heart Disease: (210.2)

Cancer: (187.1)

Stroke: (45.8)

Chronic lower respiratory disease: (41.6)

Accidents (unintentional injuries): (39.2)

Alzheimer's disease: (24.4)

Diabetes: (24.2)

Influenza/Pneumonia: (18.8)

Nephritis, nephrotic syndrome, and nephrosis: (15.0)

Septicemia: (11.4)

0   50   100   150   200   250

(b) Number of deaths for leading causes of death 2006

◀ **Figure 18.4** (a) In 1900 five of the ten leading causes of death—and three-fifths of all deaths—in the United States were infectious diseases. (b) The leading causes of death in 2006 were so-called lifestyle diseases: Heart disease, stroke, and diabetes are related in part to our diets, cancer and lung diseases to cigarette smoking, and many accidents to excessive consumption of alcohol.

essential for the formation of certain compounds the bacteria require for proper growth. But bacterial enzymes can't tell the difference between sulfanilamide and PABA because the substances are so similar in structure.

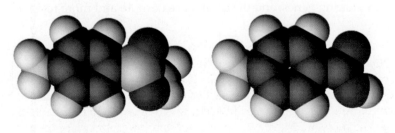

Sulfanilamide   *para*-aminobenzoic acid

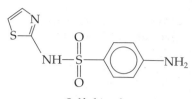

Sulfathiazole

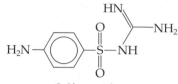

Sulfaguanidine

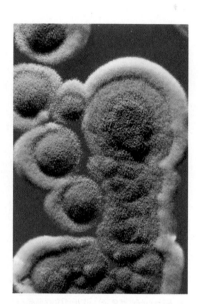

▲ Penicillin molds. These symmetrical colonies of mold are *Penicillium chrysogenum*, a mutant form of which now produces nearly all commercial penicillin.

▼ **Figure 18.5** Penicillins. A general formula where R is a variable side chain (a). Penicillin G (b) is generally considered the most effective penicillin but can only be given by injection. Amoxicillin (c) can be administered orally.

When sulfanilamide is applied to an infection in large amounts, bacteria incorporate it into pseudo–folic acid molecules that cannot act normally; hence, the bacteria cease to grow. Of the thousands of sulfanilamide analogs that have been developed and tested, only a few are used today. Two common ones are sulfathiazole and sulfaguanidine. Some sulfa drugs tend to cause kidney damage or other problems.

Penicillins

The next important discovery was that of penicillin, an antibiotic. An **antibiotic** is a soluble substance (derived from molds or bacteria) that inhibits the growth of other microorganisms. Penicillin was first discovered in 1928 but was not tried on humans until 1941. Alexander Fleming (1881–1955), a Scottish microbiologist then working at the University of London, first observed the antibacterial action of a mold, *Penicillium notatum.* Fleming was studying an infectious bacterium, *Staphylococcus aureus,* and one of his cultures became contaminated with a blue mold. Contaminated cultures generally are useless. Most investigators probably would have destroyed the culture and started over, but Fleming noted that bacterial colonies had been destroyed in the vicinity of the mold.

Fleming was able to make crude extracts of the active substance. This material, later called *penicillin,* was further purified and improved by Howard Florey (1898–1968), an Australian, and Ernst Boris Chain (1906–1979), a refugee from Nazi Germany, both working at Oxford University. Fleming, Florey, and Chain shared the 1945 Nobel Prize in Physiology or Medicine for their work on penicillin.

Penicillin is not a single substance but a group of related compounds (Figure 18.5). By designing molecules with different structures, chemists can change the properties of the drugs, producing penicillins that vary in effectiveness. Some can be taken orally; others must be injected. Bacteria resistant to one penicillin may be killed by another. Amoxicillin has a broad spectrum of activity against many types of microorganisms. It is among the top 10 most prescribed drugs with more than 34 million prescriptions in 2006. Overall, penicillins account for about half of all antimicrobial prescriptions.

Penicillin works by inhibiting enzymes that the bacteria use to make their cell walls. The bacterial cell walls are made up of *mucoproteins,* polymers in which amino sugars are combined with protein molecules. Penicillin prevents cross-linking between these large molecules. This leaves holes in their cell walls, and the bacteria swell and rupture. Cells of higher animals have only external membranes; they do not have mucoprotein walls and are therefore not affected by penicillin. Thus, penicillin can destroy bacteria without harming human cells. However, some people can't take penicillin because they are allergic to it.

In their early days, antibiotics were known as miracle drugs. The number of deaths from blood poisoning (septicemia), pneumonia, and other infectious diseases was reduced substantially by the use of antibiotics. Prior to 1941 a person with a major bacterial infection almost always died. Today, such deaths are rare except for those ill with other conditions such as AIDS. Six decades ago, pneumonia was a dreaded killer of people of all ages. Today, it kills mainly the elderly and those with AIDS. Antibiotics have indeed worked miracles in our time, but even miracle drugs are not without problems. It wasn't long after these drugs were first used that disease organisms began to develop strains resistant to them. One strain can spread antibiotic resistance to another by sharing genes, and the emergence of more and more resistant strains of

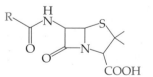

(a) General formula

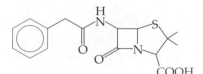

(b) Penicillin G

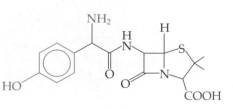

(c) Amoxicillin

bacteria is a serious threat to world health. For example, although tuberculosis was once almost completely wiped out in developed countries, drug-resistant tuberculosis is now rampant in Russia and among AIDS patients everywhere.

Another example involves erythromycin, an antibiotic obtained from *Streptomyces erythreus*. As long as it had only limited use, erythromycin could handle all strains of staphylococci. After it was put into extensive use, resistant strains began to appear. Staph infections are now a serious problem in hospitals. People who go to a hospital to be cured sometimes develop serious bacterial infections instead. Some even die of staph infections picked up at hospitals. A Centers for Disease Control and Prevention (CDC) estimate in 2007 held that staph infections cause more than 94,000 serious illnesses and about 19,000 deaths each year.

Cephalosporins

In a race to stay ahead of resistant strains of bacteria, scientists continue to seek new antibiotics. Penicillins have been partially displaced by related compounds called *cephalosporins* such as cephalexin (Keflex). Unfortunately, some strains of bacteria are resistant to cephalosporins.

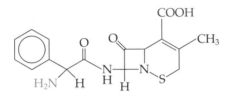

Cephalexin

Tetracyclines

Another important development in the field of antibiotics was the discovery of a group of compounds called *tetracyclines*. The first of these four-ring compounds, aureomycin, was isolated in 1948 by Benjamin Duggor (1872–1956) from *Streptomyces aureofaciens*. Scientists at Pfizer Laboratories isolated terramycin from *Streptomyces rimosus* in 1950, and both drugs were later found to be derivatives of tetracycline, a compound now obtained from *Streptomyces viridifaciens*. All three compounds (Figure 18.6) are **broad-spectrum antibiotics**, so called because they are effective against a wide variety of bacteria.

Tetracyclines bind to bacterial ribosomes, inhibiting bacterial protein synthesis and blocking growth of the bacteria. Tetracyclines do not bind to mammalian ribosomes and thus do not affect protein synthesis in host cells. Several disease-causing organisms have developed strains resistant to tetracyclines.

When given to young children, tetracyclines can cause the discoloration of permanent teeth, even though the teeth may not appear until several years later. Women are told to avoid tetracyclines during pregnancy for the same reason. This probably results from the interaction of tetracyclines with calcium during the period of tooth development. Calcium ions in milk and other foods combine with hydroxyl groups on tetracycline molecules.

Fluoroquinolones

Fluoroquinolone antibiotics were first introduced in 1986, and now make up about one-third of the $32 billion global antibiotic market. They act against a broad spectrum of bacteria and seemingly have few side effects. A major use is against bacteria with penicillin resistance. Fluoroquinolones act by inhibiting bacterial DNA replication through interference with the action of an enzyme called DNA gyrase. Humans do not have this enzyme so fluoroquinolones do not harm human cells. Because their mechanism of action is different from that of other antibiotics, it was hoped that resistance to these drugs would be slow to emerge. However, resistance to fluoroquinolones is already being seen.

3. Why won't my doctor give me antibiotics for a cold?
Colds are the result of many different viruses. Not only are antibiotics ineffective against viruses, but the indiscriminate use of antibiotics also increases the likelihood that antibiotic-resistant strains of bacteria will develop.

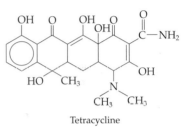

Tetracycline

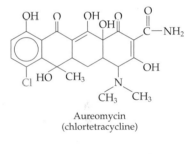

Aureomycin
(chlortetracycline)

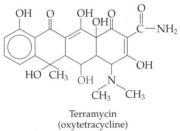

Terramycin
(oxytetracycline)

▲ **Figure 18.6** Three tetracycline antibiotics. Subtle structural differences are pointed out in color.

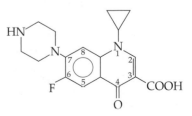

Ciprofloxacin (Cipro®)

Structure–Function Relationships

Fluoroquinolones are of particular interest to chemists and pharmacologists because many of their structural features are correlated with their activity, making possible a rational, logical approach to synthesis of new drugs in this class.

The basic skeletal structure for fluoroquinolones is

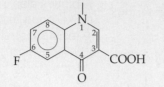

The fluorine atom at position 6 is essential for broad antimicrobial activity when taken orally. Effectiveness depends on a carboxyl group at position 3 and a carbonyl oxygen at position 4, which are responsible for binding to the bacterial DNA complex. A nitrogen-containing ring structure at position 7 or a methoxy group at position 8 broadens the range of bacteria affected. If position 7 has a bulky substituent, the drug has fewer side effects on the nervous system.

Self-Assessment Questions

1. Sulfamethoxazole is selectively toxic to bacteria because mammals can obtain a certain compound from the diet, while bacterial cells must synthesize it. That compound is
   a. *para*-aminobenzoic acid   **b.** ascorbic acid
   c. folic acid   **d.** uracil

2. Penicillins act by inhibiting the synthesis of
   a. bacterial cell walls   **b.** folic acid
   c. β-lactams   **d.** viral RNA

3. Antibiotics kill
   a. bacteria but not viruses   **b.** both viruses and bacteria
   c. neither bacteria nor viruses **d.** viruses but not bacteria

4. Taking antibiotics for a stomach virus will
   a. cure the illness
   b. increase the chance of developing drug-resistant bacteria
   c. lessen the symptoms
   d. shorten the duration of the illness

Answers: 1, *a*; 2, *a*; 3, *a*; 4, *b*

18.5 Viruses and Antiviral Drugs

More to Explore
Shnayerson, Michael, and Mark J. Plotkin. "Supergerm Warfare." *Smithsonian*, October 2002, pp. 114–126.

For most of us, antibiotics have taken the terror out of bacterial infections such as pneumonia and diphtheria. We worry about resistant strains of bacteria, but for most people these problems are not insurmountable. However, viral diseases cannot be cured by antibiotics, and viral infections from colds and influenza to herpes and **acquired immune deficiency syndrome (AIDS)** still plague us. Some viral infections—such as poliomyelitis, mumps, measles, and smallpox—can usually be prevented by vaccination. Influenza vaccines are quite effective against common recurrent strains, but there are many different strains of flu viruses, and new ones appear periodically.

DNA Viruses and RNA Viruses

Viruses are composed of nucleic acids and proteins (Figure 18.7). They have an external coat with a repetitive pattern of protein molecules. Some coats also include a lipid membrane, and others have sugar–protein combinations called glycoproteins. The genetic material of a virus is either DNA or RNA. A *DNA virus* enters a host cell where the DNA is replicated, and it directs the host cell to produce viral proteins. The viral proteins and viral DNA assemble into new viruses that are released by the host cell. These new viruses can then invade other cells and continue the process.

Most *RNA viruses* use their nucleic acids in much the same way. The virus penetrates a host cell, where the RNA strands are replicated and induce the synthesis of viral proteins. The new RNA strands and viral proteins are then assembled into new viruses. Some RNA viruses, called **retroviruses,** synthesize *DNA* in the host cell. This process is the opposite of the transcription of a DNA code into RNA (Section 16.10) that normally occurs in cells (the term *retro* implies working backward). The synthesis of DNA from an RNA template is catalyzed by an enzyme called *reverse transcriptase.* The human immunodeficiency virus (HIV) that causes AIDS is a retrovirus. HIV invades and eventually destroys *T cells,* white blood cells that normally help protect the body from infections. With the T cells destroyed, the AIDS victim often succumbs to pneumonia or some other infection.

Worldwide, as many as 33 million people were living with AIDS in 2007. More than 25 million are already dead from AIDS, and 2.1 million more died in 2007. Some 2.5 million people became infected with HIV in 2007. In the United States, an estimated 1.0–1.2 million persons are living with HIV/AIDS. Another 37,000 new HIV infections occur each year.

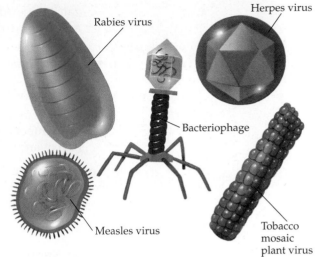

▲ **Figure 18.7** Viruses come in a variety of shapes, which are determined by their protein coats.

Antiviral Drugs

Scientists have developed a variety of drugs that are effective against some viruses. None provide cures. Structures of some common antiviral agents are given in Figure 18.8. Amantadine (Symmetrel®) is used to prevent or treat certain kinds of influenza (flu). It is given either alone or in combination with flu shots. Acyclovir (Zovirax®) is used to treat chickenpox, shingles, cold sores, and the symptoms of genital herpes.

Antiretroviral drugs prevent the reproduction of retroviruses such as HIV and are used against AIDS. Three important kinds are

- *Nucleoside analogs,* or nucleoside reverse transcriptase inhibitors (NRTIs), such as didanosine (ddI, Videx®), lamivudine (3TC, Epivir®), stavudine (d4T, Zerit®),

▼ **Figure 18.8** Some antiviral drugs. Amantadine (a) is completely synthetic. Didanosine (ddI) (b) is a nucleoside analog or nucleoside reverse transcriptase inhibitors (NRTI). Delavirdine (c) is a nonnucleoside reverse transcriptase inhibitor (NNRTI). Saquinavir (d) is a protease inhibitor. Oseltamivir (e) (Tamiflu®) can prevent influenza if it is given in early stages.

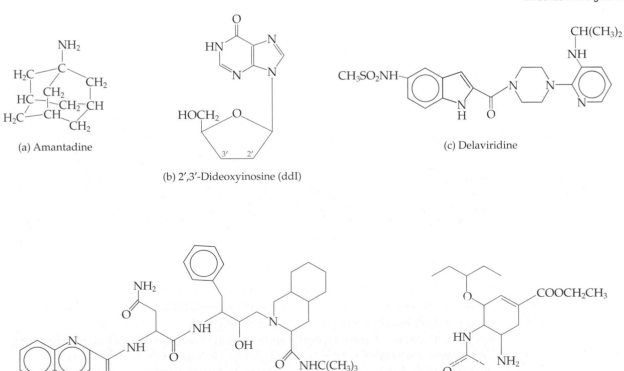

(a) Amantadine

(b) 2′,3′-Dideoxyinosine (ddI)

(c) Delaviridine

(d) Saquinavir

(e) Oseltamivir

zalcitabine (ddC, Hivid®), and zidovudine (AZT, Retrovir®). Substitution of an analog for a nucleoside in the viral DNA cripples it, slowing down its replication.

- *Nonnucleoside reverse transcriptase inhibitors (NNRTIs)* stop the reverse transcriptase from working properly to make more virus. NNRTIs include delavirdine (Rescriptor®), loviride, and nevirapine (Viramune®).

- *Protease inhibitors* block the enzyme protease so that new copies of the virus can't infect new cells. Protease inhibitors include ritonavir (Norvir®), indinavir (Crixivan®), nelfinavir (Viracept®), and saquinavir (Invirase®).

Combinations of AIDS drugs seem to be more effective than any one of them alone. For example, Trizivir® contains abacavir, lamivudine, and zidovudine. Unfortunately, these drugs are too toxic to use in quantities sufficient to stop virus replication completely.

Basic Research and Drug Development

Scientists had to understand the normal biochemistry of cells before they could develop drugs to treat diseases caused by viruses. Gertrude Elion and George Hitchings of Burroughs Wellcome Research Laboratories in North Carolina and James Black of Kings College in London did much of the basic biochemical research that led to the development of antiviral drugs and many anticancer drugs (Section 18.8). They determined the shapes of cell membrane receptors, and they learned how normal cells work. They and other scientists were then able to design drugs to block receptors in infected cells. Elion, Hitchings, and Black shared the 1988 Nobel Prize in Physiology or Medicine for this work.

Drug design in its early days was often a hit-or-miss procedure. Today scientists use powerful computers to design molecules to fit receptors, and drug design is becoming a more precise science.

Combinatorial Chemistry

For a century now chemists have synthesized one substance at a time and then tested it for biological activity. Then they made a new compound by varying the structure a bit, hoping to obtain a substance with more desirable properties and less undesirable attributes. Now many chemists use *combinatorial chemistry,* a technique in which a set of starting chemicals is reacted in all possible combinations. For example, a set of 10 compounds reacted with 10 different reagents gives 100 possible products. The products are then tested for biological activity in much the same manner. The process is highly automated.

These robotic techniques yield enormous numbers of different compounds whose properties can vary widely. Most hold little promise and are discarded or filed away. Promising substances are tested further. The very few with the most potential eventually start the long, expensive route to human testing and possible FDA approval.

Self-Assessment Questions

1. Viruses consist of a protein coat surrounding a core of
   - **a.** lipids
   - **b.** nucleic acids
   - **c.** polysaccharides
   - **d.** proteins

2. Most RNA viruses replicate in host cells by replicating RNA strands and synthesizing
   - **a.** host-cell lipids
   - **b.** host-cell proteins
   - **c.** viral polysaccharides
   - **d.** viral proteins

3. Retroviruses replicate in host cells by
  a. replicating RNA strands and synthesizing viral proteins
  b. replicating RNA strands and synthesizing host-cell proteins
  c. synthesizing DNA and forming new viruses
  d. synthesizing DNA and replicating host cells

4. What HIV enzyme catalyzes formation of a DNA copy of the viral genome inside the host cell?
  a. DNA polymerase   **b.** protease
  c. reverse transcriptase   **d.** RNA polymerase

5. The best treatment for viral diseases such as diphtheria, mumps, and smallpox is
  a. antibacterial drugs   **b.** antiviral drugs
  c. tetracyclines   **d.** vaccinations

Answers: 1, b; 2, d; 3, c; 4, c; 5, d

18.6 Chemicals against Cancer

Despite decades of "war" on cancer, it remains a dread disease. Each year in the United States, more than 1.4 million new cases are diagnosed with cancer killing more than 560,000 people. Scientists have made progress, but much remains to be done. A major problem is that the drugs that kill cancer cells damage normal cells as well, so a primary aim of cancer research is to find a way to kill cancerous tissue without killing too many normal cells. Treatment with drugs, radiation, and surgery yields a high rate of cure for some kinds of cancer. For example, with treatment prostate cancer has an overall five-year survival rate of 98.8%. For other types, such as lung cancer, the rate of cure is still quite low: The overall five-year survival rate is only 15.0%, mainly because the lung cancer is often advanced at the time of diagnosis. The five-year survival rate for all cancers is about 64%. Dozens of anticancer drugs are used widely, and this number will no doubt increase as our understanding of basic cell chemistry increases.

Cancer chemotherapy affects any body cells that undergo rapid replacement, not just cancer cells. Included are the cells that line the digestive tract and those that produce hair. Side effects of chemotherapy, therefore, include nausea and loss of hair.

Antimetabolites: Inhibition of Nucleic Acid Synthesis

An **antimetabolite** is a compound that closely resembles a substance essential to normal body metabolism and in this way interferes with physiological reactions involving it. Rapidly dividing cells, characteristic of cancer, require an abundance of DNA. Cancer antimetabolites block DNA synthesis and therefore block the increase in the number of cancer cells. Because cancer cells are undergoing rapid growth and cell division, they are generally affected to a greater extent than normal cells.

Gertrude Elion and George Hitchings patented 6-mercaptopurine (6-MP) in 1954. This compound can substitute for adenine in nucleotides, the phosphate–sugar–base unit of both DNA and RNA (Section 16.8); the pseudonucleotide then inhibits the synthesis of nucleotides incorporating adenine and guanine. This slows DNA synthesis and cell division, thus inhibiting the multiplication of cancer cells. Prior to 6-MP, half of all children with acute leukemia died within a few months. Combined with other medications, 6-MP was able to cure approximately 80% of child leukemia patients.

Two other prominent antimetabolites are 5-fluorouracil (5-FU) and its deoxyribose derivative, 5-fluorodeoxyuridine. 5-FU blocks synthesis of the thymine-containing nucleotide thymidine, which is essential for DNA replication. When given rapidly by injection, 5-fluorodeoxyuridine is metabolized to 5-FU. When infused slowly into an artery, it is converted to a fluorinated nucleotide called floxuridine monophosphate, a compound that acts to slow the division of cancer cells. Both 5-FU and 5-fluorodeoxyuridine are employed against a variety of cancers,

6-Mercaptopurine

Adenine

especially those of the breast and the digestive tract. Research over the last two decades has led to refinements in the methods of use of these drugs, increasing their effectiveness.

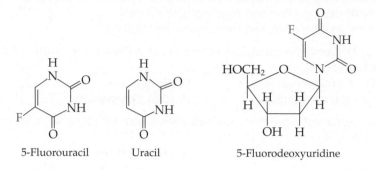

5-Fluorouracil   Uracil   5-Fluorodeoxyuridine

The antimetabolite methotrexate acts in a somewhat different manner; it is a folic acid antagonist that interferes with cellular reproduction. Note the similarity between its structure and that of folic acid. Like the pseudofolic acid formed from sulfanilamide, methotrexate competes successfully with folic acid for an essential enzyme but cannot perform the growth-enhancing function of folic acid. Again, cell division is slowed and cancer growth is retarded. Methotrexate is used frequently against leukemia. It is also used in the treatment of psoriasis and certain inflammatory diseases such as rheumatoid arthritis.

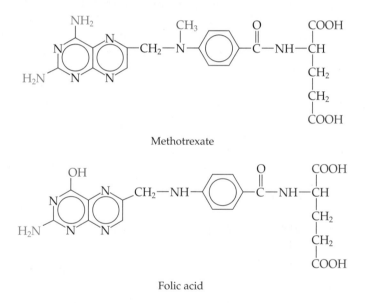

Methotrexate

Folic acid

Alkylating Agents: Turning War Gases on Cancer

Alkylating agents are highly reactive compounds that can transfer alkyl groups to compounds of biological importance. These foreign alkyl groups then block the usual action of the biological molecules. Some alkylating agents are used against cancer. Typical among these are nitrogen mustards, compounds that arose out of chemical warfare research.

The original "mustard gas" was a sulfur-containing blister agent used during World War I. Contact with either the liquid or the vapor causes blisters that are painful and slow to heal. It is easily detected, however, by its garlic or horseradish odor. Mustard gas is denoted by the military symbol H.

$$Cl-CH_2CH_2-S-CH_2CH_2-Cl$$

Mustard gas (H)

Nitrogen mustards (symbol HN) were developed about 1935. Though not quite as effective overall as mustard gas, nitrogen mustards produce greater eye damage and don't have an obvious odor. Structurally, nitrogen mustards are chlorinated amines.

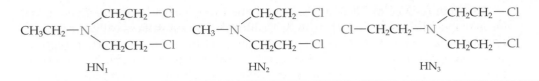

$$HN_1 \qquad\qquad HN_2 \qquad\qquad HN_3$$

Chemical Warfare

Chemical agents were used extensively during World War I. More than 30 such substances were employed, killing 91,000 and wounding 1.2 million (many of them for life). Fritz Haber supervised the release of chlorine gas in the first attack by the Germans. Adolf Hitler was among those wounded when the British retaliated with phosgene a few days later. Haber considered gas warfare to be "a higher form of killing." Fortunately, his views have not become widely accepted.

The use of chemical warfare agents was largely avoided during World War II. However, they have been used in smaller wars, such as by Iraq against Iran in the 1980s and by the Iraqi government against Kurdish rebels. They remain today as one of the *weapons of mass destruction (WMDs)* that can kill huge numbers of people and might be used by a rogue nation or a terrorist group. (The other WMDs are nuclear devices and biological agents such as anthrax and botulism.)

In cancer treatment, nitrogen mustards, which are bifunctional alkylating agents (that is, agents with two functional groups), act by cross-linking two DNA strands. Cross-linking prevents or hinders replication and thus impedes growth of a cancer. Knowledge gained through science is neither good nor evil. The same knowledge—in this case, the ability to make nitrogen mustards—can be used either for our benefit or for our destruction.

The nitrogen mustard of choice for cancer therapy is often a compound called cyclophosphamide (Cytoxan). It is used to treat Hodgkin's disease, lymphomas, leukemias, and other cancers.

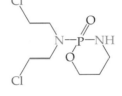

Cyclophosphamide

Cisplatin: The Platinum Standard of Cancer Treatment

The platinum-containing compound cisplatin, $PtCl_2(NH_3)_2$, is a prominent anticancer drug. Cisplatin was first synthesized in 1844. Its biological effect was discovered by accident in 1965 by Barnett Rosenberg at Michigan State University. Rosenberg was using inert platinum electrodes in an experiment designed to measure the effect of electrical currents on cell division. He found that *E. coli* were growing in length but failing to divide, reaching 300 times their normal length. After much further study, Rosenberg and his team found that the electric current had caused a chemical reaction between the platinum in the electrodes and nutrients in the solution containing the bacteria. It was found that cisplatin had been produced. Because the compound inhibited cell division, Rosenberg reasoned correctly that it might be effective as an anticancer drug. Cisplatin, an alkylat-

ing agent, binds to DNA and blocks its replication. It is widely used to treat cancers such as those of the ovaries, testes, uterus, head, neck, breast, lung, and advanced bladder cancers.

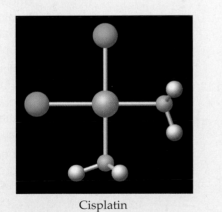

Cisplatin

Alkylating agents can cause cancer as well as cure it. For example, the nitrogen mustard HN_2 causes lung, mammary, and liver tumors when injected into mice, yet it can be used with some success in the management of certain human tumors. There is still a lot of mystery—and seeming contradiction—regarding the causes and cures of cancer.

Miscellaneous Anticancer Agents

There are many other anticancer agents. Alkaloids from vinca plants have been shown to be effective against leukemia and Hodgkin's disease. Paclitaxel (Taxol), obtained from the Pacific yew tree, is effective against cancers of the breast, ovary, and cervix.

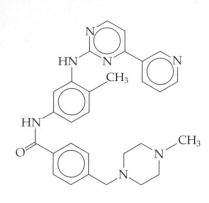

Imatinib

Imatinib (Glivec or Gleevec) is a synthetic drug particularly effective against chronic myeloid leukemia. Cell growth is controlled by substances called *growth factors,* which act by attaching to receptor proteins on the surface of certain types of cells. People with chronic myeloid leukemia produce cells with damaged receptors that send grow-and-divide signals to the cells even with no growth factor present. Imatinib inhibits the faulty receptor, preventing it from stimulating the cells to grow and divide; the cancer cells cannot multiply.

Several antibiotics have been found to kill cancer cells as well as bacteria. Actinomycin, obtained from the molds *Streptomyces antibioticus* and *S. parvus,* is used against Hodgkin's disease and other types of cancer. It is quite effective but extremely toxic. Actinomycin acts by binding to the double helix of DNA, thus blocking the replication of RNA on the DNA template. Protein synthesis is inhibited.

Chemotherapy is only part of the treatment for cancer. Surgical removal of tumors and radiation treatment remain major tools. It is unlikely that a single agent will be found to cure all cancers. Active research is underway on the mechanisms of carcinogenesis, and a better understanding will lead to better cures. Prevention of cancer is a greater hope. Deaths from smoking-related cancers account for at least 30% of all cancer deaths; reducing cigarette smoking can significantly reduce cancer deaths.

Self-Assessment Questions

1. 5-Fluorouracil inhibits the formation of nucleotides containing the base
   a. adenine   **b.** cytosine   **c.** guanine   **d.** thymine

2. 6-Mercaptopurine mimics the base adenine, forming a false nucleotide, and thus slows
   a. cell-mediated immune reactions   **b.** DNA synthesis
   c. RNA transport   **d.** viral replication

3. Methotrexate acts as a folic acid antagonist, slowing
   a. cell division   **b.** cell-mediated immune reactions
   c. transport of nutrients to the cancer cell
   d. viral replication

4. Cisplatin inhibits cell division by
   a. being incorporated in RNA
   b. binding to DNA and blocking its replication
   c. halting viral replication
   d. substituting for cytosine in a nucleotide

5. Which of the following anticancer drugs was once used as a chemical warfare agent?
   a. cisplatin   **b.** methotrexate
   c. nitrogen mustards   **d.** vinblastine

6. Cyclophosphamide acts by
   a. being incorporated in RNA
   b. cross-linking DNA chains and blocking replication
   c. halting viral replication
   d. substituting for cytosine in a nucleotide

Answers: 1, d; 2, b; 3, a; 4, b; 5, c; 6, b

18.7 Hormones: The Regulators

Before we discuss the next group of drugs, we need to take a brief look at the human endocrine system and some of the chemical compounds, called hormones, this system manufactures. A **hormone** is a chemical messenger produced in the endocrine glands (Figure 18.9). Those released in one part of the body send signals for profound physiological changes in other parts of the body. By causing reactions to speed up or slow down, hormones control growth, metabolism, reproduction, and

Major Endocrine Glands

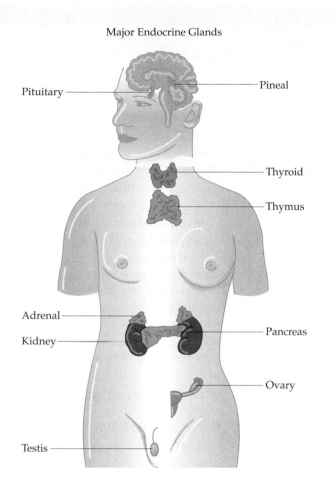

◄ **Figure 18.9** The approximate locations of the eight major endocrine glands in the human body.

many other functions of body and mind. Some of the important human hormones and their physiological effects are listed in Table 18.1 (p. 546). Still other hormones are formed in tissues of the heart, liver, kidney, gut, and placenta.

Prostaglandins: Hormone Mediators

A **prostaglandin** is a hormonelike lipid compound derived from a fatty acid. Each prostaglandin molecule has 20 carbon atoms, including a five-carbon ring. Prostaglandins act much like hormones in that they act on target cells. However, they differ from hormones in that they (1) act near the site where they are produced, (2) can have different effects in different tissues, and (3) are rapidly metabolized. Prostaglandins are synthesized in the body from arachidonic acid (Figure 18.10). There are six primary prostaglandins, and these potent biological chemicals are widely distributed throughout the body. Extremely small doses can elicit marked changes. Many others have been identified, and hundreds of synthetic analogs have been produced.

Prostaglandins act as mediators of hormone action. They regulate such things as blood pressure, blood clotting, pain sensation, smooth muscle activity, and secretion

4. Does eating lots of sugar cause diabetes? Not directly. However, high sugar intake contributes greatly to obesity, and obesity is strongly linked to Type 2 diabetes.

ANSWER

▼ **Figure 18.10** Prostaglandins are derived from arachidonic acid, an unsaturated carboxylic acid with 20 carbon atoms (a). Two representative prostaglandins are shown here: (b) prostaglandin E_2 and (c) prostaglandin $F_{2\alpha}$.

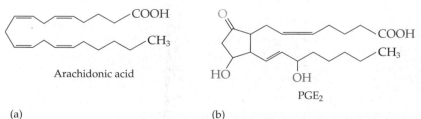

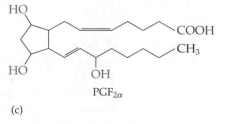

(a)   (b)   (c)

Table 18.1 Some Human Hormones and Their Physiological Effect

Name	Chemical Nature	Function(s)
Pituitary Hormones		
Vasopressin (Antidiuretic hormone)	Nonapeptide	Stimulates contractions of smooth muscle; regulates water retention and blood pressure
Oxytocin	Nonapeptide	Stimulates contraction of the smooth muscle of the uterus; stimulates secretion of milk
Growth hormone (GH) (or somatotropin)	Protein	Stimulates body growth, bone growth
Thyroid-stimulating hormone (TSH)	Protein	Stimulates growth of the thyroid gland and production of thyroid hormones
Melanocyte-stimulating hormones (MSHs)	Peptides	Controls skin pigmentation
Adrenocorticotrophic hormone (ACTH)	Protein	Stimulates growth of the adrenal cortex and production of cortical hormones
Follicle-stimulating hormone (FSH)	Protein	Stimulates maturation of egg follicles in ovaries of females and of sperm cells in testes of males
Luteinizing hormone (LH)	Protein	Stimulates female ovulation and male secretion of testosterone
Prolactin	Protein	Maintains the production of estrogens and progesterone; stimulates the formation of milk
Hypothalamic Hormones		
Corticotropin-releasing factor (CRF)	Protein	Acts on corticotrope to release ACTH and β-endorphin (lipotropin)
Gonadotropin-releasing factor (GnRF)	Decapeptide	Acts on gonadotrope to release LH and FSH
Somatostatin release inhibiting factor (SIF)	Polypeptide	Inhibits GH and TSH secretion
Thyrotropin-releasing factor (TRF)	Tripeptide	Stimulates TSH and prolactin secretion
Thyroid Hormone		
Thyroxine	Amino acid derivative	Increases rate of cellular metabolism
Parathyroid Hormone		
Parathyroid	Protein	Regulates Ca^{2+} level in blood
Pancreatic Hormones		
Insulin	Protein	Increases cell usage of glucose; increases glycogen storage
Glucagon	Protein	Stimulates conversion of liver glycogen to glucose
Somatostatin	Polypeptide	Inhibits insulin and glucagon secretion from pancreas and release of numerous gut peptides
Adrenal Cortical Hormones		
Cortisol	Steroid	Stimulates conversion of proteins to carbohydrates
Aldosterone	Steroid	Regulates salt metabolism; stimulates kidneys to retain Na^+ and excrete K^+
Adrenal Medullary Hormones		
Epinephrine (adrenaline)	Amino acid derivative	Stimulates a variety of mechanisms to prepare the body for emergency action including the conversion of glycogen to glucose
Norepinephrine (noradrenaline)	Amino acid derivative	Stimulates sympathetic nervous system; constricts blood vessels; stimulates other glands
Gonadal Hormones		
Estradiol	Steroid	Stimulates female sex characteristics; regulates changes during menstrual cycle
Progesterone	Steroid	Regulates menstrual cycle; maintains pregnancy
Testosterone	Steroid	Stimulates and maintains male sex characteristics
Pineal Hormone		
Melatonin	Amine derivative	Regulation of circadian rhythms

of substances related to reproduction. A single prostaglandin can have different, even opposite, effects in different tissues. This range of physiological activity has led to the synthesis of hundreds of prostaglandins and their analogs. Prostaglandin E_2 (PGE_2), also known as dinoprostone (Cervidil®), is used to induce labor. PGE_1 can lower blood pressure. A PGE_1 analog called misoprostol is used to help prevent peptic ulcers, a common side effect of large doses of NSAIDs. Other prostaglandins can be used clinically to relieve nasal congestion, to provide relief from asthma, to treat glaucoma, to treat pulmonary hypertension, and to prevent formation of the blood clots associated with heart attacks and strokes.

$PGF_{2\alpha}$ is used in cattle to synchronize breeding. It is also used in the artificial insemination of prize cattle. Injection of $PGF_{2\alpha}$ along with a hormone into a prize cow will induce the release of many ova, which are then fertilized with sperm from a champion bull. The developing embryos are implanted in less valuable cows, enabling a farmer to get several calves a year from one outstanding cow.

Diabetes

The pancreas produces the hormone *insulin*, with which the body increases cell usage of glucose. *Diabetes* arises when the pancreas does not produce enough insulin (Type 1) or when the insulin is not properly used by the body (Type 2). In both cases, elevated blood glucose levels are the result. Most Americans diagnosed with diabetes suffer from Type 2. Type 2 has a strong genetic component and is also closely tied to lifestyle. Poor diet, obesity, and lack of exercise all appear to be contributing factors to Type 2. Along with increases in obesity, Type 2 diabetes has become increasingly common, even among children and adolescents.

People who suffer from Type 1 diabetes ordinarily must take insulin, by injection. (Insulin taken orally is broken down by the body's digestive system before it can reach the bloodstream.) Type 2 diabetics may be able to control their disease with careful diet, exercise, and loss of weight. When these measures aren't enough, a biguanide such as metformin (Glucophage®) is often prescribed. Metformin works by reducing the production of glucose in the liver. It is one of the most widely prescribed medications, with over 25 million prescriptions annually. Compounds called sulfonylureas, which increase the production of insulin from the pancreas, are also used to treat Type 2 diabetes. If these oral drugs are not enough to maintain glucose levels for Type 2, then insulin must be administered.

$$CH_3-N-C-NH-C-NH_2$$

Metformin

A number of complications can arise from diabetes, including blindness, cardiovascular disease, kidney disease, and nerve damage. Thus, diagnosis and control of diabetes is very important. Symptoms include thirst, increased appetite, increased urination, weight loss, and blurred vision. About 24 million Americans have the disease—and almost a quarter of them aren't aware of it.

Steroids

A **steroid** is a compound with a skeletal four-ring structure. Steroids occur widely in living organisms, and not all are hormones (Figure 18.11). Cholesterol, for example, is a steroid component of all animal tissues. About 10% of the brain is cholesterol, which makes up a vital part of the membranes of nerve cells. Cholesterol is found in deposits in hardened arteries and is a major component of certain types of gallstones. Cortisol, another natural steroid, is an adrenal hormone (Table 18.1). Prednisone is a synthetic steroid used to reduce inflammation in arthritis sufferers and to treat injuries.

Many drugs, both natural and synthetic, are based on steroids, including the notorious anabolic steroids (eChapter 19), abused by athletes and body builders. Some steroid drugs are anti-inflammatories used to treat arthritis, bronchial asthma, dermatitis, and eye infections. Some are sex hormones used in formulating birth control pills (Section 18.8).

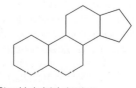

▲ Steroid skeletal structure.

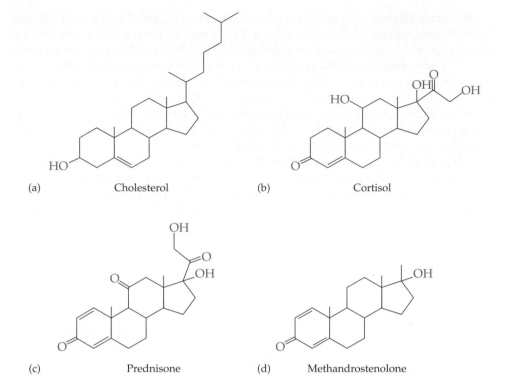

(a)   Cholesterol   (b)   Cortisol

(c)   Prednisone   (d)   Methandrostenolone

► **Figure 18.11** Line-angle formulas of four steroids. Cholesterol (a) is an essential component of all animal cells. Cortisol (b) is a hormone secreted by the adrenal glands in response to physical or psychological stress. Prednisone (c) is a synthetic anti-inflammatory substance. Methandrostenolone (d), also called Dianabol®, is an anabolic steroid (eChapter 19). Note that all have the same basic four-ring structure.

▲ Percy Lavon Julian (1899–1975) was involved in the synthesis of a variety of steroids, including physostigmine, a drug used to treat glaucoma. Untreated, glaucoma causes blindness. Julian is shown here on a 1992 U.S. postage stamp.

Serendipity often plays a role in the discovery of new drugs. Percy Julian, a grandson of slaves, was a chemist at the Glidden Paint Company doing research on soybeans when his work led to the development of new steroid-based drugs. It often happens that new drugs are discovered by chemists who are not really looking for them.

Sex Hormones

Sex hormones are steroids (Figure 18.12). Note that male sex hormones differ only slightly in structure from female sex hormones. In fact, the female hormone progesterone can be converted to the male hormone testosterone by a simple biochemical reaction. The physiological actions of these structurally similar compounds, however, are markedly different.

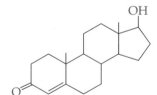

(a)   Testosterone   (b)   Estrone

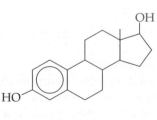

(c)   Estradiol   (d)   Progesterone

► **Figure 18.12** Line-angle formulas of the principal sex hormones. Testosterone (a) is the main male sex hormone (androgen). Estrone (b) and estradiol (c) are female sex hormones (estrogens). Progesterone (d), also produced by females, is essential to the maintenance of pregnancy.

An **androgen** is any compound that stimulates or controls the development and maintenance of masculine characteristics. These male sex hormones are secreted by the testes. In males, the pituitary hormones FSH and LH (Table 18.1) are required for the continued production of sperm and of androgens. These hormones are responsible for development of the sex organs and for secondary sexual characteristics, such as voice and hair distribution. The most important androgen is testosterone.

An **estrogen** is a compound that controls female sexual functions, such as the menstrual cycle, the development of breasts, and other secondary sexual characteristics. The two important estrogens are estradiol and estrone. These female sex hormones are produced mainly in the ovaries. Another female sex hormone is progesterone, which prepares the uterus for pregnancy and prevents the further release of eggs from the ovaries during pregnancy. A related type of compound, called a **progestin**, is any steroid hormone that has the effect of progesterone. In females, the pituitary hormones FSH and LH (Table 18.1) are required for the production of ova and of estrogens and progesterone.

Sex hormones—both natural and synthetic—are sometimes used therapeutically. For example, a woman who has passed menopause may be given hormones to compensate for those no longer being produced by her ovaries. Hormone replacement therapy provides both risks and benefits, and its use is quite controversial (see Collaborative Group Project 78).

18.8 Chemistry and Social Revolution: The Pill

When administered by injection, progesterone serves as an effective birth control drug; it fools the body into acting as if it were already pregnant. The structure of progesterone was determined in 1934 by Adolf Butenandt (1903–1995), who received the Nobel Prize in Chemistry in 1939. Other chemists began to try to design a contraceptive that would be effective when taken orally.

In 1938 Hans Inhoffen (1906–1992) synthesized the first oral contraceptive, ethisterone, by incorporating an ethynyl group ($-C\equiv CH$). (The ethynyl group is derived from ethyne, $HC\equiv CH$, also called acetylene.) However, ethisterone had to be taken in large doses to be effective and was not widely used.

In 1951, Carl Djerassi synthesized 19-norprogesterone, which is simply progesterone with one of its methyl groups missing. It was four to eight times as effective as progesterone as a birth control agent. However, like progesterone itself, 19-norprogesterone had to be given by injection, an undesirable property. Djerassi then combined the two ideas: remove a methyl group to make the drug more effective, and add an ethynyl group to allow oral administration. Djerassi synthesized norethindrone (Norlutin®), patented in 1956, which proved effective when taken in small doses (Figure 18.13). Working at about the same time as Djerassi, Frank Colton synthesized norethynodrel, another progestin, for which G. D. Searle was awarded patents in 1954 and 1955. Norethynodrel, a progestin, was used in Enovid, the first birth control pill approved by the FDA in 1960.

The progestins—norethynodrel, norethindrone, and related compounds—mimic the action of progesterone. Mestranol, a synthetic estrogen, regulates the menstrual cycle; the progestin establishes a state of false pregnancy. A woman does not ovulate when she is pregnant (or in the state of false pregnancy established by the progestin), and because she does not ovulate, she cannot conceive.

Emergency Contraceptives

Several products, called emergency contraceptives or the "morning-after pill," are available for use to prevent pregnancy after unprotected intercourse. There are two kinds of emergency contraception pills (ECPs). Some ECPs are combination pills with both estrogen and progestin—synthetic analogs of the natural substances. Others contain progestin only. Given in high doses, ECPs disrupt hormone actions on which pregnancy depends. ECPs halt development of the uterine lining and inhibit ovulation and fertilization.

▲ Carl Djerassi (b. 1923), professor of chemistry at Stanford University and president of Zoecon Corporation, Palo Alto, California, synthesized Norlutin, a progestin, in 1951.

▲ Frank Colton (1923–2003) synthesized norethynodrel, a progestin used in the first birth control pills.

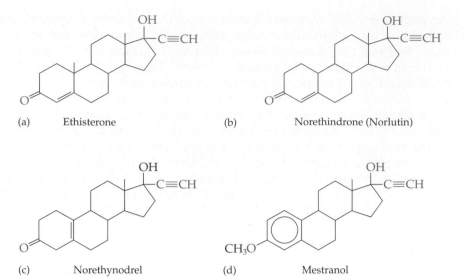

▶ **Figure 18.13** Line-angle formulas of some synthetic sex hormones: ethisterone (a), norethindrone (b), norethynodrel (c), and mestranol (d). In 1960, G. D. Searle's Enovid, which contained 9.85 mg of norethynodrel and 150 mg of mestranol, became the first birth control pill approved by the FDA in the United States. Djerassi's norethindrone and Colton's norethynodrel differ only in the position of a double bond.

(a)   Ethisterone

(b)   Norethindrone (Norlutin)

(c)   Norethynodrel

(d)   Mestranol

One brand of ECP, called Previn®, contains both an estrogen (ethinyl estradiol) and a progestin (levonorgestrel). Another ECP, called Plan B®, available over the counter, contains only levonorgestrel. ECPs are taken as two doses, 12 hours apart. They are 75–89% effective in preventing pregnancy after unprotected sexual intercourse, and they work best if taken as soon as possible after intercourse. Only one woman out of 100 will become pregnant after taking progestin-only ECPs.

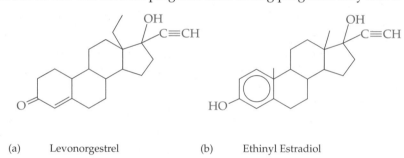

(a)   Levonorgestrel

(b)   Ethinyl Estradiol

▲ The Copper T 380A IUD (ParaGard®).

An intrauterine device (IUD), the Copper T 380A IUD (ParaGard®), is also used for emergency contraception. Emergency IUD insertion reduces the risk of pregnancy by 99.9% if inserted within three to five days of unprotected intercourse. Copper(II) ions in the IUD interfere with sperm transport and fertilization. The copper IUD probably causes an inflammation that makes the endometrium unsuitable for implantation.

ECPs are not the same as the so-called medical abortion pills such as mifepristone (also called RU-486). Mifepristone (Mifeprex®), which inhibits the action of progesterone, works after a woman becomes pregnant and the fertilized egg has attached to the wall of the uterus. It causes the uterus to expel the egg, ending the pregnancy. ECPs prevent pregnancy after sexual intercourse, whereas mifepristone ends an unwanted pregnancy at an early stage.

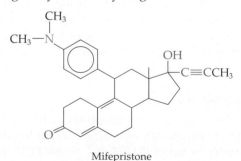

Mifepristone

People who believe that human life begins when the ovum is fertilized by the sperm equate the use of mifepristone with abortion and oppose the use of the drug. Others consider the drug just another method of birth control and see its use as being safer than a surgical abortion.

Chemical Treasure in the Rainforest

In 1940 progesterone was one of the most expensive compounds in the world. Obtained with great difficulty from animals, it sold for $200 per gram—equivalent to about $1600 in today's dollars. American chemist Russell Marker (1902–1995) wondered if progesterone could be made from a plant steroid. He took diosgenin, which occurs in tiny amounts in lily roots, and converted it to progesterone.

He then looked for a more abundant source of this plant steroid, examining more than 400 plants from all over the world. He finally identified a Mexican wild yam, *Dioscorea villosa*, that contained a significant quantity of diosgenin in its roots. He left his faculty post at Pennsylvania State University to look for more of these vines in the tropical forests of Mexico. Eventually he found a profuse growth of them and within a few months synthesized almost a kilogram of progesterone. Marker had found a commercial source not only for progesterone but for all the other steroid hormones as well. (Wild yam extracts are sold with claims that they may relieve symptoms associated with menopause, but the yam as such does not contain progesterone or anything else that would act like progesterone.)

Tropical rainforests are warm, moist, and fertile areas where there are more different kinds of plants and animals than anyplace else on Earth. These forests are living chemical factories that contain valuable and irreplaceable natural resources. Unfortunately, the world's rainforests are rapidly disappearing.

▲ The Mexican wild yam is a source of diosgenin, from which progesterone is made.

Risks of Taking Birth Control Pills

Oral contraceptives currently are used by more than 100 million women worldwide and by about 12 million women in the United States. Most women report no side effects. Of the side effects that are reported, the vast majority are minor. Some women experience hypertension, acne, or abnormal bleeding. These pills increase the risk of blood clotting in some women, but so does pregnancy. Blood clots can clog arteries and cause death by stroke or heart attack. The death rate associated with birth control pills is about 3 in 100,000, only one-tenth of that associated with childbirth. For smokers, however, the risks are much higher. For women over 40 who smoke 15 cigarettes per day, the risk of death from stroke or heart attack is 1 in 5000. The FDA advises all women who smoke, especially those over 40, to use some other method of contraception.

The Minipill

Because most of the side effects of oral contraceptives are associated with the estrogen component, the amount of estrogen in these pills has been greatly reduced over the years. Today's pills contain only a fraction of a milligram of estrogen. "Minipills" are now available that contain only small amounts of progestin and no estrogen at all. Minipills are not quite as effective as the combination pills, but they have fewer side effects.

A Pill for Males?

Why should females have to bear all the responsibility for contraception? Why not a pill for males? Many men are willing to share the risks and the responsibility of contraception, but condoms and vasectomies are the only effective forms of contraception available to them. There are biological reasons for females to bear the burden. Women are the ones who get pregnant when contraception fails, and in females contraception has to interfere with only one monthly event: ovulation. Males produce sperm continuously.

In males, the pituitary hormones FSH and LH (Table 18.1) are required for the continued production of sperm and of the male hormone testosterone. Several research groups have developed what may be a safe, effective, and reversible contraceptive for males. The pill has a progestin—the same key ingredient as is used in

More to Explore
Marzuola, C. "Male Pill on the Horizon." *Science News*, September 14, 2002, pp. 373–374.

pills for women—combined with the male sex hormone testosterone. The progestin seems to function much as it does in the women's pill; it suppresses the rate of sperm production and the quantity of sperm produced. When will such male birth control pills be available to the public? Optimistic projections estimate a wait of about three years. Others, though, say there will be no male pill in our lifetimes.

Self-Assessment Questions

1. Hormones are produced in the endocrine glands and act
   a. on the brain only
   b. on nearby cells only
   c. on sex organs only
   d. throughout the body

2. Prostaglandins are synthesized in the body from
   a. arachidonic acid
   b. ascorbic acid
   c. cholesterol
   d. progesterone

3. Which of the following is *not* a steroid?
   a. oxytocin
   b. prednisone
   c. progesterone
   d. testosterone

4. The gonads are stimulated to produce sex hormones and eggs or sperm by
   a. ACTH and LH
   b. FSH and LH
   c. LH and CRF
   d. TRF and GH

5. The "pregnancy hormone" is
   a. estradiol
   b. estrone
   c. oxytocin
   d. progesterone

6. The group that makes a birth control drug effective orally is
   a. CH_3CH_2-
   b. $CH_2=CH-$
   c. $HC\equiv C-$
   d. CH_3CO-

Answers: 1, d; 2, a; 3, a; 4, b; 5, d; 6, c

18.9 Drugs for the Heart

The heart is a muscle that beats virtually every second of every day for as long as we live. Diseases of the cardiovascular system are responsible for one-third of the deaths in the United States and more than one-fifth of deaths worldwide. Four of the top 10 prescription drugs worldwide treat cardiovascular disease. These drugs prolong life and improve its quality.

Major diseases of the heart and blood vessels include

- ischemic ("lacking oxygen") coronary artery disease
- heart arrhythmias (abnormal heartbeat)
- hypertension (high blood pressure)
- congestive heart failure

Atherosclerosis (fatty deposit buildup in the lining of the arteries) is the primary cause of coronary artery disease, which in turn causes myocardial infarction (heart attack). Thus, most drug treatments for the heart aim to increase its supply of blood (and oxygen), to normalize its rhythm, to lower blood pressure, or to prevent accumulation of lipid plaque deposits in blood vessels.

Lowering Blood Pressure

Hypertension, or high blood pressure, is defined as pressure, in millimeters of mercury (mmHg), above 140/90. Normal blood pressure is defined as less than 120/80. Hypertension is the most common cardiovascular disease, affecting more than one in four adults in the United States. Because it seldom produces symptoms, many people have hypertension without realizing it. Four major drug categories for lowering blood pressure are

- *Diuretics* [example: hydrochlorothiazide (HCT)], which cause the kidneys to excrete more water and thus lower the blood volume
- *Beta blockers* (examples: propranolol, metoprolol; see Section 18.12), which slow the heart rate and reduce the force of the heartbeat

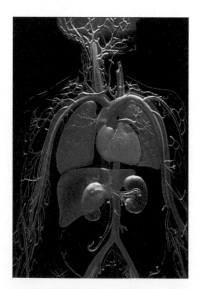

▲ The cardiovascular system. Over the course of 24 hr, the heart pumps more than 8000 L of blood through nearly 100,000 km of blood vessels.

- *Calcium channel blockers* (example: amlodipine), which are powerful vasodilators, inducing muscles around the blood vessels to relax
- *Angiotensin-converting enzyme (ACE) inhibitors* (example: lisinopril), which inhibit the action of an enzyme that causes blood vessels to contract

Studies show that diuretics, the oldest, simplest, and cheapest of the blood-pressure-lowering drugs, may also be the most effective drugs for treating hypertension in most people. However, both a diuretic and one of the other medications often are prescribed.

Normalizing Heart Rhythm

An *arrhythmia* is an abnormal heartbeat. Some arrhythmias exhibit no symptoms and are discovered only during a physical examination, but others can be life threatening. The electrical properties of nerves and muscles arise from the flow of ions across cell membranes, and drugs for arrhythmia alter this flow. There are several types of such drugs, with different mechanisms of action. These drugs tend to have a narrow margin between the therapeutic dose and a harmful amount. A change in the heart's rhythm can have serious consequences. The too-rapid heartbeat, known as *fibrillation,* is so common that defibrillator devices are now available on many airplanes and in various other public places.

Treating Coronary Artery Disease

A common symptom of coronary artery disease is chest pain called *angina pectoris.* This pain is caused by the heart getting less oxygen than it needs, usually due to partial blockage of the coronary arteries by lipid-containing plaque (arteriosclerosis). When the blockage is complete, a heart attack occurs, and some of the heart muscle dies. Medical treatment for coronary artery disease usually involves dilation (widening) of the blood vessels to the heart to increase the blood flow and slowing of the heart rate to decrease its workload and its demand for oxygen. Some of the drugs used are the same as those used to lower high blood pressure (beta blockers, for example). Several organic nitro compounds, especially amyl nitrite ($CH_3CH_2CH_2CH_2CH_2ONO$) and nitroglycerin have a long history in the treatment of angina pectoris. These compounds act by releasing nitric oxide (NO), which relaxes the constricted vessels that are reducing the supply of blood and oxygen to the heart.

Although many new drugs are now being used to treat heart failure, one that has been used for centuries still plays an important role. The foxglove plant (page 526), was used by the ancient Egyptians and Romans. The plant contains a mixture of glycosides that yield carbohydrates and steroids on hydrolysis. Hydrolysis of digoxin (one of the digitalis glycosides) yields the steroid digitoxigenin, which affects heart rhythm and strength of contractions. Digoxin is still used to treat patients with heart failure.

$$H_2C-O-NO_2$$
$$HC-O-NO_2$$
$$H_2C-O-NO_2$$

Nitroglycerin

Self-Assessment Questions

1. Which of the following is a treatment for hypertension (high blood pressure)?
   a. aspirin   **b.** a diuretic   **c.** nitroglycerin   **d.** prednisone

2. Drugs called beta blockers
   a. lessen blood flow to the kidneys
   b. lessen the damage of beta particles
   c. prevent hydrolysis of beta linkages
   d. slow the heartbeat rate

3. Drugs such as amyl nitrite and nitroglycerin act by releasing a substance that relaxes the smooth muscles in blood vessels. That substance is
   a. an amino acid   **b.** digitalis   **c.** nitric oxide   **d.** a steroid

Answers: 1, b; 2, d; 3, c

NO—A Messenger Molecule

The simple molecule NO, notorious as an air pollutant (Chapter 13), was found in 1986 to act as a *messenger molecule* that carries signals between cells in the body. All previously known messenger molecules were complex substances such as norepinephrine and serotonin (Section 18.12) that act by fitting specific receptors in cell membranes. Nitrogen monoxide, commonly called nitric oxide, is essential to maintaining blood pressure and establishing long-term memory. It also aids in the immune response to foreign invaders in the body and mediates the relaxation phase of intestinal contractions in the digestion of food.

NO is formed in cells from arginine, a nitrogen-rich amino acid (Section 16.4), in a reaction catalyzed by an enzyme. NO kills invading microorganisms, probably by deactivating iron-containing enzymes in much the same way that carbon monoxide destroys the oxygen-carrying capacity of hemoglobin (Section 13.5).

Louis Ignarro, Robert F. Furchgott, and Ferid Murad, who discovered the physiological role of NO, were awarded a 1998 Nobel Prize in Physiology or Medicine.

This award has a fortuitous link back to Alfred Nobel, whose invention of dynamite provided the financial basis of the Nobel Prizes. Nitroglycerin, the explosive ingredient of dynamite, relieves the chest pain of heart disease. In his later years, Nobel refused to take nitroglycerin for his own heart disease because it causes headaches, and he did not think it would relieve his chest pain. Murad showed that nitroglycerin acts by releasing NO.

NO also dilates the blood vessels that allow blood flow into the penis to cause an erection. Research on this role of NO led to the development of the anti-impotence drug sildenafil (Viagra®). One of the physiological effects of sildenafil is the production of small quantities of NO in the bloodstream. Related research has led to drugs for treating shock and a drug for treating high blood pressure in newborn babies.

Further research indicates that two other gases, carbon monoxide (CO) and hydrogen sulfide (H_2S), also act as messenger molecules. Research into implications of these for the development of heart medicines is under way.

18.10 Drugs and the Mind

A **psychotropic drug** is one that affects the human mind. The first drugs used by primitive peoples probably were mind-altering substances. Alcohol, marijuana, opium, cocaine, peyote, and other plant materials have been known for thousands of years. Generally, only one or two of these were used in any one society. Today thousands of psychotropic drugs are readily available. Some still come from plants, but many are synthetic.

There is no clear distinction between drugs that affect the mind and those that affect the body. Most drugs probably affect both mind and body. It can still be helpful, however, to distinguish between drugs that act primarily on the body and drugs that affect primarily the mind.

Psychotropic drugs are divided into three classes.

- A **stimulant drug**, such as cocaine and amphetamines, increases alertness, speeds mental processes, and generally elevates the mood.
- A **depressant drug**, such as alcohol, most anesthetics, opiates, barbiturates, and some tranquilizers, reduces the level of consciousness and the intensity of reactions to environmental stimuli. In general, depressants dull emotional responses.
- A **hallucinogenic drug**, such as lysergic acid diethylamide (LSD) and marijuana, alters qualitatively the way we perceive things.

We will examine some of the major drugs in each category, but first we need to look at some chemistry of the nervous system.

18.11 Some Chemistry of the Nervous System

The nervous system is made up of about 12 billion **neurons** (nerve cells) with 10^{13} connections between them. The brain operates with a power output of about 25 W and has the capacity to handle about 10 trillion bits of information. Nerve cells vary a great deal in shape and size. One type is shown in Figure 18.14. The essential parts

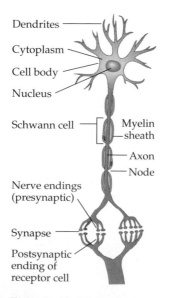

Dendrites

Cytoplasm

Cell body

Nucleus

Schwann cell

Myelin sheath

Axon

Node

Nerve endings (presynaptic)

Synapse

Postsynaptic ending of receptor cell

▲ **Figure 18.14** Diagram of a human nerve cell.

of each cell are the cell body, the axon, and the dendrites. We discuss here only the nerves that make up the involuntary (autonomic) nervous system. These nerves carry messages between organs and glands that act involuntarily (such as the heart, the digestive organs, and the lungs) and the brain and spinal column.

Axons on a nerve cell may be up to 60 cm long, but there is no continuous pathway from an organ to the central nervous system. Messages must be transmitted across tiny, fluid-filled gaps, or **synapses** (Figure 18.15). When an electrical signal from a nerve cell in the brain reaches the end of an axon of a presynaptic nerve cell, chemicals called **neurotransmitters** are released from tiny vesicles into the synapse. Receptors on the postsynaptic cell bind the neurotransmitter, causing changes in the receiving cell. After a brief interval, the neurotransmitter is released from the receptor and carried back to the presynaptic cell by substances called transporters.

There are many neurotransmitters (Table 18.2). Substances that act as neurotransmitters include amino acids, peptides, and monosubstituted amines (monoamines). For example, glutamate, the carboxylate anion of the amino acid glutamic acid, is the most common neurotransmitter in the brain. More than 50 peptides function as neurotransmitters. Messages are carried to other nerve cells, to muscles, and to the endocrine glands (such as the adrenal glands). Each neurotransmitter fits one or more receptor sites on the receptor cell. For example, nerve impulses trigger release of glutamate from the presynaptic cell. Glutamate receptors in the postsynaptic cell bind glutamate and are activated. Neurotransmitters determine to a large degree how you think, feel, and move about. If there are problems with their synthesis or uptake, things can go quite wrong. Many drugs (and some poisons) act by mimicking the action of the neurotransmitter. Others act by blocking the receptor and preventing the neurotransmitter from acting on it. The monoamines and some amine drugs are discussed in the next section.

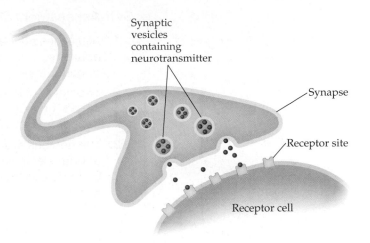

▲ **Figure 18.15** Diagram of a synapse. When an electrical signal reaches the presynaptic nerve ending, neurotransmitter molecules are released from the vesicles. They migrate across the synapse to the receptor cell, where they fit specific receptor sites.

Table 18.2	Selected Neurotransmitters*	
Transmitter Molecule	**Derived from**	**Site of Synthesis**
Dopamine	Tyrosine	Central nervous system (CNS)
Norepinephrine	Tyrosine	CNS
Epinephrine	Tyrosine	Adrenal medulla, some CNS cells
Serotonin [5-Hydroxytryptamine (5-HT)]	Tryptophan	CNS, gut
Acetylcholine	Choline	CNS, parasympathetic nerves
GABA	Glutamate	CNS
Glutamate		CNS
Aspartate		CNS
Glycine		Spinal cord
Histamine	Histidine	Hypothalamus
Adenosine	ATP	CNS, peripheral nerves
ATP		Sympathetic, sensory and enteric nerves
Nitric oxide, NO	Arginine	CNS, gastrointestinal tract

*Not all of these are discussed in this text.

18.12 Brain Amines: Depression and Mania

We all have our ups and downs in life. These moods probably result from multiple causes, but it is likely that a variety of chemical compounds formed in the brain are involved. Before we consider these ups and downs, however, let's take a look at epinephrine, an amine formed in the adrenal glands and some central nervous system cells (Figure 18.16).

Epinephrine is secreted by the adrenal glands—and is thus often called adrenaline—when a person is under stress or is frightened. A tiny amount of epinephrine causes a great increase in blood pressure, and its flow prepares the body for fight or flight. Because culturally imposed inhibitions prevent fighting or fleeing in most modern situations, the adrenaline-induced supercharge is often not used. This sort of frustration has been implicated in some forms of mental illness.

Biochemical Theories of Mental Illness

Biochemical theories of mental illness usually involve chemical imbalances of brain amines, conditions that can be improved with medication that corrects the imbalances. One such brain amine is norepinephrine (NE), a relative of epinephrine. NE is a neurotransmitter formed in the brain (Figure 18.16). When produced in excess, NE causes euphoria. In large excess, NE induces a manic state.

▼ **Figure 18.16** The biosynthesis of epinephrine and norepinephrine from tyrosine. L-Dopa, the left-handed form of the compound (see "Right- and Left-Handed Molecules" on facing page), is used to treat Parkinson's disease, which is characterized by rigidity and stiffness of muscles. Parkinson's results from inadequate dopamine production, but dopamine cannot be used directly to treat the disease because it is not absorbed into the brain. Dopamine has been used to treat hypertension. It also is a neurotransmitter; schizophrenia has been attributed to an overabundance of dopamine in nerve cells.

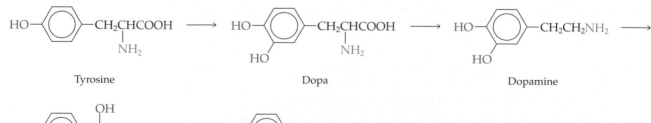

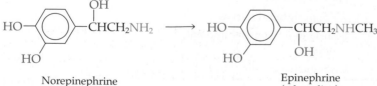

Another brain amine is the neurotransmitter *serotonin*. Serotonin is involved in sleep, appetite, memory, learning, sensory perception, mood, sexual behavior, and regulation of body temperature. This brain amine also is involved in behaviors such as eating, sleeping, and aggression. Serotonin inhibits the nervous system in ways that calm, soothe, and generate feelings of contentment and satisfaction. The main metabolite of serotonin, 5-hydroxyindoleacetic acid (5-HIAA), is found in unusually low levels in cerebrospinal fluid of severely depressed patients with a history of suicide attempts, a finding indicating that abnormal serotonin metabolism plays a role in depression. Some research suggests that a reduced flow of serotonin through the synapses in the frontal lobe of the brain causes depression.

Cerebrospinal fluid levels of 5-HIAA are also low in murderers and other violent offenders. However, they are higher than normal in people with obsessive–compulsive disorders, sociopaths, and people with guilt complexes.

Brain Amine Agonists and Antagonists in Medicine

NE and related compounds fall into several general categories. NE **agonists** (drugs that enhance or mimic its action) are stimulants. NE **antagonists** (drugs that block the action of NE) slow down various processes. Drugs called beta blockers (Figure 18.17) reduce the stimulant action of epinephrine and NE on various kinds

More to Explore
Ezzell, Carol. "Why? The Neuroscience of Suicide." *Scientific American*, February 2003, pp. 44–51.

Propranolol   Metoprolol

◀ **Figure 18.17** Propranolol and metoprolol are beta blockers, widely used to treat high blood pressure.

Right- and Left-Handed Molecules

Like people, molecules can be either right- or left-handed. Using models, we can show two structures for the amino acid alanine [$CH_3CH(NH_2)COOH$] (Figure 18.18). These structures are **stereoisomers**, isomers having the same structural formula but differing in the arrangement of atoms or groups of atoms in three-dimensional space. Models of the two stereoisomers of alanine are as alike—and as different—as are a pair of gloves. If you place one of the models in front of a mirror, the image in the mirror will be identical to the second stereoisomer. Molecules that are nonsuperimposable (nonidentical) mirror images of each other are stereoisomers of a specific type called **enantiomers** (Greek *enantios*, "opposite").

Enantiomers have a **chiral carbon**, a carbon atom to which four different groups are attached. For example, the central carbon of alanine has four groups—H, CH_3, COOH, and NH_2—joined to it. If a molecule contains one or more chiral carbons, it is likely to exist as two or more stereoisomers. In contrast, propane ($CH_3CH_2CH_3$) does not contain a chiral carbon and thus does not exist as a pair of stereoisomers.

Stereoisomerism is quite common in organic chemicals of biological importance. The simple sugars (Section 16.2) are all right-handed; 19 of the 20 amino acids that make up proteins are left-handed. The other, glycine (H_2NCH_2COOH), has no handedness.

A right-hand glove doesn't fit on a left hand, and vice versa. Similarly, enantiomers fit enzymes differently; thus, they have different effects. As an example,

consider atorvastatin (Lipitor®), the most prescribed drug in the world, used to lower blood cholesterol. Initial research in 1989 was promising but not outstanding. Later it was found that only the left-handed isomer is active and that it is much more effective when separated from its right-handed form. The development of drugs that contain only one enantiomer is becoming increasingly important in the pharmaceutical industry.

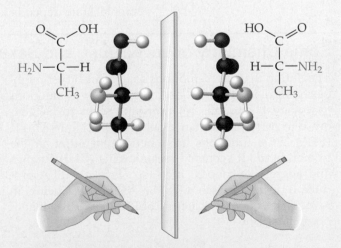

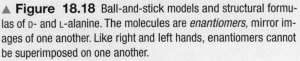

▲ **Figure 18.18** Ball-and-stick models and structural formulas of D- and L-alanine. The molecules are *enantiomers*, mirror images of one another. Like right and left hands, enantiomers cannot be superimposed on one another.

of cells. Propranolol (Inderal®) is used to treat cardiac arrhythmias, angina, and hypertension by slightly lessening the force of the heartbeat. Unfortunately, it also causes lethargy and depression. Metoprolol (Lopressor®) acts selectively on the cells of the heart. It can be used by hypertensive patients who have asthma because it does not act on receptors in the bronchi.

Serotonin agonists are used to treat depression, anxiety, and obsessive–compulsive disorder. Serotonin antagonists are used to treat migraine headaches and to relieve the nausea caused by cancer chemotherapy.

Brain Amines and Diet: You Feel What You Eat

Richard Wurtman of the Massachusetts Institute of Technology has demonstrated a relationship between diet and serotonin levels in the brain. As shown in Figure 18.19, serotonin is produced in the body from the amino acid tryptophan. Wurtman found that diets high in carbohydrates lead to high levels of serotonin. High levels of protein lower the serotonin concentration.

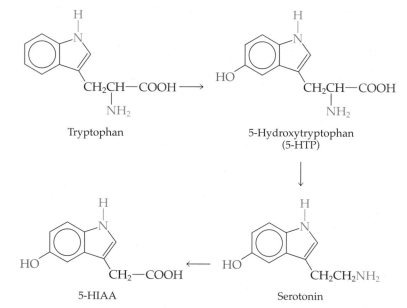

▶ **Figure 18.19** Serotonin is produced in the brain from the amino acid tryptophan. The synthesis involves several steps; some are omitted here for simplicity. 5-HIAA is a metabolite of serotonin.

Norepinephrine is synthesized in the body from the amino acid tyrosine. As noted in Figure 18.16, the synthesis is complex and proceeds through several intermediates. Because tyrosine is also a component of our diets, it may well be that our mental state depends to a fair degree on our diet.

Some Chemistry of Love, Trust, and Sexual Fidelity

The notion that such emotions as love and trust might be chemical in origin is unsettling. However, emotions that trigger romantic relationships seem to be governed in part by a substance called β-phenylethylamine (PEA). PEA functions as a neurotransmitter in the human brain. It seems to create excited, alert feelings and aggressive moods. Increased levels of PEA produce a feeling much like the feeling of "being in love." The chemical structure of PEA resembles that of norepinephrine.

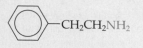

β-Phenylethylamine

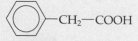

Phenylacetic acid

How much PEA does it take to get back that old feeling? Levels of PEA in the brain can be estimated by measuring levels of its metabolite, phenylacetic acid, in the urine. Low levels of urinary phenylacetic acid correlate with depression. There are no good food sources of PEA. Chocolate contains small quantities of PEA, but most of the PEA in chocolate is metabolized before it reaches the brain. PEA releases dopamine (it is a dopamine agonist), producing an antidepressant effect. Some research has shown that oral doses of PEA relieve depression in some people.

Oxytocin (Table 18.1), sometimes called the "cuddling chemical," increases the bond between lovers. It also induces contractions during childbirth and lactation afterward. Swiss scientists have shown that oxytocin increases trust in people. People playing a trust game were more trusting after taking one sniff of oxytocin.

If oxytocin equals love, vasopressin may keep couples together. Few mammals are monogamous, mating and bonding with one partner for life. The prairie vole is one example. Montane voles, a closely related species, are polygamous. Scientists have found that within 24 hours after mating, the male prairie vole is hooked for life. Postcoital production of vasopressin in the male is responsible for this monogamous behavior. Male montane voles, which have a different vasopressin receptor pattern than the prairie vole, did not react significantly to the female vole when injected with vasopressin. When given a compound to suppress the effect of vasopressin, the prairie voles lose their devotion to each other. Perhaps the solution to our high divorce rate is a few sniffs of vasopressin.

Nearly one out of every ten people in the United States suffers from mental illness. Over half the patients in hospitals are there because of mental problems. When the biochemistry of the brain is more fully understood, mental illness may be cured (or at least alleviated) by the administration of drugs or by adjusting the diet.

Self-Assessment Questions

1. Under stress, the adrenal glands release an amine called
   a. amphetamine **b.** aniline **c.** dopamine **d.** epinephrine
2. Norepinephrine is formed in the brain from the dietary amino acid
   a. asparagine **b.** glutamine **c.** tryptophan **d.** tyroslne
3. Serotonin is formed in the brain from the dietary amino acid
   a. cysteine **b.** serine **c.** tryptophan **d.** tyrosine
4. Many forms of mental illness are thought to result from abnormal metabolism of
   a. acetylcholine **b.** GABA **c.** nitric oxide **d.** serotonin

Answers: 1, d; 2, d; 3, c; 4, d

18.13 Anesthetics

An **anesthetic** is a substance that causes lack of feeling or awareness. A **general anesthetic** acts on the brain to produce unconsciousness and a general insensitivity to pain. A local anesthetic causes loss of feeling in a part of the body. Diethyl ether ($CH_3CH_2OCH_2CH_3$), the first general anesthetic, was introduced into surgical practice in 1846 by a Boston dentist, William Morton. Inhalation of ether vapor produces unconsciousness by depressing the activity of the central nervous system. Ether is relatively safe because there is a fairly wide gap between the dose that produces an effective level of anesthesia and the lethal dose. Its disadvantages are high flammability and the undesirable side effect of nausea.

Nitrous oxide, or laughing gas (N_2O), was tried by Morton without success before he tried ether. Nitrous oxide was discovered by Joseph Priestley in 1772, and its narcotic effect was soon noted. Mixed with oxygen, nitrous oxide is used in modern anesthesia. It is quick acting but not very potent. Concentrations of 50% or greater must be used for it to be effective. When nitrous oxide is mixed with ordinary air instead of oxygen, not enough oxygen gets into the patient's blood, and permanent brain damage can result.

Chloroform ($CHCl_3$) was introduced as a general anesthetic in 1847. It quickly became popular after Queen Victoria gave birth to her eighth child in 1853 while anesthetized by chloroform, and it was used widely for years. It is nonflammable and produces effective anesthesia, but it has serious drawbacks. For one, it has a narrow safety margin; the effective dose is close to the lethal dose. It also causes liver damage, is a suspected carcinogen, and must be protected from oxygen during storage to prevent the formation of deadly phosgene gas.

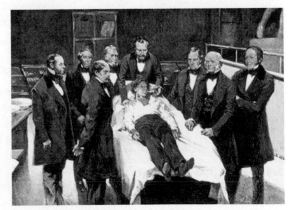

▼ This painting shows an operation in Boston in 1846 during which ether was used as an anesthetic.

Table 18.3 Inhaled Anesthetic Agents

Generic or Chemical Name	Formula	Commercial Name	Year Introduced	Currently in Use?
Diethyl ether	$CH_3CH_2OCH_2CH_3$	Ether	1842	No
Nitrous oxide	N_2O	Nitrous oxide	1844	Yes
Chloroform	$CHCl_3$	Chloroform	1847	No
Isoflurane	$CHF_2OCHClCF_3$	Forane®	1980	Yes
Desflurane	$CHF_2OCHFCF_3$	Suprane®	1992	Yes
Sevoflurane	$CHF_2OCH(CF_3)_2$	Ultane®	1995	Yes

Modern Anesthesia

Modern inhalant anesthetics include fluorine-containing compounds (Table 18.3). These compounds are nonflammable and relatively safe for patients. Their safety for operating room personnel, however, has been questioned. For example, female operating room workers suffer a higher rate of miscarriages than women in the general population.

Modern surgical practice usually makes use of a variety of drugs. Generally, a patient is given

- a tranquilizer such as a benzodiazepine (Valium or Versed; Section 18.16) to decrease anxiety
- an intravenous anesthetic such as thiopental (Section 18.14) to produce unconsciousness quickly
- a narcotic painkiller (such as fentanyl; Table 18.5) to block pain
- an inhalant anesthetic (Table 18.3) to provide insensitivity to pain and keep the patient unconscious, often combined with oxygen and nitrous oxide to support life
- a relaxant such as pancuronim bromide to relax the muscles and make it easier to insert the breathing tube

General anesthetics reduce nerve transmission at the synapses by inhibiting the excitatory effect of certain neurotransmitters on central nervous system receptors. The anesthetics bind only weakly, and it is difficult to determine their exact mode of action.

More to Explore
Perkins, Bill. "How Does Anesthesia Work?" *Scientific American*, May 2005, p. 102, and Connor, J. T. H. "The Victorian Revolution in Surgery." *Science*, April 2, 2004, pp. 54–55.

Solvent Sniffing: Self-Administered Anesthesia

Nearly all gaseous and volatile liquid organic compounds exhibit anesthetic action, which can lead to their abuse. The sniffing of glue solvents, gasoline, aerosol propellants, and other inhalants is perhaps the deadliest form of drug abuse. The dose required for intoxication often is not far from that which will stop the heart. And it is difficult to measure the dose inhaled from the plastic or paper bag normally used in sniffing. Also, as with nitrous oxide, sublethal doses can cause permanent brain damage by cutting down the oxygen supply to the brain.

Local Anesthetics

Local anesthetics are used to render a part of the body insensitive to pain. They block nerve conduction by reducing the permeability of the nerve cell membrane to sodium ions. The patient remains conscious during dental work or minor surgery.

The first local anesthetic to be used successfully was cocaine, a drug first isolated in 1860 from the leaves of the coca plant (Figure 18.20). Its structure was determined by Richard Willstätter in 1898. Scientists have made many attempts to develop synthetic compounds with similar properties. Cocaine is a powerful stimulant. Its abuse is discussed in Section 18.17.

▲ **Figure 18.20** Coca leaves contain the alkaloid cocaine, a local anesthetic and stimulant.

Table 18.4 Common Substances Used as Local Anesthetics

Substance	Duration of Action
Procaine (Novocain®)	Short
Lidocaine (Xylocaine®)	Medium (30–60 min)
Mepivacaine (Carbocaine®)	Fast (6–10 min)
Bupivacaine (Marcaine®)	Moderate (8–12 min)
Prilocaine (Citanest®)	Medium (30–90 min)
Chloroprocaine (Nesacaine®)	Short (15–30 min)
Cocaine	Medium
Etidocaine (Duranest®)	Long (120–180 min)

Because local anesthetics are weak bases (amines), they are usually used as hydrochloride salts to make them water soluble. Some of the local anesthetics (Figure 18.21) are amine-esters (cocaine, procaine, and chloroprocaine), whereas others are amine-amides (lidocaine, mepivicaine, prilocaine, bupivacaine, and etidocaine). The amine-esters cause allergic reactions in some people; the amine-amides do not. The ethyl and butyl esters of *para*-aminobenzoic acid (PABA) are topical anesthetics, applied to the skin to relieve the pain of burns and open wounds. These esters are applied as ointments, usually in the form of picrate salts.

Introduction of a second nitrogen atom in the alkyl group of the ester produces a more powerful anesthetic (Table 18.4). Perhaps the best known of these is procaine (Novocaine®), first synthesized in 1905 by Alfred Einhorn (1865–1917), who had worked with Willstätter on the structure of cocaine. Procaine takes a fairly long time to take effect, the anesthetic wears off too quickly, and it causes allergic reactions. It is no longer used in dentistry, replaced largely by the amine-amide anesthetics.

Introduced in the 1940s, lidocaine is still the most widely used local anesthetic. It takes effect quickly. When combined with a small quantity of epinephrine, it acts for an hour or more. (See Collaborative Group Project 81).

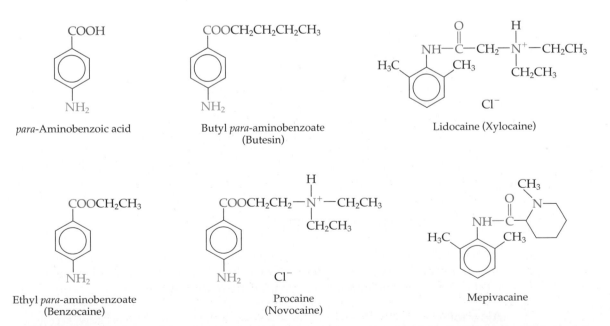

para-Aminobenzoic acid

Butyl *para*-aminobenzoate (Butesin)

Lidocaine (Xylocaine)

Ethyl *para*-aminobenzoate (Benzocaine)

Procaine (Novocaine)

Mepivacaine

▲ **Figure 18.21** Some local anesthetics. They are often used in the form of a hydrochloride or picrate salt, which is more soluble in water than the free base.

QUESTION: Which are derived from *para*-aminobenzoic acid? Which have amide functional groups? Which are esters?

Dissociative Anesthetics: Ketamine and PCP

Ketamine, an intravenous anesthetic, is called a *dissociative anesthetic* because it dissociates a person's perception from his or her sensations; it reduces or blocks signals to the conscious mind from other parts of the brain such as those that are associated with the senses. Ketamine induces hallucinations similar to those reported by people who have had near-death experiences. They seem to remember observing their rescuers from a vantage point above the scene or moving through a dark tunnel toward a bright light. Ketamine acts as an NMDA receptor antagonist. N-methyl d-aspartate (NMDA) is an amino acid derivative that mimics the action of the neurotransmitter glutamate (Table 19.2). Because ketamine acts by binding to receptors in the body, we can assume that our bodies also produce chemicals that fit these receptors. These compounds may be synthesized or released only in extreme circumstances, such as in near-death experiences.

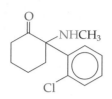

Ketamine

Ketamine is used widely in veterinary medicine. Because it suppresses breathing much less than most other available anesthetics, ketamine is used sometimes in human medicine as an anesthetic for victims with unknown medical history and for children and persons of poor health.

Phencyclidine, commonly called PCP, is closely related to ketamine. PCP was formerly used as an animal tranquilizer and for a brief time as a general anesthetic for humans. It is now fairly common on the illegal drug scene.

PCP, which is soluble in fat and has no appreciable water solubility, is stored in fatty tissue and released when the fat is metabolized, accounting for the "flashbacks" commonly experienced by users. Many users experience bad "trips" with PCP, and about 1 in 1000 develops a severe form of schizophrenia. High doses cause illusions and hallucinations and can also cause seizures, coma, and death. However, death of PCP users more often results from an accident or suicide during a "trip".

Phencyclidine
(PCP)

Self-Assessment Questions

1. Which of the following general anesthetics is nonflammable?
   a. cyclopropane
   b. diethyl ether
   c. ethylene
   d. halothane

2. A hazard of the use of nitrous oxide as an anesthetic is that it
   a. can asphyxiate if not combined with O_2
   b. has too long a recovery time
   c. induces too deep anesthesia
   d. is flammable

3. The formula for nitrous oxide is
   a. N_2O
   b. NO
   c. NO_2
   d. N_2O_3

4. Which of the following is a dissociative anesthetic?
   a. diethyl ether
   b. isoflurane
   c. phencyclidine
   d. thiopental

Answers: 1, d; 2, a; 3, a; 4, c

18.14 Depressant Drugs

In this section we explore several familiar depressant drugs. We discussed ethyl alcohol in some detail in Chapter 9, but we start with it again here because it is by far the most used and abused depressant drug in the world.

Alcohol

In nature, sugars in fruit are often fermented into alcohol by airborne yeasts, but purposeful production of ethanol probably did not begin until people began to farm. As early as 3700 B.C.E., the Egyptians fermented fruit to make wine and the Babylonians made beer from barley. People have been fermenting fruits and grains ever since, but relatively pure ethanol was first produced by Muslim alchemists who invented distillation in the eighth or ninth century.

People often think they are stimulated ("get high") when they drink, but ethanol is actually a depressant. It slows down both physical and mental activity. Nearly two-thirds of adults in the United States drink alcohol, and nearly one-third have five or more drinks on at least one day each year. There are at least 10 million alcoholics in the United States, and alcoholism is the third leading health problem, right after cancer and heart disease. Approximately 30% of current drinkers in the United States drink to excess. According to Centers for Disease Control and Prevention, excessive drinking of alcohol kills about 75,000 people each year in the United States, and underage drinking causes more than 5000 deaths of people under 21 each year. For 15- to 24-year-olds, the three leading causes of death are automobile accidents, homicides, and suicides, and alcohol is a leading factor in all three.

For some, there is a positive side to drinking alcohol. Longevity studies indicate that those who use alcohol only moderately (no more than a drink or two a day) live longer than nondrinkers. This may result from the relaxing effect of the alcohol. Heavier drinking, however, can cause many health problems, and alcoholism shortens the life span of alcoholics by 10 to 12 years.

Some Biochemistry of Ethanol

Alcoholic beverages are generally high in calories. Pure ethanol furnishes about 7 kcal/g. Ethanol is metabolized by oxidation to acetaldehyde at a rate of about 1 oz (28 mL)/hr, and a buildup of ethanol concentration in the blood caused by drinking more than this amount results in intoxication.

Oxidation of ethanol uses up a substance called NAD^+, the oxidized form of nicotinamide adenine dinucleotide, represented in the reduced form as NADH, where the H is a hydrogen atom.

$$CH_3CH_2OH + NAD^+ \longrightarrow CH_3CHO + NADH + H^+$$
$$\text{Ethanol} \qquad\qquad\qquad \text{Acetaldehyde}$$

Because NAD^+ is usually employed in the oxidation of fats, when one drinks too much alcohol, the fat is deposited—mainly in the abdomen—rather than metabolized. This results in the familiar "beer gut" of many heavy drinkers.

Just how ethanol intoxicates is still somewhat of a mystery. Researchers have found that ethanol disrupts receptors for two neurotransmitters (Table 18.2): *gamma*-aminobutyric acid (GABA), which inhibits impulsiveness, and glutamate, which excites certain nerve cells. Long-term drinking changes the balance of GABA and glutamate. Ethanol thus disrupts the brain's control over muscle activity, causing the drunk to stagger and fall. Alcohol also raises dopamine levels in the brain; dopamine is associated with the pleasurable aspects of alcohol and other drugs.

There are several theories about the cause of alcoholism. Excessive drinking over time can change the levels of some of the brain chemicals, causing a craving of alcohol to bring back good feelings or to avoid bad feelings. Studies on twins raised separately show that there is undoubtedly a genetic component. There are also emotional, psychological, social, and cultural factors. We still have much to learn about this ancient drug.

Barbiturates

A family of related depressants, the barbiturates, displays a wide range of properties. Barbiturates can be employed to produce mild sedation, deep sleep, or even death. More than 2500 barbiturates have been synthesized over the years, but only a few have found widespread use in medicine (Figure 18.22).

- Pentobarbital is employed as a short-acting hypnotic drug. Before the discovery of modern tranquilizers, it was used widely to calm anxiety.
- Phenobarbital is a long-acting drug. It is employed widely as an anticonvulsant for people suffering from seizure disorders such as epilepsy.
- Thiopental, which differs from pentobarbital only in that an oxygen atom on the ring has been replaced by a sulfur atom, is used widely as an anesthetic.

▶ **Figure 18.22** Line-angle formulas of some barbiturate drugs: (a) pentobarbital (Nembutal), (b) phenobarbital (Luminal), and (c) thiopental (Pentothal). These drugs, derived from barbituric acid, are often used in the form of their sodium salts; for example, thiopental is used as sodium pentothal.

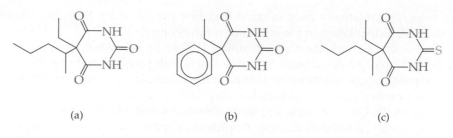

(a)   (b)   (c)

Barbiturates were once used in small doses as sedatives, the dosage generally being a few milligrams. In larger dosages (about 100 mg), barbiturates induce sleep. They were once the sleeping pills of choice but now have been replaced by drugs such as a benzodiazepine (Section 18.16) which have less potential for abuse and are less likely to be lethal. Barbiturates are quite toxic; the lethal dose is about 1500 mg (1.5 g).

Barbituric acid, from which the barbiturates are derived, was first synthesized in 1864 by Adolph von Baeyer. He made it from urea, which occurs in urine, and malonic acid, which is made from apples.

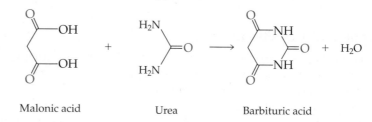

Malonic acid   Urea   Barbituric acid

The term *barbiturates*, according to Willstätter, came about because at the time of the discovery von Baeyer was infatuated with a woman named Barbara. The word comes from *Barbara* and *urea*.

Synergism: Barbiturates and Alcohol

Barbiturates are especially dangerous when ingested along with ethyl alcohol. This combination produces an effect perhaps ten times greater than the sum of the effects of two depressants. Many people have died by suicide or accident because of this combination. This effect, in which two chemicals bring about an effect greater than that of each chemical individually, is called a **synergistic effect** (Section 12.4). Synergistic effects can be deadly, and they are not limited to alcohol–barbiturate combinations. Two drugs should never be taken at the same time without competent medical supervision.

Barbiturates, like ethanol, are intoxicating, and they are strongly addictive. Habitual use leads to the development of a tolerance to the drugs, and ever-larger doses are required to produce the same degree of intoxication. The side effects of barbiturates are similar to those of alcohol: hangovers, drowsiness, dizziness, and headaches. Withdrawal symptoms are often severe, accompanied by convulsions and delirium, and can cause death.

Barbiturates are cyclic amides. The mechanism of their action is similar in some respects to that of alcohol in that they act on a GABA receptor.

Self-Assessment Questions

1. Ethyl alcohol is a(n)
   a. analgesic   **b.** depressant   **c.** narcotic   **d.** stimulant
2. When ethanol is oxidized in the body, the same substance (NAD^+) is used up as when
   a. amino acids are joined to make peptides
   b. fat is oxidized
   c. proteins are hydrolyzed
   d. sugar is oxidized

3. When two drugs in combination give a greater total effect than the sum of the effects of the two drugs acting independently, the effect is
a. activation   **b.** catalysis   **c.** enhancement   **d.** synergistic

4. Which of the following is most likely to combine with alcohol to form a lethal dose?
a. amphetamine   **b.** barbiturates   **c.** benzodiazepines   **d.** marijuana

Answers: 1, b; 2, b; 3, d; 4, b

18.15 Narcotics

A **narcotic** is a drug that produces narcosis (stupor or general anesthesia) and analgesia (relief of pain). Many drugs produce these effects, but in the United States only those that are also *addictive* are legally classified as narcotics. Their use is regulated by federal law.

Opium and Morphine

Morphine

Opium is the dried, resinous juice of the unripe seeds of the Oriental poppy (*Papaver somniferum*) (Figure 18.23). It is a complex mixture of 20 or so nitrogen-containing organic bases (alkaloids), sugars, resins, waxes, and water. The principal alkaloid, morphine, makes up about 10% of the weight of raw opium. Opium was used in many patent medicines during the nineteenth century. Laudanum, a solution of opium in alcohol containing about 10 mg morphine per mL of solution, was widely used during the Victorian era for everything from toothaches to tuberculosis. It was even used to quiet colicky babies. English romantic poets Samuel Taylor Coleridge (1772–1834), Percy Bysshe Shelley (1772–1822), and George Gordon Byron (Lord Byron, 1788–1824) were among the prominent users, as was the French poet Charles Pierre Baudelaire (1821–1867). Mary Todd Lincoln (1818–1882), wife of the sixteenth president of the United States, was prescribed laudanum for sleep problems, perhaps accounting for her anxiety and hallucinations. Paregoric, a much weaker solution containing only 0.4 mg of morphine per milliliter, was also widely used, especially for diarrhea.

Morphine was first isolated in 1805 by Friedrich Sertürner, a German pharmacist. With the invention of the hypodermic syringe in the 1850s, a new method of administration became available. Injection of morphine directly into the bloodstream was more effective for the relief of pain, but this method also seriously escalated the problem of addiction.

▲ Trade card advertising Mrs. Winslow's Soothing Syrup, an opium-containing patent medicine.

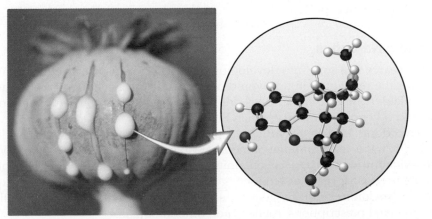

◄ **Figure 18.23** Opium poppy seed pod. The slits on the seed pod exude the resinous juice, which when dried is called opium. About 10% of the weight of raw opium is morphine.

Morphine was used widely during the American Civil War (1861–1865) for relief of pain from battle wounds. One side effect of morphine use is constipation. Noting this, soldiers came to use morphine as a treatment for that other common malady of men on the battlefront—dysentery. More than 100,000 soldiers became addicted to morphine while serving in the war. The affliction was so common among veterans that it came to be known as "soldier's disease."

Morphine and other narcotics were placed under control of the federal government by the Harrison Act of 1914. Morphine is still used by prescription for the relief of severe pain. It also induces lethargy, drowsiness, confusion, euphoria, chronic constipation, and depression of the respiratory system. Morphine is addictive, and this becomes a problem when it is administered in amounts greater than the prescribed dose or for a period longer than the prescribed time.

Codeine and Heroin

Changes in the morphine molecule alter its physiological properties. Replacement of the phenolic OH group by a methoxy (OCH_3) group produces codeine (see Additional Problem 55). Codeine is present in opium to an extent of about 0.7 to 2.5%, but it is usually synthesized by methylating the more abundant morphine molecules. Codeine is similar to morphine in its action, but it is less potent, has less tendency to induce sleep, and is less addictive. For relief of moderate pain, codeine is often combined with acetaminophen (for example, in Tylenol 2® and Tylenol 3®) or with aspirin (in Empirin® Codeine). Codeine for cough suppression is used in liquid preparations that are regulated less stringently than stronger narcotics.

Conversion of both the OH groups of the morphine molecule to acetate ester groups produces heroin (see Additional Problem 56). This semisynthetic morphine derivative was first prepared by chemists at the Bayer Company of Germany in 1874. It received little attention until 1890, when it was proposed as an antidote for morphine addiction. Shortly thereafter, Bayer widely advertised heroin as a sedative for coughs, often in the same ads describing aspirin. However, it soon was found that heroin induced addiction more quickly than morphine and that heroin addiction was harder to cure.

The physiological action of heroin is similar to that of morphine. Heroin is less polar than morphine; it enters the fatty tissues of the brain more rapidly and seems to produce a stronger feeling of euphoria than morphine. Heroin is not legal in the United States, even by prescription. It has, however, been advocated for use in the relief of pain in terminal cancer patients and has been so used in Britain.

Addiction probably has three components: emotional dependence, physical dependence, and tolerance. Psychological dependence is evident in the uncontrollable desire for the drug. Physical dependence is shown by acute withdrawal symptoms such as convulsions. Tolerance for the drug is evidenced by the increasing dosages required to produce in the addict the same degree of narcosis and analgesia.

Deaths from heroin and other narcotics are usually attributed to overdoses, but the situation is not always clear. The problem is often a matter of quality control. Street drugs vary considerably in potency, and a particular dose can be much higher in potency than expected.

Synthetic Narcotics

Much research has gone into developing a drug that would be as effective as morphine for the relief of pain but would not be addictive. Several synthetic narcotics are available. Some common ones are listed in Table 18.5. Structures of some of these and other narcotics are given in Figure 18.24.

Oxycodone is a semisynthetic opioid related to codeine (see Additional Problem 57). It is used for relief of moderate to severe pain. It is available alone and in combination with products containing acetaminophen, ibuprofen, or aspirin. A sustained-release form of oxycodone is called OxyContin®. Abused for its euphoric effects, OxyContin and its generic equivalents are widely available illegally through faked prescriptions, robbery, and other methods.

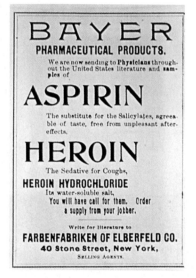

▲ Heroin was regarded as a safe medicine in 1900; it was widely used as a cough suppressant. It was also thought to be a nonaddictive substitute for morphine.

More to Explore

Nestler, Eric J., and Robert C. Malenka. "The Addicted Brain." *Scientific American*, March 2004, pp. 78–85.

Table 18.5 Narcotic Pain Medications

Substance	Duration of Action (h)	Approximate Dosage Relative to Morphine at 30 mg[a]
Codeine	4–6	200
Meperidine (Demerol®)	2–4	300
Hydrocodone	4–8	20
Fentanyl	1–2[b]	<1[b]
Oxycodone	4–6[c]	20
Hydromorphone (Dilaudid®)	4–5	7.5
Methadone	4–6[d]	(c)

[a]Oral doses for equivalent pain relief.
[b]Used in transdermal patches for chronic pain.
[c]Often dispensed in a time-release form (OxyContin®) that allows 12 h between doses.
[d]Used for treatment of heroin addiction; not usually given for pain relief.

Hydrocodone, another synthetic narcotic, is always combined with acetaminophen or another medication. Sold as HYCD/APAP (hydrocodone/acetaminophen), this combination was the second most prescribed drug in the United States in 2006.

The synthetic narcotic methadone is widely used to treat heroin addiction. Like heroin, methadone is highly addictive. However, when taken orally, it does not induce the sleepy stupor characteristic of heroin intoxication. Unlike a heroin addict, a person on methadone maintenance is usually able to hold a productive job. If an addict who has been taking methadone reverts to heroin, the methadone in his or her system effectively blocks the euphoric rush normally given by heroin and so reduces the addict's temptation to use heroin.

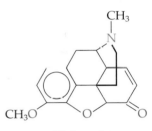

Hydrocodone

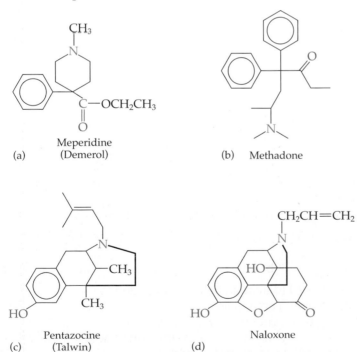

(a) Meperidine (Demerol)

(b) Methadone

(c) Pentazocine (Talwin)

(d) Naloxone

◄ **Figure 18.24** Line-angle formulas of some synthetic narcotics: (a) meperidine, (b) methadone, (c) pentazocine (Talwin®), and (d) naloxone.

Morphine Agonists and Antagonists

Chemists have synthesized many morphine analogs, but relatively few have shown significant analgesic activity, and most of those are addictive. Morphine acts by binding to receptors in the brain. A morphine receptor is shaped such that there is a flat part that binds to the benzene ring, a cavity that binds the two carbon atoms indicated by the bold line in the structure, and an anionic site that binds the nitrogen atom.

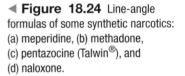

Anion site binds N

Cavity binds carbon atoms

Flat hole binds benzene ring

▲ Diagram of a morphine receptor.

A molecule that has morphinelike action is called a morphine *agonist*. A morphine *antagonist* inhibits the action of morphine by blocking the receptors. Some molecules have both agonist and antagonist effects. An example is pentazocine, Figure 18.24c, which is less addictive than morphine and is effective for the relief of pain.

Pure antagonists such as naloxone, Figure 18.24d, are used to treat opiate addicts. An addict who has overdosed can be brought back from death's door by an injection of naloxone.

Natural Opiates

Morphine acts by binding to specific receptor sites in the brain. Why should the human brain have receptors for a plant-derived drug such as morphine? Scientists concluded that the body must produce its own substances that fit these receptors. Actually, it produces several morphinelike substances, called **endorphins** (endogenous morphines). Each is a short peptide chain composed of amino acid units. Those with five amino acid units are called *enkephalins*. There are two enkephalins, which differ only in the amino acid at the end of the chain. *Leu*-enkephalin has the sequence Tyr-Gly-Gly-Phe-Leu, and *Met*-enkephalin is Tyr-Gly-Gly-Phe-Met. Other endorphins have chains of 30 or more amino acids.

Some enkephalins have been synthesized and shown to be potent pain relievers. Their use in medicine is quite limited, however, because after being injected they are rapidly broken down by the enzymes that hydrolyze proteins. Researchers have sought to make analogs more resistant to hydrolysis that can be employed as morphine substitutes for the relief of pain. Unfortunately, both natural enkephalins and their analogs, such as morphine, seem to be addictive.

It appears that endorphins are released during strenuous exercise and as a response to pain. Capsaicin (the active compound in chili peppers) can also stimulate endorphin release. Some evidence indicates that acupuncture anesthetizes somewhat by stimulating the release of brain "opiates." The long needles stimulate deep sensory nerves that cause the release of peptides that then block the pain signals for a brief time.

Endorphin release has also been used to explain other phenomena once thought to be largely psychological. A soldier, wounded in battle, may feel no pain until the skirmish is over. His body has secreted its own painkiller.

Self-Assessment Questions

1. The dried sap of the seed pods of the Oriental poppy plant is called
   a. codeine b. paclitaxel c. potpourri d. opium

2. An active ingredient in opium that is less abundant and less powerful than morphine is
   a. caffeine b. codeine c. methadone d. pentazocine

3. A semisynthetic opiate produced by adding acetyl groups to morphine molecules is
   a. heroin b. MDMA c. meperidine d. methadone

4. Methadone used to treat addiction is intended to
   a. act as an opiate antagonist
   b. detoxify the patient during withdrawal
   c. satisfy the drug craving yet leave the addict able to function in society
   d. treat the underlying causes of addiction

5. Substances produced in the body that resemble the opiates in their effects are called
   a. endorphins b. oxytocins
   c. placebos d. thyroxins

Answers: 1, d; 2, b; 3, a; 4, c; 5, a

18.16 Antianxiety Agents

The hectic pace of life in the modern world causes some people to seek rest and re-laxation in chemicals. Ethyl alcohol is undoubtedly the most widely used tranquil-izer. The drink before dinner—to "unwind" from the tensions of the day—is a part of the way of life for many people. This search for relief from stress and anxiety has led to the great popularity of antidepressant drugs. According to the Centers for Disease Control and Prevention (CDC), antidepressants are now the most common-ly prescribed drugs, with 118 million prescriptions in 2005.

One class of antianxiety drugs (anxiolytics) is the benzodiazepines, compounds that feature seven-member heterocyclic rings (Figure 18.25). Three common ones are diazepam, a classic antianxiety agent; clonazepam, an anticonvulsant as well as antianxiety drug; and lorazepam, used to treat insomnia. Antianxiety agents—sometimes called *minor tranquilizers*—make people feel better simply by making them feel dull and insensitive, but they do not solve any of the underlying prob-lems that cause anxiety. They are thought to act on a GABA receptor, dampening neuron activity in higher function parts of the brain.

What is the price of tranquility? After 20 years of use, benzodiazepines were found to be addictive. People trying to go off these drugs after extended use experience painful withdrawal.

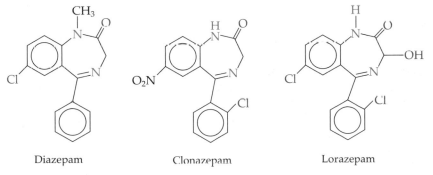

Diazepam   Clonazepam   Lorazepam

◀ **Figure 18.25** Three benzodi-azepines: (a) diazepam (Valium®), (b) clonazepam (Klonopin®), and (c) lorazepam (Ativan®).

Antipsychotic Agents

The first antipsychotic drugs—sometimes referred to as *major tranquilizers*—were compounds called phenothiazines. In 1952 chlorpromazine (Thorazine®) was ad-ministered as a tranquilizer to psychotic patients in the United States. The drug had been tested in France as an antihistamine, and medical workers there had noted that it calmed mentally ill patients being treated for allergies. Chlorpromazine was found to be helpful in controlling the symptoms of schizophrenia and truly revolu-tionized mental illness therapy.

Chlorpromazine is one of several phenothiazines that are used in medicine. Promazine (chlorpromazine without the chlorine atom) is also a tranquilizer, but it is much less potent than chlorpromazine.

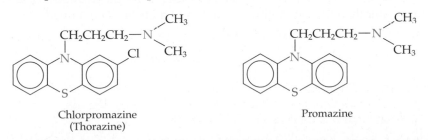

Chlorpromazine
(Thorazine)   Promazine

Phenothiazines act in part as dopamine antagonists; they block postsynaptic receptors for dopamine, a neurotransmitter important in the control of detailed motion (such as grasping small objects), in memory and emotions, and in exciting the cells of the brain. Some researchers think schizophrenic patients produce too much dopamine, whereas others think that they have too many dopamine receptors. In either case, blocking the action of dopamine relieves the symptoms of schizophrenia.

Newer, so-called second-generation or atypical, antipsychotics are now available, including aripiprazole (Abilify®), risperidone (Risperdal®), clozapine (Clozaril®), and olanzapine (Zyprexa®). Aripiprazole is a partial dopaminergic agonist. It is used to treat schizophrenia, acute manic episodes of bipolar disorder, and to treat depression. Risperidone is used to treat schizophrenia. Clozapine, the first of the atypical antipsychotics to be developed, has been shown to be the most effective drug for treating schizophrenia, but severe side effects—such as destruction of white blood cells—relegate it to use only after other antipsychotic drugss have failed. Olanzapine is used to treat psychotic disorders such as schizophrenia and acute manic episodes and to stabilize bipolar disorder.

These drugs are thought to act on a type of serotonin receptor that loosens the binding of dopamine receptors, allowing more dopamine to reach the neurons.

Antipsychotic drugs have served to reduce greatly the number of patients confined to mental hospitals. They do not cure schizophrenia but instead control its symptoms to the extent that 95% of all schizophrenics no longer need hospitalization. Patients who go off their medication relapse.

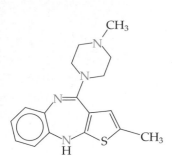

Olanzapine (Zyprexa®)

Antidepressant Drugs

The oldest class of antidepressant drugs are the tricyclic (three-ring) antidepressants. Examples are imipramine (Tofranil®) and amitriptyline (Elavil®). Note that slight changes in the structure of a molecule can result in profound changes in properties. Promazine is not very potent, but replacing the sulfur atom of promazine with a CH_2CH_2 group produces imipramine, a fairly potent antidepressant. In amitriptylene, the ring nitrogen atom also is replaced by a carbon atom. Tricyclic antidepressants block the reabsorption ("reuptake") of neurotransmitters such as norepinephrine and serotonin. Tricyclics have serious side effects and have been largely replaced by newer drugs.

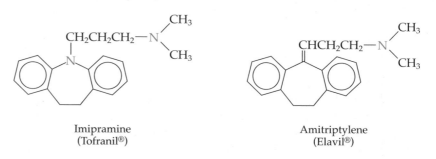

Imipramine
(Tofranil®)

Amitriptylene
(Elavil®)

Common antidepressants today include drugs called *selective serotonin reuptake inhibitors* (SSRIs). Three major ones are shown in Figure 18.26. Doctors prescribe these drugs to help people cope with a wide variety of psychological issues, including anxiety syndromes such as panic disorder, obsessive–compulsive disorders (including gambling problems and overeating), and premenstrual syndrome (PMS). The drugs work by enhancing the effect of serotonin, blocking its reuptake by nerve cells. They seem to be safer than tricyclic antidepressants and more easily tolerated.

▶ **Figure 18.26** Three selective serotonin reuptake inhibitors (SSRIs): fluoxetine (Prozac®), paroxetine (Paxil®), and sertraline (Zoloft®).

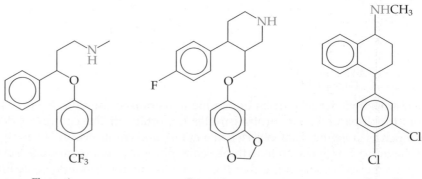

Fluoxetine

Paroxetine

Sertraline

Self-Assessment Questions

1. Which of the following classes of drugs are commonly used as antianxiety medications?
   a. amphetamines
   b. benzodiazepines
   c. opiates
   d. steroids

2. Benzodiazepine molecules feature
   a. five-membered heterocyclic rings
   b. seven-membered heterocyclic rings
   c. five-membered hydrocarbon rings
   d. seven-membered hydrocarbon rings

3. What kind of receptor do benzodiazepines act upon?
   a. acetylcholine   **b.** GABA   **c.** opiate   **d.** serotonin

4. The drug credited with greatly reducing the number of patients in mental hospitals is
   a. amitriptylene   **b.** clonazepam   **c.** chlorpromazine   **d.** diazepam

5. Phenothiazine molecules have benzene rings on each side of a heterocyclic ring that features
   a. N and Cl atoms   **b.** S and N atoms   **c.** two N atoms   **d.** two S atoms

6. Which of the following drugs is a tricyclic antidepressant?
   a. chlorpromazine   **b.** imipramine   **c.** sertraline   **d.** triclosan

Answers: 1, b; 2, b; 3, b; 4, c; 5, b; 6, b

18.17 Stimulant Drugs

Among the more widely known stimulant drugs is a variety of synthetic amines related to β-phenylethylamine (Figure 18.27). These drugs, called **amphetamines**, are similar in structure to epinephrine and norepinephrine (Section 18.12). They seem to act by increasing levels of norepinephrine, serotonin, and dopamine in the brain.

Amphetamines

Amphetamine and methamphetamine are inexpensive and have been widely abused. Amphetamine was once used as a diet drug, but it had little long-term effect. It is a standard treatment for narcolepsy, a rare form of sleeping sickness. Amphetamines induce excitability, restlessness, tremors, insomnia, dilated pupils, increased pulse rate and blood pressure, hallucinations, and psychoses.

Methamphetamine has a more pronounced psychological effect than amphetamine. Methamphetamine is made readily from an antihistamine and household chemicals. Illegal "meth labs" have been found in all parts of the country. Legal

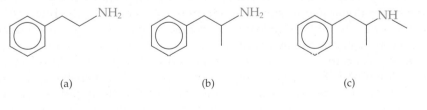

(a)   (b)   (c)

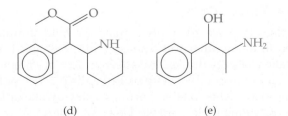

(d)   (e)

◀ **Figure 18.27**
β-Phenylethylamine (a) and related compounds. (b) amphetamine, (c) methamphetamine, (d) methylphenidate, and (e) phenylpropanolamine.

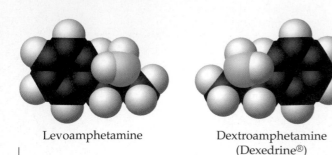

Levoamphetamine   Dextroamphetamine
                              (Dexedrine®)

Amphetamine
(Benzedrine®)

▲ Benzedrine® is a mixture of Dexedrine® and its enantiomer, levoamphetamine. As is often the case, the two isomers differ greatly in their biological activity.

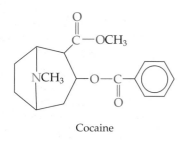

Cocaine

According to the CDC, drug overdoses killed more than 33,000 people in 2005. Contrary to popular belief, most were not caused by cocaine or heroin, but by OxyContin and other painkillers diverted into the illegal drug trade. More than 15 million Americans admitted to using prescription drugs for nonmedical reasons.

Table 18.6	Caffeine in Soft Drinks
Brand	**Caffeine***
Red Bull (8.2 oz)	80
Diet Coke	45
Mello Yello	51
Diet Pepsi	36
Dr Pepper	41
Sun Drop	63
Sunkist Orange	41
Mountain Dew	55
Pepsi-Cola	37
Royal Crown Cola	43

Source: National Soft Drink Association, U.S. FDA

* Milligrams per 12-oz serving.

restrictions on the availability of over-the-counter cold or allergy medicines has helped reduce methamphetamine manufacture in the United States. Like other amine drugs, amphetamines are often distributed as hydrochloride salts. The free-base form (the basic amine form that has not been neutralized by an acid), especially of methamphetamine, is used for smoking because they vaporize more readily than the salts. Long-term effects of methamphetamine use include severe mental problems, loss of memory, and serious dental problems.

Amphetamines are used to treat attention-deficit/hyperactivity disorder (ADHD) in children. The drug of choice is often methylphenidate (Ritalin®). Although it is a stimulant, the drug seems to calm children who otherwise can't sit still. Such use has been criticized as "leading to drug abuse" and as "solving the teacher's problem, not the kid's."

Like many other drugs, amphetamine exists as mirror-image isomers (enantiomers). Benzedrine® is the trade name for a mixture of the two isomers in equal amounts. The dextro (right-handed) isomer is a stronger stimulant than the levo (left-handed) isomer. Dexedrine® is the trade name for dextroamphetamine, a drug made of the pure dextro isomer.

Dexedrine is two to four times as active as Benzedrine. Many current drugs have dextro and levo isomers, and most of them are sold as mixtures of the isomers. The development of drugs that contain only one enantiomer is becoming increasingly important in the pharmaceutical industry. Worldwide, all five of the top-selling drugs are single-enantiomer drugs, and total sales of single-enantiomer drugs is about $200 billion annually.

Cocaine

Cocaine, first used as a local anesthetic (Section 18.14), also acts as a powerful stimulant. The drug is obtained from the leaves of a shrub that grows almost exclusively on the eastern slopes of the Andes Mountains. Many of the Indians living in and around the area of cultivation chew coca leaves—mixed with lime and ashes—because of their stimulant effect. Cocaine used to arrive in the United States as the salt cocaine hydrochloride, but now much of it comes in the form of broken lumps of the free base, a form called *crack cocaine.*

Cocaine hydrochloride is readily absorbed through the watery mucous membrane of the nose, and this is the form used by those who "snort" cocaine. Those who smoke cocaine use the free base (crack), which readily vaporizes at the temperature of a burning cigarette. When smoked, cocaine reaches the brain in 15 s. It acts by preventing the reuptake of dopamine after it is released by nerve cells, leaving high levels of dopamine to stimulate the pleasure centers of the brain. After the binge, dopamine is depleted in less than an hour, leaving the user in a pleasureless state and (often) craving more cocaine.

The use of cocaine increases stamina and reduces fatigue, but the effect is short-lived. Stimulation is followed by depression. Once quite expensive and limited to use mainly by the wealthy, cocaine is now available in cheap, potent forms. Hundreds, including several well-known entertainers and athletes, have died from cocaine overdose.

Caffeine: Coffee, Tea, or Cola

Coffee, tea, and cola soft drinks naturally contain the mild stimulant caffeine. Caffeine is also added to many other soft drinks (Table 18.6) and is the source of the "lift" of many so-called "energy drinks." An effective dose of caffeine is about 200 mg. This is about the amount in one 10-oz cup of strong drip coffee or three to five cups of tea. Caffeine is also available in tablet form as a stay-awake or keep-alert type of drug. Two brands of 200-mg tablets are No-Doz and Vivarin.

Is caffeine addictive? The "morning grouch" syndrome suggests that it is mildly so. There is also evidence that large doses of caffeine may be involved in chromosome damage. To be safe, people in their childbearing years should avoid large quantities of caffeine. Overall, the hazards of caffeine ingestion seem to be slight.

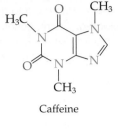

Caffeine

Nicotine

Nicotine: Going Up in Smoke

Another common stimulant is nicotine. This drug is taken by smoking or chewing tobacco. Nicotine is highly toxic to animals and has been used in agriculture as a contact insecticide. It is especially deadly when injected; the lethal dose for a human is estimated to be about 50 mg. Nicotine seems to have a rather transient effect as a stimulant. This initial response is followed by depression, but smokers generally keep a near-constant level of nicotine in their bloodstream by indulging frequently.

Is nicotine addictive? Casual observation of a person trying to quit smoking seems to indicate that it is. Consider the 1972 memorandum from a Philip Morris scientist who noted that "no one has ever become a cigarette smoker by smoking cigarettes without nicotine." He suggested that the company "think of the cigarette as a dispenser for a dose unit of nicotine."

Self-Assessment Questions

1. Several stimulant drugs are derivatives of
   a. *p*-aminobenzoic acid
   b. caffeine
   c. ketamine
   d. β-phenylethylamine

2. Dextroamphetamine is composed of molecules that
   a. differ from amphetamine by a CH_3 group
   b. are a mixture of left- and right-handed
   c. are all left-handed
   d. are all right-handed

3. Cocaine is smoked in the form of
   a. coca leaves
   b. the free base
   c. a liquid
   d. a salt

4. Compared to amphetamines, cocaine exhibits stimulation that
   a. is similar in duration and intensity
   b. lasts longer but is less intense
   c. lasts longer and is more intense
   d. is shorter but is more intense

Answers: 1, d; 2, d; 3, b; 4, d

18.18 Hallucinogenic Drugs

Hallucinogenic drugs are consciousness-altering substances that induce changes in sensory perception; they qualitatively change the way we perceive things. Common hallucinogenic drugs include natural products such as psilocybin mushrooms, peyote cactus, and *Salvia divinorum*; synthetic chemicals such as MDMA ("Ecstasy"), dimethyltryptamine (DMT), PCP, and ketamine (Section 18.13); and semisynthetic ones such as LSD and mescaline. Marijuana is sometimes considered a mild hallucinogen.

Lysergic Acid Diethylamide (LSD)

Lysergic acid diethylamide (LSD), a semisynthetic drug, is a powerful hallucinogen. Its physiological properties were discovered quite accidentally by Swiss chemist Albert Hofmann (1906–2008) in 1943 when he unintentionally ingested some LSD. He later took 250 mg, which he considered a small dose, to verify that LSD had caused the symptoms he had experienced. Hofmann had a rough time for the next few hours, experiencing such symptoms as visual disturbances and schizophrenic behavior.

LSD can create a feeling of lack of self-control and sometimes of extreme terror. The exact mechanism by which LSD exerts its effects is still unknown, but it is thought to act on dopamine receptors and to excite by increasing release of glutamate.

Lysergic acid is obtained from ergot, a fungus that grows on rye. This carboxylic acid is converted to its diethylamide derivative.

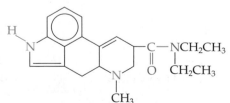

Lysergic acid diethylamide (LSD)

Clonidine, a drug used to treat high blood pressure, eases symptoms of withdrawal from nicotine, alcohol, and narcotics.

Several medical drugs are obtained from the ergot fungus. Because ergotamine shrinks blood vessels in the brain, it is used to treat migraine headaches. Ergonovine induces uterine contractions and can reduce bleeding after childbirth. Like LSD, both these compounds are lysergic acid amides.

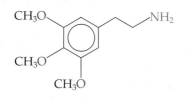

Mescaline

The potency of LSD is indicated by the small amount required for a person to experience its extraordinary effects. The usual dose is probably about 10–100 μg. No wonder Hofmann had a bad time with 250 mg. To give you an idea of how small 10 μg is, let's compare that amount of LSD with the amount of aspirin in one tablet—one aspirin tablet contains 325,000 μg of aspirin.

Plant Products

A variety of plants produce hallucinogenic compounds:

- Psilocybin, a hallucinogenic alkaloid of the tryptamine family, is found in several species of mushrooms, including *Psilocybe cubensis*. Its effects are similar to those of LSD but are of shorter duration.

- Mescaline (3,4,5-trimethoxyphenylethylamine) is a hallucinogenic drug related to phenylethylamine (compare to Figure 18.27). It is found in the peyote cactus, *Lophophora williamsii,* and other species. It is also made synthetically. Peyote is used in the religious ceremonies of some Indian tribes. The effects last for up to 12 hours.

- The herb *Salvia divinorum*, sometimes called diviner's sage, is a powerful psychoactive herb. Its main active component is a complicated compound called salvinorin A, a potent opioid receptor agonist with effect similar to those of dissociative anesthetics such as PCP and ketamine.

Club Drugs: Raves and Rapes

Club drugs are substances used by teenagers and young adults at all-night dance parties called raves. These drugs include 3,4-methylenedioxymethamphetamine (MDMA or Ecstasy), *gamma*-hydroxybutyrate ($HOCH_2CH_2CH_2COOH$; GHB), Rohypnol (flunitrazepam), ketamine, methamphetamine, and LSD. Use of such drugs can cause serious health problems, especially in combination with alcohol.

Especially troubling are some substances known as *date rape drugs*: Rohypnol, ketamine, and GHB have been used to facilitate sexual assault because they render the victim incapable of resisting. Actions occurring while under the influence of these drugs often cannot be remembered. The GHB analogs *gamma*-butyrolactone (GBL) and 1,4-butanediol ($HOCH_2CH_2CH_2CH_2OH$) have similar effects. (See Problems 55 and 56.)

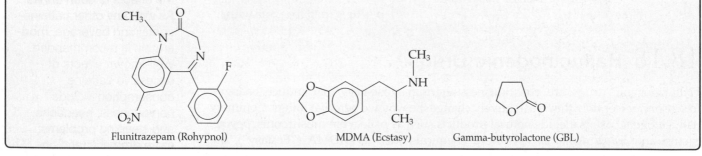

Flunitrazepam (Rohypnol)   MDMA (Ecstasy)   Gamma-butyrolactone (GBL)

More to Explore

Hogg R. C., and D. Bertrand. "What Genes Tell Us About Nicotine Addiction." *Science*, November 5, 2004, pp. 983–985.

Tetrahydrocannabinol
(THC)

Marijuana

The plant *Cannabis sativa,* commonly called marijuana, has long been useful. The plant stems yield tough fibers (hemp) for making ropes. *Cannabis* has been used as a drug in tribal religious rituals and also has a long history as a medicine, particularly in India. In the United States, marijuana is the most commonly used illegal drug.

The term **marijuana** refers to a preparation made by gathering the leaves and flower buds of the plant (Figure 18.28), which are generally dried and smoked. The principal active ingredient is tetrahydrocannabinol (THC). Although there are several active cannabinoids in marijuana, only one is shown here.

Marijuana plants vary considerably in THC potency, depending on their genetic variety. Wild plants native to the United States have a low THC content, usually about 0.1%, but some marijuana sold in North America now has a THC content approaching 6%.

The effects of marijuana are difficult to measure, partly because of the variable amount of THC in different samples. Smoking *cannabis* increases the pulse rate, distorts the sense of time, and impairs some complex motor functions. Other possible effects include a euphoric floating sensation, a feeling of anxiety, a heightened enjoyment of food, and a false impression of brilliance. Although studies have shown no mind-expanding effects, users sometimes experience hallucinations.

The long-term effects of marijuana, such as whether it causes psychotic symptoms that persist beyond temporary intoxication, are unclear. There is some evidence of increased risk of psychotic problems in people who have used *cannabis*, with the greatest risk—a 50–200% increase—in people who used *cannabis* most frequently. People who use marijuana heavily are often lazy, passive, and mentally sluggish, but it is difficult to prove that marijuana is the cause. Even if it is, the damage is less extensive than that caused by heavy use of alcohol. Some studies in animals show that high doses of THC decrease the output of gonadotropic hormones, lower testosterone levels, and hinder sperm production, motility, and viability. High THC doses also disrupt the ovulation cycle. Overall, though, the effect of *cannabis* use on fertility is uncertain.

Scientists have learned how THC acts in the brain to produce its various effects. When a person smokes marijuana, THC rapidly passes from the lungs into the bloodstream and thus to the brain, where it connects to specific sites called cannabinoid receptors on nerve cells. Some regions of the brain have many such receptors; other areas have few or none. The parts of the brain that influence pleasure, memory, thought, concentration, judgment, sensory and time perception, and regulate movement are rich in receptors.

As THC enters the brain, it activates the brain's reward system in the same way that food and drink do. Like nearly all drugs of abuse, THC causes a euphoric feeling by stimulating the release of dopamine.

The location of receptors in movement control centers in the brain explains the loss of coordination seen in those intoxicated by the drug. The presence of receptors in memory and cognition areas of the brain explains why marijuana users do poorly on tests. That few receptors are present in the brainstem where breathing and heartbeat are controlled is probably why it is hard to get a lethal dose of pure marijuana.

If THC binds to receptors in the brain, the brain must produce a THC-like substance. This substance is thought to be anandamide derived from arachidonic acid, which is the precursor of prostaglandins.

Marijuana has some legitimate medical uses. It reduces eye pressure in people who have glaucoma. If not treated, this increasing pressure eventually causes blindness. Marijuana also relieves the nausea that afflicts cancer patients undergoing radiation treatment and chemotherapy.

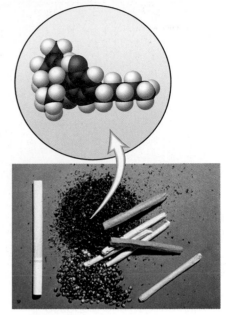

▲ **Figure 18.28** Prepared marijuana.

Anandamide

Self-Assessment Questions

1. Hallucinogenic drugs
   a. act as depressants
   b. act as stimulants
   c. cause changes in perception
   d. treat mental illness

2. A drug derived from a compound obtained from a fungus that grows on rye is
   a. anandamide   b. LSD   c. mescaline   d. psilocybin

3. LSD acts on the neurotransmitter
   a. acetylcholine   b. dopamine   c. GABA   d. serotonin

4. In the 1960s, the THC content of street marijuana was generally about 1%. Today the THC content is
   a. about the same
   b. as high as 2%
   c. as high as 6%
   d. as high as 10%

5. The THC-like substance produced in the brain is
   a. anandamide   b. an endorphin   c. LSD   d. morphine

Answers: 1, c; 2, b; 3, d; 4, c; 5, a

Chocolate contains small quantities of anandamide.

More to Explore
Nicoll, Roger A., and Bradley E. Alger. "The Brain's Own Marijuana." *Scientific American*, December 2004, pp. 68–75.

Voters in several states and cities have approved the medical use of marijuana and some other drugs. Still illegal under federal law, synthetic THC is legal by prescription for this use.

18.19 Drug Problems

The existence of illegal drugs causes enormous problems. Their marketing is covert, with sellers failing to report their income or pay taxes on it, making the illicit drug business enormously lucrative. Annual revenues for drug dealers amount to at least $400 billion, about half of which comes from the United States. Worldwide, people spend more money for illegal drugs than for food. So much unlawful wealth makes illegal drugs a dangerous business in which crimes are frequent, murders are common, and corrupt politicians abound. **Drug abuse** (using drugs for their intoxicating effects) is a serious problem, not only for the abusers but also for society in general.

The illegal drug user is the biggest loser. Street drugs are expensive, and most are addictive. Many addicted users steal in order to pay for drugs and eventually end up in jail; one study found that 16% of convicted jail inmates admitted that they committed their offense to get money for drugs. In the workplace 50 to 80% of all accidents and personal injuries are drug related. Drug users are absent from work over twice as often as nonusers, and they are five times as likely to file claims for compensation. They are a drain on their employers' income, and they often have trouble keeping a job.

A recurring problem is that illegal drugs are not always what they are supposed to be. Buyers simply have to trust the information sellers give them about the identity and quality of the products they buy. In fact, crime labs have generally found nearly two-thirds of all drugs (other than marijuana) brought in for analysis to be something other than what the dealers said they were.

Of course, there can be difficulties even with legal drugs that have been approved and tested. Problems range from faulty prescriptions written by physicians, to pharmacy errors in filling prescriptions, to patient mistakes in taking medications. Some drugs have undesirable side effects that are worse than the condition being treated. There are drugs (both prescription and over the counter) that have such similar names that they are easily confused. **Drug misuse** (for example, using penicillin, which has no effect on viruses, to treat a viral infection) is all too common. Such overuse of penicillin has led to strains of bacteria that are now immune to penicillin.

Chemistry has provided many drugs of enormous benefit to society: longer and healthier lives, relief from pain, and more. Sometimes drugs can create problems: addiction, dangerous side effects, even death. We must use them wisely.

A N S W E R

6. How should I dispose of old medicine?

The Substance Abuse and Mental Health Services Administration recommends that old medicine be mixed with coffee grounds, cat litter, or similar materials, then bagged and disposed of in the trash. This minimizes the possibility of abuse (especially if used cat litter is employed!). Alternatively, some communities are now providing periodic disposal services. There are environmental concerns about flushing medications down the sink or toilet; some medications may affect fish and wildlife even in very low doses.

Placebo and Nocebo Effects

An interesting phenomenon associated with the testing of drugs is the **placebo effect**. A placebo is an inactive substance given in the form of medication to a patient.

A common way to evaluate a new drug is to administer it to one group of patients and to give placebos to a similar "control" group. In the "double-blind" study of the sort most accepted by the medical community, neither the patients nor the doctors know who is receiving the real drug and who is receiving the placebo.

Sometimes people who think they are receiving a certain drug expect positive results and may actually experience such results, even though they have not been given the actual drug. In one study of a new tranquilizer, 40% of the patients who had been given placebos reported that they felt much better and rated the drug as highly effective. The psychological effect of a placebo can be very powerful.

The placebo effect confers health benefits from a treatment that should have no effect. Conversely, patients can experience a nocebo effect. A *nocebo* is a substance that is actually harmless but that produces harmful effects because the person involved thinks it harmful. The patient presumes a bad outcome, which then becomes a self-fulfilling prophecy. In layman's terms, this is just being "worried sick" or "scared to death." Perhaps the most extreme example is the voodoo hex, a notion so powerful that those who are "cursed" and believe in it may die of fright.

Self-Assessment Questions

1. Using an antibiotic to treat a cold is an example of
   - **a.** appropriate therapy
   - **b.** drug abuse
   - **c.** drug misuse
   - **d.** a placebo effect

2. Using the narcotic OxyContin for its intoxicating effect is an example of
   - **a.** appropriate recreation
   - **b.** drug abuse
   - **c.** drug misuse
   - **d.** legal intoxication

3. A placebo is a(n)
   - **a.** alcohol solution of an active drug
   - **b.** alcohol solution of an inactive drug
   - **c.** inactive drug that looks like real medication
   - **d.** platinum/cesium/boron drug

4. In a double-blind experiment, some patients are given the drug being tested and others are given
   - **a.** acupuncture
   - **b.** a homeopathic formulation
   - **c.** morphine
   - **d.** a placebo

Answers: 1, c; 2, b; 3, c; 4, d

More to Explore
Shermer, Michael. "What's the Harm? Alternative Medicine Is Not Everything to Gain and Nothing to Lose." *Scientific American*, December 2003, p. 50.

Critical Thinking Exercises

Apply knowledge that you have gained in this chapter and one or more of the FLaReS principles (Chapter 1) to evaluate the following statements or claims.

18.1 A television advertisement featured Dr. Robert Jarvik, inventor of the artificial heart, encouraging consumers to "talk to their doctor" about use of the cholesterol-lowering drug Lipitor. (The advertisement was pulled from the air in February 2008.)

18.2 A television advertisement urged consumers to ask their doctors if "the purple pill" might be right for them. The ads did not disclose what condition the pill was supposed to treat; they merely show young, attractive people blissfully enjoying a fantasy experience.

18.3 A nurse claims to be able to help patients by "therapeutic touch," a technique that she says allows her to sense "human energy fields" by moving her hands above the patient's body.

18.4 A television advertisement shows an actor taking a medicine that inhibits the production of stomach acid, then eating a large meal of rich, spicy food. Does the ad prove the claim of heartburn prevention?

18.5 A television advertisement shows an actor taking an antihistamine and then joyously walking though a grassy, flower-filled meadow without experiencing the usual allergy symptoms. What, if anything, does the ad prove about alleviation of allergies?

18.6 A Web article recommends *S*-adenosylmethionine (SAMe) for treatment of depression, arthritis, and liver disease.

18.7 The topical product HeadOn ("apply directly to the forehead") was introduced in television advertisements that claimed it relieved headaches, but the company, Miralus Healthcare, provided no reliable clinical data to support this claim. The product listed two active ingredients: white bryony (a poisonous vine) at 1 ppt and potassium dichromate (a known carcinogen) at 1 ppm. The company dropped all factual claims about the product after the Better Business Bureaus objected to the claim that HeadOn provided headache relief.

■ SUMMARY

Section 18.1—A **drug** is a substance that relieves pain, treats illness, or improves one's health. Many drugs are obtained from natural sources. Paul Ehrlich founded **chemotherapy**, based on the fact that some chemicals were more toxic to disease organisms than to human cells.

Section 18.2—Aspirin is a **nonsteroidal anti-inflammatory drug (NSAID)** to distinguish it from more powerful steroids. It is an **analgesic** (pain reliever), **antipyretic** (fever reducer), and **anti-inflammatory** (reduces inflammation). NSAIDs inhibit the production of prostaglandins that send pain messages to the brain and are also **anticoagulants** (inhibit clotting). Ibuprofen, ketoprofen, and naproxen are also NSAIDs. Acetaminophen is an analgesic and antipyretic but does not reduce inflammation. Many analgesic products are combinations of NSAIDs and other drugs.

Section 18.3—Cold symptoms are often treated with an **antihistamine** to relieve symptoms caused by an **allergen**. The allergen triggers the release of histamine, causing sneezing, itchy eyes, and runny nose. Colds are also treated with a cough suppressant, a nasal decongestant, and/or an expectorant to loosen mucus. None of these treatments cure a cold; they simply treat the symptoms.

Section 18.4—Sulfa drugs were the first antibacterial drugs; they mimic compounds essential for bacterial growth. An **antibiotic** is a soluble substance derived from mold or bacteria that inhibits microorganism growth. Penicillins prevent bacteria from forming their mucoprotein cell walls but do not harm animal cells, which do not have mucoprotein walls. More modern antibiotics include cephalosporins, tetracyclines, and fluoroquinolines. These **broad-spectrum antibiotics** are effective against a wide variety of bacteria. Unfortunately, antibiotic-resistant bacteria often develop after an antibiotic has been used for a while.

Section 18.5—Viral diseases cannot be cured with antibiotics. Some, such as measles, mumps, and smallpox, are usually prevented by vaccination, but vaccines are not available for all viral diseases. Herpes and **acquired immune deficiency syndrome (AIDS)** are viral infections that have no cure. Viruses are composed of nucleic acids—either DNA or RNA—and proteins. A **retrovirus** such as the AIDS virus is an RNA virus that synthesizes DNA in the host cell. Antiviral agents are used to treat some viral diseases. Drug design was often a hit-or-miss in process its early days, but now drugs can be designed to fit receptors.

Section 18.6—Most anticancer drugs are **antimetabolites**, which inhibit DNA synthesis, slowing the growth of cancer cells. Alkylating agents transfer alkyl groups to biomolecules, blocking their usual action. Nitrogen mustards, which arose from chemical warfare research, are alkylating agents.

Section 18.7—A **hormone** is a chemical messenger, sending signals for physiological changes in other parts of the body. A **prostaglandin** is a hormonelike lipid that acts as a mediator of hormone action. There are six primary prostaglandins. Some hormones and prostaglandins are **steroids**, having a characteristic four-ring structure. Cholesterol and cortisol are steroids, as are sex hormones. An **androgen** stimulates or controls masculine characteristics, and an **estrogen** controls female sexual functions. Estradiol, estrone, and progesterone are estrogens. A **progestin** is a steroid hormone that has the effect of progesterone.

Section 18.8—Birth control pills are mixtures of an estrogen and a progestin, which act by creating a state of pseudopregnancy, preventing ovulation and conception. Birth control pills are safer than childbirth for nonsmokers but much less safe for smokers.

Section 18.9—Diseases of the circulatory system include coronary artery disease, arrhythmia, hypertension, and congestive heart failure. Drugs for hypertension include diuretics, beta blockers, calcium channel blockers, and ACE inhibitors. Drugs for arrhythmia alter the flow of ions across membranes. Coronary artery disease treatment is often the same as for hypertension; however, organic nitro compounds are also used.

Section 18.10—**Psychotropic drugs** are drugs that affect the mind. They can be divided into three classes: **stimulant drugs** such as cocaine and amphetamines; **depressant drugs** such as alcohol, anesthetics, opiates, and tranquilizers; and **hallucinogenic drugs** such as LSD.

Section 18.11—The **neurons** (nerve cells) of the nervous system have tiny gaps called **synapses** between them. **Neurotransmitters** are chemicals released into the synapses during nerve transmissions that cause changes in the receiving cell.

Section 18.12—Mental illness is thought to involve brain amines such as norepinephrine (NE) and serotonin. NE **agonists** enhance or mimic the action of NE. NE **antagonists** block its action and slow down various processes. Agonist and antagonist molecules must have the proper shape. Many compounds exist as **stereoisomers**, with the same formula but different arrangements of atoms. Stereoisomers that are not superimposable on their mirror images are called **enantiomers** and usually have one or more **chiral carbon** atoms (have four different groups attached). Some brain amines appear to be affected by diet.

Section 18.13—An **anesthetic** is a depressant that causes lack of feeling or awareness. A **general anesthetic** produces unconsciousness and general insensitivity to pain. A local anesthetic causes loss of feeling in a part of the body. Modern inhalant anesthetics include fluorine-containing compounds. A variety of other drugs are used for anesthesia in surgical practice. Local anesthetics are amine-esters (example: procaine) or amine-amides (example: lidocaine). A dissociative anesthetic dissociates perception from sensation. Ketamine and PCP are dissociative anesthetics, but PCP is abused, causing illusions and hallucinations.

Section 18.14—Alcohol is the most common depressant used worldwide. It appears to disrupt two neurotransmitters that control muscle activity. Barbiturates such as pentobarbital and thiopental are used as sedatives and anesthetics. Alcohol and barbiturates together produce a **synergistic effect**, with the action of one enhancing the other; the result can be deadly.

Section 18.15—A **narcotic** relieves pain and produces stupor or anesthesia but is addictive. Opium from the Oriental poppy contains morphine, a natural narcotic. Codeine is made from morphine; it is less potent and less addictive. Codeine is often combined with other analgesics for pain relief, and it suppresses coughs. Heroin is a derivative of morphine that is much more potent and highly addictive. Several synthetic narcotics have been prepared. A morphine *agonist* has morphinelike action; a morphine *antagonist* blocks morphine receptors. The body produces morphine agonists called **endorphins** in response to extreme pain.

Section 18.16—Tranquilizers are depressants that some people use to relax. Some are addictive. Phenothiazines are one type of tranquilizer that act as dopamine antagonists and relieve symptoms of schizophrenia. Antidepressants are sometimes prescribed that enhance the effect of serotonin.

Section 18.17—**Amphetamines** are stimulants similar in structure to epinephrine. Amphetamine exists as enantiomers, one of which is much more potent than the other. Methamphetamine is even more potent than amphetamine and is easy to make. Ritalin is an amphetamine used to treat attention deficit hyperactivity disorder (ADHD). Cocaine is obtained from the coca plant. It is a local anesthetic as well as a powerful stimulant and is very addictive. Caffeine in coffee, tea, and soft drinks and nicotine in tobacco are common legal stimulants.

Section 18.18—Hallucinogenic drugs induce changes in sensory perception. The best-known hallucinogenic drugs are LSD and **marijuana**. Psilocybin from certain mushrooms and mescaline from peyote cactus are natural hallucinogens.

Section 18.19—**Drug abuse**—using drugs for their intoxicating effects—is a serious problem for abusers and society. **Drug misuse**, involving legal drugs used incorrectly or by mistake, is also a problem. Drugs must be tested properly to avoid the **placebo effect**, in which an inactive substance, given as medication, produces results in a patient for psychological reasons. The testing procedure for drugs is expensive and time-consuming.

■ REVIEW QUESTIONS

1. What is an antibiotic?
2. What is a hormone?
3. What is cortisol? Prednisone?
4. How do birth control pills work?
5. What are prostaglandins? How do prostaglandins differ from hormones?
6. Describe two medical uses of barbiturates and the relative dosage level of each. What are the hazards of long-term barbiturate use?
7. Name two dissociative anesthetics.
8. Describe a synergistic reaction involving drugs.
9. What is a narcotic?
10. Describe modern anesthesiology.
11. How may our mental state be related to our diet?
12. Are tranquilizers a cure for schizophrenia?
13. What are psychotropic drugs?
14. What are the effects of a hallucinogenic drug?

■ PROBLEMS

NSAIDs and Related Drugs

15. What is the chemical name for aspirin?
16. In what ways do aspirin and acetaminophen act similarly? Why is acetaminophen chosen over aspirin for the relief of pain associated with surgical procedures?
17. Why is aspirin chosen over acetaminophen for the treatment of arthritis?
18. What is a COX-2 inhibitor?
19. What two functional groups are present in an aspirin (acetylsalicylic acid) molecule?
20. What two functional groups are present in an acetaminophen molecule?

Antibacterial Drugs

21. What are penicillins?
22. Looking at the molecular models of sulfanilamide and *para*-aminobenzoic acid on page 535, write structural formulas for the two compounds.
23. Tetracyclines are broad-spectrum antibiotics. What does this mean?
24. What are fluoroquinolones?

Antiviral Drugs

25. Do antiviral drugs cure viral diseases? What is the best way to deal with viral diseases?
26. What are the three classes of antiviral drugs? How does each work?

Anticancer Drugs

27. List two major classes of anticancer drugs.
28. Do anticancer drugs sometimes cause cancer? Explain.

Hormones and Birth Control Drugs

29. What is an estrogen? What is an androgen?
30. What structural feature makes a birth control steroid effective orally?
31. How does mifepristone (RU-486) work?
32. What are emergency contraceptives? How do they work?

Drugs and the Mind

33. What is a psychotropic drug?
34. What is a hallucinogenic drug?

Anesthetics

35. Which of these anesthetics is dangerous because of flammability?
    a. diethyl ether
    b. halothane
    c. chloroform
36. For each of the following anesthetics, describe a disadvantage other than flammability associated with its use.
    a. nitrous oxide   b. halothane
    c. diethyl ether

Narcotics

37. How does codeine differ from morphine in its physiological effects? In what ways are the two drugs similar?
38. How does heroin differ from morphine in its physiological effects? In what ways are the two drugs similar?
39. How are endorphins related to **(a)** the anesthetic effect of acupuncture and **(b)** the absence of pain in a wounded soldier?
40. What is an agonist? What is an antagonist?

■ ADDITIONAL PROBLEMS

41. Which of the following compounds are classified as steroids?
    a. tristearin   b. cholesterol
    c. testosterone   d. prostaglandins
42. Review the "Structure–Function Relationships" essay (page 538). Then draw the structure for a fluoroquinoline drug, which should be similar to ciprofloxacin in action but does not have the large nitrogen-containing ring at position 7.
43. A patient was treated with the antibiotic ampicillin for a urinary tract infection. The daily dose was 40mg/kg divided into three doses per day. **(a)** What was the daily dose, in grams, for the 140-lb man? **(b)** The ampicillin capsules each contain 500 mg of active ingredient. How many capsules did the patient have to take every eight hours?
44. What structural feature is shared by all steroids?

45. What structural feature is shared by all tetracyclines?

46. Diphenhydramine (Benadryl®) has the chemical formula $C_{17}H_{21}NO$. Is it water soluble or fat soluble? Explain your answer.

47. Ephedrine hydrochloride has the chemical formula $C_{10}H_{16}NO^+Cl^-$. Is the compound water soluble or fat soluble? Explain your answer.

48. What is the basic structure common to all barbiturate molecules? How is this structure modified to change the properties of individual barbiturate drugs?

49. The dissociative anesthetics ketamine and PCP are classified as derivatives of cyclohexylamine. Give the structural formula for cyclohexylamine.

Problems 50–53 use the concept of an LD_{50} value. The LD_{50} of a substance is the dosage that would be lethal to 50% of a population of test animals.

50. When administered orally to rats, ketoprofen and naproxen have LD_{50} values of 101 mg/kg and 534 mg/kg, respectively. Which is more toxic? Explain.

51. When administered intravenously to rats, procaine and cocaine have LD_{50} values of 50 mg/kg and 17.5 mg/kg, respectively. Which is more toxic? Explain.

52. When administered orally to rats, aspirin has an LD_{50} value of 1500 mg/kg of body weight. If the toxicity is the same for humans, how many 325-mg tablets would kill the typical 10.0-kg child?

53. When administered intravenously to rats, the LD_{50} of procaine is 50 mg/kg body weight, and that of nicotine, 1.0 mg/kg body weight. Which drug is more toxic? What is the approximate lethal dose of each for a 50-kg human?

54. If the minimum lethal dose (MLD) of amphetamine is 5 mg/kg, what is the MLD for a 70-kg person? Can toxicity studies on animals always be extrapolated to humans?

55. As noted in Section 18.15, replacement of the phenolic OH group of the morphine molecule by a methoxy (OCH_3) group produces codeine. Draw the structure of the codeine molecule.

56. As noted in Section 18.15, conversion of both the OH groups of the morphine molecule to acetate esters produces heroin. Draw the structure of the heroin molecule.

57. As noted in Section 18.15, oxycodone is related to codeine, differing only in that the hydroxyl group of codeine has been oxidized to a carbonyl group (as in ketone synthesis, Chapter 9). Draw the structure of the oxycodone molecule.

58. Nalorphine, a morphine antagonist, is related to morphine in that the methyl group on the N atom of morphine is replaced by a propenyl group ($—CH_2CH=CH_2$). Draw the structure of the nalorphine molecule.

Problems 59 and 60 are based on the box feature "Club Drugs." You may need to review some topics in Chapter 9.

59. What kind of compound is the GHB analog *gamma*-butyrolactone (GBL)? What kind of reaction would convert GBL to GHB?

60. What kind of compound is the GHB analog 1,4-butanediol? What kind of reaction would convert 1,4-butanediol to GHB?

61. What compound does the following formula represent? To which class of anticancer drugs does it belong?

62. What compound does the following formula represent? To which class of anticancer drugs does it belong?

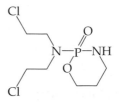

63. For the compound shown here, how will removal of the *para*-hydroxyl group affect the solubility of the molecule in (a) water and (b) fat?

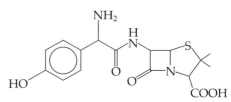

64. Which NSAID does the following formula represent? To what aromatic hydrocarbon is it related? Name the two oxygen-containing functional groups.

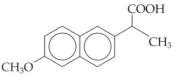

65. Chloramphenicol (Exercise 1.1A) is a broad-spectrum antibiotic that acts by inhibiting bacterial protein synthesis. It is soluble in water to the extent of 2.5 mg/mL. How would conversion of chloramphenicol (R = —H) to its (a) palmitate ester [R = —$CO(CH_2)_{14}CH_3$] and to its (b) sodium succinate (R = —$COCH_2CH_2COONa$) derivative affect its solubility? Explain.

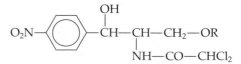

66. What type of drug is shown here? Name the three types of oxygen-containing functional groups present.

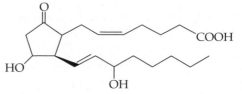

67. What type of drug is megestrol acetate (below), used for the treatment of some cancers?

a. phencyclidine
b. prostaglandin
c. steroid
d. tetracycline

■ COLLABORATIVE GROUP PROJECTS

Prepare a PowerPoint, poster, or other presentation (as directed by your instructor) for presentation to the class.

68. Prepare a brief report on one of the following combination products. Make a list of the ingredients in each and give the chemical structure, medical use, toxicity, and (if possible) side effects of each.

a. Advil® Cold & Sinus b. Aleve® Cold & Sinus
c. Dimetapp® Cold & Flu d. Theraflu®
e. Excedrin PM® f. Alka-Seltzer® Plus
g. DayQuil® h. NyQuil®
i. Excedrin® Extra Strength

For items 69–74, follow these general directions: Give the chemical structure, medical use, toxicity, and (if possible) side effects of each.

69. Anticancer drugs:
a. bleomycin b. fludarabine
c. hydroxyurea d. levamisole
e. procarbazine f. tamoxifen
g. vinblastine

70. Antihistamines:
a. azatadine b. fexofenadine
c. cyproheptadine d. cetirizine
e. dexchlorpheniramine f. loratadine
g. promethazine h. tripelennamine

71. Antibiotics:
a. vancomycin b. gramicidin
c. erythromycin d. ciprofloxacin
e. trimethoprim f. cefoxitin
g. cefoperazone h. azithromycin

72. Antiviral drugs:
a. delavirdine b. famciclovir
c. ganciclovir d. indinavir
e. ribavirin f. rimantadine
g. ritonavir h. valacyclovir

73. Opioid narcotics:
a. propoxyphene b. oxycodone
c. butorphanol d. oxymorphone
e. hydromorphone f. MPTP
g. fentanyl h. levorphanol

74. Antidepressants:
a. sertraline b. paroxetine
c. citalopram d. desipramine
e. isocarboxazid f. phenelzine
g. fluvoxamine h. nortriptyline

75. Four major drug categories for lowering blood pressure are listed on pages 552–553. Give the chemical structure, toxicity, and side effects of each of the examples listed.

76. Do a cost analysis of five brands of plain aspirin. Calculate the cost per gram of each. Compare the cost per gram of an extra-strength aspirin formulation with that of plain aspirin.

77. Discuss some of the problems involved in proving a drug safe or proving it harmful. Which is easier? Why?

78. Search the Internet for information on the use of hormone replacement therapy (HRT) in postmenopausal women. List the risks and benefits of HRT.

79. Do an Internet search to learn more about drug trials so that you can answer the following questions.
a. Why might a person who is gravely ill not want to participate in a placebo-controlled drug study?
b. What are the advantages and disadvantages of participating in such a drug study?
c. Why is it important that studies use placebos as well as the test drug?
d. Sometimes the control group receives the standard treatment for the disease instead of a placebo. Why does this happen?

80. Search the Web for information on any new drugs for treating one of the following diseases.
a. pneumonia
b. arthritis
c. cancer
d. diabetes

81. Use the *The Merck Index* or similar reference to look up the toxicities of cocaine, procaine, lidocaine, mepivicaine, and bupivicaine. Is it always accurate to compare toxicities in different animals and extrapolate animal toxicities to humans? Does the method of administration have an effect on the observed toxicity?

82. Prepare a brief report on one of the following neurotransmitters. Describe its function and site of action.
a. GABA b. glutamate
c. aspartate d. glycine
e. histamine f. adenosine

83. Prepare a brief report on thalidomide: its structure, original intended use, problems that arose, and proposed new uses. Identify the chiral carbon atom(s) on thalidomide that were the source of the problems.

Review of Measurement and Mathematics

Accurate measurements are essential to science. As we noted in Chapter 1, measurements can be made in a variety of units, but most scientists use the International System of Measurement (SI). The data they record often has to be converted from one kind of unit to another and otherwise manipulated mathematically. In this appendix, we extend the discussion of metric measurement and review some of the mathematics that you may find useful in this course.

A.1 THE INTERNATIONAL SYSTEM OF MEASUREMENT

As noted in Chapter 1, the standard SI unit of length is the *meter*. This distance was once defined as 0.0000001 of Earth's quadrant—that is, of the distance from the North Pole to the equator measured along a meridian. The quadrant proved difficult to measure accurately, and today the meter is defined precisely as the distance light travels in a vacuum during 1/299,792,458 of a second.

The SI unit of mass is the *kilogram* (1 kg = 1000 g). It is based on a standard platinum–iridium cylinder kept at the International Bureau of Weights and Measures. The *gram* is a more convenient unit for many chemical operations.

The derived SI unit of volume is the *cubic meter*. The units more frequently employed in chemistry, however, are the *liter* (1 L = 0.001 m^3) and the *milliliter* (1 mL = 0.001 L). Other SI units of length, mass, and volume are derived from these basic units. Table A.1 lists some metric units of length, mass, and volume and illustrates the use of prefixes.

A.2 EXPONENTIAL (SCIENTIFIC) NOTATION

Scientists often use numbers that are inconceivably large or small. For example, light travels at about 300,000,000 m/s. There are 602,200,000,000,000,000,000,000 carbon atoms in 12.01 g of carbon. On the small side, the diameter of an atom is about 0.0000000001 m, and the diameter of an atomic nucleus is about 0.000000000000001 m. We find it difficult to keep track of the zeros in such quantities. Scientists find it convenient to express such numbers in exponential notation.

Table A.1 Some Metric Units of Length, Mass, and Volume
Length
1 kilometer (km) = 1000 meters (m)
1 meter (m) = 100 centimeters (cm)
1 centimeter (cm) = 10 millimeters (mm)
1 millimeter (mm) = 1000 micrometers (μm)
Mass
1 kilogram (kg) = 1000 grams (g)
1 gram (g) = 1000 milligrams (mg)
1 milligram (mg) = 1000 micrograms (μg)
Volume
1 liter (L) = 1000 milliliters (mL)
1 milliliter (mL) = 1000 microliters (μL)
1 milliliter (mL) = 1 cubic centimeter (cm^3)

A number is in *exponential notation* when it is written as the product of a coefficient and a power of 10. (Such a number is called *scientific notation* when the coefficient has a value between 1 and 10.) Two examples are

$$4.18 \times 10^3 \text{ and } 6.57 \times 10^{-4}$$

Expressing numbers in exponential form generally serves two purposes.

1. We can write very large or very small numbers in a minimum of printed space and with a reduced chance of typographical error.

2. We can convey explicit information about the precision of measurements: The number of significant figures (Section A.4) in a measured quantity is stated unambiguously.

In the expression 10^n, n is the exponent of 10, and the number 10 is said to be raised to the nth power. If n is a *positive quantity*, 10^n has a value *greater than 1*. If n is a *negative quantity*, 10^n has a value *less than 1*. We are particularly interested in cases where n is an integer. For example,

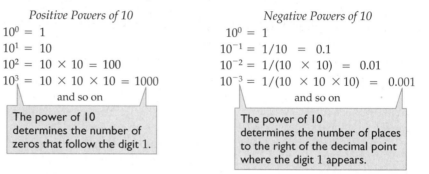

We express 612,000 in scientific notation as

$$612{,}000 = 6.12 \times 100{,}000 = 6.12 \times 10^5$$

We express 0.000505 in scientific notation as

$$0.000505 = 5.05 \times 0.0001 = 5.05 \times 10^{-4}$$

We can use a more direct approach to converting numbers to scientific notation.

- Count the number of places a decimal point must be moved to produce a coefficient having a value between 1 and 10.

- The number of places counted then becomes the power of 10.

- The power of 10 is *positive* if the decimal point is moved to the *left*.

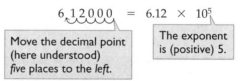

- The power of 10 is *negative* if the decimal point is moved to the *right*.

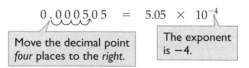

To convert a number from exponential form to the conventional form, move the decimal point in the opposite direction.

$$3.75 \times 10^6 = 3.750000$$

| The exponent is 6. | Move the decimal point *six places* to the *right*. |

$$7.91 \times 10^{-7} = 0.000000791$$

| The exponent is −7. | Move the decimal point *seven places* to the *left*. |

It is easy to handle exponential numbers on most calculators. A typical procedure is to enter the number, followed by the key [EE] or [EXP], then enter the exponent of 10. The keystrokes required for the number 2.85×10^7 are [2] [.] [8] [5] [EE] [7].

For the number 1.67×10^{-5}, the keystrokes are [1] [.] [6] [7] [EXP] [5] [±]; the last keystroke changes the sign of the exponent. Many calculators can be set to convert all numbers and calculated results to the exponential form, regardless of the form in which the numbers are entered. Most scientific and graphing calculators can also be set to display a fixed number of digits in the results.

Addition and Subtraction

To add or subtract numbers in exponential notation using a scientific or graphing calculator, simply enter the numbers as usual and perform the desired operations. However, to add or subtract numbers *by hand* in exponential notation it is necessary to express each quantity as the *same power of 10*. In calculations, this treats the power of 10 in the same way as a unit—it is simply "carried along." In the following, each quantity is expressed with the power 10^{-3}.

$$(3.22 \times 10^{-3}) + (7.3 \times 10^{-4}) - (4.8 \times 10^{-4}) =$$

$$(3.22 \times 10^{-3}) + (0.73 \times 10^{-3}) - (0.48 \times 10^{-3}) = (3.22 + 0.73 - 0.48) \times 10^{-3}$$

$$\ddagger = 3.47 \times 10^{-3}$$

Multiplication and Division

To multiply numbers by hand expressed in exponential form, *multiply* all coefficients to obtain the coefficient of the result and *add* all exponents to obtain the power of 10 in the result.

Rewrite in exponential form.

$$0.0803 \times 0.0077 \times 455 = (8.03 \times 10^{-2}) \times (7.7 \times 10^{-3}) \times (4.55 \times 10^2)$$

Multiply the coefficients.   Add the exponents.

$$= (8.03 \times 7.7 \times 4.55) \times 10^{(-2-3+2)}$$

$$= (2.8 \times 10^2) \times 10^{-3} = 2.8 \times 10^{-1}$$

Generally, most calculators perform these operations automatically, and no intermediate results need be recorded.

To divide two numbers in exponential form, *divide* the coefficients to obtain the coefficient of the result and *subtract* the exponent in the denominator from the exponent in the numerator to obtain the power of 10. In the following example, multiplication and division are combined. First, the rule for multiplication is applied to the numerator and to the denominator, and then the rule for division is used.

Rewrite in exponential form.

$$\frac{0.015 \times 0.0088 \times 822}{0.092 \times 0.48} = \frac{(1.5 \times 10^{-2})(8.8 \times 10^{-3})(8.22 \times 10^2)}{(9.2 \times 10^{-2})(4.8 \times 10^{-1})}$$

Apply the rule for multiplication to the numerator and denominator.

Apply the rule for division.

$$= \frac{1.1 \times 10^{-1}}{4.4 \times 10^{-2}} = 0.25 \times 10^{-1-(-2)} = 0.25 \times 10^1 = 2.5$$

Raising a Number to a Power and Extracting the Root of an Exponential Number

To raise an exponential number to a given power, raise the coefficient to that power and multiply the exponent by that power. For example, we can cube a number (that is, raise it to the *third* power) in the following manner.

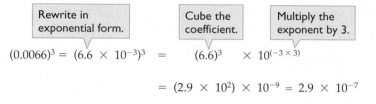

$$(0.0066)^3 = (6.6 \times 10^{-3})^3 = (6.6)^3 \times 10^{(-3 \times 3)}$$

$$= (2.9 \times 10^2) \times 10^{-9} = 2.9 \times 10^{-7}$$

To extract the root of an exponential number, we raise the number to a fractional power—one-half power for a square root, one-third power for a cube root, and so on. Most calculators have keys designed for extracting square roots and cube roots. Thus, to extract the square root of 1.57×10^5, enter the number 1.57×10^5 into a calculator, and use the $\left[\sqrt{}\right]$ key.

$$\sqrt{1.57 \times 10^{-5}} = 3.96 \times 10^{-3}$$

Some calculators allow you to extract roots by keying the root in as a fractional exponent. For example, we can take a cube root as follows.

$$(2.75 \times 10^{-9})^{1/3} = 1.40 \times 10^{-3}$$

Self-Assessment Questions

For items 1–4, express the number in scientific notation, and then select the single best answer.

1. 35,400
   a. 3.54×10^{-4} b. 35.4×10^{-3} c. 3.54×10^4
   d. 35.4×10^4 e. 35.4×10^5

2. 43,600,000
   a. 436×10^4 b. 43.6×10^{-6} c. 4.36×10^7
   d. 4.36×10^{-7} e. 43.6×10^{-7}

3. 0.000000000000312
   a. 3.12×10^{-15} b. 3.12×10^{-13} c. 3.12×10^{-12}
   d. 3.12×10^{12} e. 3.12×10^{13}

4. 0.000438
   a. 4.38×10^{-3} b. 4.38×10^{-4} c. 4.38×10^3
   d. 4.38×10^4 e. 4.38×10^6

For items 5–6, express the number in decimal (ordinary) notation, and then select the single best answer.

5. 7.1×10^4
   a. 0.00071   b. 0.000071   c. 71,000
   d. 710,000   e. 7,100,000

6. 2.83×10^{-4}
   a. 0.000283   b. 0.00283   c. 2830
   d. 28,300   e. 283,000

For items 7–14, perform the indicated operation, and then select the single best answer.

7. $(3.0 \times 10^4) \times (2.1 \times 10^5) = ?$
   a. 5.1×10^9 b. 6.3×10^1 c. 3.21×10^4
   d. 6.3×10^9 e. 6.3×10^{20}

8. $(6.27 \times 10^{-5}) \times (4.12 \times 10^{-12}) = ?$

    a. 2.58×10^{-17} **b.** 2.58×10^{-16} **c.** 2.58×10^{-59}

    d. 2.58×10^{-60} **e.** 2.58×10^{59}

9. $\dfrac{4.33 \times 10^{-7}}{7.61 \times 10^{22}} = ?$

    a. 5.69×10^{-30} **b.** 5.69×10^{-15} **c.** 5.69×10^{15}

    d. 1.76×10^{29} **e.** 5.69×10^{30}

10. $\dfrac{9.37 \times 10^{9}}{3.79 \times 10^{27}} = ?$

    a. 2.47×10^{-36} **b.** 2.47×10^{-35} **c.** 2.47×10^{-18}

    d. 2.47×10^{18} **e.** 2.47×10^{36}

11. $\dfrac{2.21 \times 10^{5}}{9.80 \times 10^{-7}} = ?$

    a. 2.26×10^{-12} **b.** 2.26×10^{-11} **c.** 2.26×10^{11}

    d. 2.26×10^{12} **e.** 4.43×10^{12}

12. $\dfrac{4.60 \times 10^{-12}}{2.17 \times 10^{3}} = ?$

    a. 2.12×10^{-15} **b.** 2.12×10^{-14} **c.** 2.12×10^{14}

    d. 4.72×10^{14} **e.** 2.12×10^{15}

13. $(2.19 \times 10^{-6})^2 = ?$

    a. 4.80×10^{-12} **b.** 4.80×10^{-6} **c.** 4.80×10^{6}

    d. 1.31×10^{11} **e.** 4.80×10^{12}

14. $\sqrt{1.74 \times 10^{-7}} = ?$

    a. 1.32×10^{-7} **b.** 4.17×10^{-7} **c.** 4.17×10^{-4}

    d. 1.32×10^{-4} **e.** 4.17×10^{-3}

Answers: 1, c; 2, c; 3, b; 4, b; 5, c; 6, a; 7, d; 8, b; 9, a; 10, a; 11, c; 12, a; 13, a; 14, c

A.3 UNIT CONVERSIONS

We discuss conversions within the metric system in Chapter 1. Because we live in a world in which almost all countries except the United States use metric measurements, we sometimes need to convert between common and metric units. Suppose that an American applies for a job in another country and the job application asks for her height in centimeters. She knows her height in customary units is 66.0 in. Her actual height is the same, whether we express it in inches, feet, centimeters, or millimeters. Thus, when we measure something in one unit and then convert it to another one, we must not change the measured quantity in any fundamental way. We can use equivalencies (given here to three significant figures; Section A.4) such as those in Table A.2 to derive conversion factors.

In mathematics, multiplying a quantity by 1 does not change its value. We can therefore use a factor equivalent to 1 to convert between inches and centimeters. We find our factor in the definition of the inch in Table A.2.

$$1 \text{ in.} = 2.54 \text{ cm}$$

Notice that if we divide both sides of this equation by 1 in., we obtain a ratio of quantities that is equal to 1.

$$1 = \frac{1 \text{ in.}}{1 \text{ in.}} = \frac{2.54 \text{ cm}}{1 \text{ in.}}$$

If we divide both sides of the equation by 2.54 cm, we also obtain a ratio of quantities that is equal to 1.

$$\frac{1 \text{ in.}}{2.54 \text{ cm}} = \frac{2.54 \text{ cm}}{2.54 \text{ cm}} = 1$$

The two ratios, one shown in red and the other in blue, are conversion factors. A *conversion factor* is a ratio of terms, equivalent to the number 1, used to change the unit in which a quantity is expressed. We call this process the *unit conversion method* of problem solving. Because we set up problems by examining the *dimensions* associated with the given quantity, those associated with the desired result, and those needed in the conversion factors, the process is sometimes called *dimensional analysis*.

What happens when we multiply a known quantity by a conversion factor? The original unit cancels out and is replaced by the desired unit. Thus, our general approach to using conversion factors is

Desired quantity and unit = given quantity and unit × conversion factors

Table A.2 Some Conversions between Common and Metric Units		
Length		
1 mile (mi) = 1.61 kilometers (km)		
1 yard (yd) = 0.914 meter (m)		
1 inch (in.) = 2.54 centimeters (cm)[*]		
Mass		
1 pound (lb) = 454 grams (g)		
1 ounce (oz) = 28.4 grams (g)		
1 pound (lb) = 0.454 kilogram (kg)		
Volume		
1 U.S. quart (qt) = 0.946 liter (L)		
1 U.S. pint (pt) = 0.473 liter (L)		
1 fluid ounce (fl oz) = 29.6 milliliters (mL)		
1 gallon (gal) = 3.78 liters (L)		

[*]This conversion is exact. The other conversions are available with more significant digits if needed; see any edition of the *CRC Handbook of Chemistry and Physics, Chemical Rubber Company.*

Now let's return to the question about the woman's height, a measured quantity of 66.0 in. To get an answer in centimeters, we must use the appropriate conversion factor (in blue). Note that the desired unit (cm) is in the numerator and the unit to be replaced (in.) is in the denominator. Thus we can cancel the unit in. so that only the unit cm remains.

$$66.0 \cancel{\text{ in.}} \times \frac{2.54 \text{ cm}}{1 \cancel{\text{ in.}}} = 168 \text{ cm (rounded off from 167.64)}$$

You can see why the other conversion factor (in red) won't work. No units cancel. Instead, we get a nonsensical unit.

$$66.0 \text{ in.} \times \frac{1 \text{ in.}}{2.54 \text{ cm}} = \frac{26.0 \text{ in.}^2}{\text{cm}}$$

Following are some examples and exercises to give you some practice in this method of problem solving.

EXAMPLE A.1 Unit Conversions

a. Convert 0.742 kg to grams.
b. Convert 0.615 lb to ounces.
c. Convert 135 lb to kilograms.

Solution

a. This is a conversion from one metric unit to another. We simply use our knowledge of prefixes to convert from kilograms to grams.

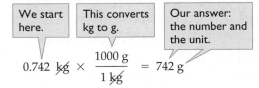

$$0.742 \text{ k\!\!\!/g} \times \frac{1000 \text{ g}}{1 \text{ k\!\!\!/g}} = 742 \text{ g}$$

b. This is a conversion from one common unit to another. Here we use the fact that 1 lb = 16 oz to convert from pounds to ounces. Then we proceed as in part (a), arranging the conversion factor to cancel the unit lb.

$$0.165 \text{ l\!\!/b} \times \frac{16 \text{ oz}}{1 \text{ l\!\!/b}} = 9.48 \text{ oz}$$

c. This is a conversion from a common unit to a metric unit. We need data from Table A.2 to convert from pounds to kilograms. Then we proceed as in part (b), arranging the conversion factor to cancel the unit lb.

$$135 \text{ l\!\!/b} \times \frac{0.454 \text{ kg}}{1 \text{ l\!\!/b}} = 61.2 \text{ kg}$$

(We give our answers with the proper number of significant figures. If you do not understand significant figures and wish to do so, they are discussed in Section A.4. In this text we simply use three significant figures for most calculations.)

■ EXERCISE A.1

a. Convert 16.3 mg to grams.   **b.** Convert 24.5 oz to pounds.
c. Convert 12.5 fl oz to milliliters.

Quite often, to get the desired unit, we must use more than one conversion factor. We can do so by arranging all the necessary conversion factors in a single setup that yields the final answer in the desired unit.

EXAMPLE A.2   Unit Conversions

What is the length, in millimeters, of a 3.25-ft piece of tubing?

Solution

No relationship between feet and millimeters is given in Table A.2, and so we need more than one conversion factor. We can think of the problem as a series of three conversions.

1. Use the fact that 1 ft = 12 in. to convert from feet to inches.
2. Use data from Table A.2 to convert from inches to centimeters.
3. Use our knowledge of prefixes to convert from centimeters to millimeters.

We could solve this problem in three distinct steps by making one conversion in each step, but it is just as easy to combine three conversion factors into a single setup. Then we proceed as indicated below.

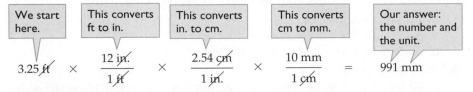

■ EXERCISE A.2

Carry out the following conversions.
a. 90.3 mm to meters   **b.** 729.9 ft to kilometers
c. 1.17 gal to fluid ounces

Sometimes we need to convert two or more units in the measured quantity. We can do this by arranging all the necessary conversion factors in a single setup so that the starting units cancel and the conversions yield the final answer in the desired units.

EXAMPLE A.3   Unit Conversions

A saline solution has 1.00 lb of salt in 1.00 gal of solution. Calculate the quantities in grams of salt per liter of solution.

Solution

First let's identify the measured quantities. They can be expressed in the form of a ratio of mass of salt in pounds to a volume in gallons.

$$\frac{1.00 \text{ lb (salt)}}{1.00 \text{ gal (solution)}}$$

This ratio must be converted to one expressed in grams per liter. We must convert from pounds to grams in the numerator and from gallons to liters in the denominator. The following set of equivalent values from Table A.2 can be used to formulate conversion factors.

$$1 \text{ lb} = 453.6 \text{ g} \qquad 1 \text{ gal} = 3.785 \text{ L}$$

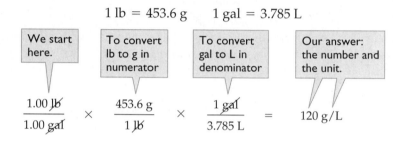

| We start here. | To convert lb to g in numerator | To convert gal to L in denominator | Our answer: the number and the unit. |

$$\frac{1.00 \text{ lb}}{1.00 \text{ gal}} \times \frac{453.6 \text{ g}}{1 \text{ lb}} \times \frac{1 \text{ gal}}{3.785 \text{ L}} = 120 \text{ g/L}$$

Note that we could also do the conversions in the numerator and denominator separately and then divide the numerator by the denominator.

$$\text{Numerator:} \qquad 1.00 \text{ lb} \times \frac{453.6 \text{ g}}{1 \text{ lb}} = 454 \text{ g}$$

$$\text{Denominator: } 1.00 \text{ gal} \times \frac{3.785 \text{ L}}{1 \text{ gal}} = 3.785 \text{ L}$$

$$\text{Division:} \qquad \frac{454 \text{ g}}{3.785 \text{ L}} = 120 \text{ g/L}$$

Both methods give the same answer to three significant figures.

■ EXERCISE A.3

Carry out the following conversions.
a. 88.0 km/h to meters per second   **b.** 1.22 ft/s to kilometers per hour
c. 4.07 g/L to ounces per quart

Self-Assessment Questions

For items 1–5, perform the indicated conversion, and then select the single best answer. You may need data from Table A.2 for some of the items.

1. 675 μs to seconds
   a. 6.45×10^{-6} s   **b.** 6.45×10^{-4} s
   c. 6.45×10^{-2} s   **d.** 6.45×10^{8} s

2. 1445 oz to grams
   a. 50.9 g   **b.** 90.3 g   **c.** 144.5 g   **d.** 4.10×10^{4} g

3. 1.24×10^{5} mm to feet
   a. 10.3 ft   **b.** 103 ft   **c.** 407 ft
   d. 4.88×10^{3} ft   **e.** 1.03×10^{4} ft

4. 12.0 fl oz to milliliters

   a. 0.405 mL   **b.** 240 mL   **c.** 341 mL

   d. 355 mL   **e.** 3.55×10^5 mL

5. 343 m/s to miles per hour

   a. 0.0592 mi/h   **b.** 12.8 mi/h   **c.** 767 mi/h

   d. 1220 mi/h   **e.** 1990 mi/h

Answers: 1, b; 2, d; 3, c; 4, d; 5, c

A.4 PRECISION, ACCURACY, AND SIGNIFICANT FIGURES

Counting can give exact numbers. For example, we can count exactly 24 students in a room. Measurements, on the other hand, are subject to error. One source of error is in the measuring instruments themselves. For example, an incorrectly calibrated thermometer may consistently yield a result that is 0.2 °C too low. Other errors may result from the experimenter's lack of skill or care in using measuring instruments. But even the most careful measurement will have some uncertainty associated with it. For example, a meter stick can be used to measure to the nearest millimeter or so. But there is no way for a meter stick to give a measurement that is correct to the nearest 0.001 mm.

Table A.3 Five Measurements of a Person's Height	
Student	**Height, m**
1	1.827
2	1.824
3	1.826
4	1.828
5	1.829
Average	1.827

Precision and Accuracy

Suppose you were one of five students asked to measure a person's height using a meter stick marked off in millimeters. The five measurements are recorded in Table A.3. The *precision* of a set of measurements refers to how closely individual measurements agree with one another. We say the precision is good if each of the measurements is close to the average and poor if there is a wide deviation from the average value.

How would you describe the precision of the data in Table A.3? Examine the individual data, note the average value, and determine how much the individual data differ from the average. Because the maximum deviation from the average value is 0.003 m, the precision is quite good.

The *accuracy* of a set of measurements refers to the closeness of the average of the set to the "correct" or most probable value. Measurements of high precision are more likely to be accurate than are those of poor precision, but even highly precise measurements are sometimes inaccurate. For example, what if the meter sticks used to obtain the data in Table A.3 were actually 1005 mm long but still had 1000-mm markings? The accuracy of the measurements would be rather poor, even though the precision would remain good.

Sampling Errors

No matter how accurate an analysis, it will not mean much unless it is performed on valid, representative samples. Consider determining the level of glucose in the blood of a patient. High glucose levels are associated with diabetes, a debilitating disease.

(a) Low accuracy   (b) Low accuracy
    Low precision   High precision

(c) High accuracy   (d) High accuracy
    Low precision   High precision

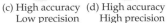

▲ Comparing precision and accuracy: a dartboard analogy. (a) The darts are both scattered (low precision) and off-center (low accuracy). (b) The darts are in a tight cluster (high precision) but still off-center (low accuracy). (c) The darts are somewhat scattered (low precision) but evenly distributed about the center (high accuracy). (d) The darts are in a tight cluster (high precision) and well centered (high accuracy).

Results vary, depending on several factors such as the time of day and what and when the person last ate. Glucose levels are much higher soon after a meal high in sugars. They also depend on other factors, such as stress. Medical doctors usually take blood for analysis after a night of fasting. These fasting glucose levels tend to be more reliable, but physicians often repeat the analysis when high or otherwise suspicious results are obtained. Therefore, there is no one true value for the level of glucose in a person's blood. Repeated samplings provide an average level. Similar results from several measurements give much more confidence in the findings. For example, a diagnosis of diabetes is usually based on two consecutive fasting blood glucose levels above 120 mg/dL.

Significant Figures

Look again at Table A.3. Notice that the five measurements of height agree in the first three digits (1.82); they differ only in the fourth digit. We say that the fourth digit is uncertain. All digits known with certainty, plus the first uncertain one, are called *significant figures*. The precision of a measurement is reflected in the number of significant figures—the more significant figures, the more precise the measurement. The measurements in Table A.3 have four significant figures. In other words, we are quite sure that the person's height is between 1.82 m and 1.83 m. Our best estimate of the average value, including the uncertain digit, is 1.827 m.

The number 1.827 has four digits; we say it has four significant figures. In any properly reported measurement, all nonzero digits are significant. Zeros, however, may or may not be significant because they can be used in two ways: as a part of the measured value or to position a decimal point.

- Zeros between two other significant digits are significant. *Examples*: 4807 (four significant figures); 70.004 (five).
- We use a lone zero preceding a decimal point for aesthetic purposes; it is not significant. *Example*: 0.352 (three significant figures).
- Zeros that precede the first nonzero digit are also not significant. *Examples*: 0.000819 (three significant figures); 0.03307 (four).
- Zeros at the end of a number are significant if they are to the right of the decimal point. *Examples*: 0.2000 (four significant figures); 0.050120 (five).

We can summarize these four situations with a general rule: When we read a number from left to right, all the digits starting with the first nonzero digit are significant. Numbers without a decimal point that end in zeros are a special case, however.

- Zeros at the end of a number may or may not be significant if the number is written without a decimal point. *Example*: 700. We do not know whether the number 700 was measured to the nearest unit, ten, or hundred. To avoid this confusion, we can use exponential notation (Section A.2). In exponential notation, 700 is recorded as 7×10^2 or 7.0×10^2 or 7.00×10^2 to indicate one, two, or three significant figures, respectively. The only significant digits are those in the coefficient, not in the power of 10.

We use significant figures only with measurements—quantities subject to error. The concept does not apply to a quantity that is

1. inherently an integer, such as 3 sides to a triangle or 12 items in a dozen
2. inherently a fraction, such as the radius of a circle equals $\frac{1}{2}$ the diameter
3. obtained by an accurate count, such as 18 students in a class
4. a defined quantity, such as 1 km = 1000 m

In these contexts, the numbers 3, 12, $\frac{1}{2}$, 18, and 1000 can have as many significant figures as we want. More properly, we say that each is an exact value.

EXAMPLE A.4 Significant Figures

In everyday life, common units are often used with metric units in parentheses (or vice versa), but significant figures are not always considered. Which of the following have followed proper significant figure usage?

a. A popular science magazine reports that "*T. rex...* gained as much as 4.6 pounds (2 kilograms) daily."

b. A urinal is rated at "1.0 gpf (gallons per flush) or 3.8 lpf (liters per flush)."

Solution

a. The quantity "4.6 pounds" has two significant figures but "2 kilograms" has only one; significant figure rules were not followed.

b. Each quantity has two significant figures; significant figure rules were followed.

▪ **EXERCISE A.4**

For which of the following is significant figure convention followed?

a. A bag of sugar says "5 lb (2.26 kg)."

b. A roll of packaging tape is labeled "1.88 in. × 54.6 yd (48 mm × 50 m)."

Significant Figures in Calculations: Multiplication and Division

If we measure a sheet of notepaper and find it to be 14.5 cm wide and 21.7 cm long, we can find the area of the paper by multiplying the two quantities. A calculator gives the answer as 314.65. Can we conclude that the area is 314.65 cm²? That is, can we know the area to the nearest hundredth of a square centimeter when we know the width and length only to the nearest tenth of a centimeter? It just doesn't seem reasonable—and it isn't. A calculated quantity can be no more precise than the data used in the calculation, and the reported result should reflect this fact.

A strict application of this principle involves a fairly complicated statistical analysis that we will not attempt here, but we can do a fairly good job by using a practical rule involving significant figures:

> *In multiplication and division, the reported result should have no more significant figures than the factor with the fewest significant figures.*

In other words, a calculation is only as precise as the least precise measurement that enters into the calculation.

To obtain a numerical answer with the proper number of significant figures often requires that we round off numbers. In rounding, we drop all digits that are not significant and, if necessary, adjust the last reported digit. We use the following rules in rounding.

- If the leftmost digit to be dropped is *4 or less*, drop it and all following digits. *Example*: If we need four significant figures, 69.744 rounds to 69.74, and to 69.7 if we need three significant figures.

- If the leftmost digit to be dropped is *5 or greater*, increase the final retained digit by 1. *Example*: 538.76 rounds to 538.8 if we need four significant figures. Similarly, 74.397 rounds to 74.40 if we need four significant figures, and to 74.4 if we need three.

EXAMPLE A.5 Unit Conversions

What is the area, in square centimeters, of a rectangular gauze bandage that is 2.54 cm wide and 12.42 cm long? Use the correct number of significant figures in your answer.

Solution

The area of a rectangle is the product of its length and width. In the result, we can show only as many significant figures as there are in the least precisely stated dimension, the width, which has three significant figures.

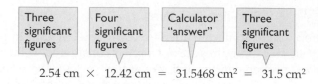

$$2.54 \text{ cm} \times 12.42 \text{ cm} = 31.5468 \text{ cm}^2 = 31.5 \text{ cm}^2$$

We use the rules for rounding off numbers as the basis for dropping the digits 468.

■ **EXERCISE A.5**

Calculate the volume, in cubic meters, of a rectangular block of foamed plastic that is 1.827 m long, 1.04 m wide, and 0.064 m thick. Use the correct number of significant figures.

EXAMPLE A.6 Significant Figures

For a laboratory experiment, a teacher wants to divide all of a 226.8-g sample of glucose equally among the 18 members of her class. How many grams of glucose should each student receive?

Solution

The number 18 is a counted number, that is, an exact number that is not subject to significant figure rules. The answer should carry four significant figures, the same as in 226.8 g.

$$\frac{226.8 \text{ g}}{18 \text{ students}} = 12.60 \text{ g/student}$$

In this calculation a calculator displays the result 12.6. We add the digit 0 to emphasize that the result is precise to four significant figures.

■ **EXERCISE A.6**

A dozen eggs has a mass of 681 g. What is the average mass of one of the eggs, expressed with the appropriate number of significant figures?

Significant Figures in Calculations: Addition and Subtraction

In addition or subtraction, we are concerned not with the number of significant figures but with the number of digits to the right of the decimal point. When we add or subtract quantities with varying numbers of digits to the right of the decimal point, we need to note the one with the fewest such digits. The result should contain the same number of digits to the right of its decimal point. For example, if you are adding several masses and one of them is measured only to the nearest gram, the total mass cannot be stated to the nearest milligram no matter how precise the other measurements are.

We apply this idea in Example A.7. Note that in a calculation involving several steps, we need round off only the final result.

EXAMPLE A.7 Significant Figures

Perform the following calculation and round off the answer to the correct number of significant figures.

$$2.146 \text{ g} + 72.1 \text{ g} - 9.1434 \text{ g}$$

Solution

In this calculation, we add two numbers and subtract a third from the sum of the first two.

$$
\begin{array}{ll}
2.146 \text{ g} & \text{three decimal places} \\
+ 72.1 \text{ g} & \text{one decimal place} \\
\hline
74.246 \text{ g} & \\
\\
- 9.1434 \text{ g} & \text{four decimal places} \\
\hline
65.1026 \text{ g} = 65.1 \text{ g} & \text{one decimal place}
\end{array}
$$

Note that we do not round off the intermediate result (74.246). When using a calculator, we generally don't need to write down an intermediate result.

■ **EXERCISE A.7**

Perform the indicated operations and give answers with the proper number of significant figures. Note that in addition and subtraction all terms must be expressed in the same unit.

a. 48.2 m + 3.82 m + 48.4394 m

b. 15.436 L + 5.3 L − 6.24 L − 8.177 L

c. (51.5 m + 2.67 m) × (33.42 m − 0.124 m)

d. $\dfrac{125.1 \text{ g} - 1.22 \text{ g}}{52.5 \text{ mL} + 0.63 \text{ mL}}$

Self-Assessment Questions

1. A measurement has good precision if it
 a. is close to an accepted standard   **b.** is close to similar measurements
 c. has few significant figures   **d.** is the only determined value

2. Following are data for the length of a pencil as measured by students X, Y, and Z. The length is known to be 11.54 cm.

	Trial 1	Trial 2	Trial 3
X	11.8	11.1	11.5
Y	11.7	11.2	11.6
Z	11.3	11.4	11.5

Which of the following best characterizes the data?
 a. X has the most precise data, Z the most accurate.
 b. Y has the most precise data, X the most accurate.
 c. Y has the most precise data, Z the most accurate.
 d. Z has the most precise data, X the most accurate.
 e. Z has the most precise data, Y the most accurate.

3. 725.8055 rounded to 3 significant figures is
 a. 725.805   **b.** 725.806   **c.** 725.81   **d.** 725.8   **e.** 726

For items 4–6, determine the number of significant figures, and then select the single best answer.

4. 0.0000073
 a. 2   **b.** 5   **c.** 6   **d.** 8

5. 0.04000
 a. 2   **b.** 4   **c.** 5   **d.** 6

6. 80.0040
 a. 2   **b.** 4   **c.** 5   **d.** 6

For items 7–16, perform the indicated calculation, and then select the single best answer using the proper number of significant figures.

7. 7.20 + 3.013 + 0.04327 = ?
 a. 10.26   **b.** 10.256   **c.** 10.2563   **d.** 10.25627

8. 4.702 − 0.4123 = ?
 a. 4.289   **b.** 4.2897   **c.** 4.29   **d.** 4.290   **e.** 4.30

9. 10.03 + 4.555 = ?
 a. 14   **b.** 14.59   **c.** 14.6   **d.** 15

10. 15.3 − 4.001 = ?
 a. 11   **b.** 11.2   **c.** 11.299   **d.** 11.3

11. 4.602 ÷ 0.0240 = ?
 a. 191   **b.** 191.7   **c.** 191.75   **d.** 191.8   **e.** 192

12. $40.625 \times 0.0028 = ?$
　　a. 0.11　　　　**b.** 0.1137　　　**c.** 0.11375　　　**d.** 0.1138　　　**e.** 0.114

13. $8.64 \div 0.1216 = ?$
　　a. 71.0　　　　**b.** 71.05　　　　**c.** 71.0526　　　**d.** 71.053

14. $(10.30)(0.186) \div 0.085 = ?$
　　a. 22　　　　　**b.** 23　　　　　**c.** 22.5　　　　**d.** 22.5388　　　**e.** 2.254

15. $(42.5 + 0.459) \div 28.45 = ?$
　　a. 1.510　　　　**b.** 1.5103　　　**c.** 1.51　　　　**d.** 1.510　　　**e.** 2

16. $(7.06 \div 0.084) - (29.6 \times 0.023) = ?$
　　a. 83　　　　　**b.** 83.3　　　　**c.** 83.4　　　　**d.** 83.369　　　**e.** 83.37

Answers: 1, b; 2, e; 3, e; 4, a; 5, b; 6, d; 7, a; 8, d; 9, b; 10, c; 11, e; 12, a; 13, a; 14, b; 15, c; 16, a

A.5 CALCULATIONS INVOLVING TEMPERATURE AND HEAT

As noted in Chapter 1, we sometimes need to make conversions from one temperature scale to another or from one energy unit to another.

Temperature

For everyday life, the following ditty will often suffice to assess the meaning of a Celsius temperature:

THE CELSIUS SCALE

Thirty is hot,
Twenty is pleasing;
Ten is quite cool,
And zero is freezing.

On the Fahrenheit scale, the freezing point of water is 32 °F and the boiling point is 212 °F, whereas on the Celsius scale the freezing point of water is 0 °C and the boiling point is 100 °C. A 100° temperature interval $(100 - 0)$ on the Celsius scale, therefore, equals a 180° interval $(212 - 32)$ on the Fahrenheit scale. From these facts we can derive two equations that relate temperatures on the two scales. One of these requires multiplying the degrees of Celsius temperature by the factor 1.8 (that is, 180/100) to obtain the degrees of Fahrenheit temperature, followed by adding 32 to account for the fact that 0 °C = 32 °F.

$$°F = (1.8 \times °C) + 32$$

In the other equation, we subtract 32 from the Fahrenheit temperature to get the number of degrees Fahrenheit above the freezing point of water. Then this quantity is divided by 1.8.

$$°C = \frac{°F - 32}{1.8}$$

The SI unit of temperature is the kelvin (Chapters 1 and 6). Recall that to convert from °C to K, we simply add 273.15. And to convert from K to °C, we subtract 273.15.

$$K = °C + 273.15 \quad \text{and} \quad °C = K - 273.15$$

Example A.8 illustrates a practical situation in which conversion between Celsius and Fahrenheit temperatures is necessary.

EXAMPLE A.8　　Temperature Conversion

At home you keep your room thermostat set at 68 °F. When traveling, you have a room with a thermostat that uses the Celsius scale. What Celsius temperature will give you the same temperature as at home?

Solution

$$°C = \frac{°F - 32}{1.8} = \frac{68 - 32}{1.8} = 20\ °C$$

■ **EXERCISE A.8**

a. Convert 85.0 °C to degrees Fahrenheit.
b. Convert −12.2 °C to degrees Fahrenheit.
c. Convert 355 °F to degrees Celsius.
d. Convert −20.8 °F to degrees Celsius.

Table A.4   Specific Heats of Selected Substances

	Specific Heat	
Substance	cal/(g °C)	J/(g °C)
Aluminum (Al)	0.216	0.902
Copper (Cu)	0.0921	0.385
Ethyl alcohol (CH_3CH_2OH)	0.588	2.46
Iron (Fe)	0.106	0.443
Ethylene glycol ($HOCH_2CH_2OH$)	0.561	2.35
Magnesium (Mg)	0.245	1.025
Mercury (Hg)	0.0332	0.139
Sulfur (S)	0.169	0.706
Water (H_2O)	1.000	4.182

Heat

In Chapter 1 we introduced the SI unit of energy, the *joule*. A joule (J) is the work done by a force of 1 newton[1] acting over a distance of 1 meter. We also introduced the more familiar unit, the calorie.

$$1 \text{ cal} = 4.184 \text{ J}$$

$$1 \text{ kcal} - 1000 \text{ cal} = 4184 \text{ J}$$

A calorie (cal) is the amount of heat required to raise the temperature of 1 g of water by 1 °C. (This quantity varies slightly with temperature; a calorie is defined more precisely as the amount of heat required to raise the temperature of 1 g of water from 14.5° to 15.5 °C.)

Some substances gain or lose heat more readily than others (Table A.4). The **specific heat** of a substance is the amount of heat required to raise the temperature of 1 g of the substance by 1 °C. The definition of a calorie indicates that the specific heat of water is 1.00 cal/(g · °C). In SI, the specific heat of water is 4.184 J/(g · °C). Note that metals all have much lower values of specific heat than does water. This means that a sample of water must absorb much more heat to raise its temperature than a metal sample of similar mass. The metal sample may become red hot after absorbing a quantity of heat that makes the water sample only lukewarm.

We can use the following equation, in which ΔT is the change in temperature (in either °C or kelvins), to calculate the quantity of heat absorbed or released by a system.

$$\text{Heat absorbed or released} = \text{mass} \times \text{specific heat} \times \Delta T$$

EXAMPLE A.9   Heat Absorbed

How much heat, in calories, kilocalories, and kilojoules, does it take to raise the temperature of 225 g of water from 25.0 °C to 100.0 °C?

Solution

Let's list the quantities we need for the calculations.

$$\text{Mass of water } = 225 \text{ g}$$

$$\text{Specific heat of water} = 1.00 \text{ cal/(g °C)}$$

$$\text{Temperature change } = (100.0 - 25.0) \text{ °C} = 75.0 \text{ °C}$$

---

[1] A *newton* (N) is the basic SI unit of force. A newton is the force required to give a 1-kg mass an acceleration of 1 m/s². That is, 1 N = 1 kg m/s². Therefore, 1 J = 1 N m = 1 kg m²/s²

Then we use the equation

$$\text{Heat absorbed} = \text{mass} \times \text{specific heat} \times \Delta T$$

$$= 225 \text{ g} \times 1.00 \text{ cal/(g °C)} \times 75.0 \text{ °C}$$

$$= 16,900 \text{ cal}$$

We can then convert the unit cal to the units kcal and kJ.

$$16,900 \text{ cal} \times \frac{1 \text{ kcal}}{1000 \text{ cal}} = 16.9 \text{ kcal}$$

$$16.9 \text{ kcal} \times \frac{4.184 \text{ kJ}}{1 \text{ kcal}} = 70.7 \text{ kJ}$$

■ **EXERCISE A.9**

How much heat, in calories, kilocalories, and kilojoules, is released by 975 g of water as it cools from 100.0 °C to 18.0 °C?

Table A.5 Some Conversion Units for Energy
1 calorie (cal) = 4.184 joules (J) (defined; exact)
1 British thermal unit (Btu) = 1055 joules (J) = 252 calories (cal)
1 food Calorie = 1 kilocalorie (kcal) = 1000 calories (cal) = 4184 joules (J)

Self-Assessment Questions

For items 1–10, perform the indicated conversion, and then select the single best answer.

1. 25 °C to kelvins
   a. −298 K   **b.** −248 K   **c.** 248 K   **d.** 298 K   **e.** 398 K

2. 373 K to degrees Celsius
   a. −100 °C   **b.** −73 °C   **c.** 0 °C   **d.** 73 °C   **e.** 100 °C

3. 301 K to degrees Celsius
   a. −72 °C   **b.** −28 °C   **c.** 28 °C   **d.** 128 °C   **e.** 374 °C

4. 473 °C to kelvins
   a. 100 K   **b.** 200 K   **c.** 300 K   **d.** 746 K   **e.** 846 K

5. 37.0 °C to degrees Fahrenheit
   a. −20.6 °F   **b.** −38.3 °F   **c.** 69.0 °F   **d.** 98.6 °F   **e.** 124 °F

6. 5.50 °F to degrees Celsius
   a. −20.8 °C   **b.** −14.7 °C   **c.** 3.06 °C   **d.** 6.31 °C   **e.** 63.1 °C

7. 273 °C to degrees Fahrenheit
   a. 152 °F   **b.** °F   **c.** 491 °F   **d.** 523 °F   **e.** 549 °F

8. 98.2 °F to degrees Celsius
   a. 38.6 °C   **b.** 54.6 °C   **c.** 72.3 °C   **d.** 156 °C   **e.** 177 °C

9. 2175 °C to degrees Fahrenheit
   a. 1190 °F   **b.** 1208 °F   **c.** 1226 °F   **d.** 3915 °F   **e.** 3947 °F

10. 25.0 °F to degrees Celsius
    a. −31.6 °C   **b.** −13.9 °C   **c.** −3.89 °C   **d.** 13.9 °C   **e.** 42.8 °C

11. Of each of the following pairs, which is the larger unit: °C or °F? cal or Cal?
    a. °C, cal   **b.** °C, Cal   **c.** °F, cal   **d.** °F, Cal

12. The temperatures, 0 K, 0 °C, and 0 °F, arranged from coldest to hottest, are
    a. 0 °C, 0 °F, 0 K   **b.** 0 °C, 0 K, 0 °F   **c.** 0 °F, 0 K, 0 °C
    d. 0 °F, 0 °C, 0 K   **e.** 0 K, 0 °F, 0 °C

For items 13–20, perform the indicated conversion, and then select the single best answer. You may need data from Table A.4 and/or Table A.5 for some of the items.

13. 0.820 kcal to calories
    a. 0.000820 cal
    b. 8.20 cal
    c. 820 cal
    d. 820,000 cal

14. 65,500 cal to kilocalories
    a. 0.0655 kcal
    b. 0.655 kcal
    c. 6.55 kcal
    d. 66.5 kcal
    e. 665 kcal

15. 0.359 kJ to joules
    a. 0.000359 J
    b. 0.00359 J
    c. 0.359 J
    d. 35.9 J
    e. 359 J

16. 0.741 kcal to joules
    a. 0.000741 J
    b. 0.741 J
    c. 3.10 J
    d. 741 J
    e. 3100 J

17. 8.63 kJ to calories
    a. 2.06 cal
    b. 20.6 cal
    c. 86.3 cal
    d. 2060 cal
    e. 8630 cal

18. 1.36 kcal to kilojoules
    a. 1.36 kJ
    b. 5.69 kJ
    c. 325 kJ
    d. 1360 kJ
    e. 5690 kJ

19. 345 cal to joules
    a. 0.345 J
    b. 1.44 J
    c. 82.5 J
    d. 345 J
    e. 1440 J

20. 873 kJ to kilocalories
    a. 0.873 kcal
    b. 209 kcal
    c. 873 kcal
    d. 2090 kcal
    e. 3650 kcal

21. How much heat in calories is required to raise the temperature of 50.0 g of water from 20.0 °C to 50.0 °C?
    a. 30.0 kcal
    b. 50.0 kcal
    c. 1500 cal
    d. 2500 kcal
    e. 6280 kcal

22. How much heat in kilojoules is required to raise the temperature of 131 g of iron from 15.0 °C to 95.0 °C?
    a. 0.058 kJ
    b. 1.11 kJ
    c. 4.65 kJ
    d. 58.0 kJ
    e. 4650 kJ

Answers: 1, d; 2, e; 3, c; 4, d; 5, d; 6, b; 7, d; 8, a; 9, e; 10, c; 11, b; 12, e; 13, c; 14, d; 15, e; 16, e; 17, d; 18, b; 19, b; 20, e; 21, c; 22, c

absolute zero The lowest possible temperature: 0 K, –273.15 °C, or –459.7 °F.

acid A substance that, when added to water, produces an excess of hydronium ion; a proton donor.

acid–base indicator A substance that is one color in acid and another color in base.

acidic anhydride A substance, such as a nonmetal oxide, that forms an acid on addition of water.

acid rain Rain having a pH less than 5.6.

acquired immune deficiency syndrome (AIDS) A disease caused by a retrovirus (HIV) that weakens the immune system.

activated sludge method A combination of primary and secondary sewage treatment methods in which some sludge is recycled.

activation energy The minimum quantity of energy that must be available before a chemical reaction can take place.

active site The region on a molecule such as an enzyme or catalyst at which reaction occurs.

addition polymerization A polymerization reaction in which all the atoms of the monomer molecules are included in the polymer.

addition reaction A reaction in which the single product contains all the atoms of two reactant molecules.

adipose tissue Connective tissue where fat is stored.

advanced sewage treatment Sewage treatment designed to remove phosphates, nitrates, and other soluble impurities.

aerobic oxidation An oxidation process occurring in the presence of oxygen.

aerobic exercise Exercise in which muscle contractions occur in the presence of oxygen.

aerosol Particles of 1 μm diameter or less, dispersed in air.

aflatoxins Toxins produced by molds growing on stored peanuts and grains.

Agent Orange A combination of 2,4-D and 2,4,5-T used extensively in Vietnam to remove enemy cover and destroy crops that maintained enemy armies.

agonist A molecule that fits and activates a specific receptor.

AIDS See **acquired immune deficiency syndrome**.

alchemy A mystical blend of chemistry, magic, and religion that flourished in Europe during the Middle Ages (500 to 1500 C.E.).

alcohol (ROH) A compound composed of an alkyl group and a hydroxyl group.

aldehyde (RCHO) An organic molecule with a carbonyl group that has at least one hydrogen atom attached to the carbonyl carbon.

aldose A monosaccharide with an aldehyde functional group.

alkali metal A metal in group 1A of the customary U.S. arrangement and in group 1 of the IUPAC-recommended periodic table.

alkaline earth metal An element in group 2A of the customary U.S. arrangement and in group 2 of the IUPAC-recommended periodic table.

alkaloid A physiologically active nitrogen-containing organic compound obtained from plants.

alkalosis A physiological condition in which the pH of the blood rises to a dangerous level.

alkane A hydrocarbon with only single bonds; a saturated hydrocarbon.

alkene A hydrocarbon containing one or more double bonds.

alkyl group (—R) The group of atoms that results when a hydrogen atom is removed from an alkane.

alkyne A hydrocarbon containing one or more triple bonds.

allergen A substance that triggers an allergic reaction.

allotropes Different forms of the same element in the same physical state.

alloy A mixture of two or more elements, at least one of which is a metal; an alloy has metallic properties.

alpha helix A secondary structure of a protein molecule in which the molecule has a spiral arrangement.

alpha (α) particle A cluster of two protons and two neutrons; a helium nucleus.

Ames test A laboratory test for mutagens, which are usually also carcinogens.

amide An organic compound having the functional group —CON— in which the carbon is double-bonded to the oxygen atom and single-bonded to the nitrogen atom.

amine A nitrogen compound derived from ammonia by replacing one or more hydrogen atoms with alkyl group(s).

amino acid An organic compound that contains both an amino group and a carboxylic acid group; amino acids combine to produce proteins.

amino group (—NH$_2$) A substituent group comprised of a nitrogen atom bonded to two hydrogen atoms.

amphetamines Stimulant drugs related to β-phenylethylamine.

amphoteric surfactant A surfactant with a hydrocarbon tail and a water-soluble head that bears both a negative charge and a positive charge.

anabolic steroid A drug that aids in the building (anabolism) of body proteins and thus of muscle tissue.

anabolism The buildup of body tissues from simpler molecules.

anaerobic decay Decomposition in the absence of oxygen.

anaerobic exercise Exercise involving muscle contractions with insufficient oxygen.

analgesic A substance that provides pain relief.

androgen A male sex hormone.

anesthetic A substance that causes loss of feeling and blocks pain.

anion A negatively charged ion.

anionic surfactant A surfactant with a hydrocarbon tail and a water-soluble head that bears a negative charge.

anode The electrode at which oxidation occurs.

antagonist A drug that blocks the action of an agonist by blocking the receptors.

antibiotic A soluble substance, produced by a mold or bacterium, that inhibits growth of other microorganisms.

anticarcinogen A substance that inhibits the formation of cancer.

anticholinergic A drug that acts on nerves using acetylcholine as a neurotransmitter.

anticoagulant A substance that inhibits the clotting of blood.

anticodon The sequence of three adjacent nucleotides in a tRNA molecule that is complementary to a codon on mRNA.

antihistamine A substance that relieves the symptoms of allergies: sneezing, itchy eyes, and runny nose.

anti-inflammatory A substance that inhibits inflammation.

antimetabolite A compound that inhibits the synthesis of nucleic acids.

antioxidant A reducing agent that is so easily oxidized that it protects other substances from oxidation.

antiperspirant A formulation that retards perspiration by constricting the openings of sweat glands.

antipyretic A fever-reducing substance.

apoenzyme The pure protein part of an enzyme.

applied research An investigation aimed at the solution of a particular problem.

aqueous solution A solution in which the solvent is water.

arithmetic growth A process in which a constant quantity is added during each growth period.

aromatic compound A compound that has special bonding and properties like those of benzene.

asbestos A group of related fibrous silicates.

astringent A substance that constricts the openings of the sweat glands, thus reducing the amount of perspiration that escapes.

atmosphere The gaseous mass surrounding Earth.

atmosphere (atm) A unit of pressure equal to 760 mmHg.

atmospheric inversion A warm layer of air above a cool, stagnant lower layer.

atom The smallest characteristic particle of an element.

atomic mass unit (u) The unit of relative atomic weights, $\frac{1}{12}$ the mass of a carbon-12 atom.

atomic number (Z) The number of protons in the nucleus of an atom of an element.

atomic theory A model that explains the law of multiple proportions and the law of constant composition by stating that all elements are composed of atoms.

Avogadro's hypothesis Equal volumes of gases, regardless of their compositions, contain equal numbers of molecules under the same conditions of temperature and pressure.

Avogadro's number The number of atoms (6.022×10^{23}) in exactly 12 g of pure carbon-12.

background radiation Ever-present radiation from cosmic rays and from natural radioactive isotopes in air, water, soil, and rocks.

base A substance that, when added to water, produces an excess of hydroxide ions; a proton acceptor.

base triplet The sequence of three bases on a tRNA molecule that determines which amino acid it can carry.

basic anhydride A substance, such as a metal oxide, that forms a base on addition of water.

basic research The search for knowledge for its own sake.

battery A series of two or more electrochemical cells; see **electrochemical cell**.

beta (β) particle An electron emitted by a radioactive atom.

binary compound A compound consisting of two elements.

binding energy Energy derived from the conversion of mass to energy when neutrons and protons are combined to form nuclei.

biochemical oxygen demand (BOD) The quantity of oxygen required by microorganisms to remove organic matter from water.

biochemistry A study of the chemistry of living systems.

biomass (in energy studies) Plant material used as fuel.

bitumen A hydrocarbon mixture obtained from tar sands by heating.

bleach A substance used to remove unwanted color from fabrics, hair, or other materials.

blood sugar Glucose, a simple sugar, circulated in the bloodstream.

boiling point The temperature at which a substance can change state from a liquid to a gas throughout the bulk of the liquid.

bond See **chemical bond**.

bonding pair (BP) A pair of electrons that comprises a chemical bond.

Boyle's law For a given mass of gas at constant temperature, the volume varies inversely with the pressure.

breeder reactor A nuclear reactor that converts nonfissile isotopes to fissile isotopes.

broad-spectrum antibiotic An antibiotic that is effective against a wide variety of microorganisms.

bronze An alloy of copper and tin.

buffer A mixture that reacts with either acid or base to keep the pH of a solution essentially constant.

builder (in detergent formulations) Any substance added to a surfactant to increase its detergency.

calorie (cal) The quantity of heat required to raise the temperature of 1 g of water by 1 °C.

carbohydrate A compound consisting of carbon, hydrogen, and oxygen; a starch or sugar.

carbon-14 dating A technique for determining the age of artifacts based on the half-life of carbon-14.

carbonyl group (C=O) A carbon atom double-bonded to an oxygen atom.

carboxyl group (COOH) A carbon atom double-bonded to one oxygen atom and single-bonded to a second oxygen atom that in turn is bonded to a hydrogen atom; the functional group of carboxylic acids.

carboxylic acid (RCOOH) An organic compound that contains the COOH functional group.

carcinogen A substance or physical entity that produces tumors.

catabolism The metabolic process in which complex compounds are broken down into simpler substances.

catalyst A substance that increases the rate of a chemical reaction without itself being used up.

catalytic converter A device containing catalysts for oxidizing carbon monoxide and hydrocarbons to carbon dioxide and reduction of nitrogen oxides to nitrogen gas.

catalytic reforming A process that converts low-octane alkanes to high-octane aromatic compounds.

cathode The electrode at which reduction occurs.

cathode ray A stream of high-speed electrons emitted from a cathode in an evacuated tube.

cation A positively charged ion.

cationic surfactant A surfactant with a hydrocarbon tail and a water-soluble head that bears a positive charge.

celluloid Cellulose nitrate, a synthetic material derived from natural cellulose by reaction with nitric acid.

cellulose A polymer comprised of glucose units joined through beta linkages.

Celsius scale A temperature scale on which water freezes at 0° and boils at 100°.

cement (Portland cement) A mixture made from limestone, clay, and sand; when mixed with water it hardens like stone.

ceramic A hard, solid product made from clay or similar materials.

chain reaction A self-sustaining change in which one or more products of one event cause one or more new events.

charcoal filtration Filtration of water through charcoal to adsorb organic compounds.

Charles's law For a given mass of gas at constant pressure, the volume varies directly with the absolute temperature.

chemical bond The force of attraction that holds atoms or ions together in compounds.

chemical change A change in chemical composition.

chemical equation A before-and-after description in which chemical formulas and coefficients represent a chemical reaction.

chemical property A characteristic of a substance that describes the way in which the substance reacts with another substance to change its composition.

chemical symbol An abbreviation, consisting of one or two letters, that stands for an element.

chemistry The study of matter and the changes it undergoes.

chemotherapy The use of chemicals to control or cure diseases.

chiral carbon A carbon atom that has four different groups attached.

coal A fossilized black rock rich in carbon.

codon A sequence of three adjacent nucleotides in mRNA that specifies one amino acid.

coenzyme An organic molecule (often a vitamin) that combines with an apoenzyme to make a complete, functioning enzyme.

cofactor An inorganic component that combines with an apoenzyme to make a complete, functioning enzyme.

cologne A diluted perfume.

compound A pure substance made up of two or more elements combined in fixed proportions.

concentrated solution A solution that has a relatively large amount of solute per unit volume of solution.

concrete A building material made from Portland cement, sand, gravel, and water.

condensation The reverse of vaporization; a change from the gaseous state to the liquid state.

condensation polymerization A reaction in which not all the atoms in the starting monomers are incorporated in the polymer because water (or other small) molecules are split out as the polymer is formed.

condensed structural formula An organic chemical formula that shows the atoms of hydrogen right next to the carbon atoms to which they are attached.

copolymer A polymer formed by the combination of two or more different monomer units.

corrosive waste A waste that requires a special container because it reacts with conventional container materials.

cosmetics Substances defined in the 1938 U.S. Food, Drug, and Cosmetic Act as "articles intended to be rubbed, poured, sprinkled or sprayed on, introduced into, or otherwise applied to the human body or any part thereof, for cleaning, beautifying, promoting attractiveness or altering the appearance."

cosmic rays Extremely high-energy rays from outer space.

covalent bond A bond formed by a shared pair of electrons.

cream An emulsion of tiny water droplets in oil.

critical mass The mass of an isotope above which a self-sustaining chain reaction can occur.

crystal A solid with plane surfaces at definite angles.

cyclic hydrocarbon A ring-containing hydrocarbon.

daughter isotope An isotope formed by the radioactive decay of another isotope.

defoliant A substance that causes premature dropping of leaves by plants.

Delaney Amendment A 1958 amendment to the U.S. Food, Drug, and Cosmetic Act that automatically bans from food any chemical shown to induce cancer in laboratory animals.

density The quantity of mass per unit volume.

deodorant A product that uses perfume to mask body odor; some claim to prevent body odor by killing odor-causing bacteria.

deoxyribonucleic acid (DNA) The type of nucleic acid found primarily in the nuclei of cells; contains the sugar deoxyribose.

depilatory A hair remover.

deposition The direct formation of a solid from a gas without passing through the liquid state; the reverse of sublimation.

depressant drug A drug that slows both physical and mental activity.

detergent A cleansing agent. Sometimes used specifically to mean a synthetic surfactant as distinguished from soap.

deuterium An isotope of hydrogen with a proton and a neutron in the nucleus (mass of 2 u).

dextro isomer A "right-handed" isomer.

dietary mineral A mineral required in the diet for proper health and well-being.

Dietary Reference Intake (DRI) A set of reference values for food and other nutrient intake that may include Estimated Average Requirements (EAR), Recommended Dietary Allowances (RDA), Adequate Intakes (AI), and Tolerable Upper Intake Levels (UL).

dilute solution A solution that has a relatively small amount of solute per unit volume of solution.

dioxins Highly toxic chlorinated cyclic compounds produced by burning wastes containing chlorinated compounds; once found as contaminants in herbicides.

dipole A molecule that has a positive end and a negative end.

dipole–dipole forces The attractive forces that exist among polar covalent molecules.

disaccharide A sugar that on hydrolysis yields two monosaccharide molecules per molecule of disaccharide.

dispersion forces The momentary, usually weak, attractive forces between molecules resulting from synchronized electron motion.

dissociative anesthetic A substance that causes gross personality disorders, including hallucinations similar to those in near-death experiences.

dissolved oxygen Oxygen dissolved in water; a measure of that water's ability to support fish and other aquatic life.

disulfide linkage A covalent linkage of two amino acid units through two sulfur atoms.

diuretic A substance that increases the output of urine.

double bond The sharing of two pairs of electrons between two atoms.

doubling time The time it takes a population to double in size.

drug A chemical substance that affects the functioning of living things; used to relieve pain, to treat an illness, or to improve one's state of health or well-being.

drug abuse The use of a drug for its intoxicating effect.

drug misuse The use of a drug in a manner other than its intended use.

elastomer A synthetic polymer with rubberlike properties.

electrochemical cell A device that produces electricity by means of a chemical reaction.

electrode A device, such as a carbon rod or metal strip inserted into an electrochemical cell, at which oxidation or reduction occurs.

electrolysis Causing a chemical reaction by means of electricity.

electrolyte A compound that, in water solution, conducts an electric current.

electron The subatomic particle that bears a unit of negative charge.

electron capture (EC) A type of radioactive decay in which a nucleus absorbs an electron from the first or second electronic shell.

electron configuration The arrangement of an atom's electrons among different energy levels.

electron-dot structure See **Lewis structure.**

electron-dot symbol See **Lewis symbol.**

electronegativity The attraction of an atom in a molecule for a pair of shared electrons in a chemical bond.

electrostatic precipitator A device that removes particulate matter from smokestack gases by forming an electric charge on the particles, which are then removed by attraction to a surface of opposite charge.

element A fundamental substance in which all atoms have the same number of protons.

emollient An oil or grease used as a skin softener.

emulsion A suspension of submicroscopic particles of fat or oil in water.

enantiomers Nonsuperimposable (nonidentical) mirror-image isomers.

end note The portion of perfume that has low volatility; composed of large molecules.

endorphins Naturally occurring peptides that bond to the same receptor sites as opiate drugs.

endothermic A process that absorbs energy from its surroundings.

energy The capacity for doing work.

energy levels (shells) The specific, quantized values of energy that an electron can have in an atom.

enrichment (food) Replacement of nutrients lost from a food during processing.

enrichment (isotope) The process by which the proportion of one isotope of an element is increased relative to those of the others.

entropy A measure of the distribution of elemental species among energy states.

enzyme A biological catalyst.

essential amino acid An amino acid not produced in the body that must be included in the diet.

ester (RCOOR′) A compound derived from a carboxylic acid and an alcohol; the —OH of the acid is replaced by an —OR group.

estrogen A female sex hormone.

ether (ROR′) A molecule with two hydrocarbon groups attached to the same oxygen atom.

eutrophication The excessive growth of plants in a body of water that causes some of the plants to die because of a lack of light; the water becomes choked with vegetation, depleted of oxygen, and useless as a fish habitat or for recreation.

excited state A state in which an atom absorbs energy and an electron is moved from a lower to a higher energy level.

exothermic A process that releases heat to the surroundings.

fast-twitch fibers The stronger, larger kind of muscle fibers that are suited for anaerobic work.

fat A compound formed by the reaction of glycerol with three fatty acid units; a triglyceride or triacylglycerol.

fat depots Storage places for fats in the body.

fatty acid A carboxylic acid that contains 4 to 20 or more carbon atoms in a chain.

first law of thermodynamics Energy is neither created nor destroyed.

flammable waste A waste that burns readily on ignition, presenting a fire hazard.

FLaReS An acronym for the rules used to test a claim: falsifiability, logic, replicability, and sufficiency.

food additive Any substance other than basic foodstuffs that is present in food as a result of some aspect of production, processing, packaging, or storage.

formula A representation of a chemical substance in which the component chemical elements are represented by their symbols.

formula mass The sum of the atomic masses of a chemical compound as indicated by the formula, expressed in atomic mass units (u).

fossil fuels Coal, petroleum, and natural gas.

free radical A reactive neutral chemical species that contains an unpaired electron.

freezing The reverse of melting; a change from the liquid to the solid state.

fuel A substance that burns readily with the release of significant amounts of energy.

fuel cell A device that produces electricity directly from fuels and oxygen.

functional group The atom or group of atoms that confers characteristic properties on an organic molecule.

fundamental particle An electron, proton, or neutron.

gamma (γ) rays Rays similar to X-rays that are emitted from radioactive substances; have higher energy and are more penetrating than X-rays.

gas The state of matter in which the substance maintains neither shape nor volume.

gasoline The fraction of petroleum containing C_5 to C_{12} hydrocarbons, mainly alkanes, used as automotive fuel.

gene The segment of a nucleic acid molecule that contains the information necessary to produce a protein; the smallest unit of hereditary information.

general anesthetic A depressant that acts on the brain to produce unconsciousness and insensitivity to pain.

geometric growth A doubling in number for each growth period.

geothermal energy Energy derived from the heat of Earth's interior.

glass A noncrystalline material obtained by melting sand with soda, lime, and various other metal oxides.

glass transition temperature (T_g) The temperature above which a polymer is rubbery and tough and below which the polymer is brittle.

global warming An increase in the Earth's average temperature.

globular protein A protein whose molecules fold into roughly spherical or ovoid shapes that can be dispersed in water.

glycogen A polymer of glucose with alpha linkages and branched chains; a storage form of starch in animals.

GRAS list A list, established by the U.S. Congress in 1958, of food additives generally recognized as safe.

greenhouse effect The retention of the Sun's heat energy by Earth as a result of excess carbon dioxide or other substances in the atmosphere; causes an increase in Earth's average atmospheric and surface temperatures.

ground state The state of an atom in which all electrons are in the lowest possible energy levels.

group A vertical column in the periodic table containing a family of elements with similar properties.

half-life The length of time required for one-half of the radioactive nuclei in a sample to decay.

hallucinogenic drug A drug that produces visions and sensations that are not part of reality.

halogen An element in group 7A of the customary U.S. arrangement and in group 17 of the IUPAC-recommended periodic table.

hard water Water containing excessive concentrations of ions of calcium, magnesium, and/or iron.

hazardous waste A waste that, when improperly managed, can cause or contribute to death or illness or threaten human health or the environment.

heat Energy transfer that occurs as a result of a temperature difference.

heat of vaporization (of a substance) The amount of heat involved in the evaporation or condensation of 1 g of the substance.

heat stroke A failure of the body's heat regulatory system; unless the victim is treated promptly, the rapid rise in body temperature will cause brain damage or death.

herbicide A material used to kill plants.

heterocyclic compound A cyclic compound in which one or more atoms in the ring is not carbon.

homogeneous Completely uniform a property of a sample that has the same composition in all parts.

homologous series A series of compounds in which adjacent members of the series differ by a fixed unit of structure.

hormone A chemical messenger secreted into the blood by an endocrine gland.

humectant A moistening agent.

hydrocarbon An organic compound that contains only carbon and hydrogen.

hydrogen bomb A bomb based on the nuclear fusion of isotopes of hydrogen.

hydrogen bond A type of intermolecular force in which a hydrogen atom covalently bonded in one molecule is simultaneously attracted to a nonmetal atom in a neighboring molecule; both the atom to which the hydrogen atom is bonded and the one to which it is attracted must be small atoms of high electronegativity, usually N, O, or F.

hydrolysis The reaction of a substance with water; literally, a splitting by water.

hydrophilic Attracted to polar solvents such as water.

hydrophobic Not attracted to water; associated with other nonpolar entities.

hypoallergenic cosmetics Cosmetics claimed to cause fewer allergic reactions than regular products.

hypothesis A reasoned guess that can be tested by experiment.

ideal gas law The volume of a gas is proportional to the amount of gas and its Kelvin temperature and inversely proportional to its pressure.

induced radioactivity Radioactivity caused by bombarding a stable isotope with elemental particles, forming a radioactive isotope.

industrial smog Polluted air associated with industrial activities, usually characterized by sulfur oxides and particulate matter.

inorganic chemistry The study of the compounds of all elements other than carbon.

insecticide A substance that kills insects.

iodine number The number of grams of iodine consumed by 100 g of fat or oil; an indication of the degree of unsaturation.

ion A charged atom or group of atoms.

ionic bond The chemical bond that results when electrons are transferred from a metal to a nonmetal; the electrostatic attraction between ions of opposite charge.

ionizing radiation Radiation that produces ions as it passes through matter.

isoelectronic Has the same number of electrons.

isomers Compounds that have the same molecular formula but different structural formulas and properties.

isotopes Atoms that have the same number of protons but different numbers of neutrons.

joule (J) The SI unit of energy (1 J = 0.239 cal).

juvenile hormone A hormone that controls the rate of development of the young; used to prevent insects from maturing.

kelvin (K) The SI unit of temperature. Zero on the Kelvin scale is absolute zero.

keratin The tough, fibrous protein that comprises most of the outermost layer of the epidermis.

kerogen The complex material found in oil shale; has an approximate composition of $(C_6H_8O)_n$, where n is a large number.

ketone (RCOR') An organic compound with a carbonyl group between two carbon atoms.

ketose A monosaccharide with a ketone functional group.

kilocalorie A unit of energy equal to 1000 cal; one food calorie.

kilogram (kg) The SI unit of mass, a quantity equal to about 2.2 lb.

kinetic energy The energy of motion.

kinetic–molecular theory A model that uses the motion of molecules to explain the behavior of matter.

lanolin A natural wax obtained from sheep's wool.

law of combining volumes The volumes of gaseous reactants and products are in a small whole-number ratio when all measurements are made at the same temperature and pressure.

law of conservation of energy The quantity of energy within the universe is constant; energy cannot be created or destroyed, only transformed.

law of conservation of mass Matter is neither created nor destroyed during a chemical change.

law of definite proportions A compound always contains elements in certain definite proportions, never in any other combination; also called the law of constant composition.

law of multiple proportions Elements may combine in more than one proportion to form more than one compound—for example, CO and CO_2.

LD_{50} The dosage that would be lethal to 50% of a population of test animals.

levo isomer A "left-handed" isomer.

Lewis (electron-dot) structure The structural formula of a molecule in which valence electrons of all the atoms are indicated by dots.

Lewis (electron-dot) symbol The symbol of an element surrounded by dots representing the atom's outermost electrons.

limiting reactant The reactant that is used up first in a reaction, after which the reaction ceases no matter how much remains of the other reactants.

line spectrum The pattern of colored lines emitted by an element.

lipid A substance from animal or plant cells that is soluble in nonpolar solvents and insoluble in water.

lipoprotein A protein combined with a lipid, such as a triglyceride or cholesterol.

liquid The state of matter in which the substance assumes the shape of its container, flows readily, and maintains a fairly constant volume.

liter (L) A unit of volume equal to a cubic decimeter.

local anesthetic A substance that renders part of the body insensitive to pain while leaving the patient conscious.

lone pair (LP) A pair of electrons in the valence shell of an atom not involved in a bond; also called a nonbonding pair (NBP).

lotion An emulsion of submicroscopic fat or oil droplets dispersed in water.

main group element In the periodic table, an element in the A groups of the customary U.S. arrangement and in groups 1, 2, and 13–18 of the IUPAC-recommended table.

marijuana A preparation made from the leaves, flowers, seeds, and small stems of the *Cannabis* plant.

mass A measure of the quantity of matter.

mass–energy equation Einstein's equation $E = mc^2$, in which E is energy, m is mass, and c is the speed of light.

mass number (nucleon number) The sum of the numbers of protons and of neutrons in the nucleus of an atom.

matter The stuff of which all materials are made; anything that has mass and occupies space.

melanin A brownish-black pigment that determines the color of the skin and hair.

melting point The temperature at which a substance changes from the solid to the liquid state.

messenger RNA (mRNA) The type of RNA that contains the codons for a protein; travels from the nucleus of the cell to a ribosome.

metabolism The sum of all the chemical reactions by which the protoplasm of an organism grows, is maintained, obtains energy, and is degraded.

metalloid An element with properties intermediate between those of metals and those of nonmetals.

metals The elements to the left of the heavy, stepped, diagonal line in the periodic table.

meter (m) The SI unit of length, slightly longer than a yard.

mica A mineral composed of SiO_4 tetrahedra arranged in a two-dimensional, sheetlike array.

micelle A spherical cluster of surfactant molecules arranged so that their hydrophilic ends all lie along the outer surface.

micronutrient A substance needed by the body in only tiny amounts.

middle note The portion of perfume intermediate in volatility; responsible for the lingering aroma after most top-note compounds have vaporized.

mineral (dietary) An inorganic substance required in the diet for good health.

mineral (geological) A naturally occurring inorganic solid with a definite composition.

mixture Matter with a variable composition.

moisturizer (skin) A substance that adds moisture to the skin or acts to retain moisture in the skin.

molarity (M) The concentration of a solution in moles of solute per liter of solution.

molar mass The formula mass of a substance expressed in grams.

molar volume The volume occupied by 1 mol of a substance (usually a gas) under specified conditions.

mole (mol) The amount of a chemical substance that contains 6.022×10^{23} formula units of the substance.

molecular mass The mass of a molecule of a substance; the sum of the atomic masses as indicated by the molecular formula, expressed in atomic mass units (u).

molecule An electrically neutral unit of two or more atoms joined by covalent bonds; the smallest fundamental unit of a molecular substance.

monomer A substance of relatively low molecular mass; monomer molecules combine to make a polymer.

monosaccharide A carbohydrate that cannot be hydrolyzed into simpler sugars.

mousse A foam or froth; a hair care product composed of resins used to hold hair in place.

mutagen Any entity that causes changes in genes without destroying the genetic material.

narcotic A depressant, analgesic drug that induces narcosis (sleep).

natural gas A mixture of gases, mainly methane, found in many underground deposits.

natural philosophy Philosophical speculation about nature.

neuron A nerve cell.

neurotransmitter A chemical that carries an impulse across a synapse from one nerve cell to the next.

neurotrophin A substance produced during exercise that promotes the growth of brain cells.

neutralization The reaction of an acid and a base to produce a salt and water.

neutron A fundamental particle with a mass of approximately 1 u and no electric charge.

nitrogen cycle The various processes by which nitrogen is cycled among the atmosphere, soil, water, and living organisms.

noble gases In the periodic table, generally unreactive gases that appear in group 8A of the customary U.S. arrangement and in group 18 of the IUPAC-recommended table.

nonbonding pair (NBP) See **lone pair (LP)**.

nonionic surfactant A surfactant with a hydrocarbon tail and a polar head whose oxygen atoms attract water molecules and make the head water soluble; bears no ionic charge.

nonmetals The elements to the right of the heavy, stepped, diagonal line in the periodic table.

nonpolar covalent bond A covalent bond in which there is an equal sharing of electrons.

nonsteroidal anti-inflammatory drug (NSAID) A milder anti-inflammatory drug as distinguished from the more potent steroidal anti-inflammatory drugs such as cortisone and prednisone.

nuclear fission The splitting of an atomic nucleus into two large fragments.

nuclear fusion The combination of two small atomic nuclei to produce one larger nucleus.

nuclear reactor A power plant that produces energy by nuclear fission.

nucleic acid A nucleotide polymer, DNA or RNA.

nucleon A proton or neutron in an atomic nucleus.

nucleon number (A) See **mass number**.

nucleotide A combination of a heterocyclic amine, a pentose sugar, and phosphoric acid; the monomer unit of nucleic acids.

nucleus Concentrated, positively charged matter at the center of an atom; composed of protons and neutrons.

octane rating The antiknock quality of a gasoline as compared to mixtures of isooctane (with a rating of 100) and heptane (with a rating of 0).

octet rule Atoms tend toward an arrangement that will surround them with eight electrons in the outermost shell.

oil (food) A substance formed from glycerol and fatty acids, which is liquid at room temperature.

oil shale Fossil rock containing kerogen from which oil can be obtained at high cost by distillation.

optical brightener A compound that absorbs the invisible ultraviolet component of sunlight and reemits it as visible light at the blue end of the spectrum.

orbital A region of space in an atom occupied by one or two electrons.

organic chemistry The study of the compounds of carbon.

organic farming Farming without synthetic fertilizers or pesticides.

oxidation An increase in oxidation number; combination of an element or compound with oxygen; loss of hydrogen; loss of electrons.

oxidizing agent A substance that causes oxidation and is itself reduced.

oxygen cycle The various processes by which oxygen is cycled among the atmosphere, soil, water, and living organisms.

oxygen debt The demand for oxygen in muscle cells following anaerobic exercise.

ozone layer The layer of the stratosphere that contains ozone and shields living creatures on Earth from deadly ultraviolet radiation from the Sun.

paint A surface coating that contains a pigment, a binder, and a solvent.

particulate matter (PM) A pollutant composed of solid and liquid particles of greater than molecular size.

peptide bond The amide linkage that bonds amino acids in chains of peptides, polypeptides, and proteins.

percent by mass The concentration of a solution, expressed as (mass of solute ÷ mass of solution) × 100%.

percent by volume The concentration of a solution, expressed as (volume of solute ÷ volume of solution) × 100%.

perfume A fragrant mixture of plant extracts and other chemicals dissolved in alcohol.

period A horizontal row of the periodic table.

periodic table A systematic arrangement of the elements in columns and rows; elements in a given column have similar properties.

pesticide A substance that kills some kind of pest (weeds, insects, rodents, and so on).

petroleum A dark, oily mixture of (mostly) hydrocarbons occurring in various geologic deposits.

pH The negative logarithm of hydronium ion concentration.

pharmacology The study of the response of living organisms to drugs.

phenol A compound with an OH group attached to a benzene ring.

pheromone A natural chemical secreted by an organism to mark a trail, send out an alarm, or attract a mate.

photochemical smog Smog created by the action of sunlight on unburned hydrocarbons and nitrogen oxides, mainly from automobiles.

photon A unit particle of energy.

photosynthesis The chemical process used by green plants to convert solar energy into chemical energy by reducing carbon dioxide.

photovoltaic cell A solar cell; a cell that converts sunlight directly to electric energy.

physical change A change in physical state or form.

physical property A quality of a substance that can be demonstrated without changing the composition of the substance.

placebo A substance that appears to be a real drug but has no active ingredients.

plasma A state of matter similar to a gas but composed of isolated electrons and nuclei rather than discrete whole atoms or molecules.

plasticizer A substance added to some plastics, such as vinyl, to make them more flexible and easier to work with.

pleated sheet A secondary protein structure characterized by antiparallel molecules with a zigzag structure.

polar covalent bond A covalent bond in which more than half of the bond's negative charge is concentrated around one of the two atoms.

polar molecule A molecule that has a dipole moment.

pollutant A chemical that causes undesirable effects by being in the wrong place and/or in the wrong concentration.

polyamide A polymer that has structural units joined by amide linkages.

polyatomic ion An ion consisting of two or more atoms bonded together.

polyester A polymer that has structural units joined by ester linkages.

polymer A molecule with a large molecular mass; formed of repeating smaller units (monomers).

polymerase chain reaction (PCR) A process that reproduces many copies of a DNA fragment.

polypeptide A polymer of amino acids, usually of lower molecular mass than a protein.

polysaccharide A carbohydrate, such as starch or cellulose, one molecule of which yields many molecules of monosaccharide(s) on hydrolysis.

polyunsaturated fat A fat containing fatty acid units with two or more carbon–carbon double bonds.

positron (β^+) A positively charged particle with the mass of an electron.

potential energy Energy by virtue of position or composition.

preemergent herbicide A herbicide that is rapidly broken down in the soil and can, therefore, be used to kill weed plants before crop seedlings emerge.

primary plant nutrients Nitrogen, phosphorus, and potassium.

primary sewage treatment Treatment of sewage in a plant with a holding pond intended to remove some of the sewage solids as sludge by settling.

primary structure The amino acid sequence in a protein or nucleotide sequence in a nucleic acid.

product A substance produced by a chemical reaction; product formulas follow the arrow in a chemical equation.

progestin A compound that mimics the action of progesterone.

prostaglandin One of several 20-carbon-atom hormone-like compounds derived from arachidonic acid; involved in increased blood pressure, the contractions of smooth muscle, and other physiological processes.

protein An amino acid polymer.

proton (nuclear) The unit of positive charge in the nucleus of an atom.

proton (H^+) The hydrogen ion in acid–base chemistry.

psychotropic drugs Drugs that affect the mind.

pure substance Also known as **substance**; a sample of matter that always has the same composition, no matter how it is made or found.

purine A heterocyclic base with two fused rings, found in nucleic acids.

pyrimidine A heterocyclic base with one ring, found in nucleic acids.

quantum An energy packet of specific size; one photon of energy.

quartz A compound composed of SiO_4 tetrahedra arranged in a three-dimensional array.

quaternary structure An arrangement of protein subunits in a particular pattern.

radioactive decay The disintegration of an unstable atomic nucleus with spontaneous emission of radiation.

radioactivity The spontaneous emission of particles (for example, alpha or beta) or rays (gamma) from the atomic nuclei.

radioisotopes Radioactive isotopes.

reactant A starting material or original substance in a chemical change; reactant formulas precede the arrow in a chemical equation.

reactive wastes Wastes that tend to react spontaneously or to react vigorously with air or water.

Recommended Daily Allowance (RDA) The recommended level of a nutrient necessary for a balanced diet.

reducing agent A substance that causes reduction and is itself oxidized.

reduction A decrease in oxidation number; a gain of electrons; a loss of oxygen; a gain of hydrogen.

replication Copying or duplication; the process by which DNA reproduces itself.

resin A polymeric material, usually a sticky solid or semisolid organic material.

restorative drug A drug used to relieve the pain and reduce the inflammation resulting from overuse of muscles.

retrovirus An RNA virus that synthesizes DNA in the host cells.

reverse osmosis A method of pressure filtration through a semipermeable membrane; water is forced to flow from an area of high salt concentration to an area of low salt concentration.

ribonucleic acid (RNA) The form of nucleic acid found mainly in the cytoplasm but also present in all other parts of the cell; contains the sugar ribose.

risk–benefit analysis An approach that estimates a desirability quotient by dividing the benefits by the risks.

rule of 72 A mathematical formula that gives the doubling time for a population growing geometrically; 72 divided by the annual rate equals the doubling time.

salt An ionic compound produced by the reaction of an acid with a base.

saponins Natural chemical compounds that produce a soapy lather.

saturated fat A triglyceride composed of a large proportion of saturated fatty acids esterified with glycerol.

saturated hydrocarbon An alkane; a compound of carbon and hydrogen with only single bonds.

science A branch of knowledge based on the laws of nature.

scientific law A summary of experimental data; often expressed in the form of a mathematical equation.

scientific model A representation that serves to explain a scientific phenomenon.

sebum An oily secretion of the body that protects the skin from moisture loss.

second law of thermodynamics The entropy of the universe increases in any spontaneous process.

secondary plant nutrients Magnesium, calcium, and sulfur.

secondary sewage treatment Passing effluent from a primary treatment plant through gravel and sand filters to aerate the water and remove suspended solids.

secondary structure The arrangement of polypeptide chains in a protein—for example, helix or pleated sheet.

sex attractant A substance or mixture of substances released by an organism to attract members of the opposite sex of the same species for mating.

shell An electron energy level; the specific, quantized energy levels that an electron can have in an atom.

significant figures Those measured digits that are known with certainty plus one uncertain digit.

silicone A polymer with a base chain of alternating silicon and oxygen atoms.

single bond A pair of electrons shared between two atoms.

SI units (International System of Units) A measuring system used by scientists worldwide; it is based on seven base quantities and their multiples and submultiples.

skin protection factor (SPF) The rating of a sunscreen's ability to limit the penetration of ultraviolet radiation.

slag A relatively low-melting product of the reaction of limestone with silicate impurities in iron ore.

slow-twitch fibers Muscle fibers suited for aerobic work.

smog The combination of smoke and fog; polluted air.

soap A mixture of salts (usually sodium salts) of long-chain carboxylic acids.

solar cell A device used for converting sunlight to electricity; a photoelectric cell.

solid A state of matter in which the substance maintains its shape and volume.

solute The substance that is dissolved in another substance (solvent) to form a solution; usually present in a smaller amount than the solvent.

solution A homogeneous mixture of two or more substances.

solvent The substance that dissolves another substance (solute) to form a solution; usually present in a larger amount than the solute.

specific heat (of a substance) The amount of heat required to raise the temperature of 1 g of the substance by 1 °C.

standard temperature and pressure (STP) Conditions of 0 °C and 1 atm pressure.

starch A polymer of glucose units joined through alpha linkages; a complex carbohydrate.

starvation The withholding of nutrition from the body, whether voluntary or involuntary.

steel An alloy of iron containing small amounts of carbon and usually containing other metals such as manganese, nickel, and chromium.

stereoisomers Isomers having the same structural formula but differing in the arrangement of atoms or groups of atoms in three-dimensional space.

steroid A molecule that has a four-ring skeletal structure, with one cyclopentane and three cyclohexane fused rings.

stimulant drug A drug that increases alertness, speeds up mental processes, and generally elevates the mood.

stoichiometric factor A factor that relates the amounts of two substances through their coefficients in a chemical equation.

stoichiometry Quantity relationships between reactants and products in a chemical reaction.

strong acid An acid that ionizes completely in water; a potent proton donor.

strong base A base that dissociates completely in water; a potent proton acceptor.

structural formula A chemical formula that shows how the atoms of a molecule are arranged, to which other atom(s) they are bonded, and the kinds of bonds.

sublevel See **subshell**.

sublimation Conversion of a solid directly to the gaseous state without going through the liquid state.

subshell A subdivision of electron energy levels in an atom; also called a sublevel.

substance See **pure substance**.

substrate The substance that bonds to the active site of an enzyme; the substance acted upon.

surface-active agent (surfactant) Any agent that stabilizes the suspension in water of a nonpolar substance such as oil.

synapse A tiny gap between nerve fibers.

synergistic effect An effect greater than the sum of the effects expected from two or more phenomena.

tar sands Sands that contain bitumen, a thick hydrocarbon material.

technology The sum total of processes by which humans modify the materials of nature to better satisfy their needs and wants.

temperature A measure of heat intensity, or how energetic the particles of a sample are.

temperature inversion A warm layer of air above a cool, stagnant lower layer.

teratogen A substance that causes birth defects when introduced into the body of a pregnant female.

tertiary structure The folds, bends, and twists in protein or nucleic acid structure.

tetracyclines Antibacterial drugs with four fused rings.

theory A detailed explanation of the behavior of matter based on experiments; may be revised if new data warrant.

thermonuclear reactions Nuclear fusion reactions that require extremely high temperatures and pressures.

thermoplastic polymer A kind of polymer that can be heated and reshaped.

thermosetting polymer A kind of polymer that cannot be softened and remolded.

top note The portion of perfume that vaporizes most quickly; composed of relatively small molecules; responsible for odor when perfume is first applied.

toxicology The division of pharmacology that deals with the effects of poisons on the body, their identification and detection, and remedies for them.

toxic waste A waste that contains or releases poisonous substances in amounts large enough to threaten human health or the environment.

tracers Radioactive isotopes used to trace movement or locate the sites of radioactivity in physical, chemical, and biological systems.

transcription The process by which DNA directs the synthesis of an mRNA molecule during protein synthesis.

transfer RNA (tRNA) A small molecule that contains anticodon nucleotides; the RNA molecule that bonds to and carries an amino acid.

transition elements In the periodic table, metallic elements in the B groups of the customary U.S. arrangement and in groups 3–12 of the IUPAC-recommended table.

translation The process by which the information contained in the codon of an mRNA molecule is converted to a protein structure.

transmutation The changing of one element into another.

triglyceride An ester of glycerol with three fatty acid units; also called a triacylglycerol.

triple bond The sharing of three pairs of electrons between two atoms.

tritium A radioactive isotope of hydrogen with two neutrons and one proton in the nucleus (hydrogen-3).

unsaturated hydrocarbon An alkene or alkyne; a hydrocarbon containing one or more double or triple bonds.

valence electrons Electrons in the outermost shell of an atom.

valence shell electron pair repulsion theory (VSEPR) A theory of chemical bonding useful in determining the shapes of molecules; it states that valence shell electron pairs locate themselves as far apart as possible.

vaporization The process by which a substance changes from the liquid to the gaseous (vapor) state.

variable A factor that changes during an experiment.

vitamin An organic compound that the body cannot produce in the quantity required for good health.

volatile organic compounds (VOCs) Compounds that cause pollution because they vaporize readily.

VSEPR theory See **valence shell electron pair repulsion theory.**

vulcanization The process of making naturally soft rubber harder by reaction with sulfur.

wax An ester of a long-chain fatty acid with a long-chain alcohol.

weak acid An acid that ionizes only slightly in water; a poor proton donor.

weak base A base that ionizes only slightly in water; a poor proton acceptor.

weight A measure of the force of attraction of Earth for an object.

wet scrubber A pollution control device that uses water or solutions to remove pollutants from smokestack gases.

X-rays Radiation similar to visible light but of much higher energy and much more penetrating.

zwitterion A molecule that contains both a positive charge and a negative charge; a dipolar ion.

Answers are provided for *all in-chapter exercises*. Brief answers are given for *selected review questions*; more complete answers can be obtained by reviewing the text. Answers are provided for *all odd-numbered Problems and Additional Problems*.

NOTE: For numerical problems, your answer may vary slightly from ours because of rounding and the use of significant figures (see Appendix).

Chapter 1

1.1 **A.** **a.** DQ would probably be small.
   b. DQ would probably be large.
   B. **a.** DQ would be uncertain.
   b. DQ would probably be large.

1.2 **A.** **a.** 1.00 kg   **b.** 179 lb
   B. **a.** 52.5 kg   **b.** 496 lb

1.3 **A.** physical: a, c; chemical: b
   B. physical: b, d; chemical: a, c

1.4 **A.** elements: He, No, Os; compounds: CuO, NO, KI
   B. 8

1.5 **a.** 7.24 kg **b.** 4.29 μm **c.** 7.91 ms **d.** 2.29 cg **e.** 7.90 Mm

1.6 **A.** **a.** 7.45×10^{-9} m   **b.** 5.25×10^{-3} s
      c. 1.415×10^3 m   **d.** 2.06×10^{-3} m
   B. **a.** 2.84×10^{-7} m   **b.** 1.19×10^{-1} s
      c. 7.54×10^5 m   **d.** 6.19×10^3 m

1.7 **A.** 500 cm^2 **B.** 1600 cm^3

1.8 **a.** 7.5×10^3 mm   **b.** 2.056 L
   c. 2.06×10^6 μg   **d.** 7.38 mm

1.9 **A.** ice   **B.** Padouk floats, ebony sinks.

1.10 **A.** 1.11 g/mL   **B.** 11.2 g/cm^3

1.11 **A.** 253 g   **B.** 819 g

1.12 **A.** 486 cm^3 **B.** 341 mL

1.13 **A.** 373 K   **B.** 195 K

1.14 **A.** 1799 kcal   **B.** 8.6 kcal

1. Science is testable, reproducible, explanatory, predictive, and tentative. Testability best distinguishes science.
2. Technological progress has enabled food production to keep pace with population growth in most of the world.
3. These problems usually involve too many variables to be treated by scientific methods.
5. Risk–benefit analysis compares benefits of an action to risks of that action.
7. A desirability quotient (DQ) is benefits divided by risks. A large DQ means that risks are minimal compared to benefits.
8. **(a)** grams;   **(b)** millimeters or centimeters
11. **(a)** applied;   **(b)** applied;   **(c)** basic
13. Penicillin has saved thousands of lives, causing harm to a very few. The DQ for penicillin is large.
15. **(a)** high;   **(b)** high;   **(c)** low
17. Food producers benefit through increased profits. The consumer assumes the greatest risk.
19. 2 kg
21. yes
23. 250 mL
25. 1.46×10^9 km^3
27. 18.7 mm
29. physical property: a, c, d; chemical property: b
31. physical change: b, c; chemical change: a
33. substance: a, b; mixture: c, d, e, f
35. homogeneous: a, d; heterogeneous: b, c
37. a substance; the law of definite proportions
39. elements: a, b, d; compound: c
41. **a.** Cl; **b.** Fe; **c.** P; **d.** Pu
43. f
45. **a.** 8.01 μg   **b.** 7.9 mL   **c.** 1.05 km
47. **a.** 0.0374 L   **b.** 1.55×10^5 m   **c.** 198 mg
   d. 1.19×10^4 cm^2 **e.** 0.078 ms
49. **a.** cm   **b.** kg **c.** dL

51. 1.00 mL; 15.3 mL
53. **a.** 1.17 g/mL   **b.** 1.26 g/mL
55. **a.** 120 g   **b.** 490 g
57. **a.** 53.1 cm^3 **b.** 18.7 mL
59.

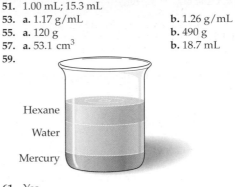

Hexane

Water

Mercury

61. Yes
63. 14.9 cm
65. 310 K
67. 38.5 kcal
69. The brass weight
71. cabbage (1.65 kg) < potatoes (~2.3 kg) < sugar (2.5 kg)
73. 5.23 g/cm^3
75. 597 g
77. 6.73×10^{-6} cm
79. 0.908 metric tons
81. 4.03 km
83. **a.** 1.3 g/cm^3; **b.** 5.4 g/cm^3; **c.** 0.688 g/cm^3; Jupiter and Earth are more dense than water, Saturn is less dense and would float (in a large enough ocean!).
85. 0.778 g/mL

Chapter 2

2.1 **A.** 1.19 g nitrogen   **B.** 63.0 g nitrogen
2.2 3 atoms hydrogen/1 atom arsenic
3. **a.** No; **b.** No.
6. A given compound always contains the same elements in the same proportions by mass. Zinc sulfide always contains the same proportion of zinc to sulfur. That proportion is 65.39 g of zinc to 32.066 g of sulfur.
9. C has 15 oxygen atoms; the initial mixture only has 14.
10. law of multiple proportions
12. That alumina is not an element.
13. law of definite proportions
15. No. The helium atoms have escaped into the air through the pores in the balloon.
17. No. In the open vessel, a gas has escaped and is not part of the weight after reaction.
19. No. They have become a part of the gas carbon dioxide.
21. Yes. Total mass before and after the reaction is the same, 1.2000 g.
23. (b)
25. 30.8 g hydrogen
27. 260 g carbon
29. C:O ratio for carbon suboxide is 1.000:0.888. Ratio of 1.332:0.888 = 3:2; 2.664:0.888 = 3:1.
31. SnO$_2$
33. The S:F ratios are (T) 1.00:2.37 and (U) 1.00:3.56. 3.56:2.37 = 1.50:1 or 3:2.
35. **a.** Yes; Dalton assumed that atoms of different elements had different masses. **b.** No; Dalton assumed that atoms of one element differ (in mass) from atoms of any other element.
37. Contradict. Dalton regarded atoms as indivisible.
39. (d)
41. No; the ratio of carbon to hydrogen is different for the different samples.
43. 0.4467 g nitrogen
45. Four hydrogen atoms
47. conservation of mass
49. 108.0 g of mercury oxide
51. 1.334:1.000 or 4:3

Chapter 3

3.1 A. 78 neutrons   **B.** 40; potassium-40
3.2 A. $^{90}_{37}X$ and $^{88}_{37}X$; $^{88}_{38}X$ and $^{93}_{38}X$ **B.** three
3.3 A. 32 electrons   **B.** 3
3.4 a. (Be) 2 2
   b. (Mg) 2 8 2; both have the same number of outer electrons.
3.5 A. a. $1s^22s^22p^5$
   b. $1s^22s^22p^63s^23p^5$ (Both have the valence configuration ns^2np^5.)
   B. a. $1s^22s^22p^63s^23p^63d^24s^2$
   b. $1s^22s^22p^63s^23p^63d^{10}4s^24p^1$
3.6 A. a. (Rb) $5s^1$ **b.** (Se) $4s^24p^4$ **c.** (Ge) $4s^24p^2$
   B. a. (Ga) $4s^24p^1$ **b.** (In) $5s^25p^1$
2. Radioactivity is spontaneous radiation from an atomic nucleus. Dalton's atomic theory stated that atoms were indestructible.
5. A and B, no; A and C, no; A and D, yes; B and C, yes
8. Cf; californium; 251 u
13. a. 2 **b.** 11 **c.** 17 **d.** 8 **e.** 12 **f.** 16
15. a. $^{40}_{19}K$, potassium-40   **b.** $^{48}_{22}Ti$; titanium-48
17. 3
19. a. 6 **b.** 16 **c.** 9
21. a. Excited state; the outer two electrons should be in the 2p subshell, which fills before the 3s.
   b. Excited state; the outer electron should be in the 2p subshell, which can hold up to 6 electrons and fills before the 3s.
   c. Incorrect; there is no 2d orbital.
   d. Incorrect; an s subshell can only hold two electrons.
23. Argon
25. Mg and Ca both have the valence configuration ns^2np^5, but Ca has one more electron shell than Mg.
27. metal, a and b; nonmetal, c and d
29. Ne and Kr
31. Four
33. Move up one atomic number for each electron added; move down one atomic number for each electron removed.
35.

Chapter 4

4.1 **a.** $:\ddot{A}r:$ **b.** $\cdot Ca\cdot$ **c.** $:\ddot{F}\cdot$ **d.** $:\dot{N}\cdot$ **e.** $K\cdot$ **f.** $:\dot{S}\cdot$
4.2 **A.** $Li\cdot + :\ddot{F}\cdot \longrightarrow Li^+ + :\ddot{F}:^-$
   B. $Rb\cdot + :\ddot{I}\cdot \longrightarrow Rb^+ + :\ddot{I}:^-$
4.3 $2\cdot Al\cdot + 3:\ddot{O}\cdot \longrightarrow 2Al^{3+} + 3:\ddot{O}:^{2-}$
4.4 **A. a.** CaF_2 **b.** Li_2O
   B. $FeCl_2$ and $FeCl_3$
4.5 **a.** K_2O **b.** Ca_3N_2 **c.** CaS
4.6 **a.** calcium fluoride   **b.** copper(II) bromide
4.7 **a.** bromine trifluoride   **b.** bromine pentafluoride
   c. dinitrogen monoxide   **d.** dinitrogen pentoxide
4.8 **a.** PCl_3 **b.** Cl_2O_7 **c.** NI_3 **d.** S_2Cl_2
4.9
   a. $:\ddot{Br}\cdot + \cdot\ddot{Br}: \longrightarrow :\ddot{Br}:\ddot{Br}:$
   b. $H\cdot + \cdot\ddot{Br}: \longrightarrow H:\ddot{Br}:$
   c. $:\ddot{I}\cdot + \cdot\ddot{Cl}: \longrightarrow :\ddot{I}:\ddot{Cl}:$

4.10 A. a. polar covalent   **b.** ionic   **c.** nonpolar covalent
   B. a. polar covalent   **b.** polar covalent
      c. nonpolar covalent
4.11 A. a. $Ca(CH_3CO_2)_2$ **b.** NH_4NO_3 **c.** $KMnO_4$
   B. a. 2 **b.** 4
4.12 a. calcium carbonate   **b.** magnesium phosphate
   c. potassium chromate   **d.** ammonium dichromate
4.13 A. a. $:\ddot{F}-\ddot{O}-\ddot{F}:$ **b.** $H-\overset{\overset{\displaystyle H}{|}}{\underset{\underset{\displaystyle H}{|}}{C}}-\ddot{C}l:$
   B. a. $\left[:\ddot{N}=N=\ddot{N}:\right]^-$ **b.** $:\ddot{O}::N:\ddot{O}:\ddot{F}:$
4.14 A. a. pyramidal   **b.** triangular
   B. a. triangular   **b.** bent (at both O)
1. Sodium metal is quite reactive; sodium ions (as in NaCl) are quite unreactive.
4. a. 2A (or 2)   **b.** 5A (or 15)   **c.** 7A (or 17)
6. N and C can form triple bonds.
7. a. $\cdot Mg\cdot$ **b.** $:\ddot{O}:$ **c.** $\cdot\dot{S}i\cdot$
9. a. $Ca^{2+} \; 2\left[:\ddot{F}:\right]^-$ **b.** $2Na^+ \left[:\ddot{S}:^{2-}\right]$
   c. $K^+ \left[:\ddot{I}:\right]^-$ **d.** $Al^{3+} \; 3\left[:\ddot{Cl}:^-\right]$
11. a. magnesium ion   **b.** sodium ion
   c. oxide ion   **d.** chloride ion
   e. silver ion   **f.** copper(I) ion
13. a. chromium(II) ion   **b.** chromium(III) ion
   c. chromium(VI) ion
15. a. Co^{2+} **b.** In^+ **c.** In^{3+}
17. a. sodium bromide   **b.** potassium chloride
   c. silver oxide   **d.** magnesium fluoride
   e. iron(II) chloride   **f.** iron(III) chloride
19. NaF, sodium fluoride, and SnF_2, tin(II) fluoride
21. a. KOH **b.** $MgCO_3$ **c.** $Fe_2(C_2O_4)_3$ **d.** $Fe(CN)_2$
23. a. sodium hydrogen sulfate   **b.** silver nitrite
   c. ammonium fluoride   **d.** copper(I) hydroxide
25. $H\cdot + \cdot\ddot{F}: \longrightarrow H:\ddot{F}:$
27. $\cdot\dot{P}\cdot + 3H\cdot \longrightarrow H:\overset{H}{\underset{}{\ddot{P}}}:H$
29. $\cdot\dot{C}\cdot + 4:\ddot{F}\cdot \longrightarrow :\overset{:\ddot{F}:}{\underset{:\ddot{F}:}{\ddot{F}:C:\ddot{F}:}}$
31. a. N_2O_4 **b.** $BrCl_3$ **c.** NI_3
33. a. carbon disulfide   **b.** phosphorus pentafluoride
   c. dinitrogen tetrasulfide
35. a. $:\overset{:\ddot{F}:}{\underset{:\ddot{F}:}{\ddot{F}-Si-\ddot{F}:}}$ **b.** $:\overset{H\;\;H}{\underset{H\;\;H}{N-N}}:$
   c. $H-\overset{\overset{H\;\;H}{|\;\;|}}{\underset{\underset{H\;\;H}{|\;\;|}}{C-N}}:$ **d.** $:\ddot{F}-\overset{:O:}{\underset{}{||C}}-\ddot{F}:$
   e. $:\overset{H\;\;H}{\underset{H}{N-\ddot{O}}}:$ **f.** $:\ddot{Cl}-\dot{S}-\ddot{Cl}:$

37.

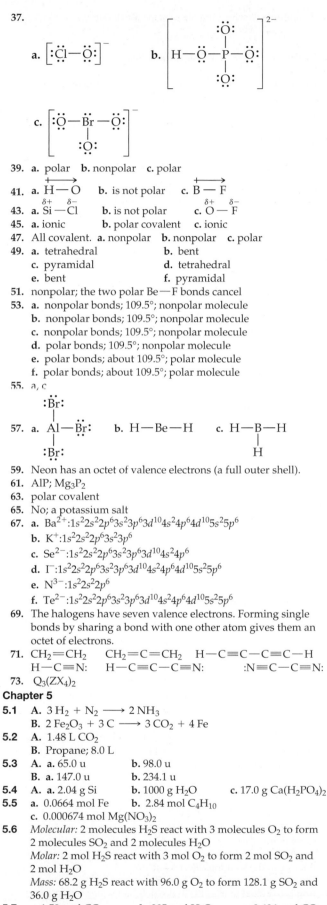

a. $\left[\,:\!\overset{..}{\underset{..}{Cl}}\!-\!\overset{..}{\underset{..}{O}}\!:\,\right]^{-}$

b. $\left[\,H\!-\!\overset{..}{\underset{..}{O}}\!-\!\underset{\overset{|}{:}\overset{..}{\underset{..}{O}}:}{\overset{:\overset{..}{O}:}{\overset{|}{P}}}\!-\!\overset{..}{\underset{..}{O}}\!:\,\right]^{2-}$

c. $\left[\,:\!\overset{..}{\underset{..}{O}}\!-\!\underset{\overset{|}{:}\overset{..}{\underset{..}{O}}:}{\overset{..}{Br}}\!-\!\overset{..}{\underset{..}{O}}\!:\,\right]^{-}$

39. **a.** polar **b.** nonpolar **c.** polar
41. **a.** H⟶O **b.** is not polar **c.** B⟶F
43. **a.** $\overset{\delta+}{Si}\!-\!\overset{\delta-}{Cl}$ **b.** is not polar **c.** $\overset{\delta+}{O}\!-\!\overset{\delta-}{F}$
45. **a.** ionic **b.** polar covalent **c.** ionic
47. All covalent. **a.** nonpolar **b.** nonpolar **c.** polar
49. **a.** tetrahedral **b.** bent
    c. pyramidal **d.** tetrahedral
    e. bent **f.** pyramidal
51. nonpolar; the two polar Be—F bonds cancel
53. **a.** nonpolar bonds; 109.5°; nonpolar molecule
    b. nonpolar bonds; 109.5°; nonpolar molecule
    c. nonpolar bonds; 109.5°; nonpolar molecule
    d. polar bonds; 109.5°; nonpolar molecule
    e. polar bonds; about 109.5°; polar molecule
    f. polar bonds; about 109.5°; polar molecule
55. a, c

57. **a.** $\underset{\overset{|}{:}\overset{..}{\underset{..}{Br}}:}{\overset{:\overset{..}{Br}:}{\overset{|}{Al}}}\!-\!\overset{..}{\underset{..}{Br}}\!:$ **b.** H—Be—H **c.** $\underset{\overset{|}{H}}{\overset{H}{\overset{|}{B}}}$ H—B—H

59. Neon has an octet of valence electrons (a full outer shell).
61. AlP; Mg_3P_2
63. polar covalent
65. No; a potassium salt
67. **a.** Ba^{2+}:$1s^22s^22p^63s^23p^63d^{10}4s^24p^64d^{10}5s^25p^6$
    b. K^+:$1s^22s^22p^63s^23p^6$
    c. Se^{2-}:$1s^22s^22p^63s^23p^63d^{10}4s^24p^6$
    d. I^-:$1s^22s^22p^63s^23p^63d^{10}4s^24p^64d^{10}5s^25p^6$
    e. N^{3-}:$1s^22s^22p^6$
    f. Te^{2-}:$1s^22s^22p^63s^23p^63d^{10}4s^24p^64d^{10}5s^25p^6$
69. The halogens have seven valence electrons. Forming single bonds by sharing a bond with one other atom gives them an octet of electrons.
71. $CH_2{=}CH_2$ $CH_2{=}C{=}CH_2$ $H{-}C{\equiv}C{-}C{\equiv}C{-}H$
    $H{-}C{\equiv}N:$ $H{-}C{\equiv}C{-}C{\equiv}N:$ $:N{\equiv}C{-}C{\equiv}N:$
73. $Q_3(ZX_4)_2$

Chapter 5

5.1 **A.** $3H_2 + N_2 \longrightarrow 2NH_3$
    B. $2Fe_2O_3 + 3C \longrightarrow 3CO_2 + 4Fe$
5.2 **A.** 1.48 L CO_2
    B. Propane; 8.0 L
5.3 **A. a.** 65.0 u **b.** 98.0 u
    B. a. 147.0 u **b.** 234.1 u
5.4 **A. a.** 2.04 g Si **b.** 1000 g H_2O **c.** 17.0 g $Ca(H_2PO_4)_2$
5.5 **a.** 0.0664 mol Fe **b.** 2.84 mol C_4H_{10}
    c. 0.000674 mol $Mg(NO_3)_2$
5.6 *Molecular:* 2 molecules H_2S react with 3 molecules O_2 to form 2 molecules SO_2 and 2 molecules H_2O
    Molar: 2 mol H_2S react with 3 mol O_2 to form 2 mol SO_2 and 2 mol H_2O
    Mass: 68.2 g H_2S react with 96.0 g O_2 to form 128.1 g SO_2 and 36.0 g H_2O
5.7 **a.** 1.59 mol CO_2 **b.** 305 mol H_2O **c.** 0.606 mol CO_2
5.8 **A.** 0.763 g O_2
    B. a. 2130 g CO_2 **b.** 2350 g CO_2

5.9 **A.** 0.00870 M **B.** 0.968 M
5.10 **A. a.** 0.0918 M **b.** 0.00756 M
    B. 0.0137 M
5.11 **a.** 673 g KOH **b.** 5.61 g KOH
5.12 **A.** 29.7 mL **B.** 6.30×10^{-4}g HNO_3
5.13 **A.** 90.7% ethanol **B.** 34.8% toluene
5.14 **A.** 2.81% H_2O_2 **B.** 51.3% NaOH
5.15 **A.** Dissolve 5.62 g glucose in enough water to make 125 g solution.
    B. Dissolve 15.6 g NaCl in enough water to make 1750 g solution.
  2. The atomic mass (O = 15.9994 u) represents the mass of one atom, which can occur in many compounds. The formula mass (31.9988 u) represents the mass of a molecule of the element as it occurs in nature—as O_2.
  4. Matter is neither created nor destroyed in a chemical reaction; thus, there must be the same number of reactant atoms as product atoms.
  7. **a.** 8 **b.** 2 **c.** 6
  9. 3 N; 3 S; 15 H; 12 O
 11. **a.** Two formula units of Na_2O_2 react with two molecules of H_2O, forming four formula units of NaOH and one molecule of O_2.
    b. 2 mol Na_2O_2 react with 2 mol H_2O, forming 4 mol NaOH and 1 mol O_2.
    c. 156 g Na_2O_2 reacts with 36 g H_2O, forming 160 g NaOH and 32 g O_2.
 13. **a.** $2K + O_2 \longrightarrow K_2O_2$
    b. $FeCl_2 + Na_2SiO_3 \longrightarrow 2NaCl + FeSiO_3$
    c. $3F_2 + 2AlCl_3 \longrightarrow 2AlF_3 + 3Cl_2$
 15. **a.** $4Fe + 3O_2 \longrightarrow 2Fe_2O_3$
    b. $CaCO_3 + 2HCl \longrightarrow CaCl_2 + CO_2 + H_2O$
    c. $C_7H_{16} + 11O_2 \longrightarrow 7CO_2 + 8H_2O$
 17.

Hydrogen gas (two volumes)   Oxygen gas (one volume)   Steam (two volumes)

 19. **a.** 82.4 L **b.** 4.46 mL
 21. 4 times
 23. **a.** 6.02×10^{23} P_4 molecules **b.** 2.41×10^{24} P atoms
 25. c
 27. **a.** 466.0 g **b.** 151.9 g **c.** 158.2 g **d.** 252.1 g
 29. **a.** 1030 g $CaSO_4$ **b.** 3.17 g $CuCl_2$ **c.** 856 g $C_{12}H_{22}O_{11}$
 31. **a.** 0.0195 mol Sb_2S_3 **b.** 0.133 mol MoO_3
    c. 3.56 mol $AlPO_4$ **d.** 0.0887 mol $Be(NO_3)_2$
 33. **a.** 16.2 mol CO_2 **b.** 11.0 mol O_2
 35. **a.** 2480 g NH_3 **b.** 193 g H_2
 37. 266 g Mg_3P_2
 39. **a.** 9.36 M HCl **b.** 0.277 M Li_2CO_3
 41. **a.** 56.0 g NaOH **b.** 34.0 g $C_6H_{12}O_6$
 43. **a.** 0.208 L **b.** 1.80 L
 45. **a.** 4.83% water by vol. **b.** 5.09% acetone by vol.
 47. Add 277 g NaCl to 3098 g water.
 49. Add 40.0 mL acetic acid to enough water to make 2.00 L of solution.
 51. b
 53. **a.** $Hg(NO_3)_2(s) \longrightarrow Hg(l) + 2NO_2(g) + O_2(g)$
    b. $Na_2CO_3(aq) + 2HCl(aq) \longrightarrow$
        $H_2O(l) + CO_2(g) + 2NaCl(aq)$
 55. **a.** yes **b.** C_2H_2
 57. 2.65×10^9 g CaO
 59. 0.418 mol H_2O_2
 61. **a.** 49.3% ethanol **b.** 0.800% acetone
 63. **a.** 1.83×10^3 g of CaC_2 required
 65. 0.00825 L (8.25 mL)
 67. No. The mole ratios remain the same; the equation is properly balanced with the coefficients divided by 2.
 69. 53,000 kg
 71. 55.5 mol H_2O

Chapter 6

6.1 400 mm Hg

6.2 **A.** 1800 mL **B.** 503 mm Hg

6.3 **A.** **a.** 2.98 L **b.** 234 L

B. $-167\,°C$

6.4 **A.** 0.179 g/L

B. 1.29 g/L; more than 7 times that of He

6.5 867 mL

6.6 **A.** **a.** 0.0471 atm **b.** 4.99 L

B. 112 L

1. Both solids and liquids are difficult to compress because the particles are close together. Liquids flow because their particles are free to move; solids have fixed shapes because their particles have fixed positions.

5. Volume is inversely proportional to pressure and directly proportional to absolute temperature.

7. b, d

9. HCl

11. No. Benzene is nonpolar; it has no dipole to attract the Na^+ and Cl^- ions of NaCl.

13. **a.** 1450 mL **b.** 3470 mmHg

15. **a.** 9120 L **b.** 1140 min (19 h)

17. 117 mL

19. 433 K (160 °C)

21. **a.** 22.4 L **b.** 22.4 L **c.** 44.8 L

23. 3.74 g/L

25. **a.** 47.5 g **b.** 66.5 g

27. **a.** decrease **b.** decrease **c.** increase

29. **a.** temperature decreases **b.** pressure decreases

31. **a.** 22.3 L **b.** 29.1 atm

33. 0.0781 mol Kr

35. 3.00 g He

37. b

39. d; a mole of each gas occupies the same volume; PF_3 has the largest molar mass.

41. 0.394 L

43. 1.29 g/L; 0.804 g/L; Moist air is less dense than dry air.

45. 1.25 atm

47. C_4H_{10}

Chapter 7

7.1 **A.** **a.** $HBr(aq) \xrightarrow{H_2O} H^+(aq) + Br^-(aq)$

b. $Ca(OH)_2(s) \xrightarrow{H_2O} Ca^{2+}(aq) + 2\,OH^-(aq)$

B. $CH_3COOH(aq) \xrightarrow{H_2O} H^+(aq) + CH_3COO^-(aq)$

7.2 **A.** $HBr(aq) + H_2O \longrightarrow H_3O^+(aq) + Br^-(aq)$

B. $HBr(aq) + CH_3OH \longrightarrow CH_3OH_2^+(aq) + Br^-(aq)$

7.3 **A.** H_2SeO_3 **B.** HNO_3

7.4 **A.** $Sr(OH)_2$ **B.** KOH

7.5 **A.** $NaOH(aq) + HC_2H_3O_2(aq) \longrightarrow NaC_2H_3O_2(aq) + H_2O$

B. $2\,HCl(aq) + Mg(OH)_2(aq) \longrightarrow MgCl_2(aq) + 2\,H_2O$

7.6 **A.** pH = 11 **B.** pH = 3

7.7 **A.** 1.0×10^{-2} M (0.010 M) **B.** 1.0×10^{-3} M (0.0010 M)

7.8 (c)

7.9 **a.** CN^- **b.** H_2O **c.** HSO_4^- **d.** H_2CO_3

1. **a.** Arrhenius: forms H^+ in water; Brønsted–Lowry: proton donor
   b. Arrhenius: forms OH^- in water; Brønsted–Lowry: proton acceptor
   c. Ionic compound from acid–base neutralization

2. No effect on litmus; no reaction with iron or zinc

4. No; a Brønsted–Lowry acid must have a hydrogen atom to donate H^+. No, but some atom in the base must have a lone pair of electrons to accept a proton.

5. A strong acid ionizes to a much greater extent than does a weak acid.

6. In acid–base chemistry the proton is a H^+ ion. In nuclear chemistry the proton usually represents a subatomic particle within a nucleus.

9. $Mg(OH)_2$ is insoluble in water so that few $OH^-(aq)$ ions form.

10. A condition in which the blood is too alkaline

13. $HIO_3(aq) \xrightarrow{H_2O} H^+(aq) + IO_3^-(aq)$

15. $LiOH(s) \xrightarrow{H_2O} Li^+(aq) + OH^-(aq)$

17. base: a, c; acid: b

19. $HCl(g) + H_2O \longrightarrow H_3O^+(aq) + Cl^-(aq)$; hydrochloric acid

21. $NH_3(aq) + H_2O \longrightarrow NH_4^+(aq) + OH^-(aq)$

23. **a.** HCl **b.** $Sr(OH)_2$ **c.** KOH **d.** H_3BO_3

25. **a.** nitric acid (an acid) **b.** cesium hydroxide (a base)
    c. carbonic acid (an acid)

27. **a.** HNO_2 **b.** H_3PO_3

29. **a.** H_2SO_4 (an acid) **b.** $Mg(OH)_2$ (a base)

31. strong base

33. weak acid

35. **a.** weak acid **b.** strong base **c.** salt **d.** strong acid

37. highest: a; lowest, b

39. **a.** $HI(aq) \xrightarrow{H_2O} H^+(aq) + I^-(aq)$

b. $LiOH(s) \xrightarrow{H_2O} Li^+(aq) + OH^-(aq)$

c. $HClO_2(aq) \xrightarrow{H_2O} H^+(aq) + ClO_2^-(aq)$

41. **a.** $HClO_2(aq) + H_2O \longrightarrow H_3O^+(aq) + ClO_2^-(aq)$
    b. $HNO_2(aq) + H_2O \longrightarrow H_3O^+(aq) + NO_2^-(aq)$
    c. $HCN(aq) + H_2O \longrightarrow H_3O^+(aq) + CN^-(aq)$

43. **a.** $KOH(aq) + HCl(aq) \longrightarrow KCl(aq) + H_2O$
    b. $LiOH(aq) + HNO_3(aq) \longrightarrow LiNO_3(aq) + H_2O$

45. $H_3PO_4(aq) + 3\,NaOH(aq) \longrightarrow Na_3PO_4(aq) + 3\,H_2O$

47. acidic: a, c; basic: d; neutral: b

49. 11

51. 1.0×10^{-12} M H^+

53. 10 and 11

55. **a.** Cl^- **b.** NH_4^+

57. $3\,HCl(aq) + Al(OH)_3(s) \longrightarrow AlCl_3(aq) + 3\,H_2O$
    $2\,HCl(aq) + Mg(OH)_2(s) \longrightarrow MgCl_2(aq) + 2\,H_2O$

59. No. Covalent OH-containing compounds do not produce OH^- ions in water.

61. $SrCO_3(aq) + 2\,HI(aq) \longrightarrow SrI_2(aq) + CO_2(g) + H_2O(l)$

63. **a.** 2 **b.** 3

65. **a.** 11 **b.** 12

67. 100 g stomach acid (500 mg HCl)

69. $HS^-(aq) + H_2O \longrightarrow H_2S(g) + OH^-(aq)$

Chapter 8

8.1 oxidation: a, b, c, d

8.2 reduction: a; oxidation: b

8.3 reduction: a; oxidation: b, c, d

8.4 **a.** oxidizing agent: O_2; reducing agent: Se
    b. oxidizing agent: CH_3CN; reducing agent: H_2
    c. oxidizing agent: V_2O_5; reducing agent: H_2
    d. oxidizing agent: Br_2; reducing agent: K

8.5 oxidation: $Al \longrightarrow Al^{3+} + 3\,e^-$; reduction: $Br_2 + 2\,e^- \longrightarrow 2\,Br^-$

8.6 **A.** half-reactions: $Fe \longrightarrow Fe^{3+} + 3\,e^-$; $Mg^{2+} + 2\,e^- \longrightarrow Mg$
    overall: $2\,Fe + 3\,Mg^{2+} \longrightarrow 2\,Fe^{3+} + 3\,Mg$
    B. half-reactions:
    $Pb \longrightarrow Pb^{2+} + 2\,e^-$; $Ag(NH_3)_2^+ + e^- \longrightarrow Ag + 2\,NH_3$
    overall: $Pb + 2\,Ag(NH_3)_2^+ \longrightarrow Pb^{2+} + 2\,Ag + 4\,NH_3$

8.7 **A.** $2\,Zn + O_2 \longrightarrow 2\,ZnO$

B. $Se + O_2 \longrightarrow SeO_2$

8.8 **A.** $2\,PbS + 3\,O_2 \longrightarrow 2\,PbO + 2\,SO_2$

B. $C_2H_5OH + 3\,O_2 \longrightarrow 2\,CO_2 + 3\,H_2O$

1. Oxidation: oxidation number increases; reduction: oxidation number decreases.

3. To allow ions to flow from one compartment to the other and, thus, keep the solutions electrically neutral.

6. Zinc; it is the anode; oxidized to Zn^{2+}.

9. Iron is oxidized to iron(III) hydroxide; salt water acts as an electrolyte.

11. Silver is oxidized by hydrogen sulfide to (black) silver sulfide; use aluminum to reduce the silver sulfide to metallic silver.

13. a. reduced (Cr^{3+} gains e^-)   **b.** oxidized (H_2O_2 loses H)
   c. oxidized (Ca loses e^-)   **d.** reduced (S_8 gains e^-)
   e. oxidized (C_2H_6O loses H)

15. a. $Mg \longrightarrow Mg^{2+} + 2\,e^-$
   b. $Al \longrightarrow Al^{3+} + 3\,e^-$
   c. $Fe \longrightarrow Fe^{2+} + 2\,e^-$; $Fe \longrightarrow Fe^{3+} + 3\,e^-$

17. a. oxidizing agent: CuS; reducing agent: H_2
   b. oxidizing agent: CCl_4; reducing agent: K
   c. oxidizing agent: C_3H_4; reducing agent: H_2
   d. oxidizing agent: Fe^{3+}; reducing agent: Ce^{3+}

19. Ag^+

21. oxidizing agent: MnO_2; reducing agent: Zn

23. a. oxidation: $Al \longrightarrow Al^{3+} + 3\,e^-$; reduction: $2\,H^+ + 2\,e^- \longrightarrow H_2$
   b. oxidation: $Mg \longrightarrow Mg^{2+} + 2\,e^-$; reduction: $Cu^+ + e^- \longrightarrow Cu$

25. a. oxidation: $2\,I^- \longrightarrow I_2 + 2\,e^-$; reduction: $Cl_2 + 2\,e^- \longrightarrow 2\,Cl^-$
   overall: $2\,I^- + Cl_2 \longrightarrow I_2 + 2\,Cl^-$
   b. oxidation: $SO_2 + 2\,H_2O \longrightarrow H_2SO_4 + 2\,H^+ + 2\,e^-$
   reduction: $HNO_3 + H^+ + e^- \longrightarrow NO_2 + H_2O$
   overall: $SO_2 + 2\,HNO_3 \longrightarrow 2\,NO_2 + H_2SO_4$

27. a. H_2CO is oxidized; H_2O_2 is the oxidizing agent
   b. C_2H_6O is oxidized, MnO_4^- is the oxidizing agent

29. Reduced; acetylene gains hydrogen.

31. MoO_3 is reduced; the reducing agent is H_2.

33. Zr was oxidized; water was the oxidizing agent

35. Nitrite ion is reduced; ascorbic acid is the reducing agent.

37. a. SO_2 **b.** $CO_2 + H_2O$ **c.** $CO_2 + H_2O$

39. Indoxyl is oxidized; O_2 is the oxidizing agent.

41. $2\,Al(s) + 3\,Cu^{2+}(aq) \longrightarrow 2\,Al^{3+}(aq) + 3\,Cu(s)$

43. $H_2S + Pb^{2+} \longrightarrow PbS + 2\,H^+$
   $S^{2-} + 4\,H_2O_2 \longrightarrow SO_4^{2-} + 4\,H_2O$

45. 10,800 L air; 2300 L O_2

47. a. V^{2+} is oxidized; it loses an electron in going to V^{3+}.
   b. Neither; there are two O atoms per N atom in both NO_2 and N_2O_4.
   c. CO is reduced; C loses O and gains H.

49. a. CaH_2 is oxidized (loses H) and Na is reduced (gains H).
   b. CaH_2 is reduced (to Ca metal) and Na is oxidized (to Na^+). The two definitions appear contradictory because the hydrogen is in an unusual oxidation state, –1. (Actually CaH_2 is reduced and Na is oxidized.)

51. $2\,Al(s) + 6\,H_2O(l) \longrightarrow 2\,Al(OH)_3 + 3\,H_2$; H_2O is the oxidizing agent and Al is the reducing agent.

53. $C_{12}H_{22}O_{11}(s) + 12\,O_2(g) \longrightarrow 12\,CO_2(g) + 11\,H_2O(l)$

Chapter 9

9.1 a. molecular: C_6H_{14}; complete structural:

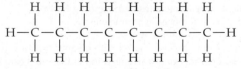

condensed structural: $CH_3CH_2CH_2CH_2CH_2CH_3$
   b. molecular: C_8H_{18}; complete structural:

condensed structural: $CH_3CH_2CH_2CH_2CH_2CH_2CH_2CH_3$

9.2 A.

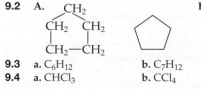

B. C_nH_{2n}

9.3 a. C_6H_{12} **b.** C_7H_{12}
9.4 a. $CHCl_3$ **b.** CCl_4

9.5 a. $CH_3CH_2CH_2CH_2OH$ **b.**

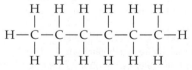

9.6 a. alcohol   **b.** ether   **c.** ether   **d.** phenol   **e.** ether
9.7 a. $CH_3OCH_2CH_2CH_3$ **b.** $CH_3CH_2OC(CH_3)_3$
9.8 a. ketone   **b.** aldehyde   **c.** aldehyde
9.9 A. a. $CH_3CH_2CH_2COOH$ **b.** CH_3CHO
   c. $CH_3CH_2COCH_2CH_3$
   B. a. $CH_3CH_2CH_2CH_2CH_2COOH$
   b. $CH_3CH_2COCH_2CH_2CH_2CH_3$
   c. $CH_3CH_2CH_2CH_2CH_2CH_2CHO$
9.10 a. $CH_3CH_2CH_2CH_2NH_2$ **b.** $CH_3CH_2NHCH_2CH_3$
   c. $CH_3NHCH_2CH_2CH_3$ **d.** $(CH_3)_2CHNHCH_3$
9.11 a. amine (NH)   **b.** amide (CONH)
   c. amine (N)   **d.** both (NH and $CONH_2$)
9.12 heterocyclic compounds: (a) and (d)

1. Carbon atoms can bond strongly to each other; can bond strongly to other elements; can form chains, rings, and other kinds of structures.
3. Isomers have the same molecular formula but different structural formulas.
4. Contains double or triple bond(s); alkenes and alkynes.
5. A set of six delocalized electrons
7. 1–4 C, gases; 5–16 C, liquids; >18 C, solids
8. All are less dense than water; hexane will float on water.
9. a. ethanol   **b.** 2-propanol (isopropyl alcohol)   **c.** methanol
10. Liver disease, nerve damage.
11. Historical use: anesthetic; modern use: solvent
13. a. alcohol   **b.** ketone   **c.** ester
   d. ether   **e.** carboxylic acid   **f.** aldehyde
15. organic: a, c; inorganic: b, d
17. a. 4   **b.** 8   **c.** 7   **d.** 5
19. a. C_9H_{20} **b.** $C_{13}H_{28}$
21. a. propane   **b.** acetylene (ethyne)   **c.** ethylene (ethene)
23. a. C_6H_{14}; $CH_3CH_2CH_2CH_2CH_2CH_3$
   b. C_8H_{18}; $CH_3CH_2CH_2CH_2CH_2CH_2CH_2CH_3$
25. a. methyl   **b.** *sec*-butyl
27. a. ethanol   **b.** butyl alcohol (1-butanol)
   c. CH_3OH **d.** $CH_3CH_2CH_2OH$
29.

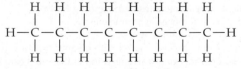

31. a. $CH_3CH_2OCH_2CH_3$ **b.** $CH_3CH_2CH_2CH_2OCH_3$
33. a. CH_3CHO **b.** HCHO
   c. propionaldehyde (propanal)
   d. methyl propyl ketone (2-pentanone)
35. a. formic acid (methanoic acid)
   b. propionic acid (propanoic acid)
   c. $CH_3CH_2CH_2COOH$ **d.** $CH_3CH_2CH_2CH_2CH_2CH_2COOH$
37. a. $CH_3COOCH_2CH_3$ **b.** $CH_3CH_2CH_2COOCH_3$
39. a. $CH_3CH_2NH_2$ **b.** CH_3NHCH_3
   c. propylamine   **d.** ethylmethylamine
41. same: a, b; isomers, c
43. a. homologs   **b.** none of these
45. a. unsaturated; alkene   **b.** saturated; alkane
47. a. ester   **b.** aldehyde   **c.** amine
   d. ether   **e.** ketone   **f.** carboxylic acid
49. a. heterocyclic, amine   **b.** not heterocyclic, cycloalkane
51. $C_{18}H_{27}NO_3$; ether, phenol, amide
53. a. $CH_3C{\equiv}CCH(CH_3)CH_2CH_2CH_3 + 2\,H_2 \xrightarrow{\text{Ni}}$
   $CH_3CH_2CH_2CH(CH_3)CH_2CH_2CH_3$
   b. $CH_2{=}C(CH_3)CH_2CH_2CH_3 + H_2 \xrightarrow{\text{Ni}}$
   $(CH_3)_2CHCH_2CH_2CH_3$
55. a. $CH_3CH_2CH_2CH_2OH$ **b.** $CH_3CH_2CH(OH)CH_2CH_3$
   c. $(CH_3)_2CHCH(OH)CH_3$ **d.**

57. isomerism
59. Equation is balanced as is. 731 g
61. **a.** dispersion and (weak) dipole–dipole **b.** hydrogen bonds. The stronger hydrogen bonds cause ethanol to remain a liquid at room temperature.
63. **a.** HO—⟨benzene ring⟩—OH **b.** O=⟨cyclohexadiene ring⟩=O

Chapter 10

10.1 $CH_2{=}CH{-}C{\equiv}N$

10.2 A.
$+CH_2CH(OCH_3)CH_2CH(OCH_3)CH_2CH(OCH_3)CH_2CH(OCH_3)+$

B. $+CH_2CH(OCOCH_3)CH_2CH(OCOCH_3)-$
$\qquad\qquad\qquad\quad{-}CH_2CH(OCOCH_3)CH_2CH(OCOCH_3)+$

10.3 a. $+OCH_2CH_2COOCH_2CH_2COOCH_2CH_2COOCH_2CH_2CO+$
b. $[OCH_2CH_2(C{=}O)]_n$

2. PVC has a Cl atom on alternate C atoms.
3. Polymerization in which all the monomer atoms are incorporated in the polymer; a double bond.
4. Teflon is like PE but with all H atoms replaced by F atoms; it is chemically quite inert.
5. Polystyrene; styrene
6. **a.** HDPE **b.** PETE
9. Synthetic fibers; they are cheaper and have a wider range of properties.
13. HDPE has linear molecules; a closely packed, fairly crystalline structure gives it greater rigidity and higher tensile strength than LDPE.
15. **a.** $CH_2{=}CHCl$ **b.** $CF_2{=}CF_2$
17. **a.** $+CH_2CH_2CH_2CH_2CH_2CH_2CH_2CH_2+$
b. $+CH_2CF_2CH_2CF_2CH_2CF_2CH_2CF_2+$
19. **a.** $+CH_2CH(OCOCH_3)CH_2CH(OCOCH_3)-$
$\qquad\qquad\qquad{-}CH_2CH(OCOCH_3)CH_2CH(OCOCH_3)+$
b. $+CH_2CH(COOCH_3)CH_2CH(COOCH_3)-$
$\qquad\qquad\qquad{-}CH_2CH(COOCH_3)CH_2CH(COOCH_3)+$
21. $CH_2{=}CH{-}CH{=}CH_2$
23. When stretched, rubber's coiled molecules are straightened. When released, the molecules coil again.
25. Polybutadiene (made from butadiene); polychloroprene (from chloroprene); styrene-butadiene rubber (from styrene and butadiene).
27. $+CO(CH_2)_2CONH(CH_2)_4NHCO(CH_2)_2CONH(CH_2)_4NH+$
29. $+CH_2(C_6H_{10})CH_2OOC(C_6H_4)COOCH_2(C_6H_{10})-$
$^{-}CH_2OOC(C_6H_4)COO+$ (where C_6H_{10} is a cyclohexane ring and C_6H_4 is a benzene ring)
31. The long chains can entangle with one another; intermolecular forces are greatly multiplied in large molecules; large polymer molecules move more slowly than do small molecules.
33. The temperature at which the properties of the polymer change from hard, stiff, and brittle to rubbery and tough; rubbery materials such as automobile tires should have a low T_g; glass substitutes, a high T_g.
35. Monomer: a; repeating unit: b; polymer: c. addition polymerization
37. $CH_2{=}CHCl$ and $CH_2{=}CCl_2$
39.

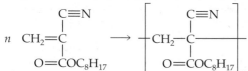

41. $\sim CH_2C(CH_3)_2CH_2C(CH_3)_2CH_2C(CH_3)_2CH_2C(CH_3)_2\sim$
43. $CH_3CH(OH)CH_2COOH$

45. **a.** HOOC—⟨naphthalene rings⟩—COOH and $HOCH_2{-}CH_2OH$
b. polyester
47. Elastomer; the golf ball
49. Carbonyl group (ketone); ethylene

51. LDPE; not crystalline, branched
53. $+NHCH_2CONHCH_2CONHCH_2CONHCH_2CO+$

Chapter 11

11.1 a. $^{226}_{88}Ra \longrightarrow {}^{4}_{2}He + {}^{222}_{86}Rn$ **b.** $^{24}_{11}Na \longrightarrow {}^{0}_{-1}e + {}^{24}_{12}Mg$
c. $^{188}_{79}Au \longrightarrow {}^{0}_{+1}e + {}^{188}_{78}Pt$ **d.** $^{37}_{18}Ar + {}^{0}_{-1}e \longrightarrow {}^{37}_{17}Cl$

11.2 A. 0.03825 mg **B.** 6.25%
11.3 A. 0.500 mg **B.** 420,000 y
11.4 A. 22,920 y **B.** 142 y
11.5 A. A neutron ($^1_0$n) **B.** Thorium-232 ($^{232}_{90}$Th)
11.6 Silicon-30 ($^{30}_{14}$Si)
11.7 Higher; the gas can be absorbed by inhalation and decay in the body.

2. (i), (b); (ii), (a); (iii), (c) 4. $^1_1H, {}^2_1H, {}^3_1H$
5. **a.** $^{59}_{27}Co$ **b.** $^{131}_{53}I$ **c.** $^{13}_{6}C$ **d.** $^{99}_{43}Tc$
6. **a.** 30 p, 32 n **b.** 94 p, 147 n **c.** 43 p, 56 n **d.** 36 p, 45 n
7. Isotopes: b, c **8.** b **9.** 157 **10.** 97
14. **a.** alpha **b.** gamma
17. Fission is the splitting of large atoms. Fusion is the joining of small atoms. Both liberate energy because in each case the nuclei formed have greater nuclear stability than the reacting nuclei (Figure 11.10).
19. **a.** $^{240}_{94}Pu \longrightarrow {}^{4}_{2}He + {}^{236}_{92}U$ **b.** $^{22}_{11}Na \longrightarrow {}^{22}_{10}Ne + {}^{0}_{+1}e$
c. $^{67}_{29}Cu \longrightarrow {}^{67}_{30}Zn + {}^{0}_{-1}e$
21. **a.** $^{179}_{79}Au \longrightarrow {}^{175}_{77}Ir + {}^{4}_{2}He$ **b.** $^{23}_{10}Ne \longrightarrow {}^{23}_{11}Na + {}^{0}_{-1}e$
c. $^{121}_{51}Sb + {}^{1}_{1}H \longrightarrow {}^{121}_{52}Te + {}^{1}_{0}n$
23. $^{99}_{42}Mo \longrightarrow {}^{99m}_{43}Tc + {}^{0}_{-1}e$
25. $^{24}_{12}Mg + {}^{1}_{0}n \longrightarrow {}^{1}_{1}H + {}^{24}_{11}Na$; sodium-24
27. $^{215}_{85}At \longrightarrow {}^{4}_{2}He + {}^{211}_{83}Bi$; astatine-215
29. $^{210}_{85}At$
31. 24.1 d
33. b
35. **a.** 3.0 μCi **b.** 0.19 μCi
37. 2 counts/min
39. 5730 y
41. 5730 y; 11,460 y
43. **a.** $^{121}_{51}Sb + {}^{4}_{2}He \longrightarrow {}^{124}_{53}I + {}^{1}_{0}n$ **b.** $^{124}_{53}I \longrightarrow {}^{124}_{52}Te + {}^{0}_{+1}e$
45. 1
47. $^{223}_{88}Ra \longrightarrow {}^{219}_{86}Rn + {}^{4}_{2}He$; $^{223}_{88}Ra \longrightarrow {}^{209}_{82}Pb + {}^{14}_{6}C$
49. 853 cm³; 131 cm³; smaller than a baseball ($\sim$200 cm³)
51. alpha particle
53. the neutron, $^1_0$n
55. **a.** 2.2×10^{13} cal; 2.2×10^{10} kcal **b.** 2.0×10^{8} bowls
57. 9.9×10^{17} atoms

Chapter 12

1. One substance enhances the effect of another; asbestos fibers and cigarette smoke.
2. Mining operations, particulate matter from crushing, high energy consumption
3. Cement mixed with sand and gravel
4. Found free in nature; easily shaped
5. Copper and tin are easier to obtain from ores than is iron.
8. The problem is keeping the metal in a usable form and not scattered in the environment.
11. Lithosphere: solid portion of Earth; hydrosphere: watery portion; atmosphere: gaseous mass surrounding Earth
13. Al, Fe, Ca; differences in atomic mass
15. Living and once-living matter
17. The 4 O atoms are also shared with neighboring Si atoms
19. The SiO_4 tetrahedra are arranged in flat two-dimensional arrays.
21. Sand (silicon dioxide), soda (sodium carbonate), and limestone (calcium carbonate)
23. Other metal oxides are substituted for silica, soda, or lime.
25. Limestone and clay
27. Iron ore or scrap iron (source of iron), limestone (removes impurities), and coke or coal (reducing agent)
29. Reduction; $Fe_2O_3 + 3\,CO \longrightarrow 2\,Fe + 3\,CO_2$
31. The furnace heats the iron ore, melts impurities as slag, and produces CO to reduce iron oxides to metallic iron.

33. Iron drawn off a blast furnace; phosphorus, silicon, and excess carbon
35. **a.** MnO_2 **b.** Si   **c.** Si   **d.** MnO_2
37. **a.** Cl_2 **b.** C   **c.** C   **d.** Cl_2
39. 114 g Cu
41. 120 million kg Al_2O_3; 250 million kg bauxite
43. 0.51 m
45. 1.7×10^7 kg ore per day
47. 282,000 cubic feet, 28 feet deep
49. In the figure, 33 mm in diameter, the lithosphere would be 0.09 mm thick, atmosphere would be about 0.26 mm thick; neither could be shown to scale properly (a thin pencil line is about 0.5 mm wide!).
51. 64.0% (paper, cardboard, yard wastes, plastics, and wood)
53. **a.** 3.4×10^9 kg   **b.** 7.7 km   **c.** 5.7×10^{10} km

Chapter 13

13.1 A. 3790 g H_2O **B.** 4940 kg CO_2
1. Refrigerant; molding of plastic foams
2. Hydrofluorocarbons; they are greenhouse gases.
5. Smoke, fog, sulfur dioxide, and particulate matter from burning of coal and other fuels.
8. No; however, some ozone-depleting gases (for example, CFCs) are also greenhouse gases.
9. Conversion of N_2 to a form usable by plants; increases the food supply.
11. Troposphere
13. **a.** $4 Fe(s) + 3 O_2(g) \longrightarrow 2 Fe_2O_3(s)$
    b. $4 Cr(s) + 3 O_2(g) \longrightarrow 2 Cr_2O_3(s)$
15. Particulate matter, sulfur oxides, carbon monoxide
17. $S + O_2 \longrightarrow SO_2$
19. $SO_3 + H_2O \longrightarrow H_2SO_4$
21. The lime reacts with and thus removes the SO_2;
    $CaCO_3(s) \xrightarrow{heat} CaO(s) + CO_2(g); CaO(s) + SO_2(g) \longrightarrow CaSO_3(s)$.
23. Hydrocarbons, NO_x, O_3, aldehydes, PAN
25. $2 NO + O_2 \longrightarrow 2 NO_2$
27. Lower operating temperature of vehicles and/or run the engine on a richer mixture (with more fuel and less air); use a catalyst for reduction of NO_x to N_2.
29. No
31. By hindering O_2 transport and thus adding to the workload of the heart
33. $O_2 + O \longrightarrow O_3$
35. Increase in skin cancer
37. Acids dissolve iron; $Fe(s) + 2 H^+(aq) \longrightarrow H_2(g) + Fe^{2+}(aq)$
39. Acids dissolve marble; $CaCO_3(s) + 2 H^+(aq) \longrightarrow$
    $H_2O(l) + CO_2(g) + Ca^{2+}(aq)$
41. Carbon monoxide from poorly ventilated heaters; nitrogen oxides from gas ranges; radon from the ground; particulates from wood burners and cigarette smoke
43. Increased risk for lung cancer; with proper ventilation
45. When CO levels exceed a maximum of 9 ppm over an 8-hour period
47. The slow warming of Earth caused by gases absorbing infrared radiation
49. Some waste heat is formed in every process (second law of thermodynamics).
51. 2.0×10^{12} t
53. Water vapor will form clouds and might even lead to cooling by reflecting sunlight back to space.
55. $2 PbS(s) + 3 O_2(g) \longrightarrow 2 PbO(s) + 2 SO_2(g); 2.67 \times 10^5$ g SO_2
57. $3 NO_2 + H_2O \longrightarrow 2 HNO_3 + NO; 2 NO + O_2 \longrightarrow 2 NO_2$

Chapter 14

14.1 A. a. 0.1 ppb   **b.** 100 ppt   **B.** 1×10^{-9} M
1. The gasoline would not dissolve; it would float.
3. Microorganisms that cause disease
4. Typhoid fever, cholera, dysentery; chemical treatment of drinking water.
5. They can leak contaminants such as gasoline and oils into underground water supplies.
7. Chlorinated hydrocarbons are unreactive and do not break down readily.

8. VOCs, MTBE, nitrate ions
9. Water expands when it freezes because it forms a rigid hydrogen-bonded structure with relatively large holes. The ice floats, protecting the deeper water from freezing.
11. 1.4×10^4 cal
13. Vaporization of water (sweat) cools our bodies.
15. Water evaporates from the skin, removing heat.
17. Dust, dissolved atmospheric gases such as carbon dioxide and oxygen, nitric acid from lightning storms; various pollutants
19. 51 mg/L
21. Acid rain and mine runoff
23. The limestone that neutralizes the acid leaves calcium ions in the water.
25. Pollution control devices on cars; power plant limestone scrubbers
27. About 1 ppm
29. Mottling of tooth enamel, possible kidney and thyroid problems
31. Sewage is collected in a pond; it removes solids that are allowed to settle.
33. Organic molecules and inorganic ions such as nitrates and phosphates
35. tertiary: a and b; secondary: c
37. $2 HNO_3(aq) + CaCO_3(s) \longrightarrow$
    $Ca^{2+}(aq) + 2 NO_3^-(aq) + CO_2(g) + H_2O(l)$
39. **a.** 14 ppb   **b.** 210 ppm
41. No. (0.009 mg Pb/L and 0.8 mg Ba/L)
43. **a.** $CHCl_3$ **b.** NaCl, phosphate ion, sand   **c.** sand
    d. NaCl and phosphate ion   **e.** phosphate ion
45. Boiling water, to boil water the molecules must be separated to relatively large distances.
47. Cl_2 is reduced; it is the oxidizing agent. SO_2 is the reducing agent.
49. a positron ($_{+1}^{0}e$)
51. The kettle with iron pellets will get hotter; Fe has a lower *specific heat* than water does.

Chapter 15

15.1 A. 58,500 J   **B.** 5.5¢
15.2 A. 91.3 kJ   **B.** 2700 kJ
15.3 A. 221 kJ   **B.** 1.43 kcal
1. Wood   2. Nuclear fusion
3. Coal; petroleum   5. fuels: a, b
7. **a.** $C_{12}H_{22}O_{11}(s) + 12 O_2(g) \longrightarrow 12 CO_2(g) + 11 H_2O(l)$
   b. $2 C_2H_2(g) + 5 O_2(g) \longrightarrow 4 CO_2(g) + 2 H_2O(l)$
9. $C(s) + O_2(g) \longrightarrow CO_2(g)$
11. $CH_4(g) + 2 O_2(g) \longrightarrow CO_2(g) + 2 H_2O(l)$
13. Yes. Both H_2 and CO are readily oxidized (will burn) forming water and CO_2, respectively.
15. A reaction goes faster at a higher temperature.
17. 4220 kJ
19. 4310 J
21. 803 kJ
23. Energy is conserved.
25. A measure of the degree of distribution of energy in a system; increased.
27. It is not possible to create energy.
29. Coal is plentiful and a good fuel, but burning it leads to significant air pollution from particulates, SO_x, and CO.
31. 41.7 years
33. 58.0 years
35. $2 C_2H_6(g) + 7 O_2(g) \longrightarrow 4 CO_2(g) + 6 H_2O(g)$
37. Ancient marine animals
39. By distillation, cracking, reforming; tetraethyllead is a highly effective octane booster but is quite toxic and deposits lead in the environment.
41. Asphalt is a residue left from petroleum distillation.
43. 19.3%
45. No; the uranium is enriched to only about 3% uranium-235; to explode, it would have to be enriched to about 90%.
47. Yes.
49. Highly reactive sodium metal must be used as a coolant; the fissile plutonium-239 could be used to make bombs.

51. Extremely high temperatures required, no physical container able to withstand reaction conditions

53. $^{232}_{90}\text{Th} + ^1_0\text{n} \longrightarrow ^{233}_{90}\text{Th}$
$^{233}_{90}\text{Th} \longrightarrow ^{0}_{-1}\text{e} + ^{233}_{91}\text{Pa}$
$^{233}_{91}\text{Pa} \longrightarrow ^{0}_{-1}\text{e} + ^{233}_{92}\text{U}$

55. 2.24×10^7 mol CH_4; 360 t CH_4

57. An electronic device that produces electricity directly from light.

59. 12 m^2; because it works only when the sun shines, energy storage is needed for nighttime and very cloudy days.

61. Plant material used directly as fuel

63. Fuel is fed into a fuel cell continuously; the electrodes serve as a surface for the chemical reaction but are not used up.

65. $C(s) + 2\,H_2(g) \longrightarrow CH_4(g)$;
$C(s) + H_2O(g) \longrightarrow CO(g) + H_2(g)$

67. Endothermic; the reaction absorbs heat, cooling the treated area.

69. 2000 watts

71. 2.35×10^6 g CO_2

73. **a.** Removal by photosynthesis   **b.** Animal respiration
c. Making plastics

75. 29 metric tons SO_2; no, there would be 27 μg SO_2/m^3

77. **a.** 85 y   **b.** 4 y   **c.** Rate of energy use will likely grow instead of remain the same, and so reserves might not last the 85 y calculated in part (a). Worldwide use is unlikely to reach the U.S. level any time soon, if ever, and so the estimate in (c) is not at all realistic.

Chapter 16

16.1 They differ only in configuration around the second carbon atom.

16.2 **A. a.** H-P-V-A   **b.** histidylprolylvalylalanine
B. a. Thr-Gly-Ala-Ala-Leu   **b.** T-G-A-A-L

16.3 **A.** 3   **B.** 24

16.4 **A.** Sugar: ribose; base: uracil; RNA
B. Neither; thymine occurs only in DNA, ribose only in RNA.

3. In every cell; muscles, skin, hair, nails

4. Polyamides; polymers of amino acids

5. All have C, H, and O; proteins also have N and possibly S.

6. The bond that joins two amino acids in a protein.

7. If the molar mass of a polypeptide exceeds about 10,000 g, it is called a protein.

9. Hydrogen bonds

11. DNA is a double helix; RNA is a single helix with some loops.

13. A process that produces millions of copies of a specific DNA sequence

15. Step 1: isolation and amplification
Step 2: gene is spliced into a plasmid
Step 3: the plasmid is inserted into a host cell
Step 4. the plasmid replicates, making copies of itself

17. A carbohydrate that cannot be further hydrolyzed; glucose, fructose, and galactose

19. Glycogen is animal starch. Amylose is a plant starch with glucose units joined in a continuous chain. Amylopectin is a plant starch with branched chains of glucose units.

21. Monosaccharides: b, d

23. lactose and sucrose

25. **a.** glucose   **b.** glucose

27. Aldehyde and alcohol (hydroxyl)

29. Aldehyde and alcohol (hydroxyl); in configuration about C-4.

31. At room temperature, fats are solids, oils are liquid. Structurally, oils have more C-to-C double bonds than fats have.

33. saturated: d; unsaturated: a, b, c

35. **a.** 18 **b.** 16 **c.** 18

37. liquid oil (right)

39. An amino group and a carboxyl group; a zwitterion is a molecule that carries both a positive and a negative charge.

41. essential: lysine and leucine

43. **a.**

$$\underset{\substack{\| \\ \text{NH}_2^+}}{\text{H}_2\text{N}-\text{C}}-\text{NHCH}_2\text{CH}_2\text{CH}_2\overset{\text{NH}_2}{\text{CHCOO}^-}$$

b.

$$\underset{\substack{\| \\ \text{O}}}{\text{H}_2\text{NCCH}_2}\overset{\text{NH}_3^+}{\text{CHCOO}^-}$$

45. **a.**

$$\text{H}_3\text{N}^+\text{CH}_2\text{CO}-\underset{\substack{| \\ \text{CH}_3}}{\text{NHCHCOO}^-}$$

b.

$$\text{H}_3\text{N}^+\underset{\substack{| \\ \text{CH}_3}}{\text{CHCO}}-\underset{\substack{| \\ \text{CH}_2\text{OH}}}{\text{NHCHCOO}^-}$$

47. aspartylphenylalanine

49. Hydrogen bonds, ionic bonds, disulfide linkages, and dispersion forces

51. DNA: a; RNA: b, c

53. ribose; uracil

55. **a.** guanine   **b.** thymine   **c.** cytosine   **d.** adenine

57. One strand will go with one daughter cell nucleus, the other strand will go with the other daughter cell nucleus.

59. **a.** DNA and mRNA   **b.** mRNA and tRNA

61. TTAAGC

63. AGGCTA

65. **a.** AAC   **b.** CUU

67. Glycogen or amylopectin

69. **a.** primary   **b.** secondary

71. **a.** pyrimidine   **b.** pyrimidine

73. **a.** purine   **b.** RNA

75. **a.** ~Thr-Ser-Met-Ala~   **b.** ~ Thr-Ser-Ala-Ala~

77. 3′-AUG-5′

Chapter 17

17.1 **A.** About 64%   **B.** About 34%

17.2 **A.** Fatty acid I
B. Fatty acid I, because fatty acid III is a cis fatty acid and not a trans fatty acid

1. energy source, thermal insulation, and protect organs

3. Vitamins are organic; minerals are inorganic.

5. Fat soluble; an excess is stored and accumulated; excess water-soluble vitamins are excreted.

9. Monosodium glutamate, a flavor enhancer

11. Vitamin E

13. The methyl ester of the dipeptide aspartylphenylalanine

16. Add nutritional value, enhance flavor or color, retard spoilage, provide texture, sanitize (and others)

17. **a.** glucose **b.** sucrose **c.** fructose

19. glucose

21. Fatty acids, glycerol, mono- and diglycerides; hydrolysis

23. Most animal fats are solids; most vegetable oils are liquids; animal fats generally have fewer C-to-C double bonds than vegetable oils.

25. An adequate protein provides all the essential amino acids in sufficient quantities; meat, eggs, milk

27. In protein synthesis, a limiting reactant is the amino acid present in a quantity so low that it limits the quantity of proper protein that can be formed.

29. All have C, H, and O; proteins also have N and S.

31. **a.** thyroid gland (regulation of metabolism)
b. hemoglobin (oxygen transport)
c. bones, teeth, blood clotting, heartbeat rhythm
d. nucleic acids, ATP (obtaining, storing, and using energy from foods), bones, teeth

33. Water soluble: pantothenic acid, biotin; fat soluble: phylloquinone

35. a, c

37. **a.** niacin (vitamin B_3) **b.** thiamine (vitamin B_1)
c. cyanocobalamine (vitamin B_{12})

39. Water soluble: b, c; fat soluble: a, d

41. **a.** improve nutrition **b.** disinfectant and preservative
c. inhibit spoilage

43. **a.** artificial sweetener **b.** improve nutrition **c.** spoilage inhibitor

45. d

47. Reducing agents, usually free-radical scavengers added to foods to prevent fats and oils from becoming rancid; vitamin E; BHT and BHA

49. About 53% from carbohydrates; 15% from fat; 20% from protein

51. About 192 kcal, about 19% from fat

53. About 100 g fat
55. a, c, e
57. about 44%
59. about 25%
61. **a.** about 53%   **b.** about 40%
63. riboflavin, niacin, and cyanocobalamin; they are water-soluble and not easily stored
65. Double bond(s) in unsaturated fats are more reactive than single bonds in saturated fats; unsaturated fats and oils would have spoiled more readily than saturated fats.
67. The thick polymeric starch is broken down to smaller, more soluble molecules by enzymes in saliva.

Chapter 18

1. A drug that kills or slows the growth of bacteria; originally limited to formulations derived from living organisms
3. Cortisol: a hormone released in the body during stressed or agitated states; prednisone: a semisynthetic hormone similar to cortisone
5. Mediators of hormone action; prostaglandins act near where they are produced; hormones act throughout the body.
6. Mild sedation (small doses) and sleep (larger doses); addiction and tolerance
7. Ketamine, PCP
9. A drug that produces stupor and relief of pain
11. Brain amines are made from dietary amino acids; for example, high-carbohydrate diets produce high serotonin levels in the brain.
12. No; they help control some symptoms.
15. Acetylsalicylic acid
17. Aspirin is anti-inflammatory; acetaminophen is not.
19. Carboxylic acid, ester
21. β-lactam antibiotics obtained from *Penicillium* molds.
23. They are effective against a wide variety of bacteria.
25. No; vaccination
27. Antimetabolites and alkylating agents
29. A female sex hormone; a male sex hormone
31. It blocks pregnancy from being established by inhibiting the action of progesterone.
33. One that changes the way we perceive things
35. a
37. Codeine is a less potent analgesic and less addictive; both are analgesics and addictive.
39. **a.** Acupuncture needles stimulate nerves that trigger release of endorphins, peptides that block pain signals.
    b. The wounds cause the body to secrete painkilling endorphins.
41. b, c
43. **a.** 2500 mg   **b.** 1.7 (about 2)
45.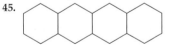
47. Water soluble; 10 C atoms with 1 O atom, 1 N atom, and an ionic charge
49.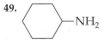
51. Cocaine; its lower LD_{50} means that less cocaine is required in a fatal dose.
53. Nicotine, because it takes a smaller amount to kill the rat. 2500 mg procaine; 50 mg nicotine.
55.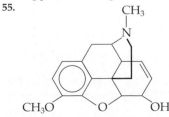
59. Ester; hydrolysis
61. 6-mercaptopurine; antimetabolite

63. **a.** Removing the polar —OH group would make the compound less soluble in water. **b.** Removing the polar —OH group would make the compound more soluble in fat.
65. **a.** The long hydrocarbon chain would make the drug much less soluble in water. **b.** The ionic and polar groups of the sodium succinate derivative would make the drug more soluble in water.
67. c

eChapter 19

19.1 **A.** 3.6 lb   **B.** 3.5 lb
19.2 **A.** 35 mi   **B.** 2250 kcal
19.3 **A.** 23.5   **B.** 174 lb
1. Meat and other animal products
3. Deplete glycogen stores and dehydrate yourself; no, it would be mostly water loss.
5. Better understanding of muscle physiology; drugs; improved equipment
7. The athlete needs more energy (more calories), best obtained from carbohydrates, especially starches.
8. Often deficient in B vitamins, iron, and other nutrients; slows metabolism, making future dieting more difficult and weight gain easier.
10. Masculinization (balding, extra body hair, deep voices) and menstrual irregularities
11. More toxic wastes that tax the liver and kidneys; such diets are also usually high in fat (artery clogging, cardiovascular disease).
13. Good vision, bone development, skin maintenance
15. Arthritis; B_6 is a cofactor for more than 100 enzymes.
17. Promotes absorption of calcium and phosphorus. The upper limit is 2000 IU; more than that can be toxic.
19. Promotes production of urine; ADH acts to retain urine.
21. Cloudy, highly colored urine
23. Muscle protein is converted to glucose.
25. 2.9 h
27. 100 g
29. 20 km
31. **a.** Cholecystokinin signals satiety.
    b. Ghrelin is an appetite stimulant.
33. 24
35. About 600
37. Actin and myosin
39. Muscle contraction; enzyme for removal of phosphate from ATP
41. Type I, aerobic; Type IIB, anaerobic
43. ATP can be generated rapidly and hydrolyzed rapidly.
45. No; the best way to build muscles is by exercise.
47. Anti-inflammatory; fluid retention, hypertension, ulcers, disturbance of sex hormone balance
49. The body releases endorphins that have much the same effect as some narcotics.
51. The "runner's high," feeling down when unable to run, having to run farther and farther to get the same high
53. Heart disease, stroke, lung cancer, emphysema, pneumonia, cancers of the pancreas, bladder, breast, kidney, and cervix
55. 2250 kcal
57. 130 kcal/day
59. **a.** 26.7%   **b.** 3.61%   **c.** Person B   **d.** 1.077 g/cm^3
61. **a.** 27   **b.** 144
63. **a.** 20 mL O_2 **b.** 1200 mL O_2

eChapter 20

20.1 **A.** $2 NH_3 + H_3PO_4 \longrightarrow (NH_4)_2HPO_4$
     B. $ZnO + H_2SO_4 \longrightarrow ZnSO_4 + H_2O$
20.2 980 mg
20.3 **A.** In 134 y; by 2142   **B.** In 89 y; by 2097
1. CO_2 and H_2O
3. Use manure from farm animals for fertilizer; rotate other crops with legumes; control insects by planting a variety of crops; minimize the use of fossil fuel.
5. Population would increase faster than the food supply unless the birthrate was controlled, poverty and war would serve as restrictions.

7. C, H, O
9. The Haber process; reacting H_2 and N_2
11. An organic (chemical sense) nitrogen fertilizer, H_2NCONH_2; ammonia and carbon dioxide
13. Phosphate rock is treated with phosphoric acid to make calcium dihydrogen phosphate; plays a role in the energy transfer processes of photosynthesis.
15. DDT is effective against many insects, but it remains toxic long after initial use.
17. They are fat soluble and become concentrated in fats moving up the food chain.
19. Mostly to bait traps and monitor infestations to determine when best to use a pesticide
21. Male insects are sterilized by radiation, chemicals, or cross breeding and then released for nonproductive breeding. Expensive and time consuming.
23. 30 g DDT
25. Addition of a constant amount each growth period; adding $10 to a savings account each week
27. 50 weeks; arithmetically.
29. 45.6 y; changes in birth rate or death rate (war, famine, . . .)
31. $6\,CO_2 + 6\,H_2O \longrightarrow C_6H_{12}O_6 + 6\,O_2$
33. (1), c; (2), a; (3), a; (4), c; (5), a
35. **a.** N, P   **b.** $2\,NH_3 + H_3PO_4 \longrightarrow (NH_4)_2HPO_4$
37. **a.** alkene, alcohol   **b.** ester, alkene, ether (epoxide)
    c. ether, alkene, ester
39. $10,737,418.24
41. $K_2CO_3 + CO_2 + H_2O \longrightarrow 2\,KHCO_3$
    $2\,KCl + 2\,H_2O \longrightarrow 2\,KOH + H_2 + Cl_2$
    $2\,KOH + CO_2 \longrightarrow K_2CO_3 + H_2O$
    not organic

eChapter 21

21.1 **a.** nonionic   **b.** anionic   **c.** anionic   **d.** cationic
1. An excellent cleanser in soft water, relatively nontoxic, derived from renewable sources, biodegradable
3. 1, a; 2, f; 3, b; 4, c; 5, d; 6, e
5. Loosens baked-on grease and burned-on food, excellent glass cleaner; vapors are irritating, toxic; don't mix with chlorine bleaches
7. Mildly abrasive and absorbs odors
9. It is softer and produces a finer lather.
11. Acid neutralizes the ionic "head."
13. Sodium palmitate and glycerol
15. $3\,CH_3(CH_2)_{10}COO^-Na^+ + HOCH_2CHOHCH_2OH$ (glycerol)
17. Precipitates hard-water ions; makes the water more alkaline
19. $PO_4^{3-} + H_2O \longrightarrow HPO_4^{2-} + OH^-$

    $CO_3^{2-} + H_2O \longrightarrow HCO_3^- + OH^-$
21. A substance added to a surfactant to increase its detergency
23. By binding Ca^{2+} and Mg^{2+} in soluble complexes
25. A surfactant molecule that carries both a positive and a negative charge
27. The positive end of the cationic surfactant would interact with the negative end of the anionic surfactant, destroying detergent action.
29. All three
31. II
33. Hexadecyltrimethylammonium chloride
35. They slowly release chlorine in water.
37. To remove paint, varnish, adhesives, waxes, and so on
39. Flammability, toxic fumes
41. A skin softener
43. OMC is insoluble in water and does not wash off as easily as PABA from sweat or while swimming.
45. Lipstick is harder, contains more wax, and is usually colored with pigments.
47. **a.** sweetener   **b.** abrasive   **c.** detergent
49. Enamel is converted from hydroxyapatite to fluorapatite, a harder substance.

51. A complex mixture of compounds used as a fragrance
53. Makes the face feel cool
55. Temporary dyes are water soluble and can be washed out; permanent dyes last until the hair is cut off or falls out.
57. A reducing agent; redox reaction
59. Oxidation
61. I, III, IV
63. I
65. $CH_3(CH_2)_{14}COOCH(CH_3)_2$
67. 11.6 g SnF_2
69. Cationic (quaternary ammonium compounds)
71. No. Brighteners convert energy in the form of ultraviolet light into visible light, but no energy is destroyed.
73. **a.** $CH_3(CH_2)_{24}COOH$ **b.** $CH_3(CH_2)_{28}CH_2OH$

eChapter 22

1. It can be poisonous in extremely large amounts.
2. People have different metabolisms and different health conditions (e.g., sugar can be more harmful to a diabetic than to a nondiabetic).
4. It is converted to fluorocitric acid that inhibits the citric acid cycle by tying up the enzyme that acts on citric acid.
6. Cigarette smoking and diet
9. b, d
11. Hydrolysis of amides; catalyst
13. A substance that blocks the transport of oxygen in the bloodstream; CO, nitrate ions
15. By converting toxic cyanide to harmless thiocyanate
17. b, d
19. By tying up sulfhydryl groups, thus deactivating enzymes
21. By chelating Pb^{2+} ions, thus enhancing their excretion
23. Sarin, tabun, soman
25. Poisons that block cholinesterase
27. By oxidizing ethanol to acetaldehyde, then to acetic acid, and finally to carbon dioxide and water
29. Cotinine is less toxic and is more water soluble (more readily excreted) than nicotine.
31. No. Some are oxidized to more toxic substances.
33. 7.94 g
35. Genes that regulate cell growth; they sustain the abnormal growth characteristic of cancer.
37. Incomplete burning of almost any organic material
39. A study of a human population to try to link human health effects (e.g., cancer) to a cause (e.g., exposure to a specific chemical); no
41. A mutagen causes mutations; a teratogen causes birth defects.
43. A waste that burns readily on ignition, presenting a fire hazard; hexane, gasoline.
45. 4.4×10^{10} molecules
47. **a.** $C_{10}H_{16}O$ **b.** 4800 mg **c.** 5 mg
49. **a.** 99.48%; % smokers in control group: 95.50%
    b. 0.9948:1 for cancer cases; 0.9550:1 for controls
    c. 0.96:1
51. Iron and zinc
53. **a.** no **b.** yes
55. The H atoms of water can hydrogen-bond to the oxygen atom of cotinine, making cotinine more readily excreted in urine.
57. **a.** 10.3 mL vodka   **b.** 15.4 mg/mL

Appendix

A.1   **a.** 0.0163 g   **b.** 1.53 lb   **c.** 370 mL
A.2   **a.** 0.0903 m   **b.** 0.2224 km   **c.** 150 fl oz
A.3   **a.** 24.4 m/s   **b.** 1.34 km/h   **c.** 0.136 oz/qt
A.4   **a.** No   **b.** No
A.5   0.12 m^3
A.6   56.8 g
A.7   **a.** 100.5 m   **b.** 6.3 L
       c. 1800 m^2 (1.80×10^3 m^2)   **d.** 2.33 g/mL
A.8   **a.** 185 °F   **b.** 10.0 °F   **c.** 179 °C   **d.** − 29.3 °C
A.9   80,000 cal; 80.0 kcal; 334 kJ

CHAPTER 1 Page xxxii: Colin Monteath\Minden Pictures. Page 1 (T): Dynamic Graphics\Jupiter Images - FoodPix - Creatas. Page 1 (B): Doris K. Kolb. Page 2: www.photos.com\Jupiter Images. Page 3 (T): Pearson. Page 3 (B): Courtesy of the Library of Congress. Page 4: UPI\Corbis\Bettmann. Page 6: James Steidl\Shutterstock. Page 8: iofoto\Shutterstock. Page 11: Corbis\Bettmann. Page 12: AP Wide World Photos. Page 13: Charles M. Duke, Jr.\NASA Headquarters. Page 14 (L & R): Tom Bochsler Photography Limited\Pearson Education\PH College. Page 17 (L): US Mint. Page 17 (ML): jmatzick\Shutterstock. Page 17 (MR): iStockphoto.com. Page 17 (R): bryngelzon\iStockphoto.com. Page 20: Richard Megna\Fundamental Photographs. Page 25: Pearson. Page 29 (L): niderlander\Shutterstock. Page 29 (R): Eric Schrader\Pearson. Page 36: Terry McCreary. Page 37 (L): Buquet\Shutterstock. Page 37 (R): David Anderson\iStockphoto.com. Page 38 (TL): Richard Megna\Fundamental Photographs. Page 38 (BL): Yana Petruseva\Shutterstock. Page 38 (R): David A. Aguilar\Harvard-Smithsonian Center for Astrophysics Dhoxax\Shutterstock. **CHAPTER 2** Page 40 (D): IBM Research, Almaden Research Center. Page 40: James L. Amos\Photo Researchers, Inc. Page 42 (L): Stamp from the private collection of Professor C. M. Lang, photography by Gary J. Shulfer, University of Wisconsin, Stevens Point. "Greece (Scott #1469);" Scott Standard Postage Stamp Catalogue, Scott Pub. Co., Sidney, Ohio. Page 42 (T): iStockphoto.com. Page 42 (D): Susumu Nishinaga\Photo Researchers, Inc. Page 43: www.photos.com\Jupiter Images. Page 44 (BL): Morozova Tatyana\Shutterstock. Page 44 (BM): Katharina Wittfeld\Shutterstock. Page 44 (BR): Richard Megna\Fundamental Photographs, NYC. Page 44 (L): Stamp from the private collection of Professor C. M. Lang, photography by Gary J. Shulfer, University of Wisconsin, Stevens Point. "Sweden (Scott #293, #295);" Scott Standard Postage Stamp Catalogue, Scott Pub. Co., Sidney, Ohio. Page 46: The Granger Collection. Page 48: Gusto\Photo Researchers, Inc. Page 51: Courtesy of the Library of Congress. Page 52: The Image Works. Page 54 (T): Harry Taylor © Dorling Kindersley; Mark Wragg\iStockphoto.com; iStockphoto.com; Tammy Peluso\iStockphoto.com. Page 54 (B): Harry Taylor © Dorling Kindersley; Michael Shake\Shutterstock; iStockphoto.com; iStockphoto.com. Page 55: Shutterstock. Page 59 (L): Richard Megna\Fundamental Photographs, NYC. Page 59 (R): Richard Megna\Fundamental Photographs, NYC. **CHAPTER 3** Page 60: Rachelle Burnside\Shutterstock. Page 62: www.photos.com\Jupiter Images. Page 63: Carey B. Van Loon. Page 65: The Burndy Library, Dibner Institute for the History of Science and Technology, Cambridge, Massachusetts. Page 66 (B): Stamp from the private collection of Professor C. M. Lang, photography by Gary J. Shulfer, University of Wisconsin, Stevens Point. "France #B76;" Scott Standard Postage Stamp Catalogue, Scott Pub. Co., Sidney, Ohio. Page 66 (T): Science VU\Visuals Unlimited. Page 72: manzrussali\Shutterstock. Page 73 (TA): Richard Megna\Fundamental Photographs, NYC. Page 73 (M): David Parker\Science Photo Library\Photo Researchers, Inc. Page 73 (B): Wabash Instrument Corp\Fundamental Photographs. Page 74: Stamp from the private collection of Professor C.M. Lang, photography by Gary J. Shulfer, University of Wisconsin, Stevens Point. "1963, Denmark (Scott #409)"; Scott Standard Postage Stamp Catalogue, Scott Pub. Co., Sidney, Ohio. Page 82 (B): Richard Megna\Fundamental Photographs. Page 82 (T): Richard Megna\Fundamental Photographs, NYC. Page 83 (T): Paul Silverman\Fundamental Photographs, NYC. Page 83 (B): Colin Keates © Dorling Kindersley, Courtesy of the Natural History Museum, London. **CHAPTER 4** Page 88 (D): Andrew Lambert Photography\Photo Researchers, Inc. Page 88: Sinclair Stammers\Photo Researchers, Inc. Page 91: Courtesy of the Bancroft Library, University of California, Berkeley. Page 93 (TA): Richard Megna\Fundamental Photographs, NYC. Page 93 (BR): Albert Copley\Visuals Unlimited. Page 94: Fundamental Photographs, NYC. Page 99: Spencer Grant\Photo Edit. Page 111: Eric Schrader\Pearson. **CHAPTER 5** Page 122: Jean-Francois Cardella\Construction Photography.com. Page 128: Stamp from the private collection of Professor C.M. Lang, photography by Gary J. Shulfer, University of Wisconsin, Stevens Point. "1956, Italy (Scott #714)"; Scott Standard Postage Stamp Catalogue, Scott Pub. Co., Sidney, Ohio. Page 129: NASA\Goddard Space Flight Center. Page 132: Fundamental Photographs, NYC. Page 140 (T): Eric Schrader\Pearson. Page 140 (B): Eric Schrader\Pearson. Page 141: Nathan Eldridge\Pearson Education\PH College. **CHAPTER 6** Page 148: NASA. Page 150 (L): Mike Liu\Shutterstock. Page 150 (R): Wikipedia, The Free Encyclopedia. Page 152: Nikita Tiunov\Shutterstock. Page 153 (L & R): Richard Megna\Fundamental Photographs, NYC. Page 156 (L): Tony Freeman\PhotoEdit. Page 156 (M): George Mattei\Photo Researchers, Inc. Page 156 (R): Liv Friis-Larsen\Shutterstock. Page 162 (L & R): Richard Megna\Fundamental Photographs, NYC. Page 164: Richard Megna\Fundamental Photographs, NYC. **CHAPTER 7** Page 170: Shutterstock. Page 172 (RA): Eric Schrader\Pearson. Page 172 (L): Lajos Repasi\iStockphoto.com. Page 173 (B): Shutterstock. Page 173 (T): Andrew Lambert Photography\Photo Researchers, Inc. Page 174: Photos.com. Page 182 (A): Richard Megna\Fundamental Photographs, NYC. Page 185: Richard Megna\Fundamental Photographs, NYC. Page 188: Creative Digital Visions\Pearson. Page 189 (T): Photos.com. Page 189 (B): Michael P. Gadomski\Photo Researchers, Inc. Page 193: Pablo Sanchez\Reuters Limited. **CHAPTER 8** Page 194: AP Wide World Photos. Page 196 (L & R): Joseph P. Sinnot\Fundamental Photographs, NYC. Page 202 (L & R): Peticolas\

Megna\Fundamental Photographs, NYC. Page 205: Creative Digital Visions\Pearson. Page 207: Walter Drake image, photographer Rich Chartier; Colorado Springs, CO. Page 208: Shutterstock. Page 209: NASA. Page 210: Richard Megna\Fundamental Photographs. Page 212 (T): Eric Schrader\Pearson. Page 212 (B): Shutterstock. Page 214 (T): Richard Megna\Fundamental Photographs, NYC. Page 214 (BL): Spencer Grant\Photo Edit. Page 214 (BR): Shutterstock. Page 215 (L & R): Spencer Grant\Photo Edit. Page 216: Shutterstock. Page 219: Shutterstock. Page 220: Shutterstock. **CHAPTER 8** Page 222: iStockphoto.com. Page 225: Shutterstock. Page 226: iStockphoto.com. Page 231 (T): Richard Megna\Fundamental Photographs. Page 231 (B): Shutterstock. Page 233: Clive Freeman\The Royal Institution\Science Photo Library\Photo Researchers. Page 235: Phil Degginger\Color-Pic. Page 239: Eric Schrader\Pearson. Page 240: Pearson. Page 241: SIU\Photo Researchers. Page 246: Richard Megna\Fundamental Photographs. Page 250: Courtesy of CEM Corporation. **CHAPTER 10** Page 260: Philippe Psaila\Photo Researchers, Inc. Page 262: Richard Megna\Fundamental Photographs. Page 264 (T): Richard Megna\Fundamental Photographs. Page 264 (B): Richard Megna\Fundamental Photographs. Page 266: Joe Appel\AP Wide World Photos. Page 267 (T): Dennis McDonald\PhotoEdit Inc. Page 267 (BL & BR): Shutterstock. Page 268 (T): Kip & Pat Peticolas\Fundamental Photographs. Page 268 (M): Shutterstock. Page 268 (B): Eckehard Schulz\AP Wide World Photos. Page 270: Wiley-VCH Verlag GmbH & Co. KGaA. Page 272 (T): Shutterstock. Page 272 (B): Shutterstock. Page 275: Eric Schrader\Pearson. Page 276: Chris Ison\AP Wide World Photos. Page 277: iStockphoto.com. Page 279: Amoco Fabrics & Fibers Co. Page 280: Marshall Troy\Evo Design. Page 284: NatureWorks LLC. Page 288: SRI International. **CHAPTER 11** Page 290: NASA. Page 293: Shutterstock. Page 303: Bettman\Corbis. Page 304 (T): Kip Peticolas\Richard Megna\Fundamental Photographs. Page 304 (B): Kip Peticolas\Richard Megna\Fundamental Photographs, NYC. Page 306: Images provided courtesy of Bristol-Myers Squibb Medical Imaging, Inc. Scans contributed by Howard Lewin, MD, Los Angeles, CA. Page 307 (L): Olivier Voisin\Photo Researchers, Inc. Page 307 (R): Wellcome Dept. of Cognitive Neurology\Science Photo Library\Photo Researchers, Inc. Page 310: ALBERT EINSTEIN and related rights TM\© of The Hebrew University of Jerusalem, used under license. Represented exclusively by Corbis Corporation Page 311 (T): UPI\Corbis\Bettmann. Page 311 (B): Bettman\Corbis. Page 314: Lawrence Berkeley Nat'l Lab. Page 315: U.S. Air Force. Page 318 (T): Corbis. All Rights Reserved. Page 318 (B): GeoEye Inc. **CHAPTER 12** Page 324: NASA Headquarters. Page 327: Shutterstock. Page 328 (R): © Dr. Richard Busch\Courtesy American Geological Institute, AGI. Earth Science World Image Bank. Page 328 (L): Peter Anderson © Dorling Kindersley. Page 329 (TL): Chip Clark. Page 329 (BL): From *Asbestos and Other Fibrous Materials: Mineralogy, Crystal Chemistry, and Health Effects* by H. Catherine W. Skinner, Malcolm Ross, and Clifford Frondel. New York: Oxford University Press, 1988. Page 329 (R): Bettmann\Corbis\Bettmann. Page 330 (TL): Andy Crump\Photo Researchers, Inc. Page 330 (BL): Lawrence Berkeley National Laboratory. Page 330 (R): Janine Wiedel Photolibrary\Alamy Images. Page 331: Shutterstock. Page 332 (T): Shutterstock. Page 332 (B): Shutterstock. Page 337 (T): Shutterstock. Page 337 (B): Robert Semeniuk. Page 337 (M): Charles D. Winters\Photo Researchers, Inc. Page 342: Photolibrary.com. Page 343: Edward Kinsman\Photo Researchers, Inc. **CHAPTER 13** Page 344: Patrick Byrd\Science Faction. Page 350: Copyright, EcoSphere Associates, Inc. 2006. Page 350 (T): NASA. Page 351: Getty Images, Inc. Page 352: Still Pictures\Peter Arnold, Inc. Page 353: AP Wide World Photos. Page 354 : Dr. Gerald L. Fisher\Science Photo Library\Photo Researchers, Inc. Page 358: EPA\Landov Media. Page 362: © Ray Pfortner\Peter Arnold Inc. Page 366: Phil Degginger\Color-Pic. Page 368: NASA. Page 372: Jonathan Patz. Page 378: John Hyde\British Antarctic Survey. **CHAPTER 14** Page 380: John Hyde\National Geographic Image Collection. Page 381: NASA. Page 382: Mike Derer\AP Wide World Photos. Page 384: Shutterstock. Page 387 (T): H.D.A. Linquist, U.S. EPA. Page 387 (M): Shutterstock. Page 387 (B): Rob Badger\Getty Images, Inc. - Taxi. Page 388 (T): iStockphoto.com. Page 388 (B): Gautam Singh\AP Wide World Photos. Page 389: iStockphoto.com. Page 390 (T): iStockphoto.com. Page 390 (B): James P. Blair\National Geographic Society. Page 392 (T): EyePress\AP Wide World Photos. Page 392 (B): Mike Segar\Reuters Limited. Page 393: U. S. Geological Survey (http://pubs.usgs.gov/sim/2005/2881). Page 396: Pavel Rahman\AP Wide World Photos. Page 398: National Institute of Dental Research. **CHAPTER 15** Page 406: Paulo Whitaker\Reuters Limited. Page 408: SOHO\LASCO (ESA & NASA). Page 408: NASA. Page 410 (T & B): Charlie Riedel\AP Wide World Photos. Page 411 (B): Tony Freeman\PhotoEdit Inc. Page 411 (TL & TR): Richard Megna\Fundamental Photographs. Page 412: Greenshoots Communications\Alamy Images. Page 413: Richard Megna\Fundamental Photographs, NYC. Page 415: iStockphoto.com. Page 416: Richard Megna\Fundamental Photographs, NYC. Page 417: Ancient Art & Architecture\DanitaDelimont.com. Page 419: © The Field Museum, Neg #GEO85637C, Chicago. Photographer: John Weinstein. Page 421 (L & R): U.S. Department of the Interior. Page 422: Shutterstock. Page 426: iStockphoto.com. Page 427: Shutterstock. Page 431: © Royalty-Free\CORBIS. Page 432 (L): U.S. Department of Energy. Page 432 (R): "Yucca Mountain, The Nevada Test Site for Terminal Isolation-Research and Development of Commercial Nuclear

*Index entries (and page numbers) in green are found in eChapters 19–22 at www.chemplace.com.